Archaeological Informatics: Pushing the Envelope CAA 2001

Computer Applications and Quantitative Methods in Archaeology

Proceedings of the 29th Conference, Gotland, April 2001

Edited by

Göran Burenhult

co-editor

Johan Arvidsson

BAR International Series 1016

2002

Published in 2016 by
BAR Publishing, Oxford

BAR International Series 1016

Archaeological Informatics: Pushing the Envelope – CAA 2001

ISBN 9781841712987 paperback
ISBN 9781407323985 e-format
DOI https://doi.org/10.30861/9781841712987
A catalogue record for this book is available from the British Library

BAR Publishing is the trading name of British Archaeological Reports (Oxford) Ltd. British Archaeological Reports was first incorporated in 1974 to publish the BAR Series, International and British. In 1992 Hadrian Books Ltd became part of the BAR group. This volume was originally published by Archaeopress in conjunction with British Archaeological Reports (Oxford) Ltd / Hadrian Books Ltd, the Series principal publisher, in 2002. This present volume is published by BAR Publishing, 2016.

BAR
PUBLISHING

BAR titles are available from:

BAR Publishing
122 Banbury Rd, Oxford, OX2 7BP, UK
EMAIL info@barpublishing.com
PHONE +44 (0)1865 310431
FAX +44 (0)1865 316916
www.barpublishing.com

Contents

Preface 1

Papers in the GIS category

Extracting "Natural Pathways" from a Digital Elevation Model 5
Applications to Landscape Archaeological Studies
Gino Bellavia

2000 Years of Town Planning in Vienna 13
Wolfgang Börner

A 5000-Years History of Settlement and Irrigation in the Murghab Delta (Turkmenistan) 21
An Attempt of Reconstruction of Ancient Deltaic System
Barbara Cerasetti

Building, building on the wall… 29
A reflection of actual building dimensions?
Dora Constantinidis

Historic landscape assessment: 35
The East of England Experience
Lynn Dyson-Bruce

GIS approach to Iberian iron age landscape in central-south Valencia region (Spain) 43
Ignacio Grau Mira

Developing an Information System for Archaeological Sites and Monuments - Data Model and Construction 49
David Haskiya

Using viewsheds wisely: developing sound methodologies from spatial analyses of megalithic monuments in western Scotland 53
Gail Higginbottom, Ken Simpson and Roger Clay

Reconstructing the seascape at the mouth of the Oder 63
Elaboration of a DBM-model based on 1912-soundings
George Indruszewski

The TimeMap Kiosk: Delivering Historical Images in a Spatio-Temporal Context 71
Ian Johnson and Andrew Wilson

The answer is blowin' in the wind. Research desires and data possibilities. 79
Hans Kamermans

A hundred years of lake contour fluctuation in the Hamun-i Helmand: A GIS based system for the study and the recovery of archaeological information in the Iranian Sistan (1899-1999) 85
Sabatino Laurenza and Sophie Pornet-Laurenza

From Stratigraphic Unit to the mouse: a GIS based system for the excavation of historical complex. The case study of Pompeii 93
Sabatino Laurenza and Cristiano Putzolu

Statistical analysis of the distribution of modern primates: a comparative approach to the spatial analysis of the Palaeolithic 105
Katharine MacDonald

The Creation and Potential Applications of a 3-Dimensional GIS for the Early Hominid Site of Swartkrans, South Africa 113
Joseph D. Nigro, W. Fredrick Limp, Kenneth K. Kvamme, Darryl J. de Ruiter and Lee R. Berger

GIS contribution to urban history and to the reconstruction of ancient landscape 125
Sofia Pescarin

GIS Modeling of the Minoan Peak Sanctuaries of East Crete 129
S. Soetens, A. Sarris, S. Topouzi and A. Tripolitsiotis

Modeling Archaeological and Historical Cognitive Landscapes in the Greater Yellowstone Region (Wyoming, Montana, and Idaho, USA) Using Geographic Information Systems 139
Thomas G. Whitley

GIS and Space Analysis in the study of the Hospitallers' fortifications in the Dodecanese 149
Nicholas Zarifis and Despina Brokou

Papers in the Virtual Archaeology category

Reconstructions of the Excavations of Two Iron Age Chariot Burials from Belgium 157
Applying Virtual Reality to Old Excavation Data
Geoff Avern

3D standards for scientific communication 163
Francesca Cantone

Acoustic reconstruction of music performance spaces using three-dimentional digital waveguide mesh models 173
Guilherme Campos, David M. Howard and Steve Dobson

A virtual journey through a Roman settlement. Aloria 177
Juan José Fuldain González

Topometrical measurements in Tiryns, Greece 181
Report on a co-operate project between physics and archaeology
Maria Shinoto, Zoltán Böröcz, Carsten Thomas, Dieter Dirksen, Joseph Maran and Gert von Bally

Advances in Geometric Modeling and Feature Extraction on Pots, Rocks and Bones for Representation and Query via the Internet 191
Utsav A. Schurmans, Anshuman Razdan, Arleyn Simon, Mary Marzke, Peter McCartney, David Van Alfen, Gram Jones, Mary Zhu, Dezhi Liu, Myungsoo Bae, Jeremy Rowe Gerald Farin and Dan Collins

Papers in the Osteology category

Neural Network Classification of Skeletal Remains 205
Suzanne Bell and Richard Jantz

Osteology for the new millennium 213
Gustav Jonsson

3-D CAT-scan: Anthropology, Archaeology and Virtual Reality 217
Niels Lynnerup

The Statistics of Archaeological Deformation Processes 221
An archaeozoological experiment
Laura Mameli, Juan A. Barcelo and Jordi Estevez

Papers in the Internet Applications and Cultural Heritage Management categories

Generating Multilingual Personalized Descriptions of Museum Exhibits – The M-PIRO Project 233
Ion Androutsopoulos, Aggeliki Dimitromanolaki, Vassiliki Kokkinaki, Jo Calder, Jon Oberlander and Elena Not

Joined up writing: an Internet portal for research into the Historic Environment ... 243
Tony Austin, Francisco Pinto, Julian Richards and Nick Ryan

The Archaeological Map of Egypt ... 253
(Archaeological Heritage Resource Management System)
Fathi Saleh and Reem Bahgat

Geographic Information Systems and Archaeology: the case of ancient Nora (Pula-Cagliari) ... 261
Anna Maria Colavitti and Giancarlo Deplano

XML Encoding of Archaeological Unstructured Data ... 267
Marco Crescioli, Andrea D'Andrea and Franco Niccolucci

Cultural landscape, computers and characterisation: ... 277
GIS-based Historic Landscape Characterisation as a tool for archaeological resource management in England
Graham Fairclough

Corinth Computer Project: Internet Education ... 295
David Gilman Romano and Nicholas L. Stapp

Desktop - Photogrammetry and its Link to Web Publishing ... 301
Günter Pomaska

A Web-based Digital Archaeological Map of Lasithi, E. Crete. ... 309
Apostolos Sarris, Katerina Bichta, Marina Giasta, Anthi Giourou, Evaggelia Karimali, Vaggelis Kevgas, Kostas Margetousakis, Eleni Peraki, Steven Soetens, Katerina Tzaneteas, SofiaTopouzi and Achilleas Tripolitsiotis

The Archaeologist Files: An approach to the digital contextualization of archaeological finds in user adaptive information systems ... 325
Isto Vatanen

Papers in the Survey and Mapping, Archaeometry, GPS and CAD categories

New Tools for Understanding Plains Indian Sites in Grasslands National Park, Canada ... 335
Gary Adams

Archaeological Survey as Optimal Search ... 341
E. B. Banning

Archaeological spatial modelling ... 351
A case study from Beagle Channel (Argentina)
Juan A. Barcelo, Ernesto L. Piana and Daniel R. Martinioni

Views on „castrum cuscin" ... 361
Archeology and archaeometry under water
Ralf Bleile, Cornelius Meyer and Burkhardt Ullrich

A PC-based stereoscopic measurement system for the generation of digital object models ... 371
Frank Boochs, Stephan Eckhardt and Ben Fischer

Infra-red (Reflectance) Spectroscopy of Ceramics from Tell Halawa, Syria ... 379
Murray Lee Eiland and Quentin Williams

Detailed Topography and Surface Survey. What is the point? Tanagra City Survey 2000 ... 385
Emeri Farinetti and Lefteris Sigalos

The application of high-resolution satellite imagery for the detection of ancient Minoan features on Crete ... 393
Lefki Pavlidis, Clive S Fraser and Cliff Ogleby

Educating the Digital Fieldwork Assistant 401
Nick Ryan and Martijn van Leusen

From Field Books to Powerbook 413
Computer Applications and the Promontory Palace of Herod the Great
James G. Schryver

Errors & Inaccuracies in Repositioning of Archaeological Sites 417
S. Topouzi, A. Tripolitsiotis, A. Sarris, S. Mertikas and S. Soetens

Three-dimensional excavation plans and 3D Studio Max 427
Experiences from the excavations of the medieval town of Naantali, Finland
Kari Uotila and Carita Tulkki

Handling digital 3-D record of archaeological excavation data 431
Mikhail Zhukovsky

Papers in the Database Applications and Statistics and Quantitative Methods categories

Approaches to petrographic data analysis using S-Plus 441
C.C. Beardah, M.J. Baxter, I. Papageorgiou and M.A. Cau

Developing an information system for archaeological sites and monuments – administration and maintenance model 449
Malin Blomqvist and Cissela Génetay-Lindholm

How can a database full of Bugs help reconstruct the climate? 453
Phil Buckland and Paul Buckland

Pattern recognition applied to Rock Art 463
Diego Diaz and Damián Castro

Creating of a database for prehistoric sites: which are the goals, the strategy and what means to put in place? 469
Pierre Corboud

The Computer Catalogue of the Kunstkammer Museum Collections and perspectives of an Internet-shared Anthropological Database 475
Yuri Osintsov, Yuri Chistov and Dmitriy Gerasimov

Reclaiming Old Data: The Wasden Site Research Project 483
E. S. Lohse and J. Anderson

The Archaeology-Palaeobotany-Palynology Database on the Palaeolithic, Mesolithic and early Neolithic sites of the Former USSR Area 491
Yuri Stepanov, Galina Levkovskaya, Mikhail Anikovitch, Nikolai Anisutkin, Elena Beliaeva, Vladimir Shumkin, Veronika Stegantzeva, Vladimir Timofeev, Galina Sinitsyna, Anastassia Bogolubova and Andrey Stegantzev

Never under-estimate the power of a model 495
Clive Robert Orton

Deriving ancient foot units from building dimensions: a statistical approach employing cosine quantogram analysis 501
Jari Pakkanen

A Genetic Algorithm problem solver for Archaeology 507
Carlos Reynoso and Edward Jezierski

Visualising the Neolithic transition in Europe 511
Thembi Russell and James Steele

Archaeological databases: what are they and what do they mean? ... 517
Robert Schlader

The Use of Permutations to Explain the Hackness Cross Tree Rune Inscription ... 521
Richard Sermon

The computer as a tool in a dialectical approach to data processing ... 527
Maria Shinoto

Papers in the Posters and Workshops categories

Presentation of a Database for funerary Analysis and Proposals for its Increments and Developments ... 543
Francesca Fulminante

A GIS Application for the study on water supply and draining system in the ancient capital cities in Japan ... 547
Akihiro Kaneda and Susumu Morimoto

Bjäre and Bowland: Computer Applications in European Pathways to Cultural Landscapes, a Culture 2000 programme ... 551
Jenny Nord Paulsson and Graham Fairclough

Ancient Mantineia's Defence Network Reconsidered Through a GIS Approach ... 559
S. Topouzi, A. Sarris, Y. Pikoulas, S. Mertikas, X. Frantzis and A. Giourou

Laser Scanning and Digital Close Range Photogrammetry for Capturing 3-d Archaeological Objects: a Comparison of Quality and Practicality ... 567
Athanasios Velios and John P. Harrison

CAA 2001 Abstracts on CD

Abstracts in the GIS category ... 1

Abstracts in the Virtual Reality and Digital Image Processing categories ... 17

Abstracts in the Osteology category ... 26

Abstracts in the Internet Applications and Cultural Heritage Management categories ... 30

Abstracts in the Survey and Mapping, Archaeometry, GPS and CAD categories ... 43

Abstracts in the Database Applications and Statistics and Quantitative Methods categories ... 51

Abstracts to workshops ... 61

Abstracts to posters ... 69

Please note that the CD referred to above has now been replaced with a download available at https://doi.org/10.30861/9781841712987_CDROM

Preface

This volume presents the proceedings of the 29th CAA conference, entitled *Archaeological Informatics – Pushing the Envelope. Computer Applications and Quantitative Methods in Archaeology*. The conference was held in Visby, on the island of Gotland, in Sweden, April 25-29, 2001. Altogether, 289 delegates from 30 countries participated in the conference. A total of 119 papers were presented, together with 19 posters and 5 workshops. In this volume, 72 papers are presented. All illustrations are printed in the volume in b/w, also those submitted as colour images, as well as on the enclosed CD-ROM disc, on which the colour images can be studied in true colours. Abstracts to all presented papers, as well as posters and workshops, are also included.

The conference was officially opened by Princess Christina, Mrs. Magnuson, in the Picture Stone Hall at the Gotland Historical Museum. The conference was introduced by an invited lecture, presented by Associate Professor Paul Sinclair and Dr. Markku Pyykönen, *Computer aided assessments of human responses and contributions to environmental change in Africa over the last 2600 years*, after which the conference was split up into parallel sessions and workshops. Titles of the sessions were:

- GIS
- Virtual Archaeology
- Osteology
- Internet Applications and Cultural Heritage Management
- Survey and Mapping, Archaeometry, GPS and CAD
- Database Applications and Statistical and Quantitative Methods

All sessions were chaired by internationally recognized experts, and those experts also acted as peer reviewers. Despite these reviewers being anonymous in the individual case, we would like to acknowledge them for their work, and also thank them for prompt and clear comments and advice to the various authors: Juan Barcelo, Martin Doerr, Ebba During, Emeri Farinetti, Ian Johnson, Hans Kamermans, Franco Niccolucci, Clive Orton, David Gilman Romano, Nick Ryan, Apostolos Sarris and Steve Stead.

We would also like to thank everyone who made the organization in Visby possible during the five hectic days in April, including individual students and the students' organization RINDI, the International Secretary, the Gotland University College technical and administrative staff, and the Historical Museum of Gotland. A special acknowledgement goes to Gustaf Svedjemo, for keeping total and successful control of the computer equipment during the conference. For the English language correction of the papers to be published, we would like to thank Mrs. Cathy Gow Sjöblom.

For generous funding of the conference, we would like to thank the European Union – the Structural Funds (Region Objective 2 Islands Sweden), the Gotland Regional Development Unit, Gotland County Council, the Municipality of Gotland, Sällskapet DBV and Trimble AB.

Göran Burenhult
Johan Arvidsson

Visby, January 2002

Papers in the GIS category

Extracting "Natural Pathways" from a Digital Elevation Model

Applications to Landscape Archaeological Studies.

Gino Bellavia
Ancient History & Archaeology
University of Birmingham
Edgbaston
Birmingham, B15 2TT
Phone: +44 24 764 16994 - E-mail: gino_bellavia@becta.org.uk

Abstract: The relationship between the distribution of archaeological sites and "natural pathways" through the landscape forms the first stage of a research programme into the relationships between prehistoric "ritual" sites in the UK and their landscape environment.

A technique has been devised to extract "natural" pathways, from a landscape Digital Elevation Model. The method is derived from GIS based hydrological modelling techniques and identifies potential multi-scale pathways across a landscape.

The "natural pathways" modelling process has been used on several areas in the UK to test the validity of the approach. Initial tests have produced promising results and indicates the usefulness and validity of the methodology. Further analysis on an area around the prehistoric site of Stonehenge hint at several possible relationships between burial monuments and "natural pathways" in the landscape.

Key words: Natural Pathways, Digital Elevation Models, Cost Surfaces, Prehistoric routes, Ritual sites.

Introduction

Overall, archaeological landscape studies have focussed on the potential relationships between sites and their landscape environment on a site by site basis, or by small-scale regional studies. While these studies may enhance the interpretation of these sites and perhaps regional patterns, they do not provide a higher level view of site characteristics. For our research, newly available data sources were to be used to facilitate a landscape analysis at the macro level. Specifically, the investigation aims to look at generalised patterns for the whole of the UK.

Initial researches into the patterns of site location focussed on relatively simple, 2-dimensional distribution analyses. These early studies generally treated the landscape as a flat plane. Recently, Digital Elevation Models (DEM) have facilitated 3-dimensional analyses. The most commonly analytically studied aspects of the landscape and sites have been intervisibility and viewshed type analyses. These techniques aim to provide an insight into site distribution and form, but may fail to take account of other, formalised aspects of the landscape, such as pathways.

It has been stated that settlements are often located at natural crossroads or thoroughfares, suggesting that before intensive road building schemes took place, sites were situated close to natural pathways through the landscape.

The formation of tracks and pathways, at any period, on a landscape, is an inevitable result of human and animal movements over time. These paths may permeate the landscape for long periods, waxing and waning in their use and significance. These pathways also provide a human structure to the landscape and allow a direct connection to the landscape and its elements. Their existence and nature should be an important element in archaeological interpretation. As Ingold (1993) states, "In short, the landscape is the world as it is known to those who dwell therein, who inhabit its places and journey along the paths connecting them".

Potential relationships between prehistoric sites and pathways have been mentioned by many authors, however the locations of these pathways are often subjectively defined. In (Llobera, 1996; ParceroOubina et al. 1998) the authors suggest that rock art and megalithic barrow sites in Iberia, are related to probable pathways. (BRADLEY et al. 1994) have looked at the relationship between potential pathways and rock art sites, however the notion of "paths" is again, subjective. Bodmin Moor and its cairns, are suggested by (Tilley, 1996), to be accessed by pathways along the main river courses and their tributaries.

(Loveday, 1998) calculated a correlation, greater than 54%, between the alignments of Roman roads and nearby double entrance henges. The author comments on the possible reasons for this correlation, which in some cases may relate to the natural topography, indicating probable pathways through the henge openings. In this scenario, the Roman roads simply follow the same terrain, as might a modern pathway. Although the alignments also occur in relatively flat terrain, which is difficult to interpret, the data set is sufficient to hint at a potential relationship between the alignments of the road and double entrance henges.

The locations of road networks are generally understood during the Roman period. Whilst a recent study[1] on the Ridgeway path in England postulates a prehistoric date for that pathway, many studies of prehistoric pathways suffer from a dearth of independent evidence. Furthermore, their dating is often concluded from the prevalence of nearby prehistoric sites.

In order to ascertain any correlation between sites and "natural pathways", a formal methodology to extract potential "natural pathways" independently from site location had to be developed.

At the scale of the analysis, discussed later, the natural topography of the UK is assumed not to have significantly changed since the prehistoric period. Therefore a modern landscape Digital Elevation Model forms the basic data set for deriving these "natural pathways".

The "natural pathways" are defined by the natural constraints of traversal across a landscape, and in general, will be routes of minimum constraint. It is important to note that the concept of natural pathways does not explicitly imply their human use as a path, but only their potential to be used as such.

Review of Path Finding Techniques

Almost all studies on the extraction of paths from a landscape relate to finding optimal least cost paths between known points across a landscape. These techniques do not possess the non-a-priori knowledge characteristics to extract natural pathways.

In (Kweon In So and Kanade Takeo, 1994) the author described a methodology using contour trees, to extract topographical features such as ridges, peaks, valleys from a contoured DEM. However, in order to extract natural pathways using this technique, we would still need to define pathways in terms of particular topographical structures. This may be relatively easy for a mountain pass, but rather more complex for a pathway that may comprise of several topographical elements such as rivers, rolling hills, and ridges.

A variety of image analysis algorithms, such as skeletization, aim to extract structure from images. These could be applied to DEMs to extract valleys and ridges, but these suffer from similar limitations to topographical feature extraction.

By far the majority of formal methods in landscape studies using GIS, have been applied to hydrology and geology. It was while studying the online documentation within Arcview GIS, relating to hydrological modelling, that a natural pathway finding methodology began to be formulated.

Description of "Natural Pathway" Algorithm

The path finding process comprises several distinct stages and is an attempt to form a drainage network of a constraint or cost surface. The complete process is illustrated by a flow diagram (figure 1). The procedure and algorithms are identical to the extraction of surface runoff characteristics used in hydrological modelling. The essential difference is the nature of the input data. Whilst hydrological analysis uses a raw DEM to produce a stream model, the extraction of "natural pathways" uses a constraint surface model as input. All of the analytical stages are cell based, as the terrain models used in this research are grid-based.

Further information on the separate stages can be found in Arcview 3.2 GIS (spatial analyst & hydrological analysis v1 extensions). These discuss a notion of flow as they relate to hydrology. This nomenclature is used here but it is useful to think of flow, in relation to "natural pathways", as equivalent to potential movements of people.

The Generation of a Constraint or Cost Surface Grid

The input to the path network algorithm is a cost grid defining the relative frictions of human traversal through each grid cell that depends on the type of landscape component in that cell.

The desired constraint surface may comprise a number of landscape and/or perceptive elements such as;

- Rivers
    - Navigable rivers
    - Rocky mountain streams
- Lakes
- Open land (default factor)
- Dense woodland
- Boggy land
- Sea
- Visual quality from a point.

The quality of the constraint surface model depends on the weighting, attributed to each of these landscape features. The model could be tailored to suit certain modes of mobility. For example, if we assume the availability of boats and ferries for river movement then this may be built into the model. A sedentary farming society may impose different constraints on the landscape which may include territorial boundaries that confine movement. The constraint surface could also be generalised to use subjective attributes such as sacred spaces, forbidden areas or landscape attractors.

The generation of a good constraint surface is the most important step in the process. The subsequent stages of the algorithm are deterministic. It should be noted that the technique relies on the relative costs between areas and so absolute measures of cost are not needed.

The cost model used in the analyses presented in this paper, uses relative terrain values, or weightings given to each landscape attribute. The effort based cost equation as outlined in (Pandolf K B et al. 1977) is not needed, however the relation of cost to slope is still deemed valid for most types of terrain. Some work by Bell, T & Wilson A (submitted for pub[2]) suggests that the tangent of the slope is a more accurate measure of traversal cost, than slope. However, the natural pathway extraction process utilises relative costs and so the form of the cost function is not crucial. This allows us to derive a simple equation. As the analysis uses a grid based DEM, the equation is described using grid nomenclature.

The grid based cost equation used to generate the cost surfaces used in the subsequent analyses, is shown below.

$$[C] = \left(\sum_{t}([G]N_t[t]+ N_t[t])\right) + \left((1-\Sigma[t])N_l(G+1)\right)$$

Where:

- $[C]$ is a grid of relative cost values.
- $[t]$ is a binary grid. Value 1 depicting areas of terrain t.
- N_t is a terrain value for terrain t.
- N_l is the default terrain value for open land based on the DEM model.
- $[G]$ is slope for each cell in the [DEM].
- $\Sigma[t]$ is the sum of all the terrain binary grids.

This particular equation is linear for each terrain factor with a gradient of G. The relative values of N_t and N_l, are important, not their absolute values. If no other constraints are used, the equation reduces to: $[C]=GN_l+N_l$

It should be noted that further work may be needed to develop better constraint models perhaps incorporating perceptive landscape attributessuch as topographic prominence (Llobera M, 2001).

The Depressionless Cost Surface Grid

This is an iterative process and involves the filling of data sinks. A sink is a cell or set of spatially connected cells whose flow direction cannot be assigned one of the eight valid values in a flow direction grid (see below). This can occur when all neighbouring cells are higher than the processing cell, or when two cells flow into each other creating a two-cell loop. As the cell size of the data increases, the number of sinks in a data set often also increases.

To create an accurate representation of flow direction and therefore accumulated flow, it is best to use a data set that is free of sinks. This ensures that sinks created by sampling errors are filled, and paths do not drain into artificial sinks. Naturally occurring cost sinks may also occur. For example, a steep circular conical hill with a flat top would be a morphological sink which if filled would result in a circular pathway at its bottom but none on its top, which could be a valid pathway. Sinks are filled to the lowest cost "pour point".

The Path Direction Grid

One of the keys to deriving the path network is the ability to determine the direction of flow from every cell in the grid. This is calculated using a flow direction algorithm (Jenson S K and J O Domingue, 1988) which outputs a grid showing the direction of flow out of each cell. The direction of flow is determined by finding the direction of best gain, that is, to the lowest cost value. If the gradient is the same to all the cells then the neighbourhood is enlarged until the steepest gradient is found. If a cell has the same change in cost value in multiple directions and is not part of a sink, the flow direction is assigned with a lookup table defining the most likely direction (Greenlee D D, 1987). The behaviour of the flow direction algorithm, at the edges of the analysis area may be chosen. Cells at the analysis boundary may de deemed to flow inwards rather than to possibly undefined cells outside the area. (See Arcview Help discussion).

This flow direction function is also used as part of the process to create the depressionless surface.

The Path Accumulation Grid

This stage creates a grid of accumulated flow each cell, by accumulating the flow from upstream cells using the method described in (Jenson S K and J O Domingue, 1988). This is determined from the flow direction. Output cells with a higher flow accumulation represent areas of the landscape that have a higher probability of being potential pathways.

The Path Network

The results of flow accumulation are used to create a path network by applying a threshold value to the path accumulation grid.

The resulting, dendritic path networks can be further processed by ordering, identical to stream ordering, using any of the published algorithms (Shreve R L, 1966; Strahler A N, 2001).

Initial Test Results

Initial tests were applied to a series of synthetic cost surfaces to test the behaviour of the technique. These clearly showed the validity of the methodology and its limitations.

In order to provide a real test of the "natural pathway" extraction process, a digital terrain model of an area of approximately 430km^2, including sections of the Chiltern Hills, was analysed. The DEM was extracted from the UK Land-Form PANORAMA data set, and consists of height values at each intersection of a 50m horizontal grid, the values have been mathematically interpolated from the contours on 1:50000 maps. DEM height accuracy is no greater than one half of the vertical interval of the source contour data, (10m). The water features were extracted from the UK STRATEGI dataset, these are at a scale of 1:250000.

The area is also crossed by a long distance footpath, the Ridgeway.

The results of the tests are illustrated in figs 2-3.

Fig. 2 is a coloured representation of a DEM for an area that covers part of the Ridgeway long distance footpath. The graticule shows the National Grid Co-ordinates. The modern route is highlighted in black. The path network, in yellow, was created from thresholding the path accumulation grid to 1000 cells. A terrain value of 10 and 1 was used for water and land respectively.

A purely visual inspection shows that the approach is reasonable. It can be seen that paths in hilly terrain generally follow ridges. It can also be seen that the methodology produces many potential path networks in relatively flat plains. This is to be expected in the absence of any constraining terrain. It is probable that in prehistory, swathes of dense woodland or wetlands existed in these flat plains near to rivers. These terrain types would be areas of high cost. An estimation of the extent and nature of woodland and any other constraints would therefore be important in order to gain a better understanding of natural pathways in otherwise flat unconstrained areas.

The very high cost attributed to water produces paths that avoid crossing rivers. This model is therefore too simplistic as some of the rivers are slow moving lower sections, (usually navigable), but serves to test the behaviour of the technique and cost surface equation.

No account of other terrain costs has been taken. The modern Ridgeway path will also follow routes of low cost and deviate from areas of high cost, which became high in modern times, such as defined rights of way, agricultural land. The modern paths cross modern river bridges, which are low cost, artificial elements of the landscape.

Fig. 3 shows the same area as fig. 2, but in this case, a terrain value of zero for water was used to produce pathways that ignore the river constraints. Note that this produces different results for paths in the relatively flat areas, as these have numerous rivers. These paths frequently cross the rivers as expected.

Discussion on Methodology

There are several factors limiting the accuracy of the paths, created by using the "natural pathway" derivation technique.

The resolution of the DEM and the accuracy of the elevation data are a major limiting factor. Coarse cells can provide only average height data, and so may hide small areas of high cost which would affect path location. This will effect the minor paths in a network more than the high thresholded segments. The effect of cell size is lessened in areas of gentle topography.

The path finding method is very sensitive to the input constraints. The accuracy of the path networks depends on a realistic model of terrain costs. This cost model could be subjective, and limited by a lack of appropriate data. As mentioned previously, in the absence of other constraints, the method is robust to the slope function used for cost.

In areas of land with visible topography, the technique will locate natural pathways as defined by the absence of any other unknown constraints. However, in flat terrain other constraints would be needed in order to provide more meaningful pathways.

The river features need to be converted to a grid with a cell size equal to the DEM cell size. The PANORAMA data used in the previous analysis has a cell size of 50m, producing, in many UK cases, an unrealistically large river. If 2 river grid cells, join at the corners, the river grid may be incorrectly crossed by pathways across the data gap. This is an unavoidable consequence of grid based schemes, which can be improved by artificially thickening the rivers. This limitation should only affect the accuracy close to rivers.

The path direction grid has only eight possible values, which is simplistic, although the digital grid based nature of the DEM defines this limitation. A possible approach to improve this is by incorporating the parsimonious techniques described in (Douglas David, 1994) as a refinement to the flow direction algorithm.

The technique does not attempt to connect natural pathways together. The located paths will generally avoid high cost areas. A long distance path may comprise segments of natural pathways combined by the crossing of relatively small, high cost areas.

The accuracy of paths close to the edges of the area being analysed depends on typical boundary effects. For the tests described in this paper, the analysis area was a subset of a larger DEM. For DEMs that include sections of coastline, the sea needs to be correctly cost modelled, or the coastline must be defined as a boundary, otherwise the sea may act as a pathway drain.

The cost grid is currently isotropic and this characteristic may need further investigation.

Natural Pathways around Stonehenge: An Initial investigation into their Relationship to Prehistoric Ritual Sites

With due consideration to the limitations of the natural pathway extraction technique, an attempt was made to investigate potential patterns and relationships between the natural pathways around Stonehenge, and the surrounding ritual monuments.

An area of approximately 30km^2 around the site of Stonehenge was analysed using the path finding method. The DEM used was the Land-Form PANORAMA data from the Ordnance Survey, as described previously. This was the only constraint data utilised in this analysis due to the lack of appropriate resolution river data. The available river data set at 1:250,000, had evident misregistration with the DEM data. It was considered reasonable to leave out the river constraints in this area as there are virtually no minor rivers around Stonehenge. Instead, the larger area is notable by the presence of a few navigable rivers including the Avon.

An archaeological site database, provided by English Heritage, was used to investigate potential relationships. The archaeological data included information on; Long Barrows, Round Barrows, Henges, Causewayed Enclosures and Ring Ditches. Basic data on the cursus monuments and the Stonehenge Palisade was also included.

The results of the analysis are shown in figs 4 and 5.

Fig. 4 shows a coloured representation of a DEM for an area centred on Stonehenge. The path network denoted by a thick yellow line, was created from thresholding the path accumulation grid to 1000 cells. This is overlaid by a thin yellow line, which is the result of thresholding the path grid to 100 cells. A terrain value of one was used for land.

As discussed previously, It is possible that in prehistory, swathes of dense woodland existed in the valleys.

Fig. 4 shows a small section of the analysis area centred on Stonehenge. The path network, denoted by a thick yellow line, was created from thresholding the path accumulation grid to 1000 cells. This is overlaid by two other yellow lines, which are the result of thresholding the path grid to 100 and 20 cells.

Discussion of Results

Some possible relationships between the prehistoric ritual sites and the natural pathways can be speculated from a purely visual analysis of the images. One of these is the possible correlation of round barrows near the paths. To test if this potential relationship was non-random, a Kolmogorov-Smirnov analysis was performed.

For the 874 round barrows in the analysis area, the difference between the cumulative area distribution and the cumulative site frequencies at several distances from the pathways was calculated. The results are shown in Table 1. The figures are based upon pathways which have a been thresholded at 20 cells.

The difference between the two distributions was found to be significant at below the 5% level. Although this result provides evidence of some relationship, it does not provide causal information. In addition the test does not allow for any information on the structural distribution of the round barrows. For example, from a purely visual analysis, some of the round barrow clusters seem to be congregated at pathway cul-de-sacs.

It could be conjectured that many of the long barrows seem to have a similar alignment to major paths. However it has been suggested that many long barrows are aligned along ridges and so this may be the cause of the visual pattern. Since slope was the only constraint used for this analysis, the natural pathway technique will locate ridges and valleys.

An initial look at the results might suggest that the henge monuments seem to be closer to the major pathways. A major pathway heads off to the Northwest from the Stonehenge area in the direction of Robin Hoods Ball, another major prehistoric site. At Stonehenge, a spur off the major pathway to the Northwest, seems to lead to Stonehenge going through the palisade. It is known that a gap exists in the palisade structure, perhaps a gateway for the pathway?

The Stonehenge Avenue, interestingly has only a vague relationship to the paths, but its changes of direction are echoed in the major path some 300m to the North.

It is interesting to note that a small path starts at Durrington Walls, leaving near the known SE entrance, and goes to Woodhenge and just beyond, joining the round barrows nearby. Some of these initial observations will form the basis of a further, in depth study, of the Stonehenge area. Further analyses will utilise the temporal dimension to investigate the existence and nature of any chronological patterns.

Future Work

As mentioned previously, one of the main limitations to the accuracy of the path network is the quality of the constraint model. Therefore, more research will be directed to improve the model with the addition of other landscape and perceptive components. These are to include;-

- Dense woodland
- River flow direction
- River type (navigable-low cost) mountain river (high cost)
- Wetlands or Bogs.
- Sea
- Visual quality: low cost attributed to cells with rich visual affordances.

The path accumulation stage can also be combined with a weighting grid. Knowledge of settlement sites could be used to produce a weighting grid to enhance the path network, although this information is scant for prehistoric Britain.

The relationship between threshold level and Kolmogorov-Smirnov test results will be investigated in more detail.

The pathway approach will be attempted with a much larger DEM of the whole UK. Potential relationships between these pathways and an archaeological database of over 20000 ritual Neolithic and Bronze age sites, will be explored.

Conclusions

From the outset, it has been our intention to be able to study the potential correlation between natural pathways through a landscape, and prehistoric ritual sites in the UK. This led to a need to develop a technique to find such pathways, independently of site location. A method to extract "natural pathways" was formulated. The nature of the techniques, based on terrain cost, produces a network of potential natural pathways. These path networks do not exclude the possibility of other, perhaps minor paths, existing between path networks or across short stretches of high cost terrain. The higher order pathways are probably more reliable as indicators of major potential thoroughfares.

The results of the initial applications suggest that research into the relationships between archaeological sites and "natural pathways" is worth developing further.

The natural pathways process, provides archaeologists with a useful tool to investigate the landscape, that may improve the interpretation of site or artifact distributions. It may also allow archaeologists to provide structural meaning to inter-site landscape studies.

Furthermore, the approach is generic and could be used for exploring any constraint surface model.

Acknowledgements

I wish to acknowledge English Heritage for providing the archaeological data. The DEM data and river data have come from the Ordnance Survey via EDINA's DIGIMAP service. I would like to thank Proffessor Richard Bradley and Dr Vince Gaffney for their enthusiastic support and the University of Birmingham for supporting my participation at CAA2001. My thanks also go to Garry Lock, Tyler Bell and the many other people who commented on the presentation at CAA2001.

End notes

[1]Bell, T. and Lock, G. Information discussed by private communications with the author.
[2]Bell, T. and Wilson, A (submitted for pub) "Tracking the Samnites: landscape and communications routes in the Sangro Valley, Italy". Extract received by private communication from Bell, T.
[3]A recent paper by Marcos Llobera in "Beyond the Map, 2000" edited by Lock, G. explained to the author by Lock, G by private communication at CAA2001.

References

Bradley, R., Boado, F.C. and Valcarce, R.F. 1994. Rock Art Research As Landscape Archaeology - A Pilot-Study in Galicia, North-West Spain. In: *WORLD ARCHAEOLOGY* **25**, 374-390.

Douglas, D. 1994. Least Cost Path in GIS using accumulated cost surface and slope lines. In: *Cartographica* **31**, 37-51.

Greenlee, D. D. 1987. Raster and Vector Processing for Scanned Linework. In: *Photogrammetric Engineering and Remote Sensing* **53**, 1383-1387.

Ingold, T. 1993. The Temporality of the Landscape. In: *WORLD ARCHAEOLOGY* **25**, 152-174.

Jenson, S. K. and Domingue, J. D. 1988. Extracting Topographic Structure from Digital Elevation Data for Geographic Information System Analysis. In: *Photogrammetric Engineering and Remote Sensing* **54**, 1593-1600.

Kweon In So and Kanade Takeo. 1994. Extracting Topographical terrain features from elevation maps. In: *Computer Vision, Graphics and Image Processing:Image Understanding* **59**, 171-182.

Llobera, M. 1996. Exploring the topography of mind: GIS, social space and archaeology. In: *ANTIQUITY* **70**, 612-622.

Loveday, R. 1998. Double Entrance Henges-Routes to the Past? In: Gibson, A. and Simpson, D. (Eds.) *Prehistoric Ritual and Religion*, pp. 14-31. Sutton]

Llobera, M. 2001. Building past landscape perception with GIS: Understanding topographic prominence. In: *Journal of Archaeological Science 28* (9).

Pandolf, K.B., Givoni, B. and Goldman, R.F. 1977. Predicting energy expenditure with loads while standing or walking very slowly. In: *Journal of Applied Physiology* **43**, 577-581.

Parcero Oubina, C., Criado Boado, F. and Santos Estevez, M. 1998. Rewriting landscape: incorporating sacred landscapes into cultural traditions. In: *WORLD ARCHAEOLOGY* **30**, 159-176.

Shreve, R.L. 1966. Statistical Law of Stream Number. In: *Journal of Geology* **74**, 17-37.

Strahler, A.N. 2001. Quantitative Analysis of Watershed Geomorphology. In: *Transactions of the American Geophysical Union* **8**, 913-920.

Tilley, C. 1996. The powers of rocks: Topography and monument construction on Bodmin Moor. In: *WORLD ARCHAEOLOGY* **28**, 161-176.

Tables

distance range to pathway (m)	cumulative area of land (m2)	area (percentage of total)	cumulative number of round barrows	round barrows (percentage of total)	difference between cumulative distributions
0-50	358369984	0.416974	354	0.405034	-0.01194
50-100	526497488	0.612595	579	0.662471	0.049877
100-150	699492496	0.813879	745	0.852403	0.038524
150-200	792272496	0.921831	828	0.947368	0.025537
Total	859455000		874		

D_{max} is Greatest cumulative difference = 0.049877

Critical value for 5% level is $1.36/\sqrt{874}$ = 0.046003

Critical value for 1% level is $1.63/\sqrt{874}$ = 0.055136

Table 1.

Figures

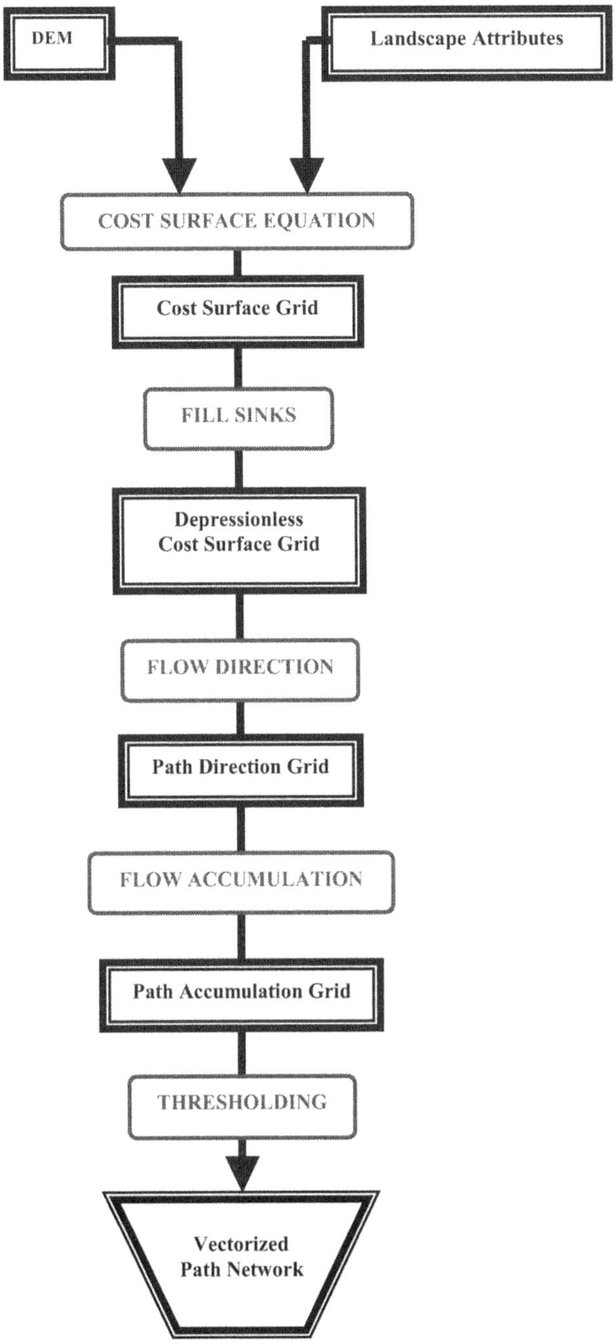

Figure 1. flow diagram of the "natural pathway" process

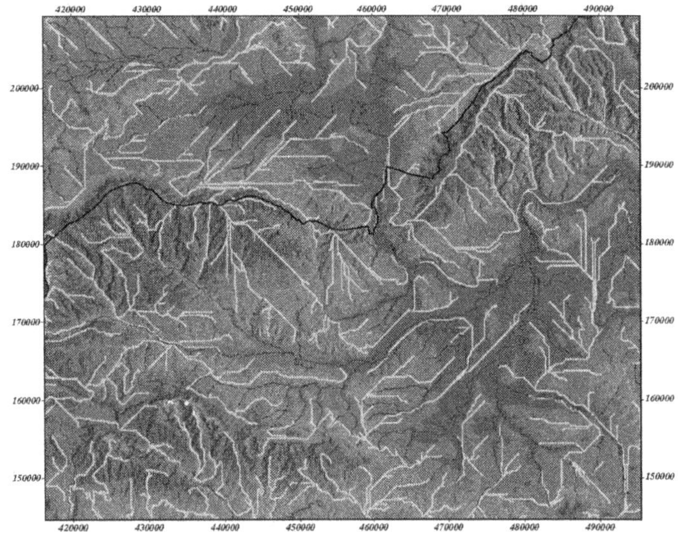

Figure 2: Natural Pathways in the Ridgeway area with river constraints

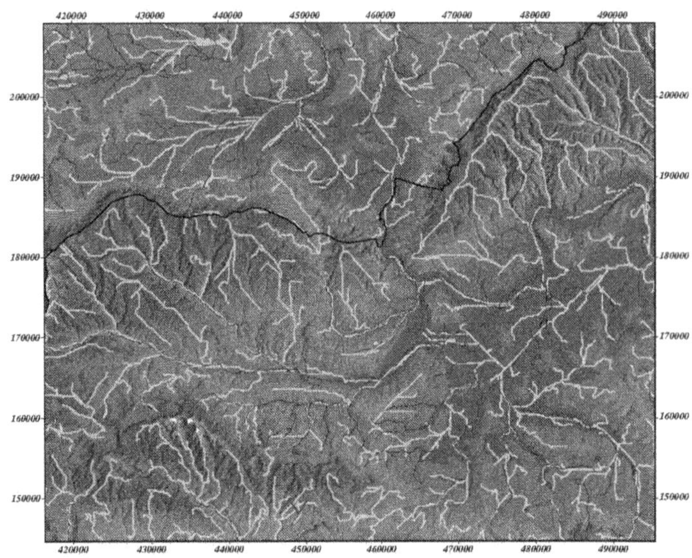

Figure 3: Natural Pathways in Rideway area, ignoring water constraints

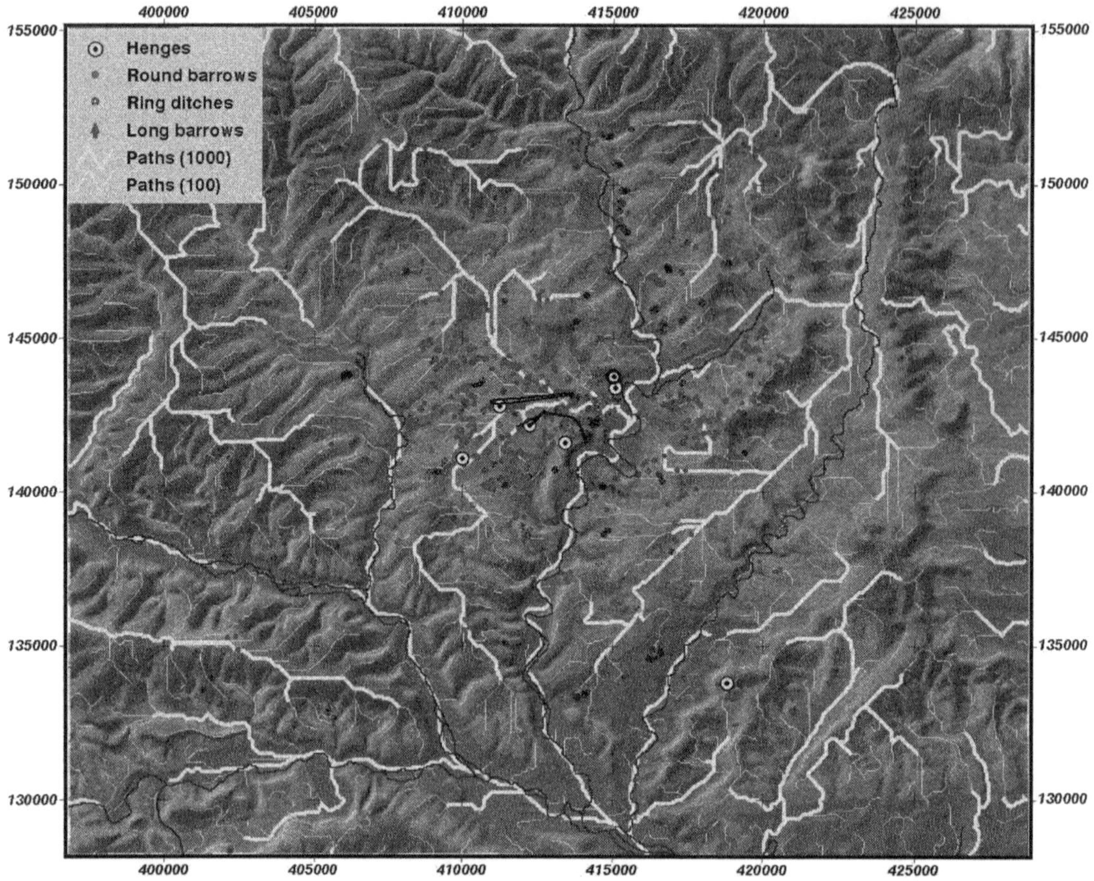

Figure 4: Natural Pathways in the larger Stonehenge area

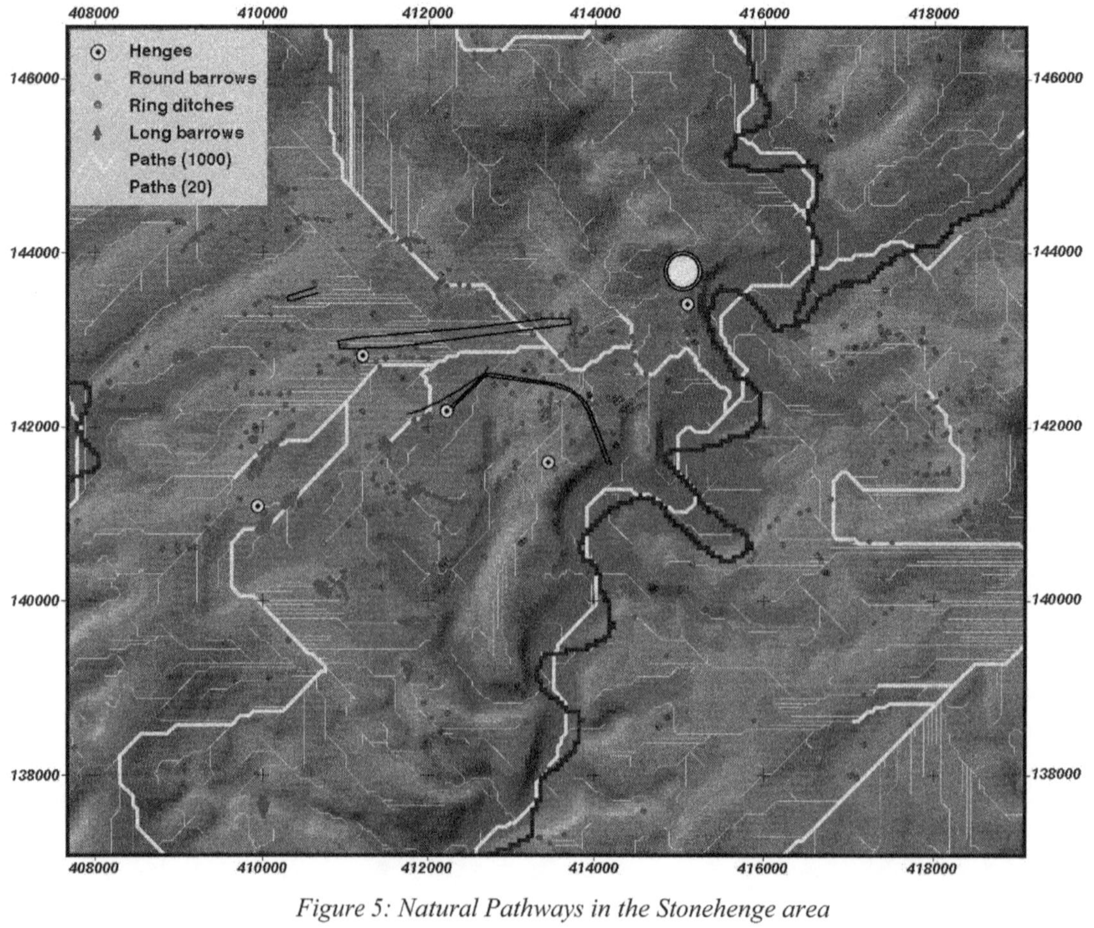

Figure 5: Natural Pathways in the Stonehenge area

2000 Years of Town Planning in Vienna

Wolfgang Börner
Geschäftsgruppe Kultur – Stadtarchäologie
Friedrich-Schmidt-Platz 5/1
1080 Vienna, Austria
E-mail: bor@gku.magwien.gv.at

Abstract: Around 2000 years ago, Roman legionnaires planned and built their military fortress upon today's city terrain according to a customary and trustworthy scheme. Due to the arrival of natives and Roman citizens, further settlements, which were constructed more or less systematically, came to exist in the whereabouts of this legionnaire fortress. On these Roman walls and on some sparse remnants from the early middle ages, the cosmopolitan city of Vienna was erected.

In cooperation with the Board for City Planning, strategies were developed to conserve the historical assets of this city for the coming generations. For this reason, a Land Registration Project for Cultural Assets, which is part of a Protective Zone Project, was elaborated for the City of Vienna. This land register has been available in the Internet since Autumn 2000 (http://service.wien.gv.at/kulturkat/). In addition to buildings worth conserving and pin-pointed protective zones, all known archeological sites of the city have been included in this GIS-application of the municipal authority.

Key words: Stadtarchäologie Wien, Cultural Heritage Land Register, GIS, Roman fortress, Gemeindeberg, Seniorarchäologie

Introduction

In a time when our cultural heritage is being destroyed by carelessness, greed, religious fanaticism and other evils at an increasingly alarming rate, it is absolutely necessary that not only the people from historic monument protection agencies and archeologists, but also city planners develop strategies to counteract these developments. To this end, it is important that these people be informed of the historical development of sites, settlements and subsequent cities. Archeology can not only be practiced in relation to a site or a time period, the entire cultural surroundings need to be observed and brought together in one unit. In this regard especially, the Stadtarchäologie Wien, in cooperation with various municipal departments of the city of Vienna, such as the EDP-department, City Surveying, City Planning and the City Archive, is attempting to pave new paths in archeology with the help of computer applications and the Internet. Only those cultural treasures that are known, can be protected.

Morphological and historical background of Vienna

Geographically Vienna can roughly be divided into 3 parts: the area north of the Danube River, the terraced landscape south of the Danube and the wooded hills of the Wienerwald in the west. The Vienna Basin provides an excellent environment for settlements and has therefore attracted settlers since prehistoric times.

The Danube served as an east-west-highway for water transport. Fords, with which the branched-out Danube could easily be crossed, provided an ideal north-south-communication.

The area south of the Danube with its terraces formed by the ice ages and rivers, in particular, was very inviting to settlers. The northern part of Vienna, the present day 20th, 21st and 22nd districts, is less favorable for settlements because of its morphological features (fig. 1). The area has no terraces and the height-differences are negligible. The constantly changing riverbed of the Danube before its regulation and the formation of new branches has largely destroyed all evidence of settlements.

It can not be said with certainty when the first settlers reached the Vienna basin, however, the first traces of settlement can be dated back to the Neolithic (Franz/Neumann 1965). Traces of early settlements can be found especially along the streams that led to the Danube River, such as the River Wien and the Ottakringer Stream, as well as in the valleys of the Wienerwald. These places were also sought out for settlement in the following time periods as well. The hills of the Wienerwald (Urban 1999), the southern region of the present day city and the region north of the Danube River were also settled (Holzer 1989).

Around 2000 years ago, Roman legionnaires planned and built their military fortress upon today's city terrain according to a customary and trustworthy scheme (Mosser 1998). There they found a relatively flat platform which was protected by brooks and rivers, and was therefore ideal for building a fort. Due to the arrival of natives and Roman citizens, further settlements, which were constructed more or less systematically, came to exist in the whereabouts of this legionnaire fortress. The civilian town (Zivilstadt) was built in today's third district (Mader 1998). Roman Villas and smaller towns could be found near further watercourses.

On these Roman walls and on some sparse remnants from the early middle ages, the cosmopolitan city of Vienna was erected.

Public Access to Heritage

The continuing development of the Internet, and associated technologies, is matched by an equally rapid growth in new business approaches to its use. In the Heritage industry this growth is dominated by a desire to make material more widely available and a corresponding necessity to find practical, efficient and usable ways to do so (Pringle 2001:329).

In the year 1993, the vice president of the United States of America, Mr. Al Gore, declared the "Super Information Highway" to an object of national concern. Only one year later, the European Commission labeled this era as the "Age of the Information Society". In summer 1999, the Austrian Government therefore introduced an initiative with the name of "Go on! Österreich ans Internet" in order to shape up the Austrian population and all Austrian enterprises in terms of these new Information-Technologies (Willenpart 2000:187).

The City of Vienna, on its part, has been attempting to reach its citizens and offer them services such as the "virtual information centers" since the year 2000 with its project "E-Vienna". An integration with cultural services and information is also planned for the near future.

It seems that we are now in a changing process comparable to the Industrial Revolution of the 19th century which had a deep impact on the economic and social structure of that period. Today, continued education and advanced training for the citizens have become a major topic for not only the different national governments but also on a European level. These new information and communication technologies are the primary cause for the momentary developments in the economy and society, as well as the means for surviving the rapid changes occurring in all branches of business. The Internet offers every organization, large and small, an opportunity to increase the dissemination of their information.

The Cultural Heritage Land Register

For the development of usable, computer-based, information systems, it is imperative that pre-implementation research is conducted into the available information, as well as the available delivery system. Understanding of any specific information set necessarily requires understanding of information at a slightly higher level, where it can best be described in terms of its part within a larger, user-centered process (Pringle 2001:329).

O´Brien and Polovina (O´Brien/Polovina 1997) give two definitions of information: the first suggests that information results from emphasizing salient features of data; the second, that information is the organization of data trends into a form that can be understood in a broader context and communicated as potential knowledge to others. In both cases it is essentially the interpretation placed on data that transforms or develops said data into information. Information by its nature, is primarily concerned with interpretation (Davies/Ledington 1991)

To provide prepared information to interested citizens, economical and political decision-makers, educators and subsequently to everyone who is interested in history and the cultural development of the City of Vienna through the ages, is the goal of the Cultural Heritage Land Register. This register of cultural treasures, which was created in cooperation with the municipal departments mentioned above, has been available in the Internet since fall 2000 (http://service.wien.gv.at/kulturkat/). The register contains information from the City Planning Department (objects worthy of protection and protected areas), the "Altstadterhaltungsfonds" (Downtown Preservation Fund) (Jugendstil buildings), as well as objects from the Stadtarchäologie Wien (archeological sites, map of the Roman legionary fortress and the Zivilstadt, and the Franziszeischer Land Register).

Technical Details of the Database for Cultural Heritage

Hardware:

- Web-Server, MapObjects-Server: NT-Workstation, Dual-Pentium, 350MHz
- Development computer MapObjects- / WEB-Application: NT-Workstation, Pentium 266MHz
- GIS-Data-Server: Unix-Workstation Bull Estrella 300 (PowerPC 604, 166MHz)
- GIS Arc/Info (GIS-Applications for processing data): Unix-Workstation Bull EscallaT (4 220MHz CPUs, 1Gb)
- ORACLE-Databank (Addresses, data concerning units such as information in order to pin-point findings and further more): Unix-Workstation Bull EscallaT (4 425MHz CPUs, 1Gb)

Software:

- Geographical Information System Arc/Info (from ESRI) for the processing of data
- MapObjects (ActiveX-Control) for the presentation of GIS data (from ESRI)
- ArcIMS as an Internet connection (from ESRI)
- VB in order to program the WEB-Application "The Cultural Heritage Land Register", using MapObjects
- ORACLE-Databank for address-search + storage of fact data of GIS data (Geometry)
- Web-Server: MS-IIS

Function of MapObjects-Server / WEB-Application:
VB-Program with embedded ActiveX-Control MapObjects and a WebControl (ArcIMS) as an Internet connection, functioning as a service (momentarily as NT but soon as Windows2000 in the "The Cultural Heritage Land Register"). WEB-Application is Frame-controlled and uses Client-side JavaScript (such as co-ordinate indication) and Server-side ASP-pages (such as the first page of the application). The fact data of the geometrical objects are to be found in the ORACLE-Databank – access to the Databank-Server possible via ODBC (SQLNet). Statistics of access are included in ORACLE .

Archeological Contents of the Cultural Heritage Land Register

The Cultural Heritage Land Register is still in development, and it should thus be noted that there are still some changes to be made regarding the application's user-interface and user-friendliness.

Archeological Content:

- Archeological sites: The archeological sites of Vienna that are linked with the so-called "site database" are marked with green triangles. A short description can be called up for each or the currently 1,375 sites which includes the address, year of the excavation/find recovery, the time period, description of the find situation and the finds. Further links are not yet available, but are being prepared.
- Maps of the Roman legionary fortress and the Zivilsiedlung: The maps of the Roman settlements, created with AutoCAD R14 and AutoCAD-Map 2000, were integrated into the application as Shape-Files. Expansion is also planned here for the near future as well. The ultimate goal is to put all of the excavation maps, fully prepared and with links to click for further information, into the Net.
- Franziszeischer Land Register: This land register, carried out by order of Emporer Franz I of Austria in 1820 (Fischer 1995), will be available for all of Vienna in digital form by the end of 2002. Its current form as a line graph will also be altered so that users will be able to click on objects in the Franziszeischen Land Register to receive information (Börner 2001:151f.).

Models

As mentioned in the section above, an expansion of the Cultural Heritage Land Register is planned and has already begun to some extent. This expansion will roughly follow two branches that will nevertheless cross paths again and again. One branch will be the intranet for the archeological colleagues of the Stadtarchäologie Wien, in which all the available information and analyses for each site will be prepared, and ultimately sent on to the second branch, the Internet. This data will then be complimented with data from other municipal departments such as the City and Provincial Archives, City Planning and data which has been gained from interested laypeople and students. The following describes three projects that, to some extent, are still in the planning phase. These projects use quite a few GIS-processes such as Distributional Analyses, Predictive Modeling und Viewshed Analyses (Constandinidis 2001).

Model 1: Citizens Write History Themselves!

For some years now, the "Senior Archeology Campaign" has belonged to the scope of responsibility of the "Stadarchäologie". This campaign was conceived in December 1994 within the framework of the research group Wiener Stadtarchäologie (Strohschneider-Laue 1998a), and was created to ensure that archaeology be available for everyone, and not just the professionals.

In contrast to the "Junior Archeology Campaign" (Strohschneider 1998b), "Senior Archaeology Campaign" is designed for those over 18 years of age. Thus the term "senior" is rather to be understood as an honorary title and not a reference to age.

These volunteer helpers support the work of archeologists through working at excavations, cleaning, labelling, and in some cases, restoring finds, (special training courses are offered at the "Volkshochschule" (community college) Meidling in the Vienna's 12th district (Kleinecke 1998)), as well as through help in the preparation of exhibits. The people who are interested and participate in the "Senior Archeology Campaign" are between the ages of 19 and 89, with the average age being between 50 and 70. The professional background and interests in archeology and history of the participants are as varied as their ages. Together with a group that is interested in the historical development of the City of Vienna, a project has been started entitled "Citizens write history themselves!" and has meet with great interest on a city political level. The above mentioned Franziszeische Land Register serves as a basis for this project. The following provides some background information as to the history of this land register.

It was under the rule of Emperor Franz I of Austria that the legislative and technical hindrances were cleared away in order to commence with big changes. One can truly say that it was due to his land tax patent from the 23rd of December 1817 that he founded the main land register of Austria. The basic ideas of this patent are still valid today. This land register, named after its originator, was also said to be a "stabile" land register because the net profit rate, which was crucial for the rating of taxation, was to be stabilized without giving consideration to higher productivity or diligence except to cases in which the fertility of the earth was destroyed by natural phenomenon. The land register was developed for the city of Vienna in the years 1819 to 1824.

Today, many churches, palaces, garden compounds, cemeteries and industrial areas can be found on the map which no longer exist.

The objects that have vanished often only remain as a street name. The question "What was here before my house was built" is only one of the many questions that this project should help to answer.

The historical development of the different districts of Vienna can be traced back by superimposing this land register onto the newly developed electronic city map of today's city. With the help of the "Senior Archaeology Campaign" and the support of the district museums, a database is being developed in which these structures are contained. According to this project, citizens interested in history will be able to gather historical data, such as literature, plans, pictures and photographs, about their home district and feed it directly into a database. The material will then be examined by historians and finally connected through the address code with the proper building site.

A simple click on a modern city map or the Franziszeische Land Register will then answer many questions concerning the historical development of the City of Vienna. It is easy to imagine the various directions in which this project can expand. It would be entirely feasible to include commercial interests as well, for example the history of traditional inns or cafés, of which quite a few still exist today, and to provide from this culturally historical link a link to the actual business homepage where the user could find addresses and even menus of the businesses. In a further developmental step, it is also planned to enlist the help of schools.

Model 2: Hidden Cultural Strata in the Area of the Roman Legionary Fortress Vindobona

A further project which is currently in the planning phase is the digital elevation model with predictive modeling characteristics of the Roman legionary fortress, which was located in Vienna's present day first district.

In this model, which was also started in the legionary fortress of Regensburg a fairly long time ago (Huber 2000), it is our aim to find out how far it is possible to make predictions concerning the structure of the Roman terrain in the region of the legionary fortress Vindobona.

The theory behind predictive modeling is that human behavior is patterned; therefore, locational behavior – that is, where sites or settlements are located throughout the landscape – should exhibit non-random tendencies. The basic theory behind the development of an archaeological location model is that if tendencies or pattern exist between site locations and one or more regionally distributed variables, then a model can be constructed through the exploitation of those tendencies (Brandt/Groenewoudt/Kvamme 1992:269).

The main aim of this archaeological topography is a reconstruction of the Roman terrain in the form of a map which is based on the determination and mapping of remaining cultural strata unto areas which have been disrupted by the construction of cellars and other forms of diggings.

Material origin:

- Site databse of the "Stadtarchäologie Wien", which among other information contains the elevation of sites.
- Cultural Heritage Database. This database also provides information as to whether a house has a cellar or not. Most of the houses in downtown Vienna have cellars, and sometimes as many as four, one over the other. Due to the lack of space before the city walls were torn down in 1860, it was necessary to store many things such as food, heating material, wine and much more in cellars. The servants also usually lived in the first level of the cellar. Many cellars were also extended, in true mining fashion, beneath the city streets. Thus it came to be that the cellars joined one another. This cellar system was used as a bomb shelter during World War II. The myth that it is possible to go from St. Stephan's Cathedral to Palace Schönbrunn is also most certainly based on the extensive cellar system.
- Building-age plans from the City Planning Department
- Drill-holes done by the Geological Department of the City of Vienna – through these drill-holes, of which there are around 1,000 from the area of the legionary fortress already, the thickness of modern and historical strata can be obtained.

The Roman surface in the area of the legionary fortress:
When settlements occupy the same area for several centuries, traces of buildings and activities may accumulate to such extent that original distinct patterns and structures become totally obscured (Holst 2001:159). Since the arrival of the Romans in Vienna's present day first district, the city has gone through many fundamental changes and many old building structures have been destroyed and rebuilt. Many of these renovation phases are undocumented. It wasn't until the 19th century that scientists became interested in the historical settlement structure of Vienna, which reaches back to the Romans. A true building boom was set off at the end of the 19th century by the fall of the city walls, which had restricted the expansion of Vienna until 1860. During this period of extensive building, many archeological remains were unearthed, and thanks to the first urban archeologist Josef Nowalski de Lila, they were in part documented (Stipanits 1998; Börner/Mosser 1996).

Initialized by the construction of new buildings, works for the channel system, underground garages and other strata disturbing measures, historically precious objects have been recovered and some arranged archeological excavations have been accomplished. The results, which have been collected in the form of lay-out plans, strata drawings, profile drawings and descriptions, excavation diaries and often enough simple sketches and draft-mappings, are being evaluated while keeping the following questions in mind:

Was the bottom of the Roman strata reached during the excavation? If the bottom stratum was reached, how far above sea level was it?

If the bottom of the last stratum was not reached, other criteria will be necessary in order to establish the bottom height none the less. According to experience, the first Roman buildings are to be found just above the upper natural strata. The upper edge of the natural strata could therefore be useful as an indicator for the bottom edge of Roman settlement. If for example the results of an excavation were a Hypocaustum, one could expect from experience that the object was built around 0.05 m deep into the natural strata.

The next step will be to synchronize the mapped objects of the legionary fortress with the patterns determined from the samples taken out of drill-holes as well as the land register for cellars. Luckily the conduit-maps are already installed in our computers, making it easy to scrutinize the information in this regard as well. With all these mappings, a digital terrain model can be established showing Vienna's Roman topography.

With the help of this model, it might indeed be possible to predict in certain circumstances whether or not planned construction work might encounter Roman strata (Börner/Öllerer 1998).

Model 3: Viewshed-Analysis of Prehistoric Settlements

In the last decades the prehistoric age has not been treated with the necessary interest. Lately the attitude has changed, even though it is still very hard to develop a settlement scheme for each period in a city as densely settled as Vienna, the federal capital of Austria. As a substitute we can, however, present the early Neolithic period settlement on the so-called "Gemeindeberg", meaning community hill, in the "Wiental", the name referring to the valley in which the river "Wien" flows into the 13th district of the city of Vienna.

Along the "Wiental", excavation findings from early history, prehistoric, Roman and middle age settlements are densely lined up one after the other. The "Gemeindeberg" (320 m above sea level but only 164 m in relative height) is part of the Jura-cliffs which suddenly appear between the regions of "Lainz" and "Ober-St.Veit" on the edge of the sandstone-zone in the 13th district. Since the sloping terrain drops towards the northwest, northeast and southwest quite steeply and the northwest hillside is flanked by the river mentioned above, this place is of special strategic value for its inhabitants. A further advantage found here is the abundance of Jasper and Flint (Chert).

The first excavation was done in the years between 1880 and 1924. It seemed then, that the complete northern hillside had been settled. The huts, which had been partially lowered into the earth itself, showed remains of one or two stake holes reinforced by several stones. The quantity of the findings too, where opulent.

This project was carried out in cooperation with the Surveyor's Department of the City of Vienna. The first step was the creation of a digital elevation model for the excavation area, were, as will be discussed later, the first difficulties arose.

Since quite a few Neolithic places have been discovered in the "Wiental", it is hoped to use the very controversial "Viewshed-Analysis" to clarify eventual correlations between the sites. This tool is closely associated with the human capacity of sight, and together with relative ease of application, results can be achieved which have clearly appealed to many different archaeological research agendas over the last decade (Tschan/ Raczkowski/ Latalowa 2000). The problem with this kind of analysis for Vienna and the surrounding Vienna Woods (including the Hallstatt and La Tène aged settlement on the hill "Leopoldsberg" including the surrounding hills), is that only very few topographical maps actually exist, and the few existing maps can only be evaluated very poorly by a GIS system such as IDRISI.

By the end of the year 2002, however, the city limits of Vienna will have been surveyed with a 3D-Laserscanner from the air (Dumfarth 2001), a process which has already proven successful during its first trials in the town of "Klosterneuburg".

3D-Laserscanning:
3D-Laserscanning functions by means of cooperation between a laser and a sensor. To describe it in general, the investigated object, such as a house, a cliff or a street, is scanned by a net of impulses from a laser. The impulses are reflected by the investigated object and immediately recorded. The position of each point of the object to the scanner is calculated by the time the signal needs and the angle of the transmitted impulses. Each second, up to 6,000 points can be surveyed allowing the recording of a large area in only a few minutes. The degree of accuracy varies by only several millimeters or centimeters. By including 4 geodetically measured coordinates, the entirety of 3D points collected can be transformed into any geodesy coordination system needed.

- maximal reach:
  2 to 350 m for natural objects with a reflectivity of approx. 50 %
  until 150 m for natural objects with a reflectivity of approx. 10 %
- minimal reach: according to type; 2 m
- Precision of the survey: 3 according to type +/- 2.5cm
- Scan density: 42.5 cm
- Velocity: up to 6,000 points per second

With this system it is also possible to fade out interfering buildings or vegetation.

Result:
With the raster data currently available, an initial analysis was made, whereby it became clear that in addition to the very incomplete topographical maps of Vienna, it would be necessary to include data from older map materials that to some extent also included topographical information and to include them as well as aerial pictures in the analysis. A truly exact analysis will not be possible until the 3D-Laserscanning has been carried out.

Prospects and Further Application Possibilities of the Cultural Heritage Database

The possibilities that the Cultural Heritage Land Register has to offer are manifold and open to virtually every area of archeology and cultural history. Subsequently, a web portal is being planned for the culture and history of Vienna. The Cultural Heritage Land Register will also serve as a basis for this portal. The number of hits in the Internet alone shows that the public is greatly interested in this application. One week after the application was brought on line, it was receiving 12,000 to 15,000 hits a week, currently the number of hits per week fluctuates between 4,000 and 8,000. The Stadtarchäologie Wien sees in this application a possibility to reach the general public, students, decision-makers, city planners and tourists and offer them and introduction to the history and culture of Vienna, thus raising the acceptance level of the population for our profession.

By incorporating modern technology such as Virtual Reality or GPS-systems, a wide range of uses are created for the application. Who would have thought even a few years ago that it would be possible to tour Vienna with a computer in their hand and be able to plan an individual sightseeing route? I am sure that not even Chenhall imagined such things when he wrote down his vision 31 years ago:

"When the anthropologist begins a research project his first step will be to query the databank, through a remote terminal (probably located in his office) in order to obtain a printed list of all references on the topic or a listing of ethnology or archaeology of a specified area, together with whatever other data he may think he will need in carrying out his project" (Chenhall 1971, 159).

This project has a long way to go and a number of as yet unknown hurdles to overcome before it is complete, but nevertheless one of our duties, in addition to our daily work, must be to keep the general population informed of the developments in our field. Information is for everyone and should be shared, but information is only interesting if it is prepared and presented correctly.

References

Börner, W. 2001. Vienna Archaeological GIS (VAGIS): A short Outline of a New System for the Stadtarchäologie Wien. In: Z. Stancic/T. Veljanovski, *Computing Archaeology for Understanding the Past (CAA 2000), Proceedings of the 28th Conference, Ljubljana, April 2000,* BAR International Series 931:149-152.

Börner, W./Mosser, M. 1996. Von der 100-jährigen Karteikarte zum Plan des Legionslagers. In: *Forschungsgesellschaft Wiener Stadtarchäologie (ed.) Workshop 1 – Archäologie und Computer:9-20.*

Börner, W./Öllerer, Ch. 1998. Der „Archäologische Kataster" im Rahmen des Schutzzonenprojektes der Stadt Wien. *FWien 1*:64-66.

Brandt, R./ Groenewoudt, B. J./Kvamme, K.L., 1992. An experiment in archaeological site location: modeling in the Netherlands using GIS techniques. In: *World Archaeology Volume 24 No. 2, Analytical Field Survey*, Routledge:268-282.

Chenhall, R.G. 1971. Computers in anthropology and archaeology. New York: *IBM Coporation Publication No. GE 20-0384*, 156-161.

Constandinidis, D. 2001. Introspective Sitescaping with GIS. In: Z. Stancic/T. Veljanovski, *Computing Archaeology for Understanding the Past (CAA 2000), Proceedings of the 28th Conference, Ljubljana, April 2000*, BAR International Series 931:165-172.

Davies, L.J./Ledington, P.W.J. 1991. *Information in Action*, Macmillian, England.

Dumfarth, E. 2001. Terrestrische 3D Laserscanning: Real World... Scan data...3D-Modell.. In: Schrenk M. (Hrsg.), *Beiträge zum 6. Symposium zur Rolle der Informationstechnologie in und für die Raumplanung*, Wien:165-169

Fischer, K. 1995. Streiflichter auf die Wien-Kartographie der letzten beiden Jahrhunderte. In: Fischer K. (ed.), *Das ist die stat Wien*, Wien, Wiener Geschichtsblätter, Beiheft 4:38-45.

Franz, L./Neumann, A.R. 1965. Lexikon ur- und frühgeschichtlicher Fundstätten, Wien.

Holst, M.K. 2001. Formalizing Fact and Fiction in Four Dimensions: A Realtional Description of Temporal Structures in Settlements. In: Z. Stancic/T. Veljanovski, *Computing Archaeology for Understanding the Past (CAA 2000), Proceedings of the 28th Conference, Ljubljana, April 2000*, BAR International Series 931:159-163.

Holzer, V. 1989. *Die latènezeitlichen Siedlungs- und Gräberfeldfunde aus Wien XXI und XXII (Leopoldau und Aspern)* [unpubl. Diss Univ. Wien]

Huber, I. 2000. *Verborgene Kulturschichten im Bereich des römischen Legionslagers Castra Regina*, Regensburg.

Kleinecke, G.. 1998. *Wenn Tonscherben wieder zu Gefäßen werden: Keramikrestaurierung mit der Unterstützung von SeniorarchäologInnen.* FWien 1:48-51.

Mader, I. 1998. *Die EDV-gestützte Erfasssung der archäologischen Fundstellen im 3. Wiener Gemeindebezirk.* FWien 1:89-97.

Mosser, M. 1998. *Das Legionslager Vindobona – EDV – gestützte Erfassung alter und neuer Grabungen.* FWien 1:74-78.

O´Brien, S./Polovina, S. 1997. Web Design and Information Architectur, *Proceedings of the Sixth International World Wide Web Conference, Santa Clara, California, USA, April 7-11, 1997.*

Pringle, M.J. 2001. Using Virtual Reality to Imporove Public Access to Heritage Databases over the Internet. In: Z. Stancic/ T. Veljanovski, *Computing Archaeology for Understanding the Past (CAA 2000), Proceedings of the 28th Conference, Ljubljana, April 2000*, BAR International Series 931:329-337.

Stipanits, U. 1998. *Über 100 Jahre handschriftliche Fundmeldungen und ihre EDV-gestützte Erfassung.* FWien 1:67-73.

Strohschneider-Laue, S. 1998a. *Über die Unbezahlbarkeit freiwilliger Helfer.* FWien 1: 38-43.

Strohschneider-Laue, S. 1998b. *Initiative Juniorarchäologie oder: Wieviel Archäologie braucht die Schule.* FWien 1:44-47.

Tschan, A./Raczkowski, Wl./Latalowa, M. 2000. Perception and Viewshed: Are the mutually inclusive? In: ####

Urban, O.H. 1999. Der Leopoldsberg. Archäologische Forschungen auf dem Wiener Hausberg. In: *Wiener Archäologische Studien, Band 2 (Hrsg. O. Harl)*, Wien.

Willenpart, B. 2000. Stadtplanung im Internet – Möglichkeiten internetgestützter Bürgerinformation und Bürgerbeteiligung in österreichischen Stadtplanungsämtern. In: Schrenk M (Ed.), *Beiträge zum 5. Symposium zur Rolle der Informationstechnologie in und für die Raumplanung.* Wien:187-192.

Figures

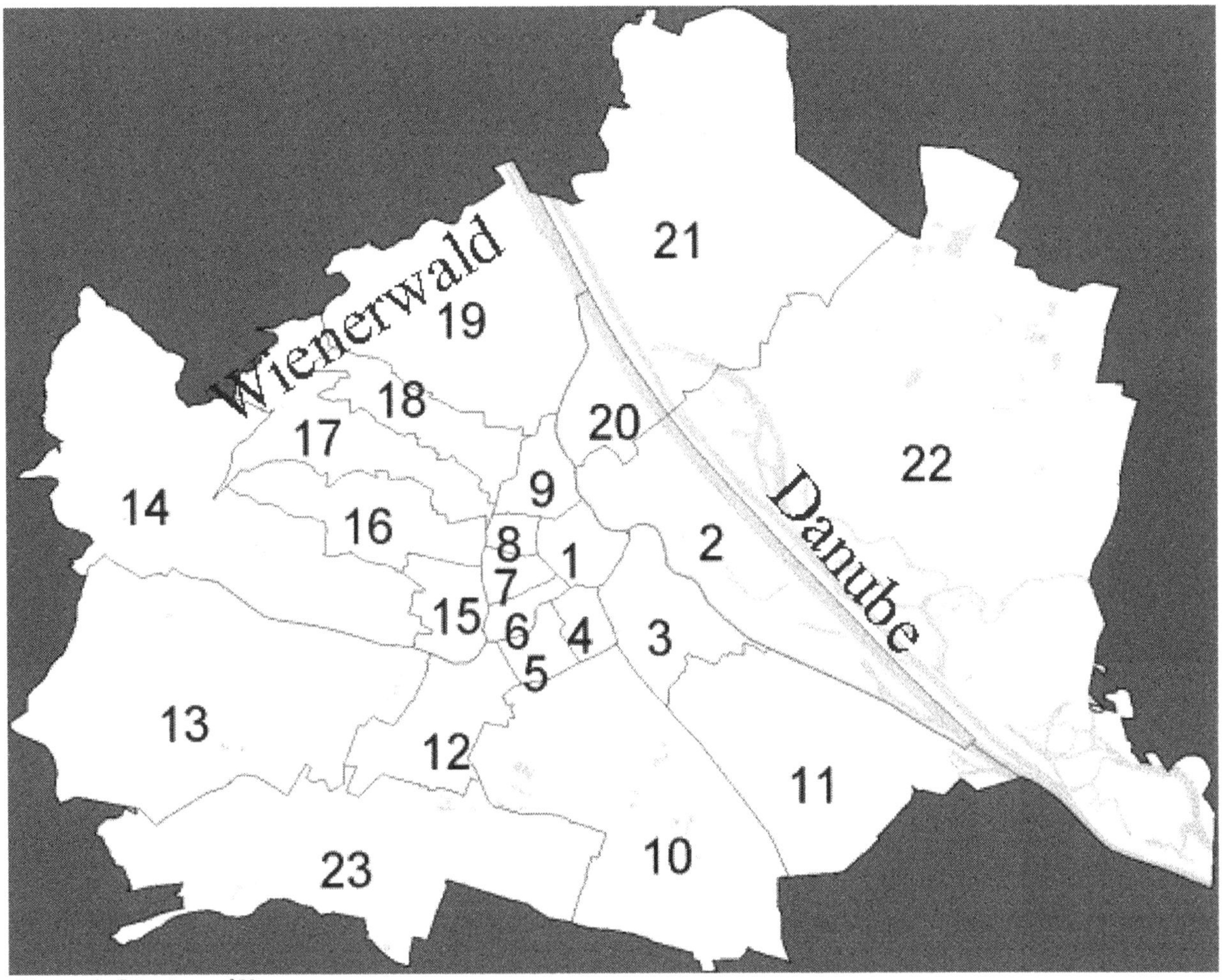

Figure 1. Districts of Vienna

A 5000-Years History of Settlement and Irrigation in the Murghab Delta (Turkmenistan).
An Attempt of Reconstruction of Ancient Deltaic System

Barbara Cerasetti
Department of Archaeology
University of Bologna
piazza S. Giovanni in Monte 2
40124 Bologna, Italy
Phone: +39 051 2097700 - Fax: +39 051 2097701 - E-mail: cerasetti@hotmail.com

Abstract: The present paper introduces the main problem of the water lack in the Murghab delta during the ancient time and the fundamental position occupied by GIS applications and Image Processing for the reconstruction of the ancient irrigation system. A vast collection of data has been gathered and organised in a Geographical Information System to prepare thematic mapping and 3-D visualisation, illustrating the settlement and the environmental history of the Murghab delta from 3000 BC to the Achaemenid period (550-300? BC).

One of the final aims of the research is to realise a Multidimensional Geographical Information System, capable of representing the dynamics of the evolution of such a complex and multitemporal-layer reconstruction of a large-scale model of geoarchaeological landscape in Central Asia

Key words: Turkmenistan – Murghab – River - Geographical Information System - Digital Elevation Model

The alluvial lowlands of Asia have long been recognised the heartland of early civilisation, expressed by high concentrations of settlements and agriculture production. The early states and the pristine agencies of political complexity developed in these vast open corridors from the economic wealth that grew throughout Chalcolithic times within the villages and kinship structure, born by the Neolithic Revolution. The first cities, where the outputs of the second leap forward in political evolution, made the social and economic breakdown of the earlier agriculture communities.

By 1966 the New Archaeology Movement advocated the priority of settlement studies for the understanding of the social and economic history (Binford, Binford 1968). The study of the human settlement provided new powerful means to read the historical fluctuation on man-made landscape (Chang 1972). The systematic study of ancient settlement in the cartographic definition was based on thematic maps, seriated according to chronological periods. Archaeological maps became an essential product to establish the quality of archaeological projects, creating a widespread demand for high definition analysis of special distribution and hierarchical rending of human settlement. The authors of the first chronological thematic maps, based on the study of ancient settlements, were Gordon R. Willey (Willey 1953) in the Virù valley in Peru and Robert Mc C. Adams (Adams 1981), working in the Diyala region in Iraq. Personal computers and specific software changed the methodology of archaeological research since the American and European specialists requested more complex elaboration and representations.

The main problem concerning the big rivers of Central Asia is the lack of sea way out (fig. 1). The deltaic branches flow across the Kara-Kum desert and the hyperarid climate causes the evaporation and the underground infiltration of the river water. In the Murghab delta the terminal lake of Dzhar, located on the north-westernmost part of the system, represents the only collecting point of the river water as represented by the Russian topographical military map[1] (fig. 2). On the '800 map (Lumsden 1885) (fig. 3) the Aina-gel, the present fossil lake on the western side of Dzhar on the 1995 maps, represents the second collecting point of Murghab water. Merv and the alluvial plain of the Murghab delta have been massively destroyed for extensive land reclaims and irrigation works from the Kara-Kum Kanal, feeding in the area the Amu-darya waters. In face of the rapid deterioration, the authorities of Turkmenistan have invited foreign projects to document the antiquities of the delta. The joint project "The Archaeological Map of the Murghab Delta (AMMD)" initiated in 1990 as a co-operative effort of Italian, Russian and Turkmen specialists[2]. A large spectrum of specialists and technicians of different nationalities and skills created a vast archive of data by intensive ground reconnaissance, low altitude flights and by adopting state-of-the-art technologies for image processing and field.

In order to preserve this unique aspect of cultural heritage of Turkmenistan, a lot of work still needs to be done in spite of the intensive research carried out in the past ten years. The urgency for the project was mainly dictated by the looming destruction of the sites for the expansion of cultivation in the northern part of the Murghab delta. Considered the more intensive series of field observations, we organized and administered the enormous quantity of cartographic and archive data so far collected by entrusting them to an interactive system like Geographic Information System (GIS). The GIS has been used like an instrument to assemble and manage the vast arrays of diversified archives to be merged in a historical analyses of settlement and population. Satellite imagery and the digitalisation of aerial photographs in general are providing another mean for the definition of ancient landscape, before and beyond direct fields of observation. Only recently, this new important instrument has been applied to a large-scale analysis in the alluvial plains of the Middle East and Central Asia.

In addition to the historical and archaeological data of Turkmenistan, the satellite survey of the region has enabled the use of new analytical methods. This type of supports is used on large scale in the archaeological project to define the ancient landscape (Gardin, Gentelle 1976; Adams 1981; Forte, Mozzi, Zocchi 1998). We therefore have directed our research towards the interpretation of the digital and cartographic data on a geoarchaeological basis. It is especially interesting in this case by using the comparison of the geomorphological landscape (fig. 4) and the situation of the ancient sites, taking the Digital Elevation Model (DEM) imagery as a background (fig. 5). The DEM is the digital representation of continuous changes of relief within space (Burrough 1986: 39; Gaffney, Stanchich 1996: 23) and any elevation variation in the territory of delta may be a trace of a palaeochannel presence. V. Castellani maintains that the irrigation system of delta has certainly changed[3]. The river cut across the plateau and left a detritus conoid, depositing again in the same cut and flowing through its deposits. It suggests the remarkable shrinkage of water supply, and the absence of river grid on aerial photographs confirms the low rain level of the delta in the past. Probably the ancient settlements stocked up the water directly from the river channels, confirming the reason of archaeological site location along the palaeochannels. In the second time the growing climate aridity and the spring dryness drastically transformed the landscape and the settlement pattern over the territory.

Considering the available information, we prepared a Vectorial Cartographic Digitalisation of historical maps on different layers. The main aim is to reconstruct the ancient irrigation system before the building of Kara-Kum canal and to obtain the original river for understanding the historical process determining the present delta. This is fundamental for a comprehension of settlement pattern distribution of archaeological sites, during the different chronological periods, into the ancient landscape. A careful analysis of topographical, geographical, historical, environmental and geological maps permitted to reconstruct the water supply history of Murghab river from late Chalcolithic to Achaemenid period.

The GIS software used in the application ArcView 3.2 allows many different data in thematic maps and complex spatial analyses, integrating information from infinite levels of subdivision recording. The transformation of archaeological map of the Murghab Delta into the GIS has presented various levels of difficulty (Cerasetti in press). The principle problem derives from the large size and high diversification of the data to be organized.

The first application of the GIS is concerning the thematic mapping of the site discoveries according to chronology and size. The set of graphs has been produced as well to provide a first level of representation for the change of population distribution (fig. 6). Massive alluvial sedimentation has buried the earliest settlement history of the Murghab lowlands. The stability of the Murghab delta has left little opportunities to detect from the surface all the Neolithic and Chalcolithic sites. In general from the middle Bronze Age onward the surface evidence represents a reliable reading-key of the archaeological situation and the data coming from the survey may be utilised with minimal constraints. As Italian specialists have well evidenced from pedological analysis and geomorphological observation, the early and middle Bronze Age original plain was contracted and cut across by the southward movement of aeolian sands that determined a chromatic change in settlement pattern. The regularly patterned settlement distribution of the early Bronze Age fertile plain changed into clusters of small oases aligned along the active watercourses (fig. 7). The relative map shows the sites of this period with the main centres located at Gonur North and Adzi Kui 8, developed in close connection with a main deltaic canal, feeding a network of natural and artificial water-courses (fig. 8). During the late Bronze Age period we record a sharp increase of the site number across the whole alluvial plain and the southwards expansion of the settlement distribution. In the following periods the settlement pattern is more coherent and the process of consolidation of a state organization is definitive. The ancient Margiana has become a region of far greater and more complex political system.

As I. Rouse said (Rouse 1972: 1), the archaeological record spatial distribution testifies the cultural and social human activities and the relations among ecological, cultural and social systems It is fundamental to understand the adaptation process into a surrounding landscape and the exploitation of available natural sources. In our case the methods of river water exploitation represent the reading key to understand every settled phenomena during Bronze and Iron Age. The site distribution maps confirm the subservience from the main water-courses from 2400 to 300 BC. The site location along the palaeochannels and the archaeological surface material allowed us to date the main water-courses inside the relative irrigation systems.

One of the main aims of our research concerns the chronological sequence reconstruction of the main irrigation systems, elaborating the data of the river morphological evolution by means a GIS application. The global visualization of the settlement pattern from middle Bronze Age to late Iron Age, connecting to the ancient deltaic systems allowed us to know the waterworks adopted in different periods for the settlement subsistence. We have reconstructed the three main systems Dzhar lake, Murghab river and Sultan-ab systems[4] (fig. 9), working approximately at the same time from the end of IV to the end of I millennium BC. The river water exploitation is a no-artificial process till Achaemenid period , characterized by the construction of the main dams. A detailed surface recognition of the area surrounding the great dams as Great Band and Sultan Band, located along the main streams of Murghab could make us able to understand better the river supply projects during the well documented historical times: these enormous waterworks were certainly supported by a central power, very well organized and capable to face the incessant problem of river regression. Mainly for this reason we dated at this period the great dams: the consolidation of central power allows economical sources to erect this type of public structures. The construction and the regular maintenance of the channels entail a great human energy employment and the surviving depends on a complex administrative and executive system. More arduous is the attempt of the main irrigation system reconstruction during the pre-historical times, considering the absence of direct sources and artificial structures. In this case the Geographical Information System has been fundamental.

At the end of IV and the beginning of III millennium BC the portion of territory between two sub-deltas of Aravalli and Merv oasis was probably occupied by alluvial swamp of Dzhar lake, 82 kilometres north-extending (fig. 10). The area is completely lacking in archaeological surface material and the middle Bronze and Iron Age sites are located along natural palaeochannel, bordering the east side of the alluvial swamp. The Dzhar lake and the surrounding landscape represented the highly important source of food, as fishes and birds, and fresh water. The Russian definition of *Dzhar*, the Persian *Jar*, means "verdant area, grassland", suggesting the lake as vital source in the ancient time and it is strictly linked with Murghab or *Morq-ab* definition (bird water), emphasizing the fundamental importance of the Murghab system[5]. The relative chronological thematic maps show that the region bordering the Dzhar alluvial swamp has been populated till the half of I millennium BC, characterized by a progressive regression of the settled area border.

The conformation and the chronology of the systems are identified according to the site distribution along the palaeochannels (fig. 11). The position of the main sites reinforces our hypothesis: the middle Bronze Age site of Gonur North is located along the same channel 2 of the final Bronze Age site of Takhirbaj 3, remaining the main water-course till the end of early Iron Age (fig. 12). The channel sides are settled less intensively during the second and the third phase of Iron Age, stopping its activity during the last phase. The water flow is going down and only 3 hectares sites last during late Bronze Age. Takhirbaj 3 is located at 20 kilometres distance from the main middle Bronze Age centre and it is gaining control over the territory during final Bronze Age. It means that during 600 years the river flow is considerably regressed and a regulating plan of the river water becomes urgent. At the beginning of the Iron Age the water decrease of channel 2 causes the advantage of channel 3 and the increasing of settlement distribution along its sides. During the second and the third phase of the Iron Age the channel 2 is practically abandoned and the channel 3 takes the upper hand, intensively exploited till the last phase. Differently from the first hypothesis[6] we suggest that the beginning of Sultan-ab system exploitation starts during the first phase of the Iron Age, confirmed by the GIS application. The main channels of Dzhar and Murghab river systems[7] are working at the same time from the middle Bronze Age to the third phase of the Iron Age: probably the intensive flow water allowed two systems working for a long time.

During the last phase of the Iron Age, with a consolidation of Achaemenid central power two impressive fortress lines were erected along the northern-eastern side of the Murghab delta (Genito 1998: 125, fig. 1) (fig. 13). Our hypothesis is that the defensive structures mainly guarded the Sultan-ab and Dzhar irrigation systems: the water lack must be the main problem for the subsistence during a period characterized by a remarkable increase of population. The problem of water sources forced enemies to try taking the control of the main courses of the Murghab river for a territorial control of the Margiana region. The more advanced line probably defended the Sultan-ab system, while the second line guarded the Dzhar lake system and Yaz-depe, the main centre of the region during the last phase of the Iron Age.

In our experiments of 3D analysis and visualization of the vectorial and raster data, we have exploited for the first time the potentialities of the software ER Mapper 6.1, realizing a 3-D model of the Murghab delta. It has been possible to visualize also different elaboration of the DEM on different colour types, producing a 3-D multilayer projection to obtain a virtual landscape as real as possible.

This multidisciplinary project implements advanced elaboration techniques both for multispectral images and for multitemporal-layer reconstruction of a large-scale model of geoarchaeological landscape in Central Asia. Soyuz KFA 1000 and ERTS-1 have been used in the first time to point out the beginning process of desertification. The ERTS, taken on 1972 only 30 years later the construction of Kara-Kum canal, testifies the regression step of the Murghab delta and the present irrigation areas (fig. 14). The CORONA 1961-1974 satellite imagery have disclosed new possibilities for the interpretation from the thematic mapping, since the major destruction of sites proceeds because of large-scale irrigation projects (fig. 15). It will also facilitate a spatial analysis between the evolution of the archaeological landscapes and the geomorphological characters of the region around Merv, opening new perspectives of research analysis in Southern Turkmenistan. In a second time we will add and overlap infinite information layers to the 3-D model and we expect that simulation will be able to advance on new pathways of higher complexity. We are still far from an exhaustive analysis of the data, but we are moving the first steps towards multimedia and 3-D publication of the elaborated data. The evolution and the integration of present applications in the GIS might represent a significant step for realizing a new 3-D addition of the digital archaeological map of Turkmenistan, containing consistent reconstruction of the ancient population.

End notes

[1] The cartographic base map of Geographical Information System of Murghab delta is represented by a mosaic of fourteen 1:100,000 topographical military maps, dated on 1995 allowed us by *Turkmenskoe Aerogeodjicheskoe Predpriyatie* of Ashgabat.

[2] The Scientific Directors of the Project are: Prof G.A. Koshelenko, Director of the Section for Classical Archaeology of the Institute of Archaeology of Russian Academy of Sciences (IARAN) in Moscow; Prof A.G. Gubaev, Dean of Faculty of Oriental Studies of State University of Turkmenistan (TSU); Prof M. Tosi, Scientific Supervisor of AMMD of Italian Institute for Africa and Orient; Prof B. Genito, Istituto Universitario Orientale of Naples.

[3] The hypothesis is a part of not published preliminary work on Murghab delta.
[4] The name of the systems are arbitrary.
[5] The translation of Persian terms of *Jar* and *Morq-ab* is a personal communication of M. Sajjadi.
[6] M. Tosi suggested that the Sultan-ab system exploitation started during the third phase of the Iron Age.
[7] The Murghab river system originated the Merv oasis delta.

References

Adams, R.McC. 1981. *Heartland of Cities. Surveys of Ancient Settlement and Land Use on the Central Floodplain of the Euphrates*. Chicago.

Binford, S.R., Binford, L.R. (eds). 1968. *New Perspectives in Archaeology*. Chicago.

Burrough, P.A. 1986. *Principles of Geographical Information Systems for Land Resources Assesment*. Oxford.

Cerasetti, B. in press. A Geographic Information System for Ancient Margiana. *Annali Istituto Universitario Orientale*.

Chang, K.C. 1972. Settlement Patterns in Archaeology. *Anthropology* 24.

Forte, M., Mozzi, P., Zocchi, M. 1998. Immagini satellitari e modelli virtuali: interpretazioni geoarcheologiche della regione del Sistan meridionale. *Archeologia e Calcolatori*, 9, 271-90.

Gaffney, V., Stanchich, Z. 1996. *GIS Approaches to Regional Analysis: a Case Study of the Island of Hvar*. Ljubljana.

Gardin, J.-C., Gentelle, P. 1976. Irrigation et peuplement dans la plaine d'Aï Khanoum de l'époque achéménide à l'époque musulmane. *Bulletin de L'École Française d'Extrême-Orient*, LXIII. Paris.

Genito, B. 1998. Trial-trench at Site No. 215. In A. Gubaev, G. Koshelenko, M. Tosi (eds), 1998, 125-35. Rome.

Gubaev, A. Koshelenko, G.A., Tosi, M. (eds), 1998. *The Archaeological Map of the Murghab Delta. Preliminary Reports 1990-95. Series Minor (Reports and Memories), volume III*. Rome.

Lumsden, P. 1885. *Countries and Tribes bordering on the Koh-i-Baba Range. Proceedings of the Royal Geographical Society and Monthly Record of Geography*. London.

Rouse, I. 1972. *Settlement Patterns in Archaeology*. Reprint 3, 1-13.

Tolstov, S.P. 1958. The Prehistoric Cultures and Primitive Irrigation Systems of Ancient Chorasmia. *Annual Report and Bulletin for 1955-56* 13, 8-36.

Trudy YuTake. 1960. *Trudy Yuzhno-Turkmenistanskoj Arkheologicheskoj Kompleksnoj Ekspedicii 10*. Ashkhabad.

Willey, G.R. 1953. Prehistoric Settlement Patterns in the Virù Valley, Peru. *Bulletin 155. Smithsonian Institution, Bureau of American Ethnology*.

Figures

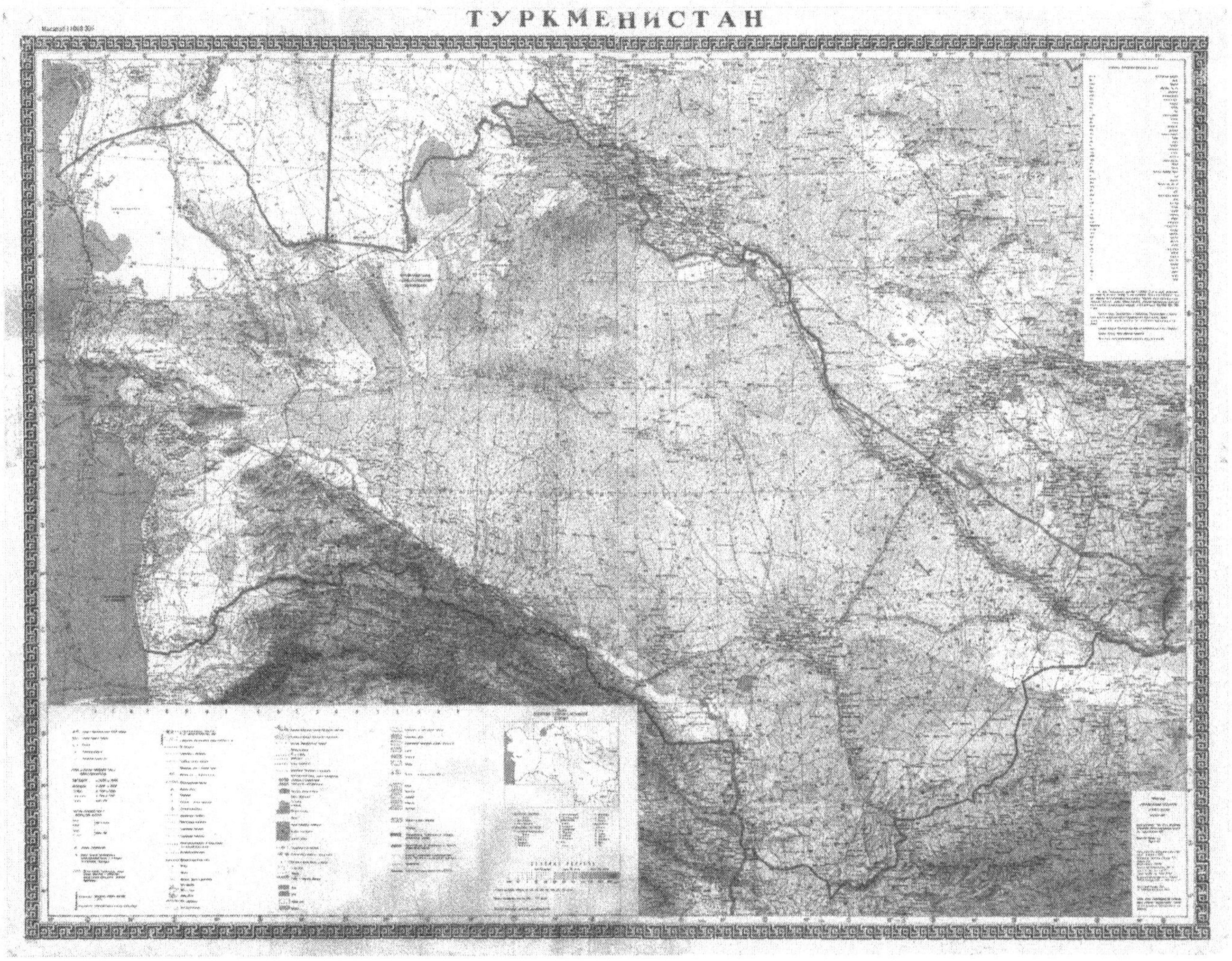

Figure 1. Map of Turkmenistan

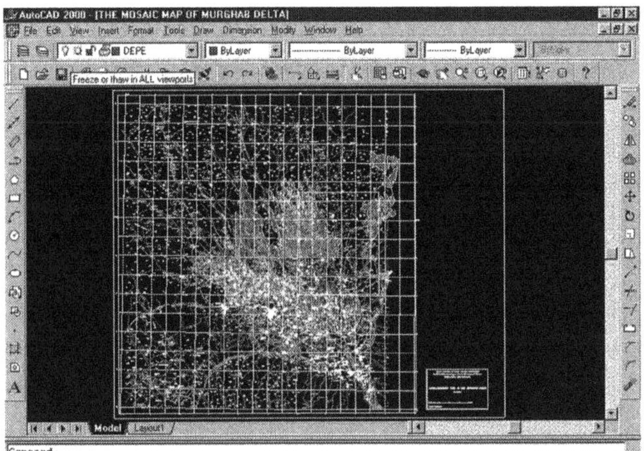

Figure 2. 1:100,1000 Map Mosaic: the Cartographic Base of Murghab Delta GIS

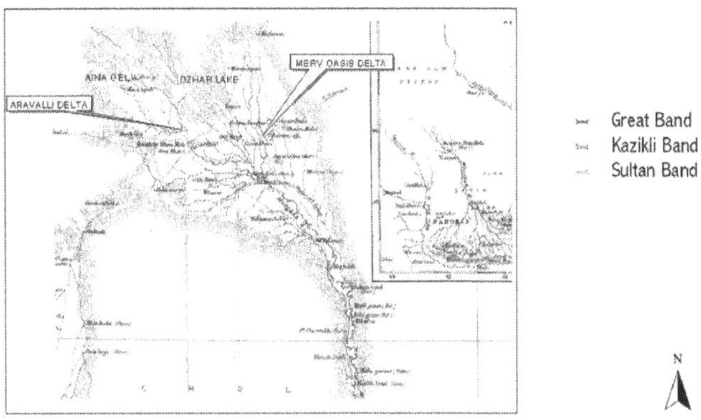

Figure 3. Map of Major-General Sir Peter Lumsden 1885

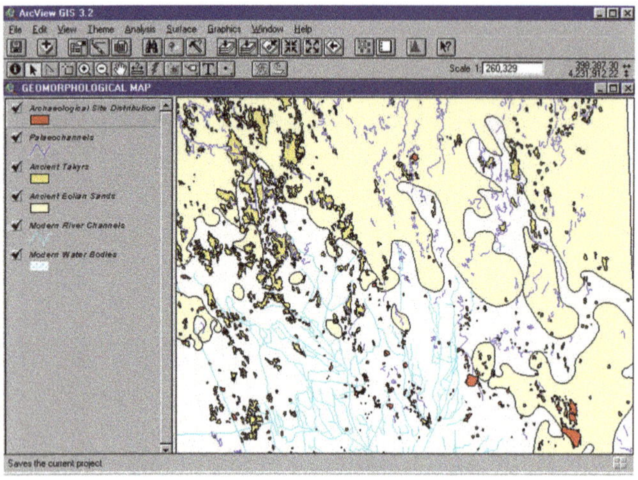

Figure 4. Geomorphological Map of Merv Oasis Delta

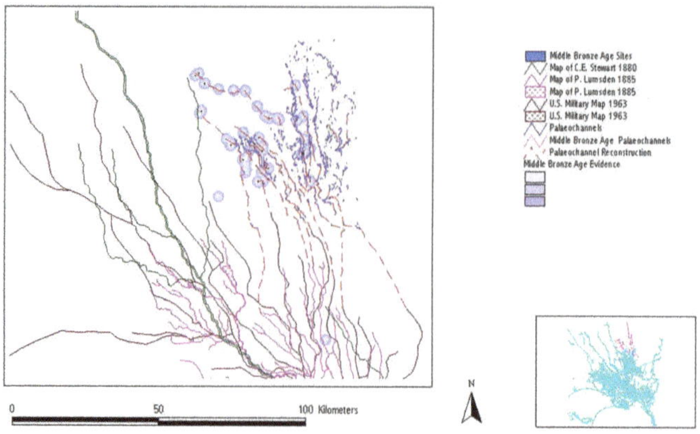

Figure 7. Middle Bronze Age Site Area Distribution

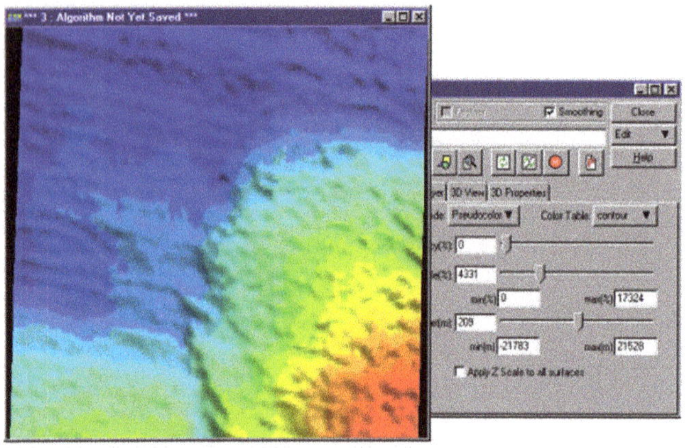

Figure 5. Digital Elevation Model of Murghab Delta: an Application of ER Mapper 6.1

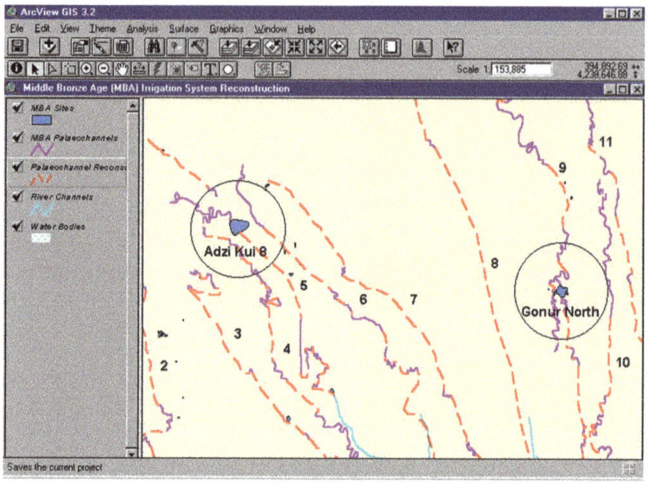

Figure 8. The Main Sites of the Middle Bronze Age

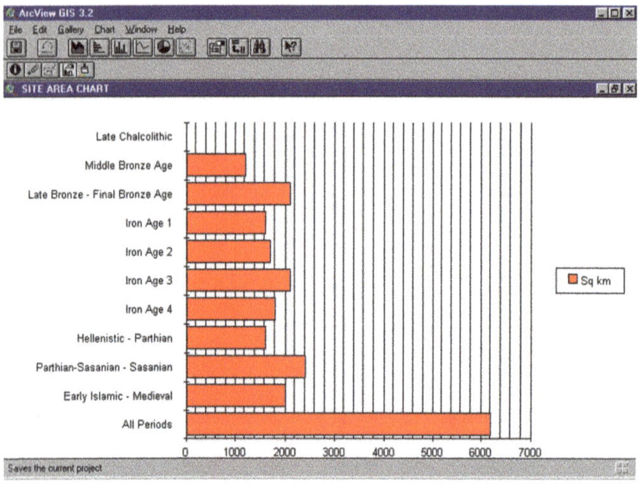

Figure 6. Classification of Settlement Size by Period

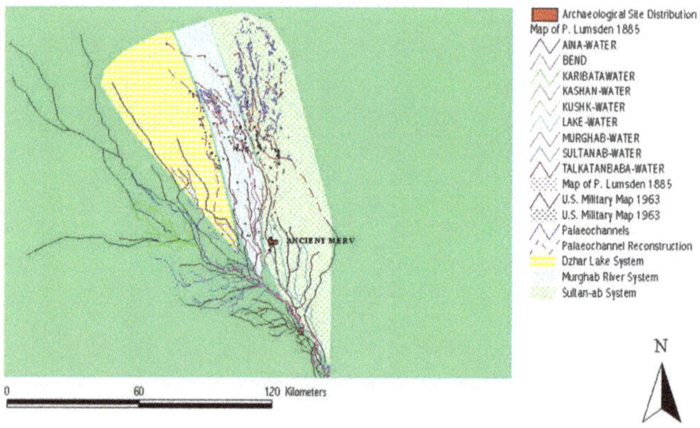

Figure 9. Main Irrigation Systems: Dzhar Lake, Murghab River, Sultan-ab

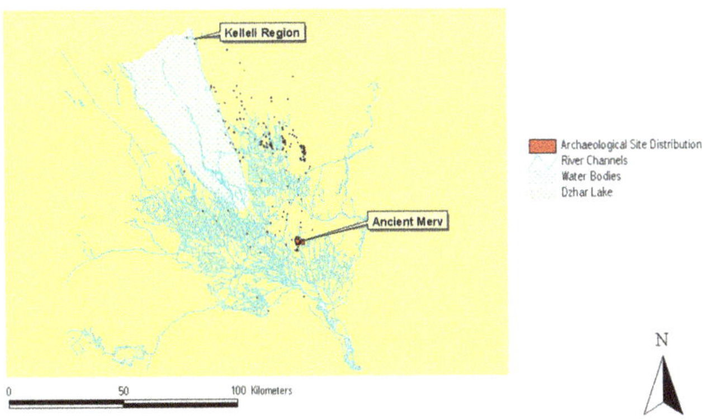

Figure 10. The Kelleli Region in relation of the Dzhar Lake

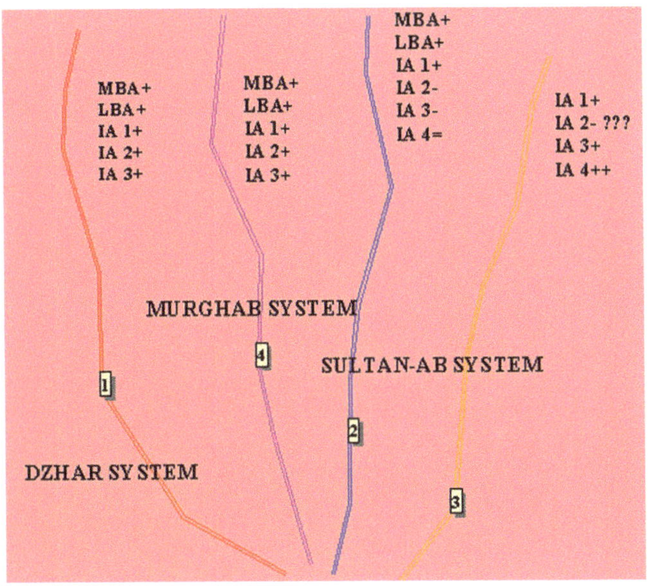

Figure 11. Chronological Sequence of Main Irrigation Systems

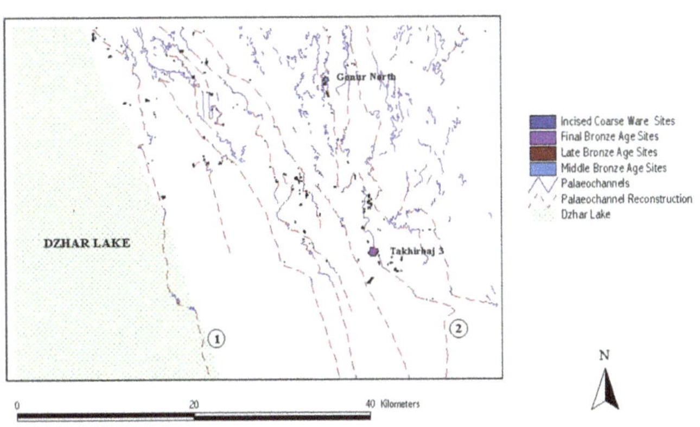

Figure 12. The Irrigation Systems of Dzhar Lake and Sultan-ab: the Sites of Gonur North and Takhirbaj 3

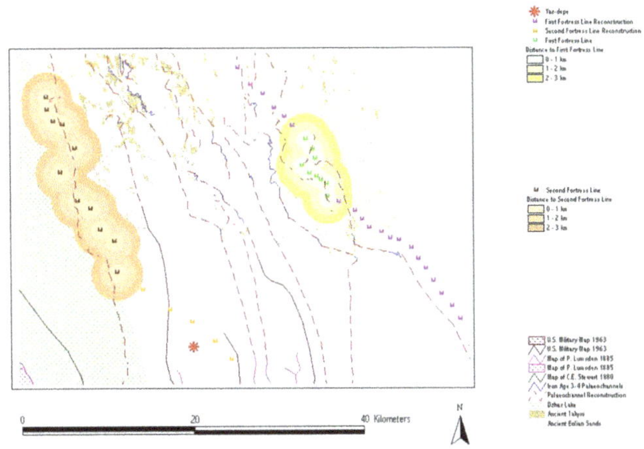

Figure 13. The Iron Age Fortress Lines on the North-easternmost Part of Murghab Delta

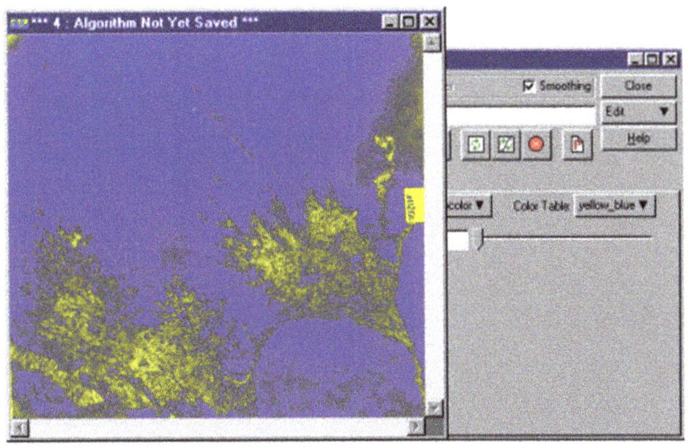

Figure 14. The ERTS-1 1972 by an ER Mapper Elaboration: the Regression Step of the Murghab Delta and the Present Irrigation Area

Figure 15. The CORONA 1961-1974 Strip Mosaic: the Murghab Delta

Building, building on the wall...
A reflection of actual building dimensions?

Dora Constantinidis
University of Athens
Thyatiron 28
Nea Smirni, 17121, Greece
E-mail: dorac@postmaster.co.uk

Abstract: By now it goes without saying that Geographical Information Systems enable archaeologists to conduct a variety of spatial analyses. Even though traditionally utilized for landscape analyses, GIS are increasingly being used for intra-site analyses as well. Another example of an intra-site analysis is presented in this paper. MapInfo, though not a fully-fledged GIS, was adequate enough to conduct a comparative analysis of the wall, to door and window ratio dimensions of actual buildings at an Aegean Bronze Age site, with ancient artists' depictions of buildings. An initial comparison of these two data sets reveals that in almost one in two instances the depictions of buildings were painted with door and window proportions equivalent to those of actual buildings.

Key words: artefeatures, spatial database, comparative analysis, ratio factors, Akrotiri wall paintings

The GIS journey continues...

As archaeologists continue on the GIS journey and move away from the usual analyses of ancient landscapes, a variety of intra-site investigations come into sight. Because archaeological data excavated at a site is defined by space, it can be mapped and accommodated quite appropriately by a GIS. Consequently there are numerous spatial analyses that can easily be performed on intra-site data once it is stored in a "GIS format". For instance there is the spatial analysis of artefact-types across a site; or the investigation of major features, and their spatial relationship to each other or to building structures at a site.

These types of intra-site GIS investigations are related to those of inter-site studies. They are related by the fact that the location of an artefact or feature (from now on referred to as "artefeature") can be used to derive distances between one artefeature to another, as is the distance between one site to another, or to some feature in the natural environment. The distances between "objects" whether a site or artefeature are measured and analysed for a variety of reasons. For instance by recording the element of distance in either scenario, the distribution of sites across a landscape can then be determined, as is the corresponding distribution of artefeatures across a site.

Moving along further with a GIS, the element of location, and hence distance, can be transposed to that of the element of dimension for each of the artefeatures found at a site. As such a GIS can then be used to analyse each artefeature not only for its relative position found at a site, but for the added dimensions of an artefeature's height, length and width. Standard databases can also cater for the storage and analysis of such dimensional factors of an artefeature. However, a GIS, when taken further to this "micro-level", is more appropriate because it can be used to centralize the details for both location and dimensions of an artefeature. This more detailed database can then be used to plot the occurrence of a particular dimension of any artefeature across a site. As such more specific and detailed distribution plots can then be performed for a site under investigation.

In effect a GIS can be used to create a "spatial database" for two major elements: those of location and dimension. In either case, by using the spatial capabilities of a GIS a variety of analyses can then be performed within the range of or in combination with each of these two "spatial" elements. For instance the element of dimension may be used for the task of formulating a pottery typology by analysing particular dimensions of pottery at a site. On the other hand, by combining the locational and dimensional data in a GIS database can allow for a detailed distribution plot of one type of artefact within a certain dimensional range to be made in relation to its location in any particular area of the site.

By travelling down another GIS path, archaeologists may arrive at an even more detailed analysis for each individual artefeature. Characteristic markings or other aspects on the surface of artefeatures can also be analysed by using a GIS in what may be termed as "surface analysis". The surface of an artefeature can be spatially analysed by using techniques developed for inter-site studies. Some sort of pattern may be determined if particular markings are distributed in a certain way across the surface of an artefeature. The GIS journey is also extending into the future with further developments, especially in the areas of Artificial Intelligence and Neural Networks (Claxton, 1995). It may then be more feasible to use a GIS to efficiently record and then use locational and dimensional data to piece back together again even the tiniest fragments of artefacts found in a room or well defined area of a site.

Only a few of all the numerous types of possible intra-site investigations that can be assisted by a GIS have been mentioned above. Research continues along the GIS road in the area of intra-site analyses. Most recently reported is a system to manage stratigraphic information and visualize excavation data at the site of Pompeii (Carafa et. al, forthcoming). The list of possible GIS analyses is confined by not only the available technology and resources, but also by the questions archaeologists ask of the excavated data, unique as it is, at each site. There are many questions that can be asked about data excavated at the Aegean Bronze Age site of Akrotiri, on Thera. This paper explores and presents an analysis of both its ancient buildings and representations of them, by using the spatial element of dimension.

As such the analysis of the buildings' dimensions is not only limited to the actual buildings themselves. It involves an analysis of the depictions of ancient buildings and actual buildings by using a GIS for the comparison of particular ratio dimensions. MapInfo was used to store dimensions of the depicted buildings of the miniature frieze found in the West House at Akrotiri and the Town Mosaic found at Knossos, on Crete. The dimensions include the length of walls, windows and doors as the artist has depicted them in the past. These measurements were then compared by a ratio factor with the measurements of buildings actually present at the site of Akrotiri on Thera.

By using MapInfo for this analysis there has been a move away from the traditional analysis of sites in ancient landscapes to a specific dimensional analysis of architecture and how it was represented in the past. The buildings that were depicted by ancient artists were scanned from photos and then digitised for analysis by MapInfo. Appropriate ratios of length were formulated and then compared with those from digitised plans of actual buildings at the site of Akrotiri. The primary purpose of this exercise was to test how accurately the depictions of windows and doors on the buildings come to portraying reality.

Aegean Bronze Age Building Depictions

From early cave paintings, and later wall paintings, to modern multi-media depictions, people have continually been seeking ways of portraying images of reality. It seems that there always has been a need to internalize, or bring indoors, external scenes by any means of representation that was and is available. Over time various depictions of the outside world have varied as to their degree of accuracy in reflecting reality. This may be due to the artists' relative skill in portraying reality, or is simply a reflection of a particular style that was chosen.

The great variety of representation techniques may lead onto questions such as what was the aim of the artist by representing images of their world in the way they did? How did this aim influence the artist's choice of style in portraying reality? Did the artist choose to depict a scene as close to reality as possible or was it more an abstraction of reality? A study of artists' depictions of reality, such as those of the Akrotiri wall paintings, can have a two-fold purpose. Firstly how accurately do the paintings reflect reality, and secondly what can they tell us about daily life? If features of the buildings in the wall paintings can be shown to approach reality it then follows that we are more likely to treat other images that are portrayed more positively regarding their reflection of reality.

The question that MapInfo assisted in answering in this instance was how close to reality were the dimensions of the depicted doors and windows of the buildings? If they were depicted close to reality then what other details can we learn from them? More light may eventually be shed on features of architectural remains that have since disappeared from a prehistoric society by a study of architectural features portrayed in images. For the purposes of this paper two different types yet related depictions of buildings were chosen: a) the depictions of buildings in the Akrotiri wall paintings were supplemented by b) the Knossos Town Mosaic.

The Akrotiri Wall Paintings

The site of Akrotiri on Thera was inhabited without a break from about the middle of the Early Bronze Age until the cataclysmic eruption of the volcano at around 1500 BC. Early on in the Late Bronze Age many of the buildings were rebuilt after an earthquake had damaged them (Doumas, 1978:777). It was during this last reconstruction phase that it can be assumed that the more recent wall paintings were applied on the newly built walls of many buildings at the site. The numerous wall paintings uncovered in the buildings at Akrotiri offer a wealth of material for investigating scenes of life in the Aegean during the Bronze Age. Televantou (1990) has carefully researched data relating to the reconstruction of the wall paintings of the West House. It appears that five distinct "towns" were depicted in the miniature frieze located in room 5 on the upper floor of the West House building. The wall paintings are in themselves a decorative architectural feature of the room yet also provide further visual clues as to the original state of the relatively well-preserved buildings at the site of Akrotiri.

The buildings in the miniature frieze of the West House present a valuable glimpse of architectural features and the past organisation of buildings in rural and town settings. The pictorial representations of the variety of scenes in the miniature wall paintings have been analysed in detail by Morgan (1988). Apart from the scenes depicting towns, others depict herding activities, battles at sea, and scenes of wildlife along a river. It appears that the artist(s) chose a single perspective from which the buildings in the town scenes were depicted. Given that the buildings would have been built on different levels, as towns were often built on an incline from the shoreline, it was difficult to represent all buildings at the same level. So the artist(s) overcame this difficulty of perspective by depicting the buildings as if viewed from a considerable distance and as if suspended in mid-air (Morgan, 1988:70).

Even based on a purely visual analysis of building types in the wall paintings of town scenes it is clear that the representation of the buildings in the miniature frieze closely reflect the architecture at Akrotiri (Morgan, 1988:70,92). Since a variety of building-styles have been depicted in the Akrotiri wall paintings it is possible to learn more about past buildings by using the depictions as evidence for architectural features that have since disappeared. The possible function of buildings may also be determined by studying them in their portrayed contexts, see Figure 1. Measurements from these depicted buildings may also reflect actual building ratios that were used at the site.

The Knossos Town Mosaic

The great array of buildings constructed at Knossos, on Crete were also decorated with wall paintings in many circumstances. However it is a collection of small individual "mosaic" pieces depicting buildings that are of more interest in this case. Most likely to have been originally used as inlays in a wooden box a number of small individual faience plaques were found at Knossos in an early Middle Minoan III deposit (Lawrence, 1983:40). Apart from depicting various scenes of people, animals and plants, most of the plaques are those of houses and towers which were all probably used in a mosaic to represent a fortified city by the sea perhaps during a seige (Robertson, 1929:20).

The houses come in an array of colours, displaying a variety of building techniques. Some of the houses are shown as though composed of large blocks of stone. Some have horizontal stripes, perhaps representing timber. Doors and windows are clearly displayed, see Figure 2., and offer complementary material to the miniature frieze at Akrotiri. However most of the plaques are partially destroyed and the restorations may not always be accurate (Lawrence, 1983:40). These depicitons of buildings were primarily selected to supplement the images of buildings available at Akrotiri because the Knossos Town Mosaic offers a very close parallel to the buildings depicted in the Akrotiri wall paintings (Morgan, 1988:68).

The Knossos Town Mosaic was selected to increase the sample of data available for analysis. This was done on the basis not only because the houses in the Town Mosaic closely parallel those in the wall paintings, but also because the actual buildings at the site of Akrotiri display strong links with Minoan styles, and hence with Crete. Hood (1990) examined the actual buildings and proposed that the Minoan style of architecture at Akrotiri suggests a strong Minoan presence at the site. Apart from the architectural elements, the numerous artefacts and pottery also attest to a strong Minoan influence on the island of Thera, or at the very least at the site of Akrotiri. Many of the internal architectural features are clearly Minoan in nature, however the extent of Minoan domination at the site remains debatable (Shaw, 1978). Regardless as to the extent and nature of the Minoan influence at Akrotiri, the clear links that this site had with Crete at the time allows for the inclusion of the Town Mosaic in the repertoire of building depictions that can be used for analysis.

Aegean Bronze Age Architecture

Buildings are one of the most fundamental aspects of a settlement. People's lives revolved in and around the buildings they created to cater for many purposes and functions. In most cases the buildings were constructed according to traditions established by either local people or ideas introduced by other cultures, or a combination of both. Styles and techniques evolved and developed into a distinguishable architecture at each site, which was influenced by the unique environment, availability of building material, prosperity and social expectations of its inhabitants. By analysing the architectural design of buildings at a site factors that determined the form and function of the buildings may be derived (Palyvou, 1990:44).

A number of still developing approaches can be used to analyse the architecture of a site. Palyvou (1990:55) has suggested that a systematic analysis of form and structure is one approach in the attempt to come to an understanding of architectural remains. Architectural design may be determined by investigating whether models, with any number of variations on a basic design, were adopted at a site (Palyvou, 1990:46). Apart from the observations that are made during the excavation process there are other possible methods for investigating architecture in its own right. For instance determining the functional purpose of a building is one aspect of investigating why or for what purposes a building was built in the past. The size, shape or form of a building may certainly be easy to determine and record, but when it comes to the reasons why a particular building displayed certain characteristics a methodology to investigate a building's function still has to be fully developed and implemented (Palyvou, 1997). As more architectural studies delve into greater detail concerning the remains of past buildings, it is more likely that new methodologies will develop to lead us into a deeper understanding of the form and function of buildings from any era in the past.

Since GIS effectively stores, manages and analyses spatial dimensions, using the capabilities a GIS has to offer can now enhance architectural studies. Most archaeological analyses of architecture are primarily descriptive, even though dimensions may occasionally be provided. No systematized standard or statistical methodology has been thoroughly developed for the study of architectural remains. It is proposed that GIS can be utilised as a tool for systematically analysing social space. Hence a GIS can be adapted to systematically approach the study of architectural remains of the past. This paper presents one very small step in this direction.

Ultimately any analysis that can lead onto new information about the original design, layout and intended purpose of a building is necessary for an all-encompassing understanding of past buildings. By representing buildings in a digitised format that can be used by a GIS program, data pertaining to dimensions such as the area of rooms, thickness of walls, length of walls, doors and windows can be easily and quickly provided. A building in its archaeological context can only be analysed for so many characteristics. For instance the West House at Akrotiri as portrayed in Figure 3., can be examined for a number of features such as material that was used, and the building techniques that were employed for its construction. Traditional "paper-based" plans of the building can also provide dimensions, however when digitised the resulting plans, see Figure 4., can then be used more efficiently for a variety of measurements to be made.

To come to an understanding of the architectural design of any remains of buildings is an important aspect of archaeological research. Palyvou (1990) has noted that many of the buildings at Akrotiri were constructed by altering basic models of design. It appears that during the Middle Bronze Age the Minoan civilisation infiltrated that of the Cyclades, especially at Akrotiri on Thera, and many elements including building techniques were adapted by the local culture (Doumas 1979:96). Many of the houses at Akrotiri reproduce Minoan architectural features on a smaller scale. Irregularity of shape, pier-and-door partitions, angled stairways leading to upper storeys, and a number of large windows, opening up, though unlike in Minoan settlements, to narrow, convoluted streets below (Lawrence, 1983:71). One of the major Minoan building characteristics, ashlar masonry, is also used at Akrotiri. Given the conglomeration of Minoan elements in the buildings at Akrotiri, local traditions may be able to be traced by not only what remains of the structures, but also by the images of buildings portrayed in the wall paintings. It is these wall paintings that this present paper has analysed to determine how accurately they reflect the proportions of windows and doors in the remnant buildings at the site.

From Depictions to Reality: A Methodology

A GIS can assist architectural analyses of ancient buildings by initially facilitating the scanning and subsequent digitisation of buildings from paper-based site plans. Then a number of analyses can be applied to these digitised plans of the buildings. However, apart from analysing only building plans from sites, wall paintings and other depictions of ancient buildings can also be analysed. Effectively two data sets are potentially available for architectural analysis: data pertaining to both real and depicted buildings. When these two data sets are merged together then a comparison of the results of related analyses can be used in order to enhance our understanding or even determine patterns of building in the past. In this case length ratios were determined from the depicted lengths of doors and windows from the Akrotiri wall paintings and Knossos Town Mosaic, and the door and window lengths from the Akrotiri building plans. A comparison was then made of these length ratios determined from the depicted buildings with those of the building plans.

Analysis can encompass a variety of approaches. The basic questions fundamental to any sort of analysis are: 1. What should be tested, and 2. How should it be tested? For the purposes of this paper the corresponding answers are: 1. How accurately the representations of buildings in the Akrotiri wall paintings and Knossos Town Mosaic reflect actual door and window dimensions, and 2. The measurements of window, door and wall lengths of actual and depicted buildings are to be compared by ratio factors. The dimensions of wall, window and door lengths are converted to L/WL and L/DL ratio factors, where L=wall length; WL=window length; and DL=door length. The analysis requires the production of two data sets that are to be used for a comparative analysis.

By digitising both site plans and depictions of ancient buildings two data sets became available in order to determine ratios and then make a comparison of these two sets of ratios to see if the artists' impressions of the buildings matches reality. However a number of assumptions have been made for this test. Firstly that the measurements from the building depictions may be a reflection of reality. That is that the depicted door and window ratio lengths relate to the ratio lengths from the plans of the actual buildings at the site. A further assumption is that the buildings that are depicted are Minoan in style and character and are closely related to the buildings actually present at the site of Akrotiri.

The two data sets were digitised and derived from a) actual plans, and b) images of buildings. This resulted in a database of measurements from the actual buildings that can be compared with a database of measurements of the depicted buildings.

Data set one: Digitised site plan

The site plan of Akrotiri on Thera was scanned and then MapInfo was used to digitise each of the buildings. The digitised image of the buildings is associated with a table of data containing dimensions of the wall, door and window lengths. These are grouped by floor level, since in some cases, the second and even third storey of a building has been preserved. However in the final analysis all window lengths were used regardless of which floor level they were located on. This was also the case for the depicted windows in the wall paintings and mosaic pieces. Effectively for this present analysis all window ratios were analysed as a unit of data with no distinction being made for floor level.

Data set two: Digitised building depictions

For the purposes of this paper the depictions were limited to the wall paintings from Akrotiri on Thera and the Town Mosaic from Knossos on Crete. MapInfo was used to digitise scanned images of buildings from the wall paintings and mosaic pieces. These digitised images then allowed for dimensions to be measured and stored in a database. Each building was measured, wherever possible, and in some cases the measurements are extrapolations of the possible original intention of the artist. Questionable restoration and even the fact that ancient artists may have made alterations to the images in an attempt to depict true-to-life scenes, meant that generalisations had to be made in some cases. That is why the depictions of the buildings were critically analysed by modern day artist and icon-painter, Stratos Kranidiotis who also assisted in the compilation of the digitised images. Mr. Kranidiotis advised what possible images the artist may have been viewing and how they may have chosen to depict them.

GIS strikes again: Results and Conclusions

Many comments have previously been made regarding the physical appearance of the depictions of buildings in the Akrotiri wall paintings and the Knossos Town Mosaic in relation to the remains of actual Minoan buildings. On the other hand this present study concentrated on a comparison of the dimensions of the depicted doors and windows with those of actual buildings. The analysis used a GIS for the easy acquisition and subsequent measurement of lengths from both the depicted and real buildings. This analysis revealed that in almost half of the cases the proportions of the windows and doorways of the depicted buildings are reflected in reality.

Even though a relatively small sample of data was available for testing, the major purpose of this analysis was essentially to demonstrate a methodology for using a GIS to store and compare data from two different sources. The two separate lists of wall/

door and wall/window length ratios of the depictions of buildings were compared with those ratios from reality. This comparison resulted in the following:

46 % of all the depictions of doors are in proportion to their corresponding door ratios as determined from the Akrotiri site plan, and

48% of all the depictions of windows are in proportion to their corresponding window ratios as determined from the Akrotiri site plan.

It is important to note that the results are limited by the relatively small sample of data that was used for analysis. So far the database of digitised depictions of buildings is limited to the Akrotiri wall paintings, and Knossos Town Mosaic. Future work is to develop a more comprehensive database and include more examples for other comparisons of buildings and depictions of buildings not only for the site of Akrotiri but with any other related and relevant sites. Another future investigation would also be to measure the height of the buildings as depicted by ancient artists and then extrapolate from the architectural remains the possible height of the original buildings.

One purpose of this paper was to illustrate another way of viewing archaeological data and then use a GIS to map and analyse spatial elements other than those used by traditional landscape investigations. Consequently another aim of this paper was to stimulate thought in order to demonstrate other ways of using GIS to map and analyse spatial elements from the various aspects of data excavated at a site. Therefore, ultimately not only landscape analyses stand to benefit by using a GIS, but each site and even each artefeature found at a site may eventually come under more analysis with a GIS. Some of the other ways of using a GIS for analysis within one site are suggested in summary to be dimensional analyses of artefeatures, plotting distributions of particular artefeatures, or even performing a surface analysis of each individual artefeature. By continuing to apply the many capabilities of GIS to various aspects of archaeological research, only time will tell what other ways of using a GIS can be thought up. As we contemplate other possibilities on the GIS journey, the sky's the limit for using a GIS in archaeological research.

Acknowledgements

I would like to thank Professor Doumas, of Athens University, for the Akrotiri material; Professor Sellis, of the National Technical University of Athens, for the MapInfo program; and Mr. Stratos Kranidiotis, artist and icon-painter, for an artist's insight on the Akrotiri wall paintings. Finally I am grateful for the computing resources provided by the Queensland University of Technology.

References

Carafa, P. et. al., forthcoming. "From Stratigraphic Unit to the mouse: A GIS based system for the excavation of a historical complex. The case study of Pompeii" in *CAA2001 Conference Proceedings*

Claxton, J.B. 1995. "Future enhancements to GIS: implications for archaeological theory." in Lock G. and Stancic Z. (eds.) *Archaeology and Geographical Information Systems: A European Perspective*. Taylor and Francis: London, pp. 335-348

Doumas, C.G. 1978. "The Stratigraphy of Akrotiri." in Doumas C.G. (ed.) *Thera and the Aegean World I*. London, pp.777-782

Doumas, C.G. 1979. "Town Planning and Architecture in Bronze Age Thera." in *150 Jahre, Deutsches Archaologisches Institut 1829-1979*. V.P. Von Zabern: Mainz, pp.95-98

Doumas, C.G. 1992. *The Wall-Paintings of Thera*. Thera Foundation: Athens

Hood, S. 1990. "The Cretan Element on Thera in Late Minoan IA" in Hardy, D.A., Doumas, C.G., Sakellarakis, J.A. and Warren, P.M. (eds.) *Thera and the Aegean World III, Proceedings of the third International Congress, Santorini, Greece, 1989*. Thera Foundation: London, pp.118-123

Lawrence, A.W. 1983. *Greek Architecture*. Revised with additions by R.A. Tomlinson, Penguin Books: England

Morgan, L. 1988. *The Miniature Wall Paintings of Thera: A Study in Aegean Culture and Iconography*. Cambridge University Press: London

Palyvou, C. 1990. "Architectural Design at Late Cycladic Akrotiri." in Hardy, D.A., Doumas, C.G., Sakellarakis, J.A. and Warren, P.M. (eds.) *Thera and the Aegean World III, Proceedings of the third International Congress, Santorini, Greece, 1989*. Thera Foundation: London, pp.44-56

Palyvou, C. 1997. "Session on The Functional Analysis of Architecture" in Hagg, R.(ed.), *The Function of the "Minoan Villa". Proceedings of the 8th International Symposium at the Swedish Institue at Athens, 6-8 June 1992*. Stockholm

Robertson, D.S. 1929. *A Handbook of Greek and Roman Architecture*. Cambridge University Press: London

Shaw, J.W. 1978. "Consideration of the site of Akrotiri as a Minoan settlement" in Doumas, C.G., (ed.) *Thera and the Aegean World I*. London, pp. 429-436

Televantou, C.A. 1990. "New Light on the West House Wall-Paintings." in Hardy, D.A., Doumas, C.G., Sakellarakis, J.A. and Warren, P.M. (eds.) *Thera and the Aegean World III, Proceedings of the third International Congress, Santorini, Greece, 1989*. Thera Foundation: London, pp. 309-322

Figures

Figure 1. Section of the miniature frieze from the West House at Akrotiri: A town scene. (From Doumas, 1992)

Figure 2. A selection of restored faince house plaques from the Knossos Town Mosaic. (From Lawrence, 1983)

Figure 3. View of the West House at Akrotiri. (From Palyvou, 1990)

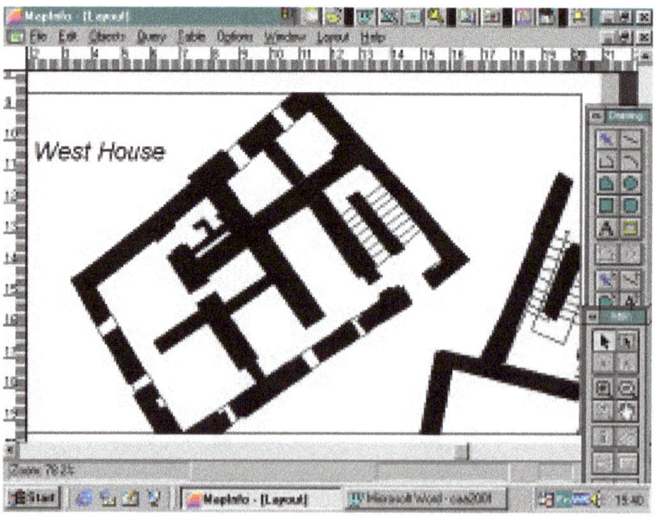

Figure 4. Digitised view of the West House at Akrotiri.(Ground and first floors are superimposed).

Historic landscape assessment:
The East of England Experience

"One of the most challenging problems for both the experimental ecologist and the theorist is the interaction of biological populations with the time dimension" (Garsd, 1984)

Lynn Dyson-Bruce
Historic Landscapes Project - East of England
County Hall
Essex County Council
Hertfordshire County Council
England
Phone: +44-1245-437618 - Fax: +44-1245-258353 - E-mail: lynn.dyson.bruce@essexcc.gov.uk

Abstract: Historic Landscape Assessment (HLA) is a new methodology being developed to assess the historic dimension of the landscape, and is based within a GIS. This will facilitate the complex analysis and representation of spatial and historical data, which will enable the 'time-depth' and complexity of the landscape to be more adequately assessed.

A new approach has been necessitated, by a combination of factors, including new legislation, land designations, and the increasing concerns about the future of our cultural landscape.

This methodology is being developed across England by English Heritage, on a county by county basis, each with a differing approach and methodology. The East of England project is unique, in that it assesses a region of political areas as a single geographic unit, with a common (and evolving) methodology.

Initial results indicate that landscape history and development is diverse and dynamic not only within political, geographic and topographic regions but between them as well, on all scales. This form of assessment is proving to have a wide range of end uses, including planning, development control, strategy, landscape and cultural resource management, research. It is currently proving to be a useful tool for landscape management.

Introduction

This paper relates the latest developments of Historic Landscape Assessment (HLA), within the project area of the East of England region. Current work has built upon the initial Historic Landscape Characterization (HLC), carried out in Suffolk, and has sought to improve the methodology in response to the diverse, complicated landscapes and methodological challenges encountered within Hertfordshire and Essex. The methodology has been improved, being more firmly based within GIS and has endeavored to make the whole process of HLA more transparent. The project plans to continue to cover the three counties of Bedfordshire, Cambridgeshire and Norfolk. Issues that will be discussed in this paper include:

- landscape
- project area
- GIS
- methodology
- interim results
- uses
- future developments

This is work in progress and until counties or topographic areas are completed, analysis, especially GIS, including spatial and 3D analysis can not be completed, neither can one fully assess historical results or landscape patterning, at this stage. All conclusions are therefore tentative and indicative of what may be revealed by the project's completion. So far it may be stated, that initial results are unexpected, exciting and increasingly reflect the rich diversity of the historic landscape

Landscape Environmental Background

In assessing landscape, which is in this case 'landscape history', it is necessary to be aware of the forces which determine how our landscape has formed, including the prevailing and interdependent influences upon such landscape development. The model below (Fig.1), attempts to chart out some of these forces, and illustrates the complexity of landscape genesis. Some of the forces are obvious, whereas others are subtle and have not always been regarded in relation with each other. For example, it is given that that underlying geology is a prime determinant of soils, but the process of soil formation is in turn affected by climate (temperature, precipitation), height, slope, drainage, etc. The resultant soils in conjunction with these factors of slope, height and climate also largely determine land capability – in other words the activities that the resultant soils will support *i.e.* agriculturally. This in turn determines farming practices, although these constraints can be modified by activities such as additional drainage, liming, fertilizing etc. These technologies are, in turn, historically specific, and are connected to particular property regimes, market access regulations and subsidies.

Furthermore, if natural forces are examined, additional factors such as vegetation, fauna etc - which are all interdependent upon each other – need to be taken into account. Lastly, in tandem with the natural forces are the anthropogenic forces, whereby human actions have manipulated the naturally occurring landscape. The landscape in itself is an historic record of these interactions. It is these anthropogenic forces and impacts on the landscape, which are part of the focus of this paper.

The results of this anthropogenic activity results in, what could be termed as 'semi-natural habitats' and 'human habitats', these are all inter-related and interdependent upon each other in a very complex series of relationships. It may be argued that due to human impact upon the landscape, that there are few 'natural habitats' left within the UK.

HLA seeks to map the key anthropogenic influences on the landscape. Results so far from Eastern England (and the various other HLA/C's already completed across the country) indicate that the history of the landscape is dynamic, diverse, complex and unpredictable. The only consistent result, appears to be that historic landscapes are diverse and dynamic. However, this factor when combined with the variations in approaches to HLA/C mapping, creates significant problems in comparing and contrasting between regions, and also some adjacent areas within regions.

To assess National, Regional and Local landscape types, an appropriate scale should be adopted, to recognize areas of diversity and rarity. For example, certain prehistoric or historic events may be strictly local, and therefore do not need to be preconceived in a national taxonomy, as they will fit into a broader grouping at the regional and national levels.

Therefore, it is proposed there should be a 'national type series' of landscape types into which these areas of study could feed, (See Fig:2). This should not deflect from other approaches and developments within an area or region, nor each area's individuality. Such a type-series should act as an 'umbrella' of generic types into which each region's diversity and uniqueness could contribute. This would result in 'national', 'regional' and 'local' sets and sub-sets reflecting an appropriate scale of landscape diversity. This is fundamental to the understanding and appreciation of our diverse and rich landscape, and to enable appropriate management of this finite cultural resource, at an appropriate scale.

Application

The East of England Project; is so far unique in being a regional project, as other counties within England are being - or have been – assessed, independently. The East of England Project is a co-ordinated approach between English Heritage (EH) and the relevant local authorities in the region. The work on this project area is in progress and includes the historic counties of: Suffolk, Hertfordshire, Essex, Bedfordshire, Cambridgeshire and Norfolk, (See Fig: 3, 4 & 5).

The experience of the project so far has shown there are distinct advantages of a single methodology being applied to a group of counties, which are being assessed as a single geographic region, even if (as is the case of East of England), it is a very recent political creation. Initial assessment and results have shown that each county has its distinctive historic landscape character and historic development, with subtle and at times dramatic variations. Broad historic landscape patterns have been identified which run across county boundaries. However there are also small pockets of unique landscapes, reflecting the diversity from one geographic/topographic region to another, and within regions, (see Fig 6).

Initial results show that HLA has proven to be a valuable management tool, by identifying and informing various issues in the East of England region, for example:

- Local, regional, historic landscape patterns
- Chronological patterns, through time and space – landscape 'time depth' and palimpsests
- Landscape management and Landscape Character Assessment
- Management of the archaeological and historical resource
- Archaeological survey
- Research objectives
- Development control in planning
- Regional and Local Planning issues and strategies

This information has already proven of use to a wide audience e.g. landscape architects and managers, archaeologists, historians, geographers, planners, etc. In the case of the East of England Project the HLA has already provided landscape managers and planners with the necessary information to make informed assessments about the impact of large development proposals on the historic landscape.

Methodology/Approach

Background

The HLA methodology originated from the seminal work carried out in Cornwall (Herring 1998), which mapped landscapes according to 'Historic Character Types' as a 'paper-based' exercise. Work in Scotland further developed the approach developing 'Historic Landuse Assessment' using a Geographic Information System (GIS), (Dyson-Bruce 1998: Dyson-Bruce *et al* 1999). Wales defined a 'Register of Landscapes' of specific or outstanding interest, (Cadw 1998). English Heritage (EH), has used a variety of paper and GIS based methodologies, to determine 'Historic Landscape Character' in the different counties, across England (Fairclough 1999).

HLA is a relatively new approach to assessing landscapes and is still under development, especially as a GIS application. However, it is the advent of GIS itself as an affordable tool within local government that has been a key driver for HLA. One of the reasons for the importance of GIS is the nature of the HLA information itself, which is spatial in four dimensions:

- horizontal space N/S – E/W
- vertical depth
- time-depth or palimpsests

The horizontal dimension forms the backdrop of this mapping project, while the elevation axis is less critical at the outset. Eventually, the topographic element should be coordinated as a part of the analytical application of the inventory. The key of HLA is the treatment of time. This together with the large quantity of complex spatial information that has to be dealt with in HLA, is ideally suited to the GIS platform.. In addition, the GIS can provide support to extend the representation of the data to include documentation concerning how the data was derived. This metadata is the core of a self-documenting reflective resource in place of the traditional simple printed map. Earlier approaches to HLA were largely based on established Landscape Character Assessment (LCA) techniques and methodology (Countryside Commission, 1997). These were paper-based, and as such resulted in a single paper map or series of simplistic maps, achieved by the aggregation of data to create thematic historic landscape types. Compared with GIS, the methodology of paper-based HLA is also relatively non-transparent. Paper-based HLA is therefore difficult to be reliably and consistently replicated. Detailed analysis or changes in the representation or update of the data are very difficult to achieve efficiently on paper maps.

However, until recently GIS has only been used as an enhanced version of the original paper-based approach to HLA, resulting in simple digital maps. Such maps may solve some of the problems of replication and representation of data, but they are lacking in inherent intelligence and metadata, that are key features of true GIS applications.

It was therefore decided that the methodology for the East of England HLA needed to be refined in response to other HLC projects, to represent the diversity of the region and also harness the capabilities of GIS. This has been achieved by creating a suite of historic 'attributes', representing historic landscape types, which may then be aggregated as required for different objectives, remits and end uses (Dyson-Bruce *et al* 1999: Fairclough 1999). HLA as a methodology has necessitated a flexible and responsive approach to record, not only the subtleties and dynamics within the landscape in an appropriate manner, but to also act as a rigorous analytical tool relating to academia, data entry, analysis and metadata.

The adapted HLA methodology used in the East of England Project is now GIS based, and may be described by the following three interactive aspects:

- HLA - creates 'historic landscape types', attributes which may be aggregated to form 'historic character areas' - *the academic, or archaeological and historical aspect.*
- GIS - which handles the data-capture process and input - *the practical aspect.*
- Metadata – this is the data concerning both of the above. It informs the complete process of data collation, ownership and map creation. This could be regarded as the *fulcrum* between academic research, and the mechanical process of that spatial data's rendition.

In assessing the historic landscape of a large region, the HLA methodology has to be, by necessity, relatively 'broad-brush' in its approach, and also largely desk based, due to restricted resources of staff and time. Only a few key and readily available sources of historical information have been used, primarily current and historic OS and 1" maps, and aerial photographs. The criteria used needed to be robust, definitive, replicable and meaningful to the data creator, the final product users and a variety of objectives. In addition the landscape types and data used, needed to be carefully selected to reflect landscape composition, in terms of its diversity, continuity and discontinuity, thereby enabling the complex concepts of 'time-depth' and 'palimpsests' within the landscape to be assessed.

The East of England HLA data has been collated at 1:25,000 (data-capture), checked and digitized at 1:10,000, using a defined series of landscape types (attributes). At the generic level, these types distinguish: field systems, woodlands, parklands, grazing, urban, industrial, extractive, military etc. (both current and relict for each type). To be recorded on the HLA, these types must still be visible within the current landscape. Finally, these types may be connected to a period or a political or economic movement, such as Parliamentary Enclosure Acts. Recording the data in this way enables it to be used flexibly in response to a variety of objectives.

The metadata has been incorporated directly within the database, using appropriate fields (See fig: 7 below). These include:

- *historic landscape type* e.g. code_c, relict_1 or Hla_code
- *data source* - the maps from which the data derived, whether current or relict e.g. date_c, date_r1
- *creator* – the person responsible for digitizing the polygons, assessing and assigning data
- *date of creation* – date data digitized
- *data owner* - Essex CC, Herts CC, Suffolk CC etc.
- *scale of data accuracy* – scale of digitizing
- *glossary of terms* e.g. code_c, code_r2.

In addition to the metadata there are databases linked to the data source that informed each polygon's creation. Their purpose is to correlate data, especially to the data sources such as the maps used to create the HLA dataset. This is to take account of the variability of the data sources. For example, the map sources:

- some are consistent across counties e.g. Ordnance Survey Landline 2000, 1st Edition
- some are consistent within a county, historic county maps e.g. Dury 1760 map for Hertfordshire; Andre & Chapman 1766 map for Essex
- and some are inconsistent within counties
- Parish based e.g. historic Tithe maps, Enclosure Maps
- Estate maps, some dating back to the 16th century

There are also links to tables of:
- definitions of the codes used in the data capture process
- a full glossary of attribute types
- counties with different but linked methodologies, i.e. Suffolk in this instance
- hyper-links to photographs, scanned images, documents,

texts, videos etc.

These databases in conjunction with the embedded metadata, will help render the methodology transparent and self-documenting for future researchers

In addition, GIS has many useful utilities and functions, which facilitate the display and analysis of this complex spatial data. In practical terms this enables a flexibility of approach. This data may then be easily analyzed, statistically and spatially, within its own dataset and in combination with others within the GIS platform. Varied outputs are easily possible e.g. maps, graphs, tables, histograms, presentations etc. In addition as an archive, the data may be 'static', providing time horizons or 'organic' allowing constant revision. This allows a far greater flexibility than any paper based methodology could achieve.

The East of England HLA: Results so far

The provisional results obtained so far are briefly discussed below.

Suffolk

Suffolk is a large rural county, and benefits from being well documented regarding its landscape development. The methodology used has been based on field morphology, with a single tier of information within the database. The types have been aggregated to form basic landscape character types, which illustrate in broad patterns the historical development of the county, e.g. pre-18th century field types, 18th century and later field types, forestry etc. (Ford, 1999).

In summary, the county has a diverse history, with a central band of 'anciently enclosed' landscape, including co-axial, sinuous, and irregular field-systems, a western area of later Parliamentary Enclosure and forestry plantations on former heaths and grazing commons, and an eastern area of sandlings and fen-lands, (see Fig: 8).

Suffolk has relatively little 20th century development compared with many counties of southern England. Urban development has primarily focused around Ipswich and Bury St. Edmunds and the county still exhibits a predominantly rural character. There is some 20th century field boundary loss, but has not been as devastating on the earlier field systems as is the case in Essex and Hertfordshire (see below). This requires further analysis, to establish if this may reflect, for example that the landscape is on a larger scale, or due to a change of landuse, (from grazing to agrarian), or CAP reforms.

Hertfordshire

Hertfordshire is in comparison a small semi-urban county (it has the highest population density of all the 'shire' counties in England) with intensive twentieth century impacts upon the landscape. Unfortunately, there has been little documentary research to facilitate the assessment of the landscape, especially with regard to field systems. The landscape is also varied with a complex sequence of landscape development, which has resulted in 'hybrids' of historic landscape types. The broad historic landscape zones seen in Suffolk do not appear to be replicated within Hertfordshire, which in comparison has a highly fragmented landscape. These factors necessitated a more analytical approach for the HLA of Hertfordshire, with more detailed research undertaken within selected parishes to inform this complex process of landscape development. In addition greater fragmentation, within the landscape required a greater emphasis on metadata. Therefore, the methodology was revised to create an 'intelligent' GIS map with multiple tiers of information, incorporating metadata.

The landscape patterns within the county are complex and show great diversity of form and function, (see Fig: 9) and are less easy to summarize than is the case with Suffolk, (see Fig. 8). The county field systems also appears to have a long 'time-depth' with the dominant types (co-axial, sinuous, irregular) often exhibiting a complex developmental morphology. A distinctive pattern is emerging, in that the distribution of certain field types appears to be mutually exclusive. For example, the co-axial and sinuous systems occur exclusively in the south of the county and the common arable fields are dominant in the north. However, the substantial but amorphous category, 'irregular' fields are distributed throughout the county. These require further analysis as to type, origin and genesis, as some may have a very different history and use.

Several other general patterns in the historic landscape of Hertfordshire (Fig: 9) are apparent:

1. The areas of Parliamentary enclosure (where the field systems have been redefined) are primarily restricted to former medieval 'common arable' fields on the Chiltern Ridge, or smaller areas of former 'common grazing' elsewhere. Otherwise, Hertfordshire has been lately enclosed by agreement, on a piecemeal basis through time, retaining most of the original infrastructure.
2. There are many survivals of 'Ancient Woodlands' (as defined by English Nature) mostly as small pockets within the matrix of the field systems. However there are, notably, no ancient woodlands remaining on the Chiltern Ridge.
3. The co-axial field systems in the south-east of the county within the Wormley Woods area, have been dated by pollen analysis to the Bronze Age, thus pushing back that part of the current landscape to a prehistoric origin. The similar co-axial areas to the west of the county may be of a similar date, thus indicating that certain areas of the southern part of Hertfordshire are prehistoric in origin, in contrast to the area to the north which appears to be mainly mediaeval in origin.
4. The county has many Parklands (with origins from the medieval period to the 20th century) of which many are now being, or have been converted, or re-used as primarily golf courses, schools, hospitals and recreational areas.
5. There have been significant impacts by 20th century developments especially in the southern half of the county, these include:

    - New Towns and 'Garden Cities' e.g. Stevenage,

Welwyn Garden City,
- gravel extraction, e.g. Lea Valley,
- modern field enclosure (associated with urban fringes),
- motorways,
- field boundary loss leading to 'prairie-style fields' i.e. large fields created from post-war field boundary removal,
- horse stud farms.

Essex

Preliminary work has shown that in some respects the historic landscape of Essex (see Fig: 10), is transitional between that of Hertfordshire and Suffolk. The broad landscape types seen in Suffolk cross over in the north and east of Essex and the fragmented landscape recorded in Hertfordshire continues across in the south and west of the county.

The field systems in Essex are, as in Hertfordshire and Suffolk of an 'anciently enclosed' landscape. Field types include irregular, co-axial and sinuous. A distinctive and extensive, rectilinear or possibly co-axial system is present in the Dengie peninsula. As in Hertfordshire, Parliamentary enclosure is not common and tends to be piecemeal, apart from the Saffron Walden area on the chalk ridge in the north east of the county.

Other interim conclusions about the historic landscape of Essex include:

1. Essex has a lower density of historic parklands than Hertfordshire
2. Gravel extraction is widespread across the south of the county, as in Hertfordshire.
3. The relict elements of the original 'ancient' forests of Epping and Hatfield have survived, but are gradually being encroached upon by modern development. These rare and important landscapes are currently the subject of a special survey within the overall project, with regard to their future management and conservation.
4. Urban development has been extensive within the county e.g. Harlow, Southend, Chelmsford, Colchester.
5. Airfields are a significant feature of the Essex lands cape and often represent the relict elements of WW II activity. Some are visible as relict elements within the field system, (preserved as field boundaries, forestry strips, chicken farms), whilst others still function as civilian airstrips. Recent and planned airport expansion at Stanstead, exacerbated by planning constraints imposed to constrain development vertically, thereby forcing all development to be spatially invasive into the surrounding landscape, also to reduce noise, obstructions and other impacts. These will all have significant impacts on the surrounding landscape.
6. The loss of field boundaries since 1950 within the county, creating 'prairie-style' fields, has been dramatic, and in some areas so devastating, as to destroy the underlying coherence and infrastructure of the parent field-systems, i.e.
   - the surrounding roads form the field boundary, all former interior field boundaries having been removed
   - in areas a contiguous landscape of 'prairie fields'.

   These constitute an area of modern cohesive landscape generated by the Common Agricultural Policy subsidies and incentive payments.
7. As with Hertfordshire, the Essex landscape exhibits an extremely fragmented character as a result of twentieth century impacts in the more southerly parts of the county, i.e. closer to the influence of London.

As the HLA in the East of England counties is completed, spatial and 3D analysis will be used to identify the subtleties of landscape change within and between counties. Topography plays a role as the analysis has demonstrated, but the exact form of the relationship is not clear and probably varies according to historical period. This will take HLA into the 4th dimension by modelling HLA data not only within landscape morphology and form but through time.

Applications of HLA

Within the East of England project HLA has already, despite being incomplete, been actively used to inform a variety of land management issues including:

- Landscape Character Analysis (LCA) recently undertaken in Hertfordshire and Essex as part of the Mineral Local Plan, has used the HLA as a key data source. The HLA has provided information on the historic dimension of the landscape in an accessible format. This has assisted the landscape architects in the process of understanding the complex issues relating to landscape value. It has also aided the appreciation of the historic environment by the general public via a public consultation process.
- HLA allows the placement of many of the 'spot site' data of archaeological sites within Sites and Monuments Records (SMR's), into their spatial context and landscape. The Hertfordshire County Archaeology Section uses the HLA on a daily basis to assess the impact of development proposals on the historic landscape.
- The HLA data has been successfully used in support of archaeological conservation at a Public Inquiry in Hertfordshire.
- HLA will provide an important element in the process of formulating Strategic Policy at county level.

It is hoped that a national historic landscape character map will be generated from all of the HLA's across the country, comparable to the National Character map produced by the Countryside Commission for England. The HLA would form a backdrop against which the work by Wrathmell and Roberts (2001), on settlement analysis of England would nest within its broader historic landscape character setting. Once completed all this information could be analyzed within GIS, in conjunction with academic research to produce invaluable results.

Conclusions

Initial results illustrate the diversity and dynamism of the history of landscape development and genesis. These have enabled us to recognize, chart and appreciate its complexity and rarity. For example several rare examples of unaltered landscapes of unenclosed common arable have been identified from the project, and extensive landscapes of co-axial fields possibly of Bronze Age date have also been mapped.

Certain parts of the landscape exhibit contiguity and continuity whereas others areas are discontinuous. It should be possible to place spot-site's within their landscapes, to inform the assessment of various issues, for example:-

- Contemporaneity of spot sites within their landscape setting
- Survival of landscape types
- Condition and quality of landscapes
- Value i.e. rarity of landscape types locally, regionally, and nationally.

It can now be recognized that the landscape is extremely dynamic through time, something we have not until now fully appreciated or been able to systematically chart, prior to GIS. These 'forces of change', may be represented as various 'engines of change', moving at different 'speeds' and 'times' throughout the region. Certain parts are changing at the speed of Formula 1 racing cars, whereas other areas are still in 'horse and cart' mode, however nothing is in neutral or in idle. In some areas these 'engines' are being re-invented, restored or replicated without being based on sound information.

Work in Hertfordshire and Essex has highlighted the need for an analytical approach to recognize and systematically represent landscape history and diversity, due to the fragmented nature of the landscape. For example, in analyzing field systems, one can not rely on field morphology alone, as shape does not necessarily indicate period of origin. Results are also beginning to challenge some of the accepted views about landscape history, e.g. field boundary loss, survival and date of field systems, in that some fields are much older than previously thought, and their morphology has in instances undergone several subtle changes.

Finally the author believes that:

- A national framework needs to be set to allow a compatibility of methodology and datacapture, within GIS. Also that national metadata standards are established, which will facilitate mutually compatible datasets. Most importantly this will facilitate replicable results with conformity of approach and methodology.
- A national framework of broad 'historic landscape types' could be established, which may act as a national umbrella, into which all the various counties could feed into and inform and provide a regional series of sets and sub-sets to reflect regional and local diversity. This could be accomplished now and still be flexible and updateable as and when necessary, all that is required are extra fields within the database, indicating National, Regional and Local types (see schema above). This would ensure a consistency of application and that results would represent landscape diversity rather than methodological differences. Thus enabling national and regional landscape models to be developed.

This is an exciting project in which to be involved, which enables the richness and diversity of our landscape history, locally, regionally and nationally, to be recorded and assessed. This information is now proving to be an effective and pragmatic management tool in a wide variety of arenas.

The HLA methodology, now incorporating metadata, facilitates all the complexity of our historic landscape to be recognized and charted for further GIS management, analysis and academic evaluation. The methodology of data capture will continue to evolve, to reflect new technological advances e.g. object-oriented databases (geodata-databases), OS Digital National Framework (DNF). The new DNF MasterMAp form the OS will allow more efficient data entry while it should also assure greater consistency with other layers, based on this polygonized data. There is no doubt that in the future, the management and analysis of spatial data lies within the GIS platform, as GIS is a powerful tool able to handle vast amounts of complex spatial data., which will reflect true landscape diversity, rather than methodological differences across England

Acknowledgements

English Heritage has been the major funding body for the project. I should like to thank Graham Fairclough (English Heritage), Stewart Bryant (Hertfordshire County Council), and Paul Gilman (Essex County Council) for their assistance and support in this project.

References

Cadw. 1998. *Register of Landscapes of Outstanding Historic Interest in Wales*, Cadw, Cardiff

Countryside Commission. 1997. *Landscape Assessment Guidance.*

Countryside Commission

Dyson-Bruce, L. 1998. *Historic Landuse Assessment*, Unpublished Report for Historic Scotland.

Dyson-Bruce, L., with Dixon, P, Hingley, R., Stevenson J. 1999. *Historic Landuse Assessment (HLA): Development and Potential of a technique for Assessing Historic Landuse Patterns. Report of the pilot project 1996-98*, Research Report, HS & RCAHMS, Edinburgh.

Fairclough, G., ed. 1999. *Historic Landscape Characterization*, English Heritage, London.

Ford, M. 1999. *Historic Landscape Characterization in Suffolk 1998-1999.* Draft Report.

Garsd, A. 1984. Spurious correlation in ecological modeling, *Ecological Modeling,* **23**: 191-201.

Herring, P. 1998. *Cornwall's Historic Landscape: Presenting a method of historic landscape character assessment.* Cornwall Archaeology Unit, Truro.

Wrathmell, S., Roberts, BK. 2001. *An Historical Settlement Atlas of England.* English Heritage, London.

Figures

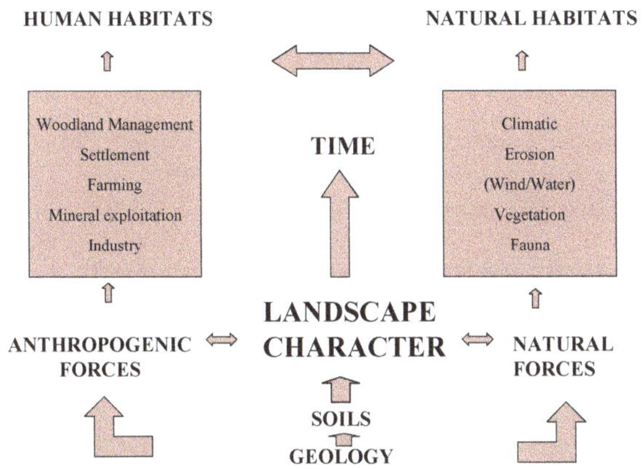

Figure 1. Landscape character in relation to natural and anthropogenic factors

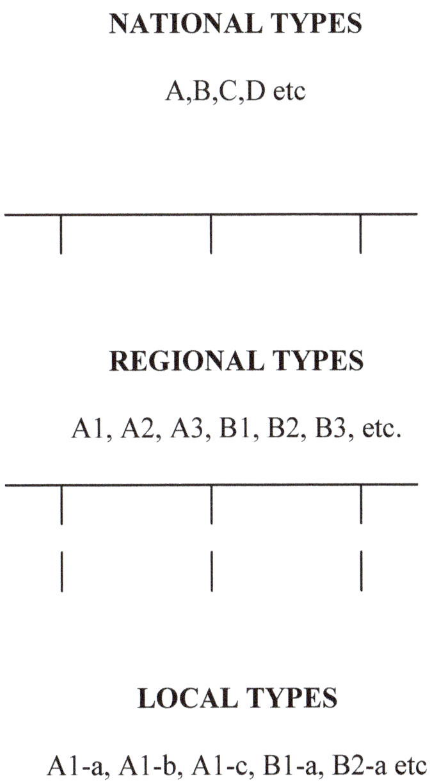

Figure 2. Proposed schema for National HLA Types.

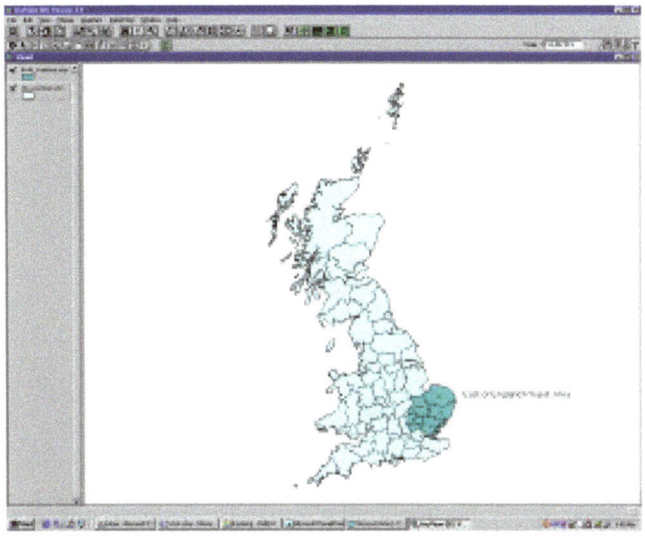

Figure 3. Map of the East of England Project Area within England

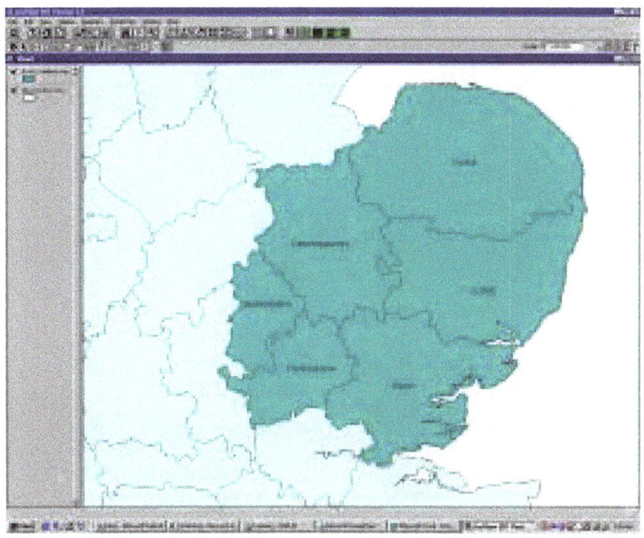

Figure 4. Detail of the Counties within the East of England Project Area.

COUNTY	COMPLETION	GIS SOFTWARE
Suffolk	1999	MapInfo
Hertfordshire	August 2001	ArcView
Essex	April 2002	ArcView
Cambridgeshire	April 2003	MapInfo
Bedfordshire	April 2003	MapInfo and Wings
Norfolk	? Date ?	? GIS ?

Figure 5. Table of Counties within East of England Project Area – Completion dates and GIS platform used.

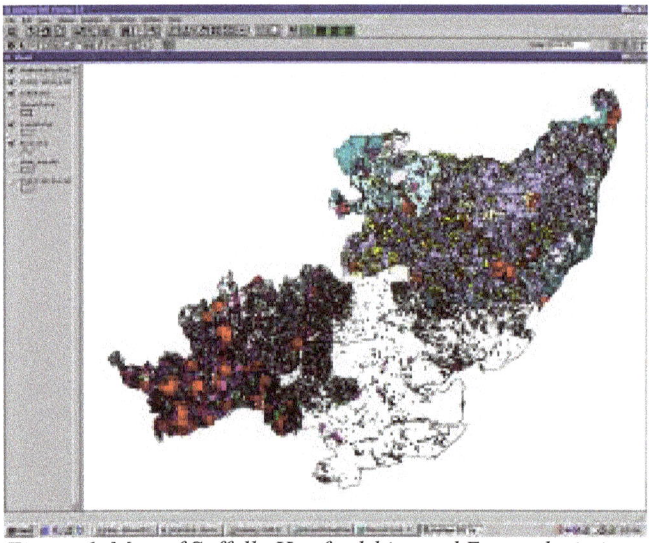

Figure 6. Map of Suffolk, Hertfordshire and Essex, depicting work in progress.

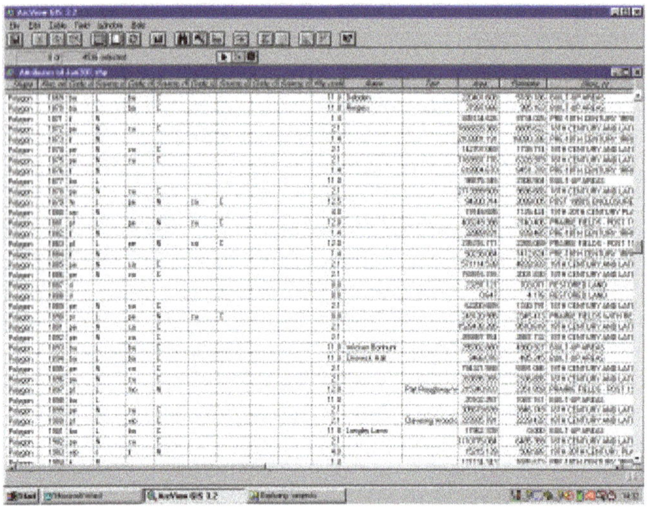

Figure 7. Sample of HLA data-base – work in progress.

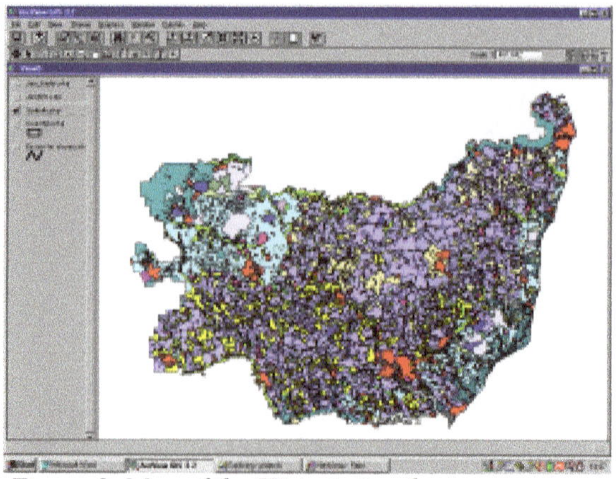

Figure 8. Map of the Historic Landscape Characterization of Suffolk

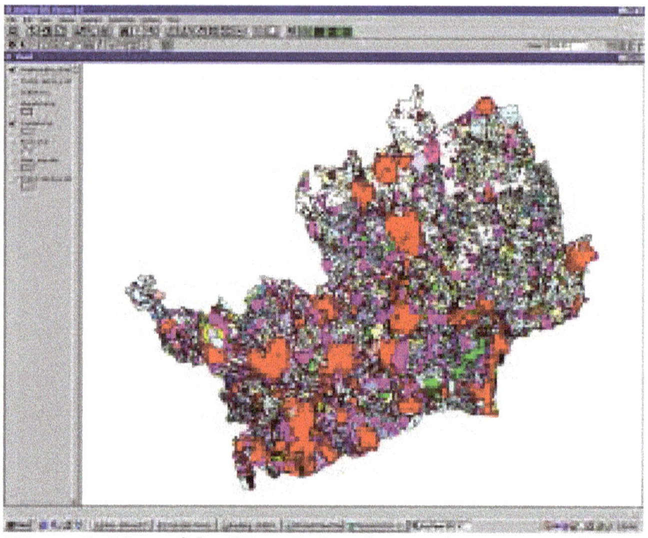

Figure 9. Map of the Historic Landscape Assessment of Hertfordshire – work in progress.

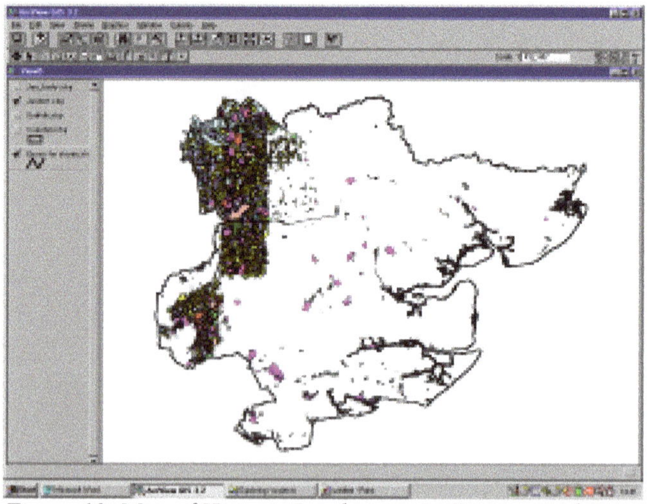

Figure 10. Map of Historic Landscapes in Essex – work in progress

GIS approach to Iberian Iron Age landscape in central-south Valencia region (Spain)

Ignacio Grau Mira
Area de Arqueologia
Facultad de Filosofia y Letras
Universidad de Alicante
Ap. Correos. 99
03080 Alicante, Spain
E-mail: Ignacio.Grau@ua.es

Abstract: This paper is an approach to landscape and territory in the central-south area of Valencia Region (Spain) through GIS techniques. Taking the information from the surveys and the revision of the known sites, an analysis of the settlement is made. We attend the dynamics of development of the Iberian oppida as the principal centres of population and political control.

Introduction

This paper shows some aspects of the recent research concerning Iberian landscapes, conduced by scholars from the Department of Archaeology of the University of Alicante (Spain). The archaeological research has been developed in the Alcoià-Comtat valleys (province of Alicante) in the central-south district of the Valencia region (Grau Mira, 1998). The aims of this research are to study aspects of settlement, landscape, and material culture among Iberian peoples.

In this region we explore the processes of spatial and societal transformations developed during the Iberian Iron Age, between the 7th and the 1st centuries BC. In Iberia this change is evidenced by the emergence of hillforts that are at the top of a hierarchical settlement pattern. These hillforts, traditionally referred to as *oppida* in Iberian research, are the principal centres of population and political control (Ruiz y Molinos, 1993). We analyse functional aspects of the Iberian hillforts and their role in regional settlement patterns using Geographic Information Systems.

Geographical background

The area of our study is L'Alcoià and El Comtat counties so-called Valls d'Alcoi, a mountainous area with peaks between the 300 and the 1200 meters, located in central Valencia Region, eastern Spain. The valleys consist of a series of folds oriented SW-NE, the typical orientation of the Betic reliefs, which define a landscape of calcareous mountains dominating narrow and irregular valleys that lead into small plains in the coast. This valleys are articulated by the river of Alcoi basin, the main valley of the area where are going to spill all the streams of the tributary valleys.

The orientation of these reliefs have a very important role in possible communication between this area and nearby region, that can only be achieved through a series of mountain corridors whose control is relevant to strategic pursuits.

The climate is classically Mediterranean with a summer drought, even temperatures in winter and maximum precipitation concentrated in the autumn months. However the broken topography modifies this general climatic pattern; for example, the northern exposed slopes will normally get more rain than southern exposed ones (Nebot *et al.*, 1993, 29).

The geological background, the climate and the broken topography have resulted in the existence of diverse types of soils in the area: in the top and slopes of the calcareous mountains we find rocket soil without agricultural possibilities. In the other hand, the better soils are located in the basins of the valleys, they are composed by alluvium and sediments deposited by the rivers in flat terrain. Middle capability soils can be found in contact areas between mountains and basin.

These natural features have led to the concentration of the settlement near the valleys, where we find the most important sites and the main ways of the area.

Procedures and methodology

We use GIS tools to analyse the structural relationships between hillforts and their surrounding area, including issues related to economic structures, control and visibility.

One of the most important points of procedure when applying GIS to landscape studies is to use the most homogeneous and precise cartography that is available. In this respect, the *Conselleria d'Obres Publiques i Urbanisme* of the Valencia Regional Government has edited many basic and thematic cartography in digital editions in the series *Sistema d'Informació Territorial*. This collection offers cartographic information in DXF and shape files for the *ArcView* program; this system allows the transfer of data with great precision.

The cartographic coverage that we used were 20 m and 100 m elevation contour lines, geology, rivers, modern roads and agriculture capability soils classification.

We produced a Digital Elevation Model of the study area generated from contour lines with a vertical interval of 20 m and covering an area of approximately 700 sq. km. With this DEM we explored cost surface analyses and viewsheds. We also examined tessellation of the landscape through Thiessen polygons.

The main analysis was carried out with GIS raster *Idrissi* and the resulting coverages were introduced into the *ArcView* program. This combination offers better possibilities of mapping, image processing and compatibility with other databases.

Settlement pattern and allocation

Different research groups have carried out excavations and intensive surveys in the study area and have uncovered the main archaeological remains. These surveys provided intensive information, which is translated into a high density of sites dating from various prehistorical periods, including those belonging to the Iberian Iron Age.

Archaeological fieldwork has provided evidence of seventy sites corresponding to the Iberian Period.

The settlement pattern in the area has two principal type of sites.

1. *Hillforts*. There are ten sites located on the hilltops. These cover approximately 2-3 hectares in area and are protected by defensive walls. These are the principal sites in the Iberian period, more commonly known as *oppida*.
2. *Farmsteads*. Sixty small sites are located at the base and on the low slopes of the valleys. Their size is typically less than a hectare, and frequently less than 5000 sq. m. These sites are interpreted as villages or farmsteads subordinate to *oppida*.

The *oppida* are generally located on summits on an approximate altitude of 800-900 m along the principal range of mountains that border the region, and around 200-300 m over the lower valleys. These locations offer defensive positions protected by the topography with good visibility of their surrounding areas. Thiessen polygons were used to assign territories to each *oppidum*, but they were first adjusted to conform to principal natural features like rivers and ridgelines (Van Leusen, 1993). We adapted the geometric lines of the theoretical Thiessen polygons to the irregular lines corresponding to natural boundaries in the proximity of these polygons. Through this approach the entire region was partitioned into territories corresponding to small valleys with boundaries recognisable in the landscape (Figure 2).

Cost surface and catchment analysis

Secondly, we employed cost surfaces analyses to define catchments encompassing to an hour's walking distance in order to explore the possibility of competition for resources. The cost surface derived catchment areas allow an examination of the economic possibilities of each *oppida* by relating the catchments with information about arable lands and soils quality. Through this analysis it is possible to quantify agricultural resources and the economic potential of each site within its theoretical territories (Figure 3).

We can conclude that it is not possible to observe similar economic strategies in relation to size and quality of land of these principal sites. However, we can detect that each *oppidum* controlled its surrounding land without frictions and competition with neighbours for land. There is an equal allocation and distribution of the *oppida* around the region. The principal interest of hillforts is oriented towards the control of medium quality soils. Land use is oriented to dry farming for cereal production, the characteristic agriculture in the Iberian period.

The large populations of the hillforts were the principal exploiters of the terrain in the region. Nevertheless the mountainous locations of the *oppida* made movement around the environment difficult, so their catchment areas were small. Therefore the exploitation of resources in these territories was completed by the subordinated farmsteads.

Communications and optimum corridor analysis

The communications between the settlements are of great importance in understanding the organisation of space. The existence of fixed pathways and communication axes were necessary to establish economic relations and to fulfill needs of political control.

In the mountainous area of study, the relief imposes some obstacles to circulation. We have carried out a GIS analysis employing the techniques of Cost Surface. These functions analyse the difficulties of travelling through a surface in relation to the characteristics of the terrain and the natural barriers. We take information to apply these techniques from a Digital Elevation Model.

We obtained optimum corridors, the routes of least resistance on wich to travel through the area. These pathways take advantage of the natural steps and the areas of lower slope. Secondly, we contrast the optimum corridors with the historically known roads in the region dating from the Middle Ages. The result shows that the traditional roads follow the same routes as natural corridors because these are the only possibilities of circulation in this mountainous region.

The layout of the natural corridors was contrasted subsequently with other layers of information about the location of the Iberian settlements. The purpose of this was to verify which of these corridors could have been used as pathways in the Iberian period. These were two principal conclusions:

1. The main pathway follows the North-South axis of the valley. Three of the most important *oppida* are located close to the road and control circulation.
2. Every *oppida* manages the secondary pathways that are present in their nearest valley. These pathways allow relations between the principal settlements.

Viewshed and intervisibility analysis

Viewshed analysis is very important to perceive territoriality and to investigate ideas about control over the landscape. We analysed the viewshed of each *oppida* through a DEM of the area as well as intervisibility between the *oppida*. Another fundamental feature of the landscape, the principal ways of travel, was also examined with respect to visibility. We developed some research similar to other GIS applications using similar technique to the analysis of Celtic roads in Burgundy (Madry and Rakos, 1996).

We can observe three principal focuses of visual control:

1. The view is oriented to the valleys below the *oppida* locations and the nearby small sites in the basin, much like GIS-based interpretations offered in other Iron Age landscapes (e.g., Lock and Harris, 1996).
2. Visibility is oriented to control the most important pathways in the area (e.g., Madry and Rakos, 1996), especially the mountain corridors. These points were very important places for achieving access to the Alcoià-Comtat Valleys.
3. Each hillfort has visual control over other *oppida*. Intervisibilty analysis allows the recognition of an important web of visual intercommunication between the *oppida* of the region. This visual connection is perhaps related to a common defensive strategy in case of incursion and war. Each *oppida* can issue the alert in case of danger.

This strategy is very important for the control of the principal North-South pathway that crosses the central part of the valley, allowing contact with other regions. This route is highly visible from the *oppida* as we can observe from the viewshed analysis. Figure 5 illustrates an example of visibility between two hillforts. The visual control of nearby valleys is clearly demonstrated, also the intervisibility with other hillforts. The control focuses on the central part of the region and the principal pathways. (Figure 5).

Final remarks

The procedures described are the basic GIS-techniques that we have employed throughout our study of the territory. This application allows us to analyse the relation between the most important Iberian settlements and their environment. Also, this system permits the management and transformation of an important volume of archaeological and geographical information.

GIS analysis illustrates the existence of early processes of proto-urbanisation and centralisation in the study area. This dynamic is shown in the emergence of several hillforts that assumed the main functions of control and exploitation of the resources in the landscape.

These hillforts or *oppida* control each one of the valleys of the region. The hill top sites manage the settlement of this environment composed of subordinate farmsteads. Also, these settlements exercise control over different roads and the defence of the territory. They provide refuge to dispersed settlements in case of danger.

This model is similar to other territorial structures in the Iberian Iron Age. These settlement patterns are characterised by the existence of hillforts controlling small territories consisting of farmsteads. These territorial structures are part of the socio-political organisation of the Iberians, characterised by the emergence of primary aristocracies and proto-state forms of development (Ruiz and Molinos, 1993).

During the Classical Iberian Period this settlement pattern changed to develop a more hierarchized model. The small valleys analysed formed part of a great territory corresponding to the complete region of L'Alcoià-El Comtat in the 3rd century BC. The Iberian town of La Serreta de Alcoi controlled this territorial structure (Olcina *et al.*, 1998) and developed its power upon the other *oppida*. This town exercised the main political, economic and social functions in the landscape.

The new territorial structure established a network of common intervisibility allowing communication inside the region in case of a hostile incursion. This control permitted a rapid evacuation of the territory and the preparation of a possible assembled reply against the invader. This change is related to the historic circumstances of the period, principally at the end of 3rd century BC when the area was severely affected by the Second Punic War. As a result of this war, the Iberian structures of settlement were transformed and the process of Romanisation developed new models of organisation.

The area analized was in the core of a major region named *Contestania* by the ancient sources. In this region the most important strategic role of the settlement pattern was related to the possibilities of communication between the north and the south of *Contestania*. The important centres of *El Tossal de Manises* (Alicante) and *La Alcudia* (Elche, Alicante) was located around 30 km to the south and *Saiti* (Játiva, Valencia) was located 25 km to the north of the area of study. The region was well connected by the principal North-South road that crosses the Alcoià-Comtat valleys. This territorial integration allowed the very important development of *Contestania* in the Iberian Period that became one of the most important territories in *Iberia*.

Acknowledgements

I am very grateful to Dr. Kenneth Kvamme for his assistance in the use of GIS applications to my research.

References

Lock, G and Harris, T.M. 1996. Danebury revisited: An English Iron Age hillfort in a digital landscape. In Aldenderf, M. and Maschner H.D.G. (eds.), *Anthropology, Space and Geographic Information Systems*. Oxford University Press, New York, pp. 214-240.

Grau Mira, I. 1998. Aproximación al poblamiento de época

Plena Ibérica (ss. IV-III aC) en la región centro-meridional del País Valenciano *Arqueología Espacial XIX-XX.* Teruel. 309-321..

Nebot, J., Massanet, C., Torró, J. 1993. *L'Alcoià y El Comtat. Guia històrica i natural.* Alcoi.

Madry, S.L.H. and Rakos, L. 1996. Line-of-sight and cost-surface techniques for regional research in the Arroux Valley. In Maschner H.D.G. (ed.), *New Methods, Old Problems: Geographic Information Systems in Modern Archaeological Research*, Occasional Paper, N° 23, Southern Illinois University, Carbondale, pp.104-126.

Olcina Domenech, M., Grau Mira, I., Moltó Gisbert, S., Reig Seguí, C., Sala Sellés, F. y Segura Martí, J.M. 1998. Nuevas aportaciones a la evolución de la ciudad ibérica: el caso de La Serreta, In Aranegui, C (ed.), *Congreso Internacional Los Iberos, Príncipes de Occidente (Barcelona, marzo 1998).* Barcelona. pp. 35-46.

Ruiz A. and Molinos, M. 1993. *Los Iberos. Análisis arqueológico de un proceso histórico.* Barcelona.

Van Leusen, P.M. 1993. Cartographic modelling in a cell-based GIS. In Andresen, J., Madsen, T. and Scollar, I. (eds.), *Computing the Past: Computer Applications and Quantitative Methods in Archeology, CAA92*, Aarhus University Press. Aarhus. pp. 105-122.

Legends

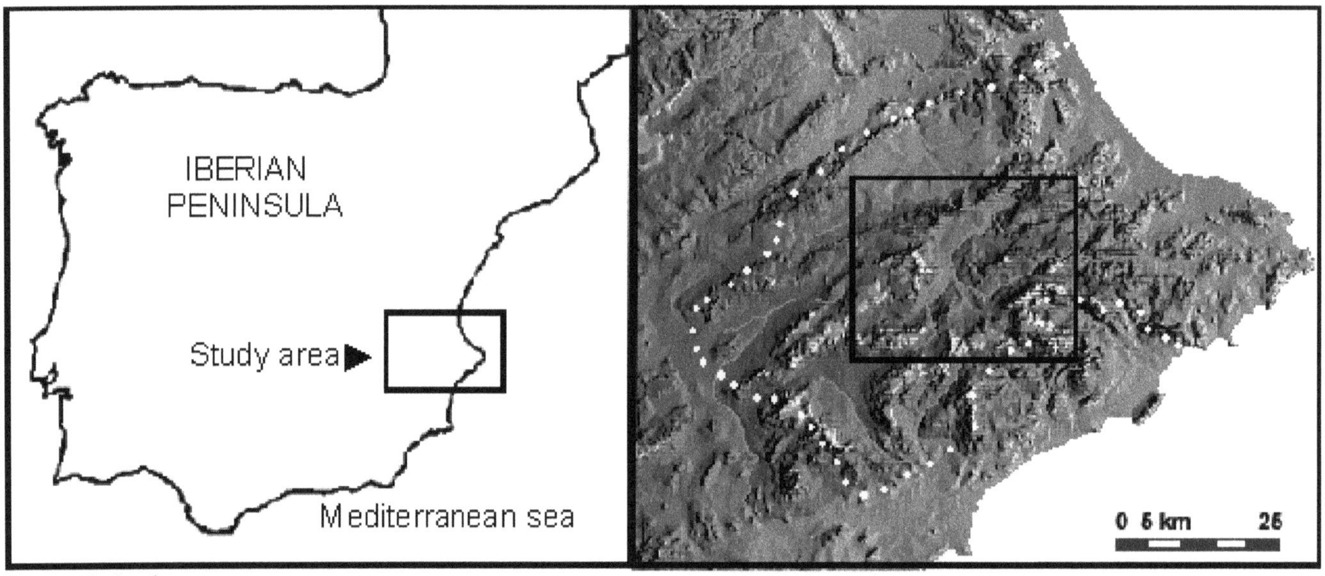

Figure 1. Study area.

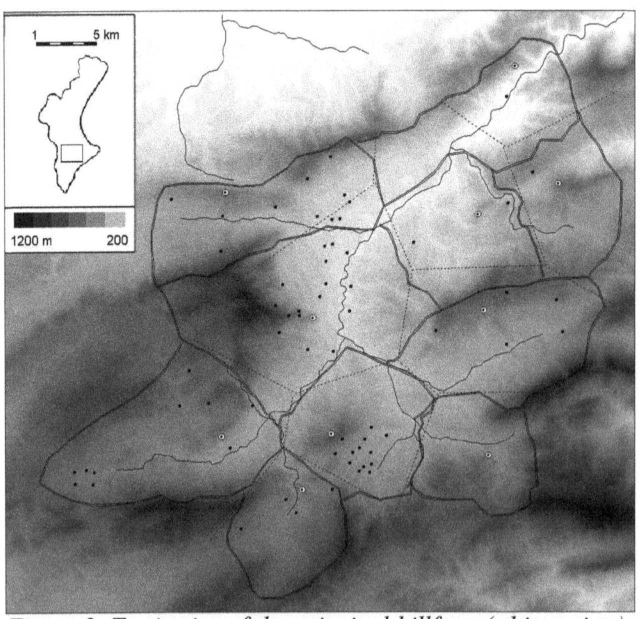

Figure 2. Territories of the principal hillforts (white points) achieved through modified Thiessen polygons conforming to landscape features

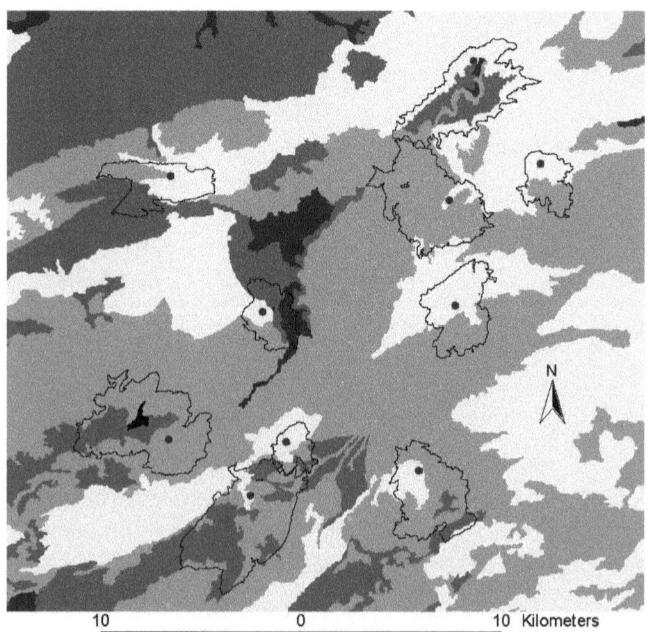

Figure 3.Cost surface derived catchment areas of the hillforts (white points) plotted over soil productivity. Dark colours are good soils; clear colours are poor soils.

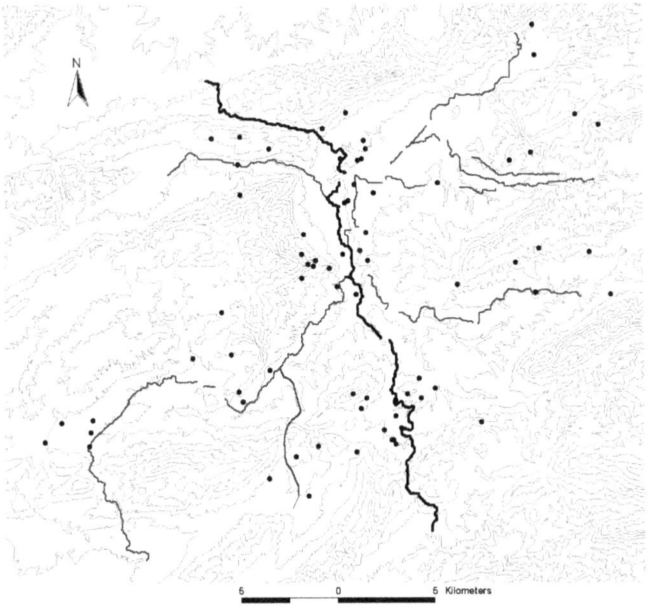

Figure 4. Settlement distribution inrelation to the principal pathways in the area.

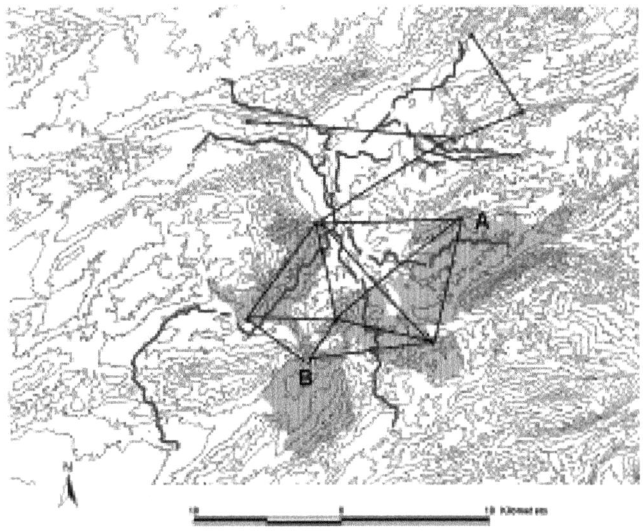

Figure 5. Visibility between two example hillforts (A and B) and the intervisibility relation of the principal hillforts. In dark grey we illustrate the principal pathways.

Developing an Information System for Archaeological Sites and Monuments - Data Model and Construction

David Haskiya
National Heritage Board
Box 5405
114 84 Stockholm, Sweden
Phone: +46 8 5191 8511 - Fax: +46 8 667 89 43 - E-mail: david.haskiya@raa.se

Abstract: The Swedish National Heritage Board is currently developing an information system for archaeological sites and monuments. The system is primarily intended for cultural resource management, forest protection management, physical planning and academic research.

The project is now in the second stage, which is composed of construction and testing of the information system and the drafting of a model for the administration and maintenance of the system.

Details are given on the modelling, design and construction of the system. Further topics are the classification of sites and monuments in the system and the use of XML for queries and data transfers.

Key words: information system, GIS, archaeological sites and monuments, data modelling, XML, meta-data

Project background

The topic of this paper is a development project where the aim is an information system for archaeological sites and monuments. The paper will elaborate on the modelling and design of the information system. It also includes a short background to the project and a brief description of Swedish cultural heritage management. The development project is carried out at the National Heritage Board, which is the central authority in Sweden for matters concerning the cultural environment and cultural heritage.

Existing information and databases

Systematic surveys and the Register of Ancient Sites and Monuments

In 1938 a nationwide survey for archaeological sites and monuments was initiated by the National Heritage Board (NHB). The reason for this survey was a government decision on mapping the landuse of the country. Archaeological sites and monuments protected by law were to be included on the Land Use Map published by the National Land Survey (NLS).

Information on the sites, in the form of geographic positions on maps and descriptions in notebooks was gathered in the Register of Archaeological Sites and Monuments (in Swedish "Fornminnesregistret"). Information from subsequent surveys, archaeological investigations and reports on finds made by private citizens, have since been added to the Register. The Register is still maintained by the National Heritage Board.

Today the Register contains information on roughly 400 000 locales each comprising one or more sites or monuments. Not all of the sites and monuments in the Register are protected by the Cultural Monuments Act; the Register also includes historically/ archaeologically interesting sites which are not protected by law.

Databases at the NHB and NLS

In the middle of the 1980´s the National Heritage Board began digitizing the descriptions in the Register. The geographical information has been, and continues to be, partly digitized by the National Land Survey (NLS).

Attribute data was stored in an Informix database, "Forn", at the NHB, but the geographical data was stored at the NLS primary databank ("Geodatabanken - alfa") for Sweden. The fact that two different applications were used to register data in two different databases has made integrated computeraided analysis and management of the data complicated and time-consuming.

In 1998, as the past efforts of the National Heritage Board were never brought to a satisfactory conclusion, the NHB was given a commission by the government to solve the problem with digital access to this kind of information. In 1999 a development project with the aim to construct a new information system was initiated. The project will end in 2002 when the information system will be gradually deployed at the National Heritage Board and at the different County-council Boards.

Modelling and design of the information system

Requirements analysis

Modelling and design of the system really began when the project team travelled around the country interviewing the potential end-users about their wishes and demands for an information system on archaeological sites and monuments. Questionnaires were also sent to all stakeholders and end-users in order to get their input.

The requirement analysis then proceeded by drafting a "requirements document" through a number of brainstorming sessions in a workgroup consisting of members of the development project and representatives from different actors in Swedish cultural heritage management.

The resulting "requirements document" deals with all aspects of the system e.g. data content, possible queries, map background and system response times. This document is at the core of all modelling and design of the system. It will also be used as the basis for the acceptance testing protocol.

The specialized data model

Modelling continued with conceptual- and data modelling – the two were initially done in parallel, but the two workgroups merged after a few modelling sessions. The resulting data model consists mainly of an object representing the archaeological site or monument. This central object has relations to the secondary objects, events and references. Much effort was put into the modelling of a classification system for the central object (see below).

The general data model

Parallel to the development of the information system for archaeological sites and monuments the NHB also initiated a project to develop an overall information technology strategy. As part of this effort different existing data models (e.g. models for archaeological excavations or historic buildings) inspired the development of a common data model for the NHB. As a result the data model for sites and monuments can now be considered as a specialization of the common data model.

Classification of sites and monuments

The sites and monuments recorded in the database will be classified according to a system devised as part of the development project. The classification-system is basically three-tiered - types of objects can belong to one or more category and features can be added to every type of object (see below for an example).

Category: "Graves"
    Type: "Cairn"
        Feature type: "Form"
            Feature value: "Long cairn"

The (meta-) database currently allows for 12 categories and 160 types of sites or monuments, but through the use of features (types and values) the 160 types can be sub-classified in a number of ways. Categories, types, feature types and feature values can easily be added or modified with the option of letting such changes affect the classification of all database-entries.

XML for queries, data and meta-data

The requirements analysis emphasized the need for a platform independent manner of data transfer between system and end-user (and also from system to system) – the system developers at the NHB quickly turned to the use of XML to satisfy these design demands.

The resulting XML-format is tailor-made to contain data and meta-data based on the NHBs common data model, but it can also be used to query the system as well as create entries in the database. The format also comprises tags referring to binary data sent in the same "package" as the XML-document.

Using GML (Geography Mark-up Language) for geographical data was considered, but the NHB decided to wait until this format is formally standardized and versioned. In the meantime geographical data will be sent as ESRI shapefiles.

Data flow

The system will exchange data mainly with the NLS, but will also receive data from the National Maritime Museum (NMM). The data received from NMM derives from their database on underwater remains. The data will be imported through a database-link (both the sending and receiving systems use the Oracle DBMS). The imported data is converted to comply with the data-model, allowing it to be sent from the data layer to application layer in the same XML-format as the other data.

The information system will automatically send the geographical data and some attribute data for all sites or monuments protected by the Cultural Monuments Act, as well the same data for a number of other types of sites or monuments traditionally printed on official maps, to the NLS (for inclusion in their Cadastral Map). The geographical data will be converted to the NLS´s proprietary data-format AutoKA-FF before transfer, but some complementary data will be sent as XML.

Landscape and cadastral data for the system will be stored on a remote server (served by the NLS) dynamically producing GIFs for the system's web-based client.

Users of the system will be able to interactively download data from the system in a number of different formats. Initially the possible formats are ESRI shapefiles, MIF/ TAB-files, DXF-files and XML-documents. These four optional formats cover the needs of the vast majority of possible users.

Finally, third-party applications can be "certified" to communicate with the NHB-system provided that they utilize the NHBs XML-format for queries and data transfers. In this way third-party applications can dynamically query the system e.g. an electronic request from a forestry company to the National Forestry Board for permission to harvest an area of forest triggers a dynamic query from the National Forestry Board's system to the NHB's system. The NHB's system then returns the data for any protected sites or monuments within the area to be harvested, which the National Forestry Boards system in turn re-packages and sends on to the forestry company.

System architecture and construction

The layers: Databases, Data layer, Application layer and Clients

As part of the NHBs information technology strategy the decision was made to implement a system platform that would be adaptable enough to handle many different types of data, uses and users. The general system architecture is as follows:

Databases containing data, geographical data and user information is stored in an Oracle DBMS. The NHB uses Oracle Spatial and Intermedia to store geographical data and images.

The data layer handles incoming XML commands and data from the application layer, converting them from XML to SQL. Database responses are in turn converted from SQL to XML and sent on to the application layer.

The application layer will handle incoming data and commands sending them on to the data layer. Responses from the data layer are converted from XML to HTML if the client is the thin web-client or just passed on as pure XML if the request comes from the client for data entry. By accessing the information stored in a user session the application layer can utilize different XSLs to produce HTML layouts in the thin web-client. In theory the application layer could also use other XSLs to produce other sorts of output e.g. PDFs or WAP-files.

In this way the system can tailor-make different outputs depending on what type of user is sending the request. A pupil in an elementary school and a professional in cultural heritage monument can thus have different layouts on their browsers with different query capabilities, receiving different system responses, and yet communicate with the same database.

Two clients for the system will be developed. The first will be a thin (web-based) client for searching and extracting data, the second a "fat" client for data entry.

The thin client will not demand any plug-ins and will run on Internet Explorer 4.01+ or Netscape 4.5+ and will allow the user to query the system either through geographical queries, attribute queries or a combination of both. Depending on the user´s rights, data can be interactively downloaded from the system in a number of different formats (see above).

Summary

In response to an increasing demand for information on archaeological sites and monuments in digital form the Swedish National Heritage Board has initiated a project aiming for the construction of an information system holding data of this kind. The system will store c. 1,000,000 data-entries and will be deployed from 2002 onwards.

The aim of the project has been to construct a system that is expandable and adaptable. In order to do this we have developed a conceptual model for archaeological sites and monuments that does not stop the results of research, or changing heritage management principles, from affecting and changing the model.

Further, the National Heritage Board has decided to construct a general system architecture that can function as a platform for many different applications, dealing with different sorts of data. The architecture designed emphasizes the separation of storing the data from presenting the data (in the clients) by using XML in communications between application and database.

Figures

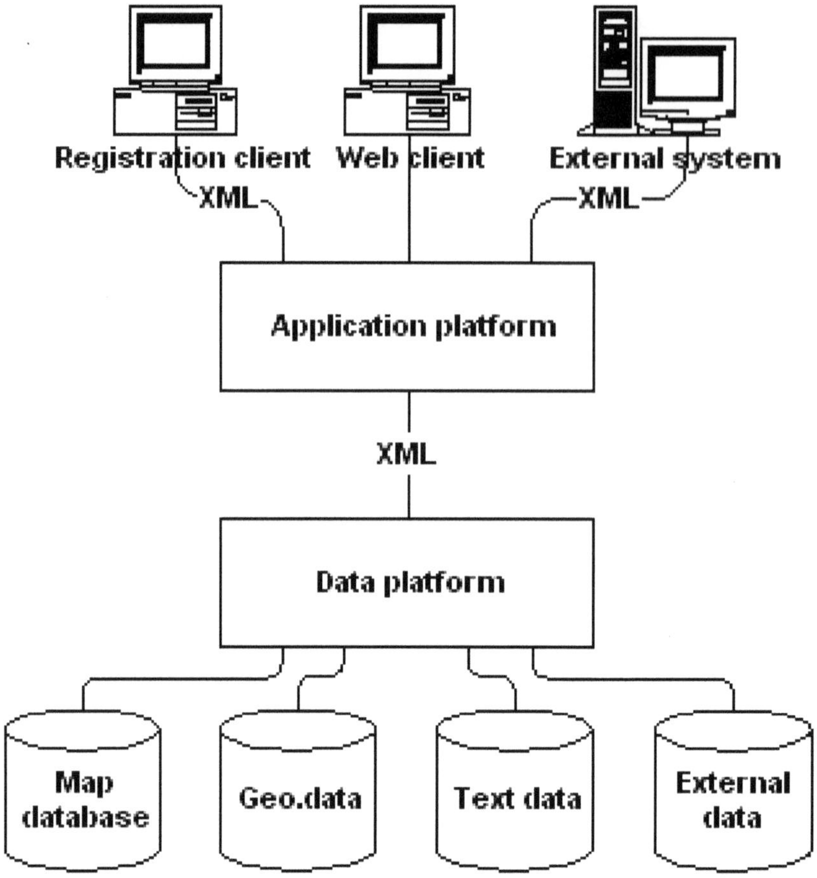

Figure 1. The architecture of the information system.

Using viewsheds wisely: developing sound methodologies from spatial analyses of megalithic monuments in western Scotland

Gail Higginbottom
Centre for European Studies and General Linguistics
University of Adelaide,
Adelaide, South Australia,
Australia. 5005.
Phone: +61 8 8303:5996 - Fax: +61 8 8303:4380 - E-mail: ghigginb@physics.adelaide.edu.au

Ken Simpson
Department of Physics and Mathematical Physics
University of Adelaide,
Adelaide, South Australia,
Australia. 5005.
Phone: +61 8 8303:5996 - Fax: +61 8 8303:4380 - E-mail: ksimpson@physics.adelaide.edu.au

Roger Clay
Department of Physics and Mathematical Physics
University of Adelaide,
Adelaide, South Australia,
Australia. 5005.
Phone: +61 8 8303:5996 - Fax: +61 8 8303:4380 - E-mail: rclay@physics.adelaide.edu.au

Abstract: In a similar way to that of Fisher et al's 1997 paper, the aim of this project is to contribute to the quantitative analysis and development of testable hypotheses concerning archaeological sites in the landscape (Fisher et al, 1997:581). The initial intention of this project was to ensure that valid and reliable outcomes regarding the original use of the free-standing megalithic monuments of western Scotland were possible through its use of appropriate spatial and statistical analyses. Whilst this objective remains the same it is no longer the sole objective. Rather, more complex theories regarding the nature of the cosmology of those who built the monuments and the possible cosmological connections between them, other monuments and the environment are considered. Based upon the initial methodologies and outcomes, further development of sound hypotheses and robust experimental designs that could be used in conjunction with GIS data and applications was assured for the more complex considerations.

This paper outlines the overall project design and demonstrates the connection between the methodologies used to date and those to come. It is believed that the project design fulfills the minimum requirements for the incorporation of systematic project design and quantitative analysis in the application of viewshed technology.

Key words: landscape archaeology, archaeoastronomy, methodology, GIS, viewshed, spatial analysis, Scotland, megaliths

Introduction

Ultimately, this study investigates the cosmology of the megalith builders of western Scotland (see Figure 1). It was decided that an investigation into territory that had previously been studied, geographically and thematically speaking, was the most sensible approach to begin with. However, it was also recognised that new methodologies were required, in order to improve the state of research into prehistoric societies and archaeoastronomy in particular. What had appeared as a single télos of the study, became twin téloi: desire for the revelation of reliable conclusions through the development of sound methodological practices.

This study began by reassessing the orientation of free-standing stone monuments and their possible associations with astronomical phenomena (originally assessed by Ruggles and contributors in 1984). Whilst investigating the possible interest of the megalith builders in astronomical phenomena it became evident that a much more complex set of behaviours (deliberate or not) might be revealed. If discovered, these could provide a deeper understanding of the cosmological issues at stake for the builders, and perhaps even those who came before and were to come afterwards. From the beginning, the project was based upon conventional scientific methods with the belief that with proper project design, and the asking of appropriate questions, one can only enhance one's knowledge of the societies one is trying to understand.

The experimental design for the expanded study relied upon the outcomes of the first two phases of the project. Due to their nature, they provided us with valid hypotheses to be tested, and directed us towards appropriate methodologies to test them with. These newer methodologies, as well as those found within the first two phases of this study, are based upon considerations current within landscape archaeology and the application of GIS. More specifically, phase three of the study is dependent upon the use or modification of GIS software (GRASS[1] and ARCINFO[2]), primarily those involving viewshed procedures. It is the intention of this paper to discuss the methodology of the project in detail.

Some current considerations

Many of the criticisms in the field of Landscape Archaeology and, to some extent Archaeoastronomy, are well rehearsed. Nevertheless, we will briefly mention some of the general issues. In this way, we may later highlight how our own research attempts rectify some of the problems still inherent in the study of archaeological sites in the landscape, especially in the application of GIS viewshed analysis. The criticisms are aimed primarily at the lack of rigorous analysis. Though the project addresses a number of critical issues in methodological approaches in Landscape Archaeology, this paper will focus upon the development of sound viewshed analysis.

What many critics of GIS analysis are looking for is the connection between visualization and statistical analyses (Lock and Harris, 1992; Kvamme 1995:7) Kvamme summarises this quest:

> simple statistics cannot convey the essence of spatial pattern in the same way that an effective graphic can. At the same time, statistical tests can inform us of the existence of (a) pattern when it is difficult or impossible to visualise, and even if we can see (a) pattern we may wish to obtain objective measures of its (existence and of its) strength. Both approaches complement each other ... (1995:7).

To create such a complementary approach between vision and statistics we must begin with the creation of testable hypotheses, for this allows for the possible derivation of firm statements of significant associations (Fisher et al 1997:582).

Testable hypotheses

The advantage of designing testable hypotheses is that we can be more certain of both our results and their interpretation. Additionally, we will gain a greater understanding of the nature of our data, our methodologies and our enquiries by the sheer fact that we have had to produce very specific questions to provide very specific answers. Added to these, it allows us to determine further hypotheses which may help to disentangle more complex issues.

To create statistically testable hypotheses, certain properties of observed or gathered data are compared to those of a hypothesised (modelled) or real population (this may be a 'parameterised' real population). These latter populations are often called 'control groups', 'expected populations/ distributions' or 'background values'. The expected distribution is often that distribution which would occur if nothing other than an interplay of chance factors were responsible for its formation. When the resultant outcome of no significant difference is found between the control and the observed data in this case it is said that the observed data are also likely to be the result of chance factors.

The creation of expected distributions of viewsheds

"There is no predetermined method for finding the statistical significance of the area visible from one location as opposed to another"(Fisher 1997:584)

In GIS the influence of landscape factors cannot really be understood from a single model. This is because the topography of the landscape varies. Thus any model or expected distribution will be "dependent on site-specific terrain" (Fisher 1997:584).

Naturally, as viewsheds are affected by terrain, we have to determine "what is the likelihood that the observed pattern of visibility is an artefact of the surrounding landscape or topography, rather than anything else?" The most accurate, but time consuming, approach would be to create viewsheds for all locations within the Digital Elevation Model (DEM) through full intervisibility analysis. From these parameters the population of the viewsheds could be calculated. A more suitable, and still very reliable, approach would be to create viewsheds from randomly sampled viewing locations. This will allow one to build a representative sample of the viewsheds of the entire landscape (Lake et al, 1998:34) from which an expected distribution can be generated. An example of this applied technique can be found in Fisher et al (1997:584). Terrain must be seen, then, as a condition or a 'background visibility property' as referred to by van Leusen (1999:§3.2).

Lack of statistical analysis

Peter Fisher et al state categorically that "(if GIS used) ... then rigorous statistical analysis should and can be applied, but usually is not" (1997: 582). Added to this is the concern of misleading interpretations and conclusions caused by the use of GIS "without (the) understanding of the underlying spatial and statistical processes" (1997: 581). If one does not understand the mechanics of either the GIS software or data there could be a 'House of Cards' effect whereby the correct procedures and statistical analyses will not be applied, results will be irrelevant and interpretations ultimately of no consequence. One might even have to ask of oneself "have I designed my hypotheses around what I thought GIS software could do?"

Viewshed studies which have employed statistical analyses include Wheatley's (1995) and Mark Lake et al's (1998) applications of the Kolmogorov-Smirnoff test, Fisher et al's use of Monte Carlo testing (1997) and, for quantitative analyses without statistical tests, Wheatley and Gillings (2000).

The following briefly outlines the outcomes of the two completed stages of this project, namely Phase 1 and Phase II.

Phase I

Rigorous statistical analysis and testable hypotheses

The following hypotheses were tested upon 125 free standing stone monuments in western Scotland using the Z^2_m tests to test for the likelihood that sites were oriented towards the same directions (see Higginbottom and Clay, 1999 for a fuller description of phase I):

(i) clustering in orientation exists in the site database of the region of western Scotland as a whole.

(ii) clustering in orientation exists in the site database of western Scotland for each of the sub-regions of Mull, Argyll, Islay, Uist, Kintyre and Lewis/Harris.

The regions included the areas of Coll, Tiree and North Argyll with Mull, Jura with Islay, and Lorn with Argyll, Uist, Kintyre and Lewis/Harris. It was an *a priori* decision to test the second set of hypotheses only if the first hypothesis could be supported.

When looking at the entire sample, some statistical evidence for the rejection of the null hypothesis (uniformity) was found (Z^2_2= 8.58, n=166, $p < .1$). The alternative hypothesis of clustering, therefore, was accepted. It was decided from this that it was worthwhile to continue the investigation of the data using regional analyses. Very significant trends were found for the regions of Uist (B = 15.93; $p < .005$) and Mull (B = 11.51; $p < .025$). Trends of minor significance were detected for Argyll (B = 8.99; $p < .1$) and Islay (B = 8.4; $p < .1$). At this level of investigation Lewis/Harris and Kintyre were the only regions that did not reveal any evidence for the rejection of the null hypothesis (uniformity).

It should be noted that phase I of this project has addressed a demand of Wheatley and Gillings to test "whether (or not) we have grouped together ... those monuments that represent the same tradition ... rather than accept it uncritically" (2000: §4.6). We have indeed found support for the likelihood of a similar tradition within the database, that of the deliberate orientation of monuments. We can now use the same database to test for more complicated behaviour, namely the orienting of sites for the purpose of indicating celestial phenomena.

Phase II Conclusions

Landscape and visions

Our results of Phase II revealed that the placement of the free-standing stone monuments in western Scottish sites included a significant consideration of the horizon for three of the regions, Argyll, Mull and Islay (Higginbottom, Smith, Simpson, and Clay, 2000). The horizon may have even been seen as an extension of the sites.[3] These regions also revealed the possible due regard given to specific celestial phenomena. What is more, the regions appeared to focus upon differing phenomena, providing evidence for regional differences within a shared, broader cultural framework (See Table 1 for specific phenomena). Added to this, the even broader cultural behaviour of the orientation of monuments is shared by a fourth area, that of Uist. Though it is not yet clear in which events or objects the sites of Uist might be interested, it is now clear that the majority of alignment forms for Uist are intersite alignments rather than intrasite alignments.

The results of the study to date make it possible to observe the connection between monument, landscape and phenomena.

Indicated directions

We can now assume that the alignment of stones within sites, and possibly between sites, is of import and that the action of looking, as well as what could be seen from the monuments, played a part in their location. It is this very evidence that allows us to apply notions of alignment, vision and visibility to our hypotheses of the next phase. Further, it gives us confidence in any future reliance upon the notion of 'line-of-site', for an interest in using monuments as the backsite and the horizon as a foresite has been verified. Finally, our choice of methodologies, namely additional cluster analyses and the application of viewshed analysis, appear appropriate for the data we have.

Phase III - Orientation and visibility analysis

Though the experimental design for this phase is complete, the application is not. The following describes in detail the methodological reasoning behind this design. To proceed with the investigation two equally important questions demand to be asked. The first is, 'is the relationship we have seen between monument, horizon and astronomical phenomena a continuous one' across the landscape? That is, are there any other areas or objects between the monuments and the indicated phenomena that participate in this relationship? Putting forward the simplest hypothesis first, then, "a spatial relationship, in the form of clustering, exists between the Ruggles sites and other Neolithic or Bronze Age objects or places" (hypothesis 1). More specifically, "there is a possible alignment of the Ruggles sites, the indicated horizon or phenomena and other Neolithic or Bronze Age features"(hypothesis 2).

The second question asks "does the connection between monument, horizon and astronomical phenomena also contain a visual component?" Might there not be a visual connection between the free-standing stones, the phenomena, the horizon and other Neolithic or Bronze age features? The hypotheses here are: "there is a significant number of other Neolithic or Bronze age sites (Group2 sites) that might be seen from the locations of the Ruggles' sites (Group1 sites) (that is, found within the non-directional viewsheds of the Group1 sites)" (hypothesis 3) and "the Group2 sites are significantly located within the Group1's directional viewsheds, the axis of which is directed by the orientation of the Group1 sites (hypothesis 4).

Finally, the investigation requires that we query the visual connection between all Group1 sites within a specific region with their visually associated Group2 sites. The question being asked is "do the Group1 monuments within the same region share any specific view of the same Group2 monuments?" The general hypothesis here states: " a significant proportion of Group1 sites share the same viewshed area(s)". More specifically: " a significant proportion of Group1 sites share the view of the same Group2 sites" (hypothesis 5 – cumulative viewshed - CVA). Naturally the issues of amount and type of Group2 sites may arise.

The first set of hypotheses (1&2) is based upon the evidence of the importance of alignments to the culture(s) of the megalith builders in western Scotland. The second group of hypotheses (3&4) is reliant upon the line of sight drawn between the monument (backsight) and the horizon points (foresight), in which an interest was also verified. The latter is, in fact, a variant of the alignment hypothesis.

Ultimately, we are trying to determine the likelihood that these other Neolithic or Bronze age sites were connected to the same cosmological belief system(s) as the free-standing stone sites that were initially assessed.

The 'secondary sites'

The list of sites that was used to determine the data for the Group2 sites came from a Royal Commission of Ancient and Historical Monuments of Scotland (RCAHMS) database that itemised all sites regardless of age in western Scotland. Neolithic and Bronze Age monuments were extracted from this. For a detailed explanation of how sites were chosen as Neolithic or Bronze age see Higginbottom, Simpson, and Clay in preparation B. The result of the application of this *a priori* method was that only "large finds", usually considered to have ritual associations, were extracted. These could be summed up as 'monuments' or 'monumental' and 'grave sites'. The actual list is: Cairn with (cinerary)urn and/or cist, Cairn or cist with cremation, chambered cairn, cup and ring marking(s), henge, ring-ditch, stone alignment, stone circle, stone setting, standing stone (RCAHMS labels). The number of these Group2 sites is 1287.

It is clear to the authors that the Group2 sites are various in age and type. Despite the fact that we are looking at site layout in relation to Group1 sites, the hypotheses do not in any way propose or rely upon the order of the sites' appearance. The order of the sites' appearance on the landscape, therefore, is not at this point accounted for. As it was clear that age could not be strongly controlled for at this stage for the entire RCAHMS data set it was decided that an in-depth study of these issues must be considered later. However, site type in itself, and as a possible key to the order of appearance of sites, will be assessed at a later stage of Phase III.

Hypotheses and the appropriate tests

For Hypothesis 1 we are asking:
(i) are there any clusterings of Group2 sites, about the Group1 sites?

For Hypothesis 2 we are asking:
(ii) are there any clusterings of Group2 sites, about the Group1 sites, in the directions indicated by the Group1 sites' orientations?

For Hypothesis 3 we are asking:
(iii) what is the percentage or proportion of Group2 sites that exist within the Group1 sites' 360 degree viewsheds? Or what is the number of Group2 sites "seen" within the hit cells versus the non-hit cells?

For Hypothesis 4 we are asking:
(iv) what is the percentage or proportion of Group2 sites that exist within the Group1's directional viewshed, as indicated by the Group1 sites' orientations?

For Hypothesis 5 we are asking:
(v)i what is the percentage of Group1 sites that share the same geographical view ?
(v)ii what is the percentage of Group1 sites that view common Group2 sites?"

Obtaining the observed data

Five sets of information are required:
(i) the orientations (azimuth values) of the Group1 sites'
(ii) the co-ordinates (eastings and northings) of the Group1 sites
(iii) the co-ordinates (eastings and northings) of the Group2 sites
(iv) a non-directional viewshed of each Group1 site.
(v) a directional viewshed of each Group1 site.

The first two sets of data were part of the original databases of Ruggles and RCAHMS. The third fourth and fifth, naturally, need to be created.

The orientations

The orientations obtained from the Group1 sites for Phase II will be applied to hypotheses 2, 4, 5. Group1 site data sets for hypotheses 2 and 4 will also be produced from the original Ruggles database but produced in the following way:

We have 276 orientations for the Group1 sites, often more than one orientation per site (n of sites=125). There are also two formats for the orientations: intersite or intrasite and one-way or two-way. From these one can code the orientations into four groups of: intersite/ one-way, intersite/two-way, intrasite/one-way or intrasite/two-way. Intersite is where the orientation is formed by the intervisibility of the two sites, and is usually "where two sites form an indication of two ranges of horizon, one in each direction"(Ruggles, 1984:66).

It was decided to use only the one-way orientations in the first instance for the following reasons:
(i) we could be sure of the intended direction that was to be sighted along;
(ii) they could be used to create future expected distribution(s), assuming that significant outcomes were obtained. We can then use all the 2-way alignments that were used in the very first set of cluster analyses in Phase I (where one of two alignments was chosen at random) and compare their distribution with that of the new expected distribution created from the 1-way alignments.

Here the expected distribution will be illustrating a situation where more than chance factors are responsible for the outcome. In relation to hypothesis 4 then, if it is found that the distributions are the same for the one-way and two-way orientations, the latter can also be said to display a significant percentage of Group2 sites in the directional viewshed areas. The testable hypothesis would be: there is no significant difference between the two distributions.

Hypothesis One

Testing hypothesis 1
To determine the occurrence, or not, of significant clusters of Group2 sites about the Group1 sites the family of tests was applied to the data for query (i), as stated under "Hypotheses and the appropriate tests".

The Z^2_m test determines whether the observed pattern of orientations is consistent with the assumption that each orientation is equally likely to be anywhere between 0 and 360 degrees, or whether that assumption can be plausibly rejected. Thus the concept of expected distribution is built into the test.

Hypothesis Two

Testing hypothesis 2
Hypothesis 2 asks 'are there any clusterings of Group2 sites, about the Group1 sites, in the directions indicated by the Group1 sites' orientations'? We already have the location of the Group2 sites that 'fall' about the Group1. From this we need to make a data cut of those Group2 sites that fall close to the orientation line of the Group1 sites. We then compare the number of these Group2 sites that fall inside the nominated band width with the number that falls outside of this band. This allow us to test whether or not there is a significant difference between the two. If there is, then it would be fair to say that there is a significantly greater number of Group2 sites that are positioned in relation to the orientation of the Group1 sites than not. The determination of this angular value or band can be found in Higginbottom, Simpson and Clay, in preparation A)

The Viewsheds

A comprehensive report of the viewshed methodology and associated theoretical bases can be found in the paper Higginbottom, Simpson and Clay, in preparation B.

The viewsheds for hypothesis 3

To obtain the viewsheds GRASS 4.3 is to be used and the function employed is r.cva. It was explained by Mark Lake to the author that the use of the cumulative viewshed (CVA) routine (r.cva) for the line-of-site (LOS) analysis was preferable as LOS (r.los) routine in GRASS truncates the height of the observer to the nearest metre whereas r.cva doesn't (personal communication).

When using r.cva for LOS assessment each site file can only have 1 set of co-ordinates which represents a single site. The procedure is to run r.cva for every site, and with the "visibility from" rather than "viewsheds of" [= -f] option chosen. The non-directional LOSs, then, were created in this manner. The directional LOS creation incorporated this technique with some additions.

The viewsheds for hypothesis 3 - directional LOS

Directional viewsheds are not possible using a single function in GRASS, yet it is essential to take account of direction in LOS calculations when assumptions or evidence for specific bearings drive the investigation. The way around this is to use r.cva for single LOS analysis as above and use the binary viewshed output as the input for r.stats. The operation of r.stats allows one to output an ascii file with the x and y co-ordinates(x3,y3) of all the 'seen' cells (non-zero data values) for each site being tested (Group1 in this case). With these data one can use trigonometrical calculations to locate the cells' positions (x3,y3) in relation to the orientation line being accounted for (in this instance it is the orientation line of the alignment produced by a Group1 site (with co-ordinates (x1,y1) and the indicated 'point' on the horizon (with co-ordinates (x2,y2)). This method was suggested to the author by Mark Lake. Alternatively, one can convert the co-ordinates of (x2,y2) and (x3,y3), in relation to (x1,y1), to azimuths. Remember also that we already have the azimuth or orientation of the line (x1y1, x2y2) for the Group1 sites.

If one chooses to use the co-ordinates, trigonometry can be used to calculate the distance and position of the seen cells from the nominated azimuths or orientations of each Group1 one site, which will allow one to get a picture of the spatial patterning of which areas were seen and where they were. One can compare this information with the co-ordinates of the Group2 sites and their distances from the bearing. This will allow one to view or calculate the number of co-incidences that have occurred between the 'seen cells' and the Group2 sites. That is, how many Group2 sites can be seen. The advantage of doing this trigonometrically is that one can actually calculate the number of sites that might have occurred within the same cell. Remembering that the Ordnance Survey data gives elevation information every fifty metres, it means our raster map is composed of 50 by 50 metre cells. The site data, however, is more detailed and it is possible to have a number of sites located within a 50-by-50 metre cell. Using something like 'r.coin' in GRASS 4.3, therefore, only allows one to readily calculate the number of cells that have coincidences (or the number of cells that have coincided with a site(s)) not the number of times there are site coincidences for the same cell(s). 'r.coin' requires that one have a raster sitemap perhaps converted by 's.to.rast'; as a result, one's raster sitemap only records absence or presence of sites.

The main reason for obtaining the x and y co-ordinates of all the seen cells is, however, that they provide data for statistical analysis. So, although getting a picture of the spatial patterning can be useful, be it via histogram or mapping, to discover whether this pattern is significant or not a more rigorous method is required to "obtain objective measures of its (existence and) strength" (Kvamme, 1995:7). Regardless of the method, an *a priori* decision must be made that allows one to choose how much of the area on either side of the indicated orientation is included in the assessment. It requires a limited decision to be made about the idea or concept of boundaries and where they are to be drawn. The reasoning behind the *a priori* decision can be found in the section below.

Creating the inclusive area for the directional viewshed

Once one has created a 360 degree viewshed, one can literally take a 'cut' from this to create a directional viewshed. The idea is to create a viewshed given the observer's location and the direction in which they are looking. This is done by taking into account all the known variables that affect visibility and vision according to the situation being investigated. The creation of the 360 degree viewshed using CVA takes care of some of the general visibility issues, based on the assumption of a clear day, apart from the curvature of the earth[4]. What is needed now, is an estimation of the horizontal visual range (areas to the left and right) a person might have when purposely directing their gaze, or looking at, a particular phenomena on the horizon. Things to take into account include horizontal visual range when focusing upon a single direction, size of the phenomena, head movement to make the edges of the phenomena the centre of vision and distance of the horizon. For a first foray into the design of a directional viewshed the range of 30 degrees was chosen, that is ± 15 degrees either side of the line of orientation (Higginbottom, Simpson and Clay in preparation B), remembering that we are not investigating what we can see overall, but what else we can look at when we are observing, or keeping in visual range/contact, the phenomena on the horizon.

Obtaining the expected data

The creation of expected distributions for viewsheds
Remembering that the influence of landscape can not really be understood from a single model we need to create a number of models or expected distributions for each separate geographical region. This will allow the determination of the influence of landscape shape upon visibility, as well as the location of a site within that landscape. Determination of influence is considered by comparing the viewshed model, which represents a pattern that occurs when chance factors are dominantly responsible, with a viewshed based upon real archaeological site data, location and landscape. To derive significant results, or arrive at significant conclusions which have archaeological relevance, the latter viewsheds must be "sufficiently different from the background visibility properties (background values or expected distributions) of the study area" (van Leusen, 1999: §3.2).

Model creation and random samples
One needs to compare the observed pattern of sites within viewsheds with four different sets of randomly generated site data for each geographical region. Three of the four sets are made using the r.random site generator option. This makes a list of locations. We will make a minimum of 2760 sets for the 360 degree, or non-directional, viewsheds for comparison with the observed sites (hypothesis 3) and at least 560 to a 1000 randomly generated sets of data for the directional viewsheds (hypothesis 4). This is in order to determine a statistical confidence level for any apparent extreme result found in the real data set. Replacement with random sampling will be allowed.

The four sets of generated data to create the expected distributions are:
Hypothesis 3
Non-directional(1): a new randomly generated set of sites (locations), where n of sites=125. The created viewsheds will be 360 degrees. This is done a minimum of 1250 times. The number of sites within and without the viewsheds will be recorded.
Hypothesis 4
Directional sites(2): (i) use one of the randomly generated site sets from Non-directional(1).
(ii) Create 56 randomly generated orientations (equalling the number of randomly generated sites) and randomly assign them to one set of the previously listed random sites. (iii) Create directional viewsheds in the manner described above (or calculate same trigonometrically) for each of the randomly generated sites and its accompanying orientation. Do steps (i) - (iiii) 560 -1000 times. The number of sites within and without the viewsheds will be recorded.
Directional sites(3): use randomly generated set of sites from Non-directional(1) above, and assign these new sites with the orientations from the original Ruggles' one-way data set (Group1 sites). The number of sites within and without the viewsheds will be recorded.
Directional sites(4): - use the original one-way Ruggles' site locations (Group1, n = 56) and randomly assign them one of the randomly generated directional orientations from Directional sites(2) 560 – 1000 times. The number of sites within and without the viewsheds will be recorded.

Note that the fourth set uses the locations of the Group1 sites but is given random orientations for their directional viewshed. Along with the first three sets, this ensures that all possible combinations of the location and orientation variables have been accounted for, further testing the hypotheses that either the location or the direction are statistically significant in the positioning of the Group2 sites in relation to the Group1 sites.

Comparing the observed and expected distributions

The general question we ask of each of our four expected distributions, is 'how many of the 560 random choices have more sites in view than that of our observed distribution"? So, if we discover that 30 expected sites have more Group2 sites in view than the observed then it is said to be significant at the 30/560 level or .048 % level. That is, we have discovered that the number of sites in the randomly generated, expected distribution are significantly less than that found in the observed distribution. Naturally we can express this in another way: the observed distribution has a significantly greater number of sites. Such a form of probability statistic is not distribution bound and is therefore suitable where we have no prior knowledge of how the expected distribution should display itself.

Hypotheses 5(i) and 5(ii) - Observed data, random samples and significance testing for CVA via r.cvs

The observed data are generated using r.cva for each of the 125 locations, but divided by region. CVA analysis differs primarily in that the generated viewshed types are of a cumulative nature (as described by Wheatley 1995). The same basic format for generating the random datasets above was used to create the random samples for CVA. Also there is only one form of random database created, a non-directional database. In this case a minimum of 1250 randomly generated sets of sites (locations) are used, where the n of sites for each set = 125, and the assessment of possible viewsheds is for a complete 360 degrees. The general format for testing is the same also.

Overview of treatise so far

This paper has so far reviewed the aims and methodology of the project. We have also presented the results of the first two phases. We now offer interpretative conclusions of these outcomes. The paper will close with remarks on the project's design

Interpretation of results to date

Both for our methodology and our interpretations we have assumed that belief systems guide, or influence, human action and thought, and that these are expressed by material culture. In our attempt to link material cultural to human action, and, finally, beliefs, we offer the following as a possible interpretation to date.

Overview

Primarily, this project has found that it was important for Neolithic and Bronze age peoples of western Scotland to purposely locate their structures to enable these sites to participate in a design of inter-relationship between themselves and the natural environment. (This raises the question whether any other recognisable entities might also be included). This suggests that it was of relevance to their belief systems to order their world in a very specific way. The patterns that we have found evidence for are 'simple' alignments. Alignments are where one or more arranged objects, or sites, are aligned with either another arranged object or site and/or a natural 'object'. What connects all these monuments is the possibility of a shared belief system across geographical space and time.

Simple alignments

In connection with these alignments is their direction and final destination. The assembling of the alignments (human action) produced distinct orientations or indicated directions to be viewed from a specific or definite place within the Landscape. For the free-standing stone monuments, in Mull, Argyll, Islay and Uist, these particular directions have been shown to be significant to the monuments as a regional group. These directions appear to be linked to celestial phenomena for the 3 regions of Argyll, Mull and Islay. To reiterate, these region titles include the areas of Coll, Tiree and North Argyll for Mull, Jura for Islay, and Lorn for Argyll. The geographical and chronological extent of these relationships, shows that the associated symbolic concepts held some weight. (that is, revealed the complexity of the relationships and the possible on-going nature of these relationships). The human action that is presented to us for each of these regions is that of forming configurations between and within structures.

From the evidence to date, Mull, Argyll, Islay, and Uist appear to be linked to the same coherent, *fundamental* system. The spatial design that they all share may well be associated with, or represent, the same or similar elementary concept(s)

Possible associated belief systems

We have seen that both the moon and the sun were important entities to the builders of many of the sites researched to date in western Scotland. The material culture has revealed this to us via the purposely created alignment indications. The strength of these indications and the geographical extent of these also emphasise the weaving and integration of an important belief system that extends through-out much of western Scotland.

What can we know about their belief system from the evidence we have so far? The lunar phenomena that were of primary focus were the major and minor standstills, and of these two, the former was predominant for Mull and Argyll and the latter for Islay. This lunar position is when the moon reaches its maximum distance north or south of the equator during its 18.6 lunar period, before it then began its march back in the other direction. The minor standstill is when the moon's path is maintained inside that of the sun's and reaches its possible maximum distance north or south of the equator whilst staying within this boundary. Argyll also has an interest in the Sun at the winter solstice and Islay has an interest around the equinox. It is possible then, that the people of these monuments had a keen understanding of the cycles of the moon and sun. It is very likely that such changeable phenomena marked out, or mirrored, the equivalent changes in their immediate worldly landscape[5], such as tides and fishing, bird and other migration, parallel human movements and plant life. The sun and the moon for such peoples, then, were likely symbols of life-giving forces. Naturally, the great phenomena of light and warmth, regardless of whether they were associated with life-giving forces or not, would be worthy of appreciation and, perhaps, adoration. The major events in these bodies' movements, such the major standstills and solstices, may have represented times of great cosmological importance and may have implied 'magical' or 'powerful' moments in time.

Summary

Through the enormous investment of time and space, and the consistent patterns over the same, that the monuments, the horizons and the associated phenomena appear to be part of the same fundamental cosmological system. This system seems to be connected to astronomical phenomena and their cycles. Added to this, regions appear to have some cultural independence, for there are variations in the astronomical system they focus upon.

Conclusion

We have above the basis for the "ultimate design" of an investigation, as explored in the introduction. Here the connection between visualization and statistical analyses is made. This connection is not only made in the final section of the analyses, where viewsheds are constructed, but from the very start of the project where the importance of vision itself is tested using other methodologies and paradigms. In this way, the reasons for using viewsheds themselves are tested for soundness and applicability. So, too, is the complimentary creation of maps of site location, ground elevation and viewsheds, once the statistical analyses are done. Other visual aids include histograms and polar plots for understanding observed orientation patterns of sites.

The results of the viewshed analyses, when complete, will provide this investigation with additional information into the cosmological systems and monuments of western Scotland.

Acknowledgements

We would like to thank most warmly the Ordnance Survey, South Hampton, UK. Firstly for the generous supply of the 1:50000 Landform® PANORAMA digital data for western Scotland, and secondly, for their continued support of the project. Ron Newbold of the Centre of European Studies is to be especially thanked for his constant encouragement and assistance. We would like to thank Göran Burenhult and the steering committee of CAA2001 for their financial support of Gail Higginbottom so that she could attend, and present at, the CAA in Gotland, Sweden. Finally we like to ackowledge the patience and kindness of Johan Arvidsson and Göran Burenhult during the preparation of this paper.

End notes

[1] See http://www.baylor.edu/grass/ for information regarding this free GIS software.
[2] See http://www.esri.com/software/arcinfo/ for details about this GIS software.
[3] See the lighter shaded area for the delineation of the furthest horizon points for all the Ruggles' sites of Figure 1. The lighter shaded area allowed us to determine which, and how much, landscape data would be required to do a full horizon analysis of all sites.
[4] We will be using the script written by Jo Wood from the University of Leicester that will allow us to take into consideration the curvature of the earth when using Grass4.3. The module is r.xy.
[5] This is not attempt to define worldly as the opposite of otherworldy, but rather as tangible and immediate.. Along with these assumptions, symbolic concepts are seen as representative constructs of these belief systems.

References

Fisher, P. , Chris Farley, C., Maddocks, A. and Ruggles C. 1997. Spatial Ananlysis of Visible Areas from the Bronze Age Cairns of Mull. *Journal of Archaeological Science*, 24, 581-592.

Higginbottom, G., Smith, A., Simpson, K. and Clay, R. 2001. Incorporating the natural environment: investigating landscape and monument as sacred space, in Martin Gojda and Timothy Darvill (eds) *One Land, Many Landscapes*. British Archaeological Reports 987, Archaeopress.

Higginbottom, G., Smith, A., Simpson, K. and Clay, R. 2000. Gazing at the horizon: sub-cultural differences in western Scotland? in César Esteban and Juan Antonio Belmonte (eds) refereed proceedings of *The Oxford VI International Conference on Archaeoastronomy and Astronomy and Culture*, 43-49. Tenerife. Organismo Autónomo de Museos del Cabildo de Tenerife.

Higginbottom, G and Clay, R. 1999. Reassessment of sites in northwest Scotland: a new statistical approach. *Archaeoastronomy*, no 17, (Journal for the History of Astronomy). S43-55.

Higginbottom, G., Simpson, K., Thornton, G. and Clay R, forthcoming. In *Creation of the sacred: form and content in the representation of cosmological landscapes*.

Higginbottom, Simpson and Clay, in preparation A. *Tradition and Culture in western Scotland: persistent perceptions of the sacred or customary behaviour?*.

Higginbottom, Simpson and Clay, in preparation B. *Designing viewshed analyses for the study of cosmological belief systems and their possible persistence through space and time*.

Kvamme, K.L. 1995. A view from across the water: the north American experience in archaeological GIS. In: G. Lock and Z. Stancic (eds), *Archaeology and Geographical Information Systems: a European Perspective*. Taylor & Francis, London, pp. 1-14.

Lake, M. W., Woodman, P. E. and Mithen, S.J. 1998. Tailoring GIS Software for Archaeological Applications: An Example Concerning Viewshed Analysis. *Journal of Archaeological Science* 25, 27-38.

Lock, G. R. and Harris, T. M. 1996. Danebury revisited: an English Iron Age hillfort in digital landscape. In M. Aldenderfer and H.D.G. Maschner (eds) *Anthropology, Space and Geographic Information Systems*: 214-240. New York: Oxford University Press.

van Leusen, M. 1999. Viewshed and Cost Surface Analysis Using GIS (Cartographic Modelling in a Cell-Based GIS II). In: J.A. Barcelo, I. Briz and A. Vila (eds), *New Techniques for Old Times*. CAA 98 – Computer Applications and Quantitative Methods in Archaeology: proceedings of the 26th Conference, Barcelona 1998. British Archaeological Reports S757, Oxford: Archaeopress, pp. 215-223.

Ruggles, C.L.N. (principal author), Appleton, P.N., Burch, S.F. Cooke, J.A., Few, R.W., Morgan, J.G. and Norris R.P. 1984. *Megalithic Astronomy: A New Archaeological and Statistical Study of 300 Western Scottish Sites*. British Archaeological Reports, 123. Oxford.

Wheatley, D.W. 1995. Cumulative Viewshed Analysis: a GIS-based method for investigating intervisibility, and its archaeological application. In: G. Lock and Z. Stancic (eds), *Archaeology and Geographical Information Systems: a European Perspective*. Taylor & Francis, London, pp. 171-186.

Wheatley, D. and Gillings, M. 2000. Vision, Perception and GIS: developing enriched approaches to the study of archaeological visibility, in *Beyond the map : archaeology and spatial technologies*. Proceedings of the NATO Advances Research Workshop on Beyond the Map: Archaeology and Spatial Technologies, Ravello, Italy, 1-2 October 1999. Gary Lock(editor). Amsterdam : IOS Press.

Tables

Region	Significant declination bin-widths in degrees	Possible astronomical phenomena	*p*	Ranges significantly avoided
Mull	¯30 -¯35 (southerly)	Lunar (Major standstill)	0.025	
	¯25 –30 (southerly)	Lunar (Major standstill)	0.095	
	25 – 30 (northerly)	Lunar (Major standstill)	0.077	
Argyll	¯20 -¯25 (southerly)	Solar (Winter solstice)	0.062	¯15 – ¯10 (southerly)
	25 – 30 (northerly)	Lunar (Major standstill)	0.026	15 – 20 (northerly)
	30 - 35 (northerly)	Unknown – but indicates densest part of Milky Way	0.002	
Islay	¯15 - ¯20 (southerly)	Lunar (Minor standstill)	0.051	¯20 -¯30 (southerly)
	¯5 - ¯10 (southerly)	Solar (flanking Equinox)?	0.035	
	0 - ¯5 (southerly)	Solar (flanking Equinox)	0.096	
	0 - 5 (northerly)	Solar (flanking Equinox)	0.095	
	15 - 20 (northerly)	Lunar (Minor standstill)	0.005	20 - 30 (northerly)

Table 1. Possible specific astronomical phenomena for which evidence has been found. Poisson statistics were used to compare the actual horizon ranges of focus with those of the expected. The expected pattern being a normal distribution within *bin-widths. Above are Poisson - distribution outcomes showing significant results of individual bin analyses, where* p is the probability *of outcome. Probabilities in bold are where* p *< 0.05, others are where* p *< 0.1.*

Figures

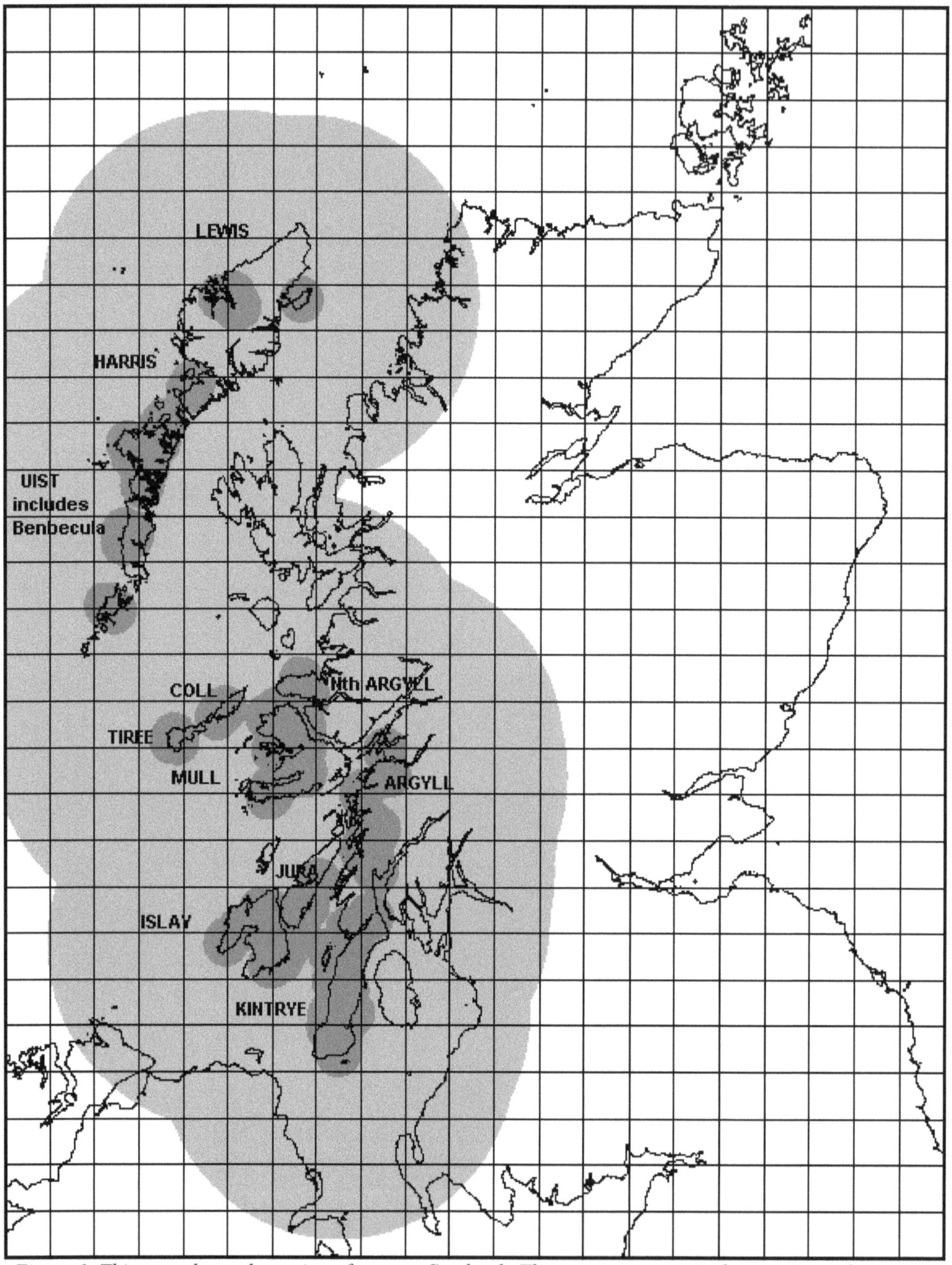

Figure 1. This map shows the region of western Scotland. The squares are equivalent in size and position to the Ordnance Survey 20 by 20 km LANDFORM® PANORAMA data tiles. The darker shaded areas are where the sites from Ruggles' data base are to be found. Map created by Andrew Smith.

Reconstructing the seascape at the mouth of the Oder
Elaboration of a DBM-model based on 1912-soundings

George Indruszewski
National Museum
Center for Maritime Archaeology
Havnevej 7
DK-4000, Roskilde, Denmark
Phone: +45 46 32 16 00 - Fax. +45 46 32 24 77 - E-mail: george@natmus.dk

Abstract: As an integral part of a 3-year research project financed by the Danish Research Foundation (DGF), the reconstruction of the early medieval seascape at the mouth of the Oder River represented a challenging task in an ever-changing environment. Besides nature-driven factors, the human factor ought be considered as one of the main agents of change active in the area, at least from the end of the 19th-century onwards. The data was processed on a GIS-based platform, and a digital model of the underwater area of the Oder Lagoon was generated from the digitising of more than 4000 soundings taken in the first half of the 20th-century. The digital bathymetric model (DBM) was used to elaborate aspect, slope, and hillshade maps of the Lagoon. Hydrological analysis was carried out on a TIN-based spatial model. The results show the potential of GIS-software in the reconstruction of past underwater areas, including the identification and matching of underwater anomalies with historical data from older cartographic sources.

Key words: Seascape, Oder Lagoon, DBM reconstruction, GIS-mapping, TIN-modeling

Introduction

During 1997 and 2000, a research project, involving institutions from Denmark, Germany, Poland, Sweden, and Norway, was carried out with the main goal of establishing a detailed view of the relationships between early medieval man, ships, and the marine environment at the mouth of the Oder River in that period. As an integral part of this project, financed by the Danish Research Foundation (DGF), the reconstruction of the early medieval seascape at the mouth of the Oder River represented one of the most challenging tasks in an ever-changing lagoon environment. Besides nature-driven factors, one had to consider, at least from the end of the 19th-century onwards, the human factor as one of the main agents of change active in the area. As a direct result of this intervention, the natural balance on both the sea side and the lagoon side of the two major islands, Usedom and Wolin, has changed drastically. Land-shaping through the building of artificial canals (the Piastowski/Kaiserfahrt canal, f. ex.), erecting protective obstacles against erosion, damming natural inlets of seawater, draining once-periodically flooded lowlands, etc. led to a radical change in the physical and geodynamic aspects of the region. Under these circumstances, the tedious work of reconstruction started as a race against all odds back in time using as much scientific knowledge as possible. Besides the use of modern prospective techniques such as seismic measurements, and geological analysis of bottom sediments, the collected data was aided by other research efforts that employed hydrodynamic modelling of water movements[1], and sediment mapping in the Oder Lagoon.[2]

A DBM-model of an early medieval landscape

The data from these sources of information help in building an approximate model of an early medieval landscape in the Oder mouth area. *ArcView* version 3.1 was the GIS-program used to construct the model. After securing the cartographic data, which varies from source to source, a digital map was constructed. The map shows the coastlines as they were measured at the beginning of the 20th c., thus before the massive anthropogenic impact on the region. Parallel with the landscape mapping, a digital bathymetric map was constructed for the Greater and the Lesser Lagoon, including the Dziwna, Swina, and the Peenestrom channels. The model took in consideration all the areas at the Oder mouth, including those heavily transformed by anthropogenic factors.[3] In this way, a complete picture was obtained by extrapolating the landscape from its actual shape back in time.

The bathymetric map is a DBM-model[4] based on floating points represented by 4232 digitized negative values (fig. 1). By comparison with modern navigation maps and with earlier models of bottom configuration in the Lagoon[5], the DBM model shows a more detailed delineation of offshore ledges both on the lagoon side and on the sea side. The Lagoon basin can be divided along the -4-3 meter isobaths into a ledge zone and the proper basin zone (fig. 2). The ellipsoidal shape of the basin strangled in the middle by the Repziner and Osiecka shoals defines the general outlook of the underwater part of the Oder mouth area.

The DBM model was used further in elaborating maps showing the underwater surfaces. The underwater surface map was produced in both monotonic and coloured versions in order to visually enhance bottom morphology (figs. 3 and 4). The comparison with modern charts shows little differentiation of bottom configuration in respect to main features such as the Wolin Shoal, the Krzecki Shoal, the Warpno Shoal, the Repzin Shoal, the Borken shoal, Göschenbrinks Flat, but the digital rendition of the underwater surface gives a more detailed overview of the bottom morphology. A reclassification of the generated surface illustrates visually that the underwater area at the Oder mouth area can be divided in nine major classes of different depth values with only four classes of depth in the Lagoon (fig. 5).

According to older cartographic sources, the Wolin Shoal (marked with the text *Das Schaar im grossen Haff ?? Fusstief?*) is known as a bottom obstruction at least since the middle of the 17th c. and this situation can be assumed for the other submerged geomorphic features. The model indicates also that the lagoon acts as a sink for heavier sediment with a higher settling velocity than that of the waterflow. This role is enhanced by the elevation in the discharge channels, which are shallower than the lagoon. Before the construction of the Piastowski/Kaiserfahrt navigational channel (that is after 1880), the mass of water had to push forward over the Wolin Shoal and into the Dziwna channel, over the Wicko Sill into the Old Swina channel, and over the Göschenbrinks shoal into the Peenestrom, in order to flow into the sea.

The analysis was carried further by calculating the aspect, the slope, and the hillshade of the underwater area. The aspect function determines the direction towards which a specific underwater slope faces. The direction is given in degrees, according to the position of the slope in relationship to cardinal points. Two aspect maps have been calculated:

- the first aspect map was calculated by using the spline interpolation method, with a cell resolution for 100 m
- the second aspect map was calculated by using the *IDW* (inverted distance weight) interpolation method, with a cell resolution of about 260-300 m
- an altered aspect was obtained by varying the brightness[6] of the second aspect map

The first aspect map shows the probable locations of underwater slopes in the Pomeranian Bay, in the lagoon and in the channels (fig. 6). The slopes, generally speaking, tend to be faced southward along the northern edge of the lagoon, while along the southern coastline, they are facing north, north-east or north-west. The Wolin shoal[7] is clearly outlined by a succession of slopes facing west on its left side, and east on its right side. A similar situation is present in the channels. On the Peenestrom and on the Dziwna, the east-facing ledges seem to cover more bottom than the west-oriented slopes from the other side of the channels. On the southern side of the lagoon, the Warpno and Repzin Shoals are shown as sloping eastward, while on the northern side, off the Karsibor Peninsula, the ledge of the Krzecki Shoal is oriented southward.

The second aspect map edited in a monotonic colour range shows more clearly the underwater morphology (fig. 7). The ridge-like shape of the Wolin shoal is more pronounced on this map. The natural trough in which the Piastowski/Kaiserfahrt navigational canal was dug is visible on this map. The most interesting feature is the circular shape shown in the middle of the Lesser Lagoon, which represents the bottom sloping in that area. This feature is enhanced when the brightness of the aspect map is altered.

The slope map shows the areas on the sea/channel/lagoon floor where there is an abrupt change in the gradient of the slope (fig. 8). The colour range can be divided in three major zones:

- the undulated zone, represented by the blue colour range
- the trough zone, represented by the white colour
- the abrasive zone, represented by the red colour range

The 'blue' zone shows that prominent features, like the Wolin shoal, the Krzecki shoal, or the Göschenbrink flat are less visible, that is their slope seems to curve up gradually without pitfalls. The same seems valid on the seaside zone. The 'white' zone shows clearly the Piastowski/Kaiserfahrt navigational channel, but also an interesting pattern off the Baltic coast of the Wolin island. This feature is indicated on modern maps with the help of contour lines, and also on the 1709-navigational chart elaborated by Nicolaas De Vries.[8] The pattern of this feature suggests geomorphologic floor changes related to the Wolinian Stage of the retreating ice mass. Another interesting feature is the white feature at the Swi touj cie-Mi dzywodzie Height, which seems to be related to an older outlet of the Dziwna in this area.

The 'red' areas represent slopes with maximum gradient that are prone to erosion and sediment transport. The Borken Shoal seems to represent such an area, and also the underwater ledge off Miroszewo, east of the Warpno Shoal. This area was recognized on land as a retreating sector, where the cliff seems to be eroded intensively.[9] Two areas off the Warpno Shoal are shown as spots with an abrupt gradient. These spots are not easily seen on the bathymetric charts, and they seem to be prime candidates to erosion and sediment transport.

The hillshade map was calculated with an azimuth of 180° and a vertical altitude, in order to enhance all the geomorphological details of the floor in the Oder mouth area (fig. 9). The altitude chosen determined an equally distributed illumination over the floor surface. The previously mentioned details are visually enhanced and spatial relationships can be seen clearly:

- the sea floor inlet/Zinnowitz-Vineta-Kosserow banks/the *Rieck* inlet
- the geomorphic postglacial features off the coast of the Wolin Island
- the track of an older Dziwna outlet in the Swietoujscie-Miedzywodzie area
- the Wolin and the Krzecki Shoals
- the Piastowski/Kaiserfahrt navigational channel
- the Repzin and the Warpno Shoals with the two abrasive areas aforementioned
- the circular floor feature in the midst of the Lesser Lagoon.
- the Borken Shoal

- the geomorphic change which divides the Achterwasser basin in a northern and a southern part

The TIN-generated surface

The analysis continued with the elaboration of a TIN model that shows the steepest paths on the floor (sea, channels, lagoon) (fig. 10).[10] The direction of the steepest paths coincides with runoffs of water if the surface would have been dry land. Here one can observe that the major collector of drainage water is the basin located in the Zalew Szczeci ski (Stettiner Haff).

Based on these results, it can be said, that the use of generated surfaces as analytic models depends to a great extent on the accuracy of the survey data. The generated model, can then be used to identify or highlight features or geophysical trends that otherwise remain hidden or overshadowed. The spatial analysis gives also the possibility to compare the generated model with older cartographic sources and seek matches of geomorphologic features. These results have been illustrated here with data from the Oder mouth area in order to show the potential brought by GIS-software to the research and reconstruction of underwater areas and past landscapes. GIS is a multifaceted research tool that can be used in different ways to visualize and analize spatially-related data sets.

Acknowledgments

My gratitude is expressed with this occasion for the help provided by Arleyn Simon, Peter McCarthney, and Charles Redman of the Archaeological Research Institute, from Tempe, Arizona in carrying out the initial steps of the spatial analysis through the use of the institute's GIS and statistical facilities. Many words of thanks also for the support and understanding Ole Crumlin Pedersen has expressed at the Center for Maritime Archaeology in Roskilde. Not the least, it should be mentioned again, that the entire work for this project was made possible through the financial support of the Danish National Foundation.

End notes

[1] Wolf, *et al.* 1998: 42-78; also, Robakiewicz,1993.
[2] Osadczuk 1999: 175-188.
[3] Lock and Harris 1996: 219, mention that 'negative areas' known to have obliterated archaeological evidence need to be integrated since they are part of the reconstructed landscape, and most important, they contain information about land development. A similar position was taken here, with respect to the underwater environment.
[4] DEM is the acronym for three-dimensional digital elevation models built with the help of cartographic software. In this case, the input values are negative, thus the change from elevation to bathymetric.
[5] Osadczuk, 1997.
[6] Minimum cell brightness was set at 80 while maximum cell brightness was set at 20.
[7] On Polish navigational charts, the shoal is represented as a continuous feature running on a north-south axis but with three different names, depending on location: at the Rów peninsula height it is named the Wolin shoal, at the Smie peninsula, it is the Pomeranian shoal, while off the Kopice cliff, it is named the Kopice shoal. In this work, the general name of Wolin shoal is used throughout to ease understanding, and also because the name represents one continuous geomorphological feature.
[8] Mingroot and Ermen 1987: 110.
[9] The erosional rate can be assessed by comparing 19th-century maps with modern maps. See Mielczarski 1987: 58-62.
[10] A TIN-based model (triangulated irregular networks) uses triangles in order to built a wire-frame representing adjoining, non-overlapping surfaces. This method minimizes the loss of information between input points, the type of surface rendering being closer to reality.

References

Wolf, T. *et al.*, 1998. Anwendung eines gekoppelten Modells im Oderhaff, in R. Lampe ed., *Greifswalder Bodden und Oder-Ästuar Austauschprozesse - Greifswalder Geographische Arbeiten* 16, 42-78.

G. A. Lock, and T. M. Harris. 1996. Danebury Revisited: An English Iron Age Hillfort in a Digital Landscape, in M. Aldenderfer and H. D.G. Maschner, eds., *Anthropology, Space, and Geographic Information Systems.* Oxford University Press.

Mielczarski, A. 1987. 250 lat antropogenicznych przemian Zalewu Szczeci skiego, *In yneria Morska* 2, 58-62.

Mingroot, van, E. and E. van Ermen. 1987. *Scandinavia in old maps and prints.* Mappamundi, Knokke.

Osadczuk, A. 1997. *Mineralogiczno-Geochemiczne aspekty wspolczesnej sedymentacji w Zalewie Szczecinskim.* Gdansk, unpubl. diss.

Osadczuk, A. 1999. An Estuary or a Lagoon? The Zalew Szczeci ski (Oderhaff) in the light of analysis of the sedimentary basin morphology and the character of present-day sedimentation, in K. Rotnicki ed., *Quaternary Studies in Poland - Special Issue 999.* Pozna , 175-188.

Robakiewicz, W. 1993. *Warunki hydrodynamiczne Zalewu Szczeci skiego i Cie nin cz cych Zalew z Zatok Pomorsk* . Gda sk.

Figures

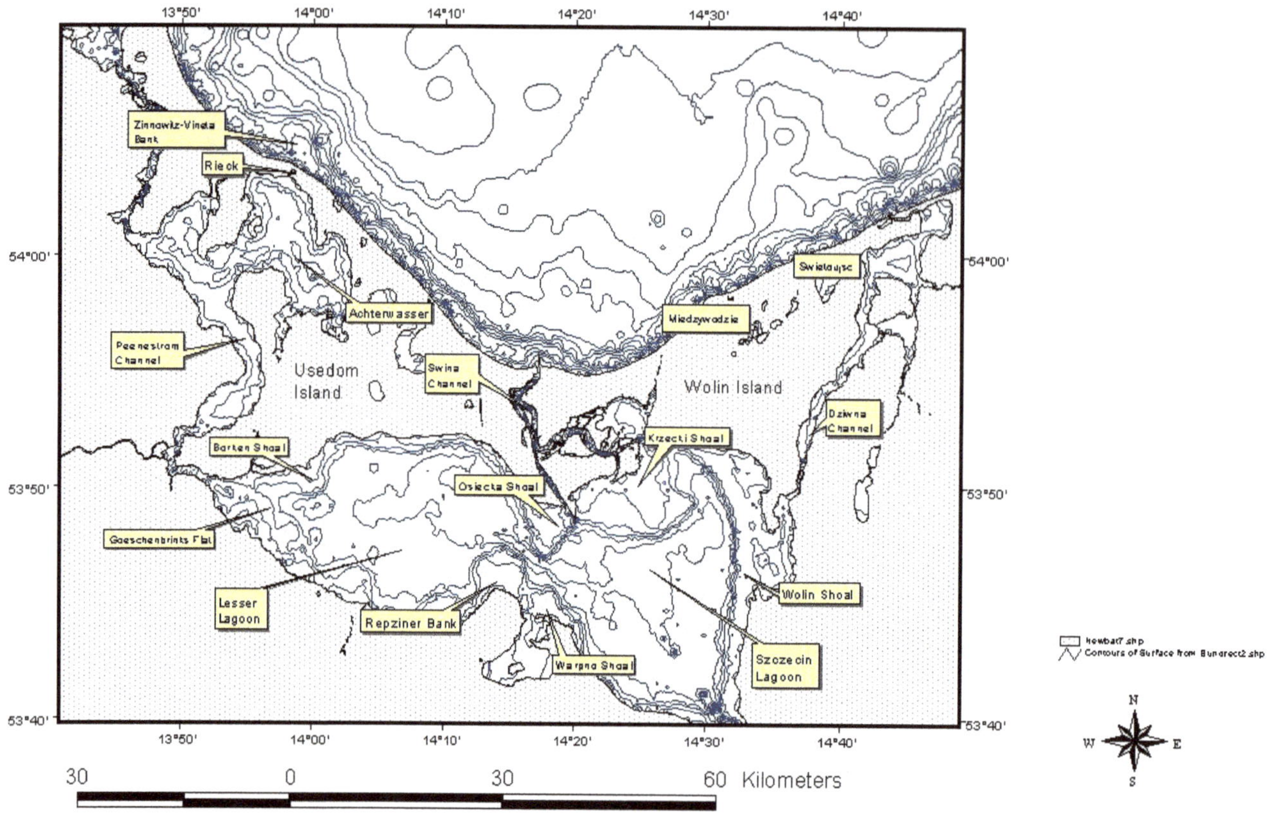

Figure 1. Isobath-contouring generated from 1912-bathymetric readings

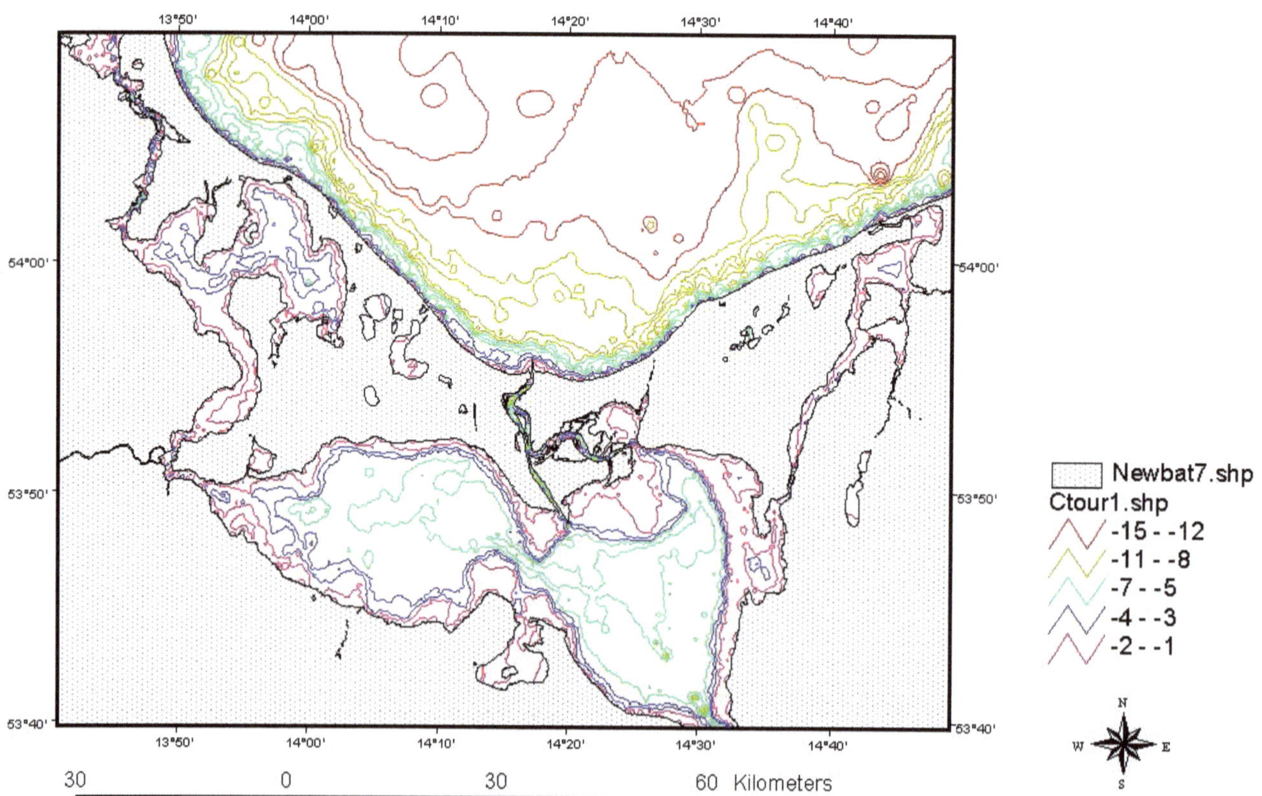

Figure 2. Graduated-colour isobath-contouring generated from 1912-bathymetric readings

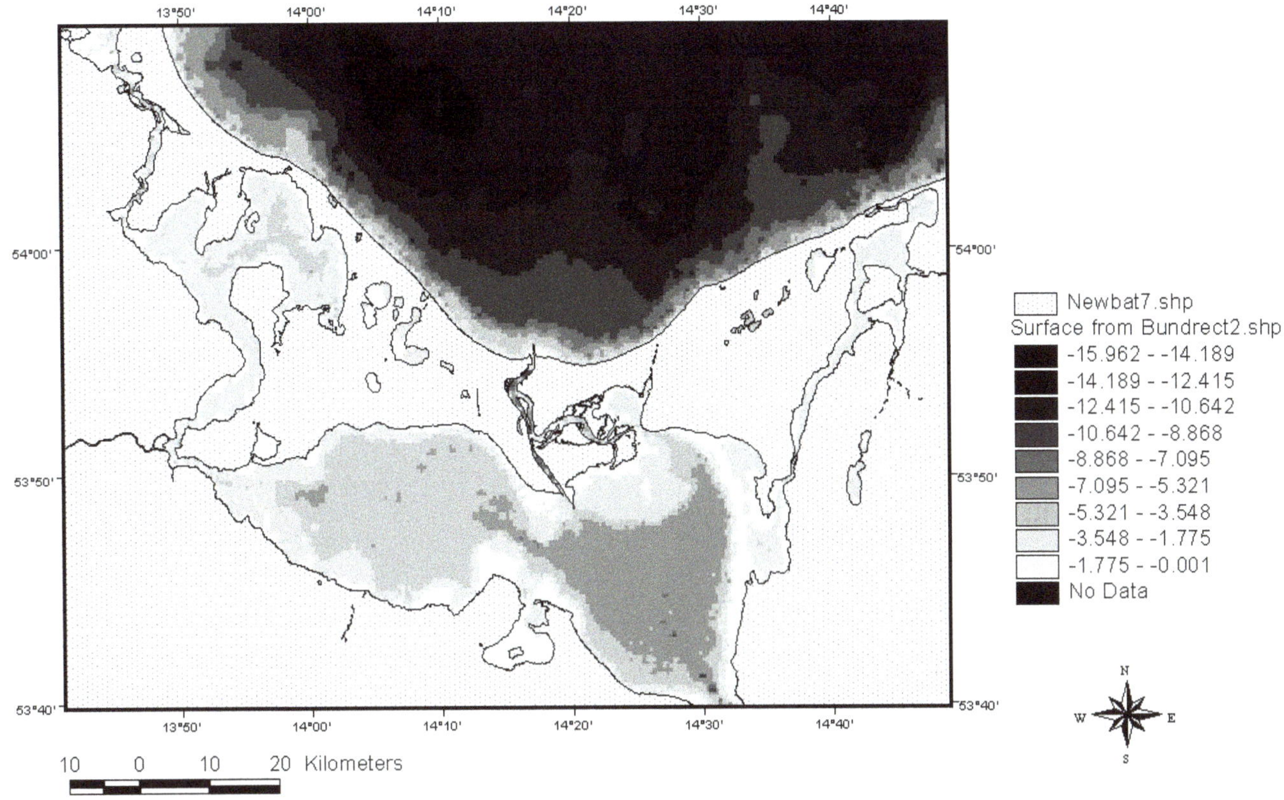

Figure 3. Underwater Surface generated from 1912-bathymetric readings

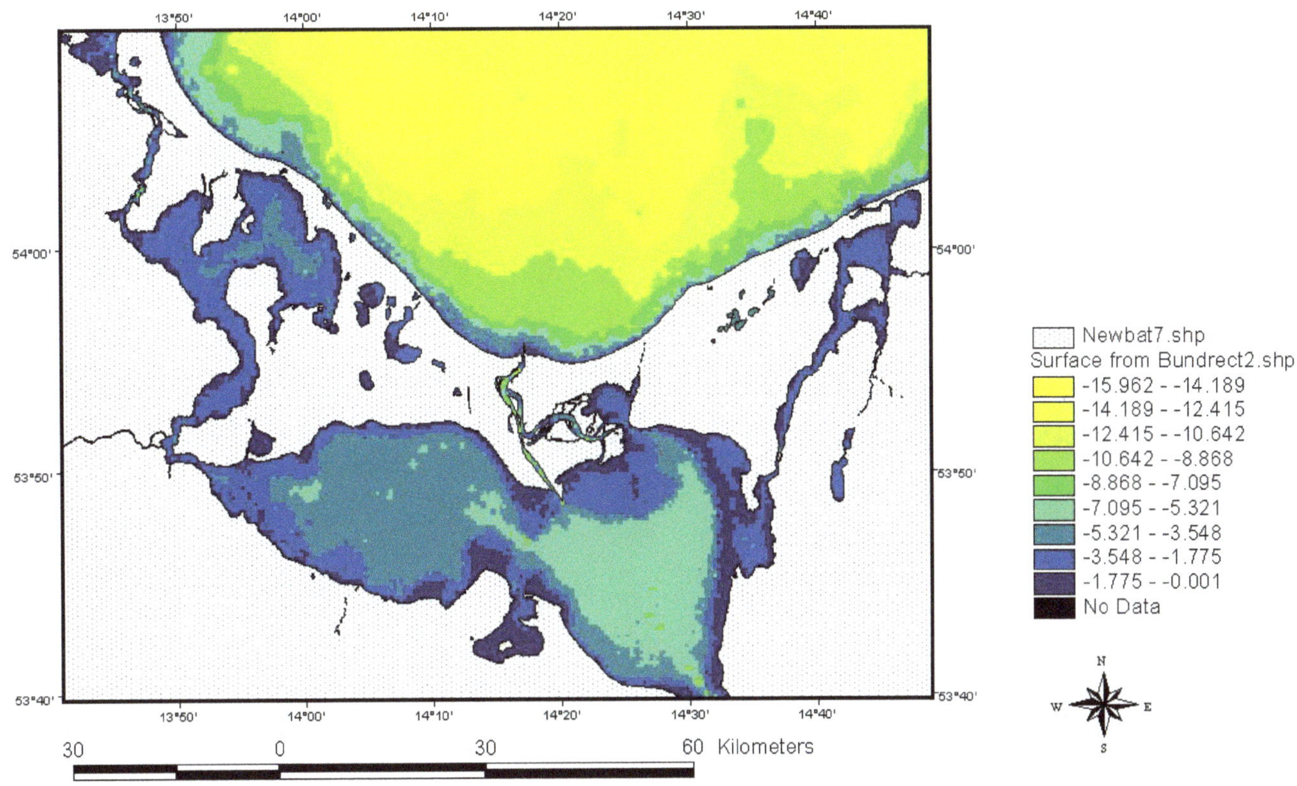

Figure 4. Graduated colour of underwater Surface generated from 1912-bathymetric readings

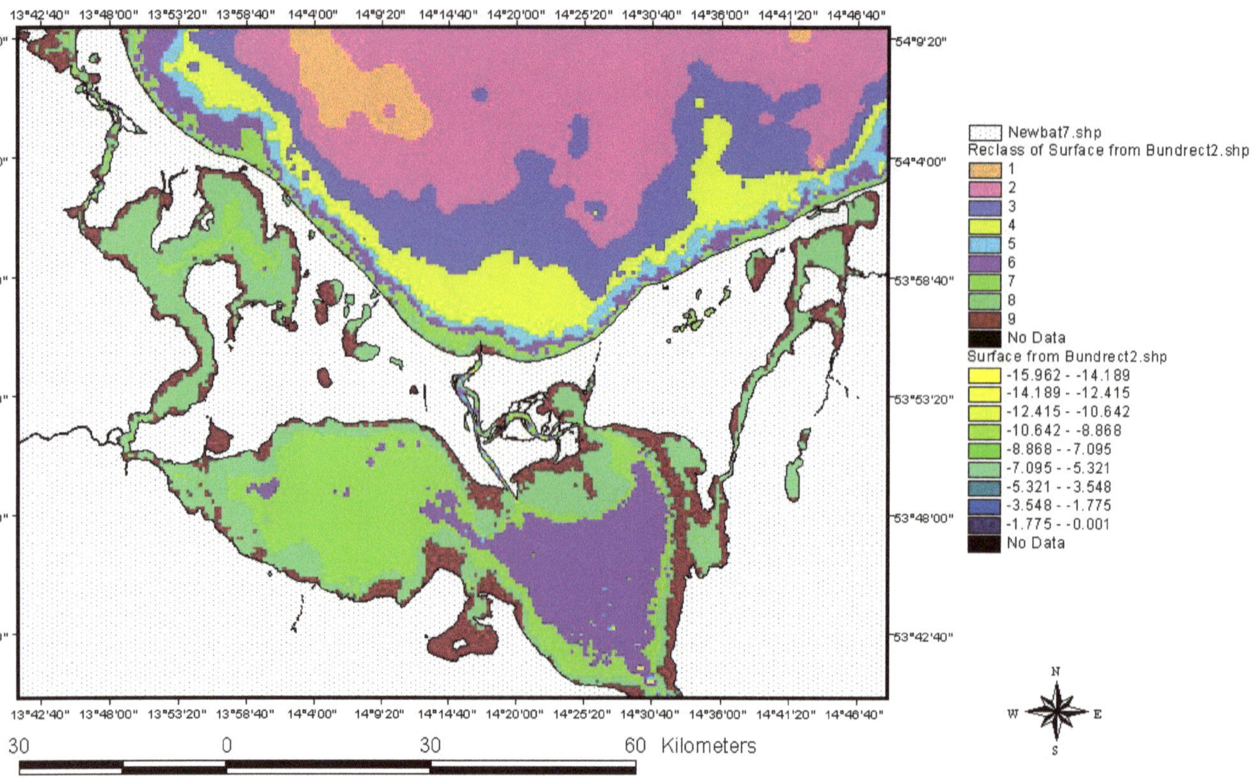

Figure 5. Reclassification of surface mapping

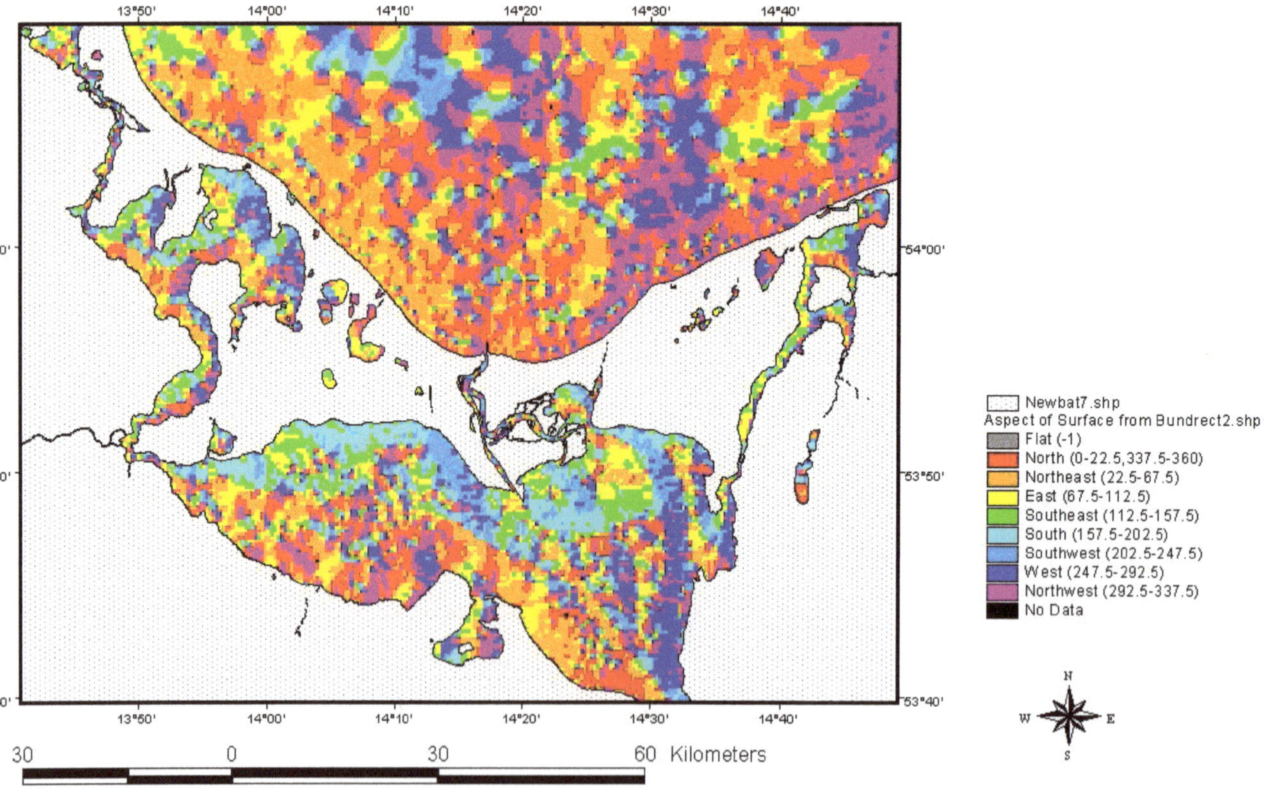

Figure 6. Aspect Mapping of generated surface

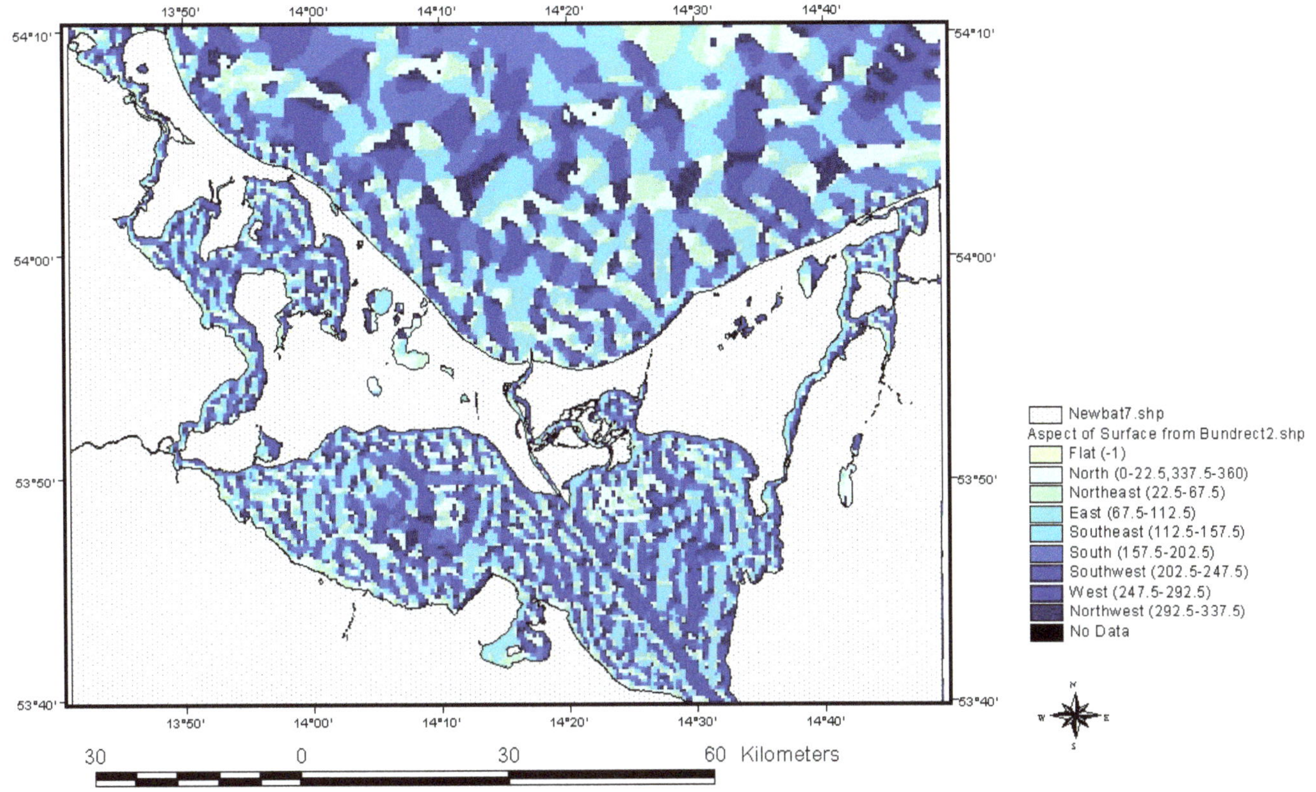

Figure 7. Second Aspect map of generated surface

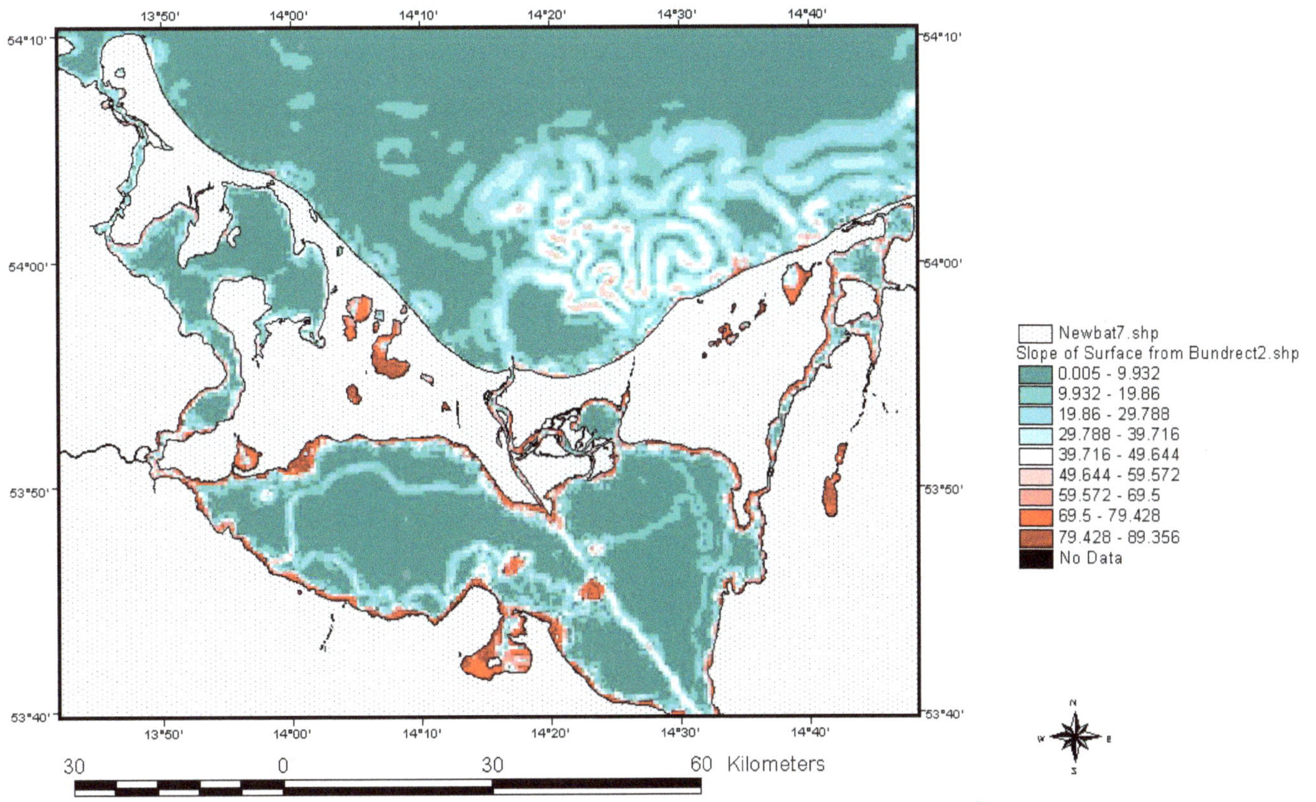

Figure 8. Slope map of generated surface

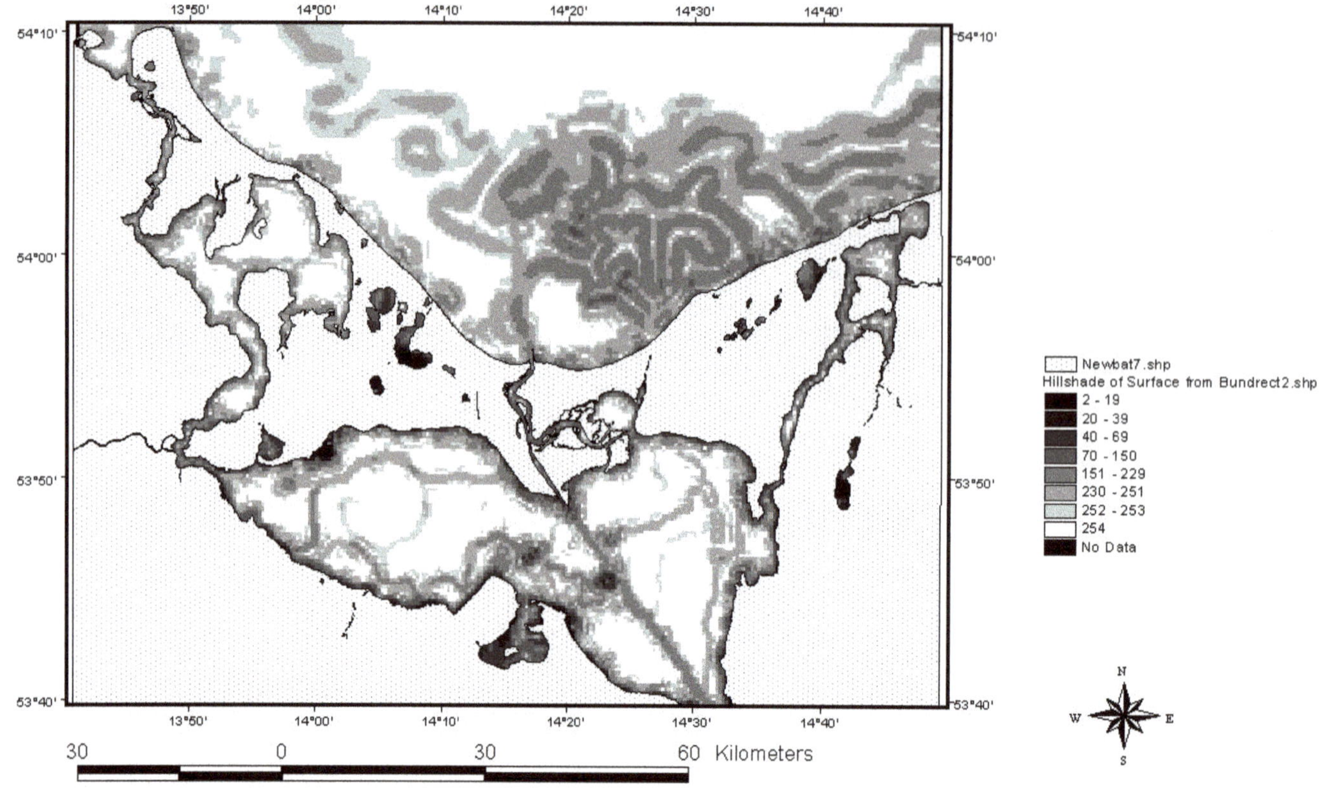

Figure 9. Hillshade map of generated surface

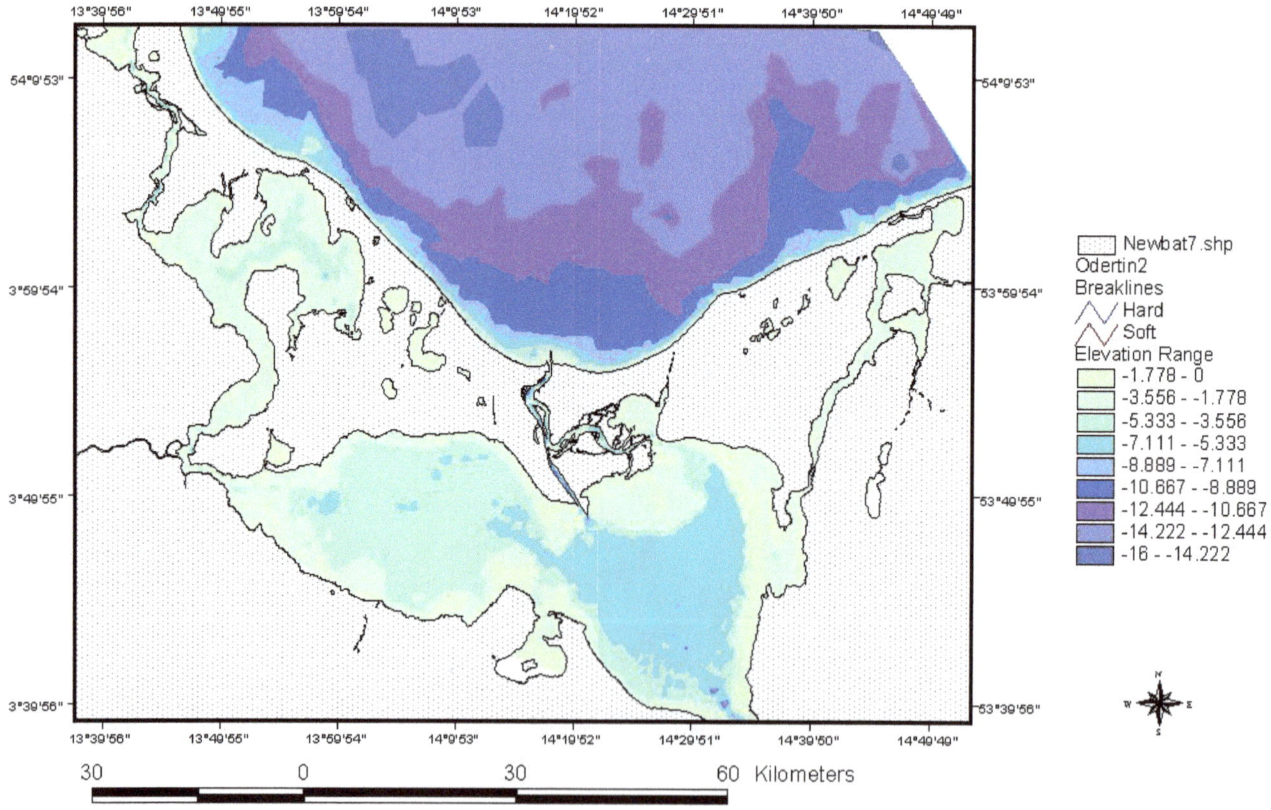

Figure 10. A TIN-based model of the Oder Mouth area

The *Time*Map Kiosk: Delivering Historical Images in a Spatio-Temporal Context

Ian Johnson
Archaeological Computing Laboratory
School of Archaeology
University of Sydney NSW 2006, Australia
E-mail: johnson@acl.archaeology.usyd.edu.au

Andrew Wilson
Archaeological Computing Laboratory
School of Archaeology
University of Sydney NSW 2006, Australia
E-mail: wilson@acl.archaeology.usyd.edu.au

Abstract: The Time*Map project of the Archaeological Computing Laboratory at the University of Sydney is investigating the use of geographic information systems (GIS) for mapping spatio-temporal data using a snapshot-transition model. The project has developed a viewer application (TMView) which will display both local and Internet-accessible datasets together in a time-enabled map window. Along with normal GIS functionality, the viewer uses time-based filtering to display time slices on demand and can generate simple map animations as AVI files. A Java map viewer for Internet-accessible datasets is under development.*

TMView has been used to implement a museum kiosk at the Museum of Sydney. The kiosk gives museum visitors access to a variety of digital resources drawn dynamically from a database. These are displayed in against series of historic maps that have been geo-registered to fit the modern map base.

Key words: Geographic Information Systems, Spatio-Temporal GIS, Time*Map, Internet, Sydney, Historic maps.*

Introduction

Our first presentation of *Time*Map[1] at CAA97 (Johnson 1999) introduced the snapshot-transition model for storing vector-based spatio-temporal data, and touched on many aspects of modelling and presenting spatio-temporal data such as fuzzy boundaries, temporal uncertainty, time-slice mapping and map animation.

While we are still grappling with the issues of spatio-temporal data modeling and temporal interpolation, we have developed a robust *Time*Map viewer application (TMView) which will display both local and Internet-accessible datasets together in a time-enabled map window. The viewer uses time-based filtering to display time slices on demand and will generate simple map animations as AVI files.

TMView was used to implement a museum kiosk – Sydney *Time*Map – at the Museum of Sydney in November 2000. We have also developed a simpler Java viewer for Internet-accessible datasets. The Java map viewer does not yet support time queries.

The TMView application – currently at version 3.0 – can be downloaded from the *Time*Map web site (www.timemap.net). The Java map viewer can be viewed on the Electronic Cultural Atlas Initiative search pages, accessible from the *Time*Map web site.

*Time*Map Windows Viewer

TMView is a Windows application written using Borland Delphi and the ESRI MapObjects GIS component[2]. TMView creates map layers on the fly by applying an SQL filter to vector datasets (stored locally as ESRI shapefiles), allowing the map to be updated as the time limits are changed. A time bar beneath the map – which can be zoomed in to individual days, months or years, or out to the full time range of the datasets displayed – allows the user to set the time limits by dragging the ends of the current time range shown by the grey bar (Figure 1).

Metadata attached to each dataset give the program information such as the way in which time fields are stored, which fields to use in labeling and how to construct URLs from attribute data to link objects on the map to database-driven web sites. Tools on the toolbar allow access to the attribute information attached to each map object, as well as linking to constructed URLs or to URLs stored directly in a data field. The key map in the top left corner can use one or more vector or raster datasets, which may or may not appear on the main map.

Internet datasets

TMView can search an Internet-accessible metadata clearinghouse developed for the Electronic Cultural Atlas Initiative (www.ecai.org). It is hoped that the clearinghouse will eventually index numerous publicly-accessible archaeological and historical datasets with a spatial and temporal component. A typical search screen in shown in Figure 2.

The clearinghouse indexes datasets accessible via the Internet and provides two sorts of metadata: *descriptive metadata* (based on Dublin Core) and *connection metadata* (information on the type of server, naming of tables and fields, passwords, georeferencing etc.). TMView uses the descriptive metadata to search for datasets by geographic area, time period or keyword, and uses the connection metadata to directly access and download all or part of a dataset for display in the map window. Datasets downloaded from the Internet are cached locally as shapefiles (for vectors) or JPG files (for rasters) and may be used offline, so that TMView MapSpaces can be packaged with TMView as CD-ROM products or transferred to a standalone computer for kiosk use.

The TMView application does not directly access datasets on Internet servers. Instead it communicates using the CORBA standard with a middle-tier application server (Figure 3) which has the necessary drivers to talk to the ECAI clearinghouse and supported data servers. The middle tier currently supports data served as JPG, GIF or shapefiles on a web server, several common SQL data servers and the MrSID image server. Other drivers are under development, including support for the Open GIS Consortium standards.

MapSpace manager

TMView has a MapSpace Manager window (Figure 4) which supports the design and saving of multi-layer maps as MapSpace files (analogous to ArcView projects or MapInfo Workspaces). MapSpaces define the datasets to be used (including non-local datasets accessible via the ECAI metadata clearinghouse), the SQL filters to be applied to create map layers, the symbology to be applied for each layer, and spatial and temporal limits. MapSpaces can be saved as local files, as files on a web server or – along with appropriate metadata – in the ECAI clearinghouse, from which they can be downloaded by other users (along with the data from datasets indexed in the clearinghouse).

Layer control

One of the key features of TMView is its handling of both time and layer display scaling. While many desktop mapping programs allow one to set the range of scales over which map layers display, most programs simply turn layers off if they fall outside the chosen scale range. It is hard to find out what layers exist and where more detailed information is available by zooming in (e.g. to specific site plans for individual archaeological sites within a city or regional map).

In TMView, layers can be set to display both within particular scale ranges and within particular date ranges using a visual metaphor (Figure 5, right). Rather than disappearing altogether, layers which lie outside their display range are collapsed to a single greyed-out line in the legend and a symbol is displayed next to them (see Figure 5 left). A clock symbol indicates that the layer lies outside the current time range for the map, a plus magnifying glass indicates one must zoom in to see the layer, and a minus magnifying glass that one must zoom out. When a greyed-out layer is highlighted in the legend, its spatial extent is flashed on the map and its temporal extent is flashed on the time bar.

The layer control dialogue also allows setting of other layer behaviours, including whether the layer should be switched on when the map is first displayed, whether it should appear in the legend, whether its symbology should be displayed in the legend and whether a temporal filter should be applied.

Java map viewer

While much of our work has gone into development of the Windows map viewer and data preparation toolkit, it became clear that the download and installation of a complex Windows application was a significant hurdle for the average user simply wishing to retrieve data in a web browser. This led us to develop a Java mapping applet, illustrated in Figure 6. The applet is an intelligent client which downloads vector and raster data and allows local zooming, panning and querying of the downloaded data. This contrasts with the approach of downloading an image generated on a server and returning to the server for each pan, zoom or enquiry. The intelligent client approach makes for a much more responsive application once the initial download has been completed.

The Java map viewer may either be pasted into a web page – in which case it will display a set of datasets specified by their ID numbers in the ECAI metadata clearinghouse – or it can be called to display the results of a clearinghouse search such as that shown in Figure 7.

In setting up the Sydney *Time*Map museum kiosk, and test datasets for the ECAI metadata clearinghouse, we have developed a number of worldwide datasets of general utility. These include a georeferenced world image (Figure 1) based on the GTOPO30 1km Digital Elevation Model (a 250Mbyte bitmap image compressed in MrSID format and served through the multi-resolution MrSID image server) and a worldwide database of cultural heritage sites (served through a MySQL database).

The worldwide database of cultural heritage sites was seeded with the UNESCO World Heritage listing and is being updated by student volunteers using a password-protected web front-end, with the aim of building a moderately comprehensive worldwide coverage of major archaeological sites.

Metdata creation tools

For TMView and the Java mapping applet to access datasets through the ECAI metadata clearinghouse, descriptive and connection metadata must be created and registered with the clearinghouse. We have developed two methods of creating metadata and registering datasets: a web-browser interface using

JavaScript and a Windows metadata editing toolkit (TME). The Windows editing program provides additional functionality such as a 'wizard' to prepare the metadata, a utility to create shapefiles from database tables and a 'data pump' to upload local datasets to an Internet-accessible server.

TME uses the middle tier application server to support registration of metadata with the clearinghouse and the 'data pump' which uploads vector GIS datasets to standard SQL data servers. TME allows point, line and polygon map data to be stored in any standard SQL data server, obviating the need for an expensive spatially-enabled SQL server or GIS data server. Effective vector GIS datasets can thus be served with freeware such as MySQL on a PC-based Linux server (MySQL is also widely available on commercial ISPs, making it a readily available method for serving spatial data without the help of a specialized GIS lab).

TME tightly controls the entry of metadata using definitions stored in the ECAI clearinghouse (cached locally so the program can be used offline). All elements are qualified by a scheme chosen from those permitted for the element, and where the scheme is a controlled list, values are chosen from a pull down menu. Mandatory and non-repeating elements are enforced. Default sets of metadata are defined to speed data entry, and metadata can be exported and imported in various formats.

Sydney *Time*Map

The Museum of Sydney describes itself as 'a multi-media, multi-disciplinary installation about the nature and narration of this place now called Sydney'. We installed the first public-access *Time*Map kiosk – termed 'Sydney *Time*Map' – at the museum in November 2000, funded by an Australian Research Council grant, in partnership with the Historic Houses Trust of NSW (parent body of the museum) and ESRI Australia. The project provides an opportunity for testing the *Time*Map methodologies within a data-rich and historically well documented environment, and for assessing user reaction.

Sydney *Time*Map uses the *Time*Map viewer, TMView, to provide a time-enabled interactive map interface to a rich and expanding collection of historical images of Sydney, presented in the context of contemporary and modern maps. The kiosk (Figure 8) consists of a Pentium III-500 with Matrox G400 dual head card, twin LCD monitors and a mouse – there is no keyboard. TMView is displayed within a wrapper written in Macromedia Director, which allows the user to choose from a menu of *Time*Map MapSpaces (left-hand screen in Figure).

MapSpaces have been created for each 20 year period since the European settlement of Sydney along with a number of specific themes such as Cadigal (the Aboriginal inhabitants of Sydney prior to the European invasion) or Archaeology (excavated sites and the NSW Heritage register). As the user rolls over the menu on the left-hand screen, appropriate images and interpretative text for the corresponding MapSpace are displayed on the right-hand screen (Figure 9).

Once launched by the Director wrapper (when the user clicks on one of the menu entries), TMView uses SQL queries against a master database of digital resources and their spatial and temporal locations to create different map layers showing image viewpoints with the direction of view (Figure 10). Users can overlay modern features (coastline, street grid) onto historical maps warped to fit modern coordinates and examine historical images from the relevant periods in context. Unlike many multimedia applications, the kiosk presents information from a continuously expanding database rather than data programmed into the system.

The kiosk application also displays points representing sites in the NSW Heritage Office site register, positioned by geocoding against an address database. As the pointer is passed over points representing sites or image viewpoints, the name of the site or title of the view is displayed. Clicking on the Heritage Office sites brings up the appropriate web-generated page from the Heritage Office site register.

The kiosk application is a valuable window into user reactions to computerized presentation of historical information, both through logging user interaction and through feedback from users. User reaction has been very positive towards the flexibility and opportunities for personal exploration that *Time*Map affords. We are using user feedback to improve the kiosk interface and develop ideas which will contribute to a full-scale exhibition at the Museum of Sydney planned for installation in 2003.

Digital image database

A common problem with smaller projects is the use of file-based data, such as Microsoft Access databases or ESRI shapefiles, which can only be shared on a local network and updated by a single user at a time. The problem is compounded if multimedia resources are stored directly in a database, resulting in unwieldy files.

While background map data for Sydney *Time*Map was stored as file-based resources (and later placed on our server and registered with the ECAI clearinghouse), the collection of many scanned historical images (along with sound and multimedia resources) by people working in two separate institutions demanded the use of a server-based approach. This collection of spatially and temporally located digital resources was, from the outset, stored as a web-accessible server database developed using freeware tools under Linux. We used MySQL as the server database and Zope and Python to provide a web-based interface.

Resources are defined as a particular source document, such as a map, photograph or painting. The resource may exist in several digital versions, such as alternative scans at different resolutions, an original scan and a cleaned up or colourized version, or a reduced resolution version of the original. The database stores the attributes of each resource, and all files which relate to that resource are identified by the database ID of the record storing the attributes, which is allocated sequentially.

Different digital versions of the same original document are distinguished by a set of standard suffixes such as _original, _cleaned etc. and may be placed in a number of directories which are identified in a configuration file. This allows us to keep the original high resolution scans, some of them up to 60 Mbytes, in a separate directory from cleaned up and resampled images of adequate quality for all normal purposes.

The web-based database access software searches all the listed directories and displays all versions of a particular resource. The resources are shown initially as thumbnails, but web resolution and original resolution images can be displayed by clicking on the appropriate links (11). All resolutions are generated on-the-fly by Python scripts which resample the images to the desired size, so that multiple resolution versions of a resource need not be stored.

The web-based software can also create web pages on the fly, using a template which defines the format of the page, the database information to be included, and the location and size of the image to be displayed. For kiosk use, a Python script generates – in three separate directories – a thumbnail, a web resolution image and an appropriate web page, again using an HTML template to determine the layout of the page and the database fields to be displayed in it. This script is used to create fixed pages which can be placed on a CD-ROM or accessed locally from a hard drive, rather than depending on an Internet connection to access the server database directly.

Figure 12 illustrates the web pages generated by TMView using the two templates in the Sydney *Time*Map kiosk application. When a user marquee-selects a number of image viewpoints on the kiosk interface, a scrollable list with basic information is generated from the thumbnails directory. When the user clicks on a single image viewpoint, or selects from the thumbnail list, the kiosk displays the web page for a single image.

Of the 1000 images included in the database, ranging from some of the first topographic sketches of the settlement to contemporary photographs, almost 400 were included in the kiosk installation. The selection was made for a variety of criteria, including aesthetic, technical and copyright issues, but most of those not used in the kiosk remain as geo-referenced images available for further research.

Issues with historical maps

The Sydney *Time*Map kiosk uses historical maps as a backdrop to contemporary images, site register data, excavated archaeological sites and other historical information. Twenty-three maps were registered, and the final display includes fourteen. These range from a 1788 plan of the encampment and fleet by John Hunter to aerial photograph mosaics from 1956, 1970 and 1994. In developing the map backgrounds we have had to solve a number of problems, both anticipated and unanticipated.

In the first place, it is essential to scan source maps at the highest reasonable resolution (e.g. 600 dpi), and this is not always possible on-site. The owners of rare maps rarely allow them off the premises, and may not be happy to allow them to be scanned or to provide space to set up a scanning facility on-site. The alternative we used in some cases was large format photography followed by scanning of the positive transparencies on a production quality scanner. In cases where we were able to scan the maps directly, we used a Abakos DeSKan Express scanning system which scans the map in controlled 180 mm strips with an overlaid graticule and stitches them together into an image of up to A0 size. We adopted a minimum scanning resolution of 300dpi and 24 bit colour.

The images generated are, of course, very large – up to 250 Mbytes - and are best archived as TIF files at original resolution, then compressed with a pyramid wavelet compression scheme such as MrSID for delivery. The MrSID compression not only reduces the image size substantially, but also speeds up image display and reduces download volumes when images are accessed over the web. Both TMView and the Java applet viewer allow download and display of lower resolution overviews and progressive update of resolution as the map is zoomed in.

Having obtained a high resolution scan of the map, the next step is geographic registration, which is normally carried out by identifying common points between the historic map and a modern map base. As maps become further removed from the present, the number of persistent points identifiable in the historic map tends to decline – street corners move, buildings are demolished, rivers wander and coastlines are built out. Furthermore, details of secondary importance to the mapmaker, such as the coastline or riverbanks, are often sketched in rather than accurately mapped.

In practice we have found that the best modern reference maps for city environments (we have worked on both central Sydney and on Parramatta) are the detailed cadastral maps. Cadastral information – property ownership – has been one of the driving forces of European large scale map making, and while coastlines, development plans and even mapped buildings are often fanciful, the basic property ownership grid of the city is faithfully represented. With suitable historical interpretation to take account of street realignment, subdivision and redevelopment it is possible to accurately locate cadastral boundary points on historical maps into a modern coordinate space.

A second method of increasing the hit rate for early maps has been to start by registering accurate topographic surveys from the mid nineteenth century, and then using those maps as a map base for registration of earlier maps. Since the structure of Sydney, and many other cities, changed more drastically between the nineteenth and twentieth centuries than throughout the late eighteenth and the nineteenth centuries, the commonality between nineteenth century and earlier maps is far greater than with modern maps.

Maps as historical documents

On some early maps of Sydney, we were unable to identify enough persistent points for registration of the whole map area. Even when adequate registration points can be identified, there is often an embarrassing lack of fit highlighted by the coastline, because the maps are simply not scale accurate. All one can do is either to fudge the registration by inventing points to squeeze recognizable features into a semblance of modern reality or, as we have chosen to do, to leave the glaring inaccuracies as witness to the historical reality of early mapmaking (Figure). A high priority of the next stage of research is develop a registration system that will systematically capture information about discrepancies in the historical maps.

The other problem which faced us with the Sydney *Time*Map kiosk was the fact that different historical maps covered different areas, were generally oriented to magnetic north rather than grid north (approximately 14 degrees rotation) and came out irregularly shaped after warping to modern coordinates. Trimmed of excess borders, we ended up with a set of ugly, inconsistent rhomboids inappropriate to the museum's sharp design principles.

To solve this problem we devised the idea of a 'map table' – a uniformly shaped rectangular area on which the scanned and warped historic map is superimposed (Figure 14) as a representation of the historic document with all its supporting material (map key, marginal labeling) and flaws (holes, torn edges). To highlight the nature of the map as a historic document, these latter features are emphasized with a slight drop-shadow.

Notes

The work described in this paper was supported by a Strategic Partnerships with Industry for Research and Training (SPIRT) grant from the Australian Research Council, in partnership with the Museum of Sydney (Historic Houses Trust of NSW) and ESRI Australia, and by funding from the Lilly Foundation and the California Digital Library through the Electronic Cultural Atlas Initiative Berkeley.

The authors would particularly like to thank Emeritus Professor Lewis Lancaster, Director of ECAI; Catherine Candee, Director of e-Scholarship at the California Digital Library; Quill Williams, ESRI Australia; A/Prof. Roland Fletcher, University of Sydney; Peter Watts and Helen Temple, Historic Houses Trust of New South Wales, and Susan Hunt and Kate Richards, Museum of Sydney; for their ongoing support and encouragement.

The Sydney City Council Archives, State Records Authority of New South Wales and the National Library of Australia kindly gave us free access to their holdings for a substantial part of the data used in Sydney *Time*Map. We would like to thank Janice Glowski of the Huntington Archive of Southeast Asian and Buddhist Art, Ohio State University, for high resolution scanning of the large format map negatives and Damian Evans, David Hobson and Bess Moylan for the painstaking work of map registration and warping.

End notes

[1] *Time*Map is a registered trademark of the University of Sydney

[2] Earlier versions of TMView used MapInfo Professional and MapX, but technical reasons led us to switch to MapObects and the open-standard shapefile format.

References

Johnson, I. 1999. Mapping the fourth dimension: the TimeMap project. In CAA97: *Computer Applications in Archaeology 1997.* BAR International Series 750. 21pp [on CD-ROM].

Figures

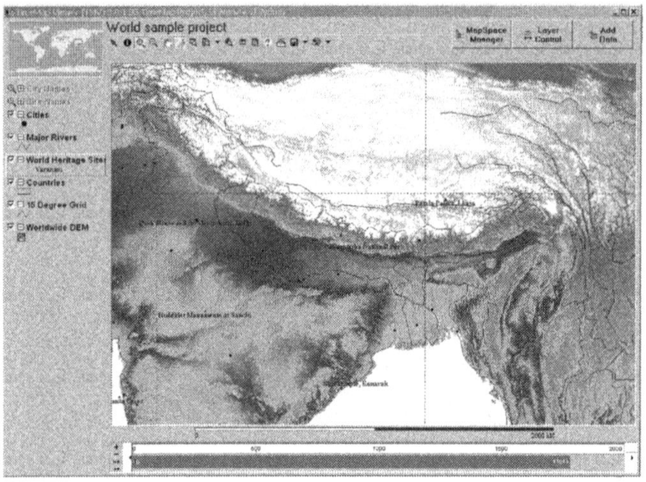

Figure 1. Time*Map viewer (TMView) map window with UNESCO World Heritage Sites superimposed on a 1km digital elevation image*

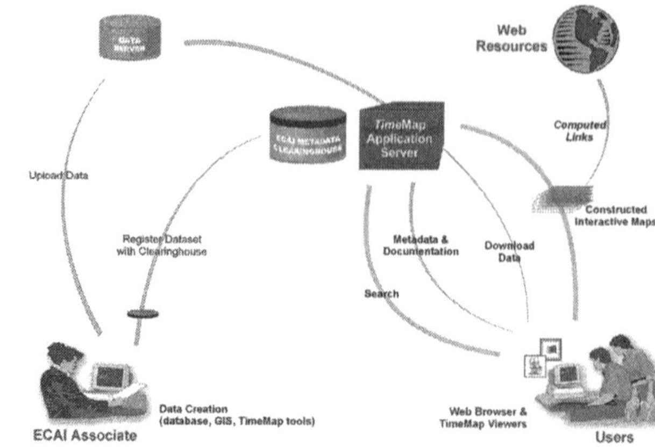

Figure 3. The relationship between TMView, the ECAI clearinghouse and the middle-tier Time*Map application server*

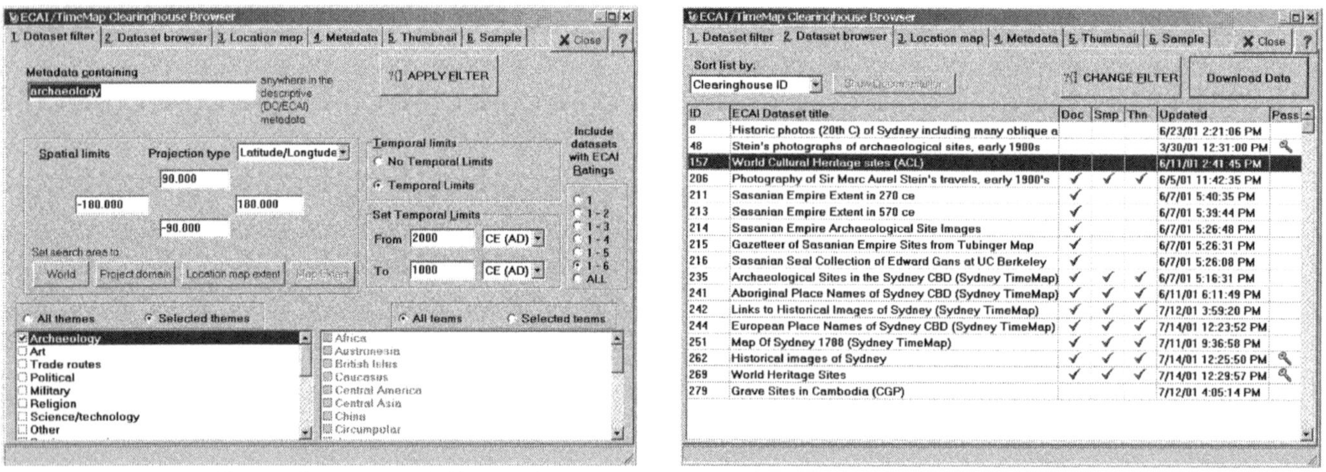

Figure 2. TMView Clearinghouse search filter and results

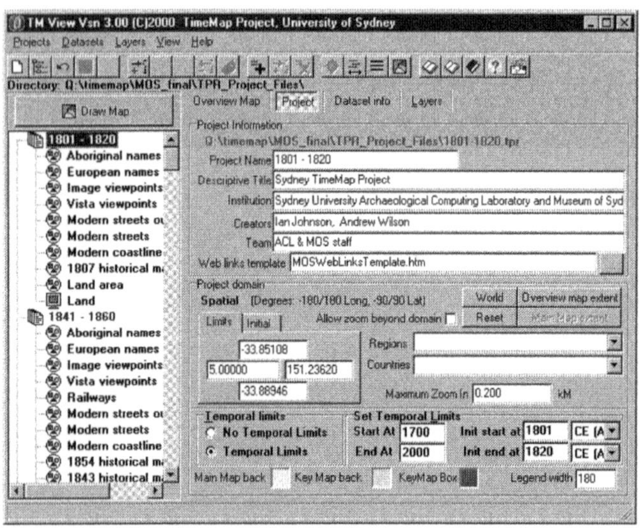

Figure 4. TMView MapSpace manager

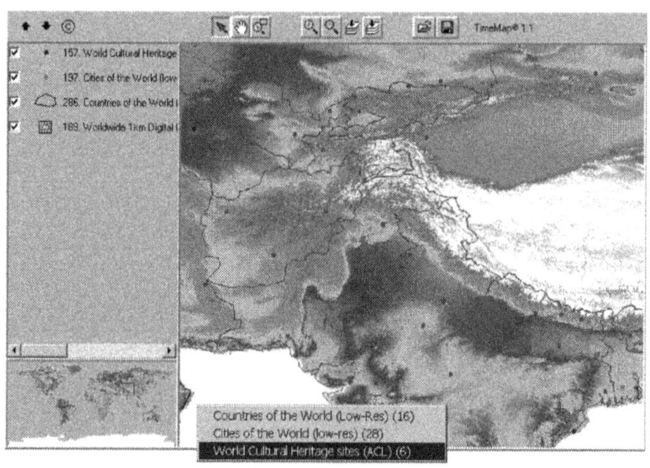

Figure 6. Java map viewer showing world cultural heritage sites (UNESCO listing) superimposed on 1km digital elevation image and world country outlines

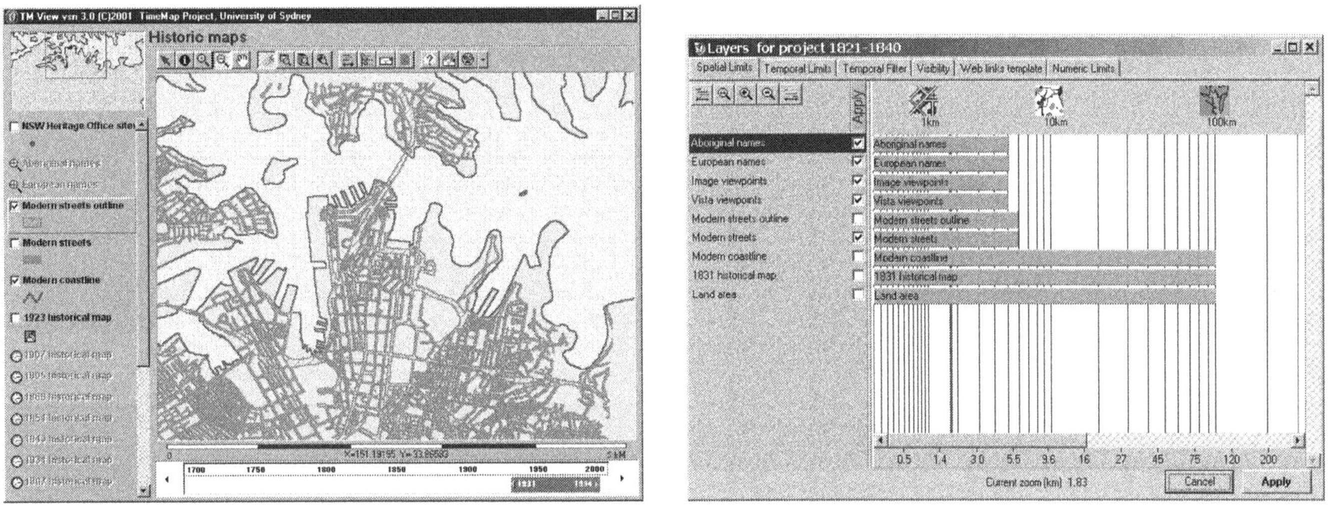

Figure 5. TMView scale visibility – map window with disabled layers and configuration dialogue

Figure 7. ECAI clearinghouse search results showing web sites, datasets and passworded datasets. Datasets are included in the map result by checking the boxes

Figure 8. The Sydney Time*Map kiosk application at the Museum of Sydney*

Figure 9: Sydney TimeMap dual screen interface – Macromedia Director wrapper

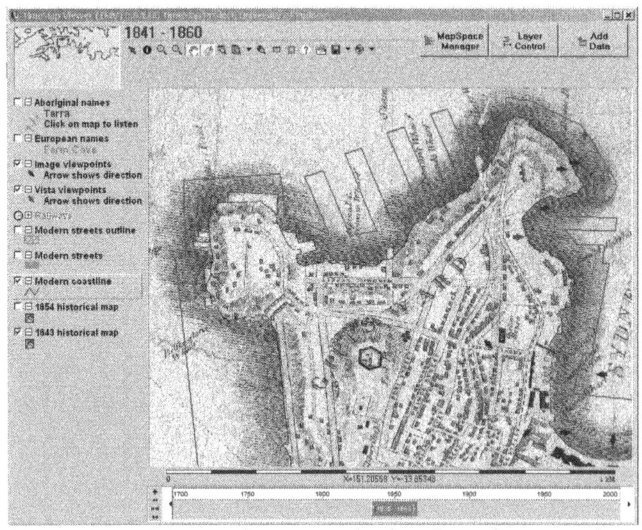

Figure 10. A historical map overlaid with modern coastline, image viewpoints and directions

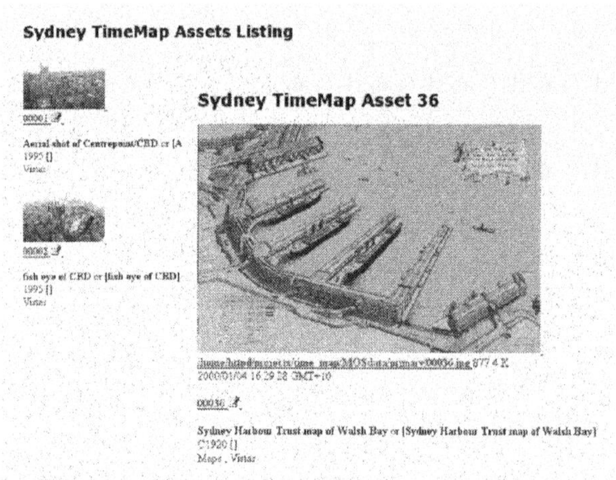

Figure 11. Web-based access to the digital image database showing thumbnail views and a higher resolution image generated on the fly

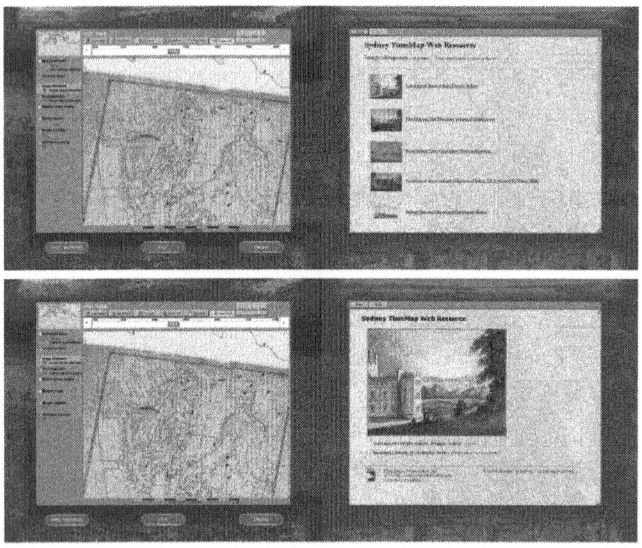

Figure 12. Web pages generated from the Sydney Time*Map kiosk application by the thumbnail template (above) and by the single item template (below).*

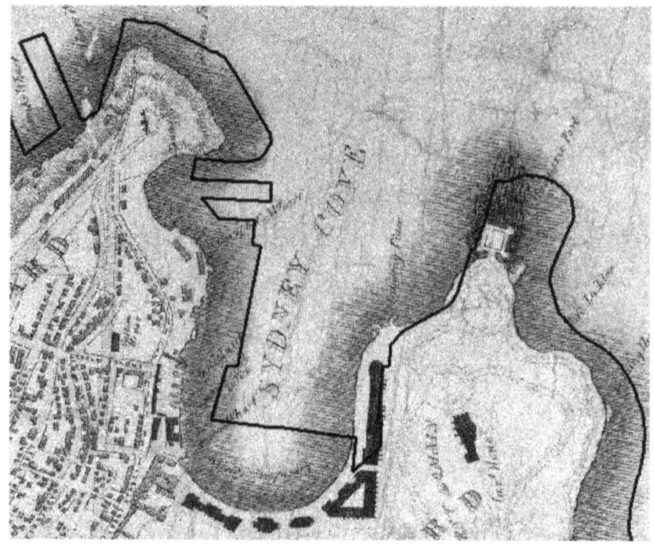

Figure 13. An 1844 map showing the effect of scale error across Sydney Cove (centre)

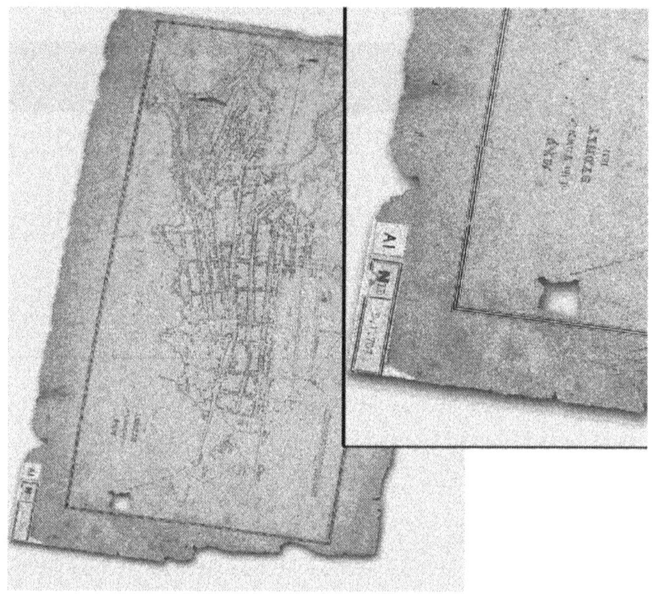

Figure 14. Map of Sydney presented as a historical document on a 'map table' display (detail of map shows use of drop-shadow to emphasise the map as a historical document)

The answer is blowin' in the wind. Research desires and data possibilities.

Hans Kamermans
Faculty of Archaeology
Leiden University
P.O. Box 9515
NL 2300 RA Leiden
Phone: +31 71 527 2385 - Fax: +31 71 527 2429 - E-mail: H.Kamermans@arch.LeidenUniv.nl

Abstract: This paper was supposed to be called "Buried by the wind. Regional analysis in Palaeolithic and Mesolithic Archaeology in the Southern Netherlands". It was supposed to examines the influence of loess deposits on the distribution of Palaeolithic and Mesolithic find spots in Limburg, the most southern region of the Netherlands, using a Geographic Information System (GIS). It was supposed to demonstrate that the archaeological visibility of Palaeolithic find spots in that part of the Netherlands is greatly hindered by the Upper Pleistocene loess cover of this part of the country. It was supposed to demonstrate also that the visibility of Mesolithic find spots is not influenced by that geological phenomenon. And it was supposed to investigate if these differences in the distribution of find spots, and the concluded differences in land use, are a result of geological processes.
But the quality of the data set did not make any of this possible.

Key words: GIS, Palaeolithic Archaeology, land use.

Introduction

The following research is part of an ongoing project about the Palaeolithic of Limburg, the southern part of the Netherlands (Kamermans & Rensink 1999). This research tries to analyse and interpret Palaeolithic and Mesolithic find spots from the Dutch loess area from a landscape perspective. One of the problems is how to evaluate correctly areas without find spots. Does this point to a selective use of specific zones by hunter-gatherers, or are these 'empty' areas a result of geological or recovery processes? In more general terms: to what degree is the observed distribution of stone artefacts across the area representative for the use of the landscape by prehistoric hunter-gatherers?

This particular part of the research started with the following question:

"We assume a different economy and a different land use between the Palaeolithic and the Mesolithic period, or at least between the Middle Palaeolithic and the Mesolithic. Can one see this in the distribution of sites in the landscape in South Limburg or is this pattern influenced by the geology? The Middle Palaeolithic sites were covered by loess during the Pleniglacial and the later Mesolithic sites were not. Are we looking at a difference in distribution or are we simply looking at moreorless the same pattern obscured by the deposition of loess. What if we use GIS and throw a loess cover over the Mesolithic sites. Would the same pattern as for the Palaeolithic emerge?"

For good research you need a good theoretical background, a good research question, a good dataset, and good tools to research your question. It looked as if all this was available for our Limburg case study.

There is no problem with the theoretical background, but what about the research question?

Research question

There is an ongoing debate about differences in subsistence strategies during the Middle and Upper Palaeolithic. What is the difference in subsistence strategy between the Neanderthals, or the Ancients as Stringer and Gamble (1993) call all pre-modern humans, and the Moderns? For instance, did Neanderthals hunt? If yes, were they general or specialised hunters? Most researchers see the change in subsistence happening between Ancients and Moderns, and considered it as part of the so-called Middle to Upper Palaeolithic "transition". Others see changes during the Middle Palaeolithic (c.f. Stiner 1994, Kuhn 1995). Differences in subsistence, means differences in human behaviour and should leave a difference in spatial patterning of the material culture in the landscape. The problem is there. Can we solve it in Limburg?

The data set

Limburg is archaeologically speaking a very well researched area. Not only for archaeology in general but especially for the Palaeolithic period. Roebroeks excavated the Bélvèdere quarry with the oldest site from the Netherlands, 250.000 years old (Roebroeks 1988). Rensink studied the Magdalenian sites in Limburg and surroundings (Rensink 1993). Amateur archaeologists are very active in that part of our country and for GIS applications Limburg is perfect; it is one of the few places in the Netherlands with some form of relief and has a great variety in landscape.

But for a GIS application you need both maps and archaeological data. For this research we used two geological maps, one soil map, one geomorphological map, two maps with the distribution of the loess, and a slope map and a slope aspect map from parts of the area. For the archaeology we used a slightly updated dataset from ARCHIS, the national Dutch sites and monuments record.

Figure 1 shows the distribution of loess and the location of the Middle Palaeolithic and Mesolithic sites. The loess map indicates areas with a loess cover of more than 1 meter thick, loam on slopes, stream sediments and areas with no loess. The loess cover was deposited during the Pleniglacial, lets say from 55.000 until 13.000 year ago. Some of the Upper Palaeolithic sites (from 35.000 until 10.000) date from during this period of deposition and some do not. Including these Upper Palaeolithic sites will complicate matters so we leave that period out of our analysis. With a visual inspection we do not see a difference in distribution.

Table 1 gives the relation between Middle Palaeolithic sites and the various deposits and shows us as many sites on the loess as expected and twice as many on the slopes as expected.

Table 2, the relation between Mesolithic sites and the various deposits, gives more or less the same picture. On the loess are as many sites as expected, on the slopes twice as many.

The Middle Palaeolithic sites on the loess are of course a problem. They date from before the loess and, for that reason, cannot be on top of the loess. The map (figure 2) confirms this.

This shows us the first problem with our data set. The maps are apparently not accurate enough although the resolution is 25 meters. The scale of the original maps is 1 : 100.000.

Another option is to use the slope map. There is a relation between loess and slope. The loess cover is still present on relatively flat surfaces. We have a slope map from part of the area, the Central Plateau. Maybe the slope map will show that the Middle Palaeolithic sites on loess are in reality lying on steep, eroded slopes (figure 3). The table (3) gives more information. One fourth of the number of sites that could be expected on the basis of chance lie on the flat area and more sites than expected lie on the slopes.

For the Mesolithic the number of sites on the flat surface is as expected and on the slopes more than expected (table 4). It looks as if the loess does influence the pattern. But there are still Middle Palaeolithic sites lying on the flat surface. So also the slope map is not accurate enough.

We know that most maps are not accurate (soil maps for instance have an average accuracy of 70% (Kamermans & Rensink 1999).

So much for GIS as a tool to solve all your problems. But is it possible to do the analysis without maps? Is there enough information in the database? If the database tells us if the site is a surface find or an excavation that will help. It does, but as I said we used a slightly updated ARCHIS dataset and we only included surface sites, so the problem remains.

A similar analysis on data collected by the Agro Pontino Project (Voorrips *et al.* 1991) utilized the field database and used maps only for illustrations (Kamermans 2000). The data that was registered in the field included not only the parent material, but also the soil type, the slope angle and the slope aspect, the geomorphology, and dozen's of other things (Voorrips *et al.* 1989). We could perform the same kind of analysis for our Limburg research but is this kind of information available in the Dutch national database ARCHIS?

There is a lot of information in the database. Administrative information about the locality, information about the landscape and information about the site and the archaeological material. There is information about the geomorphology, the geology, the texture of the soil and the situation in the landscape. But there are two problems. It is not always clear if this information is collected on site or taken from a map and the information is not always there. On the form people have to fill in when they report a new find, a suggestion is printed next to the geomorphology field, which is to take this information from the geomorphological map. In the case of the current research therefore, ARCHIS did not help very much. This is the second problem with the data set.

Conclusions

With a very straightforward application, like the one sketched above, we encountered two major problems: the digital maps and the archaeological data set. Will these problems be solved in the near future?

A new development with digital maps in the Netherlands is the AHN (Actueel Hoogtebestand Nederland), a new digital elevation database from the whole of the country made with laser altimetry (http://www.minvenw.nl/rws/mdi/geoloket/index1.html). It has a density from 1 point per 16 m². This 3D dataset is accurate enough to solve in the near future the problem of the relation between slope and loess coverage.

But the digital geological map, soil map and loess map will not become more accurate than the original maps. So these maps are not accurate enough to rely on the information. To do the analysis on the basis of a database we will need an awful lot of data in ARCHIS. We need information gathered on site about the geology, geomorphology, soils (like parent material, texture, soil type), drainage, slope angle, slope aspect, etc, etc. Only then we can do our analysis independent of our maps.

The ROB is planning a new version of ARCHIS. It will have three layers of information. One for experts, one for scientists and one for lay people. Every layer would have their own kind of information. From the documentation (Van Capelleveen *et* al. 2000) it becomes clear that the expert layer will contain mainly administrative information to manage the site. The form to register a site into ARCHIS will stay more or less the same, so it looks as if not all of the information required for a good regional analysis will be there.

The problem is highly relevant. One of the most important products of the State Service at the moment is the IKAW, the indicative map of archaeological values (Deeben *et al.* 1997). These maps are a result of predictive modelling on the basis of ARCHIS and digital maps of the physical environment. There has been a lot of criticism directed towards the first two generations of these maps (c.f. Verhagen *et al.* 2000). The new ARCHIS will play an important role in the production of future predictive maps. In my opinion it is impossible to predict site location for CRM purposes without a very detailed archaeological database. The only other solution is to rely on the unreliable digital maps.

With the general available data it was impossible to answer the simple question formulated above. The problem remains, so in the end *the answer, my friend, is blowin' in the wind.*

Acknowledgements

I would like to thank my colleague Professor Wil Roebroeks (Faculty of Archaeology, Leiden University) for the research question I could not answer and Eelco Rensink (Dutch State Archaeological Service) for the collaboration in the Limburg project. Both would have been co-author's of this paper had I answered the research question. I am very grateful to Martijn van Leusen (Groningen Institute of Archaeology, Groningen University) for making all his digital maps of South Limburg available to me.

References

Capelleveen, E.J. van, E.E.A.M.C. Kuis & J.C. Meerkerk. 2000. *Rijksdienst voor het Oudheidkundig Bodemonderzoek. Programma van Eisen ten behoeve van Europese aanbesteding vernieuwing Archis.* Twynstra Gudde, Amersfoort.

Deeben, J., D. Hallewas, J. Kolen & R. Wiemer. 1997. Beyond the crystal ball: predictive modelling as a tool in archaeological heritage management and occupation history. In: Willems, W., H. Kars & D. Hallewas (eds.): *Archaeological Heritage Management in the Netherlands. Fifty Years State Service for Archaeological Investigations.* ROB, Amersfoort, pp. 76-118.

Kamermans, H. & E. Rensink. 1999. GIS in Palaeolithic archaeology. A case study from the southern Netherlands. In: L. Dingwall, S.C. Exon, V.L. Gaffney, S. Laflin & P.M. van Leusen (eds). *Archaeology in the Age of Internet - CAA97. Computer Applications and Quantitative Methods in Archaeology.* BAR International Series 750: 81 & CD-ROM (ca. 30 p.).

Kamermans, H. 2000. Land evaluation as predictive modelling: a deductive approach. In: G. Lock (ed.). *Beyond the Map. Archaeology and Spatial Technologies.* NATO Sciences Series. IOS Press: Amsterdam. 124-146.

Kuhn, S.L. 1995. *Mousterian Lithic Technology. An Ecological Perspective.* Princeton University Press: Princeton, New Jersey.

Rensink, E. 1993. *Moving into the North: Magdalenian Occupation and Exploitation of the Loess Landscape of Northwestern Europe.* Unpublished PhD thesis, Leiden.

Roebroeks, W. 1988. From Find Scatters to Early Hominid Behaviour. A study of Middle Palaeolithic River Side Settlement at Maastricht-Belvédère (The Netherlands). *Analecta Praehistorica Leidensia* 21

Stiner, M.C. 1994. *Honor Among Thieves. A Zooarchaeological Study of Neandertal Ecology.* Princeton University Press: Princeton, New Jersey.

Stringer, C. & C. Gamble. 1993. *In Search of the Neanderthals. Solving the Puzzle of Human Origins.* Thames and Hudson: London.

Verhagen, P., M. Wansleeben & M. van Leusen. 2000. Predictive Modelling in the Netherlands. The prediction of archaeological values in Cultural Resource Management and academic research. In: Harl, O. (ed.): *Archäologie und Computer 1999. Forschungsgeselschaft Wiener Stadtarchäeologie* 4: 66-82.

Voorrips, A., S.H. Loving & H. Kamermans. 1989. Information Science in Archaeological Survey. in: T. Hackens & U. Miller (eds). *Geology and Palaeoecology for Archaeologists.* Pact 24: 189-210.

Voorrips, A., S.H. Loving & H. Kamermans (eds.). 1991. *The Agro Pontino Survey Project. Methods and preliminary results.* Studies in Prae- en Protohistorie 6. Amsterdam. 131 p.

Tables

unit	area	perc	O sites	E sites
löss cover	346.47125	50.03	35	39.03
loam on slopes	163.47375	23.61	37	18.41
stream sediments	53.19625	7.68	0	5.99
no löss	129.334375	18.68	6	14.57
	692.475625	100.00	78	78.00

Table 1. The relation between Middle Palaeolithic sites and the various deposits. Area is in km², O sites is observed sites and E sites is expected sites

unit	area	perc	O sites	E sites
löss cover	346.47125	50.03	19	22.52
loam on slopes	163.47375	23.61	24	10.62
stream sediments	53.19625	7.68	1	3.46
no löss	129.334375	18.68	1	8.40
	692.475625	100.00	45	45.00

Table 2 the relation between Mesolithic sites and the various deposits

unit	area	perc	O sites	E sites
low	61.82	78.83	2	7.88
medium	8.60	10.97	6	1.10
high	8.00	10.20	2	1.02
	78.42	100.00	10	10.00

Table 3. the relation between Middle Palaeolithic sites and slope class.

unit	area	perc	O sites	E sites
low	61.82	78.83	5	6.31
medium	8.60	10.97	3	0.88
high	8.00	10.20	0	0.82
	78.42	100.00	8	8.00

Table 4. the relation between Mesolithic sites and slope class.

Figures

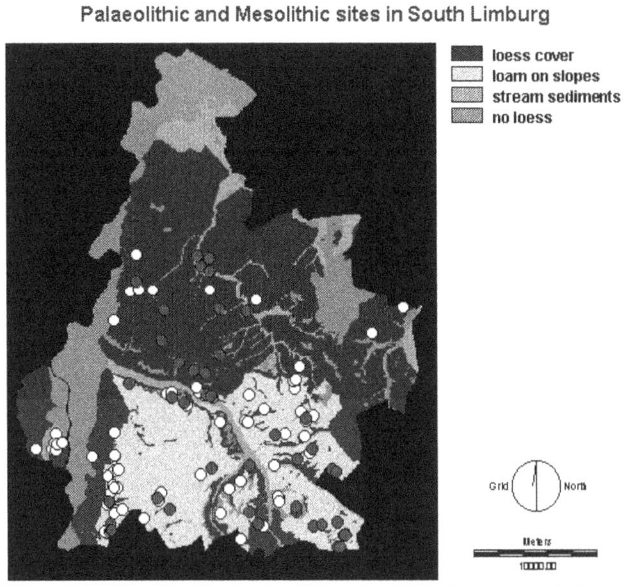

Figure 1. The distribution of loess and the location of the Middle Palaeolithic (white dots) and Mesolithic (red dots) sites.

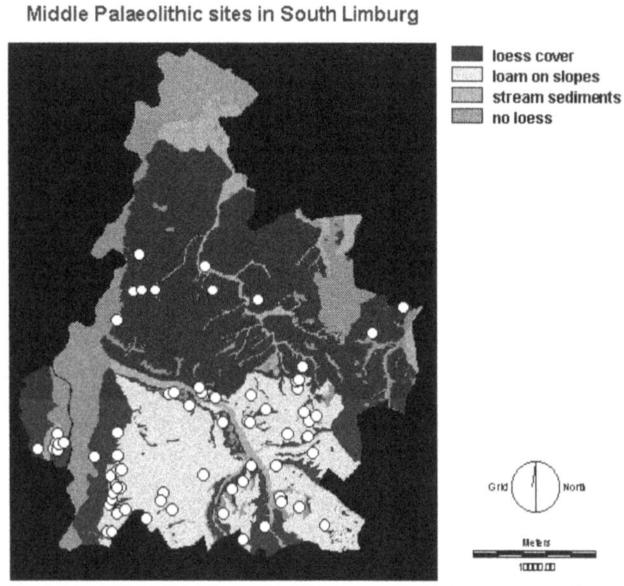

Figure 2. The distribution of loess and the location of the Middle Palaeolithic sites.

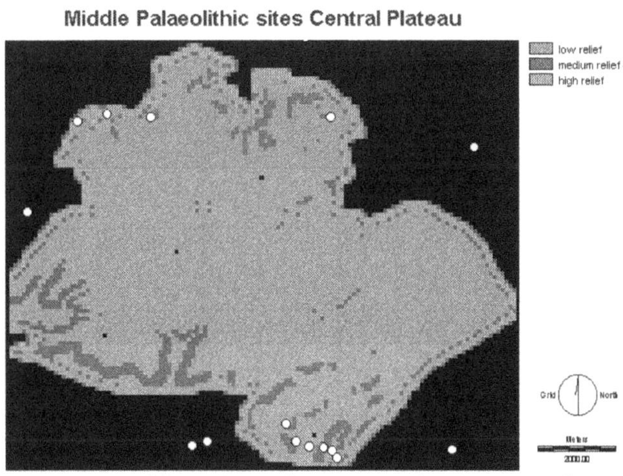

Figure 3. Middle Palaeolithic sites on the slope map from the Central Plateau.

A hundred years of lake contour fluctuation in the Hamun-i Helmand: A GIS based system for the study and the recovery of archaeological information in the Iranian Sistan (1899-1999)

Sabatino Laurenza
University of Bologna
Dept. of Archaeology, Italy
via G. Spontini 9/A 00013 Mentana (Roma), Italy
Phone: +39 06 9091843 - Mob: +39 347 5250441 - E-mail: laurenza.s@tiscalinet.it

Sophie Pornet-Laurenza
University of Bologna
Dept. of Archaeology, Italy
via G. Spontini 9/A 00013 Mentana (Roma), Italy

Abstract: The fragmentation of the contextual set of all kinds of information, which has not been preserved intact during the years, has always been a big limit for archaeological research.

Reconstruction is now realisable with a stochastic process, which gives a complete picture of the pertinent data by increasing approximations, because it is able to join together information and contexts otherwise scattered.

The recording, display, management and storage of data by computers, gave us the possibility to do complex analysis, which was previously impossible.

In this framework GIS are a set of instruments able to push archaeological data analysis further.

If we consider the spatial nature of archaeological data - the recording of the precise location of artefacts, ecofacts, architectural or stratigraphical units and sites - the GIS allows the archaeologist to match different types of geographic data (hydrography, soils, elevations, ...) in order to produce wide-scale analyses.

Key words: Historical cartography, Iranian Sistan, Lake fluctuation, GIS, Remote Sensing.

Introduction

Apart from a few and very exceptional cases, the archaeological record all over the world is too fragmentary, to return a direct vision of ancient lives and landscapes. The context is usually lost and it is left to the skills of specialists "to piece together the fragments of the past". Generations since World War II have been blessed by computer science, with an increasing capacity to combine vast archives of data. Reconstruction has been made possible, more recently, by so-called "stochastic procedures" which allow combining of very large numbers of operations, modeling the scattered fragments into true pictures of lost realities. As more data is introduced with different theoretical perspectivs, models can be constantly updated.

Geographical Information Systems (GIS) at present are the main instrument available from computer science to reconstruct past landscapes and human populations in their historical fluctuations, by combining data and information from maps, earth sciences, surveys and excavations.

The Iranian Sistan

The Hilmand Delta in Sistan represents a clear case of instability evidently connected to the fact that the basin is sharply limited to the west by a North-South barrier made by the front of the Palang Mountains and their piedmont colluvial fans of gravel sediments. The first noticeable difference between the Hilmand and the other systems, such as the Murghab or the Nile, is the fact that the densest settlement areas and the capital centres have always been, since the end of the 4th millennium BC, at the end of the delta system, to connect with the river as well as the huge terminal lake. The rapid build-up of sediments compels the waters to change their course several times, while a circular anticlockwise sequence of descending lacustrine basins connected with each other by spillways, like the Sheelag rud, drains overflows to the southernmost sections of the basin (fig. 1). Satellite imagery of southern Sistan indicates some six or seven overlapping delta fans, all dated by the archaeological sites to later historical times, spanning less than a thousand years in Islamic times. Even more unstable have been the southern limits of the Hamun, the largest and most perennial of the terminal lakes. What is left of this ancient system, mostly dating to protohistorical times, is reduced to thin *yardangs*, columnar sedimentary residues shaped by the dominant winds. Third millennium BC soils, often black from organic content, have been detected at the top of these *yardangs* about two meters above the present level. The combined action of the expanding lake and wind erosion has lowered most of the ancient surfaces, destroying all residual evidence of the Bronze Age landscapes.

The analysis of the Aeolian deflation, as a process of concentration of the objects through the effect of the lowering of the deposit layers needs to re-evaluate the surface documentation that precedes different strategies of research. Wind and water are tireless excavators, that follow scare but precise rules (fig.2). During the 70's, American researchers like C. Redman and P. J. Watson (Redman, C. R., Watson, P. J. 1971) worked on the information drawn from the dispersion of stone artefacts on the surface of prehistorical sites in oriental Anatolia. Between 1966 and 1970, measures of the Aeolian-raining erosion rates in the Oaxaca valleys in Mexico and in the Deh Luran in Iran were made and allow Kirkby and Kirkby (Kirkby, A., Kirkby M. J. 1976.) to elaborate models of geo-morphological alteration. The archaeological team of SiS, convinced by the interest that this kind of study would present in the Sistan area, couldn't apply these methods of research, as the site of SiS was much bigger than the areas studied by the Americans and the wind and water freaks are much more powerful (Tosi, M. 1985). In 1972, they applied systematic surface survey to the areas of the site where the handicraft-quarters for lapis lazuli at SiS were discovered.

Historical cartography

The pivotal position of Sistan, among Afghanistan, Baluchistan and Iran, was critical in the power struggle between the Russian and the British empires for the region of Middle Asia during most of the 19th century. This conflict has ultimately determined the present political division of the region

H. C. Rawlinson (Rawlinson, H. R. 1873.) who visited Sistan in 1872, writes that "the comparative geography of Seistan is rendered especially difficult by the constant changes in the lower course of the Helmend, changes which are common to the deltas of most rivers; but which occur with all the greater frequency when owing to the very slight difference of level, there is no defined basin to receive the waters that are discharged". It is also useful to underline the difficulties, others British officers found in defining the course of the main river - the Hilmand - from its secondary branches, and to make the difference between artificial canals dug for irrigation purposes and natural rivers beds.

Before the work of the 1899-1902 Afghan-Iranian Border Commission, under the Head of Col. M. Mac Mahon who produced the first complete and detailed maps of the area and described precisely the circular movement of the water of the river and the lakes, we can just have an idea of the variability reading the different papers of the British officers, in particular telling about their visit to the Koh-i Kwaja mount. In 1842, Captain Conolly (Rawlinson, H. R. 1873.) needs a small reed-boat to reach what he describes as an island. Thirty years later, Major Lovett (Rawlinson, H. R. 1873.) crosses the dried lake by foot to reach the same point (fig.3)

The site

Discoveries during the last thirty years in Central Asia (particularly in the Iranian area) brought to light the antiquity of the proto-urban process in that region of the world far away from the traditional civilizations (Egypt, Mesopotamia, Harappa).

The state formation process comes for the first time, between 3400 and 3000 BC in the Nile valley, in Mesopotamia, in the plateau regions of Iran, Turkmenia, and in the Indus valley thanks to the exceptional demographic increase, to the densification of settlements along the hydrographic axes, to the exploitation of the mineral resources organized following work specialization, and to the hierarchical establishment of the society. Between 3200 and 2900 BC hundreds of little settlements are abandoned, creating for the first time a precise difference between center and periphery. In such period urban centers and as Hissar, Tureng Tepe, Namazga Tepe and Mundigak, which, together with Shahr-i Sokhta (hereafter S.iS.) will represent the "Helmand civilization".

The principal settlement of Sistan, S.iS., which represents the main center of the region during the protohistorical period, is founded on the top of the terrace, at the end of the fourth millennium.

The excavations of the Italian archaeological expedition of IsIAO (ex IsMEO) recognized during ten years of research at least 11 cultural phases (also called "structural phases") of development of the city. S.iS. represents a trade center for local and extra-regional products, but also a high-specialized place of artifact production, with entire urban and extra-wall areas dedicated to these activities.

The secondary settlements (between 2 and 6 ha.) almost all belong to the protostate phase (from 2500 BC), characterized on one side by cultural development, by technological progress and by an enlargement of the settled areas; on the other side, begins the development of a cultural preservation, due mainly to the shift of the traditional commercial and cultural axes. This process increases during the second millennium for the disappearance of the urban structures in the Iranian area. The crisis of S.i.S. in the frame of the Sistan protostate system began around the beginning of the second millennium when the town becomes smaller (from 120 ha. to 5-6 ha.) (fig. 4).

Modern Cartography and Satellite Imagery

The opportunity of using historical and modern maps, and especially satellite photos now available, in such a way as to understand the interactions between man and landscape through the reconstruction of environmental changes and their consequences on the settlement in this region, can be considered an extension of the purposes of the Italian Mission in Sistan, helped by new technologies. The present work aims to enlarge the use of these new tools looking "backwards" to the management of documentation which was not originally planned.

We should underline that the present project has three aims:
1. recovery and management of large and varied archives of old heterogeneous data
2. integration of these data in a system of construction/ verification of hypothesis about the study of the material and of the environment
3. organisation of all the data collected in a GIS based system.

The use of satellite imagery begins after 1961 for military purposes. Since 1972, with the launch of Landsat-1 by the USA, it becomes available for scientists and the public. The IsIAO projects have made extensive use of this imagery for the study of ancient civilizations in Middle Asia and the Arabian Peninsula.

Our only serious possibility to understand the complex fluctuations of peoples and their socio-economic organization through time, since no occidental scientists can work in the area, rests only on the progress of satellite imagery for earth studies and related computer processing. The old surveyed data from the explorers of the 19th century and the archaeologists of the 20th will be projected on satellite images of highest resolution.

For the Archaeological Map of the Helmand Delta, currently under preparation in Rome and Bologna, a wide collection of maps and satellite images has been gathered to define altitude variations and geo-morphological conditions at the highest level. Recently NASA has released the collection of the first military images from the CORONA Program, that are particularly important because they date to the Sixties when Sistan was still very little developed (figs 5-6).

The idea of using GIS for the management, analysis and study of old and new data on Sistan will be directed towards reconstructing the landscape and environmental changes for the study of early civilizations in their territorial dimensions, beginning with Shahr-i Sokhta and the Early Bronze Age settlements. Central to the understanding of changes in the area is the erratic behaviour of the lake, whose fluctuations largely marked the wealth potentials of the region (figs7-8)

The methodology

Since 1978 the Italian Expedition of IsIAO could not continue fieldwork activities in the Sistan region, and from that moment the digital processing phases of the huge and heterogeneous collection of data began.

Until now, several tries of data recording, management, and visualization, such as a dedicated database, map digitization, and a Hypermedia system, developed in *ToolBook*, have been carried out.

The main aim of the project is to develop new methodological and technical approaches for the acquisition and the elaboration of a huge set of data, in order to create a Multimedia GIS for the publication of an archaeological landscape and sites, containing all the typical geo-spatial information.

The first step of this work was to realize a prototype simulation model, able to give to the user a complete vision of the problematic of this region and to allow him to analyze, through the computer, the data from an excavation nowadays inappropriate to support a normal management.

Thanks to the GIS-based system for the study of the excavation of Shahr-i Sokhta settlement and graveyard, the user can check the hypothesis proposed until now, but also explore some new hypotheses.

According to the available data, the methodology consists of the following steps:
- checking and analysis of the existing archives;
- codification and integration of the different tables in a relational system for the management of databases (RDBMS);
- digitization in CAD systems of the excavated inhabited areas and Necropolis phase maps, using different layers for each room or for each single grave and geocoding these maps;
- realization of GIS system for the analysis and the study of the excavated areas of the site;
- queries for the realization of phase, thematic and diachronic maps directly from the system (fig. 9)
- construction of a 3D model of the landscape and of the site of Shahr-i Sokhta (figs10-11);
- visualization of the site with excavation areas geocoded (fig.12);
- texture mapping of the satellite photo on the model of the site of Shahr-i Sokhta (fig.13).

With this approach, it would be possible to arrive, at the end of the data elaboration processes, or the first "virtual publication" of the protohistoric site. Moreover, the development of such multidimensional GIS, will make possible a multimedia publication of the digitally elaborated archaeological information, in order to give the scientific public the possibility of exploring and understanding the fieldwork data and information for the entire site complex (i.e.: a final CD-ROM).

Conclusion

It is clear that we have presented here only the methodological approach to the problem of recovery of data and information from an old excavation and survey. The first target we kept in mind was to find a system able to manage and record all the data and that allows us to study and to formulate some new interpretative hypotheses. Until now we have worked mainly on the collection and management of data, using GIS as the engine in which all the information will be stored, in order to allow the researchers to produce quickly and easily some basic thematic maps and to check all the data .

The next step will be the most interesting and exciting, as it will consist of the intra-site spatial analyses and of some analyses of the working areas definition on the site. We started some analyses and at the moment the result is very satisfying, pushing us to formulate each day new intra-site analyses.

Our wish is to achieve a Multimedia GIS-based publication of the entire *corpus* of data relative to the study of the protohistoric site of Shahr-i Sokhta by next summer, which will allow us to present new and innovative hypotheses on the difficult problematic of this region.

Acknowledgement

The authors are grateful to Prof Maurizio Tosi who provided material, data and time to make this project possible.

References

Kirkby, A., Kirkby M. J. 1976. Geomorphic Processes and the Surface Survey of archaeological sites in Semi-arid Areas. In D. A. Davidson and M. L. Schackley (eds.), *Geoarchaeology*, London, pp. 229-253.

Johnson, I., Norton, M. 1996. Archaeological Applications of GIS. In : Proceedings of Colloquium II, UISPP XIIIth Congress, Forlì, Italy

Laurenza, S. et alii. 1995. Un Sistema Ipermediale per navigare in un'antica citta' dell'Asia Mediorientale, Pubblication n.36, Aprile 1995, Collana IRSIP

Laurenza, S. et alii. 1995.. Surfing the Past: A tool for hypermedia visit of an early urban site in Middle Asia. In : Ist International Congress on : Science and Technology for Safeguard of Cultural Heritage in the Mediterranean Basin, Catania, pag.475

Laurenza, S. et alii. 1996. Sharh-i Sokhta, la Città Bruciata. In M. Forte (ed.) *Archeologia. Percorsi virtuali nelle civiltà scomparse.* pp.200-204

Laurenza, S. et alii: "Horse-Grave of Avarian Age (7th. A.D.) in S.E. Hungary: A Virtual Flight Implementing Parallel Enviroment" in Barcélo, Forte Sanders (eds.) " Virtual Reality in Archaeology". BAR International Series 843, Oxford, 2000.

Lock, G., Stancic, Z. 1995. Archaeology and Geographical Information Systems. Taylor & Francis , London

Lock, G. 2000. Beyond the Map : Archaeology and Spatial Technologies. In : Lock G. (ed) NATO Science Series, *Series A : Life sciences - Vol.321*

Redman, C. R., Watson, P. J. 1971. Systematic, Intensive Surface Collections. *American Antiquity*. 35 (3): 319-325.

Renfrew, C., Zubrow, E. B. 1994. The ancient mind. Elements of cognitive archaeology, Cambridge, 1994.

Tosi, M. 1985. Dalla "molle" Shahr-i Sokhta alla "rigida" Moenjodaro, ovvero il consiglio disatteso. In *Studi di Paletnologia in onore di Salvatore M. Puglisi.* Roma, La Sapienza, pp. 277-289.

Rawlinson, H. R. 1873. Notes on Seistàn. *RSG journal* xliii, London, pp. 272-294.

Salvatori, S. and Vidale, M. 1997. Shahr-i Sokhta 1975-1978: Central Quarters Excavations. In : Preliminary Report. Reports and Memoirs, Series Minor I, Roma: IsIAO

Vidale, M. 1995. Viaggio intorno alla mia ciotola. Evoluzione tecnologica e comunicazione non verbale in una sequenza ceramica dell'Età del Bronzo. *Annali dell'Istituto Universitario Orientale* 55, 3, *Suppl.* 84 Napoli: Istituto Universitario Orientale

Vidale, M., Tosi M. 1996. The Development of Wheel Throwing at Shar-i Sokhta. Slow and Fast Revolutions towards Statehood. *East and West* 46, Rome: IsIAO, pp. 251-269

Figures

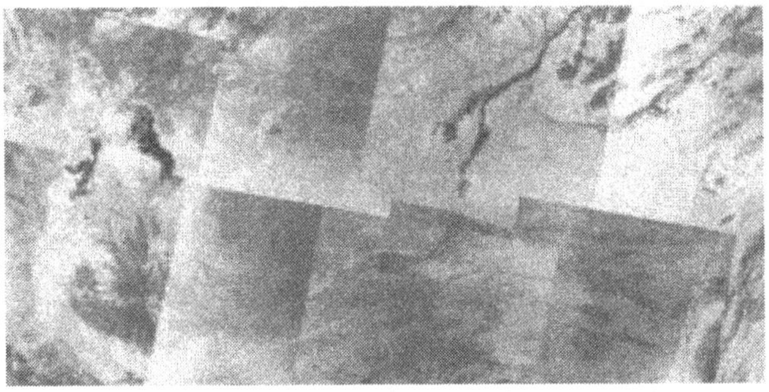

Figure 1. Helmand river view from Landsat mosaic image

Figure 2. Potsherds on the top surface of Sharh-i Sokhta

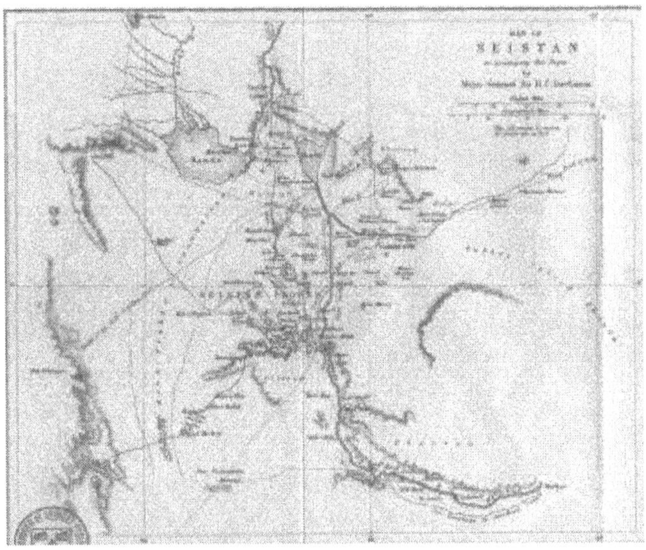

Figure 3. Historic map of Sistan by Sir H. C. Rawlinson (1873)

Figure 4. Areial photo of Eastern Residential Area of Sharh-I Sokhta

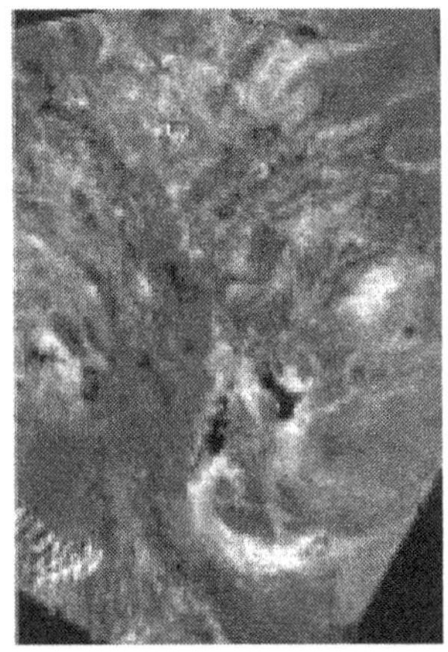

Figure 5. Mosaic of Corona satellite photos

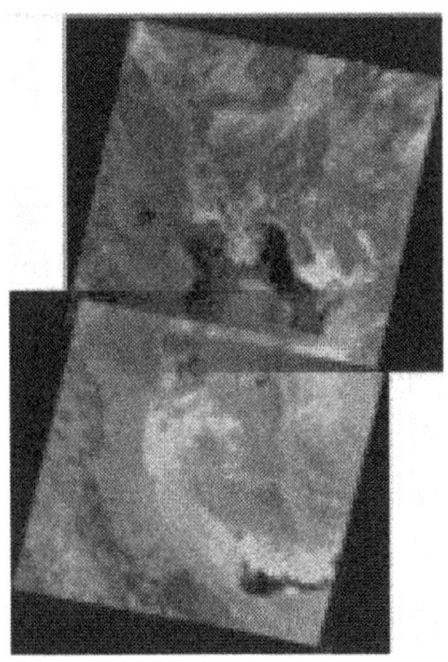

Figure 6. Mosaic of Landsat satellite photos

Figure 7. Reconstruction of Sistan base map with seasonal rivers net

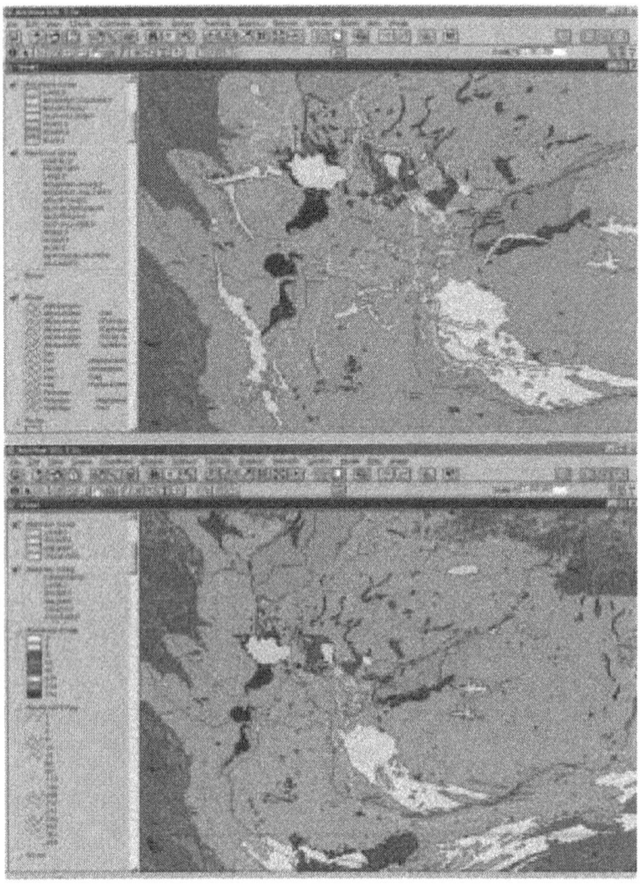

Figure 8. Overlay of vector ancient historical maps with modern reconstructed base map

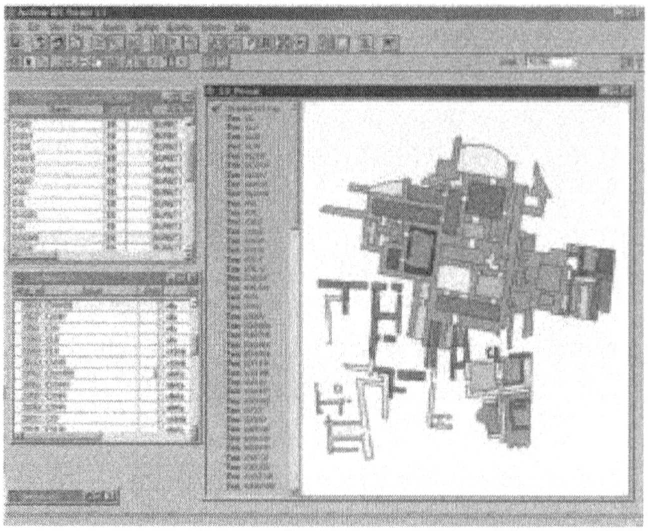

Figure 9. Queries on Burnt Building area of Sharh-I Sokhta

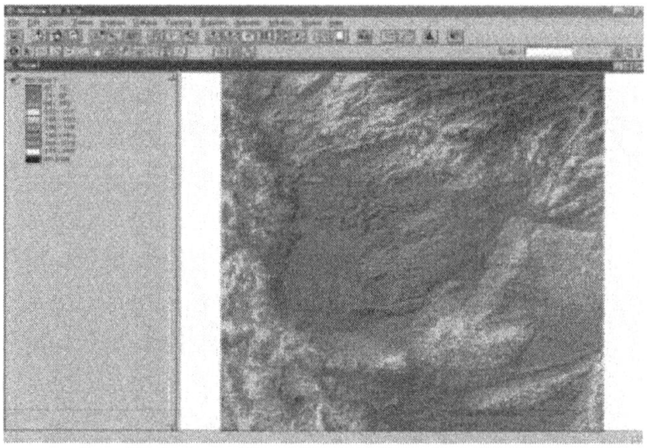

Figure10. GRID processed with 3D Analyst of Sistan basin

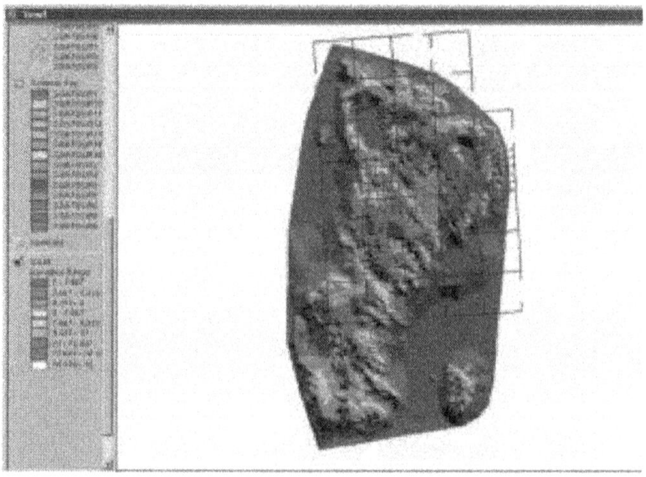

Figure11. 3D reconstruction of the site with investigated areas

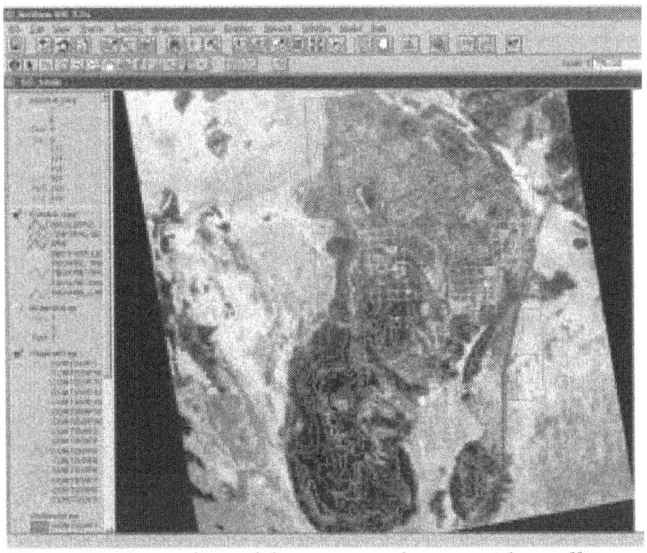

Figure12. Geocoding of the excavated areas with satellite photo

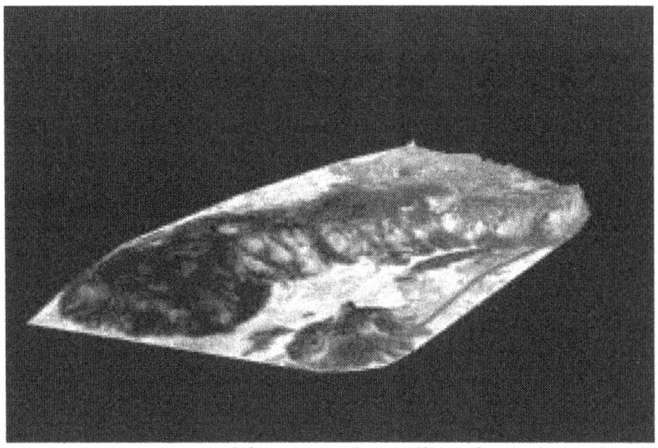

Figure13. 3D view of the site with satellite photo draped on

From Stratigraphic Unit to the mouse: a GIS based system for the excavation of historical complex. The case study of Pompeii

Sabatino Laurenza
Università degli Studi "La Sapienza"
Roma, Italy
E-mail: laurenza.s@tiscalinet.it

Cristiano Putzolu
Università degli Studi "La Sapienza"
Roma, Italy
E-mail: cputzolu@freemail.it

Abstract: The needs of archaeologists are the recording, the management and the elaboration of a huge quantity of data and information linked to different monuments which put us daily in front of difficult questions. The systematic proceedings of research increasis the quantity of information to be recorded and managed. The archaeologists need a quick and correct elaboration of collected data, in order to understand and analyse the investigated contexts.

In this framework, we developed our application, born under a MURST 1998 Project (Ministry of University and of Scientific and Technological Research). The project, started in May-June 1999 at the University "La Sapienza" of Rome, under the Scientific Direction of Prof Andrea Carandini, focused on two case studies: Pompeii and Palatino, having as its main aim the realization of a methodology and of a GIS based system for the recording and analysis of an archaeological excavation.

Briefly, the system consists in the elaboration of a desktop GIS based system helpful for the archaeologists during the excavation phases of recording and management and able to support SU forms with as much information as possible. It is mainly a system for the management and for the analysis of the stratigraphic sequence, both graphic and alpha-numeric. Moreover, a final 3-D visualisation of the excavation is displayed, broken down into its elements, Stratigraphic Units, in order also to link the information and the data coming out during the excavation to 3D SU. In this way, we hope to complete the usual graphical documentation (plans and prospects) with 3D display format, showing surfaces and volumes of excavated SU.

Key words: Stratigraphic Units, Dbase, GIS, 3D modelling, TIN.

Introduction

Archaeology is the study of the human past through material remains and traces. Archaeological remains range from complex urban sites and monumental architecture through individual dwellings to small "portable" objects such as broken tools. Archaeology is, by its nature, fieldwork based and interdisciplinary. Its scope ranges from the sciences of C14 dating and chemical analyses, through environmental and behavioural sciences, to historical and literary research and even art history. Along the way phenomenal, and ever increasing, quantities of data are collected. Far from being concerned with the individual spectacular object, out of its cultural and physical context, archaeologists attempt to piece together evidence and extract information from a myriad small details: the interrelationships between objects, the soil from which they were unearthed, the position of archaeological sites in the landscape and their relationship with the environment. The work is painstaking. To archaeologists with bigger datasets, computers have become essentials for their work.

Archaeology is a *spatial* discipline, and this is why today desktop mapping software is so important. On a Micro-scale, we use the distribution of artefacts in sites to understand the organisation of daily life. On a Macro-scale, the position of sites in the landscape and relative to one-another informs us about social organisation and economic strategies.

Archaeology uses stratigraphic methodologies, which allow us to manage wide and multistratified excavation areas and to reconstruct scientifically histories generated by the analysis of monumental complexes. The needs we have are the recording, the management and the elaboration of a huge quantity of data and information linked to different monuments which push us daily in front of difficult questions. The systematic proceedings of the research increases the quantity of information to be recorded and managed. Archaeologists need a quick and correct elaboration of collected data, in order to understand and analyse the investigated contexts.

Everyone who has directed or managed a large scale excavation project, knows how difficult it is to remain informed about daily progress by comparing the excavation diary and reports, plans, Stratigraphic Units (SU) cards, maps and sections, photos and electronic forms.

For decades directors of excavation projects have tried to put order in the wide data-set of different information, using database archives and CAD system.

In the last ten years, in Italy, we have witnessed the spread of GIS in order to record, manage and display interactively huge archaeological data-sets, with priority on spatial and geographical information.

Between 1985 and 2000, under the scientific direction of Prof A. Carandini, research on the northern edges of Palatino and on Pompeii Insulae VIII,2 and VII,9-11 carried out.

Using funding from the Italian government (Ministero per l'Università e la Ricerca Scientifica e Tecnologica, MURST) from 1999 we started planning for the realisation of a GIS-based system able to manage the entire archaeological data-set, making self operating information recording and elaboration, in order to facilitate the reconstruction of phase to phase image sets for historical interpretation.

Excavation data, archives and the GIS engine

The first target of our project was to have at our disposal, at the end of each fieldwork season, all the archives updated for all the data recorded in the field (graphic and texts) from which to extract automatically thematic diachronic plans (phase plans).

Considering that usual systems of recording archaeological excavations allow the recording of information and limited graphic elaboration, and as we want to elaborate the stratigraphic sequence for separate phases creating thematic images, as the result of the combining of graphic, spatial and written information, we decided to choose to adapt the capacity of the GIS system to our needs.

The system realised is based on a simple structure, mirroring the procedure of scientific elaboration of data collected in the field (Fig. 1). Starting from this point of view an option not to be set aside is that the system realised, based on inter-exchangeable and common file formats, which allows easy exchange with colleagues, giving us a warranty of an opening for future solutions and an insurance for the data. Looking at the future, we decided to use from the beginning the *. mdb* format for data-entry, and to use Visual Basic release 6.0 for all the scripts, programming operations and also for the realisation of the database forms. It could seem too pretentious to say that almost all software houses, after a period of transition, are waiting for a more fixed and stable release of the Win '98 operating system, to make some changes in their product, at least for the script language. It is, as everyone now knows, also the case for ESRI, which in the new release of ArcView 8.1 and of ArcInfo 8.1 changed the script language from Avenue to Visual Basic, and also with OLE/COM languages.

Considering the difficulties represented by the quantity and the diversity of data, an important step was the analysis of the existing data, both on paper (SU cards) and graphic/ topographic data (plans, maps, diagrams). So we decided to integrate the paper cards of the different samples chosen, by realising a specific *thesaurus* for each category of data, in order to realise an archive, with combo-boxes checked vocabularies that will help the user which must input data and avoiding redundancy o data and lexical problems. The same kind of analysis has been developed for the graphic documentation, analysing the existing stratum plans and beginning the vector acquisition of these.

Once shared, managed and overlapped the information on the GIS base, we could observe differences and coincidences of recorded data, and so decide when, where and how to make a more accurate and detailed level of lecture and interpretation. The main function of our GIS based system will consist on the reconstruction of the existing relationship between archaeological layers and their spatial intra site distribution. In fact, the GIS as a representation of information dropped by the physical reality, must stand out for its own capacities to collect all the objects existing in the physical world, making them suitable to the mixes that really occur in the reality.

During the vector analyses the archaeological record is known to the operator since from the first phase of the project, in such a case the computing application asks, by the beginning phase, the accurate definition of an operative route able to normalize on distinctive levels the different existing information and the final goals to reach: the reconstruction of a topographic base, the overlapping of the Stratigraphic Units in a unique file, the recording of data linked then to the spatial information.

Beginning with such methodological remarks, we planned a research route marked by the following working steps :

- Coding of the information and structuring of the field excavation data by realising of one database;
- Elaboration of *thesaurus* lexicon and of scripts of control routines in order to guide the data-entry phase;
- Digitise and transformation in vector of the cadastral and/or photogrametric plans and of all the stratum plans, each with different layers and SU;
- Geo-coding and assignment of a numeric identification key to the graphic objects which define the single SU;
- Integration of graphic and numeric/textual data, with spatial and not spatial attributes, by using a vector GIS as engine;
- Realisation of a graphical user-friendly interface (GUI) able to simplify the operations of recording and visualisation of the data in the database.

So, the work has been necessarily divided in different steps; the first one is the planning of the archives, and it took away a long time for the analysis, also to evaluate the problems connected to the data coding and to define the logical and physical structure of the system (fig. 2). Considering also that information must be organised in a way to satisfy different users at different layers, it seems to us obvious to choose a relational architecture, mirroring the elaboration of more archives connected between them by a primary key, here identified with the SU numbers.

First archive contains all the data of SU, following the form of Italian Ministry of Cultural Heritage; the second one is assigned to the recording of graphic documentation (photos and drawings), by keeping always the relational link with the primary key, the SU number; the third one, contains each object found in the context and it is also linked to the primary key. In order to make easier the data-entry for the students, we realised a database system which manages in a hierarchical mode the different tables, and that using graphic user friendly interfaces (GUI), drives the user during the choice of alternative values remotely controlled between them during the programming phase (figs 3–5).

The next steps of our project were represented by the vector transformation of the 1:20 excavation plans and by the realisation of the final GIS engine.

The vector transformation of all the excavation plans was necessary to allow an exact collocation of the excavated strata and of some finds, and for the entire visualization of the existing graphic information, without compromising the overall readability of the data. So we proceed to transform in vector the different papery *overlays*, or better defined as *stratum* plans, beginning in this way the construction of the cartographic basic system, with a basic level of plans of *stratum* at 1:20 scale.

After this step, the GIS engine realisation starts, by linking together the data-entry and the graphic overlays of each SU, giving us already the possibility to extract phase plans directly by writing a simple SQL query (figs 6-7).

At this point, once having done some joining and linking operations directly in the ESRI ArcView release 3.2a GIS, we try to extract some charts and analyses on the SU data, for example, the diagram of pottery by different periods, or queries about SSUU with some special finds (figs 8-10).

3D SSUU modelling

Established that the GIS engine works well for the management of 2D SSUU, we decided to try to accomplish 3D models of some SSUU in room VII of " Domus della Pescatrice" in order to arrive to calculate the volume of each SU and to directly reconstruct, by the computer, the excavation phases.

Thanks to our methodological approach, that appears to be soon he right one and also thanks to the module 3D Analyst release 1.0 of the ESRI ArcView release 3.2a, it has been possible to follow the excavation phases directly on the field and than to reconstruct the excavation on a Desktop Personal Computer.

The main goal of the last fieldwork campaign in Pompeii was therefore to test a new standard of documentation, consisting in a research tool.

Hardly the traditional documentation of an archaeological excavation recorded all the three dimensions of the space, in the best cases restricted to 2,5 D! The few elevations marked on the excavation plans and sections are in fact never satisfactory, because they can not represent the precise depth of the entire SU surface. Starting from this need, we elaborated a qualified strategy for 3D fieldwork data acquisition.
We used an electronic total station and a Nikon digital photo camera for the data recording, considering that the documentation strategy included the survey of control points, of borderlines and of internal surface points and the photos of each SU.

The post-processing phase is based on two data formats: the .DXF coming out from the total station and the *.tiff* digital images.

At this point, our problem was to optimise the survey proceedings and times; we didn't need a long and complex recording methodology.

If we consider the SU volumetric value as the space between its surface and the surfaces of the SU covered by it, our system to record the complete three dimension of SU could be limited to the survey of the surface of each SU. We decided in this way to document the excavation of the room using an ETS, for the recording of the contours and of the surfaces of each SU. The SU were surveyed with an average of 100 points per square meter. Other interesting problem was the borderline: in fact, the surface of SU, which we will call *upper surf*ace from this moment, touches the underlying SU surfaces, called *SU bottom*, in a portion of space limited to it. At this moment we realised that the orthorectification of the digital images of SU was necessary; once orthorectified, thanks to the four control points surveyed by the ETS, the photo was georeferred and linkable to the topographic data of SU, following the same co-ordinate system. We decided to use an ArcView extension, 3D Analyst, to build the TIN of upper surface and of bottom of each SU; as those TIN overlap on the borderline, we could see the SU as an unique solid. On the *upper surface* TIN we overlaid as texture the georeferred and orthorectified photo. In this way we obtained a jpeg photo, including the values of the elevations and two different TIN with inside the volumetric value.

Finally, the volumetric count was made by calculating the space between a before surface and an after surface, using the *cut fill* function of ARCVIEW, where with before surface we mean the until now called upper surface and with after surface we mean the bottom.

We must underline also that importing in a CAD the orthorectified and georeferred photo and points we could anyway produce a traditional plan, avoiding errors usual during the manual survey.

Because of the absence of findings and remains in the stratigraphic deposit, we couldn't make distribution analyses in this room; in fact surveying the location position of the remains in the SU deposit, we could easily visualise and analyse the primary contexts.

At the end, we want underline that often plans and sections are not enough to represent the precise and entire aspect of the investigated deposit, and they add only the physical bi-dimensional data of the SU to the Harris matrix and to the SU recorded cards.

We think that this recording methodology could provide the archaeologists a set of 3D data, including the advanced possibilities of volumetric counting (figs 11-14).

Some final considerations and future approaches

In this paper we follow almost the same pattern of some our previous papers, about the Pompeii Excavation GIS, where we presented our system as a *work in progress*, with some results already obtained but capable of many implementations.

In the present paper we tried briefly to summarise some of the results:

- The GIS database engine provides a solid archive for the thousands of US cards and the other papery documentation.
- Once in the database, it is easy to query (and to obtain answers if the queries are appropriate) this previously not nimble Leviathan.
- The digitising and geo-coding of the numerous overlays is not only a good procedure of archiving but allows the user to better visualise the relationships between the surveyed structures.
- From its beginning (i.e. without any GIS based operation but simply after the data entry of US cards) the system can fill automatically the Ministerial form for US documentation.
- After a simple query the system can provide the user with phase or period maps.
- 3D reconstructed SU with draped images give the possibility to calculate volumetric values and to visualise in a more realistic way the aspect of the excavated (i.e. destroyed) deposit.
- Even if still *in progress* our GIS already allows simple and crossed SQL queries both from the tables of the database and from the graphic representations of the structures.

But we would like to illustrate briefly the main implementations we are working for:

1. We want to extend the 3D modelling to the walls and mainly to the study of the structures, as for the reconstructed SU, making possible queries directly from the graphic interface of our GIS.
2. We are trying to realise automatically the Harris Matrix and to keep it visible and editable directly from the GIS engine.

The problems to solve for those two points are still many, even though we have already reached some good results from the research point of view.

3D managing of the Wall Stratigraphic Units[1]

Due to the scarcity of resources available this year for the entire project, we can here present only the preliminary stages of our work that, in our intentions, will push the quality of USM data managing at the same level reached for SU.

It is important here to remember that our GIS is entirely developed in ESRI ArcView[2] because we think that the winning choice for this kind of systems could be the complete intercommunication between each module[3]; that being stated 3 D Analyst 1.0 extension is, at the moment, the only available option.

If one thinks of this system as a two-faced Janus, where each object is at the same time a database record and a graphic entity, one can easily understand that the major problems come from the graphic aspect.

The planning and creation of the new database for Wall SU data-entry was made easier by the experiences gained with the previous SU one.

This doesn't mean that we duplicate the database we already have and use it for Wall SU but that now we know exactly how to plot the diagram of the relations between the tables, how to prepare the appropriate user-friendly layouts, how to control input data errors, just because we made some errors and wasted time creating the SU data-entry!

At the beginning, we hoped the same for the graphic aspect : *utinam id sit, quod spero* ! (Terentius).

But we realised immediately that ArcView 3D Analyst 1.0 doesn't allow in any *x-z* or *y-z* plane the same data processing it uses for *x-y* plane.

In fact, TINs creating method builds up a surface joining each point with the ones contiguous: stricter is the net of points, more accurate is the surface obtained.

The other factor involved in TINs creating is the *boundary*, a line that circumscribes the area where the new surface will be created. In the case of SU, the boundary lays in the x-y plane and the software have no problems in TINs creating; but working with Wall SU, the points and the boundary of the surface belong to a *x-z* / *y-z* plane polygon.

3D Analyst manages that data as if they lays in the *x-y* plane; it seems to us that the software processes only surface points with x and y coordinates that fall inside the boundary line projection on the *x-y* plane.

In fact, in our tests, we can see that there are some points involved in the process of TIN creating and others (often close to the former ones) totally ignored.

It seems at the same time that even the neighbourhood between surface points is calculated in the *x-y* plane (i.e. ignoring the elevation of each point) so that often the triangles of the surface are created from points that in reality are not so close! (figs 15-17)

Now it's important to keep in mind what we want from our system: we don't want to bring out one of the thousands of *3D Virtual Trip of Pompeii* and we don't want to digitally rebuild our *domus* case study; we just want to manage the huge mass of data coming from the excavation and gain new knowledge analysing them.

This is not a secondary aspect: if we already have Wall SU data from the database and, at the moment[4], our GIS software architecture doesn't allow a photo realistic 3D reconstruction of the walls, we can consider satisfied with the simple Wall SU extrusion.

This solution was tested and provides a valid graphic interface for Wall SU database in the GIS 3D scene.

With the extrusion of our bi-dimensional plan of Room VII we were able to extend our GIS analyses to the structures in elevation, making queries directly on 3D structures as on the 2D views. Anyway, we must recognise that in our case of study, the stratigraphic sequence of the walls is quite easy, with each Wall SU one beside the others, and that in case of a more complex sequence (with Wall SU one above the other) could be hardly visualised by a simple extrusion process (fig. 18).

The Harris Matrix

As usually said, the work of the archaeologist seems to be quite a strange activity; in fact, during a first step of his work on field he must excavate and so to remove the stratigraphy, that, in a second phase of his work he will be obliged to rebuild during the elaboration and the interpretation of data. Such a model has a its own graphic representation, the so called " Harris Matrix" or also the "stratigraphic diagram". It is necessary to resume all the existing relations, that in case of monumental complexes excavations can be also thousands, between the SU or the Actions in them identified. The layers are the most little unit of an archaeological identification and they have besides their spatial dimensions also a time (chronological) dimension. As a result Harris developed the so called " Harris Matrix" able to describe and to represent graphically the time relationships between different layers. We could imagine that the layers in a Harris Matrix are presented as a rectangle containing the layer name (or number). The relations draft as lines and the position of linked rectangles describe the type of relation. Such a graphic representation corresponds well to the common image of archaeological excavations, where it is usually expected that the most recent layer is above to the oldest one. As there are several exceptions, the relations must always be confirmed by some other observations.

From the mathematic point of view, the relation " After than" could be defined as a set partially ordered. So, we could easily define some additional rules for the relations: Relations are thoughtless, asymmetric, transitive and anti-cyclic.

For the relation "Contemporaneous to" we have different rules: These relations are reflexives, symmetric, transitive, and they are called "relations of equivalence".

We don't want here to resume the concepts of Harris Matrix but just we would like to underline that anyone which could realise a software or an application for the matrix diagram must be very careful to observe these basic rules. In order to increase the readability of the Harris matrix it will be useful to clean it from all the so called "redundant relationships". These are relations already formed by other relations.
The stratigraphic diagrams or matrix allow therefore to represent in a very schematic way on a bi-dimensional plain the 3D reality of the stratification of an archaeological deposit, and the single SU will contribute to define the phases of an historical period.

Keeping in mind all these rules and laws, we are trying from this year to realise an application able to reproduce directly on computer the Harris matrix; as all our programming language of our database is entirely made in Visual Basic release 6.0, we prefer to continue to use it also for the reproduction of SU diagram. We decided to realise an OCX applet, able to be connected directly to different database formats, i.e. dbf, mdb, etc., and that mainly will be able to read directly the data from the interesting fields and to reproduce the diagram.

Until now we have developed the OCX applet that works well for the main stratigraphic rules; we are now working in order to update and to extend the package, making possible automatically the calculation of all the redundancies (fig. 19).
Moreover, as already told before in the text, with the new version of ArcView release 8.1, we would like to make a porting of all the programming scripts made in Visual Basic directly in the GIS system, and to allow the user to have also the possibility to open the matrix during the vie of the excavation and to edit it directly by inside the system. It is hard work but we can say that is more and more possible, especially by the fact that with new Geodatabase structure of Arc View, we could establish some fixed relationships between graphical objects, in our case the SU.

Anyway, we hope to finish this application for the next CAA 2002 in order to present the full operative system applied also on other excavations (i.e. a protohistoric site).

Conclusion

At the beginning of the III millennium we can say that the Information Technologies are inside all the aspects of our work. The need to show to the others and to check ourselves directly in field the data and the information collected during an excavation is nowadays possible, thanks to the IT. Archaeologists are using the modern computer systems in the field since from the beginning of 1980, receiving a grateful help for the management and the recording of the information. Since from the '90s the diffusion on wide scale of the GIS systems give the possibility to manage not only the textual information but also the graphic aspects of it, linked directly each others and mainly to push the intra-site analyses during the same excavation phases.

The use of GIS for excavation is today a reality, also because the specialization of some archaeologists in GIS use is increasing.

Starting from this consideration we can affirm that during the realization of an excavation GIS is better to think to the archaeologist's questions than to the problems of IT experts, making to himself the most usual questions : *What is an excavation GIS for? How will it be made? Which benefits could I get from it? Could it be a tool used for analytical queries and for the making of historical and interpretative models?*

When we started, our main aim was to create a system useful

for the management and for the analysis of the excavation archaeological record; after two years, we could say that not only have we accomplished the system but also that today at the University Laboratory there is a team able to manage computers and to use it for its own needs.

Our work is based on three main and important assumptions :

- The computer is only a tool in the hands of the archaeologist;
- We don't have to approach the computer assisted systems with fear of them;
- We must be able to programming scripts, in order to create specific applications.

Our final system is not the ultimate one, but at the beginning of the new millennium we must consider that the winning philosophy of computer systems is represented by personalised solutions. We can say that our system is functional and useful for the management of different excavations, even if it would be not the absolute one.

The real problem today in Italian Universities is to decide what could be the computer knowledge degree of archaeologists. Archaeologists must know directly the recording and management processes; the computers used today give us the possibility to do it. However, it requires the diffusion in the Italian University Departments of basic computer know-how and the realisation of a theoretical and practical school. Archaeology could not survive without the use of computers, but would risk being kept out of new communication systems, which always demand complete and transparent documentation, high and fast transfer protocols, and different reading keys.

The specialists and also common people must have the possibility to access to publication and visualisation of the researches to which they are interested. If we are not able to adapt ourselves to modern communication systems and to transfer our data to different subjects, we will be out of the game and we will work only for ourselves. Archaeology is perfect for digital technology; it finds in this branch several spectacular applications, which will help us to follow modern dynamics and language based essentially on the images. The problem of image language is to keep the distance between "*false true*" and "*true false*". No reconstruction will give us the exact image of the past. This gap between virtual reality and hypotheses on the past is a constant in a inexact science as archaeology, with which the researchers use to play. The final image must increase the value of the scientific reflection and it must exist as a step of the archaeological research process.

IT therefore represents one of the new tools for archaeology and it must be used, as such, with care and attention, exactly as the trowel.

The ambitious purpose of this project, if such exists, is not to open new frontiers for the recording and the management of archaeological record, but to suggest some guidelines and methodological approaches to the excavation data management. At the end, we hope that the development and future extensions of this system could be a suggest for the institutions to begin to consider seriously those new and powerful tools for the archaeological research and for didactical purposes.

Acknowledgements

Special thanks to Prof P. Carafa for the Scientific coordination of the project. The authors would like to thanks all the students who participated to the data-entry and to the vector transformation of the SU plans.

End notes

[1] In Italy abbreviated as USM (Unità Stratigrafica Muraria, i.e. Wall Stratigraphical Unit).

[2] We developed the database with MS Access and Visual Basic and we digitised the overlays with AutoCAD R. 14 in order to provide the system with tabular and graphical data but, once created the system, we tried to exploit ArcView resources as, for example, the 3D UUSS reconstruction realised in ArcView 3D Analyst.

[3] The latest release of ESRI ArcView 8.1 seems to confirm this philosophy with implemented graphic and database (geodatabase) capabilities and the choice of Visual Basic instead of Avenue as internal programming language.

[4] We are waiting for the new 3 D extension of ESRI ArcGIS 8.1 that has been distributed in Europe only since last month.

References

Allen, K. M. S., Green, S. W., Zubrow, E. B. 1990. *Interpreting Space : GIS and Archaeology*. London: Taylor & Francis.

Alvey, B. A. P. 1993. Interpreting Archaeology with Hindsight: the use of three dimensions in graphic recording and site analysis. In E. C. Harris/m. R. Brown III/G. J. Brown (eds), *Practices of Archaeological Stratigraphy*, 218-228. London: Academic Press.

Andersen-Madsen. 1996. IDEA – the integrated database for excavation analysis. In *Interfacing the Past*, CAA 95, Annalecta Prehistorica Leidensia 28, pp. 3-14.

Barcelò J. A., Forte M., Sanders D. A. (eds). 2000. *Virtual Reality in Archaeology*. BAR International Series 843, Oxford.

Beex, W. 1995. From excavation drawing to archaeological playground: CAD applications for excavations. In : J. Wilcock/ K. Lockyear (eds.) *Computer Applications and Quantitative Methods in Archaeology 1993*, 101-108, BAR International Series 598, Oxford : Tempus Reparatum.

Biswell S., Cropper L., Evans J., Gaffney V., Leach P. 1995. GIS and excavation: a cautionary tale from Shepton Mallet, Somerset, England. In G. Lock, Z. Stancic (edds.) *Archaeology and Geographical Information Systems: a European perspective*. London, pp. 211-229.

Carafa P. 1997. What was Pompeii before 200 bc? Excavation in the House of "Joseph II" and of House of "nozze di Ercole". In

: S. Bon/R. Jones (eds.) *Sequence and Space in Pompeii*, Oxford, pp. 13-31.

Carandini A. 1995. Storie della terra, Torino, 3rd ed.

Hardy, R. L. 1988. Concepts and results of mapping in three dimensional space, *Technical Papers ACSM-ASPRS 1988* Annual Conference,2. Baltimore, MD, 106-115.

Harris, T. 1988. Digital Terrain Modelling and three-dimensional surface graphics for landscape and site analysis in archaeology. In: C. L. N. Ruggles/S. P. Q. Rahtz (eds) *Computer Application and Quantitative Methods in Archaeology 1987*, 161-170, BAR International Series 393, Oxford : Tempus Reparatum.

Harris, T. M. and Lock, G. R. 1995. Multi-dimensional GIS : exploratory approaches to spatial and temporal relationships within archaeological stratigraphy. In : *Interfacing the Past, Annalecta Praheistorica Leidensia 28.* CAA 95,

Hinshelwood-Dalla Bona. 1994. GIS and Intrasite Analysis: An Example from Northwestern Ontario. Archaeological Computing Newsletter 40, pp 12-20.

Jones, T. A., Leonard, J. F. 1990. Why 3-D modelling? *Geobyte* 5 (1), 25-26.

Laurenza, S. and Putzolu, C. 2001. Sistema Informativo Dinamico per la gestione e l'analisi dei dati archeologici : l'esempio di Pompei. In : *Usi e Consumi dell'informazione geografica 2001*, 209-217, I Quaderni di MondoGIS, Atti della Terza Conferenza di MondoGIS, Roma.

Maggiolo Schettini *et al.* 1995. Computation and representation of stratigraphic sequences in a system for archaeological data, Archeologia e Calcolatori 6, pp. 173-187.

Paradis, A. R., Becker, R. C. 1990. Interactive volume modelling-a new product for 3-D modelling. *Geobyte* 5 (1), 42-44.

Pigot, S., Hazelton, B. 1992. The fundamental of a topological model for a 4-D GIS. *Procs. 5th International Symposium on Spatial Data Handing* 2, 580-590.

Reilly, P. 1992. Three dimensional modelling and primary archaeological data. In : P. Reilly/ S. Rahtz (eds), *Archaeology and the Information Age. A Global Perspective,* One World Archaeology 21, 147-173. London : Routledge.

Rick, J. W. 1998. Total Stations in Archaeology. SAA Bullettin, 14(4) : Interface-total station.

Tucker, G. E., Lancaster, S. T., Gasparini N. M., Bras R. L., and Rybarczyk, S. M. 1999. An object-oriented framework for distributed hydrologic and geomorphic modeling using triangulated irregular networks. *Computers and Geosciences* in press.

Tucker, G. E., Lancaster, S. T., Gasparini N. M., Bras R. L., Zeidler, J. , Johnson, W., and Isaacson, J. 2000. Modelling the 3D stratigraphic context of Prehistoric sites : a new approach using process-based computer simulation. Paper presented at the 2000 SAA Conference. Philadelphia (USA).

Wunsch, G. , Arasa , E. and Perez, M. 1995. Dissecting the palimpest : an easy computer graghic approach to the stratigraphic sequence of Tunel VII site (Tierra del fuego) Argentina. SAA

Youngman, C. 1989. Spatial data structures for modelling subsurface features. In : Raper J. F. (ed.) *Three dimensional applications in Geographic Information Systems,* 129-138. London: Taylor & Francis.

Figures

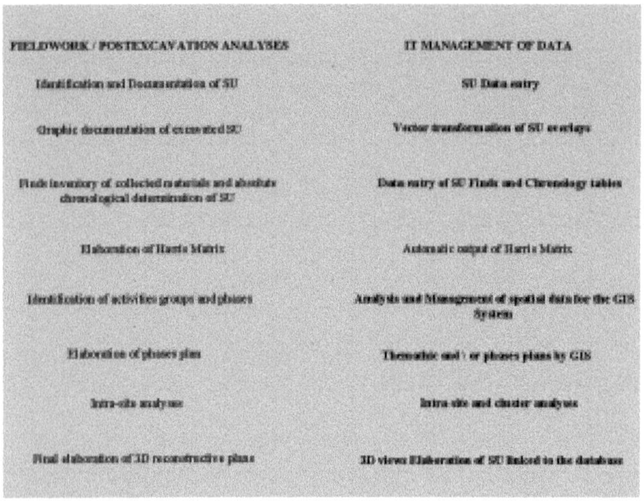

Figure 1. Fieldwork vs IT proceedings

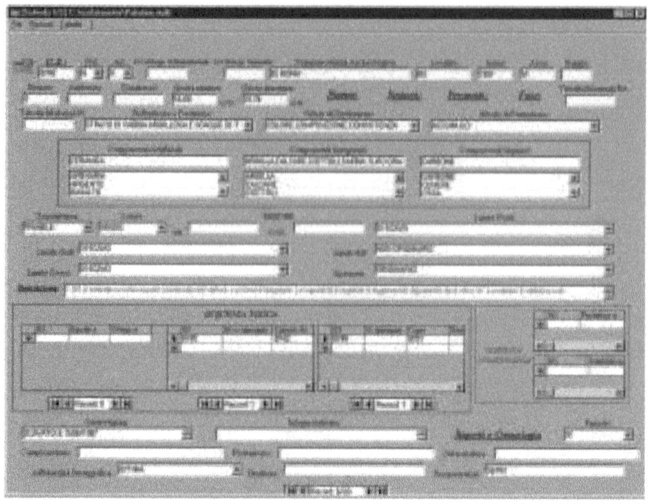

Figure 4. SU data-entry form

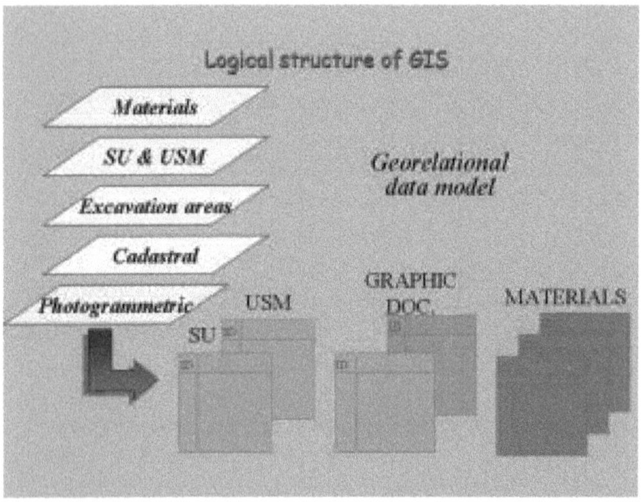

Figure 2. Logical structure of the realised GIS engine

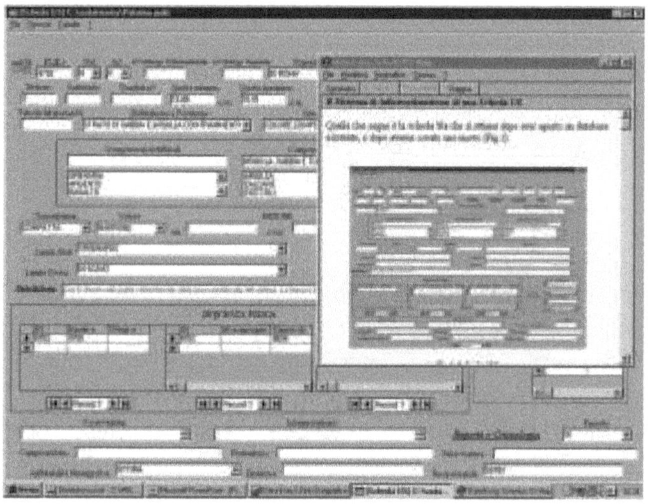

Figure 5. Help online for the dataentry system

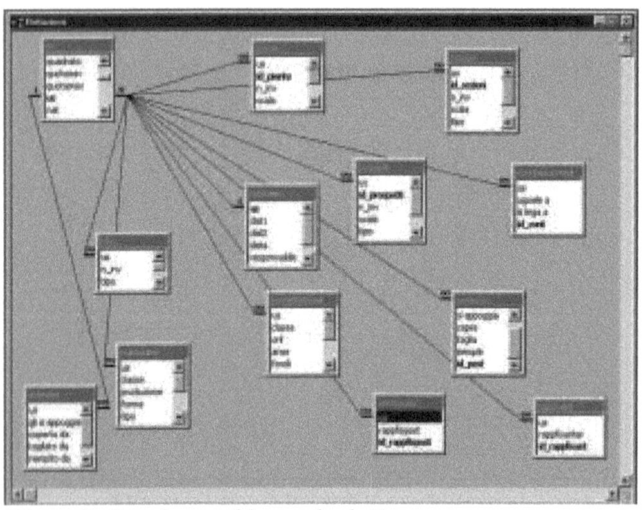

Figure 3. Database tables and relation

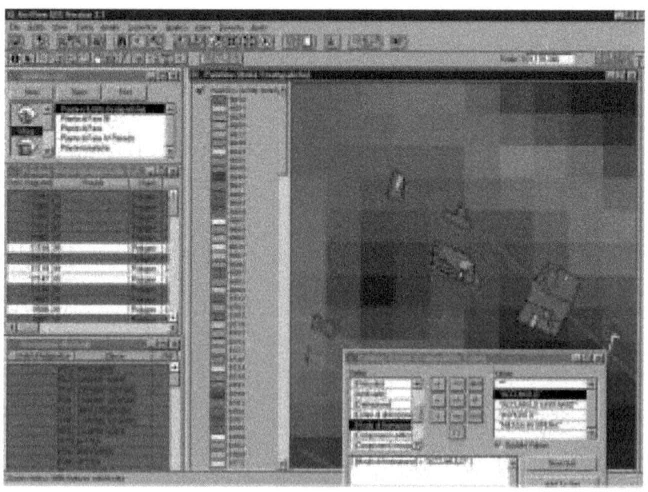

Figure 6. SQL query on formation mode of SU

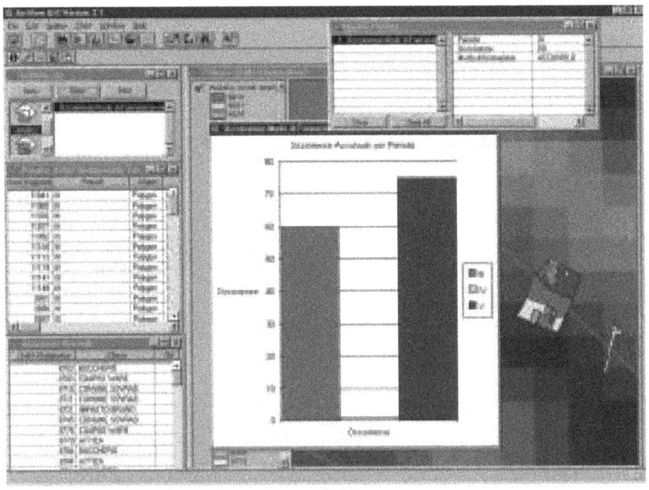

Figure 7. Graphic chart from SQL query

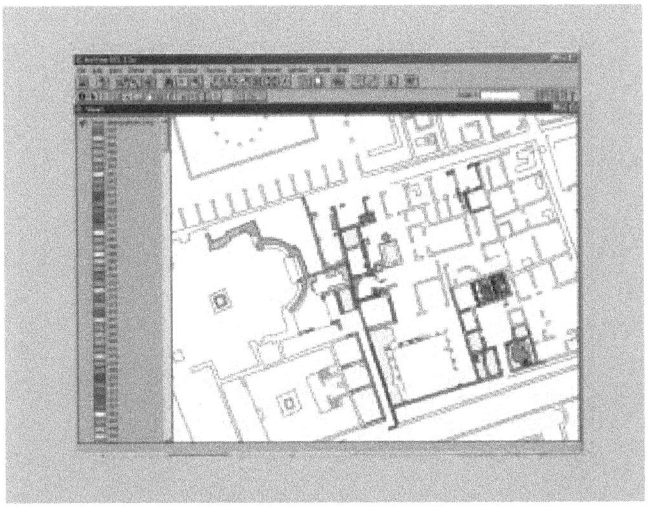

Figure 8. SU view of Pompei insula with excavated SU

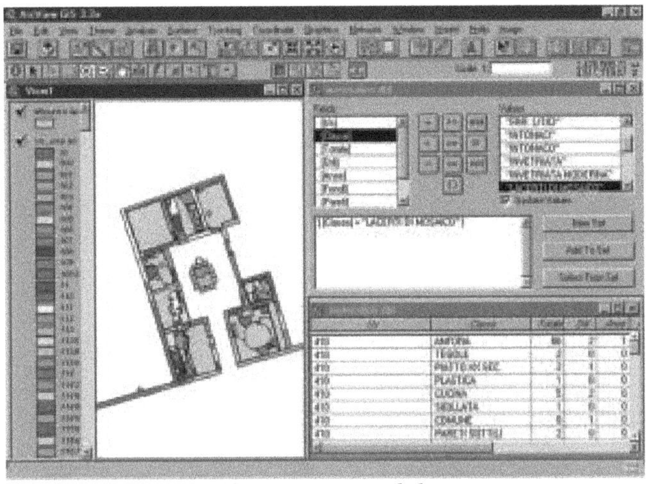

Figure 9. SQL query for mosaic tassel dispersion

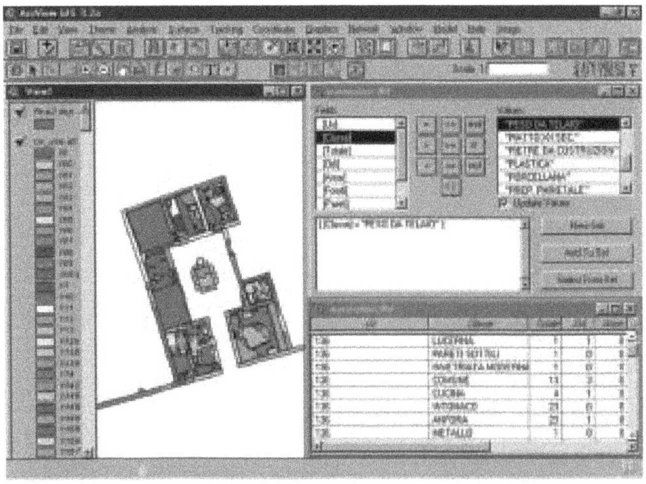

Figure 10. SQL query for the handloom weights distribution

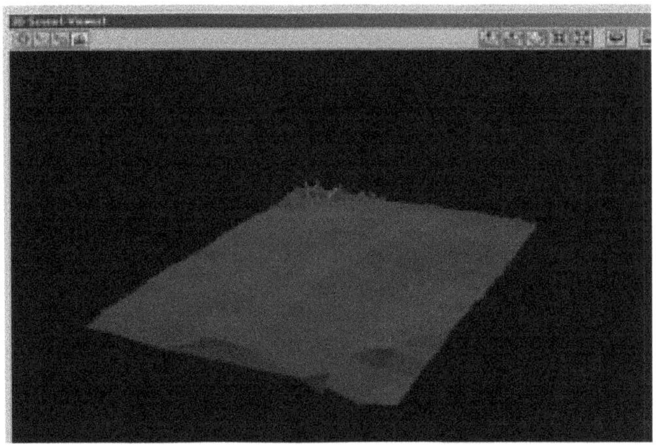

Figure 11. Surveyed points and computed TIN

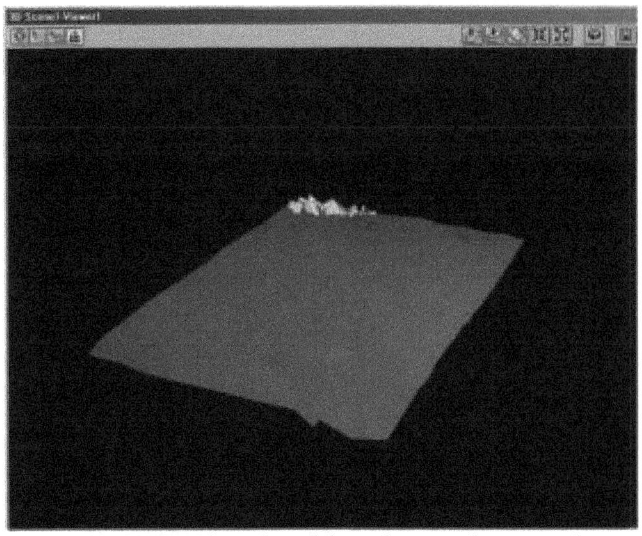

Figure 12. SU TIN based models with merged texture photos

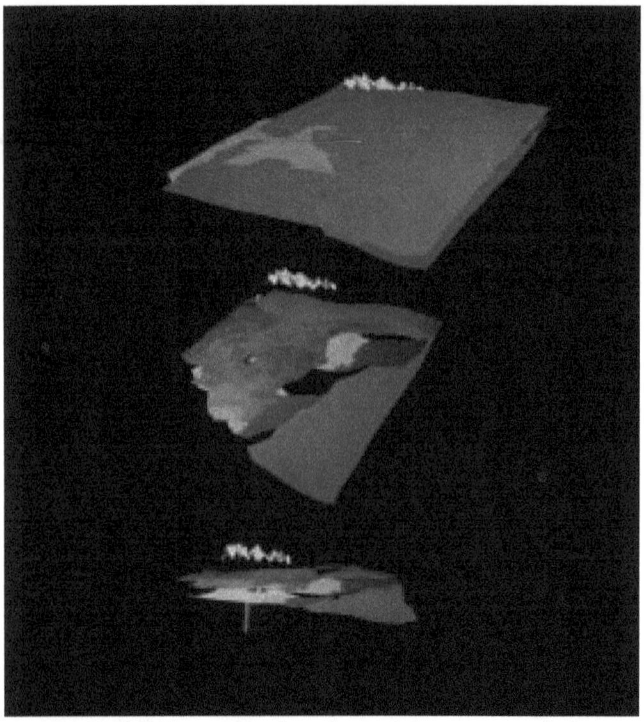

Figure 13. 3D View of archaeological deposit extruded

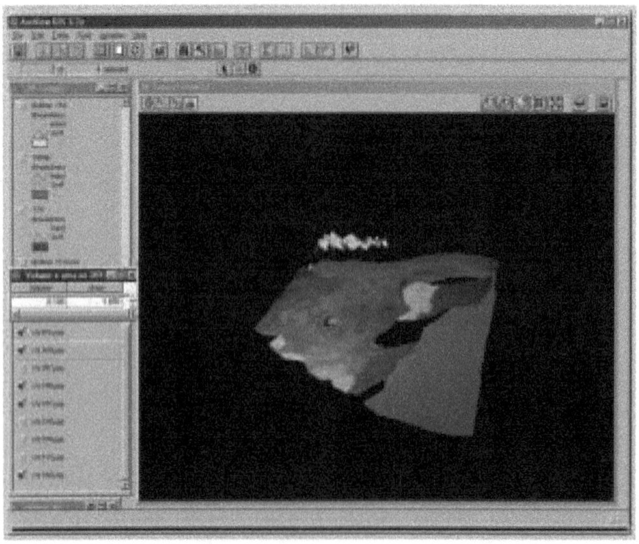

Figure 14. SU volumes count

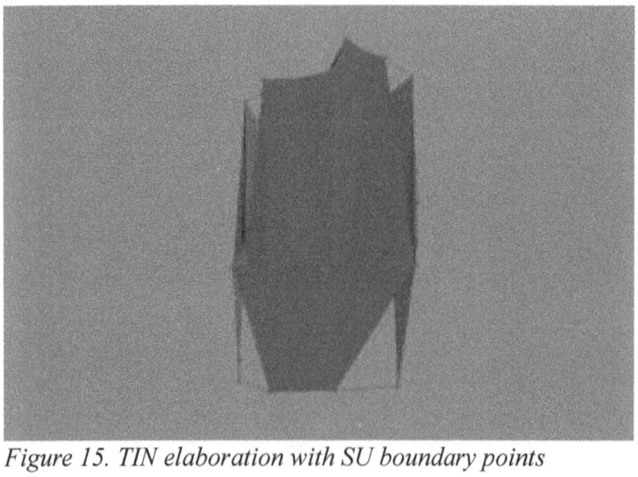

Figure 15. TIN elaboration with SU boundary points

Figure 16. TIN of SU boundary points showing points remaining outside of it

Figure 17. TIN processed as a polygon

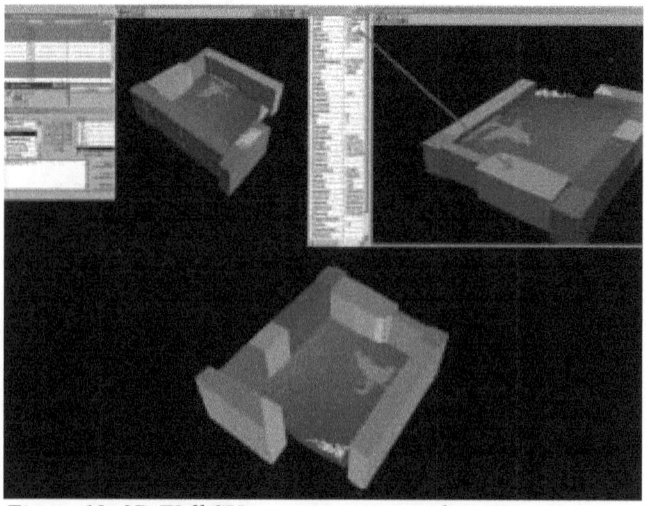

Figure 18. 3D Wall SU reconstruction and queries

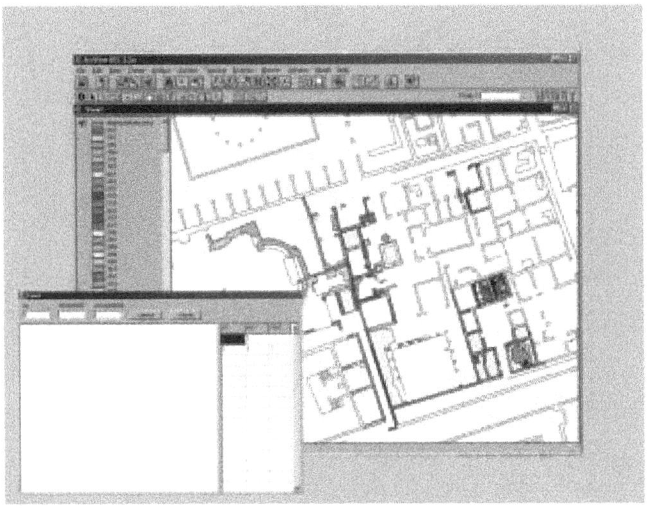

Figure 19. View of the Matrix OCX applet

Statistical analysis of the distribution of modern primates: a comparative approach to the spatial analysis of the Palaeolithic

Katharine MacDonald
Department of Archaeology
University of Southampton
Southampton, SO17 1BJ, England
Phone: +44 23 80592931 - E-mail: km@soton.ac.uk

Abstract: Palaeolithic archaeology is concerned with the analysis of hominid geographical ranges (the extent of a species distribution) as they change over time. Interpretation focuses on the cultural evolution of social and technological adaptations that permitted hominids to extend those ranges, and the environmental factors that ultimately limited them. Spatial analysis of modern primate range maps provides the opportunity to characterize the behavioural traits that favour range expansion and the environmental variables that limit it in the primate order in general. A comparative study of primate distribution in relation to physical and behavioural characteristics and environmental factors was conducted using GIS and statistical techniques. This paper will cover the methodology of comparative spatial analysis, and the potential archaeological application of the results.

Key words: Biogeography, GIS, human evolution, primates, quantitative methods.

Introduction

The overall subject of my research is to investigate large-scale patterns in the distribution of early hominid species in sub-Saharan Africa. The scientific aim is to increase our understanding of the limiting factors that constrained hominid distribution in the past, and of their effects on the probabilities of speciation and extinction. The geographical distribution of hominid species is an important factor in human evolution, and analysis of spatial patterns can help us to understand evolutionary processes affecting hominids. In addition, changes in distribution may be related to changes in behaviour which are of crucial interest to students of human origins. For example, one explanation for the first expansion out of Africa by *Homo erectus* is based on an increase in brain size and associated intensification of social processes and behavioural flexibility. Thus patterns in the geographical distribution of hominid species are of great interest to archaeologists attempting to reconstruct hominid behaviour.

In order to interpret patterns in hominid distribution, we need models of the way in which distribution processes are influenced by environmental variables and species characteristics. The science of biogeography has a theoretical framework and a methodology for interpreting spatial patterns in distribution. Modern primates provide data on distribution patterns in an order of species that shares some characteristics with the hominid lineage. In this paper I will discuss the theory, method and results of a comparative analysis of primate biogeography. The aim of this analysis is to identify characteristics shared by widely distributed primate species. The relationship between variables across a range of taxa can be used to test predictions about adaptation. Thus such a comparative analysis can increase our understanding of the evolutionary and ecological processes that have shaped modern primate distribution. It is hoped that the patterns and processes identified in this analysis can be extrapolated back in time, and evidence for or against their effects sought in the hominid fossil and archaeological record.

Discussion

Human origins and spatial distribution

Temporal variation in the geographical distribution of hominid species has implications for our understanding of evolutionary processes in human evolution. Geographical distribution has an important role in the changes leading to speciation (Foley, 1999). Geographical processes such as population dispersals and contractions, fragmentation and isolation of populations in response to the appearance of a geographical barrier, and shifting habitat boundaries result in altered selective pressures, different demographic parameters, and changes in gene flow or discontinuity (Foley, 1987). Thus the geographical distribution of early hominid species provides important evidence for the evolutionary processes that shaped human evolution.

A number of mechanisms have been proposed as driving forces in human evolution: climate change (Vrba, 1985) and variability (Potts, 1998), and competition between species (Foley, 1999). Alternatively it has been suggested that behavioural innovation may itself act as a catalyst for further evolution (Wilson, 1985). An investigation of the temporal patterns in the distribution of hominid species, and the selective pressures and adaptation associated with changes in distribution, could be used to clarify causal factors in human evolution.

There is evidence for hominid range expansion over time, possibly in the period 4-1.8 my ago among the australopithecines and early *Homo*, and certainly after 1.8 my ago with the appearance of *Homo erectus*, the first hominid species to move out of Africa. Spatial variation in hominid geographical ranges requires explanation in terms of both external limiting factors and species' characteristics.

The human brain is uniquely large among primates, and substantial increases in brain size occurred at various stages in human evolution. Potts (1998) has suggested that increasing brain size could be a variability-selected characteristic that would increase the flexibility of responses to climate change. Variability selection is a form of natural selection that arises when a period of time is characterised by increasing fluctuations in climate and environmental change (Potts, 1998). Large brain size and behavioural flexibility in other species has been linked to colonisation success (Sol and Lefebvre, 2000). Thus we might predict a general relationship between environmental tolerance, range expansion, brain size and behavioural flexibility in hominids. Furthermore, it has been suggested that behavioural innovation may act as a selective pressure (Wilson, 1985). A change in the processes affecting human evolution might also accompany brain expansion.

A number of specific changes in behaviour associated with brain expansion could have been useful in colonization. Possibilities include risk assessment via intensification of social processes (Gamble, 1993), flexible responses to environmental variability (Potts, 1998), and the ability to maintain a broader dietary niche via more innovative foraging behaviour.

According to Gamble (1993), the key to range expansion for primate species was the investment in behaviours that would not only select and alter the environments they and their offspring inhabit, but also gather knowledge about it. At a local level this behaviour would be useful in weathering climatic variability; however such an ability to gather information would also aid successful migration (Gamble, 1993). Migration is based on an assessment of risk, which can be carried out through physiological or conscious processes. The area an individual animal can assess is finite; social life provides the framework for evaluation and dissemination of information from a wider area. By increasing a population's knowledge of the environment, social learning will therefore increase chances of making an accurate assessment (Gamble, 1993). Such social frameworks would have been particularly useful as hominids entered more seasonal environments (Gamble, 1993).

However an increase in brain size has high metabolic costs. Aiello and Wheeler (1995) have suggested that these costs were met by a corresponding reduction in gut size, requiring a change to a higher quality diet. The limited modern distribution of the great apes may be due to the high dietary requirements of large brained species (Richard, 1985). One way of improving dietary quality would have been by increasing the intake of meat. It has been suggested that hominid expansion out of Africa corresponded with the opening of a new, carnivorous dietary niche (Turner, 1992). Aspects of this change in diet and the behaviour involved could have affected *H. erectus'* dispersal ability. Prey species may have provided a crucial resource in more seasonal, colder and northern environments (Turner, 1992). Finally, increased meat eating could have added to pressures favouring range expansion. Carnivorous species tend to have a lower population density than their prey, and Walker (1996) has suggested that *H. erectus* populations may have colonized new areas partly in order to meet this requirement.

Finally, there is considerable variation in mammal life histories, and humans are notable for their long period of maturation and adult lives (O'Connell et al., 1999). Attempts have been made to explain the evolution of life history traits with reference to climatic variability. According to K and r theory (MacArthur and Wilson, 1967), stable non-seasonal environments are predictable and therefore favour individuals that devote a lot of energy to a few offspring at a time. Such K selected individuals tend to invest in behaviour to buffer the effects of environmental pressure. By contrast, unpredictable environments favour individuals that produce large numbers of offspring early in life and have efficient dispersal mechanisms (r selection). Like their closest relatives (the Great Apes), the hominids display an extreme form of the K selected pattern. Data from brain and body size and dental eruption sequences indicates that longevity was increased and maturity delayed in *H. erectus* relative to the Great Ape and earlier hominid pattern (O'Connell et al., 1999). In general, the more long-lived and slow maturing species of primates are not the most widely distributed. This may be partly because populations with a few, slowly developing young are more vulnerable to the increased environmental variation found at higher latitudes.

There is some evidence that even at early periods hominids were not subject to the limitations on geographical distribution operating on primate species with similar life history and relative brain size characteristics (Jablonski et al., 2000). Ecological flexibility may have been of great importance in this pattern. However the existence of a demographic and social structure allowing a flexible reproductive rate would remove some of the constraints set by slow developing young. Flexibility in the rate of reproduction could be achieved by introducing an element of choice as to whether the mother or another member of the group took care of infants and young children. For example, according to the Grandmothering hypothesis, changes in the *H. erectus* life history pattern included increased survival of post-menopausal women, who would have been able to provision their daughters' children (O'Connell et al., 1999). This would have allowed a more flexible response to climatic variability, and could also have been useful in range expansion. The change in life history in *H. erectus* has been linked to an ability to expand their range, especially into arid areas where children could not successfully provision themselves (O'Connell et al., 1999).

Comparative studies

Biogeography is the science of the distribution of living things. Comparative, quantitative biogeographical studies reveal empirical patterns of interspecific variation in the sizes, shapes, boundaries, and internal structures of geographic ranges (Brown et al., 1996). The geographical range of a species is the total extent of its distribution. Analysis of these patterns contributes to an understanding of the historical and ecological processes that influence the distribution of species (Brown et al., 1996). Models of how these processes interact to produce distribution patterns could help us understand the mechanisms behind hominid distribution. The biogeographical literature is an important source of theory and methodology for this analysis.

Biogeographers study the distribution of individual species in detail, or look for general patterns and rules through quantitative study of large numbers of species. Comparisons among species are frequently used to test hypotheses of how organisms are adapted to their environment. The comparative method is the most general technique for asking questions about common patterns of evolutionary change (Harvey and Pagel, 1991). Hypotheses of adaptation lead to predictions of correlated evolution between the presumed adaptive character and the proposed cause (Purvis and Rambaut, 1995). Comparative studies identify evolutionary trends by comparing the values of variables across a range of taxa. The comparative method is employed here in an attempt to explain the evolutionary processes shaping modern primate geographical ranges.

Size is one characteristic of a species geographical range that appears to reflect the influences of environmental limiting factors and dispersal. Within orders or families of species, range size can vary by several orders of magnitude. Recent studies have shown that this variation is associated with variation in a number of other factors. Geographical range size decreases with decreasing latitude and decreasing elevation (Brown et al., 1996). This is known as Rapoport's rule, although the degree to which it constitutes an ecological rule is under debate (Gaston et al., 1998), and orders of species vary in the extent to which they conform. For example, the geographical ranges of African primates show a latitudinal gradient south but not north of the equator (Cowlishaw and Hacker, 1997). Species richness tends to increase with decreasing latitude, and this pattern has been linked to Rapoport's rule (Stevens, 1989). There are significant correlations between range size and both body mass and some measure of average population density (Brown et al., 1996). This pattern is complicated by the relationship between body mass and abundance. A triangular relationship between body mass, population density and range size, based on niche breadth, has been demonstrated for primates (Eeley and Foley, 1999).

In addition, the geographical ranges of primate species tend to increase with climatic variability and niche breadth (Cowlishaw and Hacker, 1997, Eeley and Foley, 1999). It has been suggested that species at higher latitudes must be able to tolerate a greater range of climatic conditions in order to persist at a site (Stevens, 1989). Ecological flexibility allows these organisms to spread across a wider range of habitats, and thus expand their geographical range. Thus Rapoport's rule (and its exceptions) can be explained with reference to the selective effects of climatic variability. In hominids, niche breadth has been linked to increased independence from environmental pressure, and lowered restrictions on ranges (Vrba, 1985).

The biogeography literature highlights a number of interesting patterns in the distribution of species. However it should be noted that there is often substantial residual variation in these analyses. Some of the exceptions to these patterns can be explained by historical factors. The current distribution of species is a product of the interaction of environmental factors and the characteristics of species over time (Brown et al., 1996). This interaction is complicated by the fact that the characteristics of species undergo natural selection from their environment (Brown et al., 1996). This observation highlights the importance of combining ecological and evolutionary perspectives on geographical distribution. Thus the combination of evolutionary and ecological theory from studies of hominid distribution and from contemporary biogeography provides a potentially comprehensive framework for the interpretation of changing hominid distribution.

Primate geographical ranges

Most non-human primates live between the Tropics of Cancer and Capricorn, and very few species occur beyond 25 degrees latitude in the north and 30 degrees latitude in the south (Foley, 1987). However primate species vary considerably in the area over which they are distributed. For example, the gelada baboon (*Theropithecus gelada*) is limited to highland Ethiopia while the vervet monkey (*Cercopithecus aethiops*) can be found throughout most of sub-Saharan Africa. Figure 1 illustrates the variation in primate distribution in Africa. The high degree of variation in distribution within the primate order requires explanation in terms of the environmental factors that limit distribution and the characteristics of species that aid dispersal.

Modern primates have often been used as a source of analogues for the behaviour of early hominid species, whether in the form of specific case studies or cross-species analysis aimed at identifying trends. As our closest relatives, primates share a number of physical and behavioural characteristics with hominid species including humans. It is likely that at least the early species of hominid had more in common with our primate relatives than with modern humans. However the period of interest in my research includes hominid species (especially *H. ergaster / erectus*) characterized by a marked increase in brain size, in tool using behaviours and use of animal resources. These adaptations could be argued to bear a closer resemblance to human behaviour. Finally, the hominids were all unique species, differing from any related species alive today. However there are a number of advantages to large scale cross species analysis as a means of investigating processes that may have been important in human evolution.

First, data is available for contemporary primates in more detail than for fossil species. Range boundary maps have been produced for most primate species, while hominid distribution must be estimated from a limited number of sites and from characteristics such as environmental tolerance. In addition, quantitative data is available on behavioural parameters for which there may not be direct archaeological or fossil evidence. Thus comparative analysis of modern primates allows us to expand our inferences about distribution and behaviour beyond what is visible in the archaeological record (Foley, 1992). Furthermore, the production of contemporary models can be used to document differences as well as similarities between the past and the present. By taking into account species with widely differing behaviour and ecological niches, comparative analysis identifies a range of possible primate strategies. The use of a cross-species, comparative analysis promotes the characterization of hominid species not as a sort of chimpanzee or a sort of human but as a unique form of life.

Hypotheses

There are a number of alternative hypotheses explaining greater environmental tolerance and range expansion in hominid species, as discussed above.

1. Behavioural flexibility or social learning favours range expansion, and high tolerance of climatic variability.
2. Range expansion is causally associated with increased meat eating.
3. Increased flexibility in hominid life history traits allowed range expansion beyond what would be predicted for a highly K selected primate.

The first hypothesis predicts covariation between behavioural flexibility or social learning and geographical range size and climatic tolerance in primates and hominids. The second hypothesis may be better tested using carnivore data, although some species of primate do occasionally eat meat. A further analysis will be required to address this question. Finally, the third hypothesis predicts a differing trend in hominids from the rest of the primates. According to this hypothesis, range size would be expected to increase with r selected life history traits in primates.

Method and analysis

A comparative study of primate distribution in relation to physical and behavioural characteristics was conducted using Geographic Information System (GIS) and statistical techniques. Geographical ranges of primates were digitized from maps, rectified and projected into equal area format. The range composition in terms of land cover and climatic variability was analysed using the statistical capabilities of GIS packages. Potential predictive variables (life history, body mass, diet, brain size) were synthesized from the literature and entered into a comprehensive database. A database of innovation frequencies was obtained from Reader (2000). Regression techniques were used to assess the relative predictive value of a comprehensive selection of variables with respect to range size and distribution. The effects of phylogenetic relationships were taken into account.

A number of questions of definition and methodology arose as part of the analysis, including the measurement of climatic variability. Climatic variation is induced by a number of climatic cycles, proceeding on different scales, from seasonal and annual up to thirty and hundred year cycles. A time series of global precipitation or temperature maps (New et al., 1999, New et al., 2000) was used as the basis for continuous measures of climatic variability within the range. This database provided monthly maps covering a thirty-year period, and the climate variability maps were based on this period in order to allow for variation caused by some of the more frequent climatic cycles.

The aim in this analysis was to measure variability at a scale that would affect individual animals. The measurements used in this analysis included spatial, seasonal and inter-annual variation. A map of mean yearly rainfall and temperature was calculated based on monthly means for 1961-90. Secondly, the temperature range for this period was used as an indicator of seasonality. Minimum and maximum yearly temperatures were calculated from monthly values and the difference taken for the final temperature range map. Finally, a map of inter-annual variability was based on variability in mean annual rainfall, as measured by the coefficient of variation. Further processing was carried out to calculate the climatic variation within each species range. Values of spatial variability tolerated by species were calculated as the coefficient of variation of mean rainfall values within the species range. The seasonality and inter-annual variability tolerated by species was calculated as mean values within the range.

The real units of geographical ranges are the complex spatial and temporal patterns in which individual organisms are dispersed over the earth (Brown et al., 1996). Any maps of geographical ranges necessarily simplify such complexity. A map of the extent of occurrence of a species represents the limits of that species distribution. This ignores smaller scale variation such as absences within the range and variation in abundance in different areas, but was deemed appropriate to the continental scale of the analysis. Primate geographical range maps were taken from Wolfheim (1983). The data from the three main geographical areas in which primates are found (Africa, South America, and Asia) were treated as separate datasets on the grounds that unique features of the continental landmass were likely to influence distribution. The Madagascar primates were excluded from the African dataset for the same reason.

Behavioural flexibility can be defined as the capacity of a species to engage in novel behaviour. In order to carry out my analysis, it was necessary to find a measure of behavioural flexibility that could be compared across species. Species vary widely in their performance in experimental tests, and it has been suggested that behavioural innovation, or opportunistic departures from the species norm, may provide an alternative measurement (Lefebvre et al., 1998). Reader (2000) has conducted a comparative study of innovation, social learning and tool use in nonhuman primates by collecting reports of such behaviour from the primate and social learning literature. This method provides a test fair to all species and data on large numbers of species by measuring the tendency to discover or learn novel solutions to environmental or social problems relevant to the animal (Reader, 2000). The database used in this analysis was provided by Reader (2000) and all frequencies were corrected for research effort using his method.

Brain size may be used as an indirect measurement of intelligence. Brain expansion is costly, so an increase in brain size (such as occurred in human evolution) must give a strong selective advantage. This advantage is generally assumed to involve increased intelligence. Corroborative evidence for this approach comes from the correlation between a number of brain measures and innovation frequency, social learning and tool use (Reader, 2000). There are a number of competing measurements of brain size available, and the choice of measurement depends on ideas about brain evolution and brain structure (Byrne, 1995). It can be argued that absolute brain size equals processing power (Dawkins, 1976). However brain size scales with body size: allometric scaling can be used to account for this relationship. Finally, the relative size of different parts of the brain also varies, and certain parts are associated with different functions. Thus a third type of measurement is based on the relative or absolute size of the parts of the brain associated with the behaviours of interest. In these analyses I used measurements based on each of these three approaches: absolute brain size, brain size relative to body size and the neocortex ratio.

Linear regression techniques were used to identify the best predictors of range size. Statistical analyses were also carried out using independent contrasts. As discussed above, comparison of continuous variables across a range of species can be used to identify covariation in characters and may suggest evolutionary trends. Possible covariation can be tested statistically using regression techniques. A strong relationship between range size and another variable may indicate that primate species that, for instance, have very long noses will tend to occupy large areas. This could be because long noses allow primates to smell each other over large areas and thus expand their ranges. It could also be because primate species with long noses are all related to a common ancestor that had a long nose and occupied a large area. Thus cross species analysis can identify spurious correlations.

To avoid this problem I used a package called Comparative Analysis for Independent Contrasts, provided by Purvis and Rambaut (1995). This calculates independent contrasts based on the comparison of pairs of species or higher nodes that share a common ancestor. The principle of this calculation is that the difference between these pairs will reflect only the changes that have occurred since they split from their common ancestor. The primate phylogeny used was a composite tree derived from 112 previously published phylogenies (Purvis, 1995).

Results

The principal findings of this study are that, as shown in Table 1, once phylogenetic relationships have been taken into account, innovation frequencies and relative and absolute measurements of brain size do not correlate with geographical range size in modern primate species. A number of significant results were obtained prior to using independent contrasts, and this suggests that the relationship between range size and behavioural flexibility is determined by phylogeny. These negative results are not entirely unexpected. The species most notable for high innovation and tool use frequency, the chimpanzee, also has a small and relatively habitat specific range. These results suggest that while behavioural flexibility is not a hindrance, other characteristics are necessary for successful range expansion. Other factors such as niche breadth are better predictors of range size (Eeley and Foley, 1999). This has interesting implications for the interpretation of hominid range expansion. While the primate analysis does not disprove the hypothesis that hominid range expansion was related to an increase in brain size and behavioural flexibility, it indicates that this is unlikely to have been part of a primate trend. If such factors were important in hominid range expansion, it is likely to have been on a different scale, or have involved behaviours absent in other primates that were particularly useful in encountering new environments.

However, some significant results have been obtained by regressing innovation and tool use frequencies on measures of spatial and temporal climatic variability for African primates (see Table 2). This is consistent with the hypothesis that more intelligent or behaviourally plastic species can tolerate greater climatic variability. Given the negative results obtained for range size this suggests that, contrary to expectations, species with a high tolerance of climatic variability do not necessarily expand their ranges. This result is particularly interesting given evidence that the habitat tolerance of hominid species increased over time. Land use studies indicate a longer distance of stone tool transport, and a wider diversity of depositional settings, as well as repeated return to particular environmental features (Potts et al., 1999). This suggests that hominids became increasingly independent of environmental constraints. Further analysis will be required to ascertain whether this increased independence or tolerance was part of a primate trend related to behavioural flexibility.

In addition, there seems to be a general pattern by which the hypothesis that behavioural flexibility allows primate species to cope with greater spatial and temporal variability within their ranges is upheld for African primates but not for South American primates. This raises interesting questions regarding differences in the evolution of primate geographical ranges in the two continents. Such variation in modern distribution patterns could be the result of differences in the historical interaction of environmental factors and species' characteristics.

The results for both range size and climatic variability are interesting in terms of predicted effects of higher levels of social learning. There is a conclusive lack of correlation with both range size and climatic variability, using species as data points and independent contrasts. This provides strong evidence against the hypothesis that higher frequencies of social learning will increase environmental tolerance and dispersal ability in primate species and hence lead to larger ranges. The relationship between social learning and environmental variability is theoretically complex. Social learning has advantages over learning independently at intermediate levels of environmental variation or where environmental variation is regular and predictable (Sibly, 1999). This could be why general intelligence as measured by innovation frequency is a better predictor of climatic variability than social learning frequencies.

Finally, there is some indication that r selected life history strategies are positively associated with larger range size. There is a significant negative relationship between geographical range size and age at first breeding (for African primates, using independent contrasts, $r = -0.522$, $n = 19$, $p<0.05$). In addition, there is a positive relationship between range size and a measurement of reproductive rate, based on variation in other life history parameters (for African primates, using independent contrasts, $r = 0.818$, $n = 14$, $p<0.01$). This conclusion is currently tentative and will need to be confirmed using a larger data set and more life history variables. R selected species are characterised by a suite of characteristics, including rapid development, frequent reproduction, shorter lifespan, and also rapid dispersal and niche breadth. Such life history traits and dispersal ability are connected to the invasion success of a species, and therefore to its range size. In addition, niche breadth covaries with geographical range size in African primates (Eeley and Foley, 1999). However such species tend not to be those with the largest brain size and greatest evidence for innovative behaviour, tool use and social learning. K selected species are likely to disperse more slowly, but may have a selective advantage once they do reach a new area (Richard, 1985). There may be two routes by which primates expand their ranges: through high levels of behavioural flexibility, allowing species to tolerate new climatic conditions; or through the fast life history and rapid dispersal characteristic of r selected species.

Conclusions

This analysis gave a number of unexpected and thought provoking results. It has been suggested that hominid range expansion could be explained by increasing brain size and associated increases in behavioural flexibility and social learning frequencies. However the primate data contradicts the predictions of this hypothesis. If these are important factors in hominid range expansion, it is unlikely to have been as part of a primate trend. A number of predictors of primate geographical range size and tolerance of climatic variability have been identified in this analysis, and still others can be found in the biogeography literature. This suggests that there may be a number of interacting characteristics determining primate range size. One reading of the results would be to suggest two alternative primate strategies involving an increase in range size. One would include a suite of characteristics including r selected life history traits: the other a high level of behavioural flexibility. A preliminary explanation for the large increase in range size relative to the primates in human evolution could be that some hominid adaptations combined these characteristics. For instance, hominid adaptations could have included a means to increase the flexibility of their behaviour and of their life history traits. Further analysis of the evidence for hominid distribution and behaviour will be necessary to evaluate this suggestion.

This approach to hominid distribution involves a combination of theory and method that has proved very productive. Biogeography provides a theoretical framework for interpreting spatial patterns in distribution, while GIS and statistical techniques provide an efficient method for carrying out large-scale spatial analyses. This larger scale approach may aid the identification of patterns and processes invisible at the site or regional level. The use of comparative material from modern primates gave access to a detailed database of distribution patterns, and allowed me to extend inferences about behaviour beyond the information available from the hominid fossil and archaeological record. In addition, the cross species approach has proved a good solution to the problem of using comparative data to interpret hominid behaviour and ecology, without assuming similarities to particular species. A number of future research directions have been made possible by this project. This combination of approaches can be applied in a further analysis of carnivore distribution in order to test hypotheses linking range expansion to meat eating. In addition, this analysis has identified patterns in primate distribution, and I have proposed some mechanisms that could explain these patterns. It is hoped that predictions based on these processes can be tested against the archaeological and fossil hominid data.

Acknowledgements

Many thanks to James Steele for ideas and criticism provided throughout this project, to Simon Reader who provided the database of innovation, tool use and social learning frequencies, and to both Simon Reader and Kevin Laland for their valuable comments on my research plans. I am supported during my PhD research by a NERC Research Studentship.

References

Aiello, L. C. and Wheeler, P. 1995. The expensive-tissue hypothesis: the brain and the digestive system in human and primate evolution. *Current Anthropology*, 36, 199-221.

Brown, J. H., Stevens, G. C. and Kaufman, D. M. 1996. The geographic range: size, shape, boundaries, and internal structure. *Annual Review of Ecology and Systematics*, 27, 597-623.

Byrne, R. 1995. *The thinking ape*, Oxford University Press, Oxford.

Cowlishaw, G. and Hacker, J. E. 1997. Distribution, diversity, and latitude in African primates. *American Naturalist*, 150, 505-512.

Dawkins 1976. *The selfish gene*, Oxford University Press, Oxford.

Eeley, H. A. C. and Foley, R. A. 1999. Species richness, species range size and ecological specialisation among African primates: geographical patterns and conservation implications. *Biodiversity and Conservation*, 8, 1033-1056.

Foley 1987. *Another unique species: patterns in human evolutionary ecology*, Longman Scientific & Technical, Harlow.

Foley, R. A. 1992. Studying human evolution by analogy, In *The Cambridge Encyclopedia of Human Evolution*, (Eds, Jones, S., Martin, R. and Pilbeam, D.) Cambridge University Press, Cambridge, pp. 335-340.

Foley, R. A. 1999. Evolutionary geography of Pliocene hominids, In *African biogeography, climate change and human evolution*, (Eds, Bromage, T. G. and Schrenk, F.) Oxford University Press, Oxford, pp. 328-348.

Gamble. 1993. *Timewalkers: the prehistory of global colonization*, Alan Sutton, Stroud.

Gaston, K. J., Blackburn, T. M. and Spicer, J. I. 1998. Rapoport's rule: time for an epitaph? *Trends in Ecology & Evolution*, 13, 70-74.

Harvey and Pagel. 1991. *The comparative method in evolutionary biology*, Oxford University Press, Oxford.

Jablonski, N. G., Whitfort, M. J., Roberts-Smith, N. and Xu, Q. Q.

2000. The influence of life history and diet on the distribution of catarrhine primates during the Pleistocene in eastern Asia. *Journal of Human Evolution*, 39, 131-157.

Lefebvre, L., Gaxiola, A., Dawson, S., Timmermans, S., Rosza, L. and Kabai, P. 1998. Feeding innovations and forebrain size in Australasian birds. *Behaviour*, 135, 1077-1097.

MacArthur, R. H. and Wilson, E. O. 1967. *Island biogeography*, Princeton University Press, Princeton, New Jersey.

New, M. G., Hulme, M. and Jones, P. D. 1999. Representing 20th century space-time climate variability. I: Development of 1902-1990 mean monthly terrestrial climatology. *J. Climate*, 12, 829-856.

New, M. G., Hulme, M. and Jones, P. D. 2000. Representing 20th century space-time climate variability. II: Development of 1901-1996 monthly terrestrial climate fields. *J. Climate*, 13, 2217-2238.

O'Connell, J. F., Hawkes, K. and Jones, N. G. B. 1999. Grandmothering and the evolution of *Homo erectus*. *Journal of Human Evolution*, 36, 461-485.

Potts, R. 1998. Variability selection in hominid evolution. *Evolutionary Anthropology*, 7, 81-96.

Potts, R., Behrensmeyer, A. K. and Ditchfield, P. 1999. Paleolandscape variation and Early Pleistocene hominid activities: Members 1 and 7, Olorgesailie Formation, Kenya. *Journal of Human Evolution*, 37, 747-788.

Purvis, A. 1995. A composite estimate of primate phylogeny. *Philosophical Transactions of the Royal Society of London Series B-Biological Sciences*, 348, 405-421.

Purvis, A. and Rambaut, A. 1995. Comparative-analysis by independent contrasts (CAIC) - an Apple-Macintosh application for analysing comparative data. *Computer Applications in the Biosciences*, 11, 247-251.

Reader, S. M. 2000. *Social learning and innovation: individual differences, diffusion dynamics and evolutionary issues*. PhD thesis. Department of Zoology, University of Cambridge

Richard. 1985. *Primates in nature*, W.H. Freeman, New York.
Sibly, R. M. (1999) Evolutionary biology of skill and information transfer, In *Mammalian social learning: comparative and ecological perspectives*, (Eds, Box, H. O. and Gibson, K. R.) Cambridge University Press, Cambridge.

Sol, D. and Lefebvre, L. 2000. Behavioural flexibility predicts invasion success in birds introduced to New Zealand. *Oikos*, 90, 599-605.

Stevens, G. C. 1989. The latitudinal gradient in geographical range - how so many species coexist in the tropics. *American Naturalist*, 133, 240-256.

Turner, A. 1992. Large carnivores and earliest European hominids: changing determinants of resource availability during the Lower and Middle Pleistocene. *Journal of Human Evolution*, 22, 109-126.

Vrba, E. S. 1985. Ecological and adaptive changes associated with early hominid evolution, In *Ancestors- the hard evidence*, (Ed, Delson, E.) Alan R Liss, New York, pp. 63-71.

Walker, A. and Shipman, P. 1996. *The wisdom of bones*, Weidenfeld and Nicolson, London.

Wilson, A. C. 1985. The molecular basis of evolution. *Scientific American*, 253, 148-157.

Wolfheim, J. 1983. *Primates of the world: distribution, abundance and conservation*, University of London Press, London.

Tables

	Innovation frequency	Tool use frequency	Social learning	Absolute brain size	Relative brain size	Neocortex ratio
Africa	r = 0.141 n = 24 p>0.1	r = -0.161 n = 24 p>0.1	r = 0.233 n = 24 p>0.1	r = 0.059 n = 33 p>0.1	r = 0.063 n = 33 p>0.1	r = 0.030 n = 12 p>0.1
South America	r = 0.397 n = 23 0.1>p>0.05	r = 0.142 n = 23 p>0.1	r = -0.087 n = 23 p>0.1	r = -0.060 n = 23 p>0.1	r = -0.255 n = 23 p>0.1	r = 0.395 n = 10 p>0.1
Asia	r = -0.390 n = 22 0.1>p>0.05	r = -0.390 n = 22 0.1>p>0.05	r = -0.017 n = 22 p>0.1	r = 0.000 n = 24 p>0.1	r = -0.387 n = 24 0.1>p>0.05	Insufficient data.

Table 1. Results of regression of innovation, tool use and social learning frequencies, and measures of brain size, on total geographical range size, using phylogenetically independent contrasts.

Dependent variable	Location	Innovation frequency	Tool use frequency	Social learning frequency
Spatial variability	Africa (n = 24)	r = 0.505 p<0.01	r = 0.633 p<0.01	r = 0.069 p>0.1
	South America (n = 23)	r = -0.192 p>0.1	r = -0.124 p>0.1	r = -0.210 p>0.1
Seasonal variability	Africa	r = 0.396 p = 0.05	r = 0.305 p>0.1	r = 0.329 p>0.1
	South America	r = -0.152 p>0.1	r = 0.022 p>0.1	r = -0.147 p>0.1
Interannual variability	Africa	r = 0.376 0.1>p>0.05	r = 0.447 p<0.05	r = -0.026 p>0.1
	South America	r = -0.312 p>0.1	r = -0.123 p>0.1	r = -0.147 p>0.1

Table 2. Results of regression of innovation, tool use and social learning frequencies on measures of climatic variability, using phylogenetically independent contrasts.

Figures

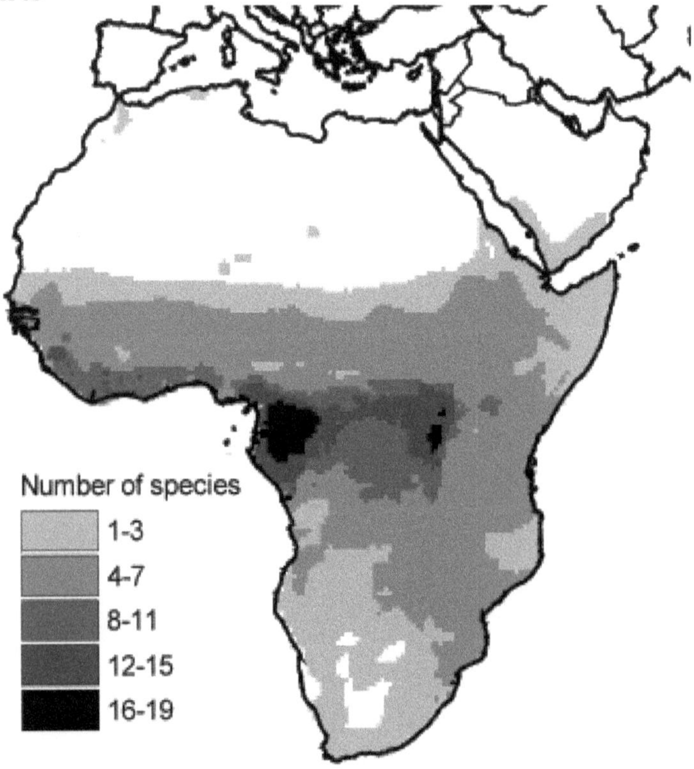

Figure 1. Primate species richness in Africa, from Wolfheim (1983).

The Creation and Potential Applications of a 3-Dimensional GIS for the Early Hominin Site of Swartkrans, South Africa

Joseph D. Nigro
Center for Advanced Spatial Technologies /
Department of Anthropology
University of Arkansas
Ozark Hall, Room 12
Fayetteville, AR 72701, USA
Phone: +1-501-582-0790 - Fax: +1-501-575-5218 - E-mail: jnigro@cast.uark.edu

W. Fredrick Limp
Center for Advanced Spatial Technologies /
Department of Anthropology
University of Arkansas
Ozark Hall, Room 12
Fayetteville, AR 72701, USA
Phone: +1-501-575-7909 - Fax: +1-501-575-5218 - E-mail: fred@cast.uark.edu

Kenneth K. Kvamme
Department of Anthropology
University of Arkansas
Old Main 330
Fayetteville, AR 72701, USA
Phone: +1-501-575-4130 - Fax: +1-501-575-6595 - E-mail: kvamme@mail.uark.edu

Darryl J. de Ruiter
Palaeoanthropology Unit for Research and Exploration
Department of Palaeontology
Bernard Price Institute for Palaeontology
Private Bag 3
University of the Witwatersrand
2050 Johannesburg, South Africa
Phone: +27-11-717-6668 - Fax: +27-11-339-7202 - E-mail: deruitd@science.pg.wits.ac.za

Lee R. Berger
Palaeoanthropology Unit for Research and Exploration
Department of Palaeontology
Bernard Price Institute for Palaeontology
Private Bag 3
University of the Witwatersrand
2050 Johannesburg, South Africa
Phone: +27-11-717-6668 - Fax: +27-11-339-7202 - E-mail: 106lrb@cosmos.wits.ac.za

Abstract: The accumulation of fossil remains, bone tools, and stone tools at the site of Swartkrans has been attributed to a variety of agents including hominin activity, carnivore activity, alluvial deposition, and gravitation. In fact, the accumulations discovered at Swartkrans and other Plio-Pleistocene cave sites in South Africa have probably resulted from a combination of these factors. In order to explore further the taphonomic nature of this deposit a Geographic Information System (GIS) was constructed incorporating all of the existing information produced from the Swartkrans excavations and a recent survey of the site. The final product represents a digital archive of Swartkrans data that allows the user to simultaneously visualize and analyze fossil, artifact, and geological data together within their spatial contexts allowing a more comprehensive investigation of the site post-excavation. The mapping and 3D reconstruction of sites such as Swartkrans present challenges when using traditional GIS approaches. These issues are addressed, as well as the potential applications of the system.

Key words: Geographic Information Systems, GIS, palaeoanthroplogy, hominin, hominid, taphonomy, fossil assemblages, intrasite analysis, Plio-Pleistocene caves, Swartkrans

Introduction

Travertine caves in the Sterkfontein Valley of South Africa's Gauteng province have yielded a rich array of Plio-Pleistocene faunal material. The site of Swartkrans, for example, has produced the single largest sample of fossils attributable to the genus *Paranthropus* in addition to numerous fossils belonging to the genus *Homo* as well as a variety of bone and stone tools. To date, approximately 350,000 pieces of fossil bone representative of both human and non-human species (Brain 1993:3), along with almost 900 stone tools and 70 bone tools, have been discovered at this site (Brain/Shipman 1993:214).

The decisive work of C.K. Brain, conducted between 1965 and 1986, has provided many important clues concerning the accumulation of the fauna at Swartkrans. Based on carnivore markings on the bones, as well as geological formation processes, Brain suggests that the fossil accumulations of Swartkrans might be best attributed to carnivore activity (Brain 1968, 1969, 1970, 1974, 1981, 1993). Recent studies have yielded an array of possible hominid-derived tools and have, by implication, raised questions concerning the formation processes responsible for their deposition (Clark 1993; Brain/Shipman 1993; Backwell/d'Errico 2001) drawn from the spatial relations of the remains at the site. In order to test the hypothesis that a variety of agents were responsible for the accumulations at Swartkrans and to extract and isolate these complex agents of accumulation, a Geographic Information System (GIS) was constructed.

In 1999 the Palaeoanthropology Unit for Research and Exploration at the University of the Witwatersrand in conjunction with the University of Arkansas' department of Anthropology and it's Center for Advanced Spatial Technologies assigned the first author to the task of developing a GIS for Swartkrans. The project arose as an exercise in applying recent technological developments to a site that was excavated at a time when GIS was either in its infancy in regards to archaeological applications or not an option at all. Swartkrans was used as the pilot site to demonstrate how this technology is capable of deriving new or enhanced information from existing data and facilitating current methods of analyses. It was chosen because of its relatively small size due and the existence of meticulously recorded fossil, artifact, and geological data collected during C.K. Brain's excavations.

The specific goals of the project were: 1) to create a system that can assist in further understanding the stratigraphy present at Swartkrans and other cave sites; 2) to create a system that can be used to further explore the relationships between the fossil and artifact distributions of these sites; 3) to realize the implications and constraints of the technology involved in creating such a system; 4) to construct a 3D model of Swartkrans in order to improve the grid-based provenience information of the finds recorded previously; and 5) to reconstruct C.K. Brain's excavation. The Swartkrans GIS was demonstrated to be very beneficial in addressing all of the goals listed above and has resulted in the creation of a digital archive encompassing all available site information. This accomplishment will allow researchers to virtually re-excavate the site an infinite number of times, a task that would otherwise prove impossible.

Provenience data of the fossils and artifacts

Palaeoanthropological exploration commenced at Swartkrans in 1948, but it wasn't until 1979 when systematic excavations were first conducted at the site. Many important specimens came to light during the first three decades of work there, including the well-known cranium of an adult *Paranthropus robustus* (SK 48). Unfortunately, all these early discoveries are lacking the vital provenience information needed to accurately reposition them in three-dimensional space (Brain 1981: 221). This is an important fact because it hinders us from producing a total reconstruction of the geological and fossil content of Swartkrans.

In 1979, this problem was remedied with the establishment of a datum, in the form of a permanent metal grid, over the Outer Cave area (fig.1). Although the main bars of the grid were constructed of rigid metal, the smaller units were made of wire and spaced every meter. Over areas representing Members (strata) 1 and 2 each grid square was divided into 50 cm^2 quadrants while subdivided even further into 25 cm^2 sub-quadrants over Member 3 due to the density of burnt bone found in twenty consecutive spits throughout a 6-m vertical profile (Brain 1993: 4). The excavation was mainly carried out in 10-cm spits in all areas of the Outer Cave, but extending as deep as 1m in certain places.

Mapping the existing site

The first stage of the project involved mapping Swartkrans as it exists today. C.K. Brain was asked to visit the site to assess if there had been any major shifts in the positions of the remaining geological structures since his first removal of excavated material, material that may have supported these structures previously. Fortunately, no major changes had taken place and so the site as it exists today reflects the site upon initial excavation in 1979 minus the excavated matrix (Brain 1999). This was useful information since the purpose of the survey was to incorporate all of the existing geology of the site into the GIS in an attempt to increase the recorded provenience of the known fossils and artifacts. For example, if a fossil was recorded to have come from the NW quadrant of the grid-based position E4S7, the accuracy of the provenience would be 50 cm^2 x 10 cm. By mapping existing structures, the accuracy of the provenience information could be increased if unexcavated material still remains in E4S7 thus decreasing the volume where the fossil was originally found to less than 50 cm^2 x 10 cm (fig.2). These structures were sometimes recorded in the excavation notes, but mapping the site filled the data gaps where representative diagrams are non-existent. In addition to increasing the provenience data of the finds, the mapping of Swartkrans provides both a three-dimensional matrix into which the excavated data can be repositioned and a complimentary source of geological information derived from the attributed survey points.

The survey was carried out using a Wild T1010™ theodolite with a Leica Distomat ™ Laser electronic distance meter attachment (fig.3). A total of 14,885 survey points were collected representing the Inner, Outer, and Lower cave systems.

The Outer Cave, now a sinkhole due to collapsed dolomite roof blocks, was the only area recorded with the intention of creating a 3D reconstruction, given available provenience information. The base perimeters of the Inner and Lower caves, however, were recorded to determine their spatial orientations within the cave network.

The survey was based on C.K. Brain's grid system to facilitate the incorporation of the data derived from his excavation notes within the 3D model. As a result, the origin (represented by the center of the grid structure) and the north line (the metal bar leading from the origin to the "Hanging Remnant", a landmark feature) remained the same as those used during the excavations (fig.4). One obstacle encountered at this point was the problem of dealing with the wire section of the grid where the wires were either sagging or gone completely. This made it impossible to map these grid sections and, as a result, each digitized excavation diagram was unable to be reprojected individually to its original location. Fortunately, the metal bars constructing the main axes of the grid were still in place and so all of the excavated squares were reprojected as one unit based on these axes.

Fifty-nine control points were set up around and within the site from where each feature was recorded (fig.5). These points were established by resecting off of two known points beginning with the origin point and the point marking the contact area where the north line touches the Hanging Remnant. Side shots were collected every 10 cm vertically while the horizontal mapping resolution ranged from between 25-50 cm depending on the degree of topographical variation of the feature. For example, a wall of breccia with small inclusions is relatively uniform so fewer points were taken to represent it, while a dolomite roof block is characterized by undulations and crevices and was respectively assigned more points due to its complexity (fig.6). These mapping intervals were chosen to be consistent with Brain's excavation levels. A line level was used as a rough guide for mapping. For areas that were inaccessible, the vertical and horizontal measurements of the theodolite were monitored and points remained unrecorded until these readings were consistent with the appropriate intervals. This degree of accuracy was sufficient enough to allow a detailed image of the terrain to be produced from the interpolation of the survey points. Attributes for the points were recorded, including feature descriptions, geological material, sediment color, degree of calcification, cave area, Member (strata), and the cardinal direction that the feature is facing. This last attribute was essential for the interpolation of the data using Voxel Analyst, our principal software, for reasons discussed below.

The survey data were stored on a memory card that was transferred daily from the total station to a Wild GIF 10 data collector. The data were then downloaded from the data collector to a Dell Inspiron 3200 laptop using a data transfer program called Convtran. The point IDs, X- coordinates, Y-coordinates, and Z-coordinates were saved in text file format. Finally, these space-delimited files were imported into MS Excel where field headings were assigned and attributes were entered manually. The application of photogrammetry was not considered for this project due to the lack of proper equipment and training and to the site being characterized by small recesses that are easier to record with a laser than with a camera. In addition, a method that can be used in dark, subterranean networks needed to be devised in order to extend this work to other cave sites.

Constructing the Inner and Lower Cave base perimeters and the excavation grid

The survey points representing the base perimeters of the Inner and Lower caves, as well as Brain's excavation grid, were imported into AutoCAD MAP R3. Each consecutive point was connected by a line using the 3D-polyline command, thus producing a three-dimensional vector representation of each feature. Upon completion these layers were saved as data interchange files (.dxf) and imported into ArcView 3.2 using the CAD Reader extension. It was essential that these layers were saved in .dxf format, as opposed to saving them as AutoCAD drawing files (.dwg) or exporting them as ArcView shapefiles (.shp) or Microstation design files (.dgn). The reason for this is that .dxf is the only format that maintains the 3D spatial relationship created by connecting two points at varying elevations when exporting them from AutoCAD MAP R3 into ArcView 3D Analyst.

Interpolation of 3D data and the limitations of convential GIS data models

A range of GIS and related software packages, including Surfer 7, ArcView 3.2, Idrisi 3.2, Slicer-Dicer and GRASS 4.0 were tested to see how their interpolation algorithms would handle point data that had multiple z values for the same x,y coordinate pair, representing the natural geological overhangs at Swartkrans (fig.7). After many attempts, it was realized that none of the interpolation methods were suitable for processing this type of data. Due to the data structures of these software packages, the resulting interpolated image contained severe errors making the 3D reconstruction of Swartkrans highly inaccurate in certain areas. Some of these software packages, such as Surfer 7, prompted the user to choose a calculation parameter, such as minimum, maximum, or average, for points possessing multiple z values. Other software packages, such as ArcView 3.2, failed to grant the user the option to select the points containing duplicate x, y coordinate values that were to be included in the interpolation. In both cases the resultant output was characterized by data spikes in places where overhangs are present at the site, such as below the Hanging Remnant (fig.8).

This is not surprising because the data model of a conventional GIS is structured in two dimensions where features are assigned x and y spatial coordinates with an attribute defining some aspect of the feature, such as soil type. Two-dimensional mapping results in planar surfaces where each data layer represents some characteristic found on a single surface. When the attribute signifies an elevation or a measured value of some sort then the data model is considered to be surficial, or representing 2.5D data. Surficial data can be visualized in three dimensions although it lacks 3D topology. A classic example of this is a Digital Elevation Model where height is considered an attribute (Raper 1989; Turner 1989). Unfortunately, up to this point archaeological applications have been limited by the constraints of these two types of data models. Yet, excavated material exists on multiple artificial surfaces overlaying one another in 3D space, making these data models insufficient. As a result, spatial modeling software was used instead of traditional GIS algorithms to produce an interpolated image containing all overhanging features. For an extensive overview and discussion concerning the progressive stages and limitations of representing multidimensional data in archaeology, refer to Harris and Lock (1996).

Constructing the 3D Model of the Outer Cave

Voxel Analyst, developed by the Intergraph Corporation, provided the solution to the problem mentioned above since it recognizes three independent axes (x,y,z). The software played a crucial role in this project and was responsible for making the 3D reconstruction of Swartkrans possible. Voxel Analyst is a data visualization and analysis tool that is designed to characterize subsurface data and to enhance the understanding of multi-dimensional data relationships within a 3D volumetric data set. A voxel, or volumetric element, is a three-dimensional version of a pixel, depicted as a cube. The voxel data model was introduced in commercial software that was developed for the purpose of oil and gas exploration It is currently used for a wide variety of applications, from reservoir engineering to mining exploration (Harris/Lock 1996; Turner 1989). Utilizing this type of software resulted in an innovative way to accurately interpolate a terrain model of an archaeological site in true 3D space.

In order to model the existing terrain at Swartkrans using Voxel Analyst, the site had to be viewed as a solid block. Thus, the solid earth as well as the void areas representing air that once contained the excavated material had to be assigned unique numeric values respectively. This allowed the software to acknowledge both features as two distinct phenomena interlocked in three-dimensional space (fig.9a). Since a reflector card was not used to collect survey points in mid-air with the laser during the initial survey, the air component was assigned using another method. Essentially all of the survey points representing the cave surface were duplicated in MS Excel and repositioned 1 cm (the distance between the center of two adjacent voxels) away from the surface in the appropriate cardinal direction in order to create the outer boundary of the air mass (fig.9b). This made it possible for the software to identify the interface between the cave structure and the air and to create a thin surface located where each voxel adjoined each other (fig.9c). This procedure eliminates the problem of spatial inexactitude that is usually inherent in voxel-based systems due to the voxel size and the lack of precise boundaries between spatial entities (Jones 1989).

Man-made features at the site, such as steps, were initially mapped and then eliminated from the survey data prior to interpolation of the natural surface. The text file representing all the survey points was brought into a text editor. A carriage return, necessary for the software to recognize the file, was placed after the last line of the data. The file was saved in .smp format and imported into Voxel Analyst. The 14,489 (14,885 minus the man-made features) points were interpolated using the Shephard's Method algorithm, an inverse-weighted approximation. This was the most appropriate interpolator since it is ideal for rough and unevenly distributed data and it satisfies the Maximum Principle in which interpolated output values will not exceed the range of the original input values (Intergraph 1994). The output represented an accurate reconstruction of Swartkrans, including all existing overhangs. The final model was exported from Voxel Analyst as a .dgn file, opened using the CAD Reader extension in ArcView 3.2, and finally converted to a .shp file.

A mistake realized in hindsight, after the survey was completed, is that no survey points were recorded on the ground level of the site since there isn't much variation in elevation. This resulted in the appearance of a "bulbous floor" in the final reconstruction of the site due to the scarcity of points representing these areas (fig.10). The ground level, fortunately, represents areas that may be excavated and recorded at a later date and so the data can always be incorporated into the GIS in the future.

Converting the excavation notes into digital format

Brain's excavations lasted from 1979-1986 and produced several volumes of excavation notes outlining in meticulous detail the geological contents, and, in some instances, the fossil content of each excavated grid square. Converting information contained in these notebooks into digital form was essential to rebuild the site as it existed prior to Brain's excavations. By doing this, the user of this system will be able to view specimens within their associated strata once again and to conduct a virtual re-excavation of the site.

The excavation diagrams were digitized using a Microscribe Digitizer, though any digitizing would have sufficed (fig.11a). Initial calibration was performed using Inscribe software to define the origin, coordinate axes and scale. Most of the diagrams were drawn on a standard form where each side of the grid-square represented exactly 80 mm (fig.11b). All of the diagrams were rescaled to convert their original dimensions to equal 1m^2. Sketched drawings were also included to provide information where standard diagrams were unavailable (fig.11c). Although these drawings are less accurate representations, it is better to have some information present in the GIS for these areas than not to have any at all. Each diagram was digitized, reprojected to its appropriate grid square in true geographic space, and stored on a unique elevation layer in AutoCAD MAP R3. The elevation layer of the diagram was assigned based on the bottom-most level of the 10-cm spit from which it was drawn. For instance, if a grid square was excavated at 80 - 90 cm below the datum, then the data in that area were assigned an elevation of -0.900 m due to the fact that each diagram was drawn after the material was removed from the grid square. Each elevation layer with associated excavation data was exported as an ArcINFO coverage and brought into PC ArcINFO to create topology. The attributes were then extracted from the excavation notes and entered in MS Excel to compile a table[1] of geological features. The digitized diagrams were finally imported into ArcView 3.2 where the attributes were linked to the appropriate feature using unique ID values selected as the join field or primary key (fig.12). These polygon themes were then clipped based on the appropriate elevation layers extracted from the 3D model. This procedure combined the new data with the old and ensured that the existing geological information of the site was taken into account before the finds were inputted into the system. At times, entities derived from the excavation note descriptions, as opposed to the diagrams or sketches, were deleted as a result of the clipping process. These features were reconstructed later using ArcView editing functions.

Creation of the fossil and artifact database

Throughout the 1999 project year, one of the authors (DJD) compiled a database of the fossils and artifacts found at Swartkrans. This task was carried out at the Transvaal Museum in Pretoria, South Africa, where all relevant information previously gathered was entered on computer using MS Access. The finds entered were only those from Members 1, 2, and 3 that were assigned accession numbers beginning with *SKX*. The resulting database contains almost 22,000 records with attributes recorded for numerous fields[2]. The fields created for the survey and excavation data were kept consistent with those listed above, when applicable.

Point representation of the fossils and artifacts in three-dimensions

Although the locations of many of the finds were recorded only to the grid system and spit (level), it seemed more beneficial to represent them as individual points within the GIS. A point theme allows the user to visualize the fossils and artifacts of interest in and about the general area of their discovery and to identify clusters of certain fossils more readily without having to deal with the confusion of having each grid square representing multiple points at a time.

Random numbers were used to place the points within the assigned grid squares by combining functions from ArcView 3.2 and SPLUS 2000, a statistical software package. To create a series of random points in specific areas of the excavation grid, a random point generator (randpts.avx)[3], downloaded from the ESRI web page[4], was employed. The extension allows the user to specify the number of random points to be created, as well as the region of interest within which the point generation will be limited. This second parameter acts as a mask. Since this extension is limited to producing points in x, y coordinate space only, this procedure was carried out on one excavation square at a time, combining all of the associated points regardless of elevation. For example, if 500 artifacts were assigned to have come from all spits in E3S2, then 500 random points were generated within this area. Since the extension is incapable of generating points with z values, SPLUS was used for this task. The number of points found in each spit, excavated in the same grid square location, was counted and the random point generator in SPLUS was used to derive random z values to represent them. These values were then added to the table containing the x, y coordinates and selected to transform the point theme into a 3D shapefile. The fossil/artifact database was then linked to these points using unique ID numbers as the join field.

The amalgamation of the data within ArcView 3D Analyst

Once these three sources of Swartkrans data were either created or converted into a usable digital format they were displayed together using the 3D Analyst extension for ArcView. All of the data was spatially sound and overlayed each other perfectly (fig.13). Now that these themes were brought into 3D Analyst, the three-dimensional capabilities of the software needed to be assessed.

Although 3D Analyst would be expected to have three-dimensional analytical capabilities based on the name, the data model of the software remains surficial. The Swartkrans data can be displayed very nicely within the module, but proper 3D analyses and functionality, such as nearest neighbor and buffering operations, cannot be conducted due to the lack of 3D topology. For example, when the user wanted to know how many fossils lie within 10 cm of a dolomite block, 3D Analyst would only output the points that are within a 10 cm area of dolomite on the same horizontal plane not taking into account the points that lie above, below, or at angles from the block. This is a serious problem since it limits the integration of these datasets to visual analysis, besides the conventional 2D and 2.5D operations.

3D buffering operation and potential applications

In an attempt to bypass this obstacle, an Avenue script was created that is successful in creating three-dimensional buffer zones. The script, however, is limited to the buffering of point features. An Avenue script (sphere3D_creator)[5] that can create a spherical graphic around a user-specified point, at a defined radius, was downloaded from the ESRI web page[6]. The script was combined with another developed independently that uses a 3D Euclidean distance calculation in order to measure the distance from a user-defined point to all other points in the dataset. The calculation is as follows: SQRT((((xcoordinate of specified point - x coordinates of all points)^2) + ((ycoordinate of specified point - y coordinates of all points)^2) + ((zcoordinate of specified point - z coordinates of all points)^2))) <= specified range represented by the radius of the spherical buffer.

The spherical graphic was included solely as a visual aid to highlight the area of interest in three-dimensional space. The degree of transparency of the sphere can be regulated allowing the user to view the points floating within it (fig.14).

Besides using 3D buffering to examine taphonomic relationships of specified skeletal remains and artifact distributions based on a number of attributes, it can be used in identifying potential specimens for bone refitting. The user can define a distance in all directions from one fossil to extract specific fossil types that are located within this distance. For example, all bovid class III metapodials can be identified within 1-m of a given specimen to see if there are any other specimens possibly representing the same individual in that area. This feature is also useful when attempting to determine whether or not a clustering of fossils represents an individual animal within the fossil deposit.

Another potential query, presently untested, would involve the conversions of all polygonal data representing the geology of the site into point data, since the 3D buffer script only works on point themes. This may enable the user to actually perform distance operations to seek associations between the fossils/artifacts and the geology in the way described above, but instead taking into account points with geological attributes. This may allow the user to overcome the obstacle stated above and discover what fossils do lie within 10 cm of a dolomite block.

Artifact and fossil distributions

It is widely accepted by researchers that the patterned spatial distributions of specimens found in these cave sites may reflect the behavior of various species while an unpatterned or scattered distribution may imply deposition by natural phenomena or episodic deposition by a variety of animals. Dolomitic caves, depending on size and location, function as lairs, shelters, or feeding places. In order to understand fossil assemblages in these caves, fossil distribution patterns must be recognized as unique indicators of animal behavior (de Ruiter/Berger 2000: 665-666) For example, hyaena dens usually consist of highly fragmented bone and scattered carcass elements (Brain 1981), while felid activity within caves often results in the presence of numerous articulated skeletons. Due to the complex nature of these cave sites, other factors such as local environmental conditions and the associated animal community need to be taken into account in order to achieve a more complete understanding of these distributions (de Ruiter/Berger 2000: 682, 684). A GIS is capable of handling complex data of these types The GIS is ideal for the examination of artifact and fossil distributions for many reasons. The ability to view and rotate a combination of elements in 3D space is the most obvious (fig.15). The query builder in ArcView can be used to extract a sample of elements restricted to a large number of associated attributes on the fly, such as the distal limb sections of all species possessing chew marks (fig.16). Informative statistics related to the fossil population and important for understanding faunal accumulations in cave assemblages (de Ruiter/Berger 2000: 683), such as Minimum Numbers of Individuals (MNI) or Minimum Numbers of Elements (MNE), can also be calculated quickly from the data. Using this system is more efficient than manually plotting or attempting to understand spatial relations of entities characterized by numerous variables based solely on the textual information recorded previously. These finds can once again be studied within their original geological context, possibly providing further insight into environmental conditions and processes responsible for the accumulations. Another benefit of archiving these data using ArcView is that statistical analyses can be performed conveniently using the SPLUS extension.

Mapping fossil and artifact densities

Density maps are useful for looking at patterns in the data by identifying high and low concentrations of specified elements (Mitchell 1999). Although this function is limited to 2D space in ArcView, it can still be useful in detecting clusters that are indicative of a vertical filtration of fossils and artifacts into the cave (fig.17). This is important because the difference in the horizontal and vertical distributions of the finds can reveal whether or not the cave represents a living floor or if the finds just fell into the cave through an opening in the surface (Clark 1993: 167).

Geological applications

Voxel Analyst can be used to further explore the interrelations of the site's geological components, such as specified pockets of material, that have been encompassed by the deposition of younger strata, and to further study the complex processes responsible for cave formation. The slicing and dicing capabilities of the software allow the user to define the yaw, pitch, and distance of a plane in order to extract strata interfaces, profiles, sections, and iso-surface values (fig.18). The excavation diagrams may also be imported into Voxel Analyst where each spit can be interpolated vertically as long as continuous numeric values are assigned to each category of geological material. Essentially, it would appear as if the user is virtually back-filling the site by converting the planimetric data to volumetric data representing geological solids. Voxel Analyst is also capable of measuring volumes and surface areas and so the amount of excavated material can easily be calculated if necessary.

Geological features, such as waterworn pebbles, can also be queried and displayed using 3D Analyst (fig.19). In this example, the output may allow one to detect paths or clusters of this material transported by water into the cave (Brain 1958). A query like this may reveal more information concerning the previous locations of cave openings that allowed transport of material to lower levels.

Summary

Attempts to easily explore and understand complex relationships between archaeological phenomena and associated strata in three-dimensional space has been hindered for more than a century, but solutions are slowly being advanced through the use of computer technology. Currently the capability to record these phenomena in 3D exists but the analysis tools essential for their interpretation do not. Multidimensional GIS would allow nearest-neighbor functions to be performed, as well as, allow the user to devise 'what if' scenarios (Harris/Lock 1996: 309). At present a GIS package that possesses a truly functional 3D data model has not been developed, limiting the archaeologist to analyze data within a vertical stack of planimetric surfaces. This was a limitation encountered at Swartkrans because the excavation material was recorded as a series of two-dimensional surfaces that had to be translated to conform to 3D space. This problem may be overcome if future excavations are recorded using the methodology described for creating the 3D model of Swartkrans.

Due to these limitations, most GIS and surface analysis applications have been carried out on a regional scale while few have been conducted at the intra-site level (Harris/Lock 1996: 307). Although these factors introduced many obstacles throughout the course of this project, the Swartkrans GIS is another step towards three-dimensional intra-site investigations of archaeological sites. The system developed is important for archiving and visualizing fossil, artifact, and geological data within their spatial contexts and providing a small set of tools for analyzing the interrelations of these 3D datasets. It still remains to be seen whether or not such a system can provide new information on the processes responsible for the accumulations found at Swartkrans and other cave sites. The answer will not be known until all data entry is complete and actual taphonomic investigations begin.

Acknowledgements

The authors would like to thank: the Palaeo-anthropology Scientific Trust for funding the project, ESRI, AutoDesk, and the Intergraph Corporation for software access and technical support, Peter Ungar, C.K. Brain, P.V. Tobias, Ron Wright, John Wilson for assistance in Avenue scripting, John Duncan for help with mathematical formulas, Brian Norris, Lewis Bowker, and Colin Menter for development of the survey design, Sohayla Hamon for the drive and inspiration needed to complete the project, Deano Stynder, Lloyd Rossouw, Alynne Bayard, Brian Culpepper, Jim Schryver, Jose Aguayo, Bruce Gorham, Terri Gorham, and the folks who reside in CAST 209 and 214.

End notes

[1] The table consisted of the following fields: ·Square, ·Level, ·Date of Excavation, ·Quadrant, ·Sub-quadrant, Member, ·Geological Material, ·Sediment Color, ·Degree of Calcification, ·Degree of Hardness, ·Presence/Absence of Granular Particles, ·Number of Stones, ·Presence/Absence of Manganese Dioxide, ·Number of Rocks, ·Presence/Absence of Decomposed Dolomite, ·Presence/Absence of Ash-like Material, ·Elevation, ·Depth of Spit, and ·Other Notes.

[2] The fields consisted of: ·Member, ·Square, ·Level, ·Quadrant, ·Sub-quadrant, ·Date, ·Order, ·Family, ·Tribe, ·Genus, ·Species, ·Body Weight Class, ·Alcelaphine/Bird Size, ·Carnivore Size, ·Element, ·Number Bone Flakes/Pebbles, ·Potential Tool, ·Burnt Categories, ·Upper/Lower Right Incisor1 (RI1), ·RI2, ·RI3, ·RI4 ·Right Canine (RC), ·Right Premolar1 (RP1), ·RP2, ·RP3, ·RP4, ·RM1, ·RM2, ·RM3, ·Left Incisor1 (LI1), ·LI2, ·LI3, ·LI4, ·Left Canine (LC), ·Left Premolar1 (LP1), ·LP2, ·LP3, ·LP4, ·Left Molar1 (LM1), ·LM2, ·LM3, ·Right Deciduous Incisor1 (Rdi1), ·Rdi2, ·Rdi3, ·Right Deciduous Canine (Rdc), ·Right Deciduous Premolar1 (Rdp1), ·Rdp2, ·Rdp3, ·Rdp4, ·Right Deciduous Molar1 (Rdm1), ·Rdm2, ·Left Deciduous Incisor1 (Rdi1), ·Rdi2, ·Rdi3, ·Right Deciduous Canine (Rdc), ·Left Deciduous Premolar1 (Ldp1), ·Ldp2, ·Ldp3, ·Ldp4, ·Left Deciduous Molar1 (Ldm1), ·Ldm2, ·Indeterminate Incisor, ·Indeterminate Canine, ·Indeterminate Premolar, ·Indeterminate Molar, ·Indeterminate Tooth, ·Teeth Portion, ·Extent of Wear, ·Diagnostic Feature 1, ·Diagnostic Feature 2, ·Diagnostic Feature 3, ·Diagnostic Feature 4, ·Diagnostic Feature 5, ·Side, ·Degree of Completeness, ·Bone Portion (Marean), ·Shaft Fragment (Villa & Mahieu), ·Specimen Physically Examined, ·Age, ·Sex, ·Surface Damage, ·Surface Damage 2, ·Surface Damage 3, ·Fracture Pattern, ·Fracture Angle, ·Fracture Outline, ·Weathering Stage, ·Shaft Circumference, ·Tooth MD Length, ·Tooth BL Width, ·Tooth Crown Height, ·Bone Maximum Length, ·Bone Maximum Width, and ·Comments.

[3] This extension was created on February 13, 2001 by Jeff Jenness, Wildlife Biologist, GIS Analyst, US Forest Service, Rocky Mountain Research Station, 2500 S. Pine Knoll Drive, Flagstaff, AZ 86001, jeffj@jennessent.com

[4] http://gis.esri.com/arcscripts/scripts.cfm

[5] This extension was created on August 7, 1999 by X. Guillotin, guillotin@esrifrance.fr

[6] http://gis.esri.com/arcscripts/scripts.cfm

References

Backwell, Lucinda, and Francesco d'Errico. 2001. From the Cover: Evidence of termite foraging by Swartkrans early hominids, in *Proceedings of the National Academy of Science* 98, pp. 1358-1363.

Brain, C.K. 1958. The *Transvaal Ape-Man-Bearing Cave deposits*. Transvaal Museum: Pretoria.

Brain, C.K. 1968. Who killed the Swartkrans ape-men?, in *South African Museums Association Bulletin* 9, pp.127-139.

Brain, C.K. 1969. The probable role of leopards as predators of the Swartkrans australopithecines, in *South African Archaeological Bulletin* 24, pp. 170-171.

Brain, C.K. 1970. New Finds at the Swartkrans Australopithecine site, in *Nature* 225, pp. 1112-1119.

Brain, C.K. 1974. A hominid skull's revealing holes, in *Natural History* 83, pp. 44-45.

Brain, C.K. 1981. *The Hunters or the Hunted?: An Introduction to African Cave Taphonomy*. The University of Chicago Press: Chicago, London.

Brain, C.K. 1993. Introduction, in C.K. Brain (ed.), *Swartkrans: A Cave's Chronicle of Early Man*, pp. 1-5. Transvaal Museum: Pretoria.

Brain, C.K. and Pat Shipman. 1993. The Swartkrans Bone Tools, in C.K. Brain (ed.), *Swartkrans: A Cave's Chronicle of Early Man*, pp. 195-215. Transvaal Museum: Pretoria.

Brain, C.K. July 29, 1999. Personal communication.

Clark, J.D. 1993. Stone Artefacts Assemblages from Members 1-3, Swartkrans Cave, in C.K. Brain (ed.), *Swartkrans: A Cave's Chronicle of Early Man*, pp. 167-194. Transvaal Museum: Pretoria

de Ruiter, Darryl J. and Lee R. Berger. 2000. Leopards as Taphonomic Agents in Dolomitic Caves - Implications for Bone Accumulations in the Hominid-bearing Deposits of South Africa, in *Journal of Archaeological Science* 27 (8), pp. 665-684.

Harris, Trevor M. and Gary R. Lock. 1996. Multi-dimensional GIS: exploratory approaches to spatial and temporal relationships within archaeological stratigraphy, in *Interfacing the Past: Computer Applications and Quantitative Methods in Archaeology CAA95 Volume II*, pp. 307-316. University of Leiden: Leiden.

Intergraph. 1994. *MGE Voxel Analyst (MGVA) User's Guide*, p. 6-14. Intergraph Corporation: Huntsville.

Jones, C.B. 1989. Data structures for three dimensional spatial information systems in geology, *in International Journal of Geographical Information Systems* 3 (1), pp. 15-31.

Mitchell, Andy. 1999. Mapping density, in *The ESRI Guide to GIS Analysis Volume 1: Geographic Patterns & Relationships*, pp. 69-85. ESRI Press: California.

Raper, Jonathan F. 1989. The 3-dimensional geoscientific mapping and modelling system: a conceptual design, in Jonathan F. Raper (ed.), *Three dimensional applications in Geographical Information Systems*, pp. 11-19. Taylor & Francis: London, New York, Philadelphia.

Turner, A. Keith. 1989. The role of three-dimensional geographic information systems in subsurface characterization for hydrogeological applications, in Jonathan F. Raper (ed.), *Three dimensional applications in Geographical Information Systems*, pp. 115-127. Taylor & Francis: London, New York, Philadelphia.

Figures

Figure 1. View of Outer Cave area of Swartkrans looking NW with metal excavation grid in place. Photo by Joseph Nigro.

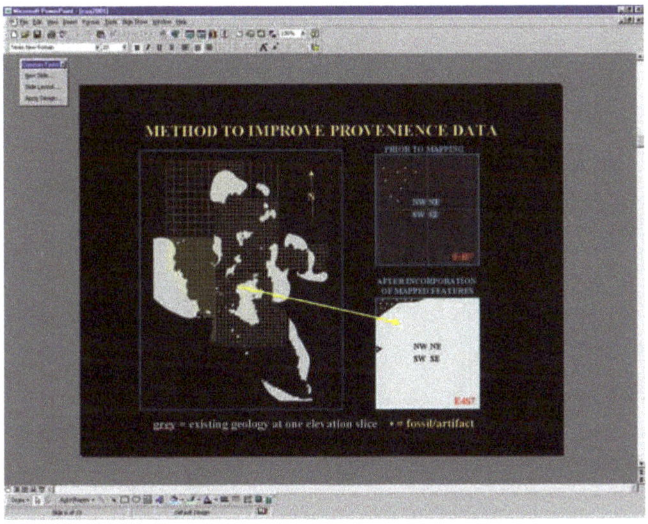

Figure 2. Arcview image depicting one elevation level of Swartkrans and the methodology used to increase the provenience of the fossils and artifacts. Image by Joseph Nigro.

Figure 3. The Wild T1010™ theodolite with a Leica Distomat ™ Laser EDM. Photo by Joseph Nigro.

Figure 4. The origin and North line of grid used during the excavations and the survey. Photo by Joseph Nigro.

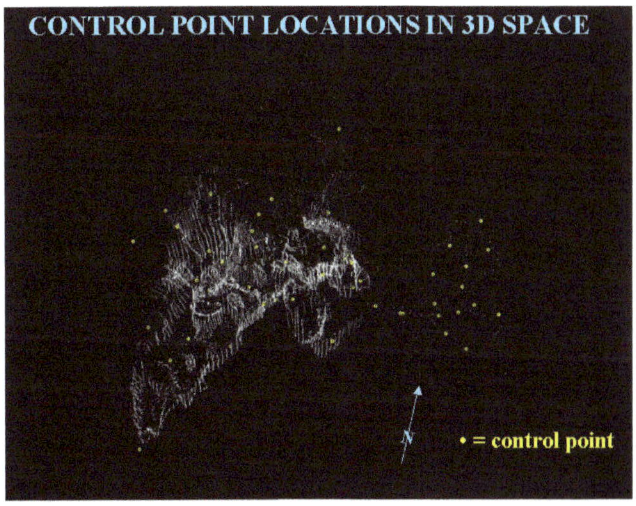

Figure 5. The control points established during the survey in 3D Analyst. The points shown in white represent the surface of the site. Image by Joseph Nigro.

Figure 6. Examples of the differences in feature topography dictating the mapping interval. Photo by Joseph Nigro.

Figure 7. Example of a natural geological overhang, represented by a portion of the Hanging Remnant. Photo by Joseph Nigro.

Figure 10. Image of Swartkrans in 3D Analyst, looking NW into the cave. The "bulbous floor" can be seen at the bottom of the image. Image by Joseph Nigro.

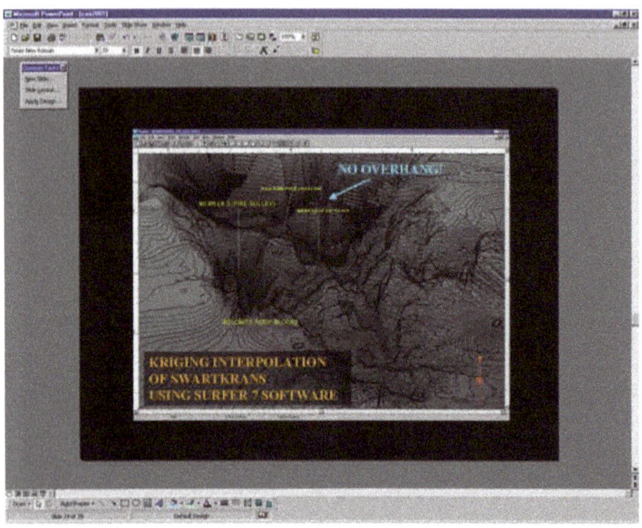

Figure 8. Interpolated image shown in 3D view using Surfer 7. The blue arrow identifies the data spikes produced. Image by Joseph Nigro.

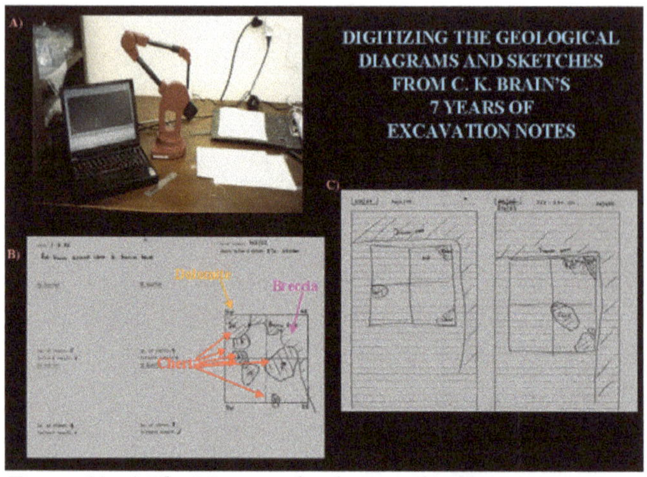

Figure 11. a) The Microscribe digitizer b) Excavation notes including a standard diagram c) Excavation sketch. Photo by Joseph Nigro. Excavation pages by C.K. Brain.

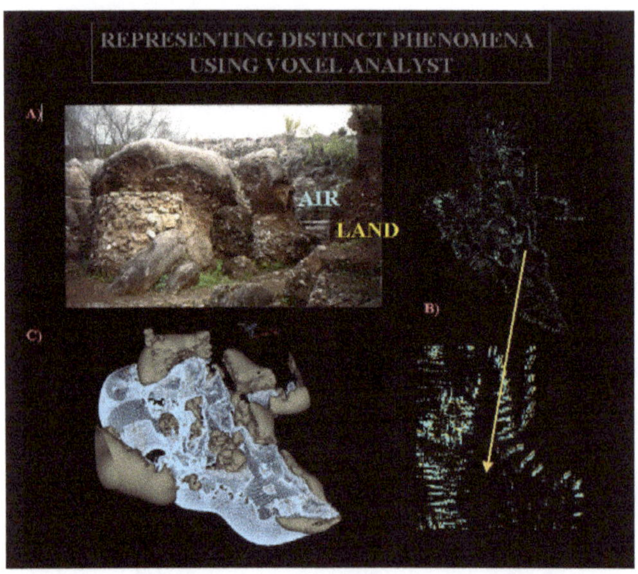

Figure 9. a) The air and land component that make up the site representing two distinct phenomena. b) The survey points in 3D Analyst with brown depicting the land and blue depicting the air c) the interpolated land and air volumes in Voxel Analyst. Photo and images by Joseph Nigro.

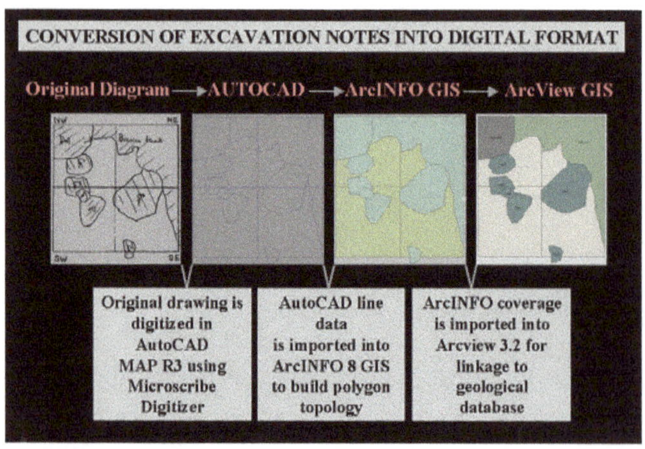

Figure 12. Procedure used to convert the excavation diagrams and sketches into digital form.

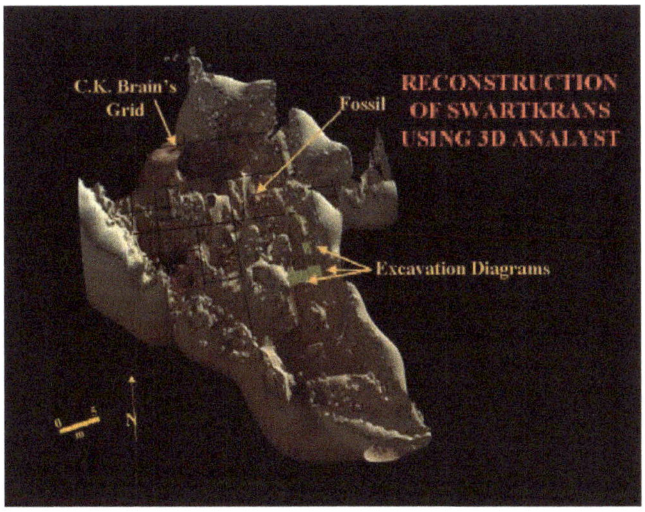

Figure 13. 3-dimensional reconstruction of Swartkrans with ancillary data in 3D Analyst. Image by Joseph Nigro.

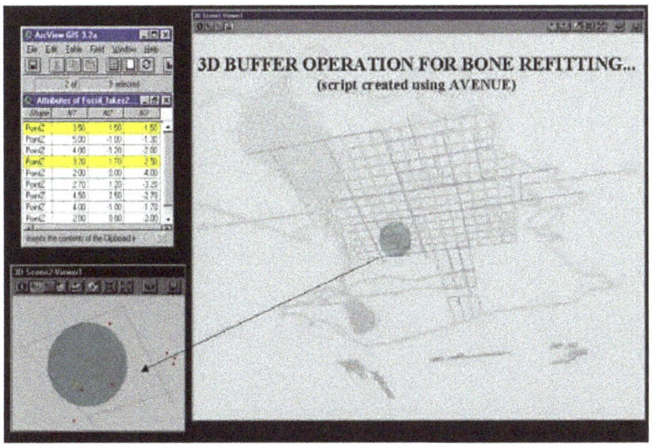

Figure 14. The 3-dimensional buffering operation in 3D Analyst. Image by Joseph Nigro.

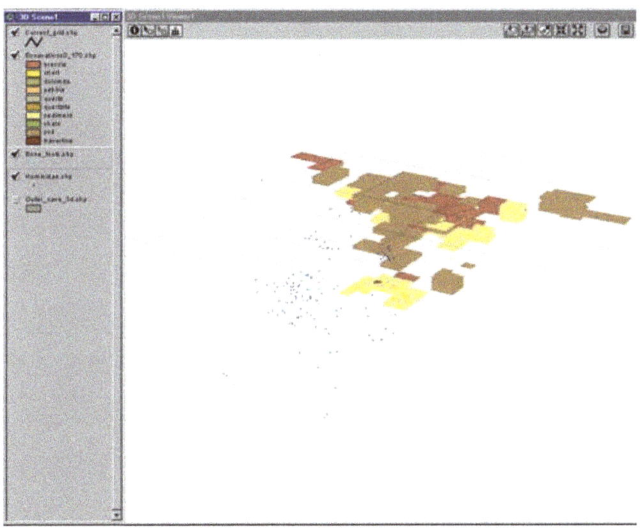

Figure 15. 3-dimensional display of geological data, stripped away to reveal the fossil hominins and the stone tools and Brain's exacavation grid in 3D Analyst. The linear feature represents Brain's excavation grid. The geological slices represent excavated material recorded at each 10 cm spit (level). Image by Joseph Nigro.

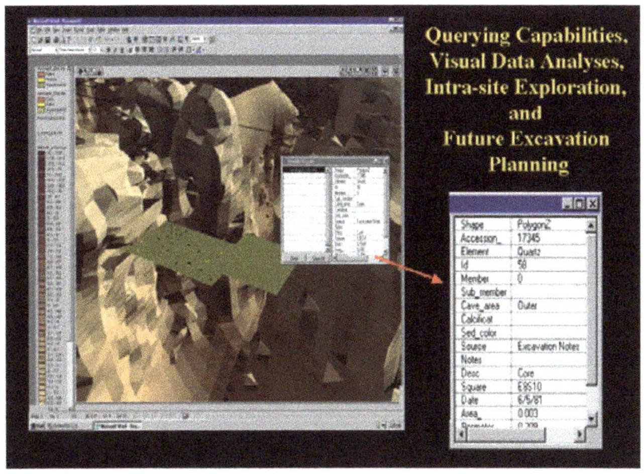

Figure 16. Image in 3D demonstrating the querying capability of 3D Analyst. The image depicts a view as though one were standing inside the Outer Cave looking at a sample of squares that were excavated in that area. Image by Joseph Nigro.

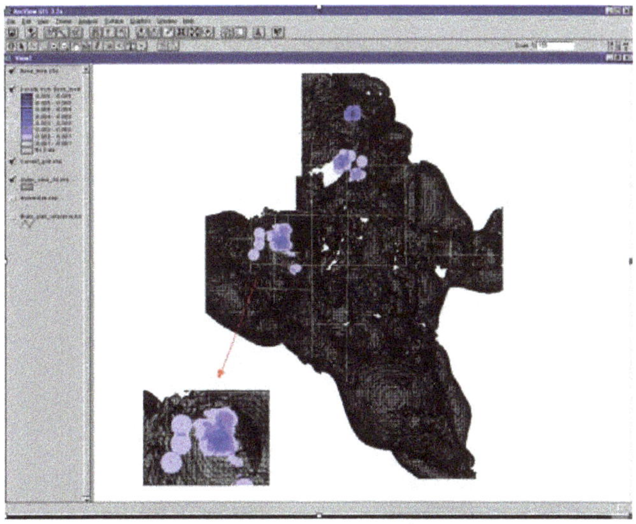

Figure 17. Plan view of Swartkrans (gray) with bone tool densities (blue) within a 25 cm² area. The individual tools are represented in red while the excavation grid is yellow. Image by Joseph Nigro.

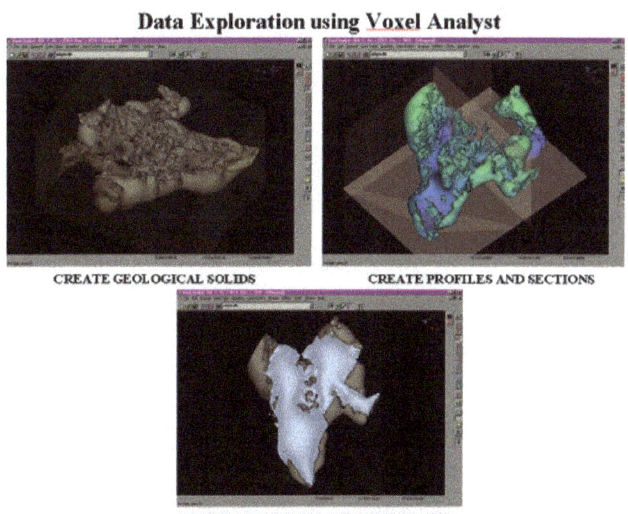

Figure 18. Possible archaeological and geological applications using Voxel Analyst. Images by Joseph Nigro.

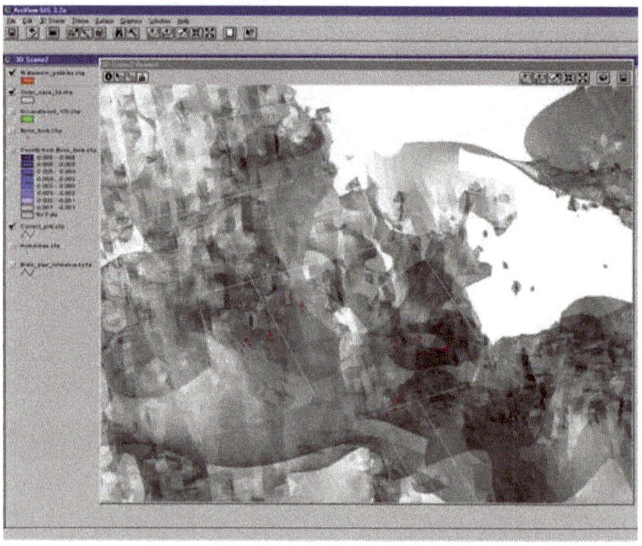

Figure 19. A sample of waterworn pebbles (red) viewed within the context of the surrounding cave (semi-transparent gray) using 3D Analyst. Image by Joseph Nigro.

GIS contribution to urban history and to the reconstruction of ancient landscape

Sofia Pescarin
University of Bologna
Dep. of Historical Disciplines
Via Tasso 9
40069 Zola Predosa (Bologna), Italy
Phone: +39-3384024402 - E-mail: pescarin@bo.nettuno.it

Abstract: This project is part of a PhD research (in History and Computing) carried out with the department of Historical Disciplines of the University of Bologna. The aim of the research is to investigate the structure of the Roman city of Bononia through the reconstruction of its surface and access to excavation data. Recognising the importance of the immersive experience, in the case of Bologna, the project attempts to answer two questions: "How did the city change from Roman period to the present?" and "What was the nature of its ancient landscape?".

Far from pretending to exactly reconstruct the ancient landscape, the reconstructed three-dimensional model tends to offer some suggestions related to distribution of urban space, in relation to its morphological structure.

Key words: GIS, Virtual reality, Ancient Landscape Navigation, Urban history, Bononia

Introduction

This work is part of a PhD research (in History and Computing) carried out with the department of Historical Disciplines of the University of Bologna.

An important research project, involving many themes related to the history of Bologna city, was started in the 1990s[1], involving mainly the department of Historical Disciplines (Prof. Francesca Bocchi) and an interdisciplinary group (CINECA Supercomputing Centre and Carpi Research Centre). The aim was to sort out the historical events that characterised the city, in term of urban structure and evolution, which led it to have its current form, through the construction of a four-dimension digital model.

The success of this project, is due mainly to the large number and accessibility of historical documents preserved in archives, as well as to the hard work of numerous historians, in meticulously analysing the information, keeping and digitising the most reliable and useful in order to reconstruct the different historical urban phases.

The city centre has been modelled in this way: modified and built following the evidence from those historical sources (documentary and iconographic) allowing the reconstruction of crucial urban phases, starting from the present, back to the 13th century[2].

A possible approach has been studied to more ancient times, but obviously obstacles were found in treating and recovering information prior to the XI Century . Historical sources are in fact quite scarce for this period and often unreliable, not allowing a proper identification and understanding of the evolution of Bologna city, between the fall of the Roman Empire and the X-XI centuries. Archaeological excavations cannot help much either in recovering the historical and urban memory.

The research project

The present research started precisely at this point - from the end of the Roman Empire, backwards - using a different approach to the mainlyarchaeological sources, but with a similar aim: the reconstruction of the ancient Roman landscape.

The first step was the development of as reliable a cartographic base scientific knowledg scientific knowledge as possible, in relation to the current form of the territory under study.

After a few attempts with Regional Technical Maps (CTR - 1 : 5000), detailed maps were chosen (1 : 2000 scale) for the city centre and less detailed for the surrounding areas (Fig. 1), used together to maintain the relationship between city and suburban zones, for both morphological aspects and human interactions.

After some verification and corrections to the digitised data, due to discrepancies between the maps at different scale and adjustment by comparison with terrain observations, a three-dimensional digital model of the territory was obtained. This model was good enough to be used as a continuous reference where necessary, particularly where documents failed to record elevation data in units above sea level.

Vectorial information, such as rivers, streets, railways and built-up areas, was overlaid on this model using ArcView GIS (Fig. 2). Aerial photographs were rectified and georeferenced in order to complete the present day 3d model.

The aim was to reconstruct a good model of the actual landscape so as to highlight the changes which occurses between Roman times and the present day[3].

In order to obtain as accurate as possible a model of Roman landscape, it was necessary to study bibliographical and archive sources in detail[4] .

Data have been systematically arranged in a relational ODBC database, consisting primarily of three tables for SITES, REMAINS and BIBLIOGRAPHY information. We now have almost 500 remains and 400 sites.

To start with, the database was interrogated for referenced points belonging to particular categories of contemporary sites. Some data couldn't be used, such as records referred to spread remains and underground remains (such as sewers, water courses and canals).

Using these points a three-dimensional model of the Roman surface has been calculated using the *krig* interpolation method.

On this new model have been placed a few themes took out from the database, through crossed-queries based mainly on the typology of the remains, their chronology and building materials. Main Roman streets (*cardines* and *decumani*) were added, as well as observations related to the presence of *domus*, suburban *villae*, industrial structures, dumping places, *amphorae* storages and building's pavements (3).

Information relating to public places was also added: the *forum*, the *basilica*, the theatre and a few temples.

Finally, remains relating to the urban water administration were placed on the surface. These included pipes (*fistulae*) belonging to Roman aqueduct and drain-pipes, and vectors representing known and hypothetical rivers[5].

Every object *geo-placed* on the model has been assigned an ID number, used as the main reference to connect with the ODBC database tables, through a *join* operation, called of in the GIS system through SQL instructions. This also allowed relations between different tables of the external database to be maintained.

The next step in the project led to a twofold direction and to a different approach to data and problems.

There's, indeed, another possible point of view, likely to be examined. Analysing the urban history going backwards - starting with known elements of modern and Medieval city - we arrived at a point were we just have spread information that cames from a few excavations. At this point lights fade away and the scene is deprived of buildings, constructions and daily life. The manuscripts are silent, leaving us with just a disturbed surface. Thanks to the reconstruction of this surface through a mathematical calculation, is it possible now to keep on working on the city, looking at the natural events and human activities that have marked it. This surface begins to be populated at *domus*, *villae*, temples, markets, theatres, industrial and commercial activities.

But the reasearch should not stop at this point. It goes on, backwards, to a time before the Romans decided to quarter the colony of *Bononia* here, before the Etruscans had left traces of their *Felsina*. It is possible to have a quick look at the large territory that the Villanovians chose as their residence in the Iron Ages[6].

Following this approach we are working in two directions: on the one hand we are now exploring the great potential of GIS, on the other developing its new potential in terms of visualisation and realtime navigation.

Analysis

GIS cannot be considered just a useful tool to visualise georeferenced two- or three-dimensional geographical data. We can experience its real value with spatial and crossed analysis of data collected in the initial phase, characterised by archival research and informatisation. This is the crucial stage, when there is direct confirmation of the validity of our project plan and the quality of the problem definition. We also have to consider that we are not dealing with simple numbers and mathematical problems, but with historical and archaeological data, sometimes so multifaceted and so difficult to interpret that there has to be room for a certain wastes to be retained, as possible, since the beginning, inside a definite range[7].

The analysis that will be carried out mostly consists of operations such as buffering, visibility, density and neighbourhood statistic, distance mapping, slope and aspect functions. The goal is to answer questions such as: the distribution of different typological elements in relation to terrain morphology and natural resources; the distribution of certain building materials; analysis - through deeper examination of relations between remains and their morphological location - of two or three macro-situations highlighted by the generation of the digital model; and analysis of the relationship of remains connected with Roman hydrographic management with soil drainage characteristics.

3D visualisation

Some analysis is related to the later phase of this project, whose final aim is the realisation of an interactive (even with data acquired) and three-dimensional navigation virtual system.

We cannot claim anything new when we talk about *archaeological landscape navigation*[8]. I think we have all witnessed years of discussion and case studies on these topics. Today, thanks to the laborious work (and information dissemination) of some researchers, it's possible to recognise that something has been achieved. Perhaps we can say that there is a common awareness of the importance of three-dimensional and immersive approaches to data. There are new opportunities for these applications together with the creation of interdisciplinary work-groups and the identification of needs and new solutions in this field.

Recognising the cognitive importance of the immersive

experience, in the case of Bologna, the project attempts to answer two questions: "How did the city change from the Roman period to the present?" and "What was the nature of its ancient landscape?".

Far from pretending to exactly reconstruct the ancient landscape, the reconstructed three-dimensional model tends to offer some suggestions related to the distribution of urban space, in relation to its morphological structure.

In this first approach the particoular software was used: *Terravista*[9]. It works with input data such as vector *shapefiles*, DEM, 3D objects and raster images (such as *geo tiff*) and it produces 3D real time navigation as output in **flt* format. Even if it cannot be considered as 'geographic' modeling software, Terravista has good characteristics of compatibility with commonly used GIS tools (ArcView, ErMapper, ...) and has the great advantage of producing perfectly georeferenced 3D scenes.

Thanks to this kind of software, a project to navigate on the actual territory (with aerial photographs) is in progress, together with the construction of a virtual model of Roman landscape displaying (and possibly interacting with) the data acquired (Fig. 5).

End notes

[1] Nu.M.E. project (New Electronic Museum for the city of Bologna)

[2] "Atlante storico delle Città Italiane": Emilia-Romagna, 2, Bologna, a cura di F. Bocchi, voll. 4, BO 1995-98; Vol. V (Atlante Multimediale di Bologna) in formato CD ROM, BO 1999; F. Bocchi, L. Calori, F. Fraticelli, A. Guidazzoli, M.Mariani, The 4 Dimensional City, Science and Supercomputing at Cineca, 1997 Report; F. Bocchi, Nuove metodologie per la storia della città: la città in quattro dimensioni, in "Medieval Metropolis", Proceedings of the Atlas Working Group, BO 1997; M. E. Bonfigli, Implementazione del progetto Nu.M.E., in "Medieval Metropolis", Proceedings, BO 1997; F. Bocchi, Medioevo Virtuale, in "Medioevo" n.11, (22) Nov. 1998; F. Bocchi, M.E. Bonfigli, L. Calori, A. Guidazzoli, M.C. Liguori, M.A. Mauri, M. Melotti & D.Vasetti; "Virtual Reality and Cultural Heritage: Some Applications"., in "Proc. of EVA 2000" , (V .Cappellini & J. Hemsley Eds.), Florence, March 2000, Pitagora Editrice Bologna, pp.162-165.

[3] For a bibliographic reference of Roman Times: Bergonzoni F., Bonora G., Bologna Romana, I, Fonti Letterarie - carta archeologica del centro urbano, Bologna, 1976; Ortalli J., Bononia romana; in 'Bologna I. Da Felsina a Bononia, dalle origini al XII secolo', "Atlante storico delle città italiane." Vol. I, a cura di F. Bocchi, Bologna, 1996; Ortalli J., Bologna città romana; in 'Atti e memorie della regia Deputazione di storia patria per le province di Romagna' vol. 47; Bologna 1997, pp. 141 – 193; Scagliarini D., L'insediamento residenziale e produttivo nel suburbio di Bologna romana; in 'Atti e memorie della regia Deputazione di storia patria per le province di Romagna.', Bologna 1969

[4] Part of the documentation was taken from Soprintendenza archives (ASAER), kindly granted by Sopritendente Dr.sa Calvani and Dr. Ortalli.

[5] Every object was always connected with its bibliographical references

[6] Forte M., La pianura bolognese nella prima età del Ferro. Note sulla topografia degli insediamenti, in "La pianura bolognese nel Villanoviano", Catalogo, Firenze, 1994, pp. 9-20; Morigi Govi C., Tovoli S., La cultura villanoviana a Bologna (IX – prima metà del VI secolo a.C.), in "La pianura bolognese nel Villanoviano", Catalogo, Firenze, 1994, pp. 46-48; Sassatelli G., Morigi Govi C., Felsina etrusca, in "Atlante storico delle città italiane." Vol. I, a cura di Francesca Bocchi, Bologna, 1996

[7] A clear example can be the case of excavation and documents without the definition of a precise dating: this led us to calculate a surface model with a generic assignment to the Roman Period (between II-I BC and III AD), considering in any case the absence of significant increase of the ground for the considered period.

[8] MOLINEAUX, B., 1992, From virtuality to actuality: the archaeological site simulation environment, In 'Archaeology and the Information Age' Edited by P. Reilly and S.Rahtz. London: Routledge, pp.312-322; FORTE M., Un esperimento di visualizzazione scientifica per l'archeologia del paesaggio: la navigazione nel paesaggio "virtuale", in "Archeologia e Calcolatori" 4, 1993, pp.137-152; BARCELO J., FORTE M., SANDERS D., 2000 (eds.), Virtual reality in archaeology, Oxford, Archeopress (BAR International Series S 843), 247-263.

[9] With the co-operation of CNR ITABC and Dr. Maurizio Forte.

Figures

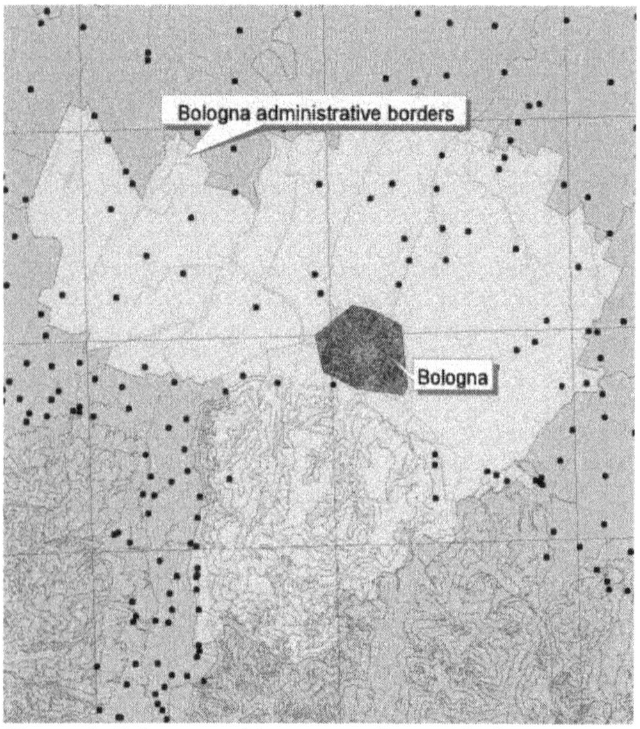

Figure 1. Bologna and its surroundings with elevation information

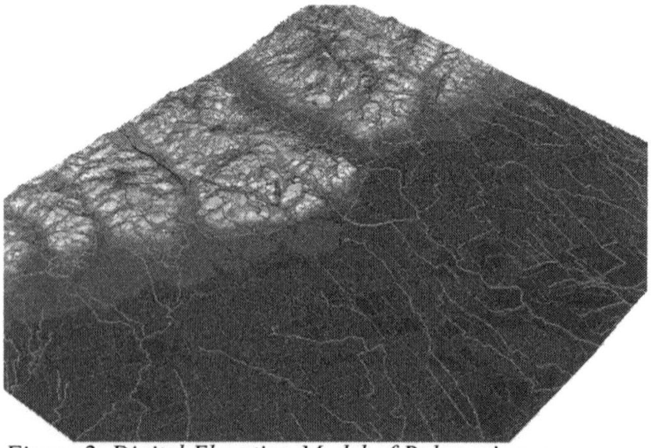

Figure 2. Digital Elevation Model of Bologna's area.

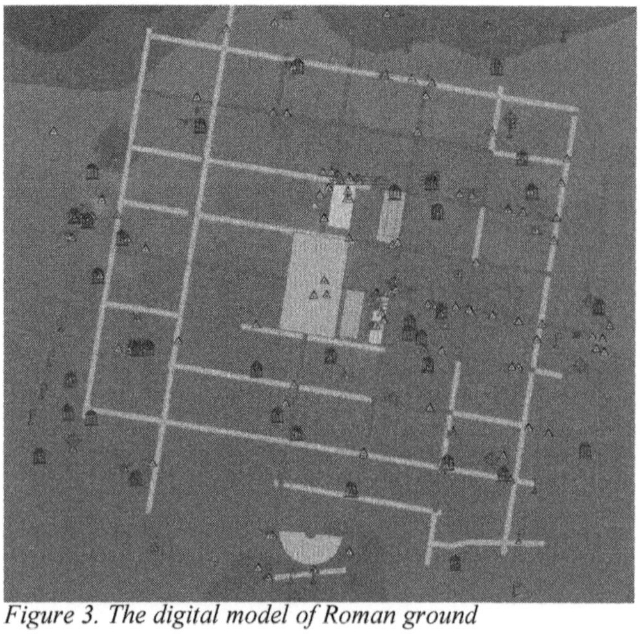

Figure 3. The digital model of Roman ground

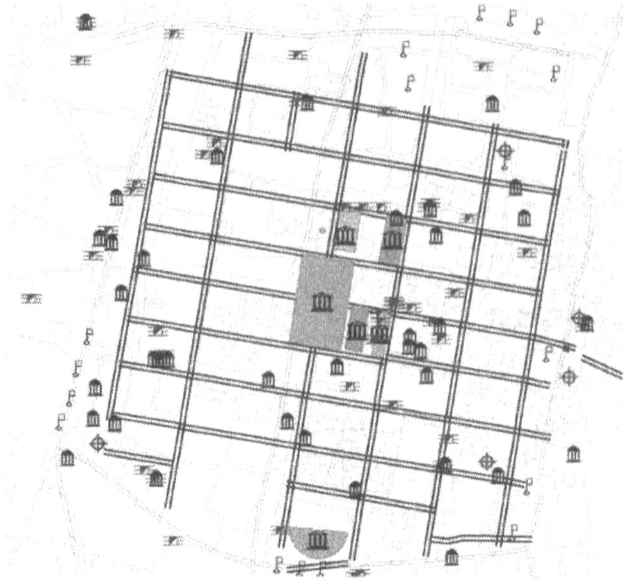

Figure 4. Themes of Roman Bononia*: mosaics (red), public buildings (blu), storages (violet), streets (brown), river (ipothetical: light blu; assumed: dark blu)*

Figure 5. A snaphot of the real time navigation in the roman virtual environment.

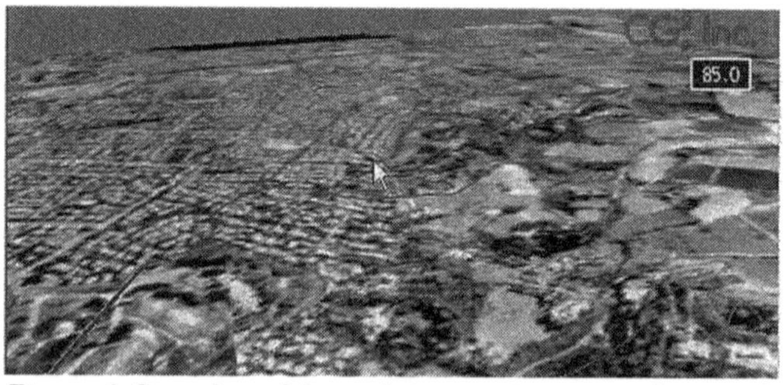

Figure 6. Snapshot of the real time navigation in the actual landscape

GIS Modeling of the Minoan Peak Sanctuaries of East Crete

S. Soetens
Laboratory of Geophysical - Satellite Remote Sensing & Archaeo-environment
Institute of Mediterranean Studies – Foundation of Research & Technology (FORTH)
Melissinou & Nik. Foka 130
PO. Box 119, Rethymnon 74100, Crete, Greece
Phone: +3-97-4238106 - Fax.: +3-831-25810 - E-mail: ssoetens@yahoo.com

A. Sarris
Laboratory of Geophysical - Satellite Remote Sensing & Archaeo-environment
Institute of Mediterranean Studies – Foundation of Research & Technology (FORTH)
Melissinou & Nik. Foka 130
PO. Box 119, Rethymnon 74100, Crete, Greece
E-mail: asaris@ret.forthnet.gr

S. Topouzi
Laboratory of Geophysical - Satellite Remote Sensing & Archaeo-environment
Institute of Mediterranean Studies – Foundation of Research & Technology (FORTH)
Melissinou & Nik. Foka 130
PO. Box 119, Rethymnon 74100, Crete, Greece

A. Tripolitsiotis
Laboratory of Geophysical - Satellite Remote Sensing & Archaeo-environment
Institute of Mediterranean Studies – Foundation of Research & Technology (FORTH)
Melissinou & Nik. Foka 130
PO. Box 119, Rethymnon 74100, Crete, Greece

Abstract: This paper presents the preliminary results of building a cultural landscape model of Minoan peak sanctuaries through a GIS approach. It is part of the wider framework project "The Topography of Power"[1] in Minoan Crete.

The reconstruction of the peak sanctuaries' landscape was based on the accurate mapping of them and their surrounding settlements through high accuracy GPS receivers and the analysis of their spatial distribution within a SPOT generated DEM. Viewsheds were exploited in order to address questions related to Minoan peak sanctuaries with respect to their topographic settings. A relational database containing the existing archaeological information was implemented to the GIS in order to provide sufficient evidence for the formation of the ritual/cultural landscape of the Minoan period.

The peak sanctuary landscape is presented in two distinctive chronological acmes, roughly Proto- and Neopalatial, with the transition from the first to the second period, and in relation to nearby settlements and burial sites.

Key words: GIS, GPS, viewsheds, Minoan, Crete, peak sanctuaries

Introduction

The phenomenon of the Cretan Bronze Age peak sanctuary is approached in terms of its definition and distribution, its history, and the various interpretations of its activities, use, and meaning.

GIS is proposed to address specific questions related to the distribution of sanctuaries, their spatial characteristics and corresponding geomorphologic and environmental settings.

The first results of this approach are presented, with some methodological comments on:

- the usage of high accuracy GPS (Global Positioning System) receivers,
- digitisation of land use and land capability maps and their related legends in a table,
- registration of all data to the same geodetic coordinate system (maps, DEM),
- creation of a relational database containing archaeological data on the sanctuaries, relating the sites, with finds, chronology and bibliography,
- viewsheds of different periods of the East Cretan peak sanctuary landscape.

These results are then evaluated to define the future prospects of the project.

General characteristics and environment

Definition

A peak sanctuary is generally characterized by its topographical location, on top or near the peak of a mountain, though reachable and within the area of human shepherding activity.

It is identified by the presence of finds indicating ritual action, such as terracotta figurines (human and animal), pottery (possibly broken on purpose) and a pebble scatter [2]. The presence of thick ash layers, architectural remains, bronze vessels, and Linear A inscriptions, is characteristic of a later phase of the peak sanctuary's history.

It has been empirically acknowledged that the topographical location of these sites was meaningful as well: "*The topography of known peak sanctuaries is the best guide in our search for new sites*" [3]. Many of the sites are located on the edge of a steep cliff, but are accessible from one slope. Although peak sanctuaries are definitely isolated from settlements, the distance to the nearby settlements seems to be important. The survey of the Atsipadhes Korakias Peak Sanctuary Project has located some small contemporary settlements in the sanctuary's "parish", but no large town or palatial villa. A common pottery type fabric at these small settlements and the sanctuary further reinforced the topographical and optical link that exists between the peak and the settlements below it. The optical link between the peak sanctuary sites themselves is even stronger [4]. For this reason proximity and intervisibility can be suggested to be some of the most promising features in a GIS approach. The previously mentioned ash feature (still under discussion) might further accentuate a possible communication role between the peak sanctuaries, as it could be a residual of fire signaling between sites.

Distribution

About 65 sites have been added to our database, including 22 sites (fig. 1, yellow with black) that conform to most of the criteria mentioned earlier. The other sites (fig.1. black) are of dubious nature: some turned out to be of a totally different period, others were later identified as settlements (with different archaeological features) or as open-air sanctuaries with different topographical nature and others again lacked the typical finds. Some may still be peak sanctuaries.

If we look at the general distribution of the discovered peak sanctuaries on Crete, a much thicker concentration is to be found in our pilot area, the East Cretan mountains.

A second concentration can be identified in central Crete, with Iuktas as the main sanctuary, the oldest, longest living and certainly the largest and most important one. In later sources, Iuktas was recognised as Zeus's tomb, and the profile of the mountain as seen from the west, does indeed resemble a human head [5]. We emphasize this feature because the recognisable shape of the mountain itself as a marker in the landscape seems like another common factor for most of the sites.

In the municipality of Rethymnon, another three certain peak sanctuaries were identified, two having common geographical and archaeological characteristics with their Eastern counterparts (Spili Vorizi and Atsipades Korakias). Further to the west some candidates were proposed, but thick vegetation and difficult access has not allowed us so far to further evaluate the peak sanctuaries of West Crete. It is one of the goals of the project to examine if these concentrations appear as the result of more intense research in the eastern part of the island or if they are indeed meaningful in the Minoan landscape.

Chronology and the peak sanctuaries in broader Minoan history

Our knowledge of the chronology of these sanctuaries is still limited, because up to very recently, hardly any detailed publication exists on the pottery. The lack of detailed publication is mostly due to the nature of the excavations: Some are more than 100 years old, while others were of a rescue nature, and hardly any detailed reports were kept. Exceptions are the publications on Atsipades, Iuktas, and Petsofas [6]. The reinvestigation of the material of most of these sites is still problematic. Nowicki noted: "We mostly deal with coarse ware, and difficult to be dated. If the site is strongly exposed and the surface eroded, potsherds are considerably washed away"[7]. And Peatfield: "...the chronology is extremely difficult to ascertain"[8]. The combined knowledge of all these reports though gives a reasonably good picture of the history of peak sanctuaries[9].

Origin: Most probably the peak sanctuaries developed in a prepalatial context. Peatfield argues that they evolved from the EM built tombs[10].

It is in this period that there is an increase in pastoral farming, reflected by large flocks of figurines in bowls, recorded on the sites. Around 2000 BC, which falls more or less in MM IA, the Near East settlements flourish and a widespread systematic trade between Crete and the East influences Minoan life [11].

MM IB-II (phase III): All peak sanctuaries are now in use, but show localised material, and a parochial character[12] (figure 2). From this period on, the Iuktas sanctuary is head of the hierarchy. This phase corresponds to the protopalatial period, and the peak sanctuary is in a period of expansion and prosperity.

Watrous suggests that due to the critical role Crete had obtained within the wider trade network in the Eastern Mediterranean, a new kind of sanctuary was initialied, namely the peak sanctuary. Some of the Petsofas figurines are Light on Dark, or Polychrome-on-Dark, and comparable to Kamares ware (MMI -II)[13]. Important for the peak sanctuary zone theory is that, in the same period, in the mountain chain west of Sitia, one can find only limited and defensible sites west of Sitia (Chamezi, Vrokastro, Katallimata, Chrysopigi, Korakia, and Tappes)[14], which were all reused in LMIIIC- Early Iron Age.

MM IIIA (phase IV): A big change on all levels can be observed from MMIII (second palace period) and onwards. Only 8 sanctuaries survive. These present recently built or at least rebuilt structures, and have much more varied and richer finds, often referred to as elite palace material (figure 3).
There is no more mention of ash-layers (possibly unstudied),

the figurines change in style (the 'regency style'), Linear A inscriptions are found on stone altars. Also in the palatial centres, the first representations of peak sanctuaries are reported, such as the Zakros peak sanctuary rhyton, the Knossos Gypsadhes serpentine vase fragments, and possibly the Petsofas terracotta model. Nowicki divides the MM III period in MMIIIA, which he recognises as the decline in numbers, and MMIII-LMI, corresponding to the neopalatial period[15]. This decline already had started in MMII. It has been frequently argued that the peak sanctuaries are here institutionalised by Knossos; a "deliberate centralisation of cult", or "part of the general process by which palatial elite used religion to maintain hierarchical position", or "they formalise in order to increase their prestige and social position"[16].

MM IIIB-LMI (phase V): After MM IIIB a general decline in use of the sanctuaries is observed, and this could be due to the natural catastrophes of this period (earthquake, Theran eruption), but even if they were not very destructive, they certainly caused a strong religious response[17]. An increase of cave sites as religious centres could present a new shift in religious practise.

LM II-III (phase VI): Only Iuktas survives the end of the second palace period up to LMII[18].

LM IIIC: The peaks are reused from LM III C and onwards as refuge settlements.

Unfortunately only three well-documented sites can be dated in such a detailed chronological scheme. In order to provide a correct representation of the peak sanctuary's chronology, and by lack of better data, the peak sanctuaries were categorized under two broader chronological periods, in which a big change can be identified, namely the protopalatial period (more specifically MM IB-II or phase III), and neopalatial period (MM IIIA or phase IV), the epoch of the so-called palatial peak sanctuaries.

Previous interpretations

After many years of research, a number of interpretations has been proposed for the function of peak sanctuaries. Most obvious is their religious character, due to the artefact assemblages which clearly indicate ritual offering. Table 2 summarizes the most popular identifications, the action that accompanies this identification, with archaeological and other arguments supporting the theory, and the hypothesized practical meaning of these sites' existence.

A GIS approach to peak sanctuaries

The implementation of GIS to the subject of Minoan peak sanctuaries offers us a way not only to analyze systematically the empirical observations, but also addresses new questions related to the broader function and distribution of peak sanctuaries, the use and perception of the Minoan landscape and the interaction of the peak sanctuaries with their surroundings.

The use of GIS

- First of all the exact mapping of these sites is essential, since already several of these sanctuaries have been destroyed, by either church construction, antennas, looting or army installations. Without having a detailed map of the location of the peak sanctuaries, it is impossible to analyse their spatial distribution, study their relation with other significant settlements and visualise them within their cultural and natural environment.
- Since a number of uncertain sites exist, the environmental characteristics of peak sanctuaries could possibly clarify their identification and improve the classification scheme.
- Analysis of the distribution of these sites is much easier and more accurate using GIS, especially if we want to compare a broad spectrum of data, such as the broader archaeological landscape, chronology, finds, geology, land use, land capability, proximity, accessibility and intervisibility.
- Cumulative viewsheds of the peak sanctuaries of different periods indicate specific tendencies related to areas of power of different categories (religious, social, economic, political, military).
- Cost distance analysis from and to the nearby settlements of the same period contributes to the study of proximity and accessibility of the peak sanctuaries.
- The creation of a predictive model might indicate candidates for undiscovered peak sanctuaries, or explain the existing gaps in the landscape, why the East has many more peak sanctuaries than the west.
- Capturing the characteristics of the peak sanctuary landscape, contributes to the whole of the Minoan land- and seascape, identifying possible humanized borders of religious influence zones, political, social or economic control areas, and the separation of the "tamed" land, such as agricultural area pasture land, settlement land, and sacred land from the "wild" area. This presumption is based on the presence of both domesticated and wild animal figurines at the peak sanctuaries.

Progress of the project and methodology

- RELATIONAL DATABASE: Basic archaeological data were collected through library research, archaeological and environmental observations made at the sites visited. The data were recorded in a relational database, including information on chronology, architecture, finds and environmental characteristics of the peak sanctuaries.

Since various opinions exist even on the general chronology of the sites, the database was formulated to include different chronologies, suggested by corresponding researchers. The database can thus be queried on the opinion of each researcher. Through lack of numerical data on the finds, the database is limited so far to the mention of absence or presence of specific types of finds. These types however can indicate in how far a site has a typical peak sanctuary artifact assembly and the way that this assemblage can be an indication of chronology[22] (see figure 4).

Problems arose in the actual site definition and the construction of the basic list of sites. As we investigated the possible presence of unknown peak sanctuaries, it was decided that sites, which were once identified as peak sanctuaries, but later erased from the list of definite peak sanctuaries, be once again included. For the purpose of chronological evolution of the peak sanctuary landscape, it goes without question that we only use certain, and dated sites (although sometimes this chronology remains vague).

Furthermore, data included in the Digital Archaeological Map of Lasithi[23] were used for chronology and typology of settlements, burial sites, "sacred" sites (not peak sanctuaries), production sites, and guard posts. The sites dated securely to either the proto- or neopalatial period, are far outnumbered by the sites dated to the full "Minoan" period and need further research.

GPS & field survey

So far, the prefecture of Lasithi has been thoroughly investigated; almost all certain peak sanctuaries were visited, with the exception of Plagia and Korfi tou Mare, located close to military installations. A few others need to be revisited, such as Etiani Kefala, since the drawing in Rutkowski's article[24] locates the peak sanctuary on the actual top of the mountain, while other reports mention that the sanctuary is about a 100m lower and further to the north[25].

Viewshed analysis took such discrepancies as far as possible into account. Coordinates of sanctuary areas were taken with **GPS** receivers, using static differential positioning. A few others in the wider region of Crete were also visited and positioned. Although difficult to determine, it was attempted to measure the extent of the peak sanctuary area, by creating polygons based on GPS measurements.

The further site distribution (of Minoan settlements, burial sites, production sites, religious sites and guard posts) was once again deduced from the Digital Archaeological Map of Lasithi, which included the pinpointing of archaeological sites of the studied area with chronology and typology. The data included excavated sites, intensive and extensive surveys, where both actual GPS coordinates and digitized survey plans were combined to achieve the fullest possible ancient site distribution.

Digitization Of Maps – Creation Of Table With Attributes Of The Polygons

Land use and Land capability maps were **digitized**, and georeferenced to Ε Γ Σ Α87 projection system (Greek Geodetic Reference System of 1987). Each polygon of those digitized maps had a property record in an accompanying table, which can be queried on any field. (such as: geological layer, physiography, soil depth, erosion, slope, artificial vegetation, aspect, capability of soil from very fertile to not cultivable at all in percentage).

Digital Elevation Model

From 1:50000 topographical maps, the coastline was created to register the Digital Elevation Model to the same projection system, namely Ε Γ Σ Α87. The DEM was constructed by a series of SPOT ORTHO images, with a pixel size of 50 x 50m. Georeferencing of the 50 x 50m pixel DEM was based on 42 reference points on the coastline of Crete, having a maximum residual of 47.431m. GPS measurements of the peak sanctuaries were also added to the DEM.

The DEM was then **resampled** with bilinear interpolation reducing the pixel size from 50x50m to 10x10m.

Pseudo Coordinates For Peak Sanctuaries

As a test, comparative viewsheds were created from the GPS points and the highest neighboring pixel (of 50 x 50m).

The viewsheds resulting from the highest pixel were much more accurate, and all further viewsheds were based on these viewpoints. The actual GPS measurements will therefore be much more valuable for analysis of aerial photographs, high-resolution satellite images (such as Ikonos), and smaller scale digitized maps (1:5000), unless if the DEM could be registered with a better accuracy.

Viewsheds

Individual viewsheds for all of the Lasithi peak sanctuaries were created, and compared to the field observations. Viewsheds were taken all from the highest pixel adjacent to the corresponding GPS points, with a maximum view distance of 25km, and at a height of 2m. The results of these viewsheds are satisfactory for general large distance observations, concerning coverage of certain areas, but not good enough for local scale landscapes. Test viewsheds and close-ups of better-known areas did not entirely cover the actual visible terrain as expected, which should cover a larger area of the fertile valleys. In order to compensate such discrepancies georeference of the DEM will be further refined by including internal reference points and using the highest-altitude pixels of the DEM, with the coordinates of known datum points. For micro-landscape analysis, it is planned to digitize some selected areas on 1:5000 maps, with 4m interval contour lines.

Furthermore, it is suggested to compare viewsheds from the highest-altitude pixel with viewsheds from the actual location of the sanctuary. A composed viewshed from the outer corners of the sanctuary would be considered as the representative view of the whole sanctuary area. As mentioned in the introduction, sanctuaries are not always on the actual top, and they could have had a directional preference in visibility. The results are hoped to clarify a difference between the religious influence zone (with possible limited visibility from the sanctuary) and the socio-economic, political or even military control areas (with the optimal visibility from the actual top).

Cumulative viewsheds of east Crete (Lasithi) and archaeological interpretations

The individual viewsheds were cumulated and evaluated by archaeological period, relevant to the peak sanctuary's history and contemporary neighboring sites. The cumulative viewsheds were based on a registered DEM with a resampled pixel size of 10 x 10m, GPS coordinates of the peak sanctuaries, relocated to the nearby highest pixel, and GPS coordinates of the main Minoan site categories.

Figure 5 is an accumulation of the viewsheds of all certain

peak sanctuaries of the Lasithi prefecture.

The first obvious observation is the high density of viewsheds in the far eastern part of the island. The highest visibility from peak sanctuaries can be found in the northeastern peninsula (Cape Sidero) and in the West Sitia mountain range. Cape Sidero might be more visible, because the topography does not obstruct visibility.

The West Sitia mountain range has the highest visibility, because these mountains are higher than the eastern ones. Remarkable though is the absence of any certain peak sanctuary in this range, but candidate sites are present (fig.1). Minoan archaeological sites are rare in this area, and this is due to either lack of archaeological investigation and/or the inaccessibility of the area itself.

The negative evidence is likewise significant: the South coast is almost invisible, and especially from Ierapetra towards the west, the area is extremely poor in visibility. This observation should be taken with precaution, since the peak sanctuary landscape to the west of our pilot area has not been added to the viewshed and once again, the eastern part is much better surveyed. The same observation has to be made for the area north of Agios Nikolaos.

Perhaps most important of all is the high intervisibility of the peak sanctuaries themselves. As already pointed out by several archaeologists (see footnote 4), the Lasithi peak sanctuaries never stand alone. The protopalatial ones have visual contact with up to six other peak sanctuaries, and eight out of twelve can be seen from at least four others. The abandonment of many peak sanctuaries in the neopalatial period (figure 6) reduces considerably their intervisibility and is rather a reflection of diminished centralization of power in this period.

Generally, it seems that the area of Sitia and especially the mountainous areas are well covered by these peak sanctuaries. This needs further verification, as field observation shows that most sanctuaries overview the fertile valleys below them.

Since all of the peak sanctuaries are protopalatial, figure 5 could be identified with the protopalatial peak sanctuary landscape. If we want to include other site categories, it has proven necessary to evaluate the numerical representation of the better-dated sites, in comparison to all Minoan sites.

Of 260 Minoan sites included here, 85 are certainly protopalatial and 75 neopalatial. Since one site can have a history in both periods, approximately one third of all sites is well dated.

Based on figure 7 the production and guard posts cannot be evaluated by period. The actual number of well-dated sites is too low and does not compare proportionally in either period. It seems relevant that of a total of 13 production sites, about half are visible in both periods from peak sanctuaries. In both proto- and neopalatial period there are only two visible production sites left.
The visibility/non visibility of all guard posts is again relatively well balanced. This is not reflected in either protopalatial, or neopalatial guard posts.

The categories settlements, "sacred" sites and of course peak sanctuaries are relatively well proportioned for further analysis of both proto- and neopalatial period, but once again the well dated site does not compare well, proportionally to the total of all Minoan sites, except for the peak sanctuaries.

The high number of visible burial sites in the protopalatial period might seem strange, but might turn out to be archaeologically relevant, since it supports the idea of the peak sanctuary's origin in the EM period in these burial sites. This is even more striking in the evolution charts, where the visibility of burial sites drops dramatically in the neopalatial period (Figure 8). Once again, the absolute number of sites is low enough for drawing valid conclusions.

It seems remarkable that even the most important sites of the protopalatial period, Petras and Agia Fotia are visible from one or two peaks at the most (field observation and confirmed by viewshed). This is reinforced in the neopalatial period, in which the number of peak sanctuaries and consequently their cumulative viewshed drops. On the other hand, the important settlement of Kato Zakros cannot be seen from any known peak sanctuary or candidate site.

Conclusive remarks

The preliminary results indicate that the project still suffers from a variety of biases on both technical and archaeological part. The quality of the DEM and positioning of GPS measurements is essential to the study of peak sanctuaries. Some of the individual viewsheds were found to be in relative contradiction with field observations. Archaeologically, a further refinement of the chronological identification of sites would improve the statistical results considerably.

However, the results have provided us with valuable information. The total intervisibility of the peak sanctuaries has been confirmed, and encourages further viewshed analysis for the peak sanctuaries of the whole island.

Some significant points have been made concerning the visual contact between peak sanctuaries and other categories of sites.

The evolution of visibility of the burial sites from peak sanctuaries further supports the theory of the peak sanctuary's origin from burial sites. The low visibility of settlements, and especially large sites, such as Kato Zakro, is rather unexpected.

If the intervisibility of the peak sanctuaries is a reflection of power, the protopalatial central authorities (religious, socio-economic, or even political) controlled a much larger area than the neopalatial ones.

The abandonment of many peak sanctuaries in the neopalatial period and their reduced cumulative viewsheds could support the hypothesis that religion has been decentralized[26].

End notes

[1] Funded by the Institute of Aegean Prehistory and directed by Prof. Jan Driessen (Université Catholique de Louvain) and Dr. Apostolos Sarris (Institute for Mediterranean Studies-F.O.R.T.H.). The project aims to further investigate the role of peak sanctuaries on the macro scale of the whole Cretan landscape, while other subprojects of "The Topography of Power" project consist in a regional survey (Tim Cunningham), a study of political structures in Neopalatial Crete (Klaas Vansteenhuyse) and an study of Minoan architecture (Jan Driessen), all of the above investigating expressions of power on different scales in the topography of Crete.

[2] Rutkowski 1986, 11; Peatfield 1992, Nowicki 1994, 34-39.

[3] Nowicki 1988, 191.

[4] Faure 1967, 118-128; Idem 1969, 190 (recognises as a possible characteristic); Peatfield 1983, 275-276 (intervisibility is characteristic of peak sanctuary); Idem 1987, 91; Rutkowski 1988, 74; Peatfield 1990, 119-120; Nowicki 1991, *passim*; Idem 1994, 34-42; Peatfield 1994, 24-25; Peatfield 1994b.

[5] Evans 1921, 153-154.

[6] Peatfield 1994b, 14 peak sanctuaries have been excavated, only three extensively published; Petsofas: Rutkowski 1991, *passim*; Iuktas: Karetsou 1974, 1975, 1976, 1977, 1978, 1979, 1980, 1981a, 1981b, 1984, 1985, e.a.; Atsipadhes: Peatfield 1992, Morris & Peatfield 1995, Nowicki 1994, 41-42, Peatfield 1991, 1992b, 1993, 1994c, 1995, 1996.

[7] Nowicki 1988, 190.

[8] Peatfield 1990,127, fn.54.

[9] Nowicki 1994, 40-41: the phases are Nowicki's proposal.

[10] Peatfield 1984, 89-90.

[11] Watrous 1994, 394-396.

[12] Peatfield 1990, 124-127; Peatfield 1994, 23.

[13] Peatfield 1987, 92, n. 22.

[14] Nowicki 1991, 144-145

[15] Nowicki 1994, 41.

[16] Peatfield 1990, 126-130.

[17] Rutkowski 1968, 163-168; Peatfield 1983, 277-278.

[18] Peatfield 1990, 131.

[19] Rutkowski 1991, 52-57; Peatfield 1992, 74-80.

[20] Peatfield 1983, 277.

[21] Faure 1969, 190; Faure 1972, 392; (on remarkable orientations); Henriksson & Blomberg 1996, 99-114.

[22] Jones 1999, *passim*.

[23] This project is available on www1.ims.forth.gr/maps/website/imslasithi2001, and was carried out by the Institute of Mediterranean Studies, under direction of Dr. Apostolos Sarris.

[24] Rutkowski 1988, 78, Pl. X.

[25] Faure 1967, 121; Peatfield 1990, 119.

[26] Knappett & Schoep 2000, 365-371, this article more or less agrees with the idea that in neopalatial period at least some part of power expression diminished and was rather decentralised than centralised.

References

Evans, A., *Palace of Minos*, I, 151-163, London 1921.

Faure, P., "Cultes populaires dans la Crète antique", *BCH* 96 (1972), 389-426.

Faure, P., "Sur trois sortes de sanctuaires crétois", *BCH* 93 (1969), 174-213.

Faure, P., "Nouvelles recherches sur trois sortes de sanctuaires crétois" *BCH* 91 (1967), 114-150.

Jones, D. W., *Peak Sanctuaries and Sacred Caves in Minoan Crete*, Jonsered 1999 (SIMA, Pocketbook 156).

Karetsou, A., *Έ ρ γ* 1985, 83-87.

Karetsou, A., *Έ ρ γ* 1984, 111-115, and *Π Α Ε* 1984, 600-614. (=*Praktika*)

Karetsou, A., "The Peak Sanctuary of Mt. Juktas", in Hägg, R., and N. Marinatos (eds), *Sanctuaries and Cults in the Aegean Bronze Age. Proceedings of the First International Symposium at the Swedish Institute in Athens, 12-13 May 1980,* 1981, 137-153.

Karetsou 1980, 1978, 1977, 1976, 1975, 1974 = Karetsou, A., *Π Α Ε* 1974, 228-239; *Π Α Ε* 1975, 330-342; *Π Α Ε* 1976, 408-418; *Π Α Ε* 1977, 419-420; *Π Α Ε* 1978, 232-258; *Π Α Ε* 1980, 337-363; *Π Α Ε* 1981b, 405-408.

Knappett, C. & Schoep, I. , "Continuity and change in Minoan Palatial power", *Antiquity* 74, Nr. 284 (2000), 365-371.

Morris, C. E. and Peatfield, A. A. D., "Pottery from the Peak Sanctuary of Atsipadhes Korakias, Ay Vasiliou, Rethymnon", *Proceedings of the 7th International Cretological Congress, Rethymnon, August 1990*, 1995, 643-646.

Nowicki, K., "Some Remarks on the Pre- and Protopalatial Peak Sanctuaries in Crete", *AEA* 1, 1994, 31-48. (=*Aegean Archaeology*)

Nowicki, K., "Some Remarks on the Distribution of Peak Sanctuaries in Middle Minoan Nowicki, K., "Investigations in the Cretan Mountains", *ArcheologiaWar* 39 (1988), 189-198.

Peatfield, A.A.D., "Crete Atsipadhes Archaeological report 1996", *AR* 1996, p. 26. (=*Annual Report of the Managing Committee of the British School at Athens*)

Peatfield, A.A.D., "Crete Atsipadhes Archaeological report 1995", *AR* 1994-1995, 1995, p. 25.

Peatfield, A.A.D., "After the 'Big Bang' - What? Or Minoan Symbols and Shrines beyond Palatial Collapse", in S. E. Alcock & R. Osborne (eds), *Placing the Gods, Sanctuaries and Sacred Space in Greece*, Oxford, 1994, 19-36

Peatfield, Alan, The Atsipadhes Korakias Peak Sanctuary Project, Classics Ireland 1994, 1. (http://www.ucd.ie/~classics/94/Peatfield94.html)

Peatfield, A.A.D., "Crete Atsipadhes Archaeological report 1994", *AR* 1993-1994, 1994, pp. 23-35.

Peatfield, A.A.D. 1993. "Crete Atsipadhes Archaeological report 1993", *AR* 1992-1993, pp. 31-32.

Peatfield, A.A.D. 1992. "Rural Ritual in Bronze Age Crete: The Peak Sanctuary at Atsipadhes", *Cambridge Archaeological Journal* 2 (1), 59-87.

Peatfield, A.A.D. 1992. "Crete Atsipadhes Archaeological report 1992b", *AR* 1991-1992, p. 27.

Peatfield, A.A.D., "Crete Atsipadhes Archaeological report 1991", *AR* 1990-1991, p. 26.

Peatfield, A.A.D. 1990. "Minoan Peak Sanctuaries: History and Society", *OpAth* 18, 117-132. (=*Opuscula Atheniensia)*

Peatfield, A., "Palace and Peak: The Political and Religious Relationship between Palaces and Peak Sanctuaries", in Hägg, Robin, and Nanno Marinatos, eds. *The Function of the Minoan Palaces. Proceedings of the Fourth International Symposium at the Swedish Institute in Athens, 10-16 June, 1984*, 1987, 89-93.

Peatfield, A.A.D. 1983. "The Topography of Minoan Peak Sanctuaries", *ABSA* 78 , 273-279. (=*Annual of the British School at Athens)*

Rutkowski, B. 1991. *Petsofas. A Cretan Peak Sanctuary, Studies and Monographs in Mediterranean Archaeology and Civilization,* I. 1.

Rutkowski, B. 1988. "Minoan Peak Sanctuaries: The Topography and Architecture", in Laffineur, R., (ed.), *AEGAEUM* 2, 71-100.

Rutkowski, B. 1986. *The Cult Places of The Aegean*, New Haven & London, 72-98. (1st edition 1972, 152 ff.)

Rutkowski, B. 1968. "The Decline of the Minoan Peak Sanctuaries", *Atti e Memorie del 1¿ Congresso Internazionale di Micenologia. Incunabula Graeca* 25:1-3, 163-168.

Watrous, Vance L., " Some Observations on Minoan Peak Sanctuaries", in Laffineur, R. and W.-D. Niemeier, eds., *AEGAEUM* 12: *Politeia: Society and State in the Aegean Bronze Age. Proceedings of the 5th International Aegean Conference / 5e Rencontre égéenne internationale, University of Heidelberg, 10-13 April 1994, Vol II.*, 393-403.

Tables

Chronological period	Nowicki 1994, phases	Major periods
(FN – EM I ?) EM II	Phase 0	Hypothetical predecessor of PS
EM II - III	Phase 1	Probably first PS (Iuktas)
EM III – MM I	Phase 2	Early **Period 1**, expansion of PS idea
MM IB - II	Phase 3	**Period 1** = Protopalatial, acme for local PS
MM IIIA	Phase 4	Abandonment of local PS serving destroyed MM II settlements, **Period 2** = Early Neopalatial, fewer PS, but richer and impressive architecture
MM IIIB – LM I	Phase 5	End of **Period 2**, cult seems more centralized towards Iuktas
LM II- III	Phase 6	Only Iuktas seems to maintain its religious function, except for shrines in defensible LM IIIC settlements, on similar topographical locations (Karfi)

Table 1. Summary of chronology

Function	Action (arguments)	Meaning
Sanctuary[1]	Offers, Sacrifice, Processions, Initiation Rites (figurines, ashlayers…)	Health for vegetal, animal and human life
Communication Centres[2]	Signaling with fire, smoke, towards settlement or other peak sanctuary (intervisibility, ashlayers)	Control over area and sea, reinforced by the divine function of the place.
Astronomical Observation Centres[3]	Observing of winter and summer solstices, spring and autumn equinoxes, moon and sun (height, topographical relative location)	Organizing of the agricultural activities, shipping season, reinforced by the divine character of the peak sanctuary

[1] Rutkowski 1991, 52-57; Peatfield 1992, 74-80.
[2] Peatfield 1983, 277.
[3] Faure 1969, 190; Faure 1972, 392; (on remarkable orientations); Henriksson & Blomberg 1996, 99-114.

Table 2. Interpretations

Figures

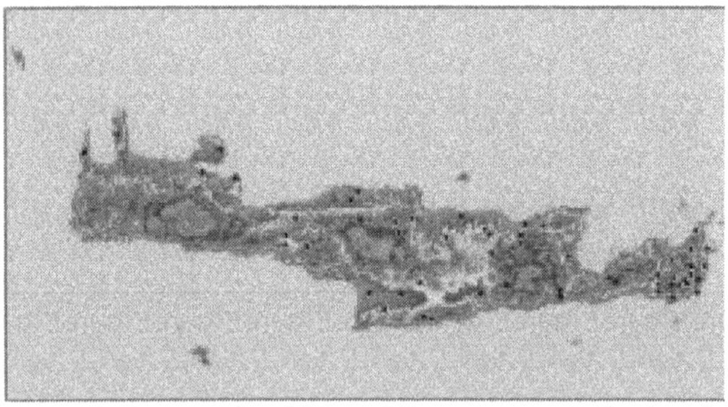

Figure 1. Topographical Map of Crete: certain (yellow) and possible (black) peak sanctuaries

Figure 2. Atzikiari – Kalamaki, protopalatial peak sanctuary

Figure 3. Petsofas, neopalatial peak sanctuary

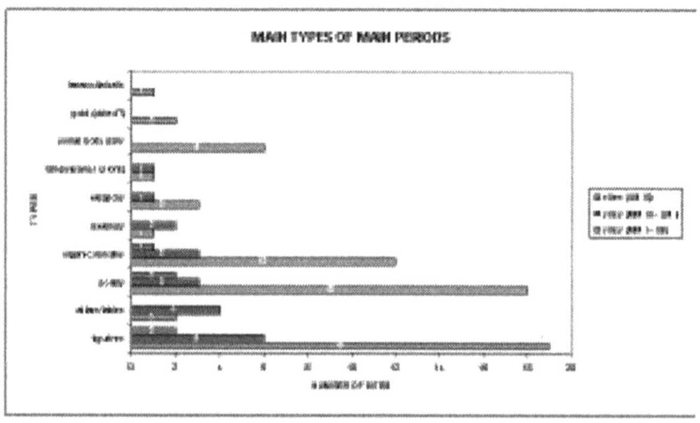

Figure 4. Main types of finds in major periods (based on Jones 1999, passim)

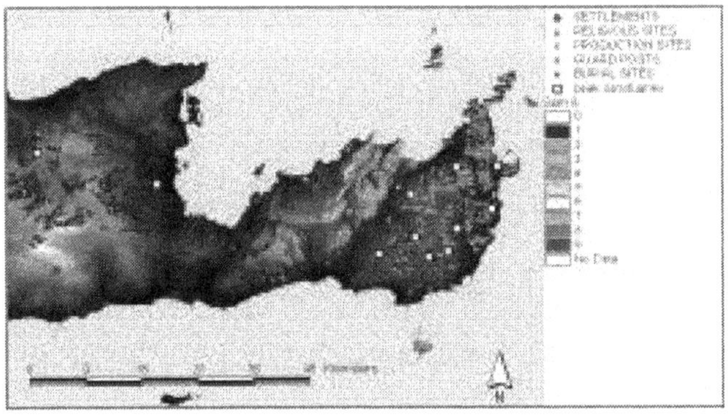

Figure 5. Cumulative viewshed of all (and protopalatial) peak sanctuaries

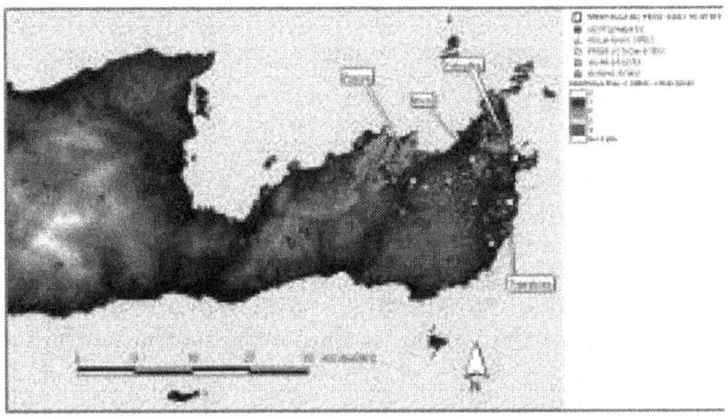

Figure 6. Cumulative viewshed of neopalatial peak sanctuaries and sites

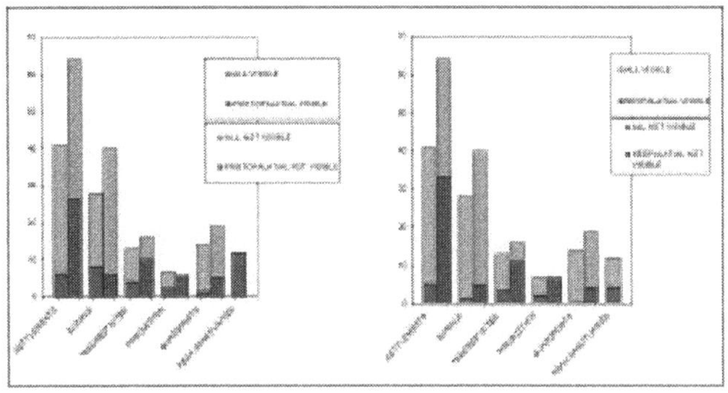

Figure 7. Visibility from peak sanctuaries: Protopalatial (left) and neopalatial (right) site categories both compared the same larger group of Minoan sites.

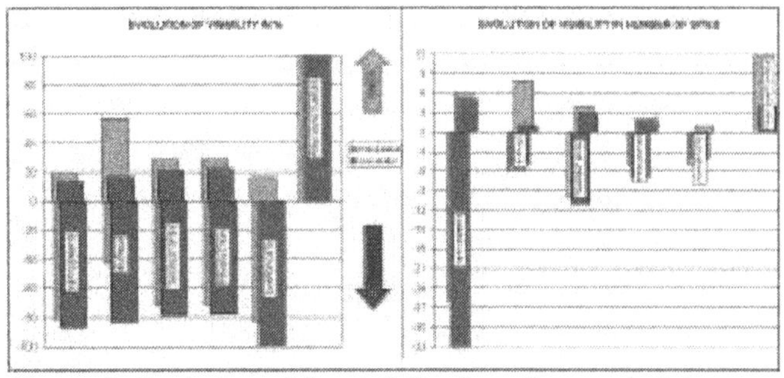

Figure 8. Evolution of the visibility from Protopalatial to Neopalatial site categories

Modeling Archaeological and Historical Cognitive Landscapes in the Greater Yellowstone Region (Wyoming, Montana, and Idaho, USA) Using Geographic Information Systems

Thomas G. Whitley
Brockington and Associates, Inc
6611 Bay Circle, Suite 220
Norcross, Georgia, USA 30071
Phone: +1-770-662-5807 ext. 13 - Fax: +1-770-662-5807 - E-mail: tomwhitley@brockington.org

Abstract: Geographic Information Systems (GIS) have been employed in the field of archaeology for more than a decade. Few archaeological studies, though, have taken advantage of the ability of GIS to provide a greater depth and understanding to issues of a theoretical or cognitive nature. In this sense GIS can be used as a tool for outlining and understanding the ways in which people have thought about and utilized their surroundings; in other words cognitive landscapes. A series of examples are provided from the Greater Yellowstone Region of Wyoming, Montana, and Idaho (USA), which illustrate the ability of an ArcView-based GIS to integrate theoretical and cognitive issues of site selection/placement into a more thorough evaluation of regional settlement. The study also incorporates structures developed in the fields of complexity research and artificial intelligence. Effective cognition-based hypotheses explaining the patterns suggested by previous research and historical documentation are then developed and evaluated.

Key words: GIS, Cognition, Modeling, Complexity, Site Selection

Introduction

Lying at the junction of Idaho, Montana and Wyoming, the Greater Yellowstone Region includes most terrain above 1500 meters in altitude between Pinedale, Wyoming and Bozeman, Montana, and between Driggs, Idaho and Cody, Wyoming. The region includes two national parks (Yellowstone and Grand Teton) as well as portions of nine national forests. Eight major mountain ranges (the Tetons, the Absarokas, the Wind River Range, The Wyoming Range, The Gros Ventre Range, the Gallatins, the Madisons, and the Beartooths) lie within the project area (fig. 1).

Research goals were initially to develop an archaeological site predictive model that could be applied to the landscape using Geographic Information Systems (GIS) technology and evaluate the results in the context of the known sites recorded in the region. The traditional means of creating and applying predictive models, however, relies on largely inductive methods where known site information is assessed on its relationship with available environmental data and a multiple regression (or some other form) of analysis is conducted to identify the predictive value of each environmental variable in an overall formula (or series of formulas).

For the Greater Yellowstone Region, a very limited amount of previous archaeological research has been carried out, and almost all of that has been in low altitude portions of the region that have undergone recent human settlement (see Whitley 2000:188-190). The database of recorded sites kept by the National Park Service does not provide an unbiased assessment of prehistoric sites from the entire region. Thus, a traditional archaeological predictive model would not be possible and more than likely would not be enlightening.

I took a different approach with the project instead. The basis of my research was focused on identifying the nature of site selection behavior. This implies that the locations of known sites were not integral to the creation of the settlement model, but instead a model was created involving the cognitive decision making process of selecting suitable site locations (site selection) and then placing sites in some of those locations (site placement). Fundamentally this involves creating a decision making model, employing some of the ways in which we model human cognition; namely analytical structures from dynamical systems research, fuzzy logic (otherwise known as approximate reasoning), and neural network analogues.

Once a suitable model of decision making was created, it could be applied to the landscape via the use of GIS analysis. The process would involve identifying some of the environmental, ecological, and sociocultural factors which were available to the region's prehistoric and historic inhabitants and are currently observable, or can be projected today. Those factors can then be referenced to the classification scheme used by the decision making model and projected as cost and benefit surfaces - reflecting the value-adding or energy expending nature of each factor across the landscape. A summary of cost and benefit surfaces will then produce an approximation of the cognitively assigned value for each land unit in the region. In other words, this is the GIS application of a cost-benefits analysis based on projections of prehistoric and Historic period cognitive decisions.

Mechanisms

To briefly address the mechanisms employed in building the model; the first consideration is: what sort of system is involved in site selection processes? The process of identifying a suitable site location involves complex highly idiosyncratic decisions which are not always based on economic principles. It cannot be conceived of as simple or static; and thereby reducible to covering law type explanations. Instead, it involves dynamical processes which are very much responsive to minor influences, and the initial conditions within which they exist. The term "dynamical" is a standard term used in Chaos Theory (more appropriately called the study of complexity) as a variant of the term dynamic - the term "dynamic" itself does not convey the highly reactive, self-organizing, and unpredictable nature of complex phenomena in its definition (cf. Gleick 1987; Waldrop 1992; Casti 1994).

Static covering law types of explanation often fail when the subject matter shows self-organizing or dynamical tendencies. This failure comes principally from the focus on inductively modeling the subject matter, through perceived rules (or laws) and attempting to predict the results. When the predictions are not met, the explanation is not forthcoming (for a discussion of the relationships between archaeological phenomena and scientific explanation see Salmon 1982). Dynamical systems research includes several different approaches for modeling extremely complex systems in a manner which does not require prediction to fulfill the goals of explanation. Dynamical systems, instead, are grounded to a much greater degree in statistical relevance and approximate rules or tendencies (e.g. causal reference type explanations - Salmon and Salmon 1979; and Salmon 1984).

An important consideration to bear in mind is that although chaos theory and dynamical systems research often deal with very complex mathematical devices, archaeological or other cultural data is not always easily reduced to mathematical relationships. As such, I do not argue that complexity-based mathematics should (or even could) be applied to all archaeological datasets. Rather, I believe that some general concepts are applicable in very real ways, particularly how we organize archaeological models and conceptualize of prehistoric cognitive processes. These concepts, though, greatly affect how we utilize GIS and other spatial types of information.

A second question about site selection decision making is: how are decisions made? This involves not only the process of decision making but the elements employed in the process. For instance, a typical pattern of evaluating surroundings, applying general knowledge to the observations, and choosing between options can be the model for the process. But it is important to understand the nature of the individuals or groups who make the observations, process the information, and choose between alternatives as well. Research into neural networks has been used to approximate the nature of decision making in artificial intelligence (e.g. Haykin 1994; Russell and Norvig 1994; Bishop 1995).

Neural networks are models which utilize traditional concepts of information flow but outline them in a dynamical manner. Modeled on the way in which our brains process information, they are essentially very complex parallel processors. The neural network of the human brain can process extremely complex tasks (such as perceptual recognition) on the order of milliseconds, whereas it can take days for a digital computer system to do the same.

The process which makes human neural networks so efficient is that the neurons process information in both parallel and hierarchical manners. They also act dynamically by learning from experience. Traditional computer systems can be programmed to carry out a task, but they will always do it in the same fashion and cannot adaptively alter their behavior. Neural networks in the brain build algorithms to carry out tasks and frequently repeated tasks are strengthened by having larger amounts of neurons dedicated to the process.

Neural network models provide an analogy for the process of site selection decision making since the dynamical cognitive rules for site placement have to be evaluated both concurrently and hierarchically. They are analogous to site placement decision making in that both neural networks and human decision making are a means for processing large amounts of information to make generalizations, classify, or evaluate. Human sociocultural systems, in general, are analogous to neural networks as well, since all individuals within a group act as "neurons" processing information and carrying out learned behavior in accordance with the needs of the group. On the individual scale, however, neural networks in our brains are exactly how this is done. In other words, neural networks and human sociocultural systems are both analogous and homologous.

Computerized neural networks have been used to identify patterns within certain types of datasets. One example from this conference is the "training" of computerized neural networks to evaluate osteological measurements to recognize distinctions within datasets of skeletal remains (Bell and Jantz 2001). For GIS applications at least one neural network extension has been developed within the ArcView environment (Looney and Yu 2000). The DataXplore module is designed to work within the Arc-SDM (Spatial Data Modeller) Extension for ArcView 3.2. It provides tools for addressing spatial data using both fuzzy clustering and neural network statistics. At the time my research was conceived and carried out there were no ArcView GIS-based extensions available that could evaluate cost-benefit surfaces using neural networks, but perhaps new tools such as Arc-SDM and DataXplore will prove useful for future research.

Another question about the process of decision making is; how are potential alternatives evaluated? If the rules of site location acceptability are conceived in terms of the surroundings, are environmental factors amenable to rigid evaluation or are approximate measures used instead? Once again, in dynamical systems research a methodology is often used which simulates approximate reasoning in complex systems: fuzzy logic.

Fuzzy logic was originally a conceptual model for mirroring human "grayscale" cognitive behavior which eventually evolved into a formal fuzzy set theory (Zadeh 1965; Yager and Filev

1994). Traditional set theory (referred to as crisp set theory) is the classification of objects, or numbers, into groups in terms of sets, unions, and intersections. Membership in any group is a black-and-white issue and most day-to-day categorizations that we use tend to follow suit. The distinction between sites and non-sites, or sterile and artifact-bearing deposits are examples.

Fuzzy set theory, however, argues that classification is not always so distinct. Many objects (or whatever we are measuring) tend toward being members of multiple groups. Or, categories tend to gradually merge, rather than be distinct. Therefore membership in any class can vary on a scale of intensity. And, in reality the differences between sites and non-sites, or sterile and artifact-bearing strata, are also not so distinct. Likewise our examination of the relationships between sites and environmental variables (such as "distance to water") should be seen as examples where fuzzy relationships are common.

If categorical distinctions are employed and we make the assumption that people were placing their sites according to a cognitive rule regarding those categories (such as "nearness to water"), then it is important to be able to grasp that the prehistoric definition of "near" will most likely be fuzzy. People are not likely to have used tape measures in determining where to pitch their tipi. The theory of approximate reasoning is integral to human cognitive processes.

In the process of site selection decision making, an inherent need to observe the surroundings and evaluate them against the needs of the group or the individual is implied. How is that evaluation carried out? Are there "rules" of acceptability that can be judged on the basis of observable criteria? In the conceptual framework of dynamical systems, attractors are used to represent states or values toward which systems tend to gravitate. In the examination of site placement dynamics it is possible to use a more literal interpretation of the terms "attractor" and "landscape" since we are actually talking about landscapes and attractive places for settlement within them (fig. 2).

Attractors are modeled in what are known as Vector/Manifold diagrams (Casti 1994). In our experience, manifolds are landscapes across which archaeological sites are placed - in more conceptual dynamical research manifolds could represent state-space (not actual physical locality). Vectors are the time sensitive paths established across the manifold (based on the rules of the system). In our experience, they equate to seasonal rounds or routes of travel.

Attractors can be seen as the representation of the observable characteristics of the rules of the system. Some represent benefits extractable from a location (lithic raw materials, utilitarian plants or hunted animals), while others represent costs accumulated while trying to extract benefits (distance from the lithic source or the terrain slope which needs to be crossed to gather plants or hunt animals).

The simplest kind of attractor is fixed at a permanent point (fig. 2a). A high quality lithic source, such as Obsidian Cliff (in Yellowstone National Park), is a good example. Sites are keyed to Obsidian Cliff to such a degree that they can be virtually indistinguishable. During the human occupation of the Yellowstone Region Obsidian Cliff has not moved, and even today it can be seen as an attractor for tourist itineraries.

A second kind of attractor is a periodic attractor (fig. 2b). An example for site placement would be seasonal, migrating resources. In the Yellowstone Region elk and bison herds spend much of the year migrating between lowland and upland areas. Likewise, some periodic attractors may actually be fixed in place, but their attractiveness as potential resources may, in fact, vary from season to season. Plants are such a resource, and Blue Camass the prime example from the Greater Yellowstone Region.

The third and most unusual type of attractor is the strange attractor (fig2c). This is the most indicative of dynamical systems. The nature of the strange attractor is that it is responsive to other systems not under study or minute effects that have major implications for the whole system. A cultural example would be the placement of sites in areas not controlled or inhabited by enemy groups. An attractive place for settlement is dependent upon the location of the enemy at any given time. In this manner it is not possible to predict where attractive places for settlement will occur without knowing all of the movements of the enemy. Archaeologically it may be extremely difficult to model for such an effect.

Bearing all of these ideas in mind, I created a decision making model for site selection that relied on a dynamical systems approach, employed neural network analogues, and classified environmental, ecological, and sociocultural factors into five fuzzy categories; *Geophysical Constraints*; *Resource-Effort Constraints*; *Cultural Biases*; *Time-Place Constraints*; and *Individual Preferences* (fig. 3).

Geophysical constraints can be thought of simply as spatial or physical barriers to site placement. Resource-effort constraints primarily reflect energy expenditure and economics. Cultural bias types of constraints are typically conscious or subconscious "taboos" for or against placement of sites in some areas for largely unspecified or strictly cultural reasons (such as a learned site placement constraint with no explicit rationale). Time-place constraints are indirect preferences for an area merely as a result of its being adjacent-to or in between consciously (or subconsciously) selected areas. Given two distant points on the landscape, for example, the likelihood of a cultural group placing their site in between the two points is greater than it would be elsewhere, if they are traveling from one to the other (such as in a seasonal round). Individual preferences are those factors which cannot be modeled or easily fit into one of the other categories. They are not predictable, often not explainable, and are much more likely to be dominating factors in situations where only a few people, or one individual, are making site placement decisions. For instance, there are countless aesthetic reasons for people to choose certain areas within which to live, or build a house. We know intuitively that they play a large role in our own lives and cannot be subsumed as mere economic concerns.

An interesting characteristic of these categories is that they are not mutually exclusive, nor are they simple black-and-white distinctions. In fact, any single variable can be seen as a member of one, two, or any number of the categories. For example, a bias against choosing to live on a steep slope may partially be a geophysical constraint (it is difficult to live on a steep slope), but may also be related to economic, or resource-effort, constraints (it is difficult to obtain water). Therefore the best way in which to conceptually model the constraint factors is with a five-dimensional matrix. Each category can be seen as one axis in the matrix with relative degrees of fuzzy membership represented along the edge. Ironically, it is not possible to depict a five dimensional fuzzy matrix in a two-dimensional medium. Therefore, fig. 3 is merely an abstraction which only conceptualizes some of the relevant characteristics. The categories are overlapping and relative membership in any class or set of classes is depicted by distance away from other categories.

Constraint factors (or rules about observable environmental variables - i.e. attractors) can be represented by points in this five dimensional categorical state-space matrix. Each constraint factor, however, plays a varying role in the site placement decision making process. Some constraints are variably more important to the cultural groups at different times, and may be highly unstable.

Each constraint factor is also linked to others. Changes in one may significantly alter the importance of similar or related constraints. For instance the bias against putting sites on steep slopes may be moderated somewhat by the preference for hunting mountain sheep (which may require a bias toward putting sites near them - on or near steep slopes). If the hunting of mountain sheep were to cease, then the steep slopes constraint might become more important and shift toward the geophysical constraint category. In fact, the Sheepeater population in Yellowstone (the Native American inhabitants during the Historic period) would have had a very different definition of "steep slopes" (since they spent so much of their time hunting mountain sheep) than perhaps the plains-dwelling Blackfeet or Crows.

Conceptualized this way, the matrix of constraint factors can be seen as a network of shifting, linked cognitive rules that each individual maintained with regard to the process of site placement decision making (in other words a neural network). Individuals who belonged to the same cultural group may have shared very similar cognitive rules but it is unlikely that they ever would be identical. The process by which such cognitive rules were employed to determine the most suitable location for a site would, however, be the same; a flow chart shows the steps involved in the process. Such steps include identifying important attractors, summarizing them and the knowledge related to them, updating stored information, and evaluating their suitability (fig. 4).

There are at least two scales at which the network of constraint factors is consulted; the regional, and the local. Regional constraints may play a larger role in the location of seasonal rounds or settlement strategies, whereas local constraints reflect more immediate necessities such as suitable slope or distance to a water source. Site placement constraint factors are going to be quite different on each of those scales. In some situations the regional goals for the group are not going to be affected by locally important factors. Rather the use of heuristic devices to assess the presence of attractors at either scale may be quite common. Given the nature of each scale, though, it is also most likely that locally important attractors will be assessed far more frequently than regional scale ones. This was the ultimate site selection decision making process which was applied to the Greater Yellowstone Region.

Regional Background

The Greater Yellowstone Region is a series of high altitude plateaus located along the continental divide, surrounded by steep mountain ranges which have always been considered largely wilderness areas. Cutting through the heart of the plateaus are several large river systems; including the headwaters of the Yellowstone, the Snake, the Madison, the Gallatin, and the Wind River.

Some areas within the region lie at lower altitudes (between 900 and 1500 meters) and their adjacent ranges create rainshadow effects. Consequently, the vegetation is drier and more arid in the northern portions of the Greater Yellowstone Region, particularly in Montana. South of Yellowstone National Park, the Teton Range includes some of the most impressive peaks in the Rockies. Rising from the Snake River floodplain at Jackson Hole, the Tetons reach over 4000 meters in altitude and present a near vertical face of approximately 1800 meters.

Feeding the Yellowstone River, America's largest alpine lake, Yellowstone Lake, lies at nearly 2400 meters in elevation. The lake is over 100 meters deep and is ringed by steep mountains, thermal basins and more than 100 feeder streams. The north end of the lake includes a complex of marshes in the Pelican Creek valley which provides ideal moose and waterfowl habitat. Similar habitats lie near the shores of Jackson Lake, at 2000 meters elevation and just east of the Tetons.

Some of the interior upland river valleys are moist alpine meadows, largely with open grazing land and bordered by mixed forests of Douglas Fir, Subalpine Fir, Lodegpole Pine, Whitebark Pine or Englemann Spruce. These meadows are ideal places to find many different kinds of edible plants, including; Blue Camass, wild onion, and Yampa. Since the fires of 1988, some of the forests have been reduced to stands of charred timber. Thick brushy undergrowth and small trees have more recently taken over many of these areas as the forests begin to rebuild themselves.

Within the geological feature known as the Yellowstone Caldera, a wide variety of thermal features can be found. Several Geyser Basins, including Midway, Upper, Lower, and Norris lie along the west side of the Grand Loop Road in Yellowstone National Park. Some of the most impressive thermal deposits are found at the Mammoth Hot Springs area.

Settlement began in the region during the early Holocene with scattered Paleoindian and Early Archaic sites. Mummy Cave is one of the oldest and perhaps most thoroughly researched sites in the region. It consists of a stratified rockshelter located along the north branch of the Shoshone River between Cody, Wyoming and the East Entrance to Yellowstone National Park. Much of what we know regarding early upland resource procurement in the region is derived from the information recovered at Mummy Cave.

A few other well stratified sites are known in the region, and they range in age from the Paleoindian through late Prehistoric periods. A large number of sites were identified along the shores of Jackson Lake, of which a small percentage were well stratified multicomponent occupations. A number of sites are also located along the margins of major creeks and rivers, and they occasionally include dateable features.

Many other types of sites are largely undateable through standard radiometric means including lithic scatters. Radiating outward from Obsidian Cliff, lithic scatters are common for miles. Similarly, in many areas of the region, deposition is virtually non-existent and isolated hearths and rock features can be found on the surface. Being repeatedly burned by forest fires over the millennia, such isolated features cannot be dated other than by diagnostic artifacts, if they are present. Isolated diagnostic artifacts are also commonly encountered. Most of what we know regarding Paleoindian settlement in the region is due to the identification of diagnostic characteristics on isolated finds of projectile points.

In the northern and eastern portions of the Greater Yellowstone Region, tipi rings are commonly encountered. Most tipi rings consist of a circle of 5 to 25 large rocks spread out in a diameter of between 2 to 6 meters. The rocks were once used to hold down the edge of hide tipi covers. When the camp was moved the rocks were left behind. Such sites rarely contain any artifact or feature remains.

Another more recent type of site encountered in the region are expedient shelters known as "war lodges". Typically associated with Crow war or hunting parties, they may have been used by a variety of different cultures and for many different purposes. They have been found throughout much of the Absarokas, and the Gallatins, as well as the nearby Bighorn Mountains. Many of these sites are currently in various states of preservation, with forest fires and tree falls occasionally destroying them. Similar types of sites include wooden corrals and traps often found in the extreme uplands and used for hunting and trapping mountain sheep or possibly deer.

The Euro-American occupation of the Greater Yellowstone Region began in the late 1800s, when trapping and mining became the main economic focus of settlements such as Cooke City, and Bottler's Ranch. Early expeditions in the 1860s and 1870s produced little in the way of archaeological deposits. Later, tourism or army based settlements and campsites produced a large number of historic structural remains and artifact dumps scattered throughout the lower altitudes. The influx of Euro-Americans and especially the first national park tourists at the end of the 1800s produced additional types of sites which are in the process of reverting back to a more natural condition. Old roads and trails are still visible in many portions of the region.

Analysis

Much of the digital GIS data was downloaded through a number of federal, state and locally sponsored web sites; including digital elevation models, polygon data and digital raster themes. Some of these websites included; the United States Geological Survey (National Spatial Data Infrastructure pages), Wyoming, Montana, and Idaho pages (state mapping programs and GIS source links), the National Park Service (for Yellowstone and Grand Teton National Parks), the Bureau of Land Management, the U.S. Forest Service (for the Shoshone, Gallatin, Targhee, Custer, Beaverhead, Caribou, and Bridger-Teton National Forests), the University of Wyoming, the University of Montana (Greater Yellowstone GIS Clearinghouse and the University of Montana Library), the Soil Conservation Service, the Fish and Wildlife Service, the ESRI homepage, NASA's Jet Propulsion Laboratory, and Microsoft's Terraserver page (for satellite imagery).

Digital themes were downloaded, translated, rectified and incorporated into ArcView version 3.1 with the Spatial Analyst extension loaded. Operating a 300 MHz Pentium II computer with 128 megabytes of RAM and a 10 gigabyte hard drive, some data layers were projected through the use of ArcView's built-in utilities (such as the hydrologic modeling or least cost path evaluations) or were written in Avenue script. All data layers were translated into a continuous surface of 90 meter grid pixels. In other words each measurable land unit represents a square 90 meters on a side. Since the time of analysis (1995 through 1997) additional data layers have become available, and many of them are at higher resolution than those used here. In the future some of this more accurate data could provide greater interpretive potential especially if used in the context of more sophisticated modeling tools (such as Arc-SDM, Geospatial Analyst, and DataXplore).

Although for most variables the raw data is a series of numeric values to three decimal points, a standardization routine was devised to incorporate the modeled relationship between energy extraction and site placement potential. This means that most variables were reclassified as spatial representations of projected cost or benefit (at each 90m grid pixel) ranging between 0 to 64 for cost variables and 0 to -64 for benefit variables. The use of negative values for benefits is due to the incorporation of least cost path analyses which automatically assume all variables are costs to be analyzed (i.e. beneficial site placement variables were modeled as negative costs). The upper/lower values of 64 and -64 reflect the incorporation of the originally nominal datasets (such as vertebrate diversity) which were translated into doubled integer classes (i.e. 0,1,2,4,8,16,32,64 or -64,-32,-16,-8,-4,-2,-1,0).

Each of the GIS data layers represent characteristic landscapes, such as typical elk/bison grassland habitat or marshy waterfowl and Moose habitat. The state of Wyoming sponsored a study in the early 1990s which looked at the distribution of 445 vertebrate species The model created by the state was translated into standardized digital data and used as a benefit surface. Cost surfaces were also produced, such as cost distance to water; where relative cost reflects the effort required to reach permanent water sources. Effort expended is moderated by distance to water and percent terrain slope.

Other cost surfaces such as the cost distance to toolstone quality obsidian were projected. Although, numerous obsidian sources have been identified in the region, especially within the Yellowstone Caldera, only a few sources, such as Obsidian Cliff, Crystal Springs, Teton Pass, Conant Creek and Grassy Lake are known to have been used. The toolstone quality sources were mapped and a cost distance evaluation was produced.

Today mountain sheep are found in a restricted environment (largely the extreme uplands, in steep rocky areas) or also in lower altitudes, along inaccessible talus slopes or wilderness areas. In the past, however, mountain sheep may have been more widespread throughout the region, including relatively dry and flat sagebrush plains, mostly with easy access to high elevation uplands. Since we have a known subsistence focus on mountain sheep (especially during the Historic period) I also included a benefit surface of potential mountain sheep habitat. Potential mountain sheep habitat was projected using a combination of elevation, preferred slope, ambient sunlight, projected snow cover, and prevailing winds.

The 15 cost and benefit surfaces (which include others such as projected seasonal availability of utilitarian and edible wild plants; and seasonal distributions of migrating ungulates) were weighted and combined into three total net cost-benefit landscapes; summer, winter, and spring/fall. No significant distinction was available to distinguish between spring and fall. Those seasons were modeled mainly as the transitions between the extremes of summer and winter. Their relative positions in the seasonal round did, however, play a part in the least cost path analyses.

The net cost-benefit landscape for summer (fig. 5a) indicates high economic benefits (darker areas equal higher benefit) for site placement in the Hayden and Lamar Valleys, areas to the north, east and south of Jackson Lake, plus much of the Middle Yellowstone, Upper Shoshone, and Clarks Fork River Valleys. As we transition into fall, high benefit areas become smaller and more dispersed, generally retreating down into the lower lying areas surrounding the Central Yellowstone Plateaus (fig. 5b). Winter shows an even greater downward migration with the highest benefit locations consisting of the Jackson Hole area, the Shoshone and Clarks Fork River Valleys, part of Lamar Valley and a few upland winter refugia (fig. 5c).

Point data was created based on the net cost benefit landscapes, modeling just those locations which for each season represent the extreme high end of the benefit scale. I then used a least cost path analysis to fit the travel vectors from summer to fall to winter to spring. Overlaid on the elevation model this shows a projected transient-seasonal residency pattern where people would use the interior of the Yellowstone Plateaus in a dispersed site pattern during the summer, aggregate sites and move downslope in the fall, and potentially spend the winter in the lower valleys surrounding the plateaus, before splitting up again and heading to the uplands in the spring (fig. 6). This data fits well with what has been archaeologically projected for Paleoindian and Early Archaic settlement patterns; periods for which a prey-based nomadism may have been the primary lifestyle.

To this point the data incorporated strictly ecological variables and assumed the prehistoric residents had good familiarity with the distribution of available resources. By including one additional attractor the picture changes dramatically. During the Historic period the Greater Yellowstone's year-round residents (the Sheepeaters) were bordered on all sides by highly mobile plains tribes or linguistically related mountain groups. Since all of the neighboring groups seasonally used interior Yellowstone resources, they would have competed with the Sheepeaters. The situation would have been similar during the prehistoric when social circumscription may have been active.

Three additional seasonal cost surfaces were used which represent the probability of encountering competing groups in each land unit based on projected locations in the lower adjacent river valleys. This was modeled according to a pattern of transient seasonality for neighboring groups. Allowances were also made for the linguistic relatedness between the Sheepeaters, the Plains Shoshoni and Bannocks in three additional cost surfaces during the Historic period (the assumption being better relations between linguistic relatives).

When the additional cost surfaces are included in the net cost-benefit seasonal landscapes the patterns of projected residency change. The summer net cost-benefit landscape shows a much reduced distribution of available high benefit areas (fig. 7a); principally being limited to the Pelican Valley, parts of the Lamar Valley, the flats south of Yellowstone Lake and some of the land along the north branch of the Shoshone River. The picture during the spring/fall shows a much wider diversity of areas but in small dispersed locations (fig. 7b). The winter landscape shows even more clearly the highly dispersed distribution of high benefit locales in the Absaroka uplands (mountain sheep habitat), the very upper reaches of the Shoshone River, a small part of the Lamar Valley and some of the potential winter refugia (fig. 7c).

A least cost path analysis between the seasonal highest benefit areas (fig. 8) indicates a permanent upland residency pattern where migration is no longer outside the high altitude region. Instead, winter locales become dispersed in either the areas of most concentrated mountain sheep habitat, or some of the upper small river valleys (with nearby access to mountain sheep habitat). Spring sees a downward migration to summer aggregated sites in those areas in which neighboring groups are not likely to be encountered. This permanent residency pattern is dependent on two conditions: first, circumscription by competing neighboring groups; and second the desire and ability to effectively exploit mountain sheep as a primary winter faunal resource.

The archaeological evidence to support the contention of year-round prehistoric residency is scant. There are very few studies which have addressed upland resource utilization, since so much of the extreme uplands is wilderness and so far has not undergone significant impact by modern land use practices. The few studies that have been done, however, support the ideas that:

1) Upland habitats were more intensively exploited than

has been typically suggested by the limited low elevation archaeological reconnaissance.

2) Such high elevation sites were not necessarily small temporary hunting or procurement sites, but may have been returned to on numerous occasions and formed a significant part of the seasonal round.
3) Year-round upland residency patterns may not have been limited to the Historic period Sheepeaters, but could have extended well into the past.
4) Upland patterns are typically associated with a reliance on mountain sheep as a primary dietary component, and specifically communal hunting in the Absarokas.

There is as of yet little evidence to indicate when in the prehistoric past a shift toward year-round upland residency may have occurred; perhaps a result of the pressures of neighboring groups. It could just as well be argued that year-round residency or seasonal exploitation requires a substantial investment in detailed understanding of the regional availability of resources. But what about cultural groups who have only a limited familiarity with a region? We would not expect that sites left behind by transient-occasional populations should accurately reflect available resource distributions, since they may have been unfamiliar with the distribution of local or regional resources. For this area we have a pretty well documented example of a transient-occasional site placement pattern: the Nez Perce.

In 1877, the Nez Perce (a Plateau tribe from western Idaho/ eastern Oregon) migrated through the newly created Yellowstone National Park while being pursued by the US Army. Without going into detail about the causes of the Nez Perce War, it is important to note that the Nez Perce had a passing familiarity with the Bannock Trail (the main east-west artery through the park at the time). A viewshed analysis of all areas immediately visible from, and within easy reach of, the Bannock Trail (in essence a cognitive landscape depiction of the net cost-benefit information available to the Nez Perce at the time of their encampment near Henry's Lake, Idaho) was created. When seen in the absence of knowledge regarding the conflict between the US Army and the Nez Perce, it is not easy to see why a transient-occasional population with limited experience in the region would choose to take an unknown route (fig. 9).

If we include a cost surface which shows the known locations of whites (known or suspected by the Nez Perce) and the risk of encountering them, the familiarity with the Bannock Trail is largely outweighed by the risk of military engagement. The net cost-benefit surface shows a high cost of travel in virtually any direction.

The Nez Perce actively enhanced their risk minimization with two strategies however. First, they engaged in sending out scouting parties with the explicit purpose of both identifying US Army and white civilian locations, and also actively misleading the US Army about the location of the main band of Nez Perce. Second, they increased their available knowledge of the region by kidnaping a number of white tourists. Several were killed but others escaped, with the exception of John Shively. His firsthand knowledge of the Mary Mountain Trail created a cognitive landscape for the Nez Perce which allowed them to make a crucial trajectory decision and minimize the risk of engaging the army along the well known Bannock Trail (fig. 10).

Since we have detailed historical information regarding the conflict we can model how the Nez Perce cognitively assessed their surroundings. But archaeologically, how would we assess the sites which were left behind by the Nez Perce if we didn't know the social context? Similarly, if strange attractors (like warfare) can so fundamentally influence supposedly simple nomadic or semi-nomadic societies, what does it mean for more complex ones? We need to consider dynamical variables in many archaeological situations and GIS is a powerful tool for hypothesizing about decision making processes.

Summary

In conclusion, the research carried out in the Greater Yellowstone Region has so far been limited to the occasional survey of areas likely to be impacted by campers, hikers, or timber harvests. Some of the most extensive work has focused on low altitude areas which have resulted in a biased interpretation of the past. By addressing social as well as ecological factors in a dynamical GIS framework, I have been able to suggest that settlement may have been more complex than was previously understood. Similarly, using GIS technology in conjunction with a theoretical model of decision making mechanisms, we can now begin to address issues of cognitive landscapes in a way which was, until recently, virtually impossible.

My approach to observing spatial patterns in the prehistoric and historic landscapes of the region has focused on providing empirical observations of phenomena that are linked to cognitive decisions. From the perspective of a European style of archaeology it may be seen as quite eco-deterministic, with such a large emphasis on resource procurement and environment. Yet from the perspective of North American archaeology it can be seen as verging on post-processual since it deals with social landscapes, cognition, and the transience of ideas. To span the ocean (so to speak) which separates these perspectives we must begin to employ GIS and related tools in a way which provides depth and meaning to our interpretations of the cognitive past, yet is grounded in a strong scientific philosophy.

References

Bell, S., and Jantz, R. 2001. Neural Network Classification of Skeletal Remains. Paper presented at the CAA 2001 Conference: Archaeological Informatics - Pushing the Envelope, Visby, Sweden.

Bishop, C. 1995. *Neural Networks for Pattern Recognition*. London and New York: Oxford University Press.

Casti, J. L. 1994. *Complexification: Explaining a Paradoxical World Through the Science of Surprise*. New York: Harper Collins.

Gleick, J. 1987. *Chaos: Making a New Science*. New York:

Penguin Books USA.

Haykin, S. 1994. *Neural Networks: A Comprehensive Foundation.* New York: MacMillen.

Looney, C., and Yu, H. 2000. Special Software Development for Neural Network and Fuzzy Clustering Analysis in Geological Information Systems. Computer Science Department, University of Nevada, Reno.

Russell, S., and Norvig, P. 1994. *Artificial Intelligence: A Modern Approach.* New York, Prentice Hall.
Salmon, M., 1982. *Philosophy and Archaeology.* New York: Academic Press.

Salmon, M. and Salmon, W. 1979. Alternative Models of Scientific Explanation. *American Anthropologist* 81:61-74.

Salmon, W. 1984. *Scientific Explanation and the Causal Structure of the World.* Princeton: Princeton University Press.

Waldrop, M. 1992. *Complexity: The Emerging Science at the Edge of Order and Chaos.* New York: Simon and Schuster.

Whitley, T. 2000. *Dynamical Systems Modeling in Archaeology: A GIS Approach to Site Selection Processes in the Greater Yellowstone Region.* Ann Arbor: University Microfilms.

Yager, R. and Filev, D. 1994. *Essentials of Fuzzy Modeling and Control.* New York: Wiley and Sons.

Zadeh, L. 1965. Fuzzy Sets. *Information and Control* 8:338-353.

Figures

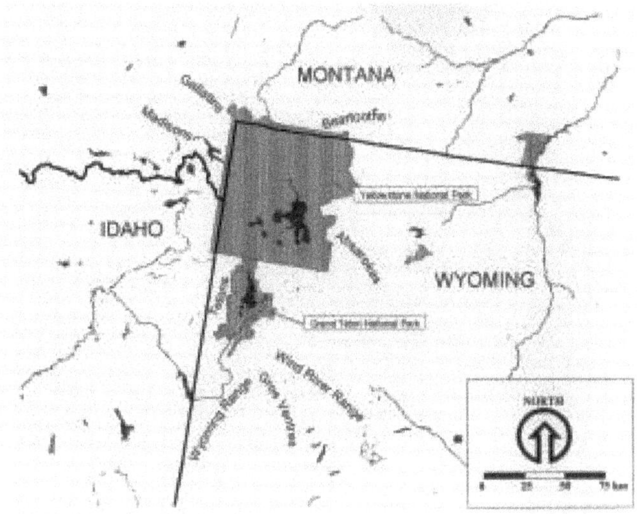

Figure 1. Location of Study Area.

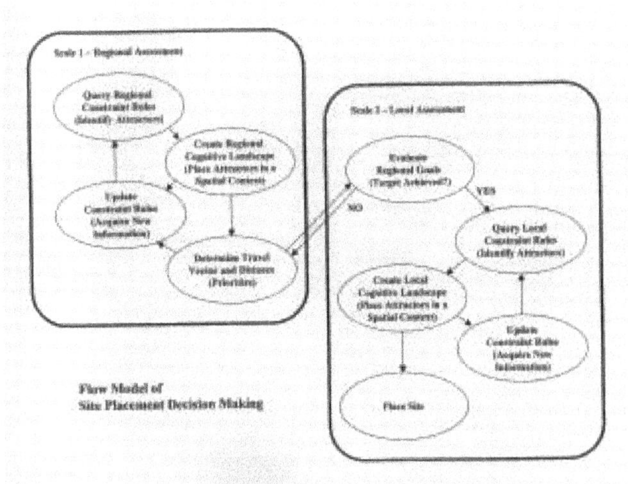

Figure 4. Process of site placement decision making.

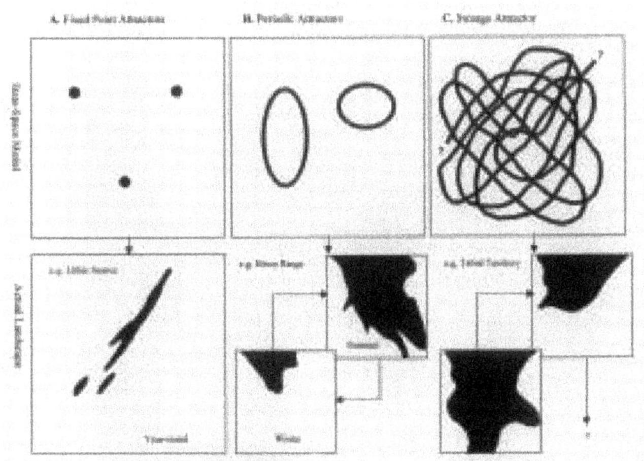

Figure 2. The three types of attractors.

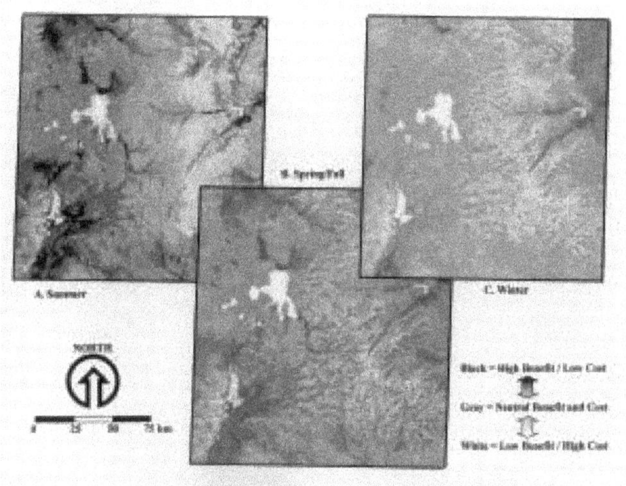

Figure 5. Seasonal cost-benefit surfaces (excluding nearest neighbor costs).

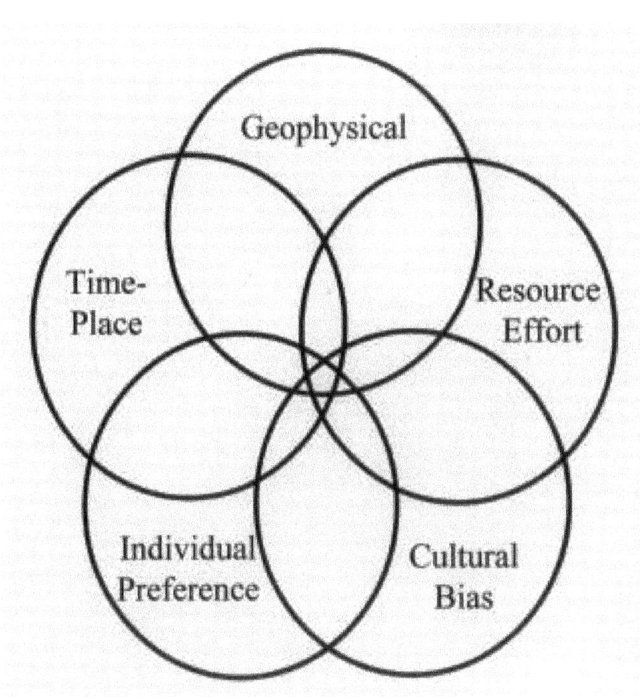

Figure 3. Matrix of site placement constraints.

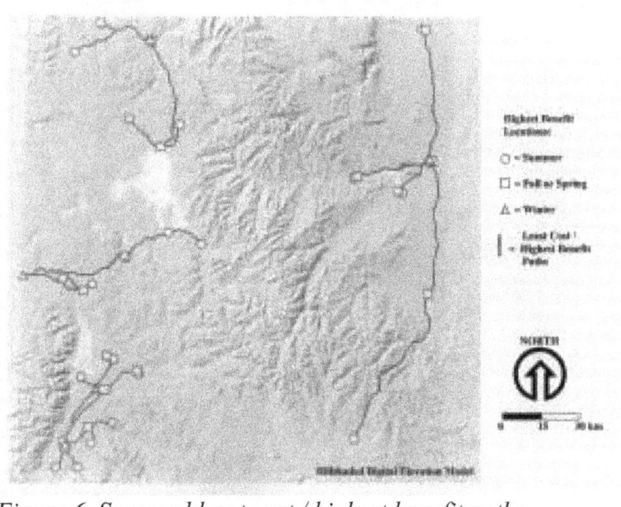

Figure 6. Seasonal least cost / highest benefit paths (excluding nearest neighbor costs).

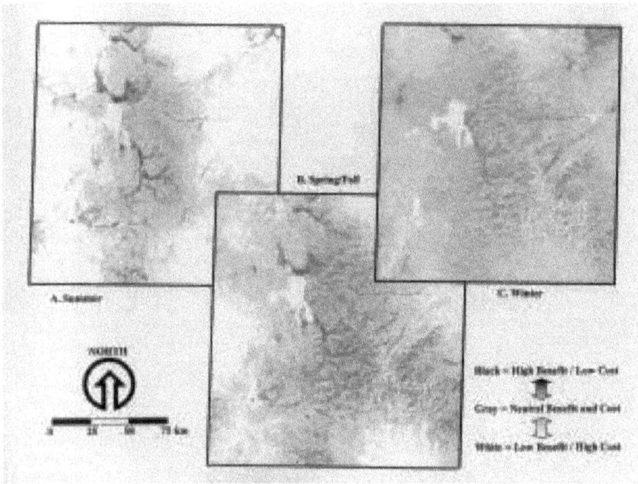

Figure 7. Seasonal cost-benefit surfaces (including nearest neighbor costs).

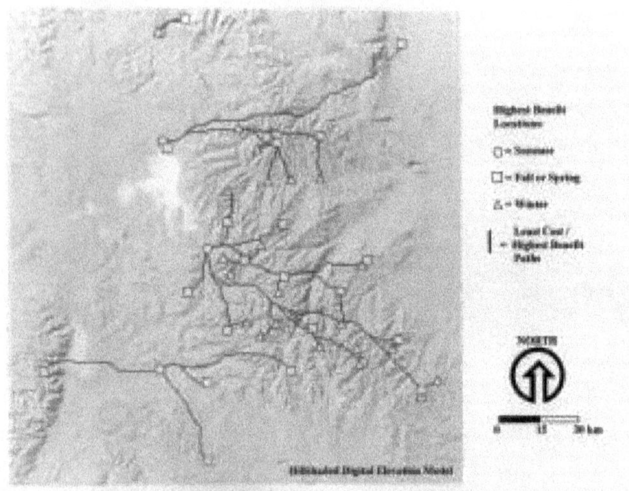

Figure 8. Seasonal least cost / highest benefit paths (including nearest neighbor costs).

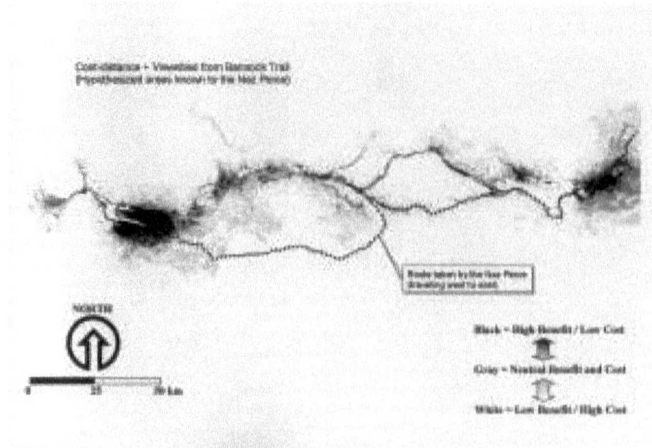

Figure 9. Cost-benefit surface (excluding cognitive variables) and route taken by the Nez Perce in 1877.

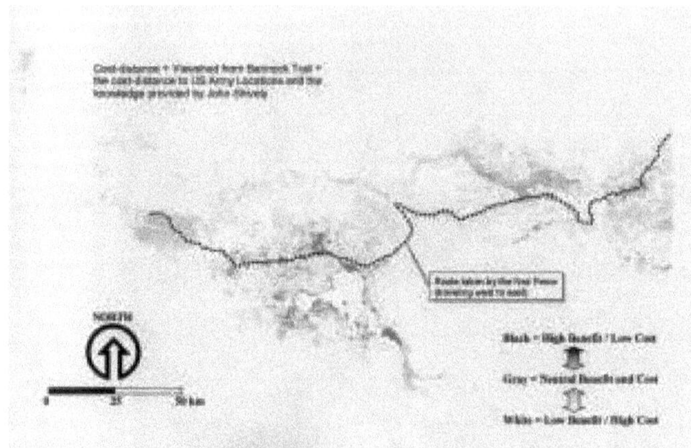

Figure 10. Cost-benefit surface (including cognitive variables) and route taken by the Nez Perce in 1877.

GIS and Space Analysis in the study of the Hospitallers' fortifications in the Dodecanese

Nicholas Zarifis
Archaeological Institute of the Dodecanese
6 Ippoton Street
85100 Rhodes, Greece
E-mail: 4eba@otenet.gr

Despina Brokou
Geodata Applications,
Fokiali Centre 2
85100 Rhodes, Greece
E-mail: geodata@tee.gr

Abstract: The Knights of St. John landed, for the first time, on Rhodes in 1306 AD and, before the end of 1309A.D., they established a strong state in the islands of the Dodecanese.

The use of G.I.S and three dimensional geography has been utilized to support the analysis of the defensive system and the evaluation of its effectiveness.

Displaying data in three dimensional perspective supports the creation of visibility surfaces, which describe the area of surveillance, in the sea or in the land, from each castle.

Analyzing visibility lines between neighbouring fortifications, or between fortifications on different islands, help us rebuild the network of communications, define the centres of administration, locate, with acceptable approximation, the places of completely demolished castles and understand better the defensive system in general.

Key words: Knights of St. John, Dodecanese, fortifications, visibility surfaces, visibility lines, network of communications.

After being forced to leave the castle of Acre, their last stronghold in the Holy Land, the Knights of the Order of St. John tried to find a suitable place from where they could continue their fight against the Islam.

They first went to Cyprus, and after a short stay they landed on Rhodes in 1306. In less than three years, that is before the end of 1309, they had conquered the island and they had established a powerful state in the Dodecanese, a state which would play a very important role for the next 213 years in the East Mediterranean area.

In times of turbulence, with the Ottoman Empire forcing its way towards the "West" and the countries of Europe defending themselves and trying to expand towards the "East", Rhodes, with its famous port, as well as the other islands of the Dodecanese, were perfectly situated to be used as an advanced military and naval base.

Facing a permanent threat from the Ottoman Empire, the Knights of the Order of St. John tried to strengthen the defence system of the islands in the best possible way. Ancient castles were rebuilt or modified. New fortresses were constructed to create a comprehensive defence network. The most advanced, for the time, structural knowledge had been applied in building the walls and towers of the town of Rhodes, the administrative centre of the Knights. Detailed orders were issued to both the Knights and the populace, in case of emergency, to move into assigned castles with their arms, livestock and harvest. Lonely watch towers had been erected along the coast to complete the surveillance and early warning system, detecting any suspicious movement of enemy vessels in the vicinity and forwarding messages to the nearby castles. More than eighty castles, fortresses and watchtowers on the islands, dating from that period, attract the admiration of visitors today. (fig.1)

On the satellite image of the island of Rhodes the various types of fortifications have been marked. Watch towers, small fortresses and large castles, all elements of a sophisticated and advanced defence system, were used to check sea-lanes, to protect the coasts and to control inner passages. (fig.2)

Not happy at all with the Knights established outside their door, Ottoman Turks had been continuously testing the effectiveness and trying to discover the weak points of the defence system with continuous raids and small-scale landings on the coasts of the islands. The knights organized their defence with the support of the local population and managed to withstand major sieges from the land or from the sea until 1522.

A systematic study of the fortresses of the Knights has been taking place in the Archaeological Institute of the Dodecanese during the last years supported by excavations, archaeological and historical research, measured drawings and detailed photographs.

Although originally designed to support decision-taking procedures, Geographic information systems and three-dimensional space analysis have proven valuable in the study of the past.

A variety of military and environmental applications as well as the rapid expansion of cellular telephony have created the need for digital terrain maps. It was only in recent years in Greece that such maps have become available at reasonable cost.

Using digital terrain maps and following a completely inverse approach, than expected, we have not been trying to find the most suitable place for the installation of a new cellular base station or antenna, or for the construction of an observation tower for forest fires. On the contrary we have tried to understand why specific places have been chosen for building castles, how they communicated with nearby fortifications, what part of the land or of the sea was under surveillance, how the defence system operated as a whole and how effective it was.

Common applications of digital terrain model analyzing software include procedures for defining the visibility surface from a specific point as well as for studying the line of sight between two points. The advantage of an analysis based on these procedures is that we are able to place an imaginary observer on the walls of castles, which are in ruinous condition today, and we can look through natural or man created contemporary obstacles in the vicinity, as in the case of the village of Kremasti, near the town of Rhodes, where modern houses have been built high and obscure vision to and from the medieval fortress.

For our study we used the digital terrain map of the islands of the Dodecanese prepared by the Greek Military Geographic Service. Using elevation contours on 20 metres intervals a TIN model has been created. On this map we placed the fortifications of the Knights and we distinguished the different types, which are watchtowers, fortresses, large castles and the administrative centre, the walled town of Rhodes.

The step of elevation contour lines of 20 metres created computing problems of two different types:

a. There have been some cases where the location of the fortress proved to be critical and minor displacement on the digital map could lead to different results in the definition of the visible surface.
b. Visibility calculations for such an extended area were time consuming, reaching the limits of both the software and the hardware at our disposition. It was thus proved necessary to divide the area of the Dodecanese into two sectors, the northern and the southern and to perform calculations on each one respectively. The division was not very easy, as we found out that the surveillance of the coastal zone of some islands was based on a combination of observation points on them as well as on nearby islands.

In all defence systems, ancient or contemporary, early warning of the presence of an enemy and close surveillance of his movements are of vital importance. In an island area, such as the state of the Knights of the Order of St. John in the Dodecanese, the enemy was expected to come from the sea. With observation points on castles and watch towers along the coasts and with visibility conditions fair (10 Km) or good (15 Km) we defined the area of the sea and of the land under surveillance and we realized that there was practically no sea lane that an enemy vessel could follow and no coastal zone that landing could take place without being noticed. (fig.3)

Defence systems depend heavily on communication connecting observation points and advanced units to the administrative centres and central units. Communication at that time was based on optical signs using flags or fires. Analyzing the lines of sight between castles we reconstructed the communication network of the Knights, a network that proved to be not only complete but also providing alternative ways for message transmission in case of destruction of a defence unit by the enemy. (figs 4-5)

The analysis of the communication network provides helpful information on how the Knights have organized their defence and expands our historical knowledge. Points with multiple connections were probably administrative centres or important military units. A dense communication network in an area indicates its military importance, as in the case of Lindos (fig.6) and of the medieval Town of Rhodes. (fig.7) At first we could not understand the presence of strong castles and numerous watchtowers in the straits between Rhodes and the islands of Chalki and Alimia. This observation seems not to be unrelated to the fact that the Ottoman fleet, before their major landing on the island of Rhodes in 1522, attacked first and destroyed the castles of Chalki and Alimia. We also know that during the Second World War the natural port of the island of Alimia was used as a major naval base for submarines. Remains of the iron net closing the entrance of the port can still be seen today. Combining these pieces of information we could possibly suggest that the Knights have also used the port of Alimia as a base for a squadron of their fleet. The dense communication network could thus be explained by the necessity to detect, as soon as possible, any enemy vessel approaching the naval base. It is thus understood why the Ottomans, not wanting to have the squadron of the navy of the Knights at their back during the landing, decided to attack its base in advance. (fig.8)

The importance of communication between nearby castles can also help us to define with better accuracy the position of fortresses that don't exist any more. For example we know that in Paradisi area, near the airport of the island of Rhodes, on a hill called "Koufa" there was a fortress of the Knights not existing any more. For years it has been thought that this castle was built somewhere on the top of the hill. The natural shape of this hill, having two peaks, makes it impossible for the castle to have visual contact with both the nearby castles of Fanes and Kremasti. Analyzing visibility we could suggest that the castle was built lower on the hill, at the intersection of the visibility surfaces of the adjacent castles. This location proved to be a more favourable place as from there the Knights could more easily protect the coast and control the land passage leading towards the Town of Rhodes. (fig.9)

Our enthusiasm on the results of the applications of three dimensional space analysis and of Geographic Information Systems, in the study of the fortifications of the Knights lead us to another research. We considered as possible areas for enemy landing the sandy shores with low inclination of the soil in the coastal zone. Then we described as easily defensible area a zone of 5Km around each castle or fortress. Combining these coverages we located the weakest points in the protection of the coastal zone. This approach from the point of view of the Geographic Information Systems appears interesting, but it was misleading, as it was wrong from the historical point of view. At this time the navy did not have the means to take into consideration the combination of sandy beaches and low soil inclination for choosing a landing position. Their interest was to land as close as possible to the capital, the town of Rhodes, to be able to move their army and heavy weapons easily.

The fortresses of the Knights in the Dodecanese attract the interest of the visitor as important structures of military architecture. They also attract the interest and the admiration of the researcher who approaches them using modern tools of space analysis, as belonging to a comprehensive defence network, covering completely the area to be protected and having an effective and sophisticated communications network.

References

Brockman, E. 1969. *The two sieges of Rhodes 1480-1522*, London

Luttrell, A. 1992. *The Hospitallers of Rhodes and their Mediterranean World*, London 1992

Ormsby, T., Jonell, A. 1999. *Extending ArcView GIS*, Redlands

Pavlides, V. 1999. *Rhodes 1306-1522: A Story*, Rhodes

Poutier, J.C. 1989. *Rhodes et ses Chevaliers*, Liban

Sire, H.J.A. 1994. *The Knights of Malta*, New Haven and London

Figures

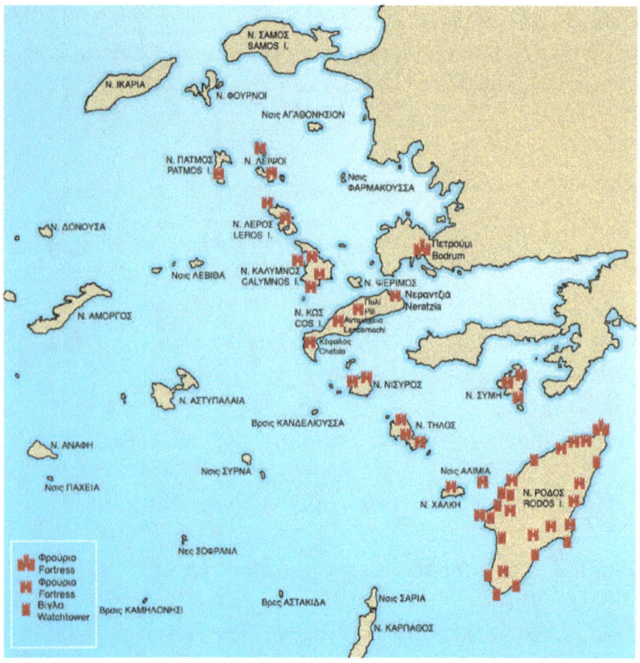

Figure 1.

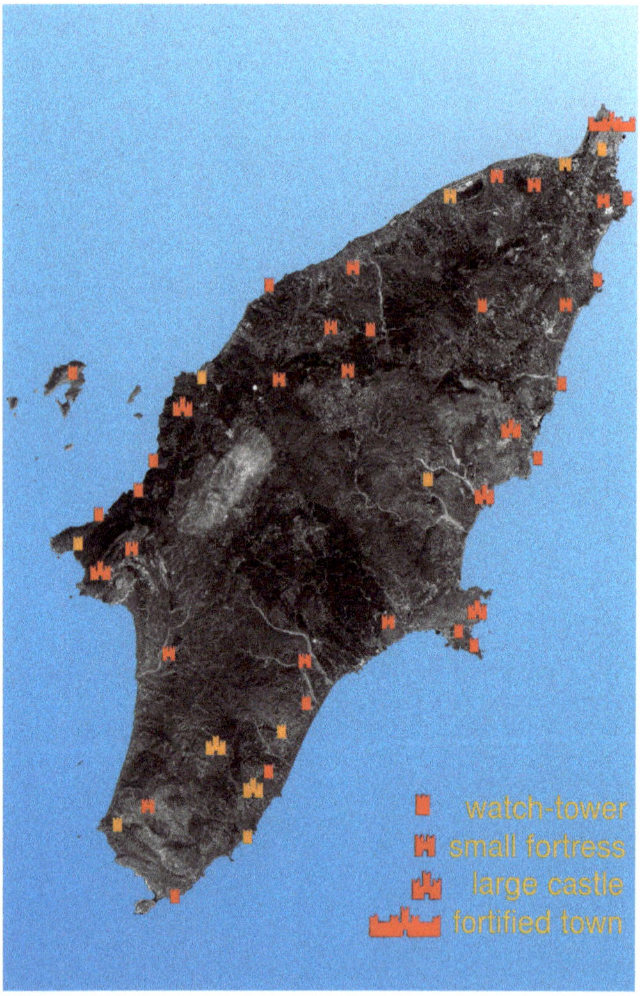

Figure 2.

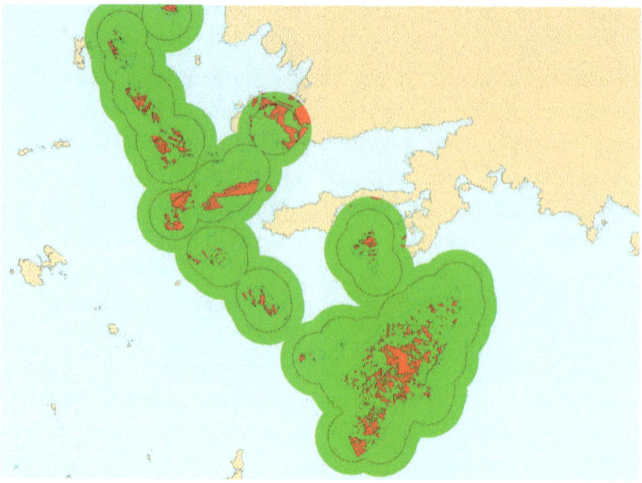

Figure 3.

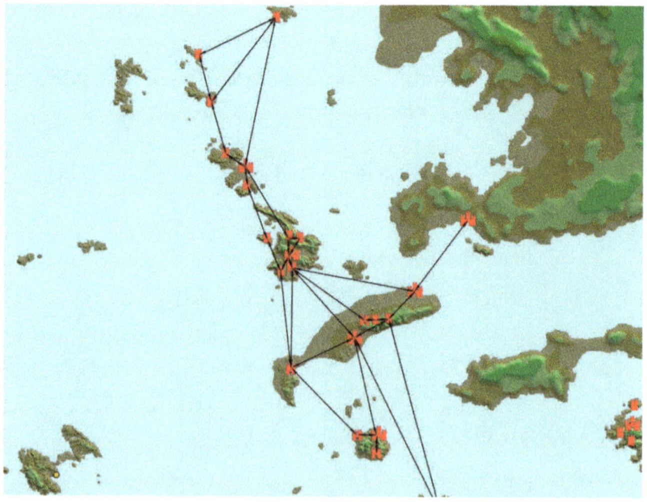

Figure 4.

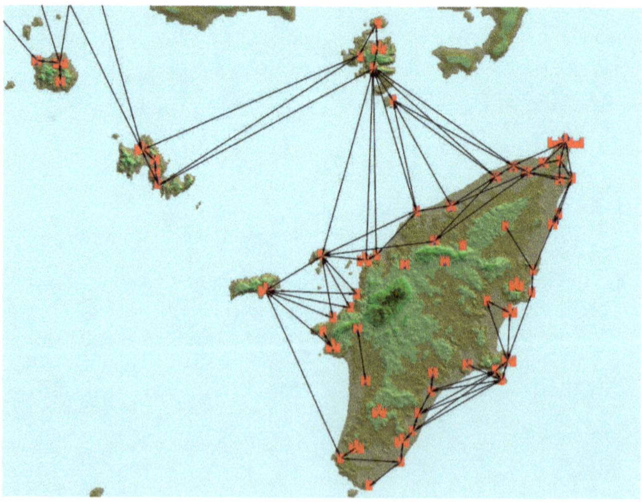

Figure 5.

Figure 6.

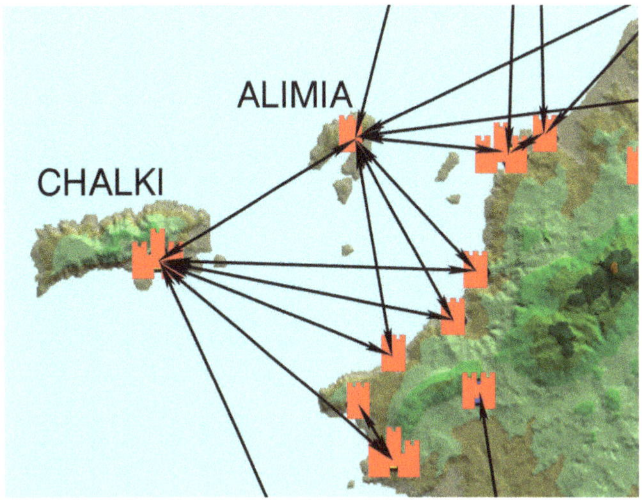

Figure 8.

Figure 7.

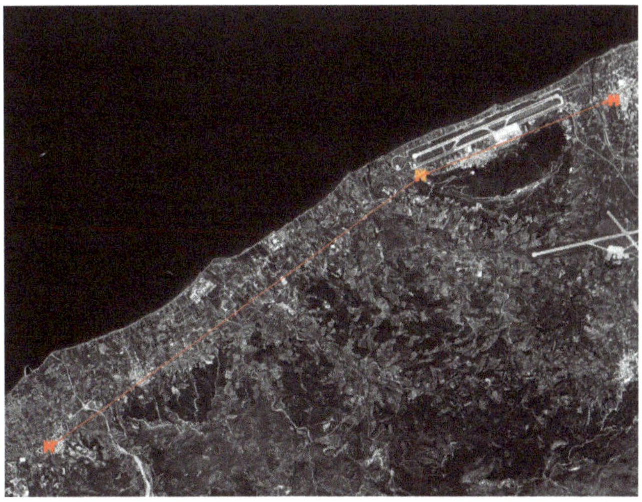

Figure 9.

Papers in the Virtual Archaeology category

Reconstructions of the Excavations of Two Iron Age Chariot Burials from Belgium

Applying Virtual Reality to Old Excavation Data

Geoff Avern
Bramble Cottage
Winkfield Row
Bracknell
Berkshire, RG42 6NE, UK
Phone: +44 01344 885535 - E-mail: oxynoe@madasafish.com

Abstract: While Virtual Reality modelling in archaeology is commonly used for creating, visualising and interpreting hypothetical reconstructions of buildings or complexes, the author has applied the technique to the excavation records of two Iron Age chariot burials, or tombes-à-char, from Belgium.

A number of Iron Age chariot burials have been excavated near the town of Neufchâteau in the Ardennes region of southern Belgium but, unlike the hundreds of excavated chariot burials from the Marne region of France, the two considered below contained many small pieces of mineralised wood which may provide clues to the form of the wooden superstructure of the vehicle.

The aims of the following reconstructions are, firstly, to demonstrate the potential of applying current 3D modelling techniques to old excavation records and, secondly, to exploit the enhanced visualisation capabilities to search for information which may shed new light on the construction of La Tène chariots.

Key words: Virtual Reality, VR, chariot burial, tombe-à-char, modelling, excavations.

Introduction

The author's doctoral research (Avern, 2001 - unpubl. thesis) concentrates on the creation of better excavation records by use of high resolution digital recording methods. One of the numerous benefits of having a 3D model as a record of an excavation is the comparative ease by which even complex 3D arrays of features and finds may be visually interpreted. This point was demonstrated over a decade ago in the pioneering work by IBM UK Scientific Centre and Southampton University on the Hamwic pit project (Reilly,1989) where a model of a Saxon rubbish pit and its contents allowed researchers to peruse and manipulate the visual data, helping them to draw conclusions on the formation of the deposits.

This paper deals with 3D modelling, but not from a current or recent excavation. Instead, we aim to demonstrate its applicability to old excavations by simple 3D modelling based on old excavation records. Our second aim is to exploit the advantages of visualisation in 3D to search for answers to a specific archaeological question. We used 3DStudioMax to create 3D models of two Iron Age chariot burials from southern Belgium which were excavated in 1967 (Bonenfant, 1967) and 1969 (unpublished) by Professeur P.-P. Bonenfant (Service des Fouilles, Université Libre de Bruxelles, Belgium).

Iron Age Belgium

The Ardennes, in south-eastern Belgium, lies between the two major La Tène centres of western Europe, *viz.* the Marne (Champagne), France, and the Hunsrück-Eifel, Germany. Grave goods excavated from cemeteries around Neufchâteau are typically Marnian, while those from burials near Houffalize display stylistic affinities with the Hunsrück-Eifel (as described by Cahen-Delhaye, 1974a). Thus the Belgian Ardennes seems to be a frontier zone in the early La Tène (fig. 1).

At least nine chariot burials from the Belgian Ardennes have been discovered and excavated, all in the vicinity of Neufchâteau, that is, from the Marnian sector, and all display typically Marnian characteristics.

Of the more than 130 La Tène chariot burials excavated in the Marne proper, the author knows of none which have yielded any wood finds. Attempts at physical reconstructions of La Tène chariots, such as Furger-Gunti (1991), have been based largely on metal finds which, while helpful for interpreting wheels, axles, articulations and even suggestions of suspension, shed very little light on the nature of the superstructure of the chariots.

The chariot burials from southern Belgium, however, have frequently contained fragments of wood, either carbonised (Cahen-Delahye, 1974b) or, in the case of the two burials considered here, mineralised by the oxidation products of the iron to which they are attached. Bonenfant (pers. comm.) recorded the 3D coordinates of all finds of his 1967 and 1969 excavations with the idea that a 3D reconstruction might be possible in the future, enabling a visual interpretation of the spatial relationships between the small finds, which might shed light on the wooden structure, particularly the superstructure, of Marnian La Tène chariots.

This presentation, then, shows the author's 3D reconstructions made from Bonenfant's plans, profiles and finds registers.

The Method of Reconstruction

It should be noted that the two models created in this exercise differ from the more typical Virtual Reality reconstructions in that "realism", in terms of complex morphologies and naturalistic colours and textures, has not been pursued. Consequently, the modelling process was relatively simple and fast, making it, in the author's opinion, a valuable tool for visualisation and interpretation of the original excavations. The procedure is given below in a series of steps.

1. The plans were scanned and the resultant files were "cleaned-up" using various image processing routines, annotated (some colour-coding, scale bars, etc) and saved as TIFF files (fig.2). (In the case of the 1969 excavation, the 5 plans from different levels of the excavation were also registered and merged into a single image.)
2. In 3DStudioMax a box was created whose top surface had the same aspect ratio as the TIFF image of the scanned plan. The scanned plan is then mapped onto the top surface of the box (fig.3). In the "Top" viewport (there are various "viewports" in this software which allow you to display different views of the model simultaneously) the box simply looks like the scanned plan. Simple objects were created from the software's range of primitives; cylinders (squat, like hockey pucks) for small finds, boxes (elongate and thin) for the iron wheel hoops, pyramids (long and flat) for spearheads, etc and then were colour-coded (purple for wheel hoops, blue for other iron, ochre for ceramics, green for wood, yellow for bronze). In the Top viewport each object, as it is created, is placed over the relevant small find on the plan, i.e. in its correct X/Y position (fig.4).
3. With multiple viewports open on screen, objects can be tilted/rotated to their correct orientation, or bent to their correct shape (e.g. the fragments of the wheel hoops). More complex shapes are created, such as pots (by creating a profile and "lathing" it at the appropriate radius) and the fragmented ring from a horse bridle (creating a torus and cutting it into pieces).
4. With all objects in their correct X/Y positions, the Front viewport is chosen, and a box is ticked to restrict subsequent movement of objects to the Z-axis. Each object is, in turn, selected and raised to its correct elevation (fig.5).
5. The top surface of the box is "sculpted" to the shape of the burial pit. When the box was constructed, for simplicity we nominated that it should have only one segment in X, Y and Z, giving it only 4 vertices per face. At this point in the modelling, we modify the box by increasing the number of segments in the X and Y directions to 30, i.e. the top and bottom surfaces of the box are defined by 31 x 31 (= 961) vertices. After changing to wireframe model and restricting movement to the X/Y plane, many of the vertices are repositioned along the boundary and contours of the grave cut (fig.6a, b). In the Top viewport all vertices along one contour are selected and in the Front viewport, after confining movement to the Z-direction, they are moved, *en masse*, to their appropriate elevation (fig.6c). The process is repeated for all contours.
7. Changing back to the rendered model, the mapping of the plan onto the top surface of the box is disabled and a colour or texture chosen to replace it. Lights are created and positioned to illuminate the model.
8. While the software permits us to navigate around the model, we may also choose to create an animation for presentation as an .avi file.

Results

The accompanying images are taken from the .avi files made for presentation at CAA2001. The animation sequences 1967.avi and 1969.avi can be found on the accompanying CD-ROM.

1967 excavation

Figure 7 shows the model in plan view. Figure 8 shows the model in perspective view.

1969 excavation

Figures 9, 10 & 11 show the model in plan, front elevation and perspective views.

Interpretations

On viewing elevations from the rear of the chariot burials (with surrounding soil base made semi-transparent) (figs 12 & 13) it is, unfortunately, immediately apparent that we will not be able to deduce anything of the superstructure of the chariots. For the 1969 excavation (fig. 13), we are able to judge the height of the chariot axle from the remains of the left wheel hoop (in purple). It is then quite apparent that the finds of mineralised wood (in green) are either from this level or immediately above it, probably corresponding to the axle and platform of the chariot. Only one piece of wood, possibly two, is found at a higher level which might be attributed to the superstructure.

Similarly, for the 1967 excavation, all wood finds are restricted to levels which are lower than where we would expect to find *in situ* superstructure.

We can, however, make some other observations. In figure 13 the remains of both iron wheel hoops are slightly skewed to the left (except for the most elevated piece of the left wheel). The right example of the paired iron rods (in front of ceramic vessel) is angled instead of lying flat (as is its opposite), and the left bias of the scatter of small iron and ceramic pieces all suggest that the chariot is not in its original position but has been "pushed" to the left.

Close examination of the models reveal that a number of small finds form "alignments". In the 1967 excavation model (fig.14) we can see 3 fragments of wood in a line and lying obliquely on the curved edge of that part of the grave cut which accepts the right wheel.

In the 1969 excavation model (fig.15) we can see a similar alignment of 3 wood fragments and an adjacent cluster of a number of iron remains which appear to form either a couple of alignments in close proximity or which perhaps belonged to a larger discrete mass.

Conclusions

By foregoing realism, the models shown here were simple and relatively quick to create in 3DStudioMax.

Both the 3D models and the ability to easily manipulate their orientation permit excellent visualisation of the spatial relationships between the finds, themselves, and the grave cut. They quickly and clearly showed that we would not be able to answer our specific questions on the nature of the superstructure of La Téne chariots.

Alignments and clusters of finds which did not follow the normal drawing axes may well have been apparent in the original excavation but are difficult to perceive from subsequent plans and elevations. A number of such alignments were easily identified from our models.

The use of 3D modelling on these old excavation data have allowed us to "revisit" the excavation and view it as the excavator was originally able to and further, in ways and orientations that were impossible at the time of the excavation. The benefits of such visualisation are strikingly evident when a final model is viewed side-by-side with the original plans from which it was constructed. Given the relative simplicity and speed of the modelling we recommend this technique for reviewing and reinterpreting old excavation data whenever elevation or height data accompanies the plan drawings.

References

Avern, G.J. 2001. *High-resolution computer imaging in 2D and 3D for recording and interpreting archaeological excavations.* Unpublished doctoral thesis, Université Libre de Bruxelles, Belgium.

Bonenfant, P. 1967. Langlier (Massul): Tombe à char de La Tène I. *Archéologie. 2*: 74-76.

Cahen-Delhaye, A. 1974a. Fouilles dans les Tombelles de La Tène en Ardenne (1970-1972). *Archaeologia Belgica.* No.166: 7-23.

Cahen-Delhaye, A. 1974b. Necropole de La Tène I à Hamipre, Offaing. I. Trois Tombes à Char. *Archaeologia Belgica.* No.162: 5-48.

Cahen-Delhaye, A. 1978. Tombelles de l'Age du Fer en Ardenne. *Archaeologicum Belgii Speculum. VIII.* Bruxelles. p.18, fig.10.

Cahen-Delhaye, A. 1984. Fouilles récentes dans les fortifications de l'Age du Fer en Belgique. (in) *Les Celtes en Belgique et dans le Nord de la France: Les fortifications de l'Age du Fer.* Eds A. Cahen-Delhaye, A. Duval, G. Leman-Delerive, P. Leman. Actes du sixième colloque tenu à Bavay et Mons. Association Français d'Etude de l'Age du Fer. p.152, fig;1.

Furger-Gunti, A. 1991. The Celtic War Chariot: the Experimental Reconstruction in the Schweizerisches Landesmuseum. (in) *The Celts.* Ed. Bompiani. Milan. pp.356-359.

Reilly, P. 1989. Data visualisation in archaeology. *IBM Systems Journal, 28* (4): 569-79.

Figures

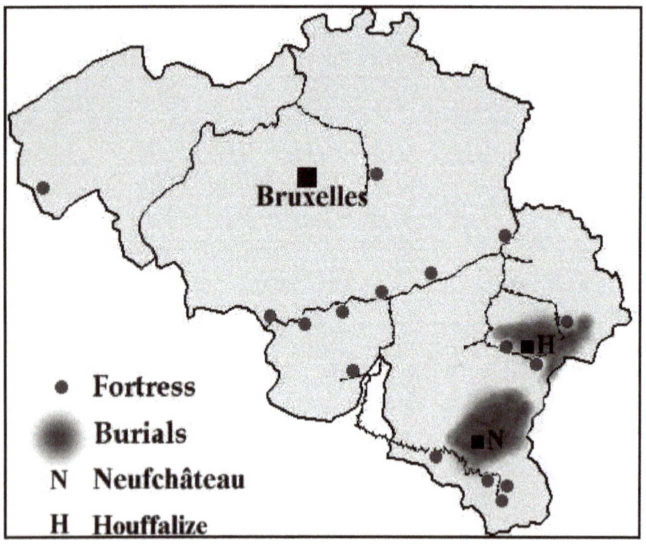

Figure 1. Belgium in the Iron Age (modified after Cahen-Delhaye, 1978 and 1984).

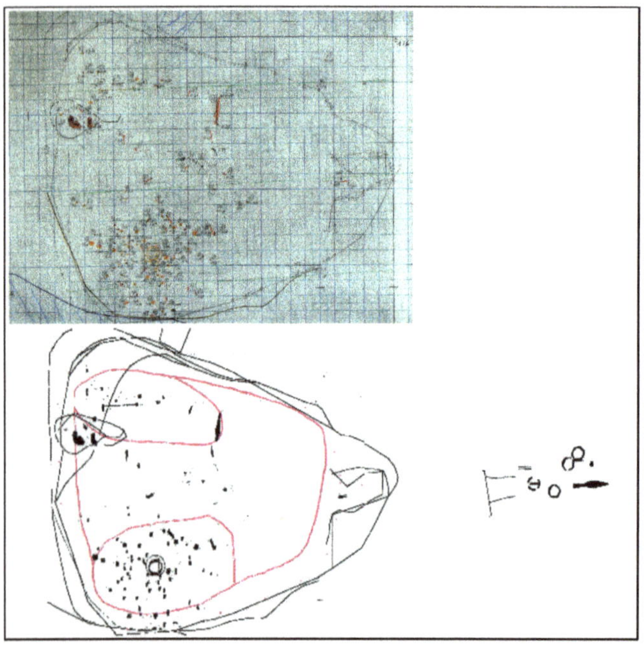

Figure 2. (above) part of the original plan of the 1967 excavation and (below) the plan after image processing, merging of multiple plans and saving as a TIFF file.

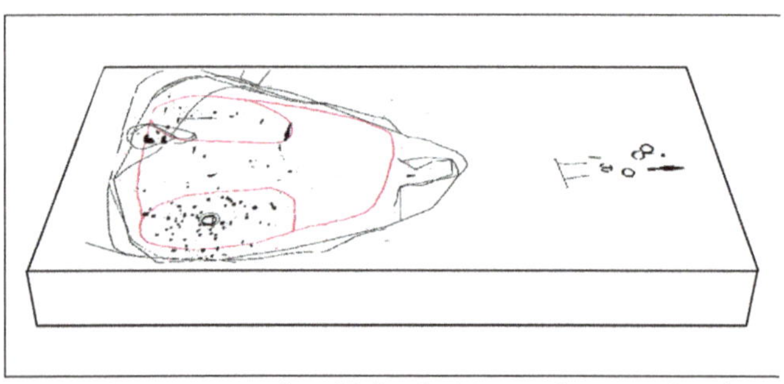

Figure 3. Box created in 3DStudioMax and textured with TIFF file of the 1967 excavation plan.

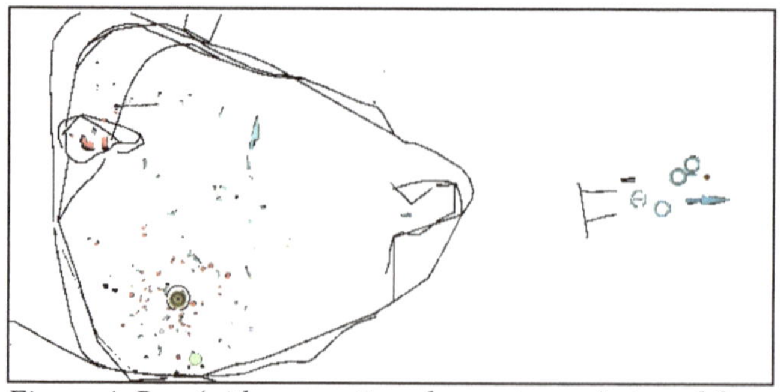

Figure 4. Box (with excavation plan as texture) viewed from above with modelled finds in correct X/Y positions.

Figure 5. Front view of box showing finds lifted to their correct elevation.

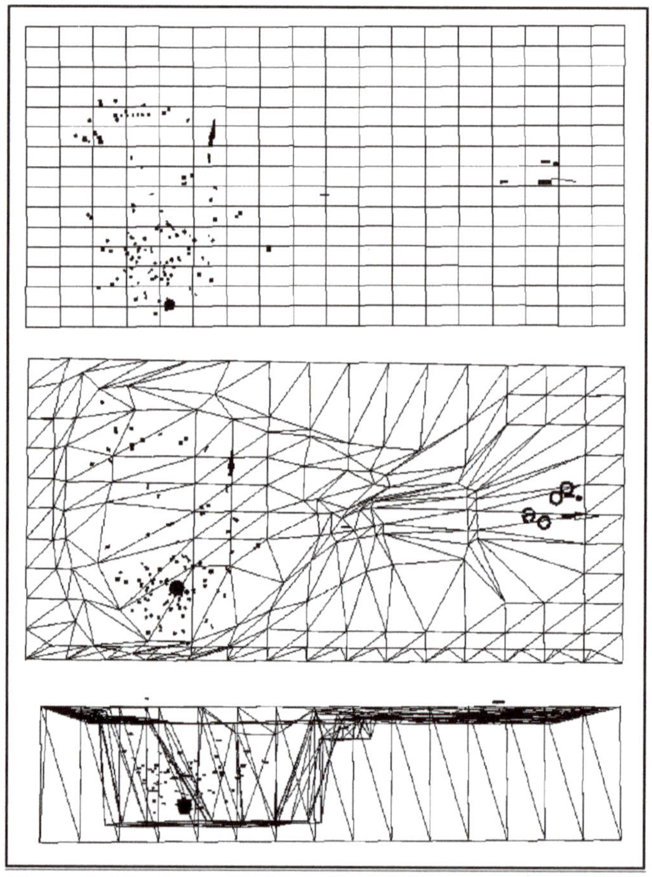

Figure 6. Wireframe models - (above) Top view, vertices in normal position; (centre) top view, vertices repositioned to follow contour lines on excavation plan; (below) front view, vertices lowered to correct elevation, creating the grave cut. (Note: these figures are from an early attempt using fewer segments than in the final models)

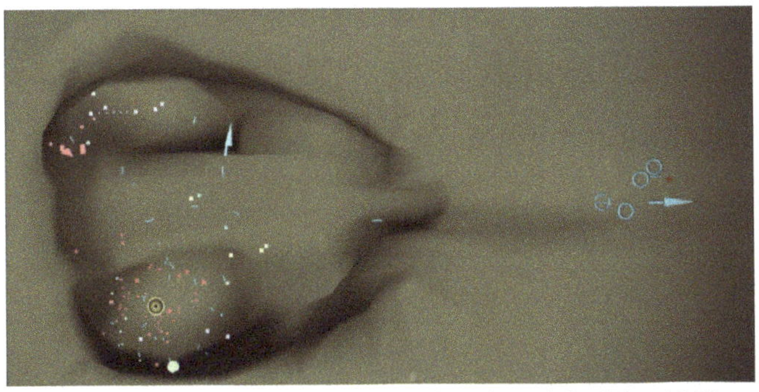

Figure 7. Top view of the final rendered model of the 1967 excavation.

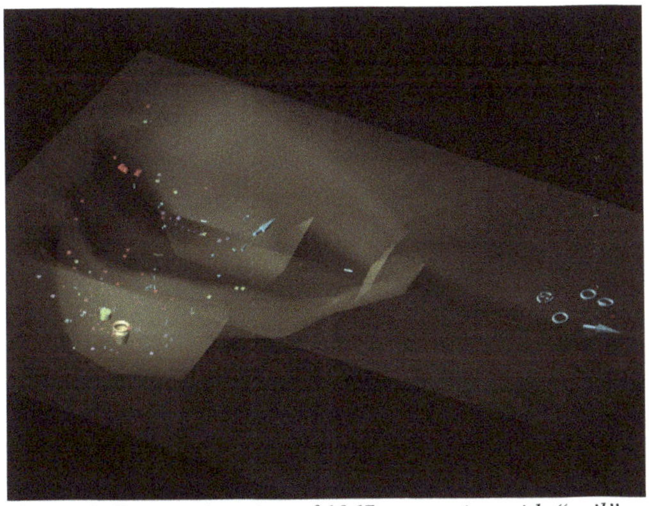

Figure 8. Perspective view of 1967 excavation with "soil" base made 30% opaque.

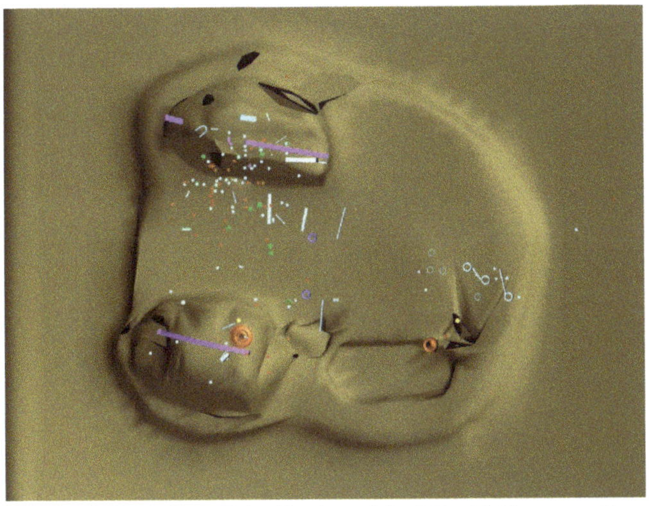

Figure 9. Top view of the final model of the 1969 excavation.

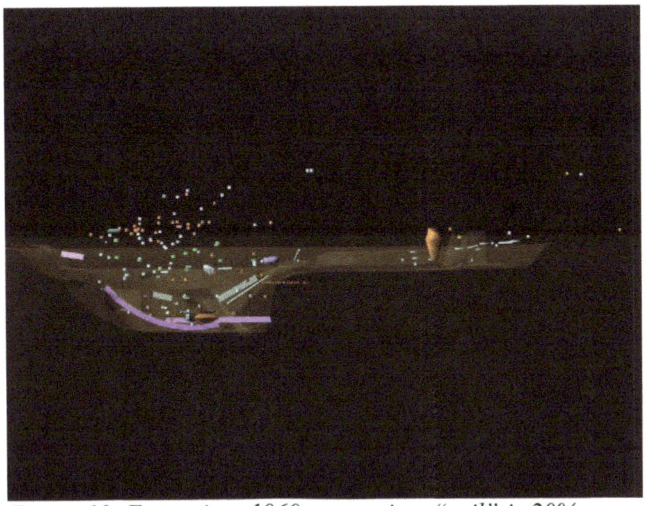

Figure 10. Front view, 1969 excavation, "soil" is 20% opaque.

Figure 11. Perspective view (elevated, rear, right), 1969 excavation, "soil" 20% opaque.

Figure 12. Left view (= rear elevation), 1967 excavation.

Figure 13. Left view (= rear elevation), 1969 excavation.

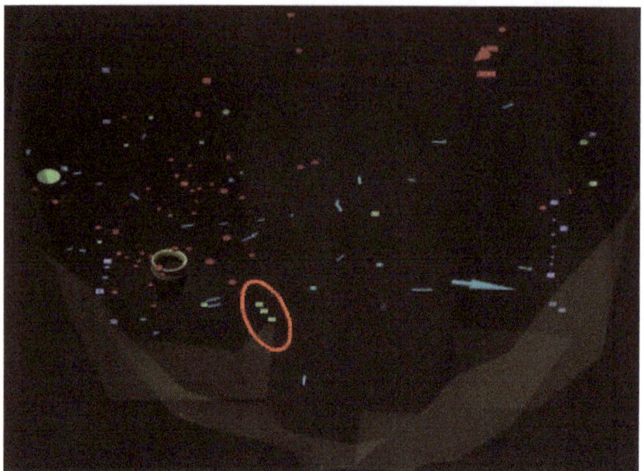

Figure 14. Elevated front view, 1967 excavation; red circle indicates alignment of wood fragments.

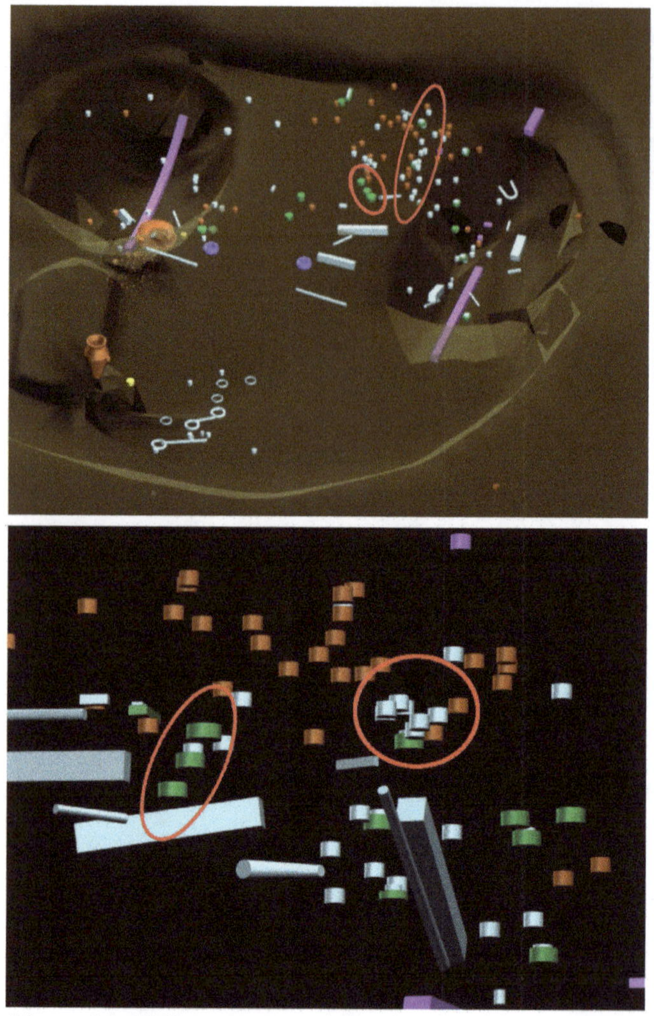

Figure 15. 1969 excavation; (above) right-of-front, elevated view with alignments of wood fragments and iron remains circled; (below) close, right-of-front, low view showing circled wood and iron alignments.

3D standards for scientific communication

Francesca Cantone
Dipartimento di Discipline Storiche "E. Lepore"
Università degli Studi di Napoli "Federico II"
Via Castagnola 30
80046, San Giorgio a Cremano
Napoli, Italy
Phone/Fax: +39 81 483047 - Mob: +39 347 8344222 - E-mail address: francesca@digitale.it

Abstract: This paper aims at verifying some of 3D modeling and VR visualization techniques in archaeology and cultural heritage management and at testing their use for scientific communication. The monument chosen for this research is the eighteenth century great staircase of SS. Niccolò and Cataldo Monastery in Lecce, that was modeled by means of CAD, VRML, X3D forthcoming standard.

In this way, it was possible to examine every method's advantages, evaluating the economic and intuitive use of WYSYG *tools versus the dimension of generated files, and the possibility of expansion and of updates. A particular focus was made on under-definition X3D standard, and on its possibilities in archaeological applications, in particular to build up 3D-DBMS, Virtual Reality scenes and Augmented-Reality applications.*

Key words: X3D, XML, VRML, Virtual Reality, 3D, Internet, Scientific Communication, Augmented Reality

Introduction

This research aims at applying and verifying some of 3D modeling and VR visualization techniques in archaeology and cultural heritage management and at testing their use for scientific communication.

It is well known that lately 3D models and virtual reality visualizations of artifacts, sites, environments and archeological, historical and artistic monuments have been increasingly used, mostly over popularization and didactics (for educational use see Mitchell – Economou 2000; for use of VR to give a public access to archaeological data see Pringle 2000; Bonfigli – Guidazzoli 2000 propose the use of VR for increase the knowdledge of the History of Bologna; Frischer – Abernathy - Favro - Liverani - Blaauw 2000 show various kinds of scientific, didactical, educational disseminations of a scientific model of Santa Maria Maggiore), but, recently, also in scientific applications. For instance Forte 1993 develops an application of scientific visualization for the topographic exploration of Reno Valley next to Bologna; Ozawa 1993 uses VR techniques for the 3D reconstruction of ancient Japanese villages of Yoshinogari in IV-VI centuries DC; Gottarelli 1995:101 analyses methods and techniques for 3D record of archaeological data, and outlines a virtual model of the archaeological excavation; Carafa – Laurenza – Putzolu 2000 propose a GIS based integration of 3D data in archaeological record and documentation.

An important laboratory for scientific visualization is CINECA VIS.I.T laboratory, a supercomputer sistem for developing real time 3D experiences, and to focus visual components of Cultural Heritage information (Guidazzoli 2000). An other important Laboratory for visual information for archaeological research is the SHAPE Lab, that has developed the ARCHAVE sistem, a software sistem that utilizes VR as an Interface to access scientifical data (Vote - Acevedo– Laidlaw – Joukowsky 2001). The Virtual Model is specifically defined an "extension of the record" in the Virtual Vilars Project. In the general project of record, documentation and reconstruction of an Iron Age Fortress in Catalogna, the authors state that "The three-dimensional and individualized restoration of the different features must also constitute an instrument for the study and typological analysis of the aforementioned structures" (Junyent – Lores 2000:228).

Goodrick – Harding 2000 use VR to solve a specific archaeological problem: the authors demonstrate the orientation of the site of Thornborough creating a virtual model of the site and of the stellar constellations.

An interesting application of VR to the analysis of pictorial information of the "Cristo Sole" Sepulcher in Vatican Necropolis is shown in Forte - Tilia - Bizzarro - Tilia 2001.

In the above applications VR tools are considered as a kind of archaeological and historical documentation, in addition to traditional two-dimensional ones.

This use of VR has lead to pay more attention on methodological aspects.

A first definition of virtual archaeology is given by P. Reilly in 1990 (Reilly 1991), who stresses the concept of "virtual" as related to a replica, or model, whose realism depends on the quality and quantity of data included. This paper opened a debate up to date till today, with interesting consequences both on methodological and technical aspects.

This concept is investigated in Sarti-Forte 1995 and it seems to be accepted in Forte 1997; Gillings and Goodrick (Gillings-Goodrick 1996; Gillings 1999; Gillings - Goodrick 2000) analyse the various techniques grouped under the term "VR" in archaeological studies. The authors consider the imprecision of the term itself a cause of the use of various alternative notions and techniques. So they emphasize the need for a framework for archaeological VR applications. Gillings analyses the problem of an "uncritical" enthusiasm in archaeology for VR, that caused a lack of interest for methodological issues, such as the definition of the actual meaning of the term VR for archaeologists and again the relationship between the reality and the virtual model. The authors come to the conclusion that VR has to stress the process more than giving passive images to be really useful for archaeological research and to help the data interpretation. With this more critical approach Gilling proposes the incorporation of VR into everyday archaeological study (see before, note 2) (Gillings 2000). One of the steps to reach this goal is considered the "development of an archaeological-VRML in its own right"(Gillings-Goodrick 1996, 2.6).

Barcelò 2000 deals again with the relationship between reality and model, and comes to the conclusion that "Visualization can be defined as the mapping of abstract quantities into graphical representations (geometric representations of lines, surfaces and solids) as an aid in the understanding of complex, often massive numerical representations of scientific concepts or results". He points out that "future advancement of Virtual Reality techniques in archaeology should not be restricted to "presentation" techniques, but to explanatory tools", thus envisaging the need for integrated visualization and data management tools. Analysis of present and possible or future applications in archaeological research are outlined in Borra 2000; Forte 2000a; Forte 2000b; Forte - Beltrami 2000.

Frischer – Niccolucci – Ryan – Barcelò 2000 point out a history of the diffusion of VR and of its applications in cultural heritage sites representation. The authors come to the conclusion that much of the past failures in CVR were caused by a lack of scientifical contents of high-tech experiments. They think that in the future VR will be much more present in archaeological research and practice, in collaborative and interactive worlds, but that this will be possible only by stressing the importance of scientific contents, by using a "philological" approach and by creating a set of standards that will allow interopeability, testing and validation.

Sanders 2000a - 2000b describes the status of archaeological uses of VR in archaeology, also envisaging in the future the use of multi-user, interactive worlds. Particular attention is paid on the need for shared communication standards, focusing on the role of VRML for developing scientific libraries, interactive worlds, data dissemination on Internet.

Zuchovsky 2001 deals again with the concept of realism, and proposes to refer the realism to the integrity, homogeneity, volume and quality of initial data, rather than on the realism of the VR final model, that will never be "real".

To sum up one of the key issues for the development of future archaeological VR applications seems to be the definition of widely accepted, flexible and transparent standards that allow the archaeologists to manage their data with no technical intermediation and to develop specific set of archaeological components.

Therefore, the first purpose of this research is testing and using VR as a tool for archaeological and historical scientific documentation and communication.

The monument chosen for this research is the SS. Niccolò and Cataldo Monastery in Lecce, in the South of Italy (fig. 1): a complex building, characterized by an interesting stratification in building phases through centuries (Photos by author, if not specified. For the historical and artistic aspects of Saints Niccolò and Cataldo Monastery see Wackernagel 1911; Infantino 1973; De Leo 1978; Bucci Morichi 1983; Fagiolo Cazzato 1984; Calò Mariani 1985; Storia 1993; Calò Mariani 1994; Kemper 1994; Pepe 1994; Vetere 1994; Palumbo 1996; Pellegrino-Vetere 1996.).

In this construction, the eighteenth century great staircase was selected (fig. 2).

So this analysis purposes are: 1 - the great staircase structural study; 2 - building stratification observation and recognition; 3 –VR tools applied test.

SS. Niccolò and Cataldo Monastery in Lecce

As we said, this research focuses on a part of SS. Niccolò and Cataldo Monastery in Lecce, the eighteen-century great staircase and the barrel vaults system that carries it.

The Monastery built by Tancredi, earl of Lecce in 1180, situated out of Lecce, had various building phases, rebuildings and restorations, until today, when it is the Department of Medieval and Contemporary History and of Cultural Heritage Studies of Lecce University.

In 1494 king Alfonso D'Aragona committed the monument to Monte Oliveto monks, who deeply changed its structure and decoration.

After this classicist moment, new architectonic and decorative manieristic changes were made in XVII sec., and in XVIII century, when the great staircase was built up, dated between 1733 and 1738.

After 1870, the monastery received poor and mad people, with very bad consequences for decoration and architectural features of the building.

During XX century, many restorations took place, sometimes deeply changing monument features and, finally, in 1985 the building was conceded to Lecce University, which made it respectable and usable.

The last restoration also involved the eighteenth century great staircase, examined in this paper, showing the old front door on the surface of the wall, where banister leans.

The eighteeth century great staircase: I PHASE methodology and analysis steps

In the first phase of this research, the great staircase was examined following these steps: 1-Analysis and measurement; 2-Drawing; 3-Photographic documentation; 4-Graphical restitution; 5-3D modeling in AutoCAD and exportation in VRML; 6-3D modeling in VRML.

Firstly the attention was focused on VRML (for VRML standard see Colombo-Marana 1997:145; Colombo-Marana 1999; Hartman-Wernecke 1996. Some uses and possible perspectives of VRML in archaeology are shown in Gillings-Goodrick 1996; Gillings 1999.), because since 1994 this language has been more and more used, to become a universal standard for virtual reality, mainly because it is free, platform independent, expansible, interactive, hyper medial and so on (theorical and methodological aspects in Ronchi 1994; technical issues in Hartman-Wemwcke 1996; Brutzman 1998; http://www.vrml.org/; http://www.lucia.it/vrml; http://www.web3d.org).

In order to test VRML standard advantages for our purposes it was decided to create the virtual scene of the eighteenth century great staircase by using different tools and techniques. The research about our tools for reconstructing, and about how they differently work and benefit our research seemed to be one of our most interesting analysis fields. So the VRML scene was created by means of: manual editing; WYSYG editing; exporting a 3D model made with AutoCAD.

In this way, it was possible defining every method's advantages, evaluating WYSYG tools economic and intuitive use versus generated files dimension, and expansion and updates possibility.

The great staircase was reconstructed, in order to make a 3D structural model to visualize and study building architectural features. The model of Saints Niccolò and Cataldo staircase aims at helping to understand the structure of the building: so high resolution model was not needed, nor particular lights, surface accuracy, detailed decorations modeling (for higher detail level models or lighting, surface and texture problems see Gottarelli 1995:75; Meucci-Buzzanca 1996; Kantner 2000; Lucet 2000; Pollefeys - Proesmans - Koch - Vergauwen -Van Gool 2000). Average measurements were taken where repetitive elements were found, such as staircase steps (the same process is applied in Medrano Marqués-Dìaz Sanz-Tramullas Saz 1991:287-288; Daniels 1997; Lucet 1997; Huggert – Chen Guo - Yan 2000). So the model here presented is a structural model (the word "structural" is not used like Daniels does in Daniels 1997, 3. 1., where the author refers it to the distinction between perception and structural models).

It was important to keep in mind our study purpose since this first step, in order to plan the right data set to use. Soon it was clear that notwithstanding the fact that the traditional 2D graphical documentation, plan and section have been considered sufficient to record a 3D object, such as a building or an archaeological excavation, 3D reconstruction needs a much greater amount of data and measures than the traditional 2D model, and consequently conveys richer information.

So, the 3D wire-frame model was made with AutoCAD (on the use of AutoCAD for archaeological applications see Buzzanca-Giorgi 1996; Messika 1996; Wood 1996; Daniels 1997; Drap – Hartmann –Virnich – Grussenmeyer 2000; Martens – Legrand - Legrand – Loots – Waelkens 2000) (fig. 3). Likewise, in order to test VRML benefits, the monument virtual scene was created by using various tools: exporting a model from AutoCAD to VRML; using a WYSYG VRML interface; hand-writing the .wrl file (fig. 4) (for more informations on this phase of this case-study see Cantone 2000, under press). The main results are: 1 - exporting an AutoCAD 3D model in VRML produces a light file, interactive, and easily usable on the web. But, analyzing the source code of this file, it is clear that this way it is obtained a single Indexed face set node, which is a polygon with the external surface of the whole model as external area. 2 - direct drawing in VRML, using a graphical WYSYG interface. That way we created a virtual world similar to all appearance to the world exported from AutoCAD. But it is just the source code that is completely different. This time in the code we can easily recognize every part of the "building": every wall, every vault has its own node, easily visible in the code, especially by using VRML possibility to name each element. Moreover, this code makes the most of some powerful features of VRML, such as the primitives. This is a double benefit: a - the files is lighter; b – it is possible updating and changing the virtual world every time you want, modeling also only one node - i.e. architectonic element. Nevertheless the GUI seems not completely able to use powerful VRML features, particularly to lighten files: it does not use, for instance, keywords DEF and USE: the result is a still heavy and not clean code. 3 - manual editing. Thus it is obtained a clean and light code, but of course much less easily and quickly. As already noticed, there are many differences between 3D VRML modeling tools, so it seems important to chose the way to model virtual words every time, evaluating how each method works, and which are its benefits for the specific research purposes. In this way, we could test each method advantages and limits, in particular evaluating quickness and intuitive editing versus generated files dimension and updates possibility.

I PHASE: Results

The analysis made this way brought out many observations on the monument studied, which was recorded and examined by means of the photographic and graphical documentation collected.

3D model interactive visualization enhanced the monument analysis enabling a better and quick test of the data set collected. In particular it was possible analyzing: 1-the great staircase architectural features; 2-the building phases stratification.

This shows that a monument VRML model is an additional tool to record, document, and study data, in order to achieve powerful and useful benefits. Indeed VRML allows: immersivity; individual interaction with the model; updating - work in progress; networking.

Individual interaction with the model enables much more completely and directly data studying, communicating and transmitting, also in modeling still existing monument reconstruction. Moreover, constant updating possibility make our graphical information more flexible and useful than traditionally.

The same features enhance networking and interacting during the whole research, introducing interesting and useful changes in the methodology of our work.

II PHASE: X3D

During the first phase of our research, we could notice VRML standard powerful possibilities. However, it was possible too noticing some limits in this standard use.

First of all, probably the inadequacy itself of efficient standardized, and widely known GUI, limited, somehow, an intensive application of VRML as a common tool for archaeological documentation.

Furthermore, VRML is a complex and heavy language. This feature has two consequences: 1-it is difficult developing error-free applications; 2- it is necessary downloading heavy client-plug in (> 3Mb), and this dramatically reduces Internet users who can visualize a VRML scene (about 10% Internet users). Moreover, VRML expansions are possible only by using PROTO (http://www.web3d.org/technicalinfo/specifications/vrml97/index.html).

These limits are remarkably strong for archaeology and cultural heritage management study purposes: for instance, it is not easy making VRML 3D model interact with other applications and, first of all, with DBMS (but this feature is under development: http://www.web3d.org/TaskGroups/dbwork).

These are some of the reasons that induced Web3d consortium to develop X3D, a new and powerful language, which is likely to be the 3D standard for Next Generation Web (Locatelli 2000; http://www.web3d.org/TaskGroups/x3d/faq/index.html) (fig. 5).

X3D stands for Extensible 3D, a next-generation, extensible, XML-compliant 3D graphics specification that extends the capabilities of VRML 97. The name X3D was chosen to indicate the integration of XML.

In short, X3D main goals are:

1. Integration with XML (X3D expresses VRML 97 capabilities using the Extensible Markup Language);
2. Componentization. (X3D enables a lightweight core that can be easily extended to provide new functionality);
3. Extensibility. (X3D uses components to add fundamentally new nodes and corresponding implementation code to the core).

The integration with XML is a powerful feature that enables easy pages re-hosting, better pages technology integration, and, finally integration in Next Generation Web, which will be based more and more on XML.

Componentization feature means that the new standard has a light core, in which only 23 nodes have been included: Anchor ; Appearance ; Background; Color; Coordinate; CoordinateInterpolator; DirectionalLight; Group; ImageTexture; IndexedFaceSet; IndexedLineSet; Inline; Material; NavigationInfo; OrientationInterpolator; PointSet; PositionInterpolator; Shape; TextureCoordinate; TimeSensor; Transform; Viewpoint; WorldInfo.

This core can be extended with Components, sets of related nodes. Various Components constitute a Profile, i.e. subset of related Components that provide a set of task-specific functionalities. Different Profiles are useful to fulfill different users needs: Base, Core e Full are examples of Profiles.

Some extensions have been already integrated, and, particularly interesting for archaeological applications purposes, also GEO VRML extension (http://www.ai.sri.com/geovrml).

For these reasons it seemed useful applying this new technology to our study, so the great staircase was modeled in X3D, by using X3D-edit, a scene-graph editor, developed by the same working group that is defining the standard.

X3D-Edit uses the XML tagset defined by the X3D Compact Document Type Definition (DTD) <x3d-compact.dtd> in combination with Sun's Java, IBM's Xeena XML editor and an editor profile configuration file <x3d-compromise.profile>.

X3D-edit processes X3D scene graph using different XLS to generate .html, .wrl files and, possibly, other kinds of outputs, such as .pdf and so on (fig. 6). A X3D browser is being developed too by the same group, called XJ3D, tested under Linux, Solaris and soon Win 32 (fig. 7)(http://www.web3d.org/TaskGroups/x3d/faq/index.html).

Use of X3D makes it possible to figure out more structured data managing, integrating a XML DBMS with the X3D model. Besides, the generated core is extremely readible and transparent, allowing a flexible management and quick control of data. Moreover, XML enables using output independent data: it is possible using data collected and analized in various ways: from the site, with PDA, to the study, with 3D DBMS, to the dissemination. Furthemore, this standard is lighter than VRML, so we can figure out a dramatic increasing in data availability.

X3D Code specimen

```
<?xml version="1.0" encoding="UTF-8"?>

<!DOCTYPE X3D PUBLIC "http://www.web3D.org/
TaskGroups/x3d/translation/x3d-compromise.dtd"
                          "file://localhost/C:/
www.web3D.org/TaskGroups/x3d/translation/x3d-
compromise.dtd">
"file://localhost/C:/www.web3D.org/TaskGroups/
x3d/translation/x3d-compromise.dtd">
```

The file heading informs the browser the file language: xml, version 1.0. The encoding used is the one defined by utf8 abbreviation, the transformation format UCS (Universal Multiple-

Octet Coded Character Set) at 8 bit, ISO/IEC 10646-1:1993 Information technology , including also not ascii characters.

This part indicates too the location of document type definition.

```
<X3D>
```

This is the beginning of the X3D code.

```
<Header>
<meta content="monastery.xml" name="filename"/>
<meta content="monastero dei santi niccol&#242; e
cataldo, lecce" name="description"/>
<meta content="francesca cantone" name="author"/>
<meta content="march, 2001" name="created"/>
<meta content="http://www.archeologia.net/
monastery.xml" name="url"/>
<meta content="X3D-Edit, http://www.web3D.org/
TaskGroups/x3d/translation/README.X3D-Edit.html"
name="generator"/>
</Header>
```

The file header includes meta tags, with various additional informations about the file, the author, the date of creation, etc.

```
<Scene>
<NavigationInfo type="&quot;EXAMINE&quot;
&quot;ANY&quot;"/><!-
      Scene graph nodes are added here -><Group>
      <children>
        <Shape>
          <appearance>
            <Appearance>
              <material>
                <Material ambientIntensity="0.033"
                          diffuseColor="0.0 0.0 1.0"
emissiveColor="0 0 0"
                                   shininess="0.100"
specularColor="0.8 0.8 0.8"/>
              </material>
              <texture>
                <ImageTexture url="ultimo.jpg"/>
                </texture>
        </Appearance>
       </appearance>
       <geometry>
         <IndexedFaceSet ccw="true" convex="true"
          coordIndex="0 1 2 -1 &#10;
1 3 2 -1 &#10;
...
(skipping coordinates)
...
6 22 28 -1 "
          creaseAngle="1.222" solid="true">
         <coord>
              <Coordinate point="6.04496 -56.09990 -45.89980&#10;
                           37.14500 -56.09990 -54.89980&#10;
2.07639 -56.09990 -45.89980&#10;
...
(skipping points coordinates)
...
25.14500 -67.90000 -18.89980&#10;"/>
           </coord>
         </IndexedFaceSet>
       </geometry>
     </Shape>
    </children>
   </Group>
      <Group/>
       <Shape>
         <appearance>
           <Appearance/>
         </appearance>
       </Shape>
       <Group/>
       <Shape>
         <appearance>
           <Appearance>
             <material>
               <Material/>
             </material>
             <texture>
               <ImageTexture url="ultimo.jpg"/>
             </texture>
           </Appearance>
         </appearance>
```

These code lines describe various aspects of a geometrical form: its appearance, material, color, shape. The geometry is defined by the series of vertex, and by their coordinates. Then a texture is applied to the geometrical form. The ambient illumination and features are described too. This example refers to a part of a wall of the staircase.

```
...
(skipping code)
...

         <Viewpoint DEF="vp" description="vp"
fieldOfView="0.785398"
       jump="true" position="0 0 300 "/>
   </Scene>
</X3D>
```

There is a definition of the viewpoint, then the end of the scene and of the X3D file.

Perspectives

Besides the advantages described before, the language structure itself seems to be useful for archaeological applications: there is a standard core, which can be extended with specific profiles. Therefore, an archaeologist can trust on a worldwide diffused language and in the same time can expand it to fulfill his research needs. Besides X3D code is flexible and transparent, so it can be easy for archaeologists themselves to control and update their data with no intermediation.

X3D seems helpful in various steps of archaeological research, from data-entry to data managing and to dissemination and publication.

In data collecting, it is possible to get a richer documentation, integrating 3D model in DBMS.

In data analyzing, X3D allows networking, simulations, building up augmented reality (for an introduction to the use of Augmented Reality in archaeology see Barcelò 2000; Ryan 2000) and achieving cognitive enhancement (Forte 2000a; Forte, M.- Beltrami, R., 2000 for the concept of VR in archaeology as related to the Cognitive Archaeology). Furthemore, GEOVRML profile integration enables using and analysing geo-referenced data, such as maps and 3-D terrain models.

Finally, in data publishing, this standard allows not only data web publishing, but also various outputs and visualizations generation from the same data set.

To sum up it seems clear that X3D will enhance the advantages of the present VRML standard.

It promises wider data dissemination, easier and more effective use of task-specific profiles, immediate interaction with other applications and DBMS.

So X3D seems to be a powerful tool, useful for several applications: excavation data 3DBMS, eventually also making the most of on site data acquisition devices; studying buildings, artifacts, sites, environments, making territorial analysis, etc.

Obviously all that will be possible after this definition phase. It is easy to suppose that the next step will be the developing of GUIs, that may be useful to archaeologists to manage their data themselves.

At the same time, in hardware developing, XML-based devices are going to be worked out, enriching possibilities in data entry, managing, dissemination.

One of the most important fields of research for archaeologists will be developing archaeological profiles and archaeological primitives libraries, to make the most of this powerful standard.

Acknowledgements

The author wants to acknowledge those who helps and sustains her in her research work: Prof. Giovanna Greco (Dipartimento di Discipline Storiche "E. Lepore", Università degli Studi di Napoli "Federico II"); Prof. Franco Niccolucci (Università degli Studi di Firenze).

References

Barcelò, J.A. 2000. *Visualizing what might be: an introduction to virtual reality techniques in archaeology.* In Barcelò – Forte – Sanders 2000:9-35.

Barcelò, J. A. - Forte, M. - Sanders, D. H. 2000. *Virtual Reality in Archaeology*, BAR Int. series 843, Oxford.

Bonfigli, M.E. - Guidazzoli, G. 2000. *A www virtual museum for the history of a city.* In Barcelò – Forte – Sanders 2000:143-147.

Borra, D. 2000. *La modellazione virtuale per l'architettura antica. Un metodo verso l'isomorfismo percettivo.* Atti del I Workshop nazionale di archeologia computazionale – Napoli-Firenze 1999, Archeologia e calcolatori 11, Firenze:259-272.

Bucci Morichi, C. 1983. *Lecce. Monastero dei Santi Niccolò e Cataldo.* In AAVV. Restauri in Puglia 1971-1973, Bari: 428-434.

Brutzman, D. 1998. *The Virtual Reality Modeling Language and Java.* Communications of the ACM, 41, 6, June 1998:57-64 http://web.nps.navy.mil/~brutzman/vrml/vrmljava.pdf

Buzzanca, G. – Giorgi, E. 1996. *Come usare AutoCAD e vivere ugualmente felici (L'ennesimo sistema per la raccolta dei dati storico-conservativi).* Archeologia e Calcolatori 7:907-916.

Calò Mariani, M.S. 1985. *Insediamenti Benedettini in Puglia per una storia dell'arte dall'XI al XVIII secolo: Catalogo della Mostra*, Bari.

Calò Mariani, M.S. 1994. *Dal chiosco alle corti.* In Vetere 1994:661-734.

Carafa, P. – Laurenza, S. – Putzolu, C. 2000. *Stratigraphic excavation from the field to the computer: the Pompeii Prototype.* In Niccolucci 2000a.

Colombo, L. – Marana, B. 1997. *La documentazione geometrica di opere monumentali in siti web*, Centro di Ricerche Informatiche per i Beni Culturali VII, n. 1-2:141-152.

Colombo, L. – Marana, B. 1999. *Standard di comunicazione visiva nel World Wide Web*, Centro di Ricerche Informatiche per i Beni Culturali IX, n. 1:115-127.

Daniel, R. 1997. *The need for the solid modeling of structure in the archaeology of buildings.* Internet Archaeology, 2 (http://intarch.ac.uk/journal/issue2/daniels_index.html) (20/04/01).

De Leo, P. 1978. *Le carte dell'abbazia dei Santi Niccolò e Cataldo in Lecce*, Lecce.

Drap, P. - Hartmann-Virnich, A. – Grussenmeyer, P. 2000. *Photogrammetric stone-by-stone survey and archaeological knowledge. An application on the romanesque priory church Notre-Dame D'Aleyrac (Provence, France).* In Niccolucci 2000a.

Fagiolo, M., Cazzato, V. 1984. *Lecce*, Lecce.

Forte, M. 1993. *Un esperimento di visualizzazione scientifica per l'archeologia del paesaggio: la navigazione nel paesaggio virtuale.* Archeologia e Calcolatori 4:137-152.

Forte, M (ed). 1997. *Virtual Archaeology: Great Discoveries Brought to Life Through Virtual Reality*, Thames and Hudson, London.

cognitive interactions and virtuality. In Barcelò – Forte – Sanders 2000:247-259.

Forte, M. 2000b. *Archaeology and Territorial "Musealisation": the virtual-cognitive reconstruction of archaeological landscapes.* In Niccolucci 2000a.

Forte, M.- Beltrami, R. 2000. *A proposito di virtual archaeology: disordini, interazioni cognitive e virtualità.* Atti del I Workshop nazionale di archeologia computazionale – Napoli-Firenze 1999, Archeologia e calcolatori 11, Firenze: 273-300.

Forte, M. – Borra, D., 2000. *The Estense Castle of Ferrara*

(Italy): Multimedia Project and Virtual Reconstruction. In Barcelò – Forte – Sanders 2000:237-245

Forte, M. – Tilia, S. – Bizzarro, A. – Tilia, A., 2001. *3D Visual Information and GIS Technologies for Documentation of Paintings in the M Sepulcher in the Vatican Necropolis*, Stancic, Z. Veljanovski, T. (ed.), Computing Archaeology for understanding the past, CAA 2000, Oxford:25-31

Frischer, B. – Abernathy, D., Favro, D., Liverani, P., Blaauw, S., 2000. *Virtual Reality And Ancient Rome: Rome Reborn and the Santa Maria Maggiore Project.* In Barcelò – Forte – Sanders 2000:155-161

Frischer, B. – Niccolucci, F. – Ryan, N. – Barcelò, J. A., 2000. *From CVR to CVRO: The past, present, and future of cultural virtual reality.* In Niccolucci 2000a

Giuliani, C.G., 1990. *L'edilizia nell'antichità*, Roma

Gillings, M., 1999. *Engaging Place: a Framework for the Integration and Realisation of Virtual-Reality Approaches in Archaeology.* In Archaeology in the Age of Internet. CAA 97, BAR International Series 750:247-254

Gillings, M., 2000. *Plan, elevations and virtual worlds: the development of techniques for the routine construction of hyperreal simulations.* In Barcelò – Forte – Sanders 2000:59-69

Gillings, M. – Goodrick, G., 1996. *Sensuous and reflexive GIS: exploring visualization and VRML*, Internet Archaeology 1, (http://intarch.ac.uk/journal/issue1/gillings_toc.html) (20/04/01)

Goodrick, G. – Gillings, M., 2000. *Constructs, simulations and hyperreal world: the role of Virtual Reality (VR) in archaeological research.* In Lock, G. – Brown, K., 2000 (ed.). On the Theory and Practice of Archaeological Computing, Oxford:41-58

Goodrick, G. - Harding, J., 2000. *Virtual Reality at the neolithic monument complex of Thornborough, North Yorkshire.* In Barcelò – Forte – Sanders 2000:115-119

Gottarelli, A., 1995. *La modellazione tridimensionale del documento archeologico: livelli descrittivi e processamento digitale.* Archeologia e Calcolatori 6:75-103

Guidazzoli, A., 2000. *Experiences of Immersive Graphics for Cultural Heritage.* In Niccolucci 2000a

Hartman, J. , Wernecke, J., 1996. *The VRML 2.0 Handbook: Building Moving Worlds on the Web*, Silicon Graphics

Huggert, J. – Chen Guo-Yan, 2000. *3D interpretative modeling of Archaeological sites/a computer Reconstruction on Medieval Timber and Earthwork Castle.* Internet Archaeology, 8 (http://intarch.ac.uk/journal/issue8/huggert_toc.html) (20/04/01)

Infantino, G.C., 1973. *Lecce sacra*, Bologna

E. Junyent, E. – Lorés, J., 2000. *Virtual Reality as an extension of the archaeological record. reconstruction of an Iron Age Fortress: Els Vilars (Arbeca, Catalonia, Spain).* In Barcelò – Forte – Sanders 2000:225-231

Kantner, J., 2000.*Realism vs. reality: creating virtual reconstructions of prehistoric architecture.* In Barcelò – Forte – Sanders 2000:47-51

Kemper, D., 1994. *SS. Niccolò e Cataldo in Lecce als ein Ausgangspunkt für die Entwicklung mittelalterlicher Bauplastik in Apulien und der Basilicat*, Worms

Locatelli, I., 2000. *X3D: il nuovo standard per il 3D in rete. Processo di standardizzazione e prime implementazioni,* (http://eureka.lucia.it/vrml/it2000/atti/materiali.html) (20/04/01)

Lucet, G. L. , 1997. *The Virtual Restitution of the Mesoamerican site of Cacaxtla: a methodological point of view.* Archeologia e calcolatori 8:129-138

Lucet, G., 2000. *Visualization of virtual environments of ancient architecture: the problem of illumination.* In Barcelò – Forte – Sanders 2000:87-95

Martens, F. – Legrand, P. - Legrand, L. – Loots, L. – Waelkens, M., 2000. *Computer aided design and archaeology at Sagalassos: methodology and possibilities of 3D computer reconstructions of archaeological sites.* In Barcelò – Forte – Sanders 2000, pp. 205-211

Mediano Marqués, M. – Dìaz Sanz, M. – Tramullas Saz, J., 1991. *Reconstitución del edificio monumental de Contrebia Belaisca (Botorrita, Zaragoza).* Applicaciones Informaticas en Arqueologia, Complutum 1:281-292

Messika, N. R., 1996. *AutoCAD for Archaeology. A New Era for Archaeology.* Archeologia e Calcolatori 7:951-954

Meucci, C. – Buzzanca, G., 1996. *Simulazione e/o seduzione (la rappresentazione mediante modelli di reperti, relitti, oggetti ed altro).* Archeologia e Calcolatori 7:973-982

Mitchell, W. L. – Economou, D., 2000. *The Internet and Virtual Reality in Heritage Education: more than just a technical problem.* In Barcelò – Forte – Sanders 2000:149-153

Niccolucci, F., 2000a. *Proceedings of the VAST Euroconference, Arezzo, Novembre 2000.* BAR International Series, in press

Niccolucci, F., 2000b. *Virtual archaeology vs virtual reality.* In Niccolucci 2000a

Ozawa, K., 1993. *Reconstruction of Japanese ancient tombs and villages.* In Andresen, J-Madsen, T.- Scollar, I (ed), Computing the past. Computer Applications and Quantitative Methods in Archaeology, Oxford

Palumbo, P. 1996. *Storia di Lecce*, Bari

Pellegrino, B. - Vetere, B. 1996. *Il Tempio di Tancredi. Il monastero dei Santi Niccolò e Cataldo in Lecce*, Lecce.

Pepe, A. 1994. *La cultura architettonica fra età normanna e aragonese*. In VETERE 1994.

Pollefeys, M.-Proesmans, M.-Koch, R., Vergauwen, M.-Van Gool, L. 2000. *Acquisition of detailed models for virtual reality*, In Barcelò – Forte – Sanders 2000:71-77.

Pringle, M. J. 2000. *Using Virtual Reality to Improve Public access to Heritage Databases over the Internet*. In CAA 2000 abstracts (http://www.zrc-sazu.si/caa/abstracts.htm) (20/04/01).

Reilly, P. 1990. *Towards a virtual archaeology*. In Lockyear, K. –Rahtz, S. (ed), Computer Applications in Archaeology 1990, Oxford: British Archaeological Reports (Int. Series 565):133-139.

Ronchi, A. M. 1994. *Virtualità reale*, Centro di Ricerche Informatiche per i Beni Culturali IV, n. 1:7-31.

Ryan, N. 2000. *Back to reality: Augmented Reality from field survey to tourist guide*. In Niccolucci 2000a.

Sanders, D. H. 2000a. *Archaeological publications using virtual reality: case studies and caveats*. In Barcelò – Forte – Sanders 2000:37-45.

Sanders, D. H. 2000b. *Virtual archaeology and museums: where are the exhibits?*. In Niccolucci 2000a.

Sarti, A. – Forte, M. 1995. *Tra Archeologia e Realtà Virtuale: verso l'Archeologia Virtuale*. Archeologia e Calcolatori 6:105-118.

Terras, M. 1999. *A Virtual tomb for Kelvingrove: Virtual Reality, Archaeology and Education*, Internet Archaeology 7, (http://intarch.ac.uk/journal/issue7/terras_toc.html) (20/04/01).

Storia. 1993. AAVV, *Storia di Lecce: dai Bizantini agli Aragonesi*, Bari.

Vetere, B. 1994. *Storia di Lecce. Dai Bizantini agli Aragonesi*, Lecce.

Vote, E. L. - Acevedo, D. – Laidlaw, D. – Joukowsky, M. S. 2001. *ARCHAVE: A Virtual Environment for Archaeological Research*. In Stancic, Z. Veljanovski, T. (ed.), Computing Archaeology for understanding the past, CAA 2000, Oxford:313 - 315.

Wackernagel, M. 1997. *Die Plastik des XI. und XII. Jahrhunderts in Apulien*, Leipzig.

Wood, J. 1996. *Digital data and the conservation of the historic built environment*. Archeologia e Calcolatori 7:901-905.

Zuchovsky, M. 2001. *Virtual 3D Reconstruction of the Kiafar Site, North Caucasus, Russia*. In Stancic, Z. Veljanovski, T. (ed.), Computing Archaeology for understanding the past, CAA 2000, Oxford:297-301.

http://www.ai.sri.com/geovrml (20/04/01)

http://www.lucia.it/vrml (20/04/01)

http://www.vrml.org/ (20/04/01)

http://www.web3d.org/(20/04/01)

Figures

Figure 1. SS. Niccolò and Cataldo Monastery – from Pellegrino - Vetere 1996

Figure 2. the part of the building in which the eighteen century staircase is built:view from internal court

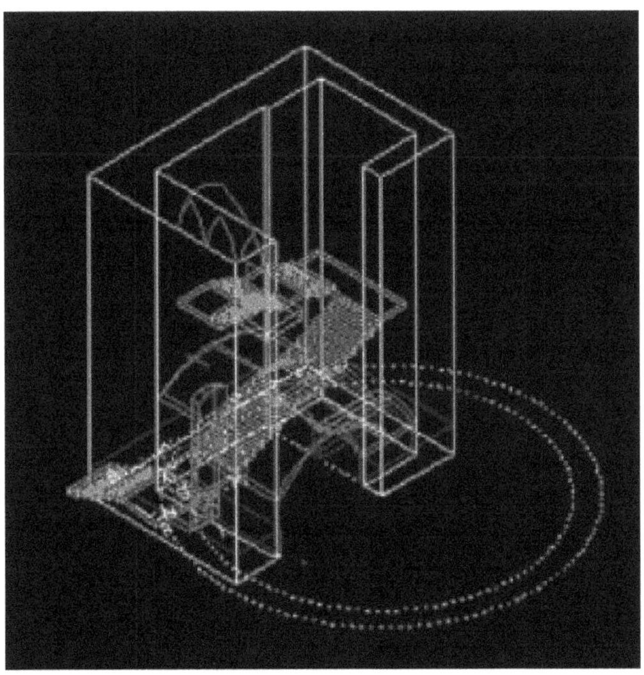

Figure 3. The wire frame model: isometric view

Figure 4. The VRML scene: front view

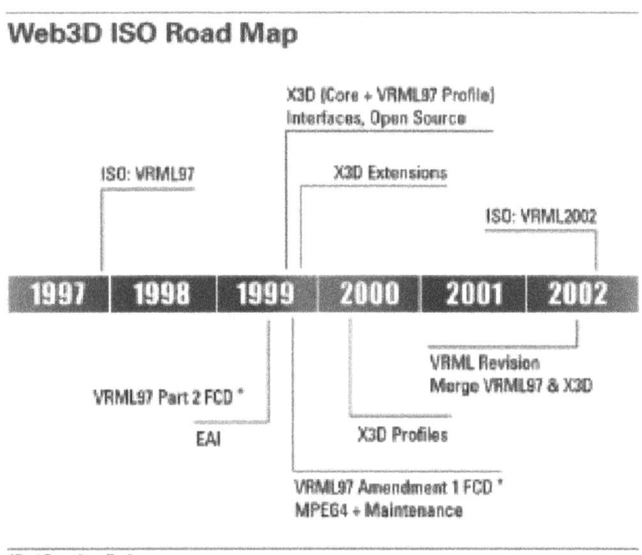

Figure 5. The road-map for the consortium to develop X3D standard

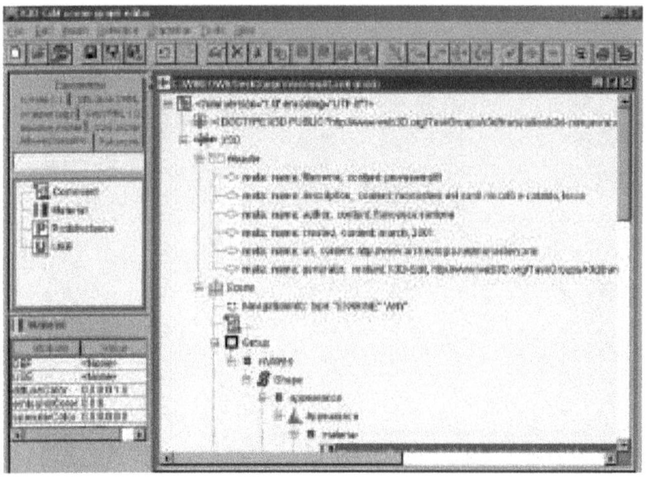

Figure 6. X3D-edit

Figure. 7 The pretty-print generated applying a XLS

Acoustic reconstruction of music performance spaces using three-dimentional digital waveguide mesh models

Guilherme Campos
Departments of Electronics
University of York
Heslington
York, YO1 5DD, UK

David M. Howard
Departments of Electronics
University of York
Heslington
York, YO1 5DD, UK

Steve Dobson
Departments of Archaeology
University of York
Heslington
York, YO1 5DD, UK

Abstract: The concept of auralisation and its importance in room acoustic simulation are briefly discussed and an overview is given of the various acoustic modelling techniques available. In particular, 3-D digital waveguide mesh physical modelling is discussed in more detail. It offers a number of implementation advantages and is deemed capable of meeting the accuracy requirements of Acoustic Reconstruction applications. A complete room acoustic modelling system is proposed, including a CAD front-end for room configuration.

Introduction

Room Acoustics is concerned with how the physical characteristics of a room (dimensions/shape/ surfaces) influence sound perception. For given positions of sound source and listener, this influence is fully described by the Room Impulse Response (*RIR*). The RIR can be used to synthesise the acoustic stimuli induced at the listener's ears by any given sound source. These can be listened to over headphones, in what is known as *auralisation* (figure 1).

Only recently has auralisation become a practical possibility, with the development of sophisticated RIR measurement techniques and advances in digital signal processing. Auralisation with measured RIR can be very useful (e.g in listening tests for concert hall subjective evaluation surveys, design and test of loudspeakers and sound diffusion systems) but is obviously restricted to the study of existing rooms. In Auditoria Design or Acoustic Reconstruction applications, the room under study does not actually exist. Methods of obtaining the RIR *directly* from a knowledge of its physical characteristics are therefore required.

Acoustic reconstruction is relevant to the fields of music history, musicology and ethno-musicology with obvious application in the context of authentic, or historically accurate musical performances, normally associated with the use of period instruments. The study of the inter-relationship between the evolution of the acoustics of music performance spaces and the development of musical styles and instruments over the centuries is another possible application.

Outside the musical/musicological sphere, the investigation of the acoustic behaviour of ancient buildings, particularly prehistoric structures, can contribute to a more complete understanding of their function and significance. Many archaeological sites, notably Mayan ruins and northern European megalithic monuments, feature striking acoustical phenomena.

Acoustic modelling and accuracy requirements

The problem of calculating a RIR amounts to solving the sound wave equation for a particular room under analysis. Because analytical solutions can be obtained only for very simple geometries and idealised boundary conditions, *acoustic models* have to be utilised. Analog modelling, based on indirect RIR measurement using ultra-sounds to excite a scale-model, is still very much in use. However, as in other fields of engineering, it is being gradually superseded by digital computer modelling based on numerical methods.

Models directly based on the wave equation are normally called *physical models*. Excessive computational loading has been their main drawback for room simulation. To circumvent this problem, simplified approaches to the analysis of sound propagation have been proposed, namely geometric models (*ray-tracing, mirror-image source method*) inspired by optics. In their standard formulations, reflection and absorption are the only modelled phenomena. Sound waves are treated like rays, based on the assumption that the room surfaces are relatively flat and their dimensions large when compared to the sound wavelength. This corresponds to the limiting case of very high frequencies. These simplifications allow relatively fast calculations but severely restrict the model's validity, making it largely dependent on the architectural details of the room.

In Auditoria Design, it is generally the case that before the acoustic behaviour of the building can be considered, many of its architectural features are already fixed. This, together with the fact that modern architecture does tend to favour large flat interior surfaces, is the main reason why software packages based on geometric modelling, although far from ideal, are considered acceptable.

In contrast, whenever absolute rather than comparative appreciation is intended, as is the case in the acoustic reconstruction of music performance spaces, approximate models are not suitable. Such applications call for the maximum possible level of accuracy, limited only by human auditory capabilities. Physical modelling is arguably the best approach in this context, since all wave propagation phenomena are taken into account.

3-D digital waveguide mesh modelling

Digital-waveguide mesh modelling was originally developed for musical applications. One- and two-dimensional versions have been successfully applied to sound synthesis and simulation of acoustic instruments.

Application to room acoustics requires 3-D meshes of suitably high density. The volume under analysis is spatially discretised as a grid of regularly spaced *nodes* interconnected by bidirectional *delay units*. Travelling waves can be simulated with this arrangement. Under the computation algorithm employed, the model implements a standard finite-difference time-domain approximation to the solution of the wave equation. *Air nodes* represent the propagation medium and can be assigned special functions, namely *mesh excitation* (modelling sound *sources*) and *sound output* (modelling *receivers*). *Boundary nodes* model the surfaces delimiting the space. Their scattering parameters vary according to the physical properties of the corresponding material.

Simulation is achieved by injecting an anechoic sound signal at a *source* node and recording the resulting acoustic pressure variation at the node(s) specified as receiver(s). The errors associated with the discretisation process, in particular frequency dispersion, can be reduced by increasing the density of the mesh.

Various 3-D mesh topologies have been proposed in the literature. Lossless models have been implemented using two of these topologies, illustrated in figure 2. Initial objective and subjective validation tests have confirmed their ability to accurately predict the low frequency room mode distribution and provide appropriate sound localisation cues.

Various model refinements are possible. For example, new techniques have recently been introduced to improve the dispersion error characteristics of the rectilinear mesh. It is also possible to model frequency-dependent air absorption using digital filters. These can be easily fitted into the waveguide mesh modelling framework – an important advantage. Integrated in boundary node transfer functions, they can also allow frequency and surface orientation dependent boundary conditions, making it possible to simulate the acoustic behaviour of surface materials more accurately. The total *RIR* computation time is given, in *s*, by

$$T_M \quad \left(\frac{1}{c}\right)^3 .V.RT_{60}.f_S^4.t$$

where c is the medium's sound propagation speed (ms^{-1}); V is the volume of the room (m^3); RT_{60} is the reverberation time @ -60 dB; f_S is the audio sampling frequency (Hz); t is the average computation time per model node per iteration (s) and ± is an adimensional factor equal to 2 and $\sqrt{3}$ respectively for the rectilinear and tetrahedral meshes.

For example, it takes approximately 68h to obtain a 2s impulse response at 22050Hz for a room of 110.6m^3 using a rectilinear model running on the fastest single-processor platform tested. Tetrahedral models can be expected to be more than twice faster, but a typical music performance space can be hundreds of times larger and have longer RT_{60}. Moreover, in order to attain the desired level of perceptual accuracy it may be necessary to use higher f_S (i.e. denser meshes) and/or implement at least some of the model refinements mentioned before, only possible at the expense of t. Practical application to the study of real rooms therefore requires a very significant performance improvement. This seems achievable in the near future, as the models lend themselves extremely well to large-scale parallelisation. The results obtained so far indicate that the computational overhead due to communication between blocks operating in parallel is very low and, more importantly, can be made independent of their total number.

3-D CAD-based model configuration

A complete room acoustic simulation system is proposed in figure 3. Its main element (block 5) is the 3-D digital waveguide modelling program. Two ways of obtaining a digital representation of the room to be modelled can be employed. A commercial 3-D CAD package (block 2) provides a framework for regular-shaped rooms. However, it is impractical for rooms with irregular boundaries (e.g. pre-historic chambers), for which the ideal solution is an automatic spatial data acquisition system (block 1). Systems of this kind have been used in Archaeology as well as contemporary architectural and medical applications. In view of recent developments in precision portable laser scanners, it is likely that affordable, accurate systems for room interiors will become available in the near future.

An automated procedure to convert between the formats of CAD file and model mesh with user-defined meshing parameters (block 4) is currently being developed. It is based on the rasterisation of horizontal room 'slices' at regularly spaced elevations. The resulting 2-D node grids are then assembled into the appropriate 3-D format for physical modelling. It will be possible to extract the information directly from files produced by commercial 3D CAD software packages. It may be necessary to go through an additional stage of drawing preparation (block 3), to make sure an enclosed space is correctly defined, with all its boundaries characterised acoustically, and to position sound source(s) and receiver(s).

Summary and Future Work

3-D waveguide mesh models can provide accurate auralisation as required for the acoustic reconstruction of music performance spaces. Their practical applicability is currently restricted to very small-sized rooms but can be significantly extended in the near future by a combination of mesh topology optimisation, introduction of new algorithm refinements and large-scale parallelisation. A crucial element in this process is the development of a general tool for model validation reliant upon accurate 3-D surveys and *in situ* experimental measurements. For this purpose, an automated mesh configuration procedure using data from standard 3-D CAD models is being developed. This will allow easy integration of data from advanced room surveying techniques, for example laser scanning. Model validation case studies are being selected among English Heritage historic sites in Yorkshire, such as Helmsley Castle

Throughout much of the current work in archaeology, concerning computer visualisation of buildings, there have been strong emphases on finding methods to achieve the most photorealistic representations possible. Whilst GIS enables landscape data to be rigorously explored, the built structure itself remains a relatively underused resource. An holistic reconstruction, therefore, might contain a building's acoustic properties and character in addition to its visual properties and therefore help to reveal alternative interpretations of its form and structure. Experimental archaeology has explored the acoustic properties of buildings bit is obviously not applicable to the virtual structure and moreover, does not generate the kind of data necessary for the generation of computer-generated sound models.

Acknowledgements

The first author is sponsored by the Portuguese Government through *FCT - Fundação para a Ciência e Tecnologia* (programme *PRAXIS XXI*). Special thanks to Paul Bryan (English Heritage) for his kind collaboration on this project.

References

Barron, M. 1993. Auditorium Acoustics and Architectural Design, London, E & FN Spon.

Beranek, L. B. 1954. *Acoustics*, New York, McGraw-Hill.

Campos, G., Howard, D. M. 2000. On the Computation Time of Three-Dimensional Digital Waveguide Mesh Acoustic Models, *Proceedings of the 26th EUROMICRO Conference*, Maastricht, Holland, 5-7 Sept., II, pp 332-339.

Campos, G., Howard, D. M. 2000. A Parallel 3D Digital Waveguide Mesh Model with Tetrahedral Topology for Room Acoustic Simulation, *Proceedings of the COST-G6 DAFx Conference*, Verona, Italy, 7-9 Dec., pp 73-78.

Van Duyne, S. A., Smith III, J. O. 1996. The 3-D Tetrahedral Digital Waveguide Mesh with Musical Applications, *ICMC Proceedings*, pp 9-16.

Van Duyne, S. A., Smith III, J. O. 1993. Physical Modeling with the 2-D Digital Waveguide Mesh, *ICMC Proceedings*, pp 40-47.

Fontana, F., Rocchesso, D. 1998. Physical Modelling of Membranes for Percussion Instruments, *Acustica – acta acustica*, 84, pp 529-542.

Hidaka, T., Beranek, L. L., Masuda, S., Nishihara, N, Okano, T. 2000. Acoustical design of the Tokyo Opera City, *Journal of the ASA*, 107 (1), January, pp 340-354.

Huopaniemi J., Savioja, L., Karjalainen M. 1997. Modeling of reflections and air absorption in acoustical spaces: a digital filter design approach, *Proc. of the IEEE Workshop on Applications of Signal Processing to Audio and Acoustics (WASPAA'97)*, Session 5.2.

Jahn, R. G., Devereux, P., Ibison, M. 1996. Acoustical resonances of assorted ancient structures, *Journal of the ASA*, 99, No. 2, February, pp 649-658.

Kuttruff, H. 1973. *Room Acoustics*. London, Applied Science Publishers.

Savioja, L.; Karjalainen, M., Takala, T. 1996. DSP Formulation of a Finite Difference Method for Room Acoustics Simulation, *Proceedings of the IEEE Nordic Signal Processing Symposium (NORSIG'96)*, Espoo, Finland, 24-27 Sept., pp 455-458.

Savioja, L., V. Välimäki. 1999. Reduction of the Dispersion Error in the Interpolated Digital Waveguide Mesh using Frequency Warping, *Proc. of the IEEE Int. Conf. Acoust., Speech, Signal Processing (ICASSP'99)*, Phoenix, Arizona, USA, 15-19 March, pp 973-976.

Figures

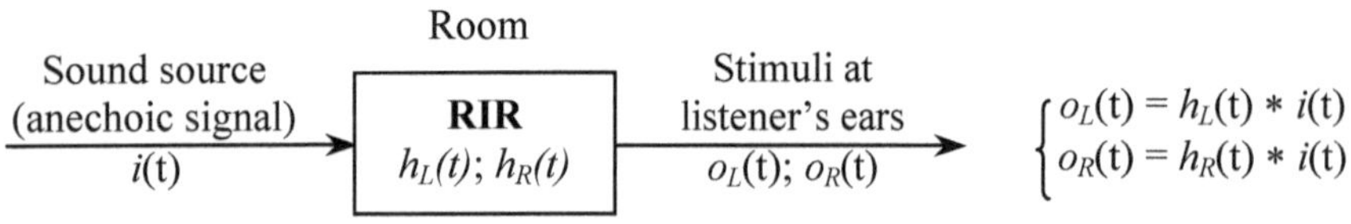

Figure 1. Auralisation ['' represents convolution; $h_L(t)$ and $h_R(t)$ stand for left and right ear RIR, respectively]*

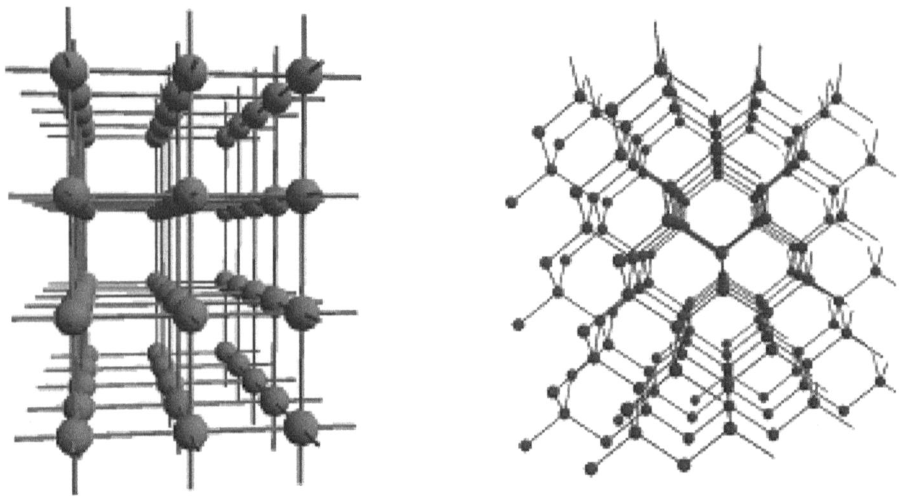

Figure 2. 3-D mesh topologies: Rectilinear (left) and tetrahedral (right)

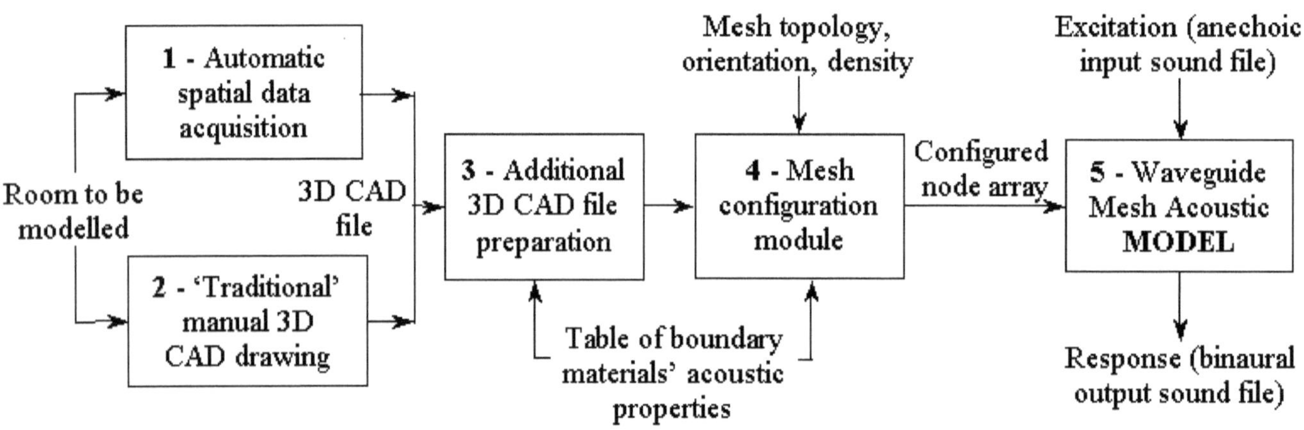

Figure 3. A 3-D waveguide-mesh modelling package with 3-D CAD front-end for room definition

A virtual journey through a Roman settlement. Aloria

Juan José Fuldain González
TEMPUS 3D
Paulino Mendivil no 5
1 Izq. Las Arenas (Getxo)
48930 Vizcaya, Spain
Phone: +94 480 15 55 - E-mail: fuldain@tempus3d.com

Abstract: This article presents the virtual reconstruction of a portion of a Roman Settlement found in Aloria (Alava – Spain). Such reconstruction was carried out using the 3DMAX program. The purpose of the reconstruction is to serve as a graphic base to the ultimate publishing of the memories of the excavation, the monographic exhibition about such settlement that will be soon held at the Archaeological Museum of Alava, as well as the making of a video for the exhibition and its later visualization on the net. It is this latest part, the visualization of the reconstruction on the net under WRML format at http://tempus3d.com, that the article is about.

Key words: virtual reality, virtual archaeology. WRML/VRML. Reconstruction. Education. Roman villa. 3D Modelling.

The virtual reconstruction of archaeological settlements is lately becoming the easiest and most attractive method of transmitting the results of the archaeological interventions to the general public because they are easily understood by everybody. On the other hand, the process of virtual reconstruction of any settlement arises new questions about the openings, roofing, etc, that archaeologists were not used to answer so far, making the analysis of the structures found deeper.

The reconstruction of the Roman settlement of Aloria has an special relevance given the fact that it has been covered, thus being impossible to be visited anymore.

The Roman settlement of Aloria is located in the Orduña valley (Fig.1), where the Nervion river is born and whose riverbed has always been used as a communication route between the Alavan plains and the Cantabric Sea. As a proof of its importance as a crossroad Orduña held the customs house for centuries just because most of the merchandises coming from Castilla had to pass through that strategic spot.

We clearly find three phases of occupancy, all of them dated from the Roman period (Fig.2). In the beginning, in the early 1st century AD, five enclosures were built. During the second phase of the settlement, around mid 2nd century AD, two of the enclosures were enlarged with tiled roofs, one of which led to a cesspool built with a restrain wall, creating a sort of *impluvium*. A sewer was also built to carry the water from the *impluvium*. Besides, four new enclosures were built, three of them were attached to the previous five, and the streets were paved. The last phase took place during the early 4th century AD. The enclosures built so far were falling into disuse and in ruin so a new one was built apparently performing the same functions as the previous ones.

The remains found in Aloria seem to be the stables, warehouses and workshops of a small *villa*. Such *villa* must have been placed a few metres uphill. However the extraction of soil from the place where the *villa* supposedly was, only left a big amount of debris all over the area. Out of the enclosures found in Aloria, none of them is a dwelling in itself. There are two stables, two warehouses, one forge, some attached tiled roofs and a barn in late Empire period. The remains of cow are numerous, as well as those of horses buried in one piece. There are also many slags as a result of a second treatment of metal in the forge. So far no remains have been found to make us think of a first transformation of iron in Aloria, basically because the ovens found were designed to work the metal, probably imported, and were not foundry ovens by themselves. Besides there are no veins nearby and there are not any foundry slags either.

The virtual journey in WRML is just a part of the graphic documentation and virtual reconstruction carried out in this settlement. The virtual reconstruction and the drawing of the plans (Fig.3) were possible thanks to the help and financing received from the Basque Society of Studies. Such reconstruction has several goals, being one of them the recreation of this virtual journey and its consequent release through the web. The reconstruction will as well be used as illustrations in the memories of the excavation and also in the monographic exhibition about the Aloria settlement that will be inaugurated in a coming future at the Archaeological Museum of Alava (Fig.4, 5 & 6). This exhibition will show a brief video focused on the 3D reconstruction of the hypothetical appearance of the buildings. The three phases previously mentioned, 1st, 2nd and 4th century AD, are explained thoroughly in the video.

The planimetric drawing of the settlement was recreated by recording the surface on video and then it was digitized. Afterwards, with the help of a topographic plan and applying the Airphoto program a photo mosaic was created in order to faithfully represent the structures and trace them.

The recreation of the 3D reconstruction was fully made with the 3D Max program (Fig.7). Then it was exported as WRML to be visualized in the web. The 3D program is also being used in the videos and illustrations of the exhibition at the Archaeological Museum of Alava and in the Memories of the Excavation (Fig.8).

The WRML virtual journey can be found in the web at: http://www.tempus3D.com, this journey is the 2nd phase of the settlement, 2nd century AD, the largest of the three.

First the geographical environment of the settlement was created (Fig.9). To do so, the Orduña valley was 3 dimensionally modelled using the level lines taken from the army topographic service map, scale 1:25,000, with a 20 metres equidistant between curves.

Once the topographic environment was finished the structures were lifted up (Fig 10). This is what you will find in the Internet address cited above. The virtual journey allows visitors to hang around by themselves or through an already determined path, following previously fixed points of view. Visitors can walk out of the predefined path anytime to look around by themselves having the choice of coming back to the journey anywhere.

The use of this kind of constructions is nowadays unquestionable, for it enables the general audience to be attracted to and understand the archeological findings which, otherwise, are only accessible to a few experts in this field. It becomes even more necessary when the referred findings are currently inaccessible or have disappeared, as it is the case of Aloria. The fact that the audience is able to understand and enjoy the archeological exhibition of a settlement with no flamboyant materials is greatly facilitated by 3D technologies. These design technologies allow for photographic-quality illustrations and videos to be made, as well as free visits to the settlement, which stimulate the curiosity of the visitors.

The WRML format allows us to publish the reconstruction on the web, being thus potentially accessible to a large audience at a low cost. Due to the reduced size of the files, WRML format is widely used on the web. The reduced size of the files comes as a result of the 3D environment reduction, in WRML there are no shadows nor light effects and the textures must be flat (2 dimensional) photographs. To sum up, WRML format provides us with a wide and cheap broadcast but it carries with it a great loss in quality. The connections to the net and the zipping programs are getting better and better day by day so it will soon be possible to load photorealistic 3D environments on the web and they will be easily accessible to everyone.

References

Adam, J.P. 1996. *La Construcción Romana, Materiales Y Técnicas*. Editorial de los Oficios. León.

Barceló, J.A. 2000. *Visualizing What Might Be: An Introduction To Virtual Reality Techniques In Archaeology*. BAR International Series 843 200. CAA 1999. Oxford. Pp. 9-37.

Choisy, A. 1873. *L'art De Bâtir Chez Les Romains*.Librairie Générale De l'architecture Et Des Travaux Publics Ducher Et Cie. Paris.

Fernández, P. A. 1993. *Arquitectura Y Urbanismo En La Ciudad Romana De Juliobriga*. Universidad de Cantabria. Santander.

Fernández, M. C. 1982. *Villas Romanas En España*. Complutum. Madrid.

Forte, M. 2000. *About Virtual Archaeology: Disorders, Cognitive Interactions And Virtuality*. BAR International Series 843. CAA 1999. Oxford. Pp. 247-260.

Frischer, B.; Favro, D.; Liverani, P.; de Blaauw, S.; Abernathy, D. 2000. *Virtual Reality And Ancient Rome: The UCLA Cultural VR Lab. Santa Maria Maggiore Project*. BAR International Series S843. Virtual Reality in Archaeology. Oxford. Pp 155-162.

Fuldain, J.J. 2000. *Computer Aided Drawing Of Archaeological Material*. Archaeological Computing Newsletter, number 55. The institute of archaeology. University of Oxford. Oxford. Pp. 7-15.

Pimentel, K., Teixeira, K. 1994. *Virtual Reality. Through The New Looking Glass*. U.S.A.

Prieto, J.J. 2000. *Virtual Reality And Reconstruction Of Archaeological Places: A Practical Task, "El Castillar" Of Mendavia (Navarra)*. BAR International Series 843 200. CAA 1999. Oxford. CDR.

Sanders, D.H. 2000. *Archaeological Publications Using Virtual Reality: Cases Studies And Caveats*. BAR International Series 843 200. CAA 1999. Oxford. Pp. 37-47.

Vitruvio, M.L. 1997. *Los Diez Libros De Arquitectura*. Obras Maestras Iberia. Barcelona.

Figures

Figure 1. *Aloria's location.*

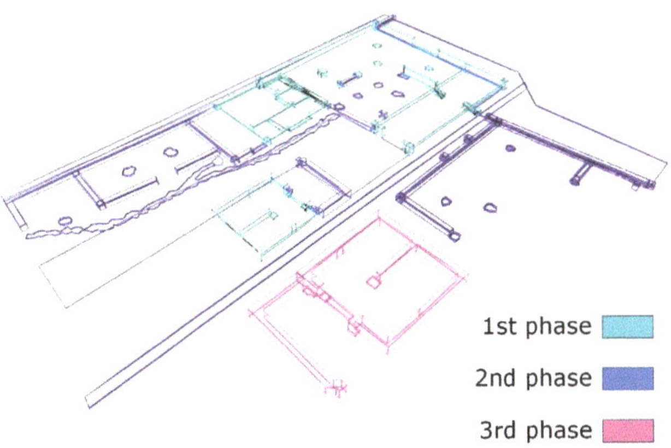

Figure 2. Phases of occupancy.

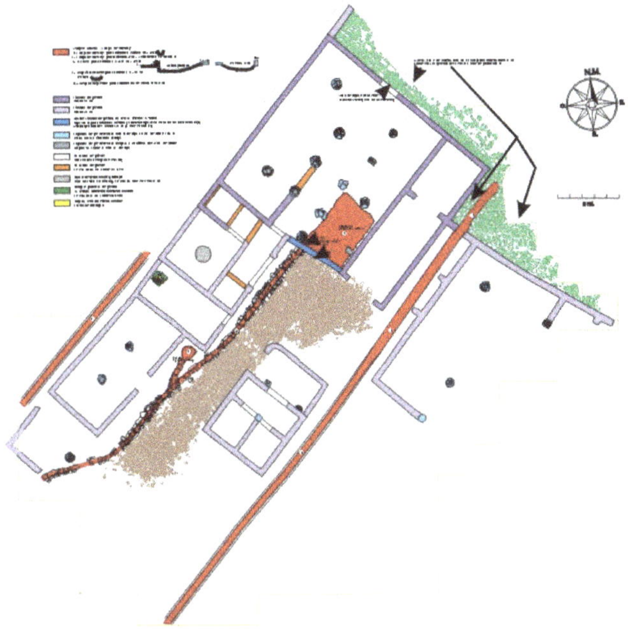

Figure 3. General plan of Aloria.

Figure 4. Inside the warehouse.

Figure 5. General view.

Figure 6. Enclosures M, W, Y.

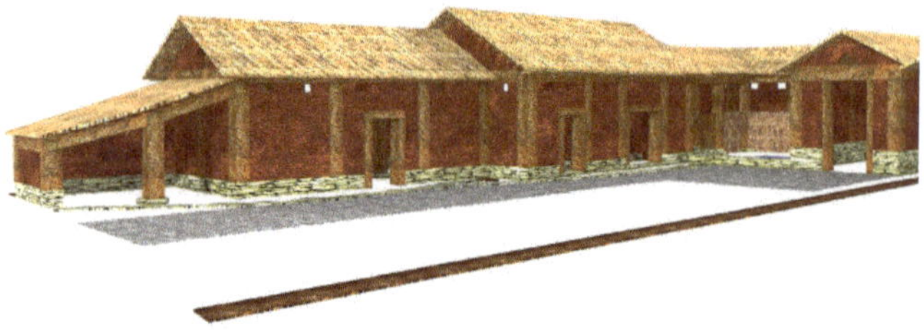

Figure 8. Enclosures M, W, Y, at the memories of the excavation

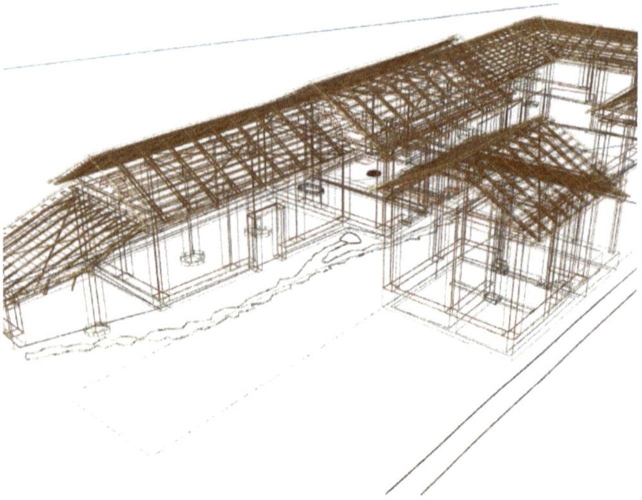

Figure 7. Modelling the reconstruction.

Figure 9. Geographical envoriment three dimensionally modelled.

Figure 10. Final aspect of the virtual reconstruction.

Topometrical measurements in Tiryns, Greece
Report on a co-operate project between physics and archaeology

Maria Shinoto
Institut für Ur- und Frühgeschichte
Marstallhof 4
69117 Heidelberg, Germany
http://www.uni-heidelberg.de/institute/fak8/ufg
E-mail: maria@shinoto.de

Zoltán Böröcz
Institut für Experimentelle Audiologie
Labor für Biophysik
Universtität Münster
Robert-Koch-Str. 45
48129 Münster
http://hobbes.uni-muenster.de/ExpAud/
E-mail: borocz@gabor.uni-muenster.de

Carsten Thomas
Institut für Experimentelle Audiologie
Labor für Biophysik
Universtität Münster
Robert-Koch-Str. 45
48129 Münster
http://hobbes.uni-muenster.de/ExpAud/
E-mail: thomaca@gabor.uni-muenster.de

Dieter Dirksen
Institut für Experimentelle Audiologie
Labor für Biophysik
Universtität Münster
Robert-Koch-Str. 45
48129 Münster
http://hobbes.uni-muenster.de/ExpAud/
E-mail: dirksen@gabor.uni-muenster.de

Joseph Maran
Institut für Ur- und Frühgeschichte
Marstallhof 4
69117 Heidelberg, Germany
http://www.uni-heidelberg.de/institute/fak8/ufg
E-mail: m17@ix.urz.uni-heidelberg.de

Gert von Bally
Institut für Experimentelle Audiologie
Labor für Biophysik
Universtität Münster
Robert-Koch-Str. 45
48129 Münster
http://hobbes.uni-muenster.de/ExpAud/
E-mail: bally@uni-muenster.de

Abstract: The cooperation between archaeologists from Heidelberg and physicists from M•nster started in 1999 in order to develop an application based on optical systems of phase measuring profilometry for non contact acquisition of 3D coordinates from objects of various sizes. 3D coordinates are stored in file formats that can be processed for 3D mapping, interactive 3D analysis, video output and two dimensional documentation.

Already existing devices had to be adapted to transportation and to the situation on site and in the museum. A modular software with parallel and automized data processing as well as automatic matching of point clouds from single measurements with special reference cubes was developed. Due to automation, the measuring process became quicker and flawless, and the amount of data to be stored was reduced dramatically. The modular software design is open to additional, independent features.

This paper presents the steps of development and three examples of different scale: A small find, a burial during excavation and a large stair case from the site of Tiryns.

Key words: phase measuring profilometry, photogrammetry, 3D, fringe projection technique, digital image processing, Tiryns, small finds, architectural remains

Project outline

The Laboratory for Biophysics in the department of medicine at the university of Münster is focused on research in optics and has a tradition in cooperation with social sciences (e.g. Dirksen et al., 2000.; Dirksen, von Bally, Bollmann, 2000). A few years ago cooperation with archaeological institutions began, and in 1999 the archaeological institute (Ur- und Frühgeschichte) at the university of Heidelberg joined a project to create a portable system for non-contact 3D measurement of archaeological finds and sites of various sizes on the basis of photogrammetry (Böröcz et al., 1999, Böröcz et al., 2000). Münster already had some experience with a similar, non-portable system that had previously been developed for medical applications.[1]

The task of the archaeologists was to check the developments in Münster for their archaeological relevance, organize test measurements of sites during excavation and of archaeological remains and small finds. They had to take part in the test measurements to evaluate the applicability for archaeologists. Test measurements took place in two campaigns in Tiryns, Greece. Tiryns is a site in the Argos plain near Nafplion. Although it bears remains from many archaeological periods, it is best known for its Bronze Age remains on a hill rock and in the surrounding plain which are of great importance in the Aegean world beside those in Mycenae, Pylos and Theben. Tiryns had been excavated in several campaigns between 1884 - 1888 by H. Schliemann and W. Dörpfeld, between 1905 and 1929 by W. Dörpfeld, K. Müller and G. Karo, and by K. Kilian between 1976 and 1985 and it has been on the UNESCO World Cultural Heritage List since 1999. Since 1994 new excavations of the Institut für Ur- und Frühgeschichte, Heidelberg on behalf of the German Archaeological Insitute in cooperation with the Fourth Ephorate of the Greek Antiquity Service are going on in the lower citadel (Unterburg) and the northeastern lower town (Maran 2000). The project team could carry out measurements of the cyclopic walls, of architectural remains, of the excavation site and of small finds in the storehouse.

In spring 1999 the team worked on portability and outdoor applicability of the system and focussed on basic archaeological problems. The system should be stable rather than elegant or fast in this project phase. It was tested on site during the excavation campaign in Tiryns in summer 1999. Analysis of the experiences from the test phase lead to a new design of many modules - hardware, software and additional instruments. In the following year until the next excavation campaign, the team tried to realize at least part of the ideas. These were tested successfully in summer 2000, which means that only slight changes to the ideas from 1999 have to be implemented now, development can go on as designed in autumn 1999.

Technical Background

Physics

The basis of the system is a topometric sensorhead consisting of two CCD cameras, and a fringe projector in the centre. During the measuring process a sequence of phase shifted quasi-sinusoidal fringes from a slide made of metal coated glass substrate is projected onto the object and registered as stereo images by the CCD cameras. After calculating the phase distribution from the digitized images, the stereoscopic images are evaluated by photogrammetric techniques, and a 3D coordinate is calculated for each valid pixel. The achievable measurement accuracy depends on the triangulation angle (the angle between the cameras), the image field size, as well as on the number of camera pixels. With a triangulation angle of 40 degrees and an image diagonal of approximately 20 cm for small finds for example, the height resolution is situated at less than 50 micrometers, the lateral at approximately 200 micrometers.

Using the obtained coordinate points - the point cloud - the object surface is reconstructed by triangulation, that is by covering it with a grid of triangles. As the investigated objects are quite complex, an approach is used which does not make special assumptions about the surface structure, it iteratively connects neighbouring points, only controlled by preset angular and distance constraints. It allows automatic processing - including the computation of range images - of large numbers of samples and single measurements (Dirksen, Kozlov, von Bally 1997, Dirksen et al. 2000).

To match sample point clouds from different angles into one 3D model, at least three identical reference points have to be identified in the first point cloud as well as in the second. These reference points can be part of the object itself or can be added into the object artificially before measuring. Artificial reference points can be designed for automatic identification by the software.

Outline of the measurement procedure

At the beginning of each measurement process, the camera lenses and their triangulation angle are chosen according to the size of the object and the measurement accuracy to be achieved. Cameras and slide projector are positioned on the sensorhead and calibrated before the measuring session starts (fig. 1).

For measuring, the slide projector throws patterns of light and shadow on the object (fig.3). The use of white light projection requires darkness or at least a subdued background illumination in order not to distort the projection. In a museum this is usually no problem, on site, however, measurements have to be made during night or with e.g. a tent. Fig. 2 shows the adjustment of lightning before projecting the fringe pattern that are recorded by the two video cameras. Colours in the false colour image inform about best lighting conditions. In some cases, measurements have to be done twice because the CCD cameras lack a sufficient dynamic range.

After acquiring the 3D coordinates, the reference points have to be identified. If artificial reference points are used, they have to be deleted from the point cloud since they do not refer to the real object. The next measurement can take place, it has to include the reference points of the previous measurement to be adjusted to the next point cloud (fig. 4).

To check the results, all point clouds can be matched from time to time and evaluated in a viewer on the computer screen (fig. 5) or by the software.

Many of these steps can be automized, which was one of the main tasks of the developing team. They shall be introduced later.

Hardware equipment

The project started with this hardware equipment: The main processor was a Pentium II with 266 MHz and 256 MB of RAM storage (PS/2 EDO-RAM), had 10 and 4 Gigabyte hard discs. After each measurement session, backup CDs had to be burnt. Different camera lenses are used for objects of different size, focal length used on site most often is 6mm, for small finds 12 mm, rarely 25 mm, all Pentax/Cosmica. The camera is XC-75CE from Sony, with one colour channel of 8 bit depth and a solution of 768 by 572 pixels.

The basic measurement software that runs the projecting device and cameras (ATOS, Advanced Topometric Sensor) was provided by the company GOM, Braunschweig, Germany. The software was partly replaced by new modules developed in the course of the project (OSCAN Toolbox, Object Scanner Toolbox). Both work with Linux as Operating System.

In the first measurement compaign the computer equipment was completely busy during repositioning of the sensorhead. The developement of OSCAN provided automation of time-consuming manual parts of the original measurement process. Therefore in the second campaign the attended time for computing the 3D-information for each measurement was greater than the time for repositioning the sensorhead. Implementation of new computer hardware will increase the measurement process by a factor of 3 and more, depending on the find.

Reference points, reference tables and reference cubes

The project started with reference points that were individually attached to the object. They had to be identified by mouse clicks and named in a dialogue box. For calculating the 3D coordinates of the object itself, the area of these additional points had to be cut out before calculating the point cloud. For automatic identification and naming of reference points, reference tables were developed, which worked reliably already during the first test phase. For the next test phase, these were further improved to reference cubes (fig. 6).

Advantages of the cube are manyfold: Instead of placing several refence tables facing to different directions, it only needs one cube in a certain area because the software already knows the identity of reference points and the 3D coordinates on all sides of the cube. The design allows the automation of several steps during the measurement.

The three circles form the reference points. The binary code differs on each side of the cube and on each cube, the software can identify the reference points automatically. The size of the cube differs according to the size of the measurement area. Sometimes, it is necessary to measure an interesting spot more detailed. For this case, the small pattern can serve as reference points.

The cubes can be dismantled, so they need to be calibrated before usage, another part of the OSCAN toolbox. Small cubes for small finds were very stable, they are not dismantled and can be used without calibration for several weeks and even after long transports. The reference cubes are filed to patent by the German Patent Office Munich, AZ 100 50 892.8 1 2.

Simplification of the measurement process

Fig. 7 shows the steps of measurement between calibration of the cameras and storing the data. The steps displayed in the light grey area refer to the object or take place in the site itself, those in the dark grey area are steps of control on the computer screen. The most important improvement in the light grey area are the reference cubes, the most important improvement in the dark grey area is automation. It increases measurement speed by a factor of 4 to 5, not to mention handling errors that can be avoided.

In addition to automation, improvements in the dark grey area of the diagram refer to data management. The original system stored every acquired coordinate point for each single measurement. It was impossible during the measurement session to check the overlap between single measurements, thus, many coordinate points were stored more than once. From time to time, the measurement process was interrupted to check the point cloud in a viewer. OSCAN now calculates the overlap and deletes coordinates which are stored more than once after each measurement automatically without using the viewer. The user is always informed whether there is too much overlap, so the sensorhead can be focused on a new part of the object instead of repeating measurement of the same part from a different angle. Reduction of data by avoiding redundant measurements and by deleting reduntant coordinate points is another achievement of OSCAN.

The preceptions from the campaign in the year 2000 lead to a new graphic user interface for a more friendly use with additional functions for documentation of the measurement and analysis. Thanks to the modular software design, additional modules can be developed without detailed knowledge of the internal software structures.

Fig. 8 shows two examples of the new user interface. OSCAN can handle several objects measurings at the same time. Each object is stored in a file structure as a "project", the four most important projects can be accessed immediately with the button on the control dialogue box. The dialogue box in fig. 8b has to be checked after each single measurement: As soon as the software has identified the reference points, the numbers are shown on the screen in order to check them. With a click on the button "accept" the next steps go on automatically as shown in fig. 7.

Samples of application

Figs. 9, 10, 11 show examples of measurements of objects with different size, table 1 compares effectiveness and speed of measurement. Fig. 9 shows a small find from Tiryns. Cross sections can substitute drawings; cross cuts from cuneiform tables, seals or cast objects are particularly interesting for archaeological analysis, since they show details in the range of several micrometers. Measurement of this sample took place in the second campaign with a reference cube.

Fig. 10 shows a burial from the lower town. Measurement took place during the second campaign, but since reference cubes could not be positioned in the steep enclosure, reference tables were used instead. Their design was improved exactly like that of the reference cubes, so the whole measurement process was automized.

Fig. 11 shows a staircase from the citadel (Mittelburg) that was measured in the first campaign. With a length of more than 8 m it is an extraordinarily large object.

Each measurement session took place under different conditions. Table 1 shows, how much the improvements between first and second campaign served a quick and efficient measurement: For the burial, almost all of the acquired coordinate points were used, whereas for the staircase in the first year measurement almost two thirds of the acquired points were still reduntant. The table presents the original measurement time for the staircase and a time estimation calculated from the parameters of the automized procedure of the second year. With the new hardware equipment, the measuring time can be further improved. The effectiveness for small finds seems to be low on first sight, but the object covered only about 10% of the area the cameas could capture. The redundancy of coordinate points on the object itself was only about 10%, which is even better than the results from the burial.

Discussion

Non contact acquisition of 3D coordinates on the basis of photogrammetry is a well established technique and already employed for documentation of archaeological finds. This project developed a high-precision and easy-to-use system that cannot only be applied to smaller objects, but also in the field of excavation, to scan sites and large objects like walls and other architectural remains. The robust equipment can be transported and used anywhere. The results are as accurate as those of established indoor solutions for smaller objects. 3D data of this kind are inevitable for VR-site documentation that may play a more important role in future archaeological publications, but as well for better methods of analysis. Further modules for presentation, documentation and analyses can be developed as additional modules (add-ins) to OSCAN.

End notes

This and similar techniques are being applied in several interdisciplinary projects concerning archaeology and related fields, some of them have been introduced in this and preceding CAA conferences. More projects and literature are introduced in the Internet, this is an arbitrary selection:
http://www.brunel.ac.uk/project/murale/home.html
http://www.prip.tuwien.ac.at/Research/ArcheologicalSherds/lit.html
http://www.research.ibm.com/pieta/index.htm
http://www.research.ibm.com/pieta/pieta_refs.htm
http://3dk.asu.edu/newhome.html
http://3dk.asu.edu/DOCUMENT/archives/publication/publication.html
http://www.i3mainz.fh-mainz.de/institut/personal/boochs/
http://www.i3mainz.fh-mainz.de/db/publikationen_thema?utab=autoren&thema=Boochs.

Acknowledgements

The project was supported by the Ministry of Education and Research of the Federal Republic of Germany, under contract number 03VB9MU2-0. We would like to thank the Fourth Ephorate of the Greek Antiquity Service for the support during the campaign as well as for the permission to publish some of the data.

References

Böröcz, Z., Thomas, C., Dirksen, D., von Bally, G. 1999. *Optical acquisition and evaluation of range data for analysis and documentation of archaeological samples*. Conference

Proceedings. EVA Europe _99 Berlin. ISBN 3-9807029-0-1.

Böröcz, Z., Thomas, C., Dirksen, D., von Bally, G., Rieckhoff, s., Fleischer, F., Teegen, W., Funke, P., Freitag, K. Rummel, N., Maran, J., Shinoto, M., Papadimitriou, A. 2000. *Automation of topometric 3D-acquisition of complex surfaces using optimized reference cubes*. Elektronische Bildverarbeitung & Kunst, Kultur, Historie. Proceedings der EVA Europe 2000 Berlin. Electronic Imaging & the Visual Arts. ISBN 3-9807029-2-8.

Dirksen, D., Kozlov, Y., von Bally, G. 1997. *Cuneiform surface reconstruction by optical profilometry*. 257-259, D. Dirksen, G. von Bally (eds.), Optics within Life Sciences (OWLS IV). Optical Technologies in the Humanities. Heidelberg: Springer.

Dirksen, D., Freitag, K., Böröcz, Z., Thomas, C., Fell, M., Ruffing, K., Funke, P., von Bally, G. 2000. *Dreidimensionale optische Vermessung von Inschriftenträgern für die digitale Analyse und Dokumentation*. 66-81, M. Hainzmann, Ch. Schäfer (eds.), Computer und Antike, 5, Alte Geschichte und neue Medien. St. Katharinen: Scripta Mercaturae.

Dirksen, D., von Bally, G., Bollmann, F. 2000. *Automatic acquisition and evaluation of optically achieved range data of medical and archaeological samples*. 147-150, C. Fotakis, T. Papazoglou, C. Kalpouzos (eds.), Optics Within Life Sciences (OWLS V). Biomedicine and Culture in the Era of Modern Optics and Lasers. Heidelberg: Springer.

Maran, J. 2000. *Tiryns. Mauern und Paläste für namenlose Herrscher*. 118ff., Archäologische Entdeckungen. Die Forschungen des Deutschen Archäologischen Instituts im 20. Jahrhundert.Mainz: Zabern.

Tables

	small find	**burial**	**staircase**	**staircase (calculated)**
time	0.5h	4h	12h	4h
single measurements	8	35	200	97
max. achievable pixels	3,500,000	15,000,000	88,000,000	43,000,000
pixels calculated	300,000	11,000,000	34,000,000	34,000,000
effectiveness	9% (90%)*	73 %	39 %	77 %
average distance of pixels	0.2mm	1-2mm	5mm	5mm

Table 1. Comparison of measurement parameters from the samples.
** The small find covered approximately 10 % of the area that was measured due to the measuring field size of the cameras.*

Figures

Figure 1. Equipment for measuring on site: Sensorhead with cameras and slide projector in its centre (front) and computer with frame grabber on top of a transportation box (back).

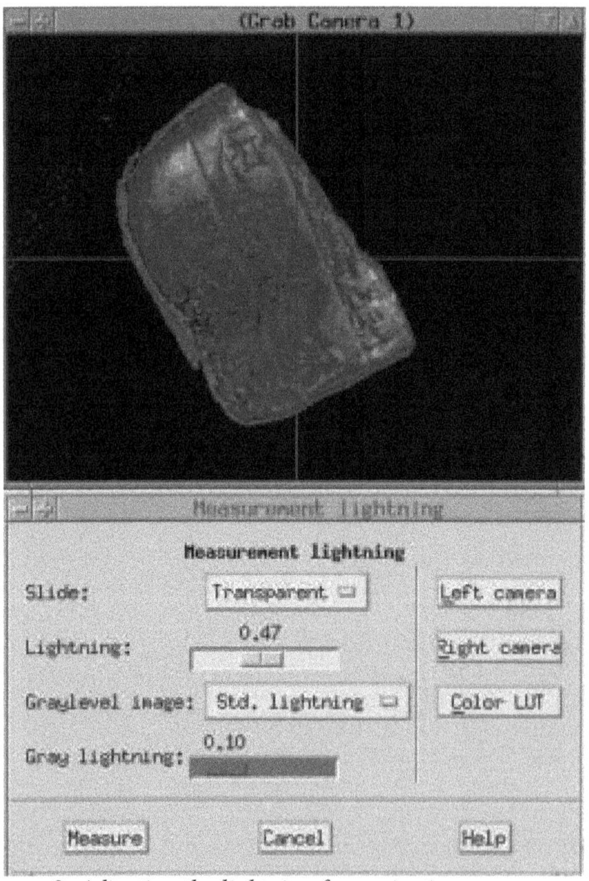

Figure 2. Adjusting the lightning for projection.

Figure 3. Projecting fringes for measurement. Back right: Reference cube for small finds.

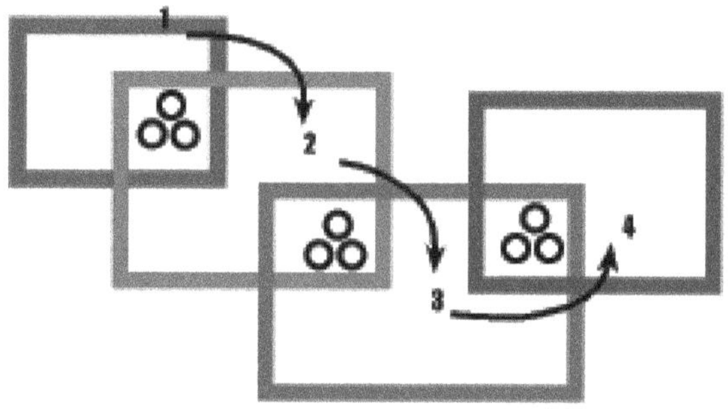

Figure 4. Covering large surfaces with reference points. Measurement steps onwards, always covering three previous and three new reference points.

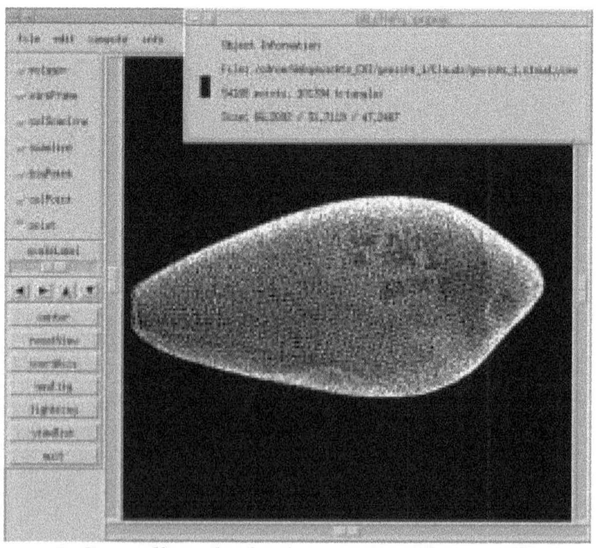

Figure 5. Controlling the final point cloud from single measurements on the screen. The viewer plots e.g. 55,000 3D-coordinates out of several million acquired and provides zooming etc.

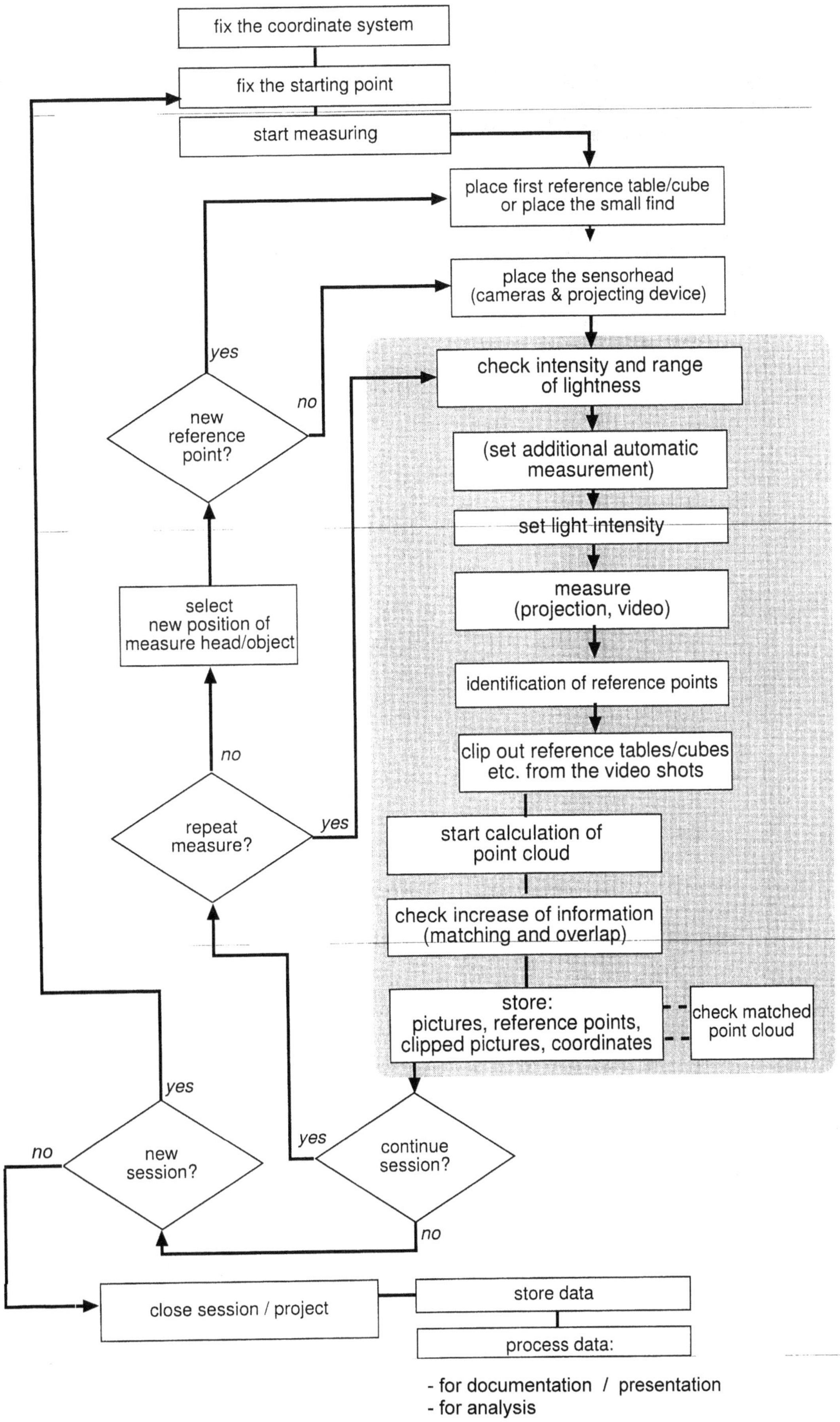

Figure 7. Steps of the measurement procedure.

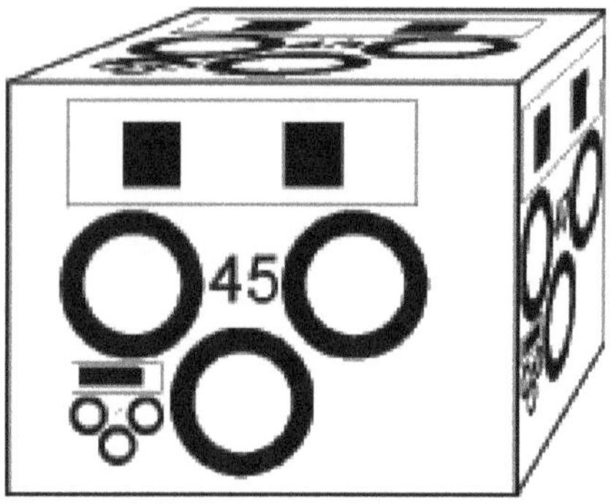

Figure 6. Reference cube.

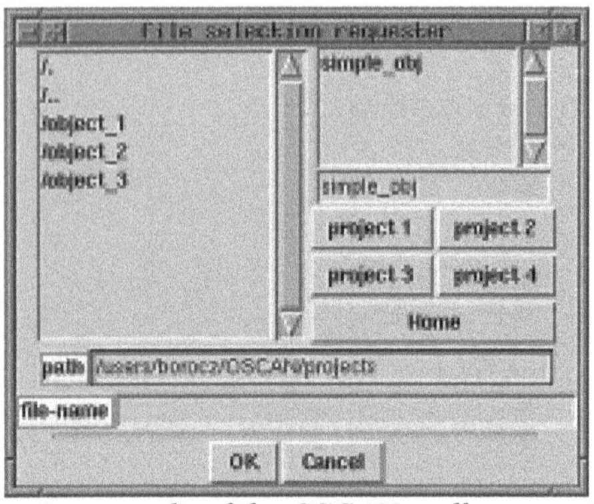

Figure 8a. An example of the OSCAN toolbox user interface. Management of measuring projects from a central control dialogue box.

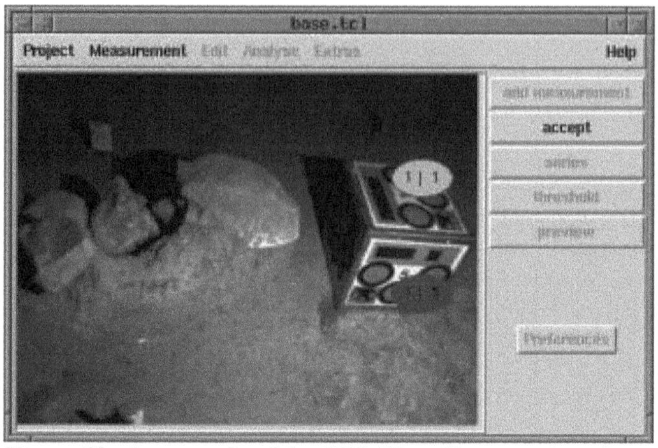

Figure 8b. An example of the OSCAN toolbox user interface. Checking the automatic identification of reference points on different sides of a reference cube in a graphic on screen.

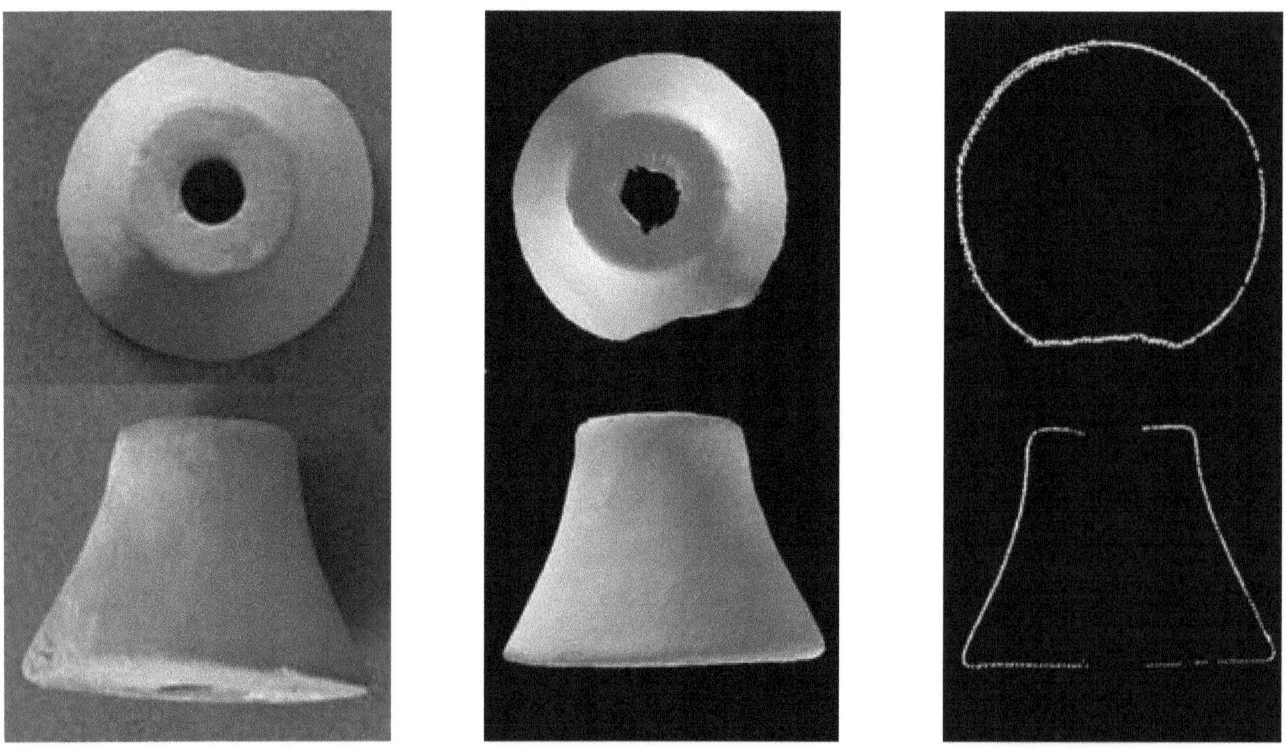

Figure 9. Small find from Tiryns, stone. (a) Photograph, (b) triangulated model, (c) cross section of the point cloud.

10a

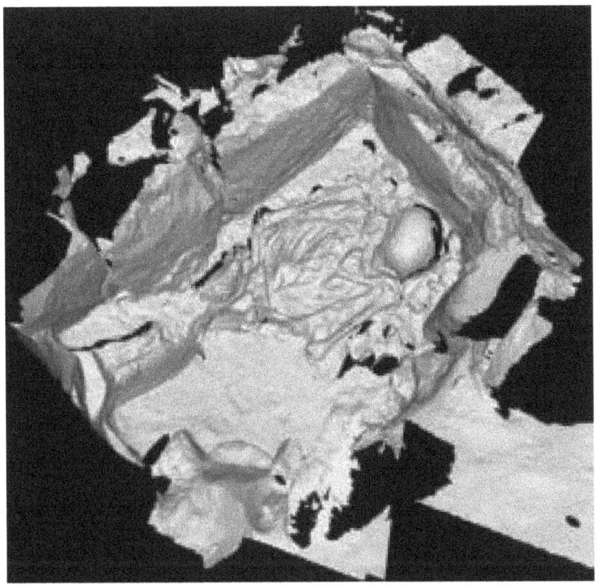

10b

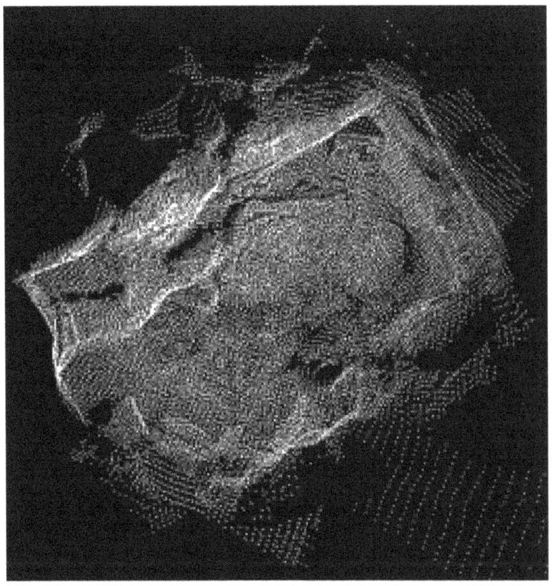

10c

Figure 10. Burial in the lower town, Tiryns. (a) Photograph, (b) point cloud, (c) triangulated model.

11a

11b

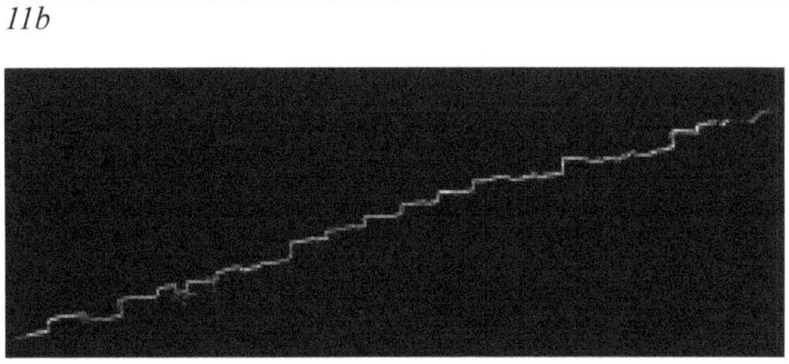

11c

Figure 11. Staircase in the Mittelburg, Tiryns. (a) Photograph, (b) triangulated model, (c) cross section of the point cloud. A video (9d) is provided as a download available at www.barpublishing.com/additional-downloads.html.

Advances in Geometric Modeling and Feature Extraction on Pots, Rocks and Bones for Representation and Query via the Internet

Utsav A. Schurmans
Department of Anthropology
University of Pennsylvania
325 University Museum
Philadelphia, PA 19104-6398, USA
Phone: +1-215-898-7461 - Fax: +1-215-898-7462

Anshuman Razdan
Partnership for Research in Stereo Modeling
Arizona State University
Tempe, AZ 85287-5106, USA
Phone: +1-480-965-5368 - Fax: +1-480-965-8692
E-mail: razdan@asu.edu

Arleyn Simon
Department of Anthropology
Arizona State University
Tempe, AZ 85287-2402, USA

Mary Marzke
Department of Anthropology
Arizona State University
Tempe, AZ 85287-2402, USA

Peter McCartney
Center for Environmental Studies
Arizona State University
Tempe AZ 85287-2402, USA

David Van Alfen
Partnership for Research in Stereo Modeling
Arizona State University
Tempe, AZ 85287-5106, USA

Gram Jones
Partnership for Research in Stereo Modeling
Arizona State University
Tempe, AZ 85287-5106, USA

Mary Zhu
Partnership for Research in Stereo Modeling
Arizona State University
Tempe, AZ 85287-5106, USA

Dezhi Liu
Partnership for Research in Stereo Modeling
Arizona State University
Tempe, AZ 85287-5106, USA

Myungsoo Bae
Partnership for Research in Stereo Modeling
Arizona State University
Tempe, AZ 85287-5106, USA

Jeremy Rowe
Information Technology
Arizona State University
Tempe AZ 85287, USA

Gerald Farin
Computer Science and Engineering Department
Arizona State Unversity
Tempe AZ 85287-5406, USA

Dan Collins
Institute for Studies of the Arts
Arizona State University
Tempe AZ 85287-3302, USA

Abstract: This paper outlines progress on the 3D Knowledge and Distributed Intelligence project (3DK) at Arizona State University. Three of the six 3DK pilot projects involve archaeological and biological material, namely ceramic vessels, lithic artifacts, and bones. These projects are: (1) "3D Morphology of Ceramic Vessels" which aims to learn about vessel uniformity and proportionality as indicators of developing craft specialization and complex social organization among prehistoric cultures; (2) "Lithic Refitting" which intends to develop the algorithms required to (partially) automate the refitting process through 3D scanning and surface modeling; and (3) "3D Topography of Joint Surfaces" which investigates the ability of human ancestors to make tools and walk upright. This paper discusses underlying research aspects common to all three pilot projects: data acquisition, modeling, feature recognition, database construction, and development of a visual query interface aimed at making 3D models of these artifacts available online for query.

Key words: 3D; modeling; query; archiving; feature extraction; ceramics; lithic refitting; joint surfaces

Introduction

Recent theoretical and technological breakthroughs in mathematical surface modeling and data-capturing techniques present the opportunity to advance 3D knowledge into new realms of cross-disciplinary research. The focus of PRISM's 3D Knowledge (3DK – http://3dk.asu.edu/) project is to research and develop the acquisition, representation, query and analysis of 3D information within a distributed environment such that researchers in many fields can exploit the full potential of 3D objects and phenomena.

The understanding of 3D structures is essential to many scientific disciplines, including archaeology and physical anthropology. For example, (1) archaeologists study the 3D form of Native American pottery to characterize the development of prehistoric cultures; (2) Lithic researchers conjoin the remains of individual lithic reduction sequences to understand the development of stone tool technology over time; and (3) Physical anthropologists compare 3D joint surfaces of human and nonhuman primate bones and study the effects of surface area, curvature, and congruency on the mechanics of a joint to reconstruct manipulative and locomotor behavior of early humans. To date, the methods that anthropologists use to measure, record, and classify 3D data such as calipers, photographs, drawings, and the use of classification schemes that reduce objects to simplified descriptors are inadequate for accurately representing such complex data. Within the past decade, improved technologies for capturing 3D objects have emerged, offering the possibility to obtain and store accurate mathematical models of objects in digital form.

At the heart of the 3D-knowledge environment are the kernel, the user interface, and the data archiving (Figure 2). The system enables user-access to archived 3D data at multiple levels of abstraction. The power resides in its ability to accept queries posed in a variety of forms and at high levels of abstraction and return the results as visualizations of the 3D objects and phenomena. Queries take the form of text, numerical data, visual primitives, and hand sketches. Input to the 3D Knowledge kernel is made through a variety of methods and agents including object segmentation, feature extraction, feature analysis and correspondence, indexing, and relationship to other objects. The 3D knowledge extraction phase of the project is the most challenging one to implement. The filtering of the representation from raw data to knowledge in the 3D Knowledge Kernel comes from the joint interaction of archaeologists, computer scientists and engineers to enable definition and query of features. To facilitate use among a broad range of researchers, instructors and students, the 3D Kernel visual query interface is accessible via the World Wide Web at http://3dk.asu.edu. Further, this research aims to develop a universal vocabulary of cognitive pattern primitives (CPPs) and an algebra (Feature Algebra) for defining complex features that cross disciplinary boundaries. We are building upon such feature extraction and representation methods in order to create a robust 3D-knowledge environment.

To effectively conduct research using 3D data, one is required to go through a number of steps irrespective of the particular application (Figure 1). This paper focuses specifically on these steps and applies them to the pilot projects involving pots, lithics, and bones.

Pilot projects

3D Morphology of Ceramic Vessels

The aim of the prehistoric ceramic vessel pilot project is to learn about vessel uniformity and variation with respect to function as indicators of developing craft specialization and complex social organization among prehistoric Native American cultures. To accurately capture the complex curvatures and vessel form and size variation, the use of metric rulers and visual inspection of 2D profiles is inadequate. Therefore, scalable, visual, and quantitative comparisons of curvatures have been developed for the study of prehistoric pottery traditions.

Based on 3D scanned data, 2D measurements (height, rim diameter, minimum diameter, maximum diameter) and 3D measurements (area, volume, volume of wall) are generated, and a measure of symmetry is developed (Figure 3). The pilot project uses a control collection consisting of two Mexican flowerpots, two Mexican mold-made piñata pots, and three hand-built Tarahumara jars to verify the validity of the formulas generated. The two prehistoric collections studied consist of a total of 87 Native American vessels from the Tonto Basin in Arizona, and the Casas Grandes in northern Mexico.

The control study and the two sets of prehistoric vessels are used to generate a quantitative analysis of vessel uniformity and symmetry. The automated measurements are important improvements that facilitate the quantitative assessment of ceramic uniformity and standardization, both of which are indicators of the development of craft specialization, differentiation of labor, and the development of complex forms of social organization among prehistoric cultures.

Lithic Tool Manufacture and Refitting

The goal of the lithic refitting pilot project is the reconstruction of stone tool manufacturing activities and related cultural behaviors within prehistoric archaeological sites by identifying sequences of conjoining lithic artifacts. Refitting lithic artifacts by manual trial and error is an accurate, but highly labor-intensive, method that requires the entire sample of artifacts to be present in a single lab. These conditions are not always possible given various antiquities restrictions. Automated 3D surface matching of conjoinable artifacts will dramatically enhance the efficiency of this valuable analytic method and extend the scope of searches to collections from different sites. Lithic refitting, or assembling the pieces of stone, broken apart by prehistoric people, has proven very useful in our attempt to better understand the prehistoric past. The technique is especially useful in technological studies, taphonomic research, and spatial analysis. The 3DK project is developing software that would reduce the time cost of refitting and facilitate development of an electronic database storing 3D images of lithic collections available for study by researchers all over the world. Theoretically, there are two ways to go about refitting. One starts from the core to which the last flakes removed from that core can be fitted (Figure 4). Once fitted, the refit set is regarded as the new core to which still other artifacts can be fitted. The advantage of this approach is that the negative scar on the core fits perfectly with the ventral surface of the detached flake, allowing us to look for a perfect match between two surfaces. However, this approach starts from the unwarranted assumption that all pieces will be complete and even more problematic, that all pieces fitting in the original block are present at the site. Therefore a second approach is needed, one in which we can, not only search for perfect matches, but also search for partial fits between two surfaces.

For this study we are currently using experimental reduction sequences as a test to evaluate and refine matching algorithms. Once the algorithms are developed we will proceed and use the technique on WHS 623X, an Ahmarian (Upper Paleolithic) site from the Wadi al-Hasa in West-Central Jordan. At this small knapping station some refitting has been done, allowing us to evaluate the manual refitting results against computer based refitting results.

3D Topography of Joint Surfaces

The objective of the "3D topography of joint surfaces pilot project" is to further our understanding of the abilities, limitations, and adaptations of our early human ancestors to make tools and walk upright. Developing biomechanical models of manipulative and locomotor behavior using 3D osteological data does this. The use of calipers and visual inspection to capture the complex curvatures of 3D joint surfaces and to control for body size differences in cross-species comparisons is inadequate. We expect to overcome this by developing scalable, quantitative models of reciprocal wrist, hand, and knee joint surfaces that will allow for comparative quantitative analysis of the effect of surface area, curvature, and congruency on joint mechanics in extant and fossil apes and humans.

Since the project inception, more than 600 bones representing the wrist and hand joints of humans, chimpanzees, gorillas, orangutans, and gibbons have been digitized to create a database that will eventually include approximately 1000 bones representing the wrist, hand, and knee joints of humans and apes. With an aim to better understand the functional morphology of joint surfaces, the segmentation of features from a bone that are of particular interest to a physical anthropologist, such as joint surfaces, may be automated, avoiding manual digitization of such features that is both time consuming and labor intensive. The surface areas and curvatures of joint surfaces and the congruencies between reciprocal joint surfaces are then quantified and analyzed. Using this digitized osteological data, combined with musculoskeletal data from cadavers and manipulative and positional behavioral data, static and dynamic models can be built to analyze the mechanics of manipulative and locomotor behavior.

Common elements

This paper focuses on the aspects common to the three pilot projects outlined above. All of these projects deal with digital 3D information and the development of a database structure to make 3D models of artifacts available online for query. These two elements constitute an intricate part of the 3DK research. In the following sections we will discuss five aspects crosscutting discipline specific applications. These are: (1) the data acquisition process, (2) modeling of point cloud data, (3) segmentation and feature extraction, (4) database storage, and finally (5) online display and query. These five aspects make this project truly an interdisciplinary research endeavor as the results of the project will provide contributions in the disciplines involved, such as computer science (specifically computational geometry), engineering, archaeology/anthropology, information technology, and database development research.

Data Acquisition

The 3DK project currently uses two laser scanners for data acquisition: a Cyberware Model 15 scanner and a Cyberware Model 3030RGB/MS scanner. The former is a mobile scanner particularly useful for smaller objects (Figure 6), while the latter is a large stationary scanner, which allows for the capture of large objects. Both scanners are equipped with turntables on which the objects to be scanned are mounted. The objects are scanned horizontally; subsequently the turntable rotates by a user-defined number of degrees, after which the scanners captures more information and so on. After this initial data capture, the resulting scans are manually edited and automatically merged, after which the object is repositioned to scan the bottom and top of the artifact. Finally, the orientations are merged to form the complete 3D image. Different types of objects require more or less scanning time. For example particular types of stone are more or less easily captured depending on the degree of translucence and luster of the artifact as well as the sharpness of the edges. In addition, complex artifact topographies require additional scans and orientations for an adequate 3D representation. As a result scanning time varies considerably depending on the size and complexity of the object ranging from under 15 minutes to two hours for scanning and data cleanup. Resolution and accuracy of the scans are a result of the distance between the scanning head and the object. The large scanner is therefore less accurate than the small scanner. For specifics on the accuracy and resolution of the laser scanners that we use, please refer to http://www.cyberware.com. Resulting from the data capture are sets of ordered point clouds as well as, in the case of the large scanner, a separate data file for color.

Modeling

When an object is scanned using a laser digitizer, several tens or hundreds of thousand point coordinates are generated, each representing a location on the object. This collection of points is referred to as a 'point cloud'. It has no structure and is simply a file containing x ,y, and z values. In order to work with this point cloud data, it has to be structured or modeled. The most common method used is to triangulate the point cloud data points. The result of this process is a collection of triangles having the digitized points as vertices (Figure 7). Each triangle 'knows' about its neighbors, which is the structure that allows fast processing of the geometry represented by the triangulation. It is important to stress that the same set of vertices or data points could have several triangulations. The emphasis therefore, should be on the vertices themselves, rather than the 'surface' as represented by the triangles.

3D triangular meshes are currently a popular surface modeling primitive, used extensively to represent real world and synthetic surfaces in computer graphics. In order to generate meshes, real world objects are often either digitized, as described earlier, or sampled as a volume from which a surface is extracted. Alternately, the points, and their connectivity, can be generated using computer software programs, either interactively (eg using a surface design tool) or automatically from mathematical functions.

Since triangulation is only a modeling primitive, additional modeling steps are necessary. Figure 8 shows the different elements of the geometric modeling module of the 3D kernel subdivided in surface and volume models. Surface contour shape information plays an important role in the analysis of archaeological material. To extract this information modeling is necessary to study the profile curve of a ceramic vessel, or the cross section of a bone, for example. To represent profile curves of archaeological vessels, we use two kinds of models: chain codes and NURBS (Non uniform Rational B-Splines) (Farin 1996; Razdan et al 2001).

Chain codes can be used to extract a 2D profile curve of a vessel. A cutting plane is projected through the 3D mesh and intersection points with the mesh are extracted and ordered in the chain code. These points are then connected to form the profile curve. After chain points are extracted, NURB curves are fitted to the chain using least squares approximations. The result is a profile curve from which appropriate features such as corner points, end points, and symmetry between both halves of the profile curve can be readily computed.

If an object is inherently smooth, it is advantageous to represent it by a smooth data format. The most widely used formats are NURBS. A NURB data set consists of control points arranged in a grid of points, roughly representing a given shape. NURB forms provide a compact data representation making it easy to calculate the curvatures of objects to extract their essential shape characteristics.

Feature Extraction and Segmentation

The next step, after the scanning and initial modeling of the artifacts, is the extraction of appropriate feature data for study. Indeed, the features are the primary goal of the subdiscipline-specific research. Although features such as corner points on a pot, joint surfaces on a bone, and flake negatives on lithics are diverse sets, they can be extracted using a common set of algorithms.

Extraction of features from objects and the use of these features to search for particular shapes involve multiple levels of analysis and extraction algorithms. However, as a common denominator, we assume that features are built from cognitive pattern primitives (CPP) that are essentially the geometrically meaningful feature building blocks. Examples of CPPs include curves, for example a part of a profile curve, and surfaces such as segments of the ventral surface of a lithic. Building features from these primitives requires a construction method, which allows these CPPs to be arranged and connected in various ways and for these connections and arrangements to be described mathematically. A set of CPPs and a set of operators to connect and arrange these CPPs (Figure 9) can then define features. Extracting information from modeled data can take on various forms depending on the desired information. We separate features into four categories: point, curve, surface and volume data. Point, curve, and surface features will be described using examples from the pilot projects. While we have not advanced our archaeological investigations into the realm of volume, such applications are being developed for our pilot projects dealing with biological cells and may be readily incorporated into archaeological research at a later time.

Extracting point features: corner points

Using the chain codes and NURB curves described earlier, we can extract profile curves on ceramic vessels (Razdan et al 2001). These profile curves have a number of points of interest to archaeologists such as corner points. Corner points can be described as an abrupt change in the orientation of a vessel wall, or a distinct angle in the joining of vessel parts such as the neck and body. For the automatic extraction of such points on a profile curve there is a need to adequately describe this feature mathematically. Hence, corner points are points of maximum curvature change on a profile curve and are extracted by comparing curvature values for all points on the profile curve. Comparison of the corner points identified mathematically (Figure 10) correspond very well with the locations of these points as identified visually by archaeologists.

Extracting curve features: profile curves

The profile curves contain various other sources of valuable information . One may be interested in the symmetry between left and right side of the vessel wall of the profile curve. This information is extracted into a signed curvature plot (Figure 11) and provides a valuable tool for the evaluation of vessel symmetry (Bae 1999). Further, we can generate an ideal model from the profile curve and evaluate how the real pot deviates from this model. These elements can provide information on the production of ceramic vessels and quantify the degrees of standardization within vessel classes. Finally, it should be stressed that although cutting planes to extract 2D information from a 3D model have here been applied to vessels only, the same procedure can be readily used to extract cross sections of bones.

Extracting surface features: facets on stone artifacts

A third and final level of feature extraction treated in this paper deals with surfaces. Surfaces form one of our major emphases as they allow us to retrieve 3D information rather than reducing a 3D phenomenon to 2D information. To extract surface information, the complete object has to be segmented into archaeologically and geometrically meaningful surfaces (Figure 12). Examples are the different facets on a lithic artifact such as the ventral surface, the partial dorsal surfaces, and the heel. These form the building blocks in any effort to refit these rocks into a complex 3D jigsaw puzzle.

Segmentation refers to the problem of extracting features or regions of interest from surfaces. Archaeologists observe and define features typically using subjective morphological definitions with implicit mathematical ramifications. The challenge, as in the corner point example is to make the criterion explicit. For example, ridges of high curvature bound facets of lithic artifacts.

Our first progress on segmentation applies to triangular mesh surfaces. Working with Anshuman Razdan and Gerald Farin, Sandeep Pulla used absolute curvature estimates along with the basic principles of a watershed algorithm (Mangan & Whitaker 1999; Pulla 2001; Pulla et al nd) to identify regions of interest and similarity on the surfaces of bones, lithics and pots. The problem of segmenting features, however, is not easily resolved. Often the definition of "meaningful" regions is application dependent, user dependent, or both. Owing to such dependency, decisions in segmentation are often ambiguous. We have, therefore, sought a solution that is unambiguous and consistent by way of its mathematical basis. "The key is to find the natural segmentation of a surface, rather than that forced by an arbitrary approximating scheme" (Fan et al 1987). However, fully automated segmentation is far from satisfying in many realistic situations. Some tweaking of internal methods and various tolerance values is therefore allowed in order to generate results that "best" suit a user's needs. (Pulla 2001)

Based on the idea of a watershed as used in geography, in which a watershed forms the dividing line between drainage basins, the computerized watershed segmentation scheme defines minima to which water would flow from the peaks surrounding that minima. By contrast to geography where this principle uses elevation in relation to neighboring areas as the defining characteristic for subdivision, this application uses absolute curvature allowing the recognition of peaks as well as valleys as the separating lines between watersheds (Figure 13). It was found that absolute curvature aided by smoothing equations yield superior results to Gaussian curvature. Applications of this method are particularly useful in segmenting facets on a lithic artifact (Figure 12). These regions in turn form the basis for the search for matches between stone artifacts.

The first step in the segmentation process is the computation and storage of curvature at each vertex or data point in the original point cloud. The curvature calculation for each point is based on a patch of nine or more points, around a particular vertex and the vertex itself (Pulla 2001). Next, absolute curvature minima are selected and form the "sink-holes" for the individual regions. Subsequently, vertices are assigned to a specific minimum by determining the way the imaginary water would travel (down the steepest drop in absolute curvature in this case). The initial regions are typically too numerous to be useful. Therefore, we increase the "watershed depth" to merge adjacent similar regions (Figure 14). Making the action of determining the watershed depth user defined we ensure flexibility to meet individual research needs. While significant depths that allow for the recognition of flake scars are typically found at similar watershed depths, these depths tend to be different for other application such as recognizing the joint surfaces on bones or the parts of a pot such as its neck, body, and base (Figure 15).

Data Storage

Having extracted a series of defining features from archaeological data, the next step is to organize the information to facilitate storage, query, and interaction with this information (Rowe 2001; Razdan et al 2001). An archiving and indexing system was developed and maintained to accurately describe and consistently search and retrieve shape data. Initial efforts to develop general index categories and feature descriptors are based on data from the pilot projects. While the data storage and query steps are similar across pilot projects, there are application specific differences across the pilot projects. Success will require reliable feature extraction, descriptions of features, shape algebra transformations, and indexing using an Object Data Base Management System (ODBMS).

To provide efficient access to text, volume, and surface information the project team developed a detailed and consistent strategy to catalog and organize the data. Significant research effort was devoted to identify and build upon the descriptive and cataloging standards that have traditionally been used to describe artifacts and on emerging metadata standards. For example, for the ceramic vessels, standards were adapted to describe surface models and derived mathematical descriptive data were analyzed to develop a vocabulary for the mapping of shapes, transition points, and critical data elements to identify and describe the vessels by shape and feature (Figure 16).

XML (eXtensible Markup Language) schemas were developed to provide a structure to organize, catalog, and describe the data elements. These schemas include application specific categories and are organized to differentiate general information (eg provenience, time period) and 2D information (eg vessel height, blade length) from 3D information. The 3D information data structures at the top level house the original binary data files, modeled files, and archaeological features. One level down the hierarchy, cognitive pattern primitives (CPPs) store the basic shape data that form the building blocks out of which meaningful archaeological features are constructed. The goal is to use the shape grammar and shape algebra to provide a natural language syntax to write a shape as a construct of tokens (CPPs) and combination rules. To link this spatial, modeled, and constructed data to descriptive data in existing databases, the schema provides a master identification number for use as the key to link data elements for a given artifact. This model is consistent with Dublin Core cataloging structures and emerging data standards within the discipline. This design permits the query engine to link to additional databases by developing an XML schema to map appropriate data fields between the databases such as a master ID#. Ideally, this process will become simpler as standards are developed for data definition and database design. In the meantime, this technique is scalable and provides the capability to add additional datasets to the query process as a proof of concept.

Query

A final element common to the three pilot projects is the query of information. We address the development of the interface, which is a Graphic User Interface (GUI), and the query process itself. The query provides access to a broad range of application specific features and allows the individual researcher the freedom to extract whatever data is relevant to his or her study. The first query interface developed by the project team supports interaction with the ceramic vessel data. Interaction with the bone surface data will be similar to the ceramic data and the capabilities of this interface will grow to incorporate differences and add appropriate tools for interaction. The lithic component addresses a much more complex interaction and will add pattern matching and fitting to the palette of tools available to researchers.

The Visual Query Interface
The visual query system permits users to input, analyze, refine and limit searches via interaction with a Graphic User Interface (GUI). The initial query request can be made in a variety of modes including text, vector graphics, and interactive 2D and 3D models. The user can manipulate and resubmit the query image in real time to refine searches. The interface includes interactive surface and volume visualization capabilities and quantification tools to extract curvature, volume, scale, and linear dimensions for each data set.

The query process combines researcher input via a sketch-based interface with the ability to search by traditional text and metric data. For example, representative vessel shapes can be selected and modified from the supplied palette or a freeform profile sketch can be drawn in the interface window. Input fields for text and numeric fields support parallel query of descriptive and derived data within the databases.

The research intent for the interface is to enable the highest degree of interactivity possible with complex 3D data while avoiding steep user learning curves. This means understanding not only the science and art of interface design, but developing an appreciation of the cognitive complexity of the user's dynamic relationship to the databases. Web designers, archaeologists, and general users are currently evaluating the prototype interface design. As additional models, such as lithic and bone data become available, the interface and related tool sets and capabilities will continue to evolve.

The query process
Behind the interface is a process that extracts key elements (features defined by the CPPs) and manages submission of the queries to the related databases (Figure 17). After the search information has been entered and the query is submitted to the server, a Java program extracts features from the sketch. A CGI (Common Gateway Interface) program parses, manages, and executes searches in the linked databases. The program currently matches text, numeric, and calculated vessel data. The capacity to more abstractly search and match by pattern within the binary data is being explored and will be added as the project continues to develop.

The results of the search from the linked databases are compiled and tabulated. A Java Server Program subsequently sends the processed search results to the client as an XSL (Extensible Stylesheet Language) file. The XSL file returned to the user browser contains thumbnail images and descriptive data for the top five vessels in the search results. The XSL file provides the data needed to view detailed information for any of the five vessels without the need of further interaction between the server and a particular user's computer.

When the search results are returned from the server, the client browser extracts and displays thumbnail and summary data from the XSL file. This information is extracted and displayed on the initial search result screen. The user can then select one of the thumbnails from the display. This action prompts the browser client to extract and display a manipulative 3D model of the selected vessel and detailed descriptive data. Currently, modification of the query sketch requires returning to the initial search result or new search screen. We are currently exploring how to bookmark or store search results to create a traceable research trail over time or multiple searches in a manner that are transportable by the user but does not require storage on the query server.

Discussion

We hope to have demonstrated that advancing archaeological analysis into the 3D domain can yield significant contributions to archaeology. While much of the research presented in this paper is ongoing and at an early stage of development, 3D data does permit us to attack a multitude of questions using a much refined means of analysis. Simple measurements on actual objects or images and drawings derived from these artifacts using calipers and/or rulers represent crude approximations at best and will not allow us to address issues such as vessel symmetry and uniformity adequately. Although advancing into the analysis of 3D digitized data can yield important contributions this move does not come without cost. Indeed, specialists outside archaeology are needed in the development of software and a distributed environment. This software and online interface development need to happen in close collaboration with archaeologists. Currently interdisciplinary research is a popular means of pooling resources and working toward similar ends, but the ability to work in a close collaboration with others outside of one's own discipline and reach an appreciable understanding of the knowledge and goals of co-collaborators from other disciplines poses serious challenges.

Additional costs such as the availability of technological equipment and artifact scanning time further limit the application of this type of research. It is for this reason that it is absolutely crucial that available information, software tools, and scan data are widely distributed. We are trying to achieve just that by making data available online for query and analysis. In doing so, much effort is going toward making the interface platform independent and data structures flexible to permit the addition of external data. It is also important that both raw and specialized data are available to allow researchers to tackle diverse sets of questions, rather than limiting the effort to questions deemed important by the archaeologists associated with a specific project.

While the costs are substantial, they are far outweighed by the potential advances that may be gained from this kind of research. This is particularly true when we consider that future advances in technology will ever expand research possibilities. Further, digital data will make it possible to preserve vulnerable information long term. This can be illustrated by the fact that numerous collections of human bones, ceramic vessels, and other Native American artifacts are often required to be returned to Native American tribes for reburial under the provisions of the Native American Graves Protection and Repatriation Act (NAGPRA). Digitally preserving these artifacts allow us to store and study this material long after its reburial. Further this type of data prevents the need to excessively handle fragile remains and facilitates the study of diverse sets of collections without continuously having to travel from museum to museum to study scattered bits of relevant material.

Conclusion

The interdisciplinary 3DK project at ASU provides ways to bring 3D research in archaeology into a new realm of analysis. By studying digital models of artifacts and bones acquired with recently developed and improved laser scanning technology and integrating the research with current problem sets in numerous disciplines, such as computer science, engineering, and information technology, truly interdisciplinary research beneficial to all disciplines involved becomes possible. Further, making the raw data as well as the developed research tools available online for query and analysis allows researchers from around the world to assess findings and conduct additional or alternative research on the artifacts. With the advent of rapidly developing technology and the broadened use of the World Wide Web, these steps appear as a logical stage of development in archaeology.

Acknowledgements

This has been a truly interdisciplinary project and the successes result from a dynamic interaction between researchers in a broad range of disciplines at Arizona State University. This work was partially funded by the National Science Foundation: 3D Knowledge: Acquisition, Representation & Analysis in a Distributed Environment (IIS-9980166), and support from the Arizona State University Vice Provost for Research, Vice Provost for Information Technology, and Deans of the Colleges of Engineering and Applied Sciences, Liberal Arts, and Fine Arts. Project team members and additional information are available at the Partnership for Research in Stereo Modeling (PRISM) site: http://3dk.asu.edu. We also would like to thank Andria Johnson for editorial assistance. Her help much improved this document.

References

Bae, M. 1999. *Curvature and Analysis of Archaeological Shapes*. MS Thesis, Arizona State University.

Fan, J-T., G. Medioni & R. Nevatia. 1987. Segmented Descriptions of 3D Surfaces. In: *IEEE Journal of Robotics and Automation*, 3, 5: 527-538.

Farin, G. 1996. *Curves and Surfaces for Computer Aided Geometric Design*. Academic Press.

Mangan, A. & R. Whitaker. 1999. Partitioning 3D Surfae Meshes Using Watershed Segmentation. In: *IEEE Transactions on visualization and Computer Graphics*, 5, 4: 308-321.

Pulla, S. 2000. *Feature Extraction Using Watershed Algorithm*. MS Thesis, Arizona State University.

Pulla, S., A. Razdan & G. Farin, nd. Improved Curvature Estimation for Watershed Segmentation of 3-Dimensional Meshes. Submitted to: *IEEE Transaction on Visualization and Computer Graphics*. Available at: http://3dk.asu.edu/DOCUMENT/archives/publication/publication.html

Razdan, A., D. Liu, M. Bae, M. Zhu, G. Farin, A. Simon & M. Henderson. 2001. Using Geometric Modeling for Archiving and Searching 3D Archaeological Vessels. *CISST 2001*, June 25-28. Available at: http://3dk.asu.edu/DOCUMENT/archives/publication/publication.html

Rowe, J. 2001. A Model Digital Library for 3D Pottery Data. *Coalition for Networked Information Meeting April 8-10*. Available at: http://3dk.asu.edu/DOCUMENT/archives/publication/publication.html

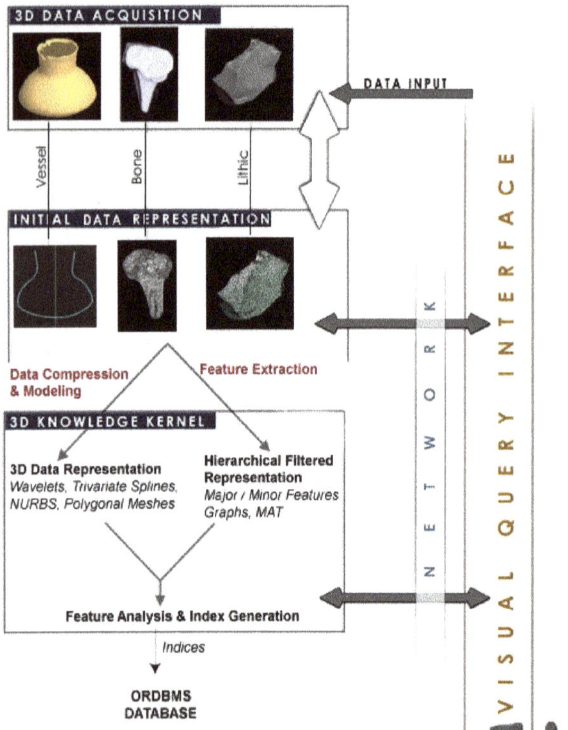

Figure 1: Data flow, knowledge abstraction and query in the 3DK project.

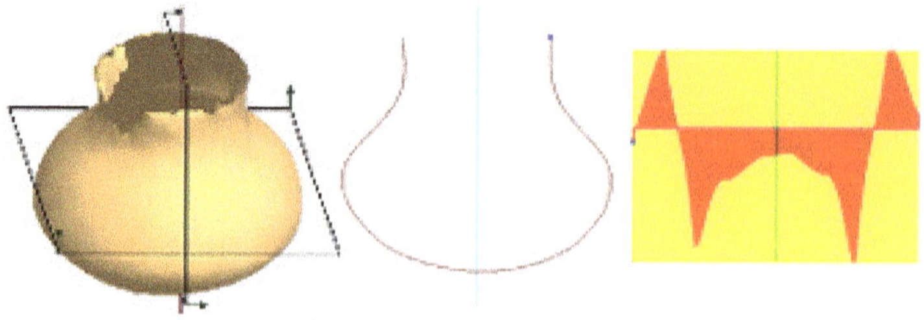

Figure 3: Views from the 3D interface developed for feature identification and profile graphing of prehistoric ceramic vessels.

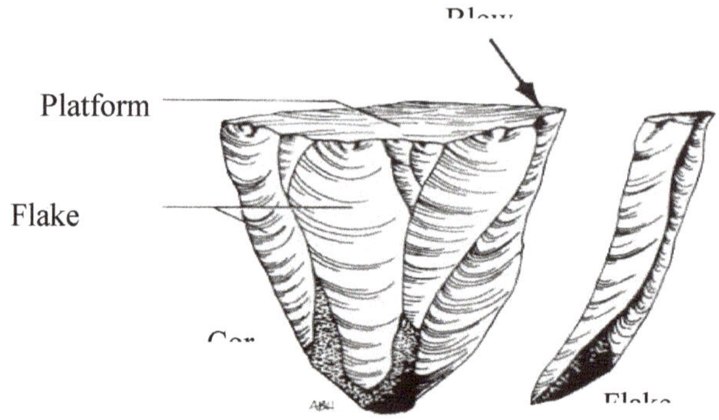

Figure 4: A flake detached from its core (Whitaker 1994)

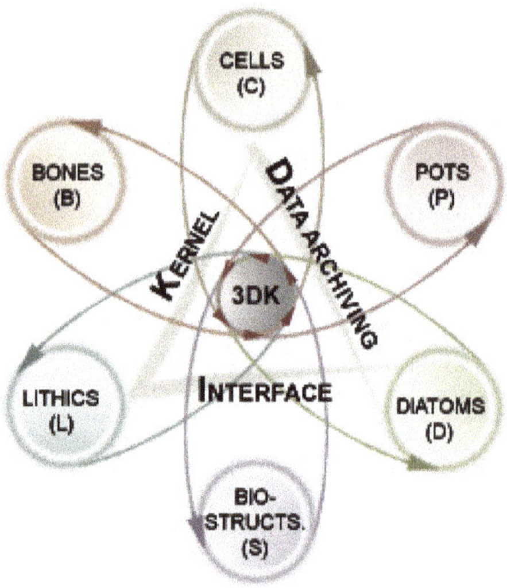

Figure 2: Pilot projects of the 3DK project and their relation.

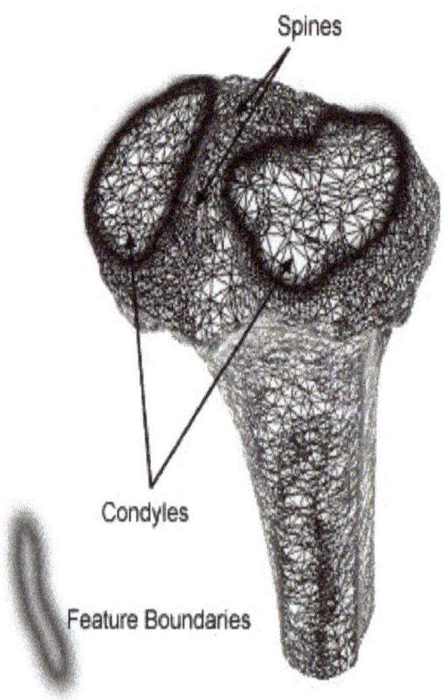

Figure 5: 3D scan of a human tibia illustrating the proximal articular surface for the knee joint, feature boundaries are outlined.

Figure 6: The small M15 Cyberware scanner. The width is about 75 cm.

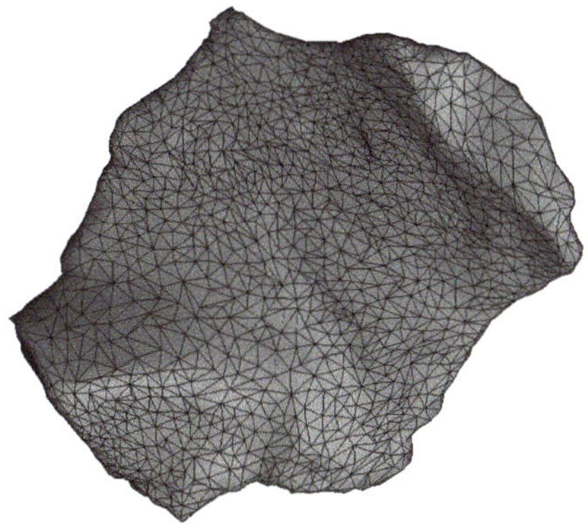

Figure 7: A triangulated, highly decimated point cloud data set representing the ventral surface of a stone flake.

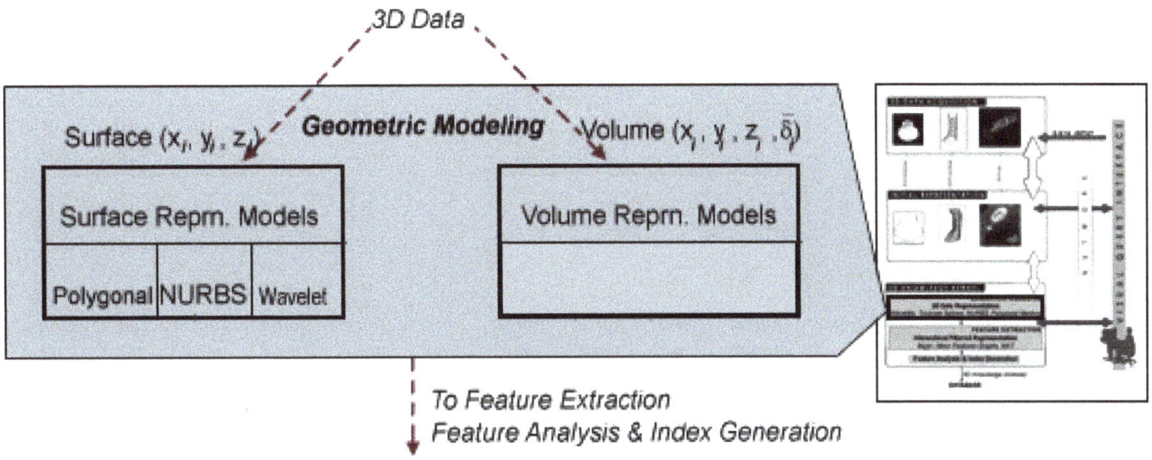

Figure 8: The Geometric Modeling module of the Kernel

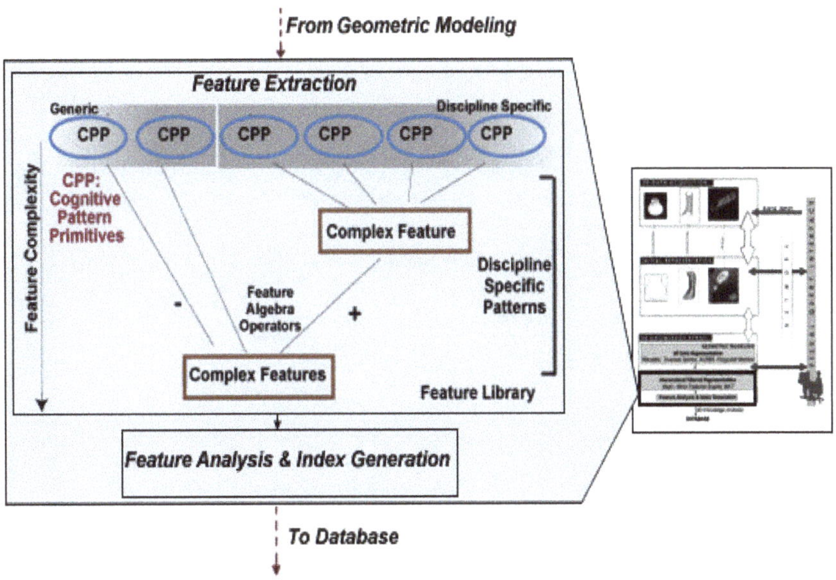

Figure 9: The Feature Extraction, Analysis and Index Generation module of the Kernel

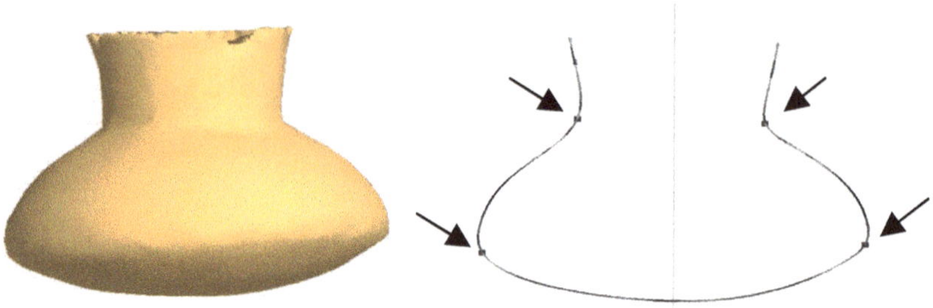

Figure 10: Original digitized pot (left) and the derived profile curve and corner points indicated by arrows on the right.

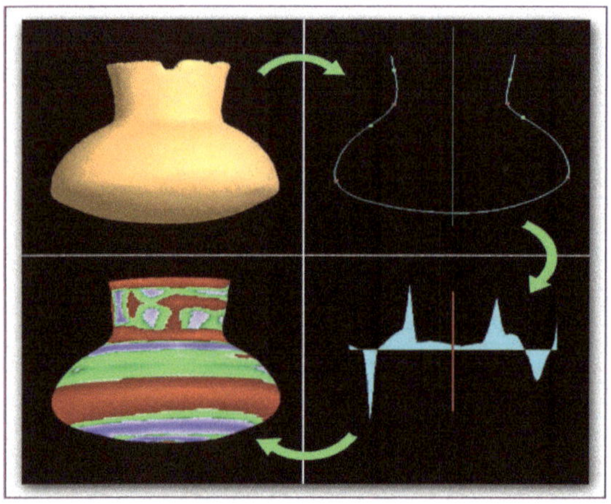

Figure 11: A Ceramic vessel, its profile curve and curvature plots reflecting the skill level of the vessel maker.

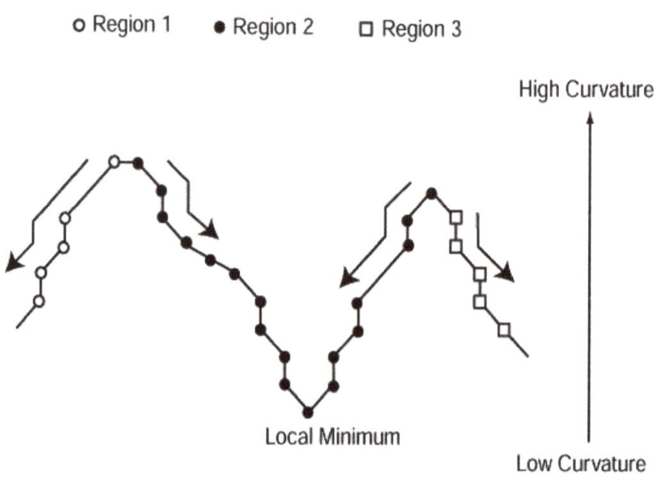

Figure 13: Curvature based segmentation

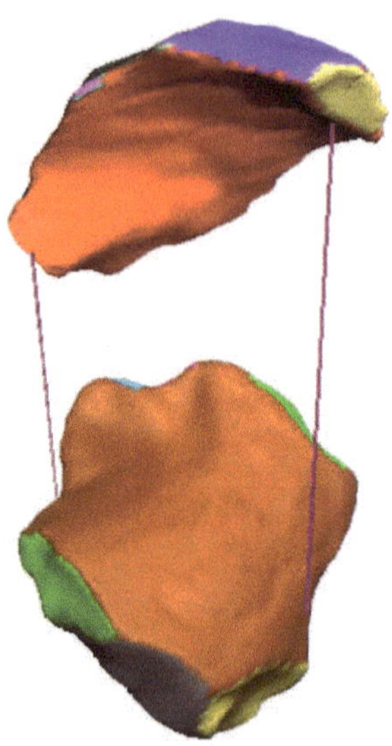

Figure 12: A fitting core and flake, each segmented using the watershed algorithm

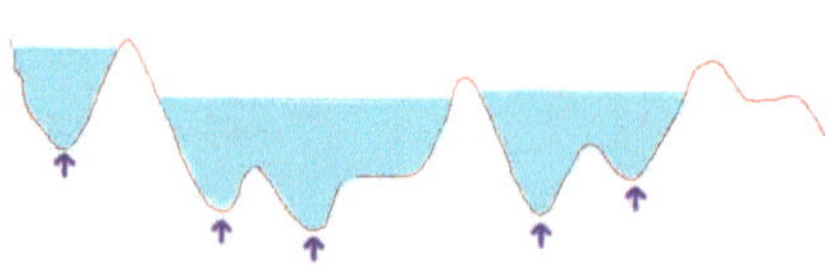

Figure 14: Merging similar regions by increasing the water depth.

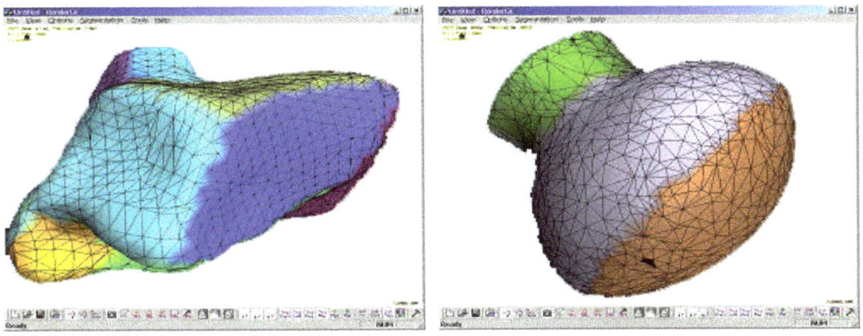

Figure 15: Triangulated surface meshes from a bone (left) and pot (right), segmented using the watershed algorithm based on absolute curvature.

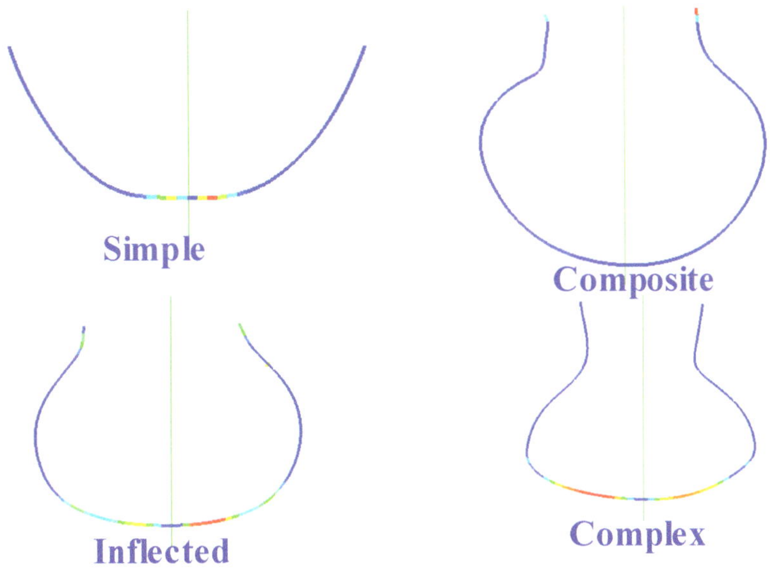

Figure 16: Types of base shapes for vessels.

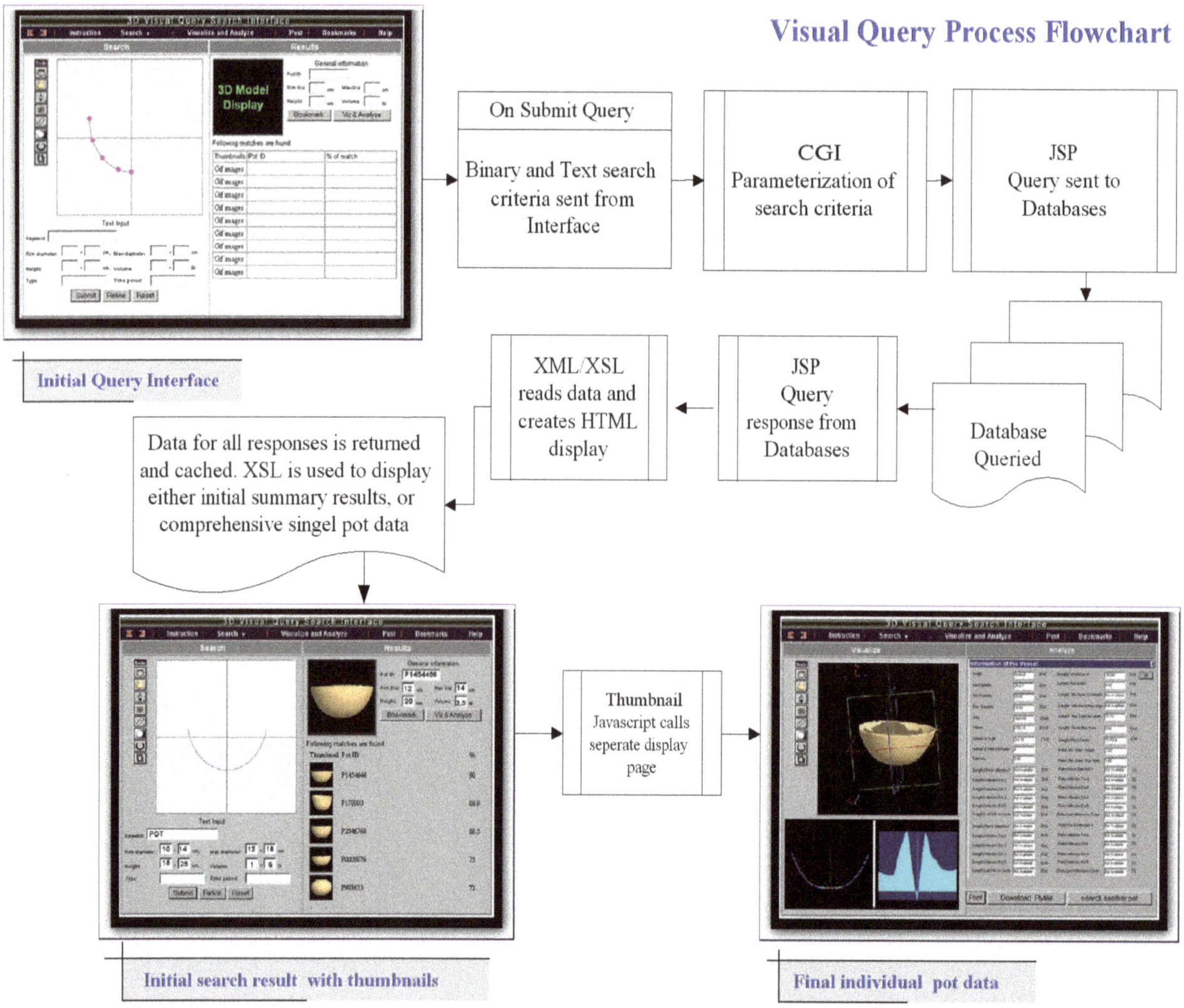

Figure 17: Figure showing the data flow in the 3DK Visual Query Process

Papers in the Osteology category

Neural Network Classification of Skeletal Remains

Suzanne Bell
Department of Chemistry, Eastern Washington University
226 Science Building
Cheney, WA, USA
Phone 509-359-7932 - Fax 509-359-6973 - E-mail: Suzanne.Bell@mail.ewu.edu

Richard Jantz
Department of Anthropology
University of Tennessee, Knoxville, USA

Abstract: Neural networks are data analysis tools well suited to archaeological applications. With both supervised and unsupervised architectures available, networks are flexible and complement existing multivariate analytical techniques. In the present application, networks were used to classify skeletal remains based on common osteological measurements. Both a contemporary (mid-1800s to 1970) and an older database (~1600-1800) were investigated to learn if networks could classify based on several different criteria including race, sex, time frame, phase, and date of birth. Use of genetic algorithms and variable contribution measures allowed an evaluation of relative importance and predictive power of variables retained in the models. Results were compared with traditional multivariate techniques and while performance was not, in general, dramatically different, interesting and potentially significant differences in variable contribution measures were noted. This emphasizes the value of networks as complimentary tools for statistical analysis of archaeological data.

Key words: osteological analysis, neural networks, genetic algorithms, cranial measurements.

Introduction and Background

Multivariate statistical techniques have been used to classify skeletal remains based on age at death, sex, and race. In the present example, results from such statistical analyses were compared to and contrasted with results obtained using back propagation neural networks coupled to genetic algorithms. Two different data sets, one an archaeological and the other a forensic, were used to evaluate the feasibility of classifying remains based on skeletal cranial measurements. Although specific applications and models varied, each study had four overriding goals:

1) To determine if a neural network model could successfully predict the attribute in question.

2) To determine which variables were the most and least important for classification and prediction.

3) To evaluate trends and relationships among and between these critical variables

4) To compare the results with those obtained using traditional multivariate analysis

Although neural networks have becoming an increasingly common tool in business, engineering, and sciences, few archaeological applications have been reported (for example, Barcelo 1993 and 1997, Bell 1998, Reeler, 1999). While a detailed description of network theory, design, and applications is beyond the scope of this paper, general references and texts are available in most university libraries. A good overview is presented by Wasserman (Wasserman 1989).

Several characteristics lend networks to complex problems as often encountered in physical anthropology and archaeology. Networks are capable of uncovering intricate, non-linear relationships among variables that might be overlooked or missed by statistical techniques such as principle component analysis. Networks do not require that the underlying variable distributions be normal. Finally, a variety of network types are now available in commercial PC-based software, providing the investigator with a number of options in both supervised and unsupervised learning techniques.

Datasets and Methods

Two databases were examined, both containing cranial measurements exclusively. A contemporary database, spanning the years AD 1850-1970, was derived from anatomical and forensic investigations. This database was comprised of measurements from 885 individuals classified by sex and race (whites versus blacks). Most of the individuals with 19th century birth dates were obtained from the Terry and Todd anatomical collections, while those with 20th century birth dates were obtained from the Forensic Anthropology Data Bank (Ousley and Jantz 1997) Fifteen cranial measurements were reported for each. The second collection was archaeological, representing 559 individuals from various sites occupied by the Arikara Indians located in the Middle Missouri region of South Dakota (Key 1983 and McKeown 2000). Years represented ranged from AD 1600-1817 and data was collected from 25 excavations. For each individual, 66 cranial measurements were reported.

The archaeological data was classified by phase, time frame, and bank of the river on which the site was located. For time frame, the identifier "Arikara" refers to the historic and protohistoric sites, while "EC" refers to the prehistoric. A summary is presented in Table 1. Although the data in the forensic database was evenly distributed, this archaeological database was not. When broken down by time frame, 392 of the skeletons were Arikara while 158 were EC. Similar imbalance was seen based on riverbank, with 403 skeletons representing the left bank and 147 representing the right bank. Further complications were introduced by the uneven distribution of site dates over time. For example, with right and left bank, only data from the right bank was dispersed enough over time to allow for an investigation of how cranial measures evolved. When considering phase, imbalances were even more significant. As a result, models created to predict phase were limited to distinguishing the Le Beau 3 phase from all others. While such imbalances in data distribution do not preclude statistical or neural network evaluation, it is important to keep it in mind when interpreting the models.

Neural network analysis was conducted using two software packages: Neurogenetic Optimizer (NGO, Biocomp Systems, Redmond, WA USA) and Statistica Neural Networks (Statsoft, Tulsa, OK USA). Both of these packages utilize genetic algorithms for network optimization and variable selection. Multivariate statistical analysis was conducted using Statistica. The databases were converted to Excel and comma-delimited text file format for network analysis.

Neural networks are layered structures of interconnected mathematical processing elements linked together as shown in figure 1. Note that the first layer consists of one processing element per input variable and that these processing elements do not perform any mathematical operations. Rather, they pass the value of the variable on to each element in the hidden layer. Nodes in the hidden and output layers perform the mathematical operations illustrated in figure 2. Each of these nodes has associated with it a set of weights (equal to the number of inputs) that are initially set to small random values. The processing element performs two operations. First, the input values are multiplied by their corresponding weight and the value summed. Second, this sum is passed onto a transfer or "squashing" function the compresses the value between 0 and 1 or –1 and +1, depending on the transfer function selected. This is where the analogy to the nervous system can be seen; the processing element, also called a neuron, will either "fire" or not fire (be activated or not) based on the sum of the inputs. A common feature of most transfer functions is the abrupt transition between a relatively large value (analogous to a nerve cell activating) and a relatively small value (analogous to an inactive neuron). If the node is in the hidden layer, the output from the transfer function is passed to the next layer.

The size of the hidden layer is determined by optimization experiments but in general, the smaller this layer is, the better the model will generalize. Although any number of nodes can be used in the output layer, in the present case, all networks models utilized one. Outputs in these models could be continuous (such as year of birth for the forensic database) or discrete (such as which bank of the river the individual was recovered from). In such cases, a binary encoding was used. For example, 1 would encode for a skeleton recovered from a site on the left bank while 0 would encode for a skeleton recovered on the right bank. Numerical results from the output node were scaled accordingly between 0 and 1.

Training of a neural network involves iterative adjustment of the weights until all training data presented to the network results in a correct response within the error threshold specified by the user. Individually, inputs (here a set of cranial measurements for one individual) are presented to the network and calculations are performed. The result is compared to the desired value and the resulting error is back propagated through the weights and used to make corrections. In practice, error is usually accumulated over one complete pass through the training data (commonly called an epoch) before weights are adjusted. This process is repeated through hundreds or thousands of epochs until training is complete. Progress is monitored not by the training data itself, but by a representative cross section removed prior to training. The training process continues for a set number of epochs (for example 100), and then the test data is run through the network and an error calculated. Training is resumed for another 100 epochs and the test is repeated. Training is halted if one of three conditions is met: 1) test performance meets the user-defined threshold; 2) test performance is no longer improving; or 3) the maximum number of epochs has been reached.

Note that the training and test data are both involved in the training process and as such do not provide an independent evaluation of network performance. For this, a separate set of validation data is needed, which consists of a representative cross-section removed from the original dataset before training. The performance of a network can be reliably measured using this validation data, and this is the metric reported throughout this paper. Figure 3 summarizes how data was divided, and how these subdivisions were used to generate the five replicates for each model constructed. Each replicate run represents a reshuffling of training, test, and validation data and these replicates were used to gauge the reproducibility and uncertainty (at the 95% confidence interval) of the models.

Until recently, optimization of the many parameters and settings associated with a given neural network model, as well as the selection for variables to include, has been problematical. Settings such as number of hidden layers, number of nodes in the hidden layer, transfer functions, learning rates, etc. all must be determined, yet little more than experience and rules-of-thumb existed for guidance. In the last ten years, an increasing number of neural network packages have incorporated genetic algorithms, which have greatly increased the utility of the technique. To implement a genetic algorithm optimization, the first step is to create a chromosome (figure 4), which is a binary encoding of all the parameters necessary to construct a neural network. This can include type of network, number of hidden layers, number of hidden nodes per layer, etc., but also includes instructions as to which variables are to be included. Genetic algorithm optimization (figure 5) begins with the generation of a pool of such chromosomes, typically 30-50. For each generation, a network is built according to the instructions encoded on each chromosome and the network is trained long enough to gauge its fitness relative to other networks. Fitness takes into account accuracy and size; smaller networks tend to generalize better and are favored. At the end of the training and comparison step, the fittest chromosomes (the ones generating the best performing networks) are allowed to pass to the next generation. The candidate pool is brought back to original size by allowing the fittest chromosomes to mate and mutate, and by adding additional, randomly generated chromosomes into the pool. The cycle repeats until one of three conditions is met: 1) no further improvement in fitness is occurring; 2) a user-defined level of performance is achieved; or 3) the maximum number of allowed generations has elapsed. At the end of this process, the algorithm will have converged on an optimal solution; however only replicate analyses (as described previously) will verify that the solution is reproducible and reliable. All network models discussed in this paper were obtained using genetic algorithm optimization as described here.

Because the outputs of all the networks were binary, the issue of cutoff or threshold values must be addressed and is best illustrated through an example. With the archaeological database, one set of models was constructed to predict bank of origin based on cranial measurements. Thus, each set of measurements was associated with either the right bank (coded as 1) or the left bank (coded as 0) of the Missouri river. Accordingly, the value of the output node varied continuously from 0 to 1. A decision must be reached as to what error threshold will be accepted as a correct response. Commonly, values of 0.51 and above are considered to be 1, while values of 0.49 or less are considered to be 0, and the cutoff value in this case is 0.50. Clearly, this is a generous threshold and may or may not be reasonable based on the problem at hand. Rather than set an absolute threshold to start with, results were evaluated for three thresholds: 0.90, 0.75, and 0.60. In all cases, the metric used to monitor network performance was the percentage of validation cases that were correctly classified by the model based on the threshold value selected.

Once working models were constructed and optimized, the next step was to determine which cranial measurements were the most important, how much each contributed, how they were related to each other, and how they were changing over time. Terms such as variable contribution, sensitivity, weight, and importance are often cited in the literature, but there are no universally accepted definitions. Within this study, the methods of calculating variable contribution were used as defined below:

1) Sensitivity: Determined using the trained network and validation data. The sensitivity of a variable is calculated by dithering its value +/- 10% of the original and averaging the change in network output across all validation cases. This measurement, while not truly multivariate, helps to identify variables than can have a large impact on network output.

2) Number of occurrences: Each network model was built using five replicates as discussed above. Accordingly, the genetic algorithm was repeated five times as well. A variable that is important to achieving optimal network performance should be included in the model more often than variables that are not important. Accordingly, the number of occurrences of a variable (0-5) is one measure of its importance. Moreover, this metric is multivariate in that the genetic algorithm will converge on an optimal variable combination.

It is important to note that a variable can be important and not sensitive and vice versa. Consider a variable that is included by the genetic algorithm in most of the models, but small changes in that variable do not result in large changes in network output. However, removal of this hypothetical variable results in dramatic degradation of network performance. Such a variable is an example of an important variable that is not sensitive.

Both of these criteria (sensitivity and number of occurrences) were combined in a quantity called the variable score calculated as follows:

Score = (0.33(normalized sensitivity) + 0.67(normalized number of occurrences)) * 100

The weighting factors were assigned somewhat arbitrarily but reflect the greater weight attributed to the multivariate measure. Since networks excel at finding complex, nonlinear, multivariate relationships among variables, the metric used to gauge variable importance should be slanted in favor of the multivariate measurement.

Results and Discussion

Results for the archaeological database are summarized in figures 6-8. Not surprisingly, under the most stringent threshold (0.90), none of the models predicted their target characteristic as well as chance. However, at a threshold of 0.75, the models were able to predict time frame slightly better than chance (looking at the average % correctly classified). At the 0.60 threshold, all three characteristics were predicted better than chance (even considering uncertainties) and the difference in performance between networks predicting time frame and bank were statistically significantly. Across all models and all thresholds, there were no statistically significant differences when males and females were considered separately. Given the nature of the predictions and the inevitable uncertainties and inconsistencies that arise from different investigators collecting data, it was assumed that the 0.60 threshold was the most reasonable one from which to draw conclusions. Thus, results suggest that the river served was an effective barrier between sites and peoples since cranial measurements were sufficient to distinguish them. Certainly, other factors play a role and investigation is continuing.

Once models were optimized, the next step was to evaluate variable contribution and importance. Using the variable score defined above, the ten most important variables for each of the three models (predict time frame, predict bank, and predict phase) were determined across five replicate networks. These variables were then examined to evaluate changes over time and to study the relationships between them. The ten most important variables for each model are summarized in Tables 2,3, and 4. For predicting time frame, the most important cranial measurement is Nlh, followed by a group of similar scores for Ocs, Aub, Ocr, Sos, Asb, Mls, and Xcb. The trend listed in the table is how the measurement appears to change over time and is discussed in detail below. Note that Ocs was also among the top ten for predicting phase. For predicting time frame, Occ was the most important variable with the next closest (Avr) nearly ten points lower. Finally, for predicting phase, Mdb was even more dominant than Occ was for predicting time frame. However, since the imbalanced dataset limited what could be done with phase, this finding has limited application.

With important variables identified, the next step was to study how these variables are changing and how they are related to each other. Since the dates associated with the skeletons were discrete and not continuous, it was necessary to group individuals into time spans as shown in Table 5. This grouping was used for plotting purposes only. Examples plots are presented in figures 9-11, with examples of increasing, decreasing, and indeterminate patterns. Several difficulties were encountered in this phase of the work. First, given the imbalance in the data set (as per Table 1), it was difficult to obtain a representative number of individuals in many of the time spans for a given subset. Thus, while the networks were able to create good predictive models, and important variables were clearly identified, the exact nature of the relationships among these cranial measures cannot be deciphered with sufficient confidence to draw firm conclusions.

Analysis of the Forensic Database

Neural network models were also applied to a contemporary database comprised of an anatomical database from the nineteenth century combined with a forensic database from the twentieth. While the detailed results of this investigation are pending, a comparison of results with those obtained above is useful. The primary goal of this study was to determine if a network model could predict race, sex, and the year of birth based on cranial measurements. Previous work with multivariate statistics and regression models resulted in a mean square error of ~ 25 years, with a 95% confidence interval of +/- ~ 50 years.

Neural network models were constructed using the general techniques previously described with network performance and variable importance measured as before. Optimized models were able to predict race from cranial measurements ~86-92% of the time while models to predict sex were correct ~81-88% of the time. Models to predict year of birth with an uncertainty of ~ 20 years, better than obtained with regression. Interestingly, the genetic algorithm retained most variables in the models used to predict race and sex while fewer variables were retained for models used to predict year of birth. This finding suggests that information regarding sex and race is widely dispersed across the cranial measurements while information on year of birth is concentrated in fewer variables.

For predicting sex, Zyb was most important followed by Nph. For predicting race, the order was Bnl, Bpl, Nph, Nlh, and Nlb. For year of birth models and across all race and sex divisions, Bbh, Zyb, Aub, and Pac were the most important. With the interesting exception of Aub, these variable rankings coincide with results from multivariate statistics, in which Aub was not flagged as important. In the network models, Aub appeared to be sex-linked and more useful for predicting year of birth for females as opposed to males. In males, results suggested that Aub was a sensitive variable, but not an important one, a distinction that was discussed above.

Summary and Conclusions

The above studies illustrate the value of neural networks in archaeological and osteological applications, particularly when combined with traditional statistical methods. Networks excel at identifying complex, multivariate, and nonlinear relationships among variables and often identify important variables that other techniques such as regression analysis or principle component analysis overlook. However, because of the complex nature of the variable interactions, it is sometimes difficult to completely characterize these relationships. In the present case, much of this difficulty arises not from the networks, but from the nature of the data itself. Nonetheless, neural networks, particularly using up-to-date PC software and genetic algorithms, represent a valuable addition to the mathematical toolbox available to archaeologists and physical anthropologists.

Acknowledgements

Dr Bell's work was funded by the US National Science Foundation, POWRE program award.

Dr. Jantz' work on the forensic anthropology data bank was supported by a grant from the National Institute of Justice.

References

Barcelo, JA. 1993. Back-propagation algorithms to compute similarity relationships among archaeological artifacts, in *Computer Applications and Quantitative Methods in Archaeology, 1993*, BAR International Series, Oxford.

Barcelo, JA and Faura, JM. 1997. Time series and neural networks in archaeological seriation. An example on early pottery from the near east, in *Compuer Applications and Quantitative Methods in Archaeology 1997*, BAR International Series, Oxford.

Bell, S. 1998. *Artificial Neural Networks as a Tool for Archaeological Data Analysis*, Archaeometry 40(1), 139-151.

Key, P.J. 1983. *Craniometric Relationships Among Plains Indians: Culture-Historical and Evolutionary Implications, Report of Investigations, No. 34*, University of Tennessee, Knoxville.

McKeown, A.H. 2000. *Investigating Variation Among Arikara Crania Using Geometric Morphology, Doctoral dissertation*, University of Tennessee, Knoxville.

Ousley, SD and Jantz, RL. 1997. The Forensic Data Bank: Documenting skeletal trends in the United States. In: *Forensic Osteology*. 2nd ed. (Ed: Reichs, KJ) Charles C. Thomas, Springfield, IL, 441-458.

Reeler, C. 1999. Neural networks and fuzzy logic analysis in archaeology, in *Computer Applications and Quantitative Methods in Archaeology*, BAR International Series (supplementary), 750 (1998), 150.

Wasserman, PD. 1989. *Neural Computing: Theory and Practice*, New York: Van Nostrand Reinhold.

Tables

Table 1. Distribution of archaeological data

Time frame	Phase	River Bank	Period	n
Arikara	Arikara	Right	1817	52
Arikara	Bad River 1	Right	1705-1707	10
Arikara	Bad River 2	Right	1766-1768	39
Extended Coalescent (EC)	La Roche	Both	1612-1663	158
Arikara	Le Beau 1	Left	1687-1726	21
Arikara	Le Beau 2	Both	1700-1766	50
Arikara	Le Beau 3	Both	1688-1728	220

Table 2.Ten most important variables for predicting bank

Cranial measurement	Score	Trend[1]
Nlh	87	Increasing
Ocs[2]	80	Indeterminate
Aub	78	Indeterminate
Ocr	77	Decreasing
Sos	76	Indeterminate
Asb	76	Decreasing
Mls	75	Increasing
Gol	71	Decreasing
Xcb	70	Increasing
Fmb	62	Indeterminate

1: Trend over time; see text for discussion
2: Important variable in more than one model (predict bank/phase/time frame)

Table 3.Ten most important variables for predicting time frame

Cranial measurement	Score	Trend[1]
Occ	100	Indeterminate
Avr	91	Indeterminate
Frs	88	Increasing
Sts	78	Decreasing
Mdb	78	Decreasing
Frf	77	Indeterminate
Nlb	77	Indeterminate
Jub	76	Increasing
Frc	71	Indeterminate
Iml	71	Indeterminate

1: Trend over time; see text for discussion

Table 4. Ten most important variables for predicting phase.

Cranial measurement	Score
Mdb[1]	100
Prr	75
Ocs	72
Wmh	71
Nar	62
Nph	62
Wcb	61
Osr	61
Lar	61
Pac	60

1: Important variable in more than one model (predict bank/phase/time frame)
Note that trend is not included since there were not enough representative measurements from different groups

Table 5: Grouping of time spans

Period	Midpoint used for graphing
1612-1688	1650
1700-1708	1704
1725-1728	1726
1766-1777	1766
1817	1817

Figures

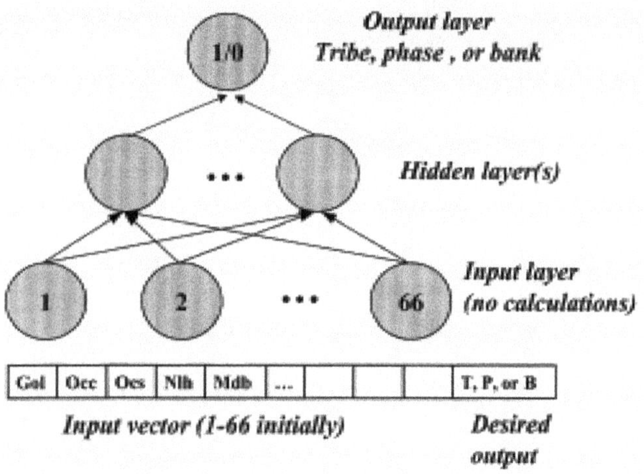

Figure 1. Back propagation neural network.

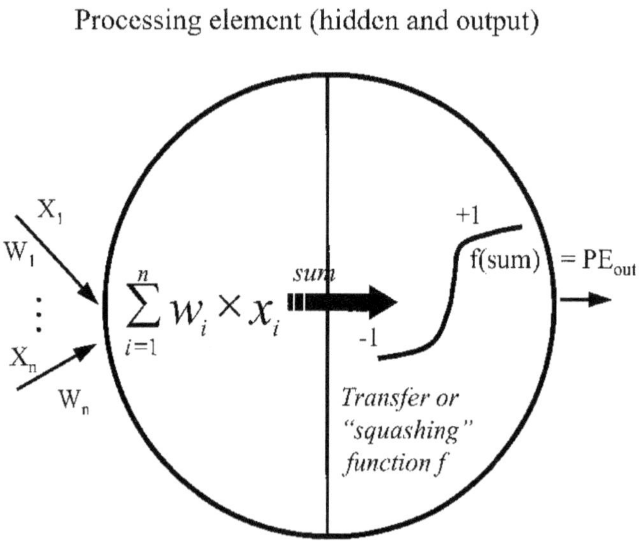

Figure 2. Processing elements, hidden and output layers.

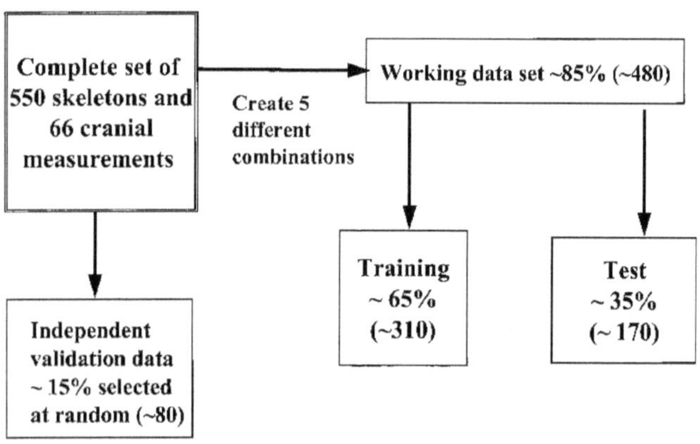

Figure 3. Division of data for network training and validation.

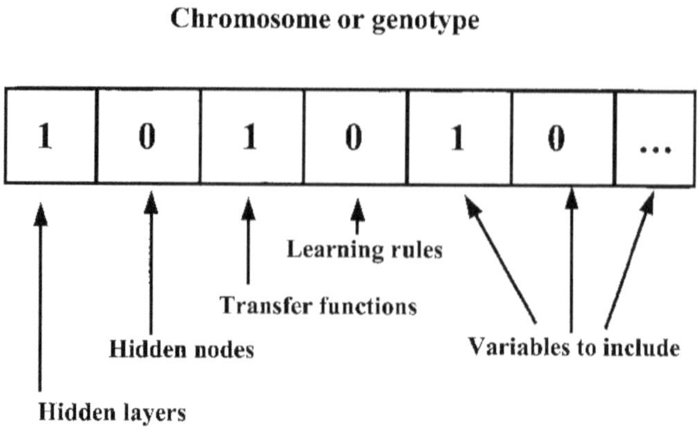

Figure 4. "Chromosome" used for genetic algorithms.

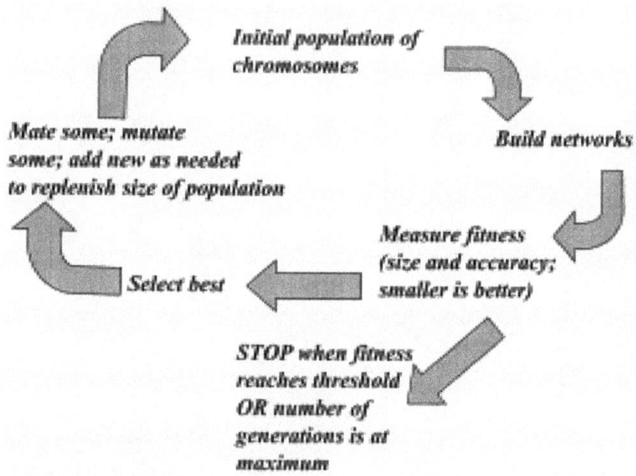

Figure 5. Application of a genetic algorithm to neural network optimization

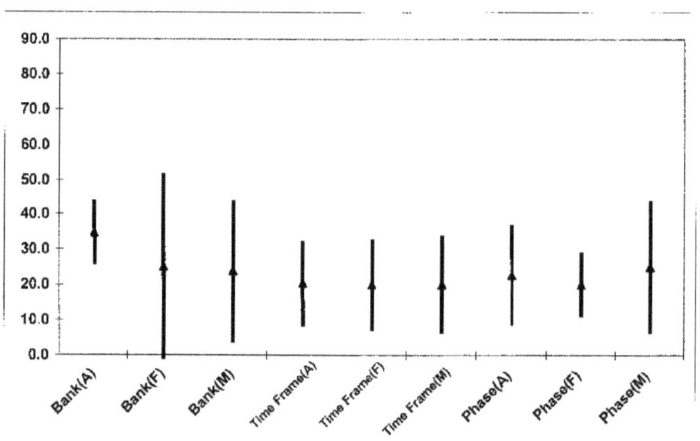

Figure 6. Summary of results, archaeological database, 0.90 threshold. From left to right, results are reported for models that predict bank (left vs. right); time frame (Arikara vs. EC); and Phase (Le Beau 3 vs. all others). Data is reported for all data (A), females only (f), and males only (m).

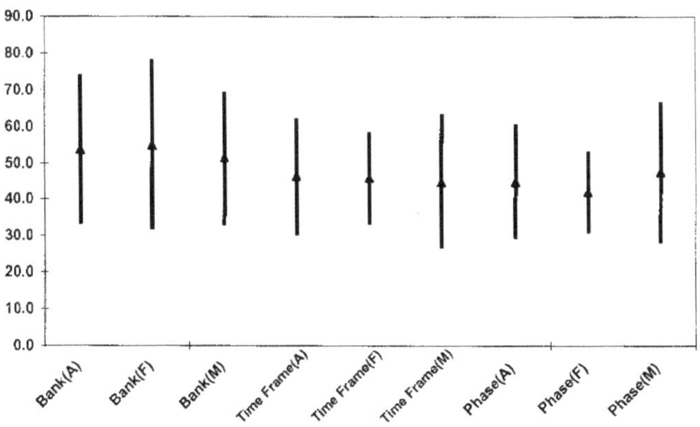

Figure 7. Summary of results, archaeological database, 0.75 threshold. Labels as in Figure 6.

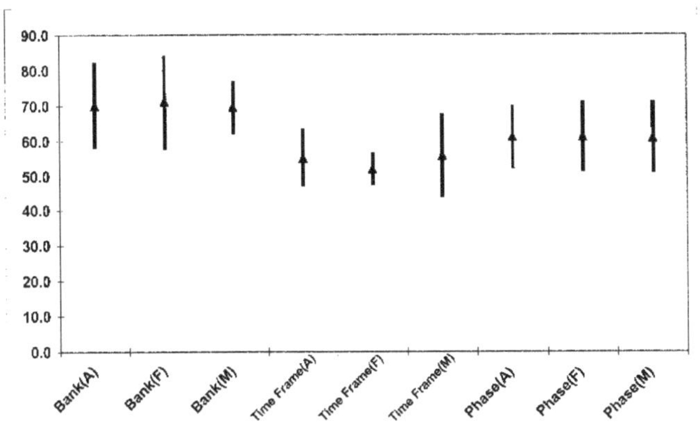

Figure 8. Summary of results, archaeological database, 0.60 threshold. Labels as in Figure 6.

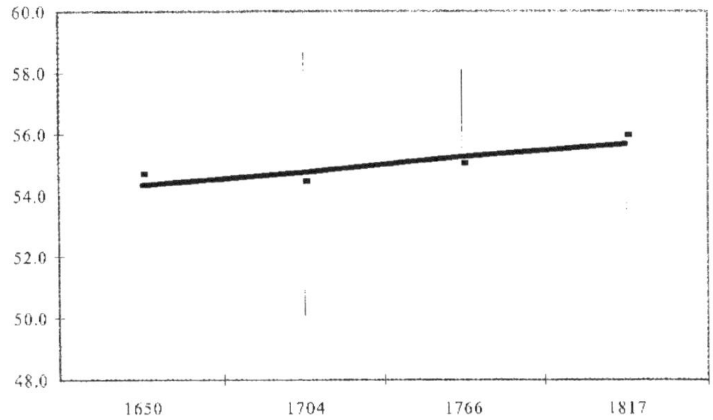

Figure 9. Example variable plot, Nlh (males only, right bank), predicting bank model. Note that the trend is generally increasing, although with the large uncertainties (+/-1 standard deviation), such observations are not conclusive. The x-axis is the midpoint date as presented in Table 5 and the y axis is the mean cranial measurement for that time period. The line is the best fit regression through the average.

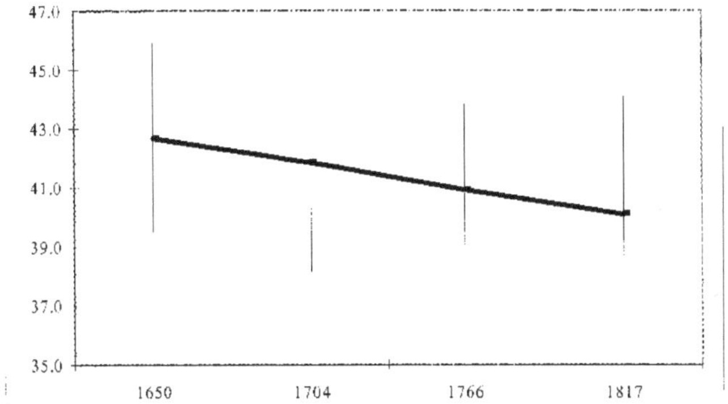

Figure 10. Example variable plot, Ocr (males only, right bank), predicting bank model. This is an example of a variable showing a general decrease. The same plot for the corresponding females (not shown) also shows a general decrease but it is not as uniform or smooth.

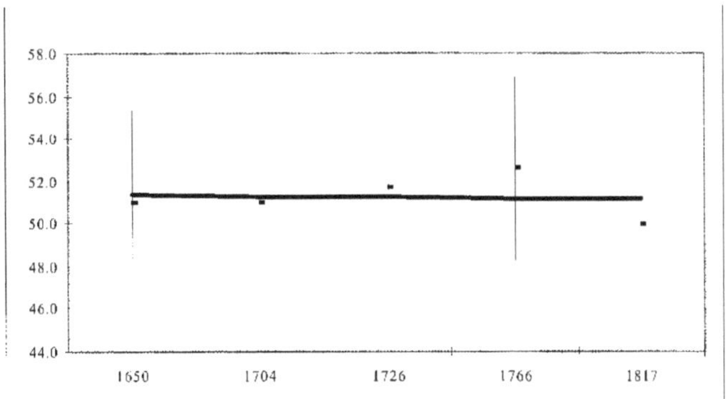

Figure 11. Examples variable plot, Frf (all), predicting time frame model. This is an example of a trend that was labeled as indeterminate.

Osteology for the new millennium

Gustav Jonsson
Gotland University College
Cramérgatan 3
621 67 Visby, Sweden
E-mail: gustav.jonsson@hgo.se

Abstract: This presentation will tell you about a way to make the osteologists work more efficient and timesaving. The lecturer has constructed a database for osteological analyses, OsDat. The "OsDat" helps you to make the calculations like MNI and MNE automatically and makes your diagrams and presentations in reports neater and gives the reader a better overlook than traditional presentations. You can also upgrade your results with just a "click", no further calculations or consideration is needed. This lecture will take you from the basics in database construction via logical operations such as MNI and MNE and ending in the final report.

Key words: Archaeoosteology, Quantification, Database

Introduction

I have called this lecture "osteology for the new millennium". And it is a new era, now we have got computers to help us in our work. In an osteological analysis we use a lot of numbers and we do a lot of calculation. For an example, the calculation of the "minimum numbers of individuals" or the "minimum number of elements". In this lecture I will try to describe an osteological database, OsDat, this is a database that can be used as it is or as a model when you want to construct a database for a specific material.

An osteologist´s work can be divided into three parts. First the osteological handicraft, then the part when the results will be transformed to understandable data. For last we have the interpretation part. A database will mostly be a tool for the second part of the work. If we look at a few typical osteological analyses we can see that all of them have tables that look rather the same. In this program you can get some of these tables just with a "click" on a button. And if you later on want to add a few other bone fragments, you don't have to do any calculation at all, you just press the same button and the computer will make a new updated set of tables. This will have two major effects on your work. First of all it will shorten the time of the work, which means that you will do the analyse faster or that you got more time for the interpretation.

When to create a program like Osdat you must have a few things in mind. First you must do the program as easy as possible, otherwise nobody will ever use your program in the work. Another criterion is that the program must be competent. This picture shows the meaning of OsDat, we have to translate bone fragment into binary code and form the binary code to an understandable picture. The program must in the same time be very complicated and easy, a paradox as it seems.

Today we have many different databases in use for osteological materials. Some of them are more famous than others. Many of them have only one user. One of the first attempts is the basic code presented by Klein and Cruz-Uribe in "the analysis of animal bones from archaeological sites" (Klein & Cruz-Uribe 1984). This was in fact many programs, one for registry of bones and one for the teeth and one for getting the results and so on. One of the more famous for human materials is "Standardized Osteological database (SOD)" which was created after the pattern that was presented in Standards for data collection from human skeletal remains (Bukistra J. E. & Ubelaker D. H. 1990, SOD). A third example is the Bonestack presented by Milles (1995). The big difference between these three programs is that the two programs for animal material was based upon the bone fragments and the SOD was for whole skeletons.

The user interface

At first look the interface in OsDat is rather the same as that you find in Microsoft Access. This is because I have been working with this system and because this program is well known and therefore the user will feel at home. The first window that you see is the main menu, from this you got three choices. First you can choose "registry", in this window the fragments can be described and recorded. The second choose is the "look up" part and it is a help window for looking up information such as when a specific bone will be fused or in which biotope a specific species is found. The last choice to do is the "quantification". Here are, for now, seven functions that all creates it's own specific table.

Register

This is the core of the program. Here we find all information about every bone fragment. Every fragment will be one post in the database, with a few exceptions. If two fragments can be put together they can be recorded as one fragment but then the number of fragments should be set to two instead of one. The same technique can be used for recording all unidentified fragments. Just let "species" and "element" be empty and set the number to the number of unidentified fragments.

Every fragment shall be coded by the same technique as the one presented by Lepiksaar 1988 (Lepiksaar J. 1988). The fragment will be divided into different parts or zones. Zone 1 and 6 in the long bones are the epiphyses. Other types of element like scapula or coxae will be coded differently. At lest 51% of the zone must be represented for be recorded.

For the fragments from the skull they all shall be recorded as "cranium" in the field "element". Then the fragment can be described in detail under the "skull parts". This is so that you won't mess up the quantification.

Tables

The tables are MNI, MNE, Nisp, Measurements, Stage of fusion, Bone modification. I will return to some of these functions and what the technique behind them looks like later. The inspiration to which tables I should create and the design of them I have got from the ones presented by Ekman (1973).

The database

Now, let us take a step down and look behind the shell. The OsDat is a relation database and normalized (Connolly T. & Begg C. 1999) to get a faster and more easily worked program. If we look on a sketch over the structure we se that the database is a lot of tables linked to each other. This is in order to save space on the hard drive and to facilitate the process of changing the information. For an example, if we want to change the name of "Bos taurus" to only "Bos" we just have to change this in one post and every other post reference to this post will be changed at an instance.

Functions

Minimum number of elements (MNE) & individuals (MNI)

The function minimum number of elements sorts all posts by species, element, side and part of the element and returns the number of elements and age of the element. The function will check which zone or part of the element that is most frequent for each element. The MNI-function does rather the same thing but it takes the maximum value of young and adults from both sexes for every species. (Table 1)

Measurements

This function sorts all measurements by species and element. For every group (species/element) it presents the maximum, minimum, average, standard derivation and how many elements that were measured. (Table 2)

Bone modification

Bone modification is presented in a table. The modification is sorted by species, element and type of modification and where on the bone element it appears and how many bone fragments that have this change.

Nisp

This is only a sorting by species and for each species presents a number of fragments and the summary of weight for each species. In the end of the list it presents the total identification grade of the material. (Table 3)

Stage of fusion

This table shows how many fragments you have in the material sorted by the stage of fusion. All the parts are counted separately so that one fragment can and will be counted for its proximal part as well for it's distal part. All fragments are sorted by species, age of fusion and part of element. Then the function will take a look at the stage of fusion of the fragment and add one to one of the groups "infant", "open", "closing" or "closed". (Table 4)

References

Bukistra J. E. & Ubelaker D. H. 1990. *Standards for data collection from human skeletal remains.* Arkansas.

Connolly T. & Begg C. 1999.*Database systems. A practical approach to design, implementation, and management.* Harlow.

Ekman J. 1973. *Early mediaeval Lund – the fauna and the landscape.* Lund

Klein R. G. and Cruz-Uribe K. 1984. *The analysis of animal bones from archaeological sites.* Chicago.

Lepiksaar J. 1988. *Om den tafonomiska förlustens betydelse vid kvantitativ analys av antropogena tanatocoenoser. I: Faunahistoriska studier tillägnade Johannes Lepiksaar.* Lund.

Milles A. 1995. The bonestack: a stack for old bones. In *Computer applications and quantitative methods in archaeology 1994*/edited by Jeremy Huggett & Nick Ryan. BAR International series, 600. Oxford

Standardized Osteological Database (2001-07-31)
http://www.cast.uark.edu/~shelley/sod/

Tables

Species	Bone element	Nisp	MNE
Sus scorfa dom.			
	Cranium	3	2
	Femur	2	2
	Humerus	5	3
	Mandibula	1	2

Table 1. Example of the Nisp/ MNE table.

Species	Element	Measurement	n	min	max	avarage	st.dev
Bos taurus	MT	GL	3	210	250	223,00	15,13

Table 2. Example of the Measurement table.

Species	n	Weight (g)
Bos taurus	31	0
Equus caballus	4	0
Ovis ammon	3	0
Sus scorfa dom.	13	0
Grade of identification:	37%	

Table 3. Example of the Nisp table.

Bos taurus

Age of fusion	Bone/ part	Inf.	Open	Closing	Closed	Nisp
42 - 48 mon.	Coxae/ -	0	0	0	0	1
42 - 48 mon.	Femur/ Dist.	1	0	0	3	8
42 - 48 mon.	Femur/ Prox.	1	0	0	2	8
18 - 24 mon.	Phalanx 1/ Prox.	0	0	0	1	1
18 mon.	Phalanx 2/ Prox.	0	0	0	1	1

Table 4. Example of the Age of fusion table.

Figures

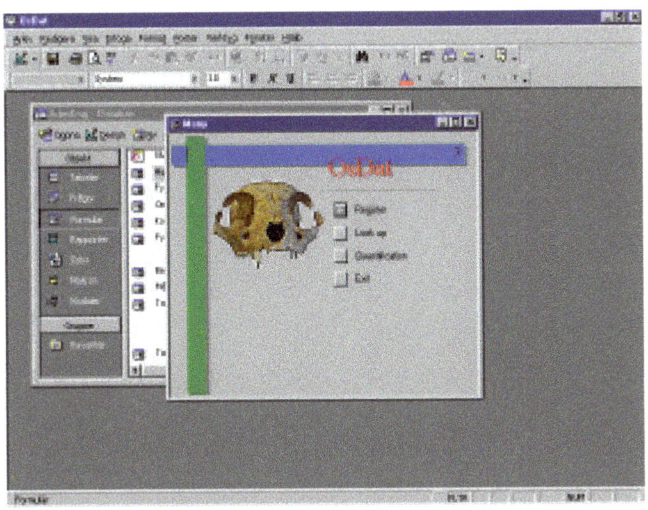

Figure 1. The main menu

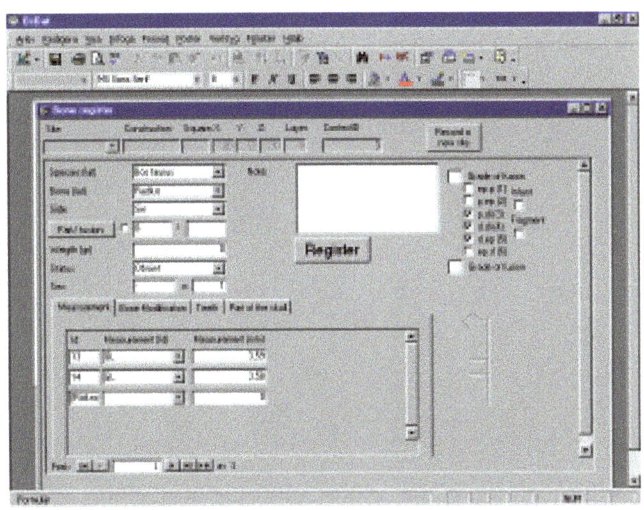

Figure 2. The register form.

3-D CAT-scan: Anthropology, Archaeology and Virtual Reality

Niels Lynnerup
Laboratory of Biological Anthropology
University of Copenhagen
The Panum Institute
Blegdamsvej 3
DK-2200 Copenhagen, DENMARK
Phone +45 3532 7239 - Fax +45 3532 7215 - E-mail: n.lynnerup@antrolab.ku.dk

Abstract: The use of CAT-scanning and three-dimensional image rendering is a powerful tool in the documentation, analyses and exhibition of human remains. Human remains may be CAT-scanned to ensure a complete documentation of the specimens. Furthermore, such documentation is easily stored and shared on computers, allowing researchers access to otherwise non-accessible materials. CAT-scanning is also an important method for osteological analyses, yielding cross sectional and composite data of bones.

Most importantly perhaps, CAT-scanning allows visualisation of hidden structures, a case in point being mummy studies, were skeletal structures may be visualised without opening the mummy wrappings. Furthermore, we have been able to visualise non-excavated skeletal structures, thus pointing to a wholly new avenue of archeological and preservational methods.

Finally, using so-called stereolithography, we have produced 1:1 models of CAT-scanned structures directly from the computer visualisations, detailing structures at a 1mm resolution. This has some very evident uses for exihibits.

Key words: CAT-scan, 3-d image, visualisation, stereolithography, human remains

Introduction

The introduction of CT-scanning by Hounsfield in the early 1970s represented a major improvement within the field of radiology. Unlike conventional X-ray images, the regions of interest in CT-scannings are presented without disturbing superimpositions of juxtapositional structures. The special characteristics of the CT-images allow three-dimensional (3-D) reproductions of the objects examined. Furthermore, the continuos refinements of equipment and techniques have made the production of plastic models a realistic and promising option, with interesting prospects for the future in a wide range of fields of anthropology.

A 3-D analysis starts with a CAT-scan of the item, done however with a close scan slice setting, preferably 2 mm or less. CAT-scanners are usually confined to hospitals, but are gradually becoming available at bigger research institutions. The Department of Anthropology of the National Museum of Natural History, the Smithsonian Institution, is an example of the latter.

After acquisition of the images, these are usually transferred to workstations or powerful PCs, equipped with software, which allows reading of these files and editing of the single images, and finally 3-D rendering. The 3-D renderings may further be used to build solid models by the technique called stereolithography (see below).

We will illustrate the various applications of these methods by some examples of the analyses we have carried out using these methods.

Documentation: King Gorm the Old of Denmark

Documentation may in many ways be viewed as the most basic use of these techniques. CAT-scanning and 3-D rendering allows items to be "virtually" accessible at any workstation. More importantly, it means that a complete documentation exists of the items, literally covering every square millimetre, including accurate representation of internal structures. This level of documentation may be especially important when the items in question are very valuable or when they may not be accessible at all times.

When excavation was undertaken in the Viking Age Jelling Church, the archaeologists found a crypt containing a skeleton underneath the church. The research team thought, after thorough investigation, that it was possible they could be the bones of King Gorm the Old (died 958/959 A.D.), transferred from one of the tumuli outside the church. Last year (2000) the bones, after being CAT-scanned, were replaced in the crypt as part of a ceremony that also marked the introduction of Christianity into Denmark. Thus, although the bones themselves are no longer immediately available for further investigation, the research team could at any time "assemble" the image sections and make 3-D reconstructions to serve as a basis for, e.g., determination of bone size and morphology (fig. 1).

Anthropological analysis: The skull from Hedegård, Denmark

CAT-scan may of course also be used directly for detailed analyses. The Hedegård skull was found by radiocarbon dating to be almost 8000 years old, making it one of the oldest human remains found in Denmark. When it was analyses at our lab, we noticed a couple of indentations of the skull surface, probably the result of healed fractures. CAT-scan allowed us to section the skull "virtually" at these lesions, thus allowing us a more detailed view of the bony changes (fig. 2). Indeed, using the 3-d rending techniques, we can section, i.e. cut, the specimen at any angle we wish.

Non-invasive and non-destructive: Mummy mismatch.

The founder of the Ny Carlsberg Glyptotek Museum in Copenhagen, Denmark, Carl Jakobsen, like many wealthy men of his time, had acquired an Egyptian mummy complete with sarcophagus. The names of two male priests were inscribed on the sarcophagus, which was said to date from circa 1100 B.C. It was decided to "look inside" virtually, by use of 3-d CAT-scan. Special focus was put on the head and hip region, and by examining the pelvis of the visualised remains inside the embalming material; it was possible to say with relative certainty that the mummy was a female (fig. 3). This may not be so surprising, as we know there was once a brisk trade in mummies and sarcophagi, a veritable "souvenir industry", in which mummies and sarcophagi were mixed up in the effort to ensure that the customer got a "full set".

Exhibition purposes: The face of the Mummy

Another mummy from the collections of the New Carlsberg Glyptotek Museum is decorated with the portrait of a young man, presumably depicting the embalmed individual. The portrait is considered one of the finest of its kind in the world, and naturally the question arose whether the portrait really bore a resemblance to the now mummified person. After the mummies skull had been rendered in 3-D, a solid 1:1 plastic model was made using stereolithography. The basis of stereolithography is the process of polymerization, i.e., the formation of larger molecules from the bonding of many, identical oligo- or monomers. This is achieved by guiding a laserbeam into a vat of liquid, i.e. monomeric, plastic resin. The laser generates heat when it strikes the monomers, which then bond, i.e. polymerise, and thus become solid. The laser beam is in turn guided by the 3-dimensional item rendered in the computer. In this fashion a polymeric model of the mummy skull was produced. This skull model was then used by forensic anatomical artists to produce a face of the deceased (fig. 4). These reconstructions in turn served as exhibition objects, allowing the public to compare the mummy portrait with a likeness of the real once-living person.

Non-invasive archaeology: Excavating without excavating

When the Danish Khapur Expedition undertook the excavation in 1990-1995 of an approximately 7000-year-old Sumerian settlement, Tell Mashnaqa in Syria, the archaeologists found numerous graves. The bones in the graves were, however, very badly preserved, and those that could be uncovered were very fragile. To get round this, the archaeologists came up with a technique for casting the entire cephalic region of a grave in plaster. Using this technique the grave could be lifted out in its entirety and flown to Denmark for CAT-scanning. The scan showed, among other things, an intact mandible that could be visualised in 3-D by the computer. Using lasers and a highly photosensitive plastic polymer – a technique called stereolithography – the computer's 3-D images were rendered as a solid model. Thus an accurate model could be produced on an not yet excavated item (fig. 5).

Not just bones

It should be noted that while the above cases all pertain to bones, as this is our field of research, the techniques are not thus limited. Indeed, wooden objects, clay and stone, also show up on CAT-scanning, and may be likewise visualised (fig. 6). This may have some implications for archaeologists working with such materials and objects. Metal more or less blocks X-rays completely, thus also disturbing other materials in an item, but fine metal objects may be probably be visualised also.

Computers and software

The computers used to run the software for these kinds of analyses used to be high-end workstations and machines capable of running UNIX. However, given the rapid development of machine power, the software now runs on ordinary, albeit powerful, PCs.

The software in question are usually specific kinds of software made for medical imaging, ranging from low cost (Digital Dr. ®) to high end (Materialise®) products. A product such as Mimics by Materialise® furthermore has the possibility of formatting the data to allow stereolithographical modelling.

Stereolithographical modelling is made possible by machines built for industrial purposes (as a whole, the field is called Rapid Prototyping, as the idea is to allow for a rapid generation of solid, 1:1 models of products, ranging from valves to mobile phones). As such, these machines are very expensive, but can be accessed through industrial research centres. In the above-presented cases, all models were made by the Danish Institute of Technology, which carries our research of such innovative techniques for the benefit of industrial purposes.

References

Hjalgrim, H., Lynnerup, N., Liversage, M., Rosenklint, A. 1995. Stereolithography: Potential applications in anthropological studies. In *American Journal of Physical Anthropology, 97*: 329-333.

Lynnerup, N., Hjalgrim, H., Rindal, L., Gregersen, H., Thuesen, I. 1997. Non-invasive Archaeology of Skeletal Material by CT Scanning and Three-dimensional Reconstruction. In *International Journal of Osteoarchaeology, 7*: 91-4.

Figures

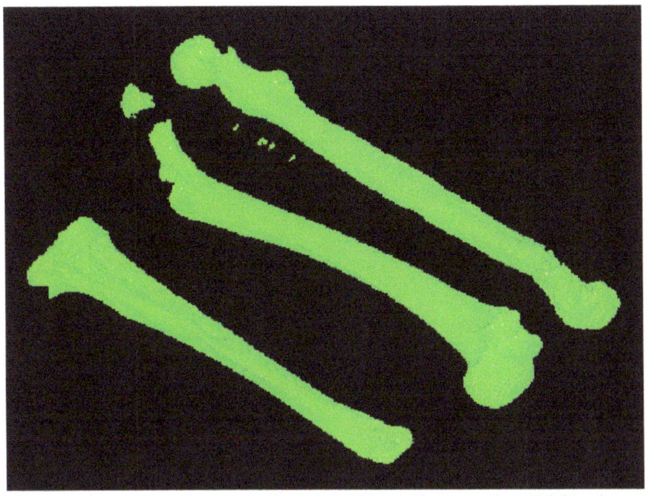

Figure 1: 3-D visualization of King Goem the Old´s long

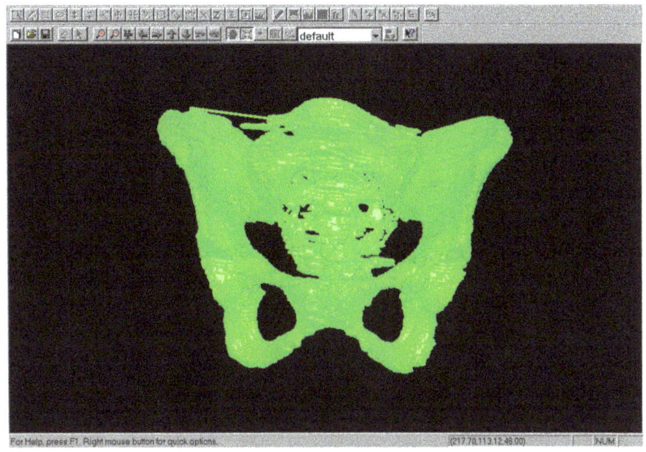

Figure 3: 3-D visualization of bone structures of an Egyptian mummy. The hip morphology indicates that the mummy is of a female.

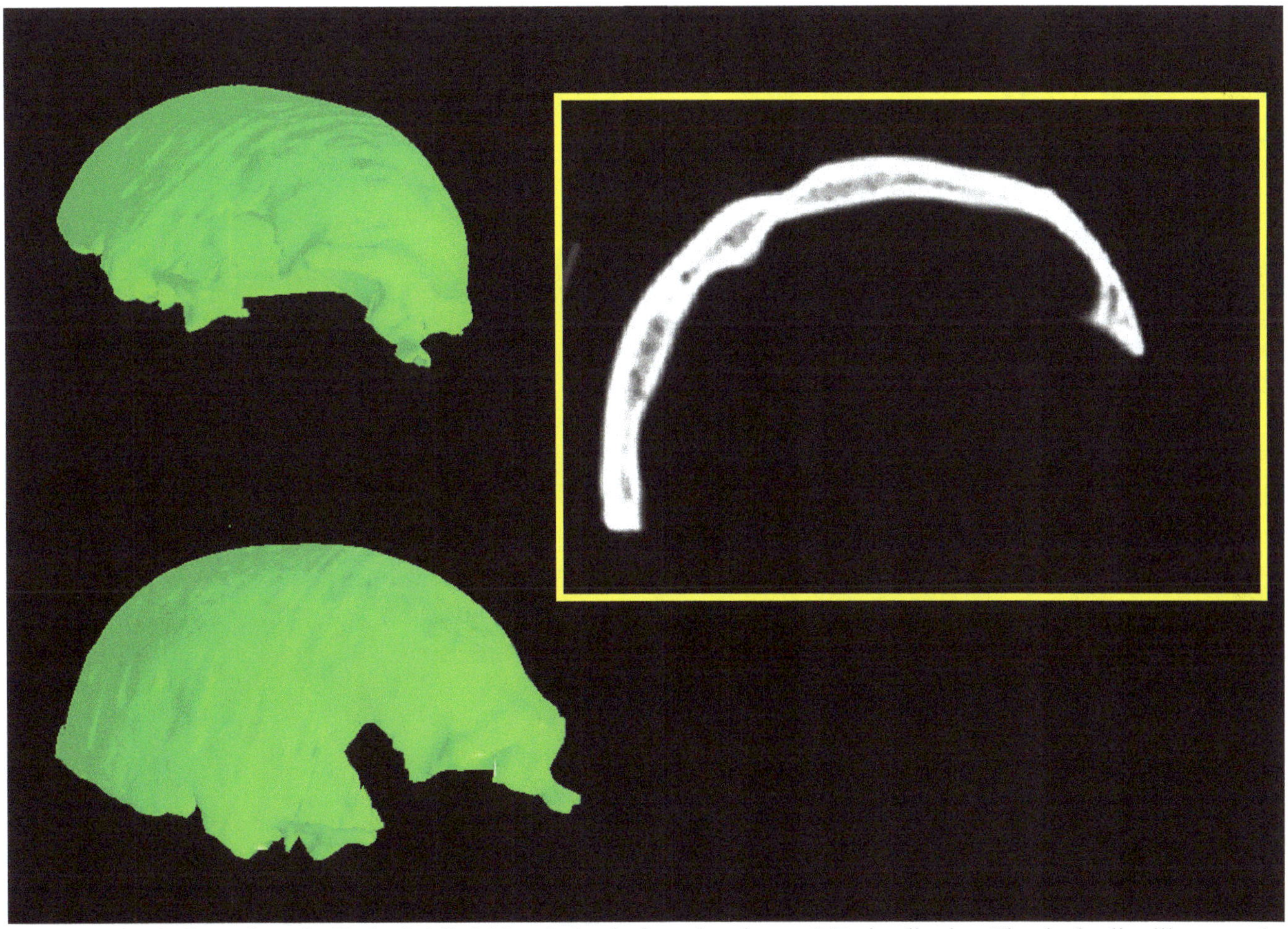

Figure 2: The skull cap from the Hedegård find. Note lesion in frontal region on 3-D visualisation. The single slice illustrates the lesion on a "virtually" cut surface.

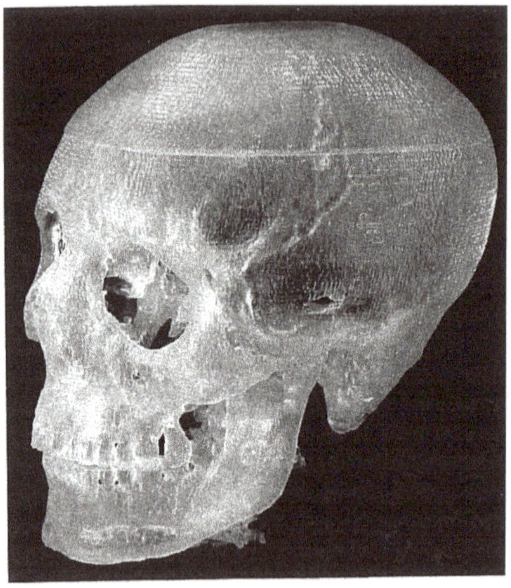

Figure 4: Stereolithographical model of mummy skull.

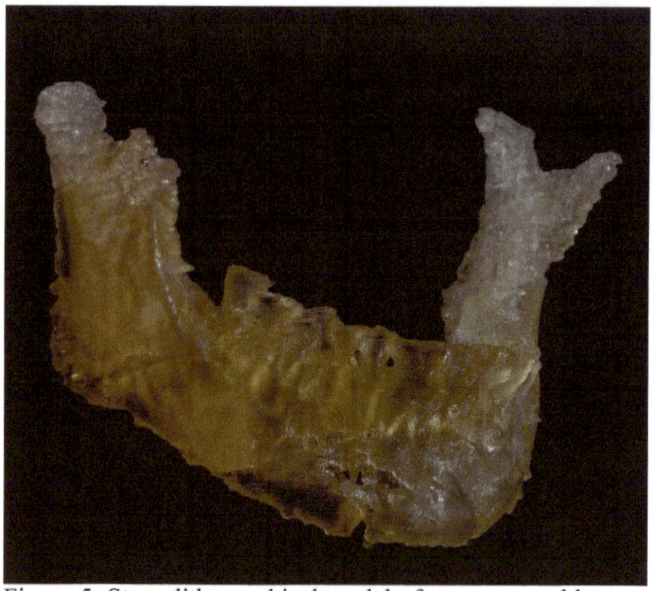

Figure 5: Stereolithographical model of un-excavated human mandible.

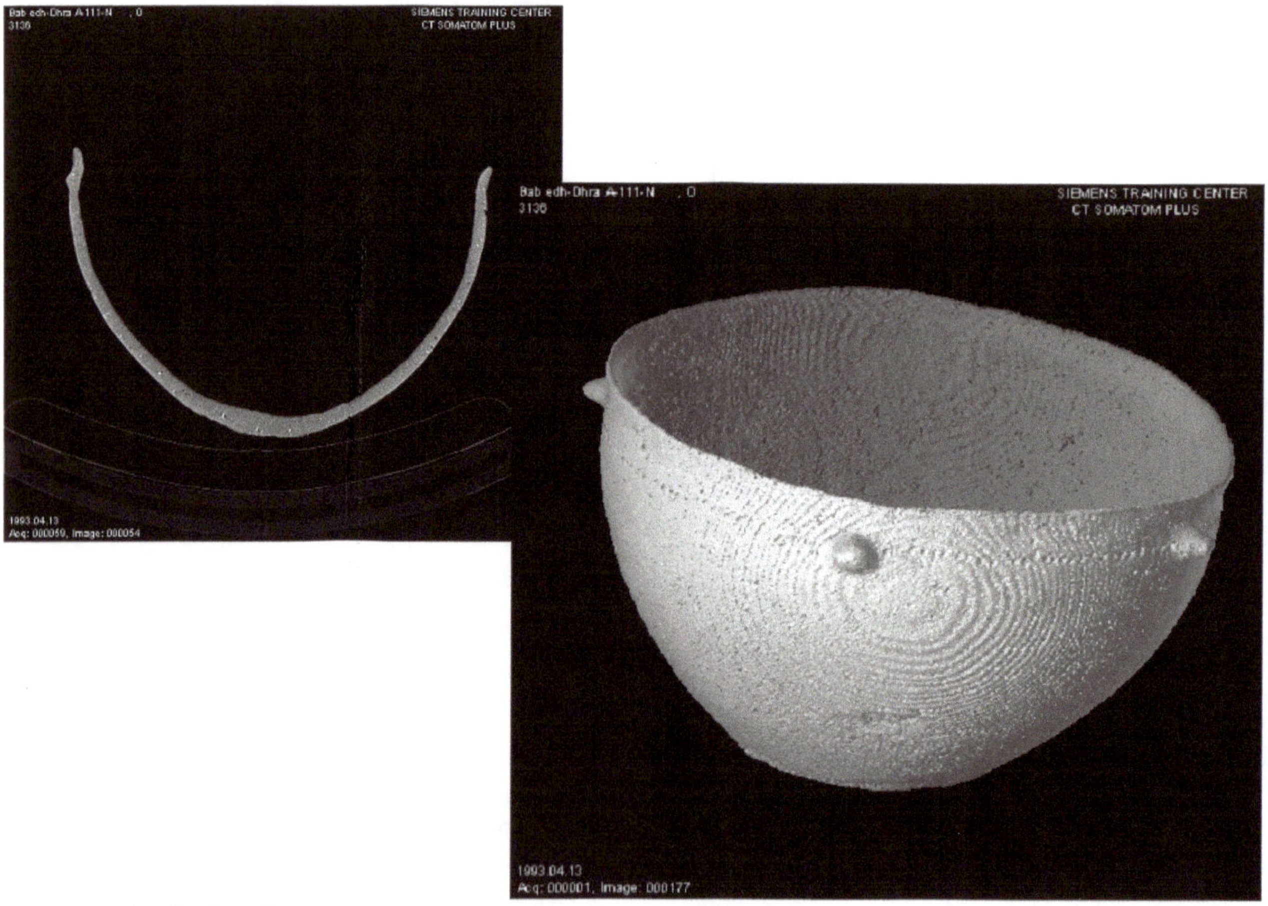

Figure 6: 3-D visualisation of pottery ware.

The Statistics of Archaeological Deformation Processes
An archaeozoological experiment

Laura Mameli
Divisió de Prehistòria
Facultat de Lletres. Edifici B
Universitat Autònoma de Barcelona
E-08193 Bellaterra, Spain
E-mail: laura.mameli@campus.uab.es

Juan A. Barcelo
Divisió de Prehistòria
Facultat de Lletres. Edifici B
Universitat Autònoma de Barcelona
E-08193 Bellaterra, Spain
E-mail: ilphg@blues.uab.es

Jordi Estevez
Divisió de Prehistòria
Facultat de Lletres. Edifici B
Universitat Autònoma de Barcelona
E-08193 Bellaterra, Spain
E-mail: ilphd@blues.uab.es

Abstract: The quantification of faunal bone assemblages from archaeological sites has a long tradition. Nevertheless, in most cases those studies are somewhat "passive", as if the bones have always been in those conditions, waiting for the archaeologist. There are many taphonomic studies to solve these problems, but in most occasions, taphonomic results seem totally unrelated to archaeological research, as if the natural factors were studied independently from social factors.

In this paper we present a case study where the natural formation process of bone assemblages is experimented. Contrary to the usual view, when wild animals (scavengers) are the causal agent of the assemblage, the archaeologically observable consequence is not an accumulation of bones, but a considerable spread of them. As a result, bones enter into the archaeological record as individual items, without an understandable spatial pattern.

We have studied 30 animal carcases scavenged by foxes in Tierra del Fuego (Argentina). A general statistical analysis of those examples is presented in order to describe an important "archaeological de-formation process". The goal of the study is to discuss how to analyse de-formation process of the archaeological record, and how they affect the quantification of faunal remains in archaeological sites.

Key words: Zooarchaeology, Spatial Archaeology, Formation Process, Correspondence Analysis, Disturbance, Post-Depositional alteration

The Deformation of Archaeozoological Assemblages

Archaeologists traditionally have drawn their inferences about past behaviour from dense, spatially discrete aggregations of artefacts, bones, features, and debris. They have traditionally assumed that the main agent responsible for creating such aggregates was only human behaviour. Even though nowadays most archaeologists are aware of natural disturbance process and the complexities of archaeological formation, archaeological contexts are still usually viewed as a deposit or aggregate of items, which are part of single depositional events. We usually speak of human behaviour being fossilised in archaeological accumulations or deposits, as if the materials of social action through time were only accumulative.

In the case of faunal processing, the "supposed" main observable consequences of social action are bone accumulations. The underlying idea seems to be that the bones of the creatures under consideration have survived since deposition to a more-or-less similar extent, and that the relative abundance of species is representative of that originally deposited. However, the accumulation of animal bones in the archaeological record not always is the result of purposeful human activity. An aggregation of bones may not reflect past human social action, but rather post depositional processes. Loss, discard, reuse, decay, and archaeological recovery are numbered among the diverse formation processes that in a sense, mediate between the past behaviours of interest and their surviving traces (see among others Meadow 1976, Hassan 1987, Schiffer 1987). Most post-depositional processes make archaeozoological assemblages more amorphous, lower in elements density, more homogeneous in their internal density, less distinct in their boundaries, and more similar (or at least skewed) in composition. Those processes have the effect of disordering artefact patterning in the archaeological record, and increasing entropy. Furthermore, some post-depositional disturbance process may increase the degree of patterning of artefact disturbances, but towards natural arrangements (Ascher 1968, Carr 1984). Consequently, determining whether the various frequencies of items in an assemblage or deposit have resulted from undisturbed deposition, differential distribution, differential preservation is the problem (Brain 1980, Lyman 1987).

Archaeological assemblages should be regarded as aggregates of individual elements, which interact with various agents of modification in statistical fashion, with considerable potential for variation in the traces they ultimately may show. Cowgill (1970) proposed a preliminary solution: we have to recognize three basic populations (in the statistical sense):

(1) events in the past,
(2) material consequences created and deposited by those events, and
(3) artefacts that remain and are found by the archaeologist ("physical finds").

By stressing the discontinuities, Cowgill states for viewing formation process as agents of bias within a sampling framework. At the beginning, material items are organized in the archaeological record in a way coherent with the resource management strategies and social practices that generated them. Once the location of social action was left, those remains were subject to bio-geologic forces, which introduce a new material organization. This new patterning of social material remains is opposite to the original pattern, and consequently increases entropy (des-organization, chaos, and ambiguity), until the original patterning become unrecognisable. Each population is then a potentially biased sample drawn from the previous population that was itself a potentially biased sample. We may view these discontinuities as sampling biases in the sense that what we recover and observe does not proportionately represent each aspect of the antecedent behaviour.

From our dialectic point of view, changes and transformations in the original patterning of activity sets are not a simple accumulation process from low entropy sets (primary deposition) to higher entropy patterns (disturbed deposits), but a non-linear sum of quantitative changes, which beyond a threshold, produce a qualitative transformation. A depositional set may be thought of as a mathematical set, the organization of which is the end product of structural transformations operating upon a previously structured set. In this sense, the occurrence of specific formation process is determined by specific causative variables. That means, that depending of the degree of entropy, the transformed archaeological set is not necessary a random sample of the original population. That is, the difference between a depositional and an activity set is based on a deep qualitative discontinuity generated by the aggregate of minor quantitative modifications (Estevez 2000).

The main point is not the recovery of the "social action direct effect" by reversing the formation process of "depositional sets" (for definitions, see Carr 1984, Urbancyk 1986). Rather, the processes responsible for generating organizing, preserving, and presenting the archaeological record should be viewed simply as a dialectic formation processes. Attention should be drawn to the dynamic life history of archaeological remains and the processes of different temporal frequency on the ultimate position, content, and pattern of archaeological remains. This perspective provides a strong antidote to the facile "reconstruction of culture" by "correcting" for apparent disturbances or distortions.

Substraction as an Archaeological Formation Process

In most archaeological bone assemblages, for any given species the frequencies of different skeletal elements show at least some significant departures from the frequencies in which they would be represented in complete skeletons. Modifications in original skeletal frequencies may appear: as a result of human actions, as a result of subtraction by animal scavenging, or as a result of differential preservation and recovery (Klein 1980, Monahan 1998, Bartram & Marean 1999, Estevez 2000).

There is a long discussion whether the difference between primary deposition of animal bones and the recovered bones may bear directly on the distinction of hunting from scavenging at archaeological sites. Since the days of Efremov (1940), archaeologists are looking for regular relevant linkages –signature criteria- between static attributes of the archaeological record and their dynamic causes and associations. A signature criterion is a criterion that is constant and unique and that discriminates one modifying agent or set of agents from another. The idea of some authors is to create a dictionary of material consequences of human hunting and animal scavenging to be able to "predict" the structure of the assemblages produced by a specific action. These predictions should be tested experimentally, that is, by evaluating the predictions in light of modern assemblages known to have been formed by the process stipulated in the predictions (Gifford 1981, Binford 1981, Schiffer 1987, Dominguez-Rodrigo 1999). For instance, scavenged assemblages appear to be without some bones selected by size, weight and shape according to the particular size, mechanical capacity, and foraging range of the scavenger. Recognizable signatures therefore can be characterized by specific ranges of taphonomic loose, generally defined through upper size and weight limits (Marean & Bertino 1994, Stahl1996). For instance, an over-abundance of cranial and distal limb elements of middle-to-big sized game would be characteristic of butchery sites, whereas the over-abundance of proximal limbs would be a definitional feature of consumption sites (Binford 1981, Stahl 1999). That means that a characteristic "human" pattern of disarticulation would be the refuse of lower legs because they lack sufficient meat to make them desirable. The heavier and less nutritious portions of a carcass (the axial skeleton) tend to remain at animal death sites, while the lighter and more nutritious portions of a carcass (the appendicular skeleton or limbs) tend to be transported more extensively away from animal death sites.

But nothing is so easy. The actual combination of those variables related to causal processes that could have given rise to specific deposits is nearly infinite, and so one cannot expect to find many simple correspondences between a priori lists of evidences and the characteristics of specific deposits. Bone subtraction also appears under the form of differential preservation. Under a given destructive regime, the individual parts will survive in proportion to their robusticity. Survival may be correlated with the compactness of the bone, expressed as specific gravity (Brain 1980). Skeletal parts of high utility tend to be low in density, and then those bones are not resistant to damage (Lyman 1987, Rogers 2000): an assemblage dominated by parts of low utility is also likely to be dominated by parts that are dense and therefore resistant to scavenger gnawing. Such an assemblage could have been produced either by selective transport or by selective destruction of low-density parts

One can hardly argue that uniformitarian principles may be formulated concerning the social scope of human communities, given the profoundly varied, and specific exploitation strategies of resources by different societies (Hassan 1987, Castro et al. 1993, Marciniak 1999). There are many actions and processes, both social and natural having acted during and after a primary cause, and also primary causes act with different intensities and in different contexts, in such a way that effects may seem unrelated with causes.

The fact that we cannot predict the degree a bone assemblage has been scavenged, does not mean we cannot analyse an animal carcass as a by-product of a series of social actions and which has been altered by other processes (or the reproduction of the same actions at the same place). In most real cases, we should speak about multiple causes and complex causal relationships, rather than indeterminism or intrinsic randomness. In this paper we are interested in measuring the probability an archaeological assemblage of animal bones can be "deformed" by the action of scavenging. We think that the aggregate of quantitative modifications experimented by a carcass (in content and spatial distribution) can produce a significative qualitative change (a bone assemblage). Scavenging should be considered as a sequence of modifications which convert an animal carcass into a disintegrated set of bones. It is no more an animal, but it contains some distorted elements of what once was (palimpsest). Given the probabilistic nature of causal relationships, we cannot assert that, the survival parts of a skeleton will follow an entirely predictable pattern if the destructive influences are known (Brain 1980, p. 117). That means, that simple documentation of frequencies of disarticulated and articulated joints in an assemblage may not permit the inferential identification of social action before/after/ in absence of subtraction by scavenging.

We have designed a series of controlled observations in order to be able to calculate the probability relationship between the disturbance effect and the composition and spatial pattern of bone remains. That is, causal significance of scavenger activity for bone assemblages composition and spatial patterning corresponds to the difference that the presence or absence of scavenger activity makes on the features of bone assemblages. In terms of Stochastic Interaction, the probability of existence for scavenger activity and the features of bone assemblages are determined reciprocally. Therefore, changes in the probability of scavenger disturbance determine changes in the probability of spatial patterning of bones and changes in the probability of spatial patterning determine changes in the probability of scavenger activity. We want to test the hypothesis that an increasing skeletal disorganization in terms of taphonomic loose suggests a more complex archaeological formation history.

Tierra del Fuego. An Archaeozoological Case Study

We have studied 30 carcases of "guanaco" (Lama guanicoe, a comparable middle sized herbivore) scavenged by foxes in Tierra del Fuego (Argentina). During three years we have taken measurements from animal carcasses produced by a catastrophic natural death in 1995. We have yet published some preliminary results using traditional frequency approach (Mameli & Estevez 1999, Estevez & Mameli 2000).

In this study we have used the following controlled variables: presence/absence of bone elements in each carcass, the quantity of bitted bones in each carcass, and the Euclidean distances between bones in each carcass, according to the longitudinal axis of orientation (from head to tail).

The Statistics of "Deformation" Process: general patterns

Archaeological deposits (bone and artefact) are usually described and analysed using global attributes:

- the average density of attributes,
- the form of arrangement of artefacts within it –clustered, random, or systematically spaced- independent of dens ity (Carr 1984).

In this paper, we have followed this approach. As a preliminary stage we have produced some qualitative statistics of the general pattern of scavenging, as observed in our data. Different carcasses were observed, and the evidences of scavenging on any kind of tissue were annotated.

Taphonomists have paid little attention to the statistical problem. They draw inferences from bone counts to develop general patterns of scavenging, transport and attrition, relying almost exclusively on measures of association between the frequencies of body parts (cf. Binford 1981, Davidson & Estevez 1986, Blumenshine 1988). Our preliminary analysis has been a frequency analysis of scavenging as observed in controlled situations. The goal is to record the relative frequency with which different body parts occur in modern carcasses lying on land surfaces, which permits the construction of a statistical model of disarticulation.

To our surprise, the observed statistical pattern of scavenger disturbance does not coincide exactly with the standard decay order of natural joint disarticulation of large ungulates (Hill 1979), probably because scavenging is not always related to disarticulation, but with edible tissues subtraction. Hill studies have shown that there tends to be a pattern in the sequence of natural skeletal disarticulation from forelimb-body disarticulation until cervical vertebrae breaking. Disarticulation of Mandible and Skull arrives only in fourth place, after forelimb-body, caudal vertebrae-sacrum and Scapula-proximal humerus. Disarticulation of carpals and tarsals is even later. In our case (see Figure 2), softer parts (anus, belly) where the first ones to be scavenged until its total consumption. Mandible and Skull of the animal is scavenged in full well before the rest of the carcass. In general, the less scavenged parts during the first year are limb bones, thorax, neck, lumbar and sternum. In the second year, those neglected parts are scavenged in greater proportion, probably because the skin of the animal preserved edible tissue in those parts. During the third year, many elements of the carcasses are hardly visible, and nearly all remaining parts are equally scavenged.

The very process of Scavenging is hardly visible in the archaeological record, because it affects softer tissues. The only indirect evidences we have are those bitted bones in the archaeological record. That is to say, from the total amount of bones in a carcass, only the quantity of bitted bones can be used as evidence of the intensity of scavenging (Mameli & Estévez, 1999, Estevez & Mameli 2000). The obvious question is then the relationship between scavenging and morphological damage discovered on the bone surface.

If we consider now the number of bones with traces of animal scavenging (bitten or gnawed bones)(Figure 3), the results of our controlled observations suggest that cranium is the only damaged skeletal part in the first year, which coincide with the fact of preliminary scavenging on less skin protected parts. During the second year, skulls are still being damaged, together with scapulae, pelvic bones and ribs. Long bones, although scavenged show very few evidences of surface modification, concentrated on epiphyses. In the third year, vertebrae are the most damaged bones.

There is also a strong temporal dynamics, which prevent a proper description of the scavenging process from archaeological remains. If we consider that archaeological record is very similar to our last year observations –when trapping has already begun-, then we see that most bones are bitten in a similar relative frequency, with the exception of zigopodia, sternum, ribs. But to affirm that those parts have been less scavenged than others would be a misleading conclusion, as the results for first and second year suggest.

To go beyond this frequency description we need multivariate techniques to disentangle the multiplicity of effects produced by different post-depositional processes. We have studied through Correspondence Analysis the presence/absence of observed evidences for scavenging.

A Bi-dimensional factorial solution accounts for a very low proportion of total variance (42%). That means that we cannot speak about a general pattern of variation which can explain observed variation among carcasses. 87 scavenging events have been observed, and it seems impossible to infer a regular pattern (see Figure 4). Where the anus or belly have been scavenged, other body parts have not evidences of being disturbed or modified as a result of scavenging. In the opposite, where limb bones or autopodia have been scavenged, there are no evidences of disturbance in the softer parts.

When we consider all observations made during the three years of the experiment, the only pattern which emerges is that of the influence of the softer parts scavenging, instead of disarticulation.. In figure 4, it is interesting to observe the differences and similarities between head parts (cranium, mandible, mouth, even neck) and limb bones (limbs and feet).

Differences between the theoretical pattern (Hill 1979) and the observed one in our case study is probably the result of the characteristic of grey foxes scavenging (little canids) and the anatomical features of guanaco, with very robust bones and a thick skin. We have observed that before disarticulation, little canids ravage soft tissues from the head of the animal (eyes, tongue, etc.) or the belly. Only when other disturbance process have acted, meat is easier to scavenge. This is why forelimb bones are mostly scavenged at later moments.

The Statistics of "Deformation" Process: Temporal Dynamics

In most taphonomic analysis of scavenged carcasses, temporal dynamics are only partially taken into account. We think that it is impossible to understand scavenging as a disturbance process, if we do not characterize it as a dialectical process, where a non-linear aggregation of quantitative changes, beyond a threshold, produce a qualitative transformation. That means that a single description of some events of scavenging is not enough for understanding the process. Scavenging is a dynamic process, introducing quantitative modifications which eventually produces a deep discontinuity.

Why carcasses with softer parts being scavenged have no evidences of modification in other body parts? Because scavenging is not a static event, but a continuing process. That means, what is not scavenged at first, will be ravaged later. We have controlled three temporal stages for each case, and we have documented a strong temporal component

An opposition between first year observations, and the rest characteristically dominate dimension 1, which accounts for 29% of total variance (see Figure 5). If we combine these results with those obtained in preceding chapter, we can arrive at the conclusion that softer parts are much scavenged during the first year that later in the process. It is interesting that we cannot separate the last two stages on the basis of softer parts scavenging.

Let we process the results differentiating the effects for each temporal stage.

Variability among carcass during the first year is very great (Figure 6a). A Bi-dimensional Multiple Correspondence Analysis only accounts for 36%. The main aspect of this variability is the relevance of low neck. That means, carcasses with evidence of scavenging at the level of the low neck (4 items), are significatively different from the others. In general, the same body part seems to be scavenged differently in different carcasses. During the second year (Figure 6b), the perceived regularity only accounts for 25% of total variance, but events are totally different. Softer parts have disappeared, and the carcasses become disarticulated. As a consequence, the same body part is processed in the same way in different carcasses. As the time goes, the effects of scavenging are more similar, and we can define some redundancy patterns. During the third year (Figure 6c), most body parts have disappeared, and foxes scavenge all remains they can find. The overall similarity increases, because the Multiple Correspondence Analysis accounts for 40 % of total variance.

It is highly significative that the important differences between forelimbs and hindlimbs, as between left and right during the first year of observations (Figure 6a), diminish during the second year (Figure 6b). That means that extremities seem to have been scavenged differently in different carcasses at the beginning of the process, tending towards an increasing similarity. We can explain all this temporal dynamics suggesting that the more scavenged the bones, the more disturbance, but also, the more similarity between body parts. The primary deposition has been totally altered, but similar bones are scavenged in similar ways, when carnivores have less edible choice, than when carcasses are complete.

We have also studied the influence of the intensity of scavenging. Figure 7 plots factor scores for individual carcasses against a measure of intensity (number of bones displaced or modified). In all cases, the most disturbed carcasses (cases where intensity values are the highest) are usually at the center of the distribution, what suggests the high similarity between them. Most cases of low intensity appear everywhere in the plot, suggesting the low degree of global similarity between the best preserved carcasses. This pattern is clearer at later stages than an the beginning of the scavenging process.

Consequently, we cannot conclude that the only effect of scavenging is to increase entropy. In fact, although disorder increases, the general similarity also increases. This final patterning of animal bones at butchery sites seem to be very different to the original pattern, which becomes unrecognisable. The more disturbances scavenging has produced in the past, the less possibilities of differential re-scavenging remains in the present. Consequently, as the process continues, scavenging concentrates in those bone which survive, appearing some hints of regular pattern at a later stage. In any case, this pattern does not go beyond 40% of total variance.

Temporal dynamics of scavenging are not characterised by a simple accumulative process from low entropy sets (death animals) to higher entropy patterns (scavenged carcasses), but it should be defined as a non-linear aggregate of quantitative changes, which beyond a threshold, produce a qualitative transformation (an archaeological deposit). We have not identified any regular pattern that could differentiate scavenging from other social actions, but a general pattern towards increasing qualitative deformation. This result is in strong opposition with usual approaches trying to identify natural disturbance processes using single identifiers.

The Statistics of "Deformation" Process: Spatial Dynamics

In our case study, we have discovered evidence for a relationship between the intensity of scavenging, the distance of transported elements and the relative frequency of lost material: although the intensity of scavenging diminish as the time goes, the spatial disturbance (distance) increases .

These graphs show the spatial distribution of skeletal parts of some carcasses analysed in our case study. Spatial coordinates have been measured in reference to local reference points (a theoretical line along the orientation axis of the carcass). Given that results are always expressed in the same scale, they can be compared, although points in the graphs cannot be interpreted as individual comparisons. Graph coordinates should be seen as non-normal transformations of original coordinates, preserving ordinal differences and similarities. In Figure 8, a 3 meters zoom window from the original location has been selected. It is important to realize that the 3 meters zoom window only includes a subset of bones from the third year. We have given emphasis on the examination of the original deposition area and how it is transformed through time, rather than a general exam of the total distribution (it can arrive until 30 meters).

We see that differences are not greater between the second and third year, than between the first and second. That means, scavenging during the second year is still produced on relatively dense carcasses. Dispersal seems to be a consequence of long scavenging and not a direct consequence of a single event animal action on carcasses.

We obtain a new test of the previous hypothesis: as the time goes, the density of locations diminishes and the spread of bones increases. That is, although scavenging is less intense after the first year, occasional disturbance produces much more spatial dispersal than original scavenging.

Our analysis pretends to examine if the number and nature of bones in one location have anything to do with characteristics in a neighbour location through the definition of a general model of spatial dependence. We usually assume that "everything is related to everything else, but near things are more related than distant things" (Tobler's law). This assumption is based on the Neighbourhood Principle, which relates the intensity of influences converging to a single location from the spatially neighbouring locations. This axiom is not always true, specially when we deal with natural disturbance processes.

What we are looking is whether the location of individual bones or skeletal parts is homogeneous or heterogeneous in the area defined by the natural disturbance process. The effects of scavenging as a disturbance process should be explained in terms of the "influence" an action performed at a location has over all locations in the proximity. A formation or de-formation process can generate some material effects (quantitative and/ or qualitative changes) around it, or it can prevent that effects of other processes leave evidences in the same vicinity. Some of the processes acting in the vicinity of the location increase the probability of some material (observable) effects and decrease the probability of others. The approach here relies on a prior hypothesis of spatial smoothness, which considers that two neighbouring observations are supposed to have been more likely originated from the same group than two observations lying far apart. (Barcelo & Pallares 1998).

This can be easily computed by estimating the spatial probability density function associated to each location. Given that locations are defined bi-dimensionally, we can calculate an interpolated surface representing the form of a probability density distribution for two continuous random variables, Cartesian co-ordinates x and y. The idea is to estimate this 2D dimensional density function, given a sample of known locations, by estimating the density in that area, the relative frequency of all observations falling in a given interval is counted. We use Kernel estimation techniques for this task (Baxter and Beardah 1997, Delicado 2000). These techniques are characterised by the use of a weight function (the kernel function) that permits give more mass to observed data near the point when the function is estimated.

A preliminary exam of point clouds graphs shows that as the time goes, the material effects of scavenging (spatial disturbance) increases, the density of bone locations diminishes and the probability of inferring the correct place for the original deposition of the carcass also diminish. We have translated this information using a kernel density approximation. The results for carcass number 1 appear in Figure 9.

Contours in Black corresponds to the probability distribution of first year observations, those in grey to the second and third year. That is, when animal bones begin to be integrated into what will become an archaeological record (after trapped in soil) we have a disturbed spatial distribution characterised by its lower density. Given that that distances among bones have increased, the probability of inferring the original placement of the carcass diminish: the increase of entropy (disorder) is the cause of an increase in the ambiguity of the archaeological record. The main effect of scavenging is then a negation of Tobler's Law: in an altered context, near skeletal parts are not more related than distant parts. In some cases distant skeletal parts can even be more related than near bones. The question is then if we can discover any traces of the original deposition even after post-depositional processes.

We have calculated a general spatial density model for all carcasses in the case study.

We use contour maps, to represent spatial densities. Probability maps should be considered a visual model of locational features, and not an explanation of spatial causality. Contour maps as those presented here are a graphical convention for showing changes in the probability of action as a function of disturbances generated by scavenging at neighbouring locations. Figure 10a shows a density map corresponding to undisturbed or only partially disturbed carcasses: all bones are in the "correct" place according to the anatomy of the dead animal, and according to the cause of death. As the time moves on, spatial disturbance increase, and the density of bones diminish. Fig.10c shows the global density of bone placement after three years of continuing scavenging. It remains a core area of more dense findings, corresponding to the precise location of the original action (animal death). It is at this level when we can affirm that the higher the density of findings, the higher the probability of the original action. We are not calculating the probability of scavenging, but evaluating the possibilities to infer the original action in the palimpsest generated by post-depositional processes. Results are clearer if we use a 3D representation of the same kernel filter.

Conclusions

Usually, archaeologists assume that prehistoric social practices, including procurement, butchering, storage, cooking and disposal will produce faunal assemblages distinct from those generated by natural processes (catastrophic death, for instance). However, animal bones found in archaeological conditions have usually been exposed to a long scavenging process, and characteristics of the original deposition are no more visible. We have shown that scavenging increases entropy, and we have also shown that the patterning at the end of the disturbance process may be explained in probabilistic terms.

In this paper we have argued that archaeological assemblages should be regarded as aggregates of individual elements, interacting with various agents of modification in statistical fashion, with considerable potential for variation. As a result, what we recover and observe does not proportionately represent each aspect of the prior behaviour. It is not yet possible to explain scavenging in terms of a regular subtraction process characterized by a logarithmic reduction of the amount of bones to be scavenged. Consequently it does not exist a simple rule able to separate the effects of scavenging from primary deposition. Each bone assemblage is a potentially biased sample drawn from an original population that was itself a potentially biased sample. We should view these discontinuities not only as sampling biases, but as a discontinuous, and non-linear process: *an increasing skeletal disorganization in terms of taphonomic substraction and spatial disturbance is related with complex site formation histories.*

Therefore, element-abundance data cannot be used in archaeology in a simple way to investigate social practices, like butchery, carcass-transport decisions, human nutritional needs, activity specialization, discard, rubbish formation, and so on, without considerations of the specific formation process. Interpreting the content and frequency of an archaeological assemblage must be grounded in an understanding of both the social and natural events that have influenced the presence/ absence, alteration, and displacement (relative to it as a primary site of production, use or discard) of its individual components and of the assemblage as a whole.

References

Ascher, R. 1968. Time's arrow and the archaeology of a contemporary community. In *Settlement archaeology*, Edited by K.C. Chang, pp. 43-52. National Press Books, Palo Alto.

Barcelo, J.A., Pallares, M. 1998. *Beyond GIS: The archaeology of social spaces*. Archeologia e Calcolatori Num. 9, pp. 47-80

Bartram, L.E., Marean, C.W. 1999. Explaining the Klasies Pattern: Kua Ethoarchaeology, the Die Kelders Middle Stone Age Archaeofauna, Long Bone Fragmentation and Carnivore Ravaging. In *Journal of Archaeological Science 26*, pp. 9-29

Baxter, M., Beardah,C.C. 1997. Some Archaeological Applications of Kernel Density Estimates. In *Journal of Archaeological Science*, 24, pp. 347-354.

Binford, L.R. 1981. *Bones. Ancient Man and modern myths*. Academic Press.

Blumenshine, R.J. 1988. An Experimental Model of the Timing of Hominid and Carnivore Influence on Archaeological Bone Assemblages. In *Journal of Archaeological Science 15*, pp. 483-502

Brain, C.K. 1980. Some criteria for the recognition of bone-collecting agencies in african caves. In *Fossils in the Making. Vertebrate Taphonomy and Paleoecology*. Edited by A.K. Behrensmeyer and A. Hill University of Chicago Press, pp.107-130.

Carr, C. 1984. The Nature of Organization of Intrasite Archaeological Records and Spatial Analytic Approaches to their Investigation. In *Advances in Archaeological Method and Theory. Vol. 7*. Edited by M.B. Schiffer. New York, Academic Press, pp.103-222.

Castro, P.V., Lull, Mico, R. 1993. Arqueología: algo más que Tafonomía. In *Arqueología Espacial vol. 16-17*, pp. 19-28.

Cowgill, G.L. 1970. Some sampling and reliability problems in archaeology. In *Archéologie et Calculateurs. Problèmes Semiologiques et Mathematiques. Colloque International du CNRS*. Editions du CNRS, Paris, pp. 161-175.

Davidson, I., Estevez, J. 1986. *Problemas de Arqueotafonomía. Formación de yacimientos con fauna*. Quaderns, 1985-1986, Homenaje Dr. J.M. Corominas. Pp. 67-87.

Delicado, P. 2000. Statistics in Archaeology: New Directions. In *New Techniques for Old Times. Computer Applicxations and Quantitative Methods in Archaeology*. Edited by J.A.Barcelo, I.Briz and A.Vila. Oxford, ArcheoPress (BAR Int. Series, S757)

Dominguez-Rodrigo, M. 1999. *The Study of Skeletal Part Profiles: An Ambiguous taphonomic tool for archaeology*. Complutum 10, 15-24.

Efremov, I.A.. 1940. *Tafonomy: a new branch of Paleontology*. Pan American Geologist 74, pp. 81-93.

Estevez, J. 2000. *Aproximación dialéctica a la Arqueología*. Revista Atlántica No. 3. Cádiz (Spain).

Estevez, J., Mameli, L. 2000. Muerte en el Canal: experiencias bioestratinómicas controladas sobre la acción sustractora de cánidos. In *Archaeofauna, vol. 9*, pp. 7-16.

Gifford, D.P. 1981. *Taphonomy and Paleoecology: A Critical Review of Archaeology's Sister Disciplines. Advances in Archaeological Method and Theory. Vol. 4*. Edited by M.B. Schiffer. New York, Academic Press, pp. 365-438.

Hassan, F.A. 1987. Re-Forming Archaeology: A Foreword to Natural Formation Processes and the archaeological Record. In *Natural Formation Processes and the archaeological Record*. Edited by D.T.Nash and M.D. Petraglia. Oxford, British Archaeological Reports S352, pp. 1-9.

Hill, A.P. 1979. Butchery and natural disarticulation: an investigatory technique. In *American Antiquity 44*, pp. 739-744.

Klein, R. 1980. The interpretation of Mammalian Faunas from Stone-Age Archaeological Sites, with Special Reference to Sites in the Southern Cape Provincia, South Africa. In *Fossils in the Making. Vertebrate Taphonomy and Paleoecology*. Edited by A.K. Behrensmeyer and A. Hill University of Chicago Press, pp.223-246.

Lyman, R.L. 1987. Archaeofaunas and Butchery Studies: A Taphonomic Perspective. In *Advances in Archaeological Method and Theory. Vol. 10*. Edited by M.B. Schiffer. New York, Academic Press, pp.249-337.

Mameli, L., Estevez, J. 1999. Procesos Post-deposicionales. Un caso de experimentación tafonómica en un área arqueológica. In *Reunión de Experimentación en Arqueología*. Edited by L. Mameli and J. Pijoan. UAB-Prehistory Series. "Treballs d'Arqueologia", Special Issue.

Marciniak, A. 1999. Faunal Materials and Interpretive Archaeology. Epistemology Reconsidered. In *Journal of Archaeological Method and Theory No. 6*, (4), pp. 293-320

Marean, C.W., Bertino, L.W. 1994. Intrasite spatial analysis of Bone: subtracting the effect of secondary carnivore consumers. In *American Antiquity 59(4)*, pp. 748-768.

Meadow, R.H. 1976. *Methodological concerns in Zoo-Archaeology. IX Congress UISPP*, pp. 110-123.

Monahan, C.M. 1998. The Hadza Carcass Transport Debate Revisited and its Archaeological Implications. In *Journal of Archaeological Science 25*, pp. 405-424.

Rogers, A.R. 2000. On equifinility in Faunal Analysis. In *American Antiquity 65(4)*, pp. 709-723.

Schiffer, M.B. 1987. *Formation processes of the Archaeological Record*. Albuquerque, University of New Mexico Press.

Stahl, P. W. 1996. The Recovery and Interpretation of Microvertebrate Bone Assemblages from Archaeological Contexts. In *Journal of Archaeological Method and Theory vol. 3 (1)*, pp. 31-75

Stahl, P.W. 1999. Structural Density of Domesticated South American Camelid Skeletal Elements and the Archaeological Investigation of Prehistoric Andrean Ch'arki. In *Journal of Archaeological Science 26*, 1347-1368.

Figures

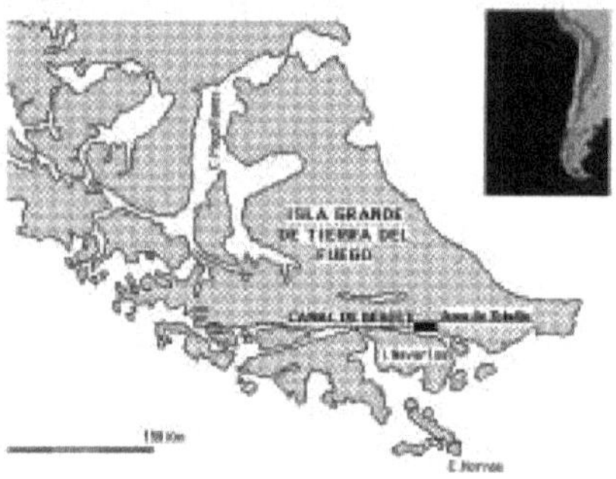

Figure 1. Study Area

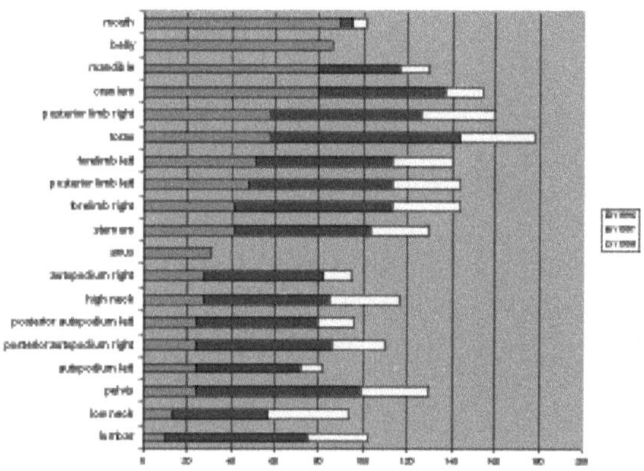

*Figure 2. Presence/absence Data. Preliminary Qualitative Statistics. Body parts representation * year*

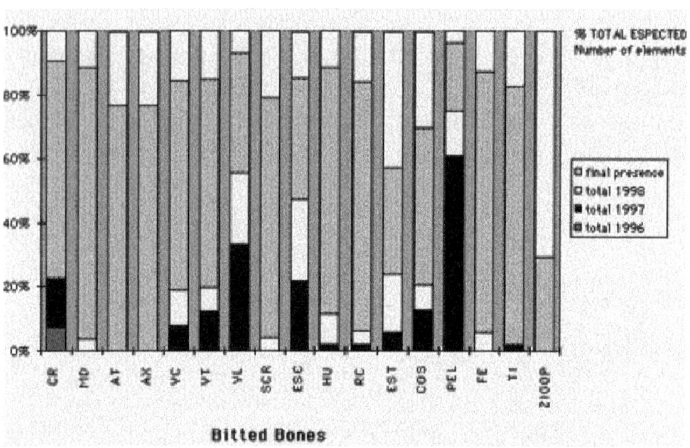

*Figure 3. Bitted Bones data. Preliminary Qualitative Statistics. Bitted bones representation * year*

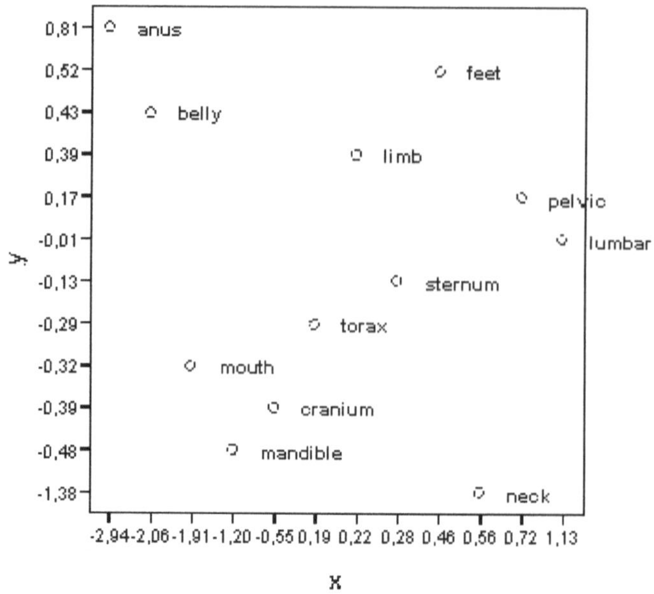

Figure 4. Presence/Absence Data. Correspondence Analysis of all carcasses during 3 years of observation. Variables Plot.

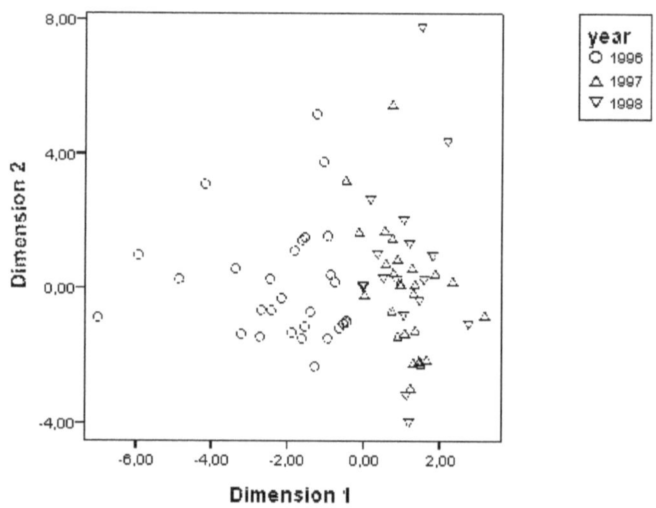

Figure 5. Presence/Absence Data. Correspondence Análisis of all carcasses during three years of observation. Individual Scores with temporal values stratification

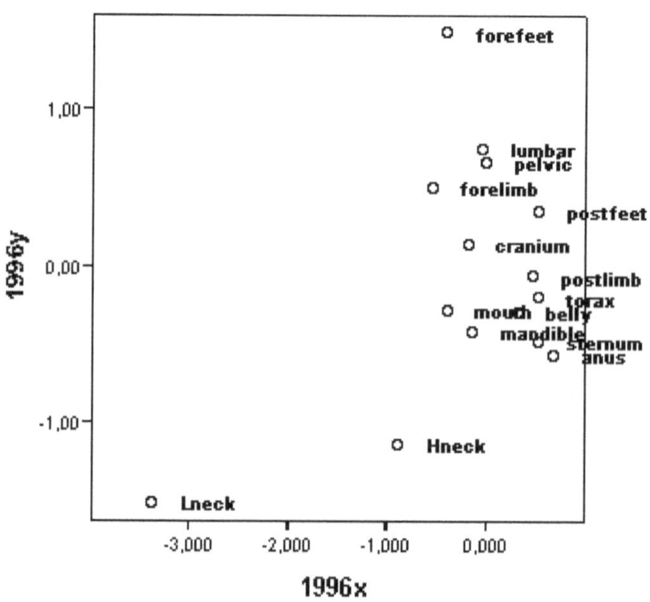

Figure 6a. Presence/Absence Data. Correspondence Analysis. First year of observations. Variable Plotting

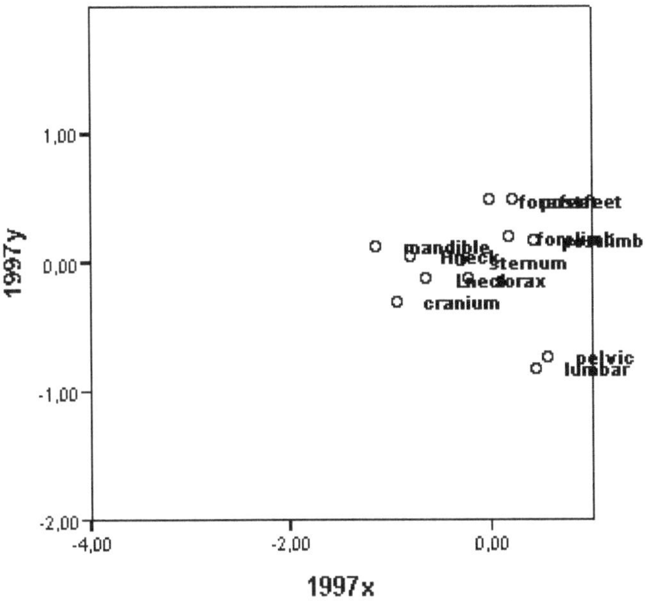

Figure 6b. Presence/Absence Data. Correspondence Análisis. Second year of observations. Variable Ploting

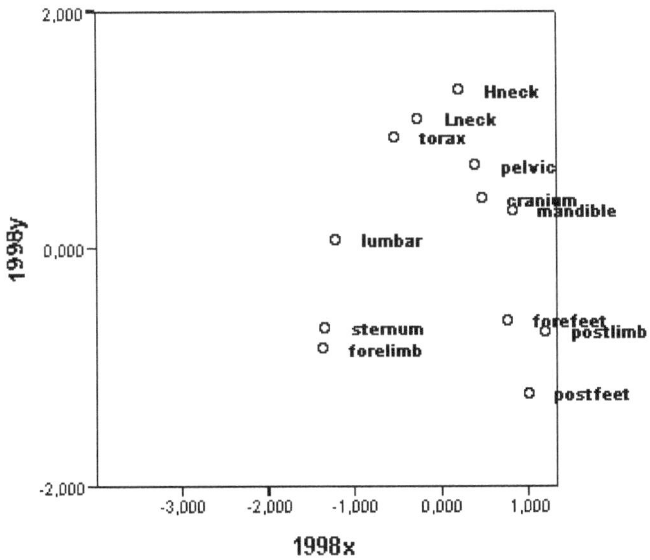

Figure 6c. Presence/Absence Data. Correspondence Analysis. Third year of observations. Variable Plotting

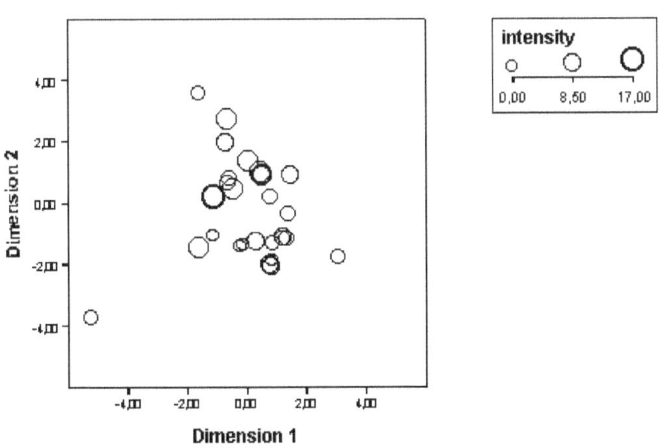

Figure 7a. Presence/Absence Data. Correspondence Analysis Individual Scores for each carcass during the first year of observations

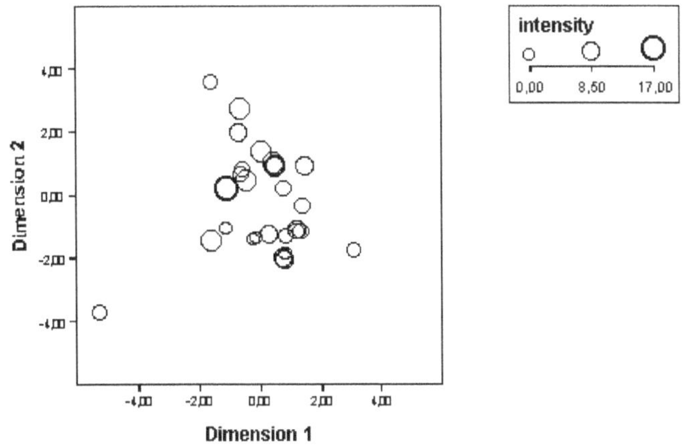

Figure 7b. Presence/Absence Data. Correspondence Analysis Individual Scores for each carcass during the second year of observations

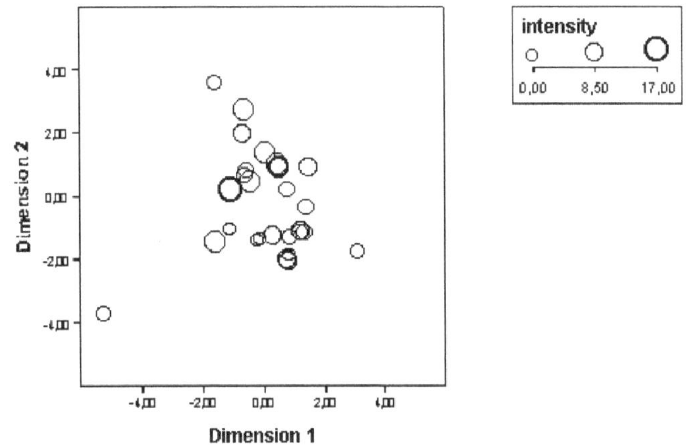

Figure 7c. Presence/Absence Data. Correspondence Analysis Individual Scores for each carcass during the third year of observations

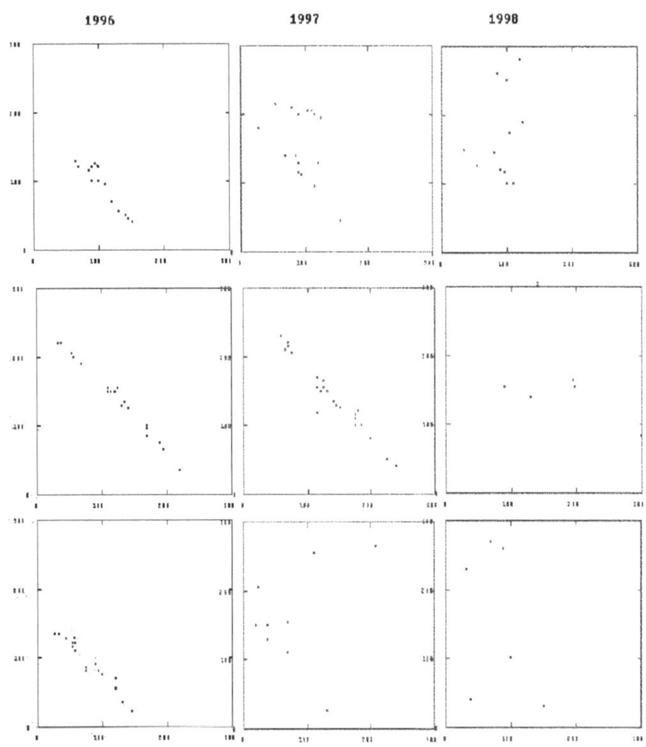

Figure 8. Spatial distribution of bones in three different carcasses

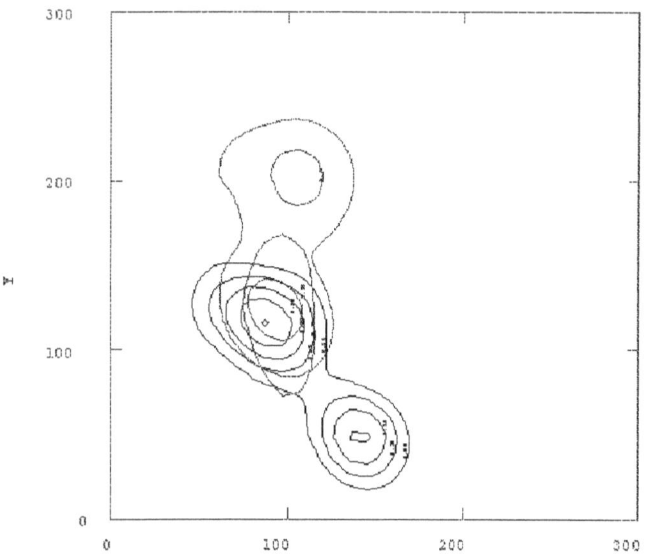

Figure 9. Kernel Function interpolation on point cloud for Carcass No. 1

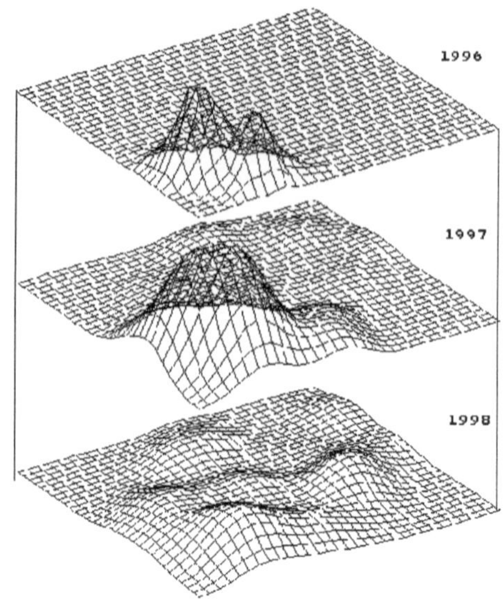

Figure 11. A 3D view of Figure 10

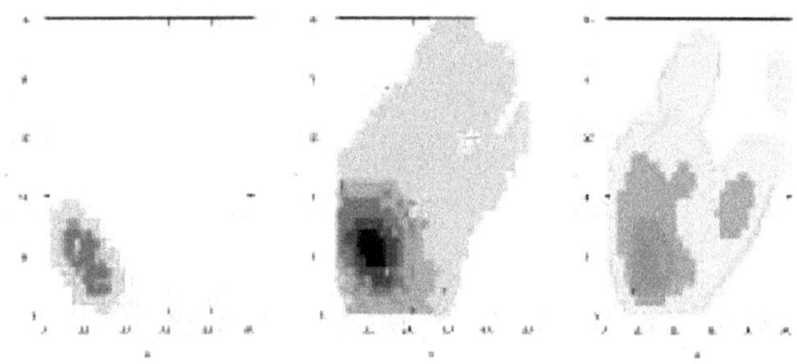

Figure 10. Kernel density maps for all carcasses in the sample

Papers in the Internet Applications and Cultural Heritage Management categories

Generating Multilingual Personalized Descriptions of Museum Exhibits – The M-PIRO Project

Ion Androutsopoulos
Software and Knowledge Engineering Laboratory
Institute of Informatics and Telecommunications
National Centre for Scientific Research (NCSR) "Demokritos"
P.O. Box 60228, GR-153 10 Ag. Paraskevi, Athens, Greece
E-mail: ionandr@iit.demokritos.gr

Aggeliki Dimitromanolaki
Software and Knowledge Engineering Laboratory
Institute of Informatics and Telecommunications
National Centre for Scientific Research (NCSR) "Demokritos"
P.O. Box 60228, GR-153 10 Ag. Paraskevi, Athens, Greece
E-mail: adimit@iit.demokritos.gr

Vassiliki Kokkinaki
Foundation of the Hellenic World
38 Poulopoulou St., GR-118 51 Athens, Greece
E-mail: vkokk@fhw.gr

Jo Calder
Division of Informatics, University of Edinburgh
2 Buccleuch Place, Edinburgh EH8 9LW, UK
E-mail: J.Calder@ed.ac.uk

Jon Oberlander
Division of Informatics, University of Edinburgh
2 Buccleuch Place, Edinburgh EH8 9LW, UK
E-mail: J.Oberlander@ed.ac.uk

Elena Not
Cognitive and Communication Technologies Division,
IRST, Istituto Trentino di Cultura,
via Sommarive 18 - Povo, 38050 Trento, Italy
E-mail: not@irst.itc.it

Abstract: This paper provides an overall presentation of the M-PIRO project. M-PIRO is developing technology that will allow museums to generate automatically textual or spoken descriptions of exhibits for collections available over the Web or in virtual reality environments. The descriptions are generated in several languages from information in a language-independent database and small fragments of text, and they can be tailored according to the backgrounds of the users, their ages, and their previous interaction with the system. An authoring tool allows museum curators to update the system's database and to control the language and content of the resulting descriptions. Although the project is still in progress, a Web-based demonstrator that supports English, Greek and Italian is already available, and it is used throughout the paper to highlight the capabilities of the emerging technology.

Introduction

In recent decades, there has been significant effort in museums, galleries, and other "memory institutions" to make their collections more accessible to the public. A main target is to advance beyond museum tags, by providing more information in a coherent and pleasant way that treats the exhibits not only as interesting material objects, but also as carriers of valuable information for the understanding of the past [Dallas 1993]. Collections available in electronic form, for example via the Internet or on CDs, are also increasingly common, and can be accessed by very wide audiences, augmenting the reach of traditional forms of presentations. As the audience of the collections grows, however, so does the need for mechanisms to tailor the available information to different target groups, including visitors speaking different languages, of different ages, and with different backgrounds and levels of expertise.

At the physical premises of museums and galleries, visitors can be accompanied by curators or professional guides, who use their experience to present information which is both interesting to the particular visitors and important from an educational point of view, often in the visitors' native languages. When visitors interact with electronic collections, however, curators and guides cannot usually be present, and information is more difficult to tailor. While it is possible, for example, to prepare in advance written or recorded descriptions of the exhibits for various ages or levels of expertise, this requires additional effort and increases the maintenance cost; and the cost increases significantly if the descriptions are to be translated into many languages. Furthermore, in settings where electronic visitors are allowed to view the exhibits in any order and revisit the collections several times, static pre-written descriptions or recordings are problematic, as they do not take into account what the visitor has already seen, and how long ago. For example, suppose a visitor sees several pieces of work by some sculptor, and later approaches yet another statue by the same artist. Repetition of information about the sculptor will then be redundant, unless the lapse in time is significant. It would be more constructive to draw the visitor's attention to differences between the current and the previous statues, or to convey new information, for example, about the creation period of the exhibit. The use of analogies and comparisons enhances informal learning, and turns the museum visit into a more fruitful experience [Hooper-Greenhill 1994]. This calls for ways to store information in a language- and visitor-independent manner, rather than pre-written or pre-recorded texts, and for mechanisms to convert dynamically this information to personalized textual or spoken descriptions of exhibits.

The M-PIRO project aims to develop technology that will address these needs. Using natural language generation, user modeling, and speech synthesis techniques, M-PIRO is developing technology that will allow information about exhibits to be stored in a database and to be converted into high-quality textual or spoken personalized descriptions in several languages (currently, English, Greek, and Italian) for presentation over the Web or in virtual reality environments. It is important to note that the descriptions are not written or recorded in advance, but are generated automatically from small fragments of texts and database fields that contain symbolic codes representing, for example, particular statues, sculptors, styles, etc. The same database is used to generate descriptions in all the languages, dramatically reducing translation costs and guaranteeing consistency across languages. Furthermore, an authoring tool is being developed to enable museum curators to update the source information, and control the language and contents of the resulting descriptions. As the descriptions are generated dynamically, they can be tailored in various respects, including the facts they convey, the language style and vocabulary, the length of the descriptions, and the hyperlinks suggested for further explorations.

M-PIRO builds upon the ILEX natural language generation system [Oberlander et al. 1998], which was originally used to produce dynamically exhibit descriptions for a Web-based electronic gallery of 20th century jewellery.1 M-PIRO extends ILEX's technology by incorporating improved multilingual capabilities, high-quality speech output, authoring facilities, extended user modeling mechanisms, as well as a more modular core generation engine. Furthermore, M-PIRO incorporates and extends machinery from the HIPS project [Benelli et al. 1999; Not & Zancanaro 2000], which allows pre-existing texts to be merged more naturally with dynamically generated descriptions; this is useful in cases where a piece of information is too difficult to be generated dynamically. We shall return to these points in following sections. Although M-PIRO is still in progress, a large-scale demonstrator is already available, and it will be used throughout this paper to demonstrate the capabilities of the emerging technology.

The remainder of this paper provides a general overview of M-PIRO, attempting to avoid technical details and focusing on the functionality that the project provides to curators and other museum staff. Section 2 presents the overall system architecture. Section 3 highlights the process that generates dynamically exhibit descriptions from the database. Section 4 presents the authoring tool that is under development. Section 5 concludes and discusses future work plans.

System architecture

Figure 1 shows M-PIRO's system architecture in abstract terms. In Web-based environments, visitors select exhibits to be described by clicking on thumbnail images, as shown in Figure 2, which is a screenshot from the current M-PIRO demonstrator. (The user interfaces shown in this paper are interim versions, which will be improved in later stages of the project.) In virtual reality presentations, exhibits are selected by approaching them. Once an exhibit has been selected, the system retrieves from the database all the information that is relevant to the exhibit, and using natural language generation [McDonald 2000; McKeown 1995; Reiter & Dale 1997, 2000] and speech synthesis techniques [Dutoit 1997] it produces an appropriate textual or spoken description. Figure 3 shows an English description generated by the current demonstrator, and Figure 4 shows the same description in Greek. The structure of the database and the generation process will be discussed further in Section 3.

The generation and synthesis components employ a variety of language resources; for example, lexicons and grammars of the supported languages, rules specifying which words can be used to express each concept, etc. Most of these resources are domain-independent; i.e., the system can be used with exhibits of different kinds, for example frescoes or jewels rather than statues, without modifying them. Some of the language resources, however, are domain-dependent, and need to be tuned to each collection domain, as part of the authoring process, to be discussed in Section 4.

A user model is also consulted during the generation of the descriptions. This is a collection of user preferences, such as the preferred language and length of exhibit descriptions, and records showing which exhibits the visitor has seen and what the system has told the visitor about them. These records allow the system to avoid repeating information that has already been conveyed, and to compare the current exhibit to previous ones [Milosavljevic & Dale 1997; Milosavljevic & Oberlander 1998; Milosavljevic 1999]. In Figure 3, for example, the description reminds the user that both the current and the previous exhibit were created in the archaic period, helping the visitor build a more coherent view of the collection. Following ILEX, the user model also contains scores indicating the educational value of each piece of information, as well as how likely it is for users of a particular type to find the information interesting [Mellish et al. 1998a; Oberlander et al. 1998]. For each exhibit, the database typically contains more information than can be expressed in a description of reasonable length. The system attempts to convey only facts that have not been expressed in the past, and among those, it focuses on facts of high interest and educational value. (This process will be discussed in more detail in Section 3.) The "say more" button (cf. Figure 3) allows the visitor to receive more information about the selected exhibit, until all the relevant facts in the database have been exhausted. This adheres to the principle that all the available information should ultimately be available to all the visitors. Additional user modeling mechanisms are being developed in M-PIRO to allow the system to adjust its vocabulary and language complexity according to the user type. With children, for example, it may be preferable to use shorter sentences and more common words (e.g., "shows" rather than "depicts") compared to texts for adults, and some domain experts may prefer the telegraphic style of traditional museum labels, rather than full text.

One of M-PIRO's most ambitious goals is that domain experts, such as museum curators, should be able to update the system's domain-dependent knowledge without the intervention of language technology experts. That is, it will be possible for domain experts, called authors, to configure M-PIRO for a new collection, and to inspect or modify the system's knowledge about a collection via an authoring tool (Figure 1), to be discussed in Section 4. Although some training and familiarity with computers will still be required, this constitutes a major advance compared to most natural language generation systems, where porting the system to a new domain requires programming and language technology expertise.

Dynamic generation of exhibit descriptions

Let us now examine more closely the process of generating exhibit descriptions from the database, starting from the structure and contents of the database. An entity-relationship database model is assumed; i.e., the database is taken to hold information about entities (e.g., statues, artists) and relationships between entities (e.g., the artist of each statue). Entities can be both concrete and abstract objects (e.g., historical periods or styles), and they are organized in a hierarchy of entity types. In the domain of M-PIRO's current demonstrator, for example, the basic entity types include – among others – the types "exhibit", "historical-period", "place", and "person". The "exhibit" entity type is further subdivided into "vessel", "statue", and "coin", as shown in the left pane of Figure 5, and "statue" is further divided into "kouros" and "imperial-portraits". Each entity is declared to belong to a particular entity type; for example, "exhibit 2" in Figure 5 is declared to be a "kouros" and, therefore, also a "statue" and an "exhibit", while "archaic-period" is a "historical-period". To make the system easier to use, we have opted for a single-inheritance hierarchy; i.e, each entity type can have only one direct super-type, and, excluding inheritance, each entity can belong to only one entity type. In fact, the core system can also handle multiple inheritance, but multiple-inheritance is not available to the users of the authoring tool, for the sake of simplicity.

Relationships between entities are expressed using fields. At each entity type, it is possible to introduce new fields, which then become available to all the entities of the type and its subtypes. For example, the "statue" type in Figure 5 introduces the field "sculpted-by"; consequently, all the entities of this type, including entities of type "kouros" and "imperial-portrait", will carry this field. The "creation-period" field is inherited from the "exhibit" type, and is, therefore, also available with non-statue exhibits, i.e., entities of type "vessel" and "coin".

The fillers of each field must be entities of a particular entity type. In Figure 5, the fillers of "sculpted-by" are declared to be entities of the type "sculptor", while the fillers of "creation-period" are required to belong to the type "historical-period". The latter allows entities like "archaic-period" and "classical-period" to be used as values of "creation-period". The "set-valued" option in Figure 5 allows a field to be filled by multiple fillers of the specified type; in the "previous-locations" field, this allows us to enter more than one previous locations of the exhibit.

Fields are also used to express attributes of entities, for example, their names or their dimensions. A number of built-in data-types are available, like "string" and "date", and these are used to specify the allowed values of attribute-denoting fields. In Figure 5, the "exhibit-depicts" and "exhibit-purpose" fields are declared to be string-valued attributes. They are intended to hold short canned texts in the three languages, describing what the exhibits depict and their purposes, as this information is typically too difficult to express using automatic text generation. Larger, paragraph-long canned texts can be associated with particular entities or entity types via the "stories" tab of Figure 5. In Figure 3, the sentence "It is decorated with a wedding scene: two newlyweds... and friends" is a canned string stored as the value of "exhibit-depicts"; the rest of the text is generated dynamically.

Following ILEX, M-PIRO associates with each fact in the database an interest, an importance, and an assimilation score (not shown in Figure 5) per user type [Mellish et al. 1998a; Oberlander et al. 1998]. The interest score shows how likely it is for a visitor of a particular type to find the fact interesting. Domain experts, for example, may be interested to see references to published articles that discuss the selected exhibit, while casual visitors would probably find such information uninteresting. The importance score, on the other hand, shows how important it is for the museum to convey the fact to each visitor type, the idea being that some uninteresting facts may, nonetheless, be important to convey. Finally, the assimilation score shows the extent to which the fact can be assumed to be known to the user, either from general knowledge or through previous interaction with the system. After being set by the author to an initial base value, the assimilation score changes dynamically during the interaction to help the system avoid repeating information the user already knows. The three scores are used during the first stage of the generation process, called content selection, to select the facts (relations or attributes) that the exhibit description should convey (Figure 6). Ideally, the facts must all be related to the selected exhibit, they must not have been expressed, they must be both interesting and educationally important, and they must not exceed a maximum number of facts per description [O'Donnell 1997]. Future Web-based M-PIRO demonstrators will allow the visitor to adjust the maximum number of facts interactively, effectively allowing the visitor to select the desired length of the descriptions.

The next stage in the generation process is document planning (Figure 6). This outputs the overall document structure, which specifies, for example, the desired sequence of the facts in the generated description and their rhetorical relations; for example, whether a fact amplifies or contrasts another one [Hovy 1993; Mann & Thompson 1988]. M-PIRO has inherited from ILEX a variety of domain-independent document planners [Mellish et al. 1998a,b], which are being extended to allow the authors to specify domain-dependent schema-like planning rules [McKeown 1995] to capture the particular structural characteristics of museum descriptions. Museum labels, for example, typically start with information about the type and creation period of the exhibits. The curator of a collection of coins may wish to specify that descriptions should then proceed with a description of what the two sides of the coin depict, followed by information about the material and style, with bibliographical references given a much lower priority. In many descriptions, they might not be presented at all. While some aspects of content selection and document planning may in general be language-dependent, in the context of museum collections we have so far managed to produce good-quality descriptions in all three languages (English, Greek, and Italian) using ILEX's content selection and document planning processes, which were originally developed for English. In contrast, the next two stages, micro-planning and surface realization, are language-specific, and separate resources had to be developed for each language.

Micro-planning specifies in abstract terms how a fact can be expressed as a phrase in each language; for example, which verb to use, in what voice and tense, and which participating entities of a relationship should be expressed as subject and object. In the case of the "sculpted-by" relationship of Figure 5, micro-planning could specify that the verb "to sculpt" should be used in English, and that the verb should be rendered in passive past form, with the subject expressing the statue and the agent expressing the sculptor. This would give rise to sentences like "This statue was sculpted by Polyklitus". Micro-planning specifications of this form are provided for each database field and language, using authoring facilities that will not be covered in this paper. Ongoing work investigates how multiple micro-planning specifications per field can be exploited (e.g., allowing both "to sculpt" and "to create" to be used in "sculpted-by" in both passive and active forms) to allow for greater variety in the resulting descriptions, and to tailor the vocabulary and style of the descriptions according to the visitor type. Micro-planning also includes processing steps that determine which facts can be aggregated in a single sentence (e.g., "This vase dates from approximately 550 BC and was found in Attica" rather than "This vase dates from approximately 550 BC. It was found in Attica."), and what type of referring expression should be generated for each entity (e.g., "Doryphorus", "this statue", or "it" may be more or less appropriate in the context of previous sentences); these steps normally do not require any input from the authors, and will not be discussed further.

The last generation stage, surface realization, is responsible for producing the final textual form of the descriptions. This includes producing the appropriate word forms (e.g., verb tenses) based on the document specifications output by micro-planning, placing the various constituents (e.g., subject, verb, object, adverbials) in the correct order, accounting for number and gender agreement, etc. Surface realization is based on large-scale systemic grammars [Halliday 1985; Teich 1999], one for each supported language, that capture the necessary linguistic information. While the grammars are domain-independent, the lexicons that they employ are to a large extent domain-specific, and they need to be tuned when the system is ported to a new domain; this will be discussed further in Section 4. M-PIRO employs ILEX's grammar of English, which is in turn based on the WAG system [O'Donnell 1994]. The Greek grammar was constructed by taking the English grammar as a starting point, and gradually modifying the elements of the grammar where the two languages differ [Bateman et al. 1991; Bateman 1997; Dimitromanolaki et al. 2001; Kruijff et al. 2000]. The Italian grammar was developed in a similar manner, starting from a grammar for Spanish [Morales-Cueto 1998].

Once a textual description has been generated, it can also be passed to a speech synthesizer to produce a spoken form. This is particularly useful in virtual reality tours, where dynamic visual content may make textual descriptions impractical. M-PIRO is developing high-quality synthesizers for all three languages, aiming to exploit additional speech-related markup that can be added to the textual descriptions. Unlike many 'text-to-speech' applications, the fact that the texts are generated by computer means that significant additional information can be provided to the synthesizer. Factors such as phrasal boundaries, rhetorical relations between phrases, and whether some item of information has been previously expressed can then be taken into account in producing synthesized speech [Theune et al. 2001]. This approach is expected to lead to improved prosody, adding to the acceptability of the system in real-usage scenarios.

Interactive symbolic authoring

Let us now examine more closely M-PIRO's authoring tool (see Figure 1). The tool is intended to be used at two stages: *domain authoring* and *exhibit authoring*. Domain authoring provides the general information about the domain, which includes the available entity types, their fields, and domain-dependent language resources, while exhibit authoring allows particular entities (e.g., exhibits, artists) to be entered into the database and their fields to be filled. Although the authoring tool is designed to be used by domain experts with no language technology expertise, some general training on the use of the tool is still required at both stages. Most of the routine authoring in a museum is expected to concern exhibit authoring, which is easier to master. It is, thus, possible to train only a few museum staff members as domain authors; they will be responsible for the initial configuration of the system and "advanced" modifications, such as the addition of new types of exhibits, or the tailoring of the domain-dependent linguistic mechanisms. The task of entering and maintaining the information about individual exhibits can then be assigned to a larger number of exhibit authors, who may have received briefer training on the use of the authoring tool.

Domain authoring starts with the construction of the hierarchy of entity types and the definition of the fields that will be available at each type, as discussed in Section 3. For each entity type, the domain author then specifies one or more nouns that can be used to refer to the entities of the type; for example, the noun "statue" and its Italian and Greek equivalents "statua" and " μ ", respectively, can be used to refer to entities of type "statue". The authoring tool forces the vocabularies of the supported languages to be kept aligned, guaranteeing that equivalent texts can be generated in all languages. Morphological components are also provided, that save the author from having to enter all the forms of each noun. This is particularly important for highly inflected languages like Italian and Greek, as demonstrated in Figure 7, where the system has generated automatically all the forms of the Greek noun " μ " from its base form. Each entity type inherits the nouns that have been associated with its super-types, and thus nouns need only be associated with the most general types they can be used with. In a similar manner, facilities are provided that allow the domain author to specify which verbs can be used to express each relationship or attribute, as well as the tense and voice of the resulting phrase, and other micro-planning specifications (Section 3).

A final stage of domain authoring, shown as *document planning* in the top menu of the authoring tool in Figure 7, is the specification of the document structure, which includes the desired sequence of the facts in the generated descriptions and their rhetorical relations (Section 3). This aspect of authoring will not be discussed further in this paper, as related work in M-PIRO is still in progress. It is expected that future versions of the authoring tool will also allow the domain author to specify user types (e.g., child, casual adult visitor, archaeologist) and assign different values to user modeling parameters, such as the interest and importance scores of the facts, depending on the user type.

Once the general information about the domain has been provided, exhibit authors can start entering information about the exhibits. This amounts to inserting entities in the appropriate entity types, and filling in their slots with appropriate symbolic codes. This is illustrated in Figure 8, where information about a kouros has been inserted (cf. Figure 5). To capture default information about all the entities of a type, *generic entities* can be introduced. For example, the "creation-period" field of the "generic kouros" entity in Figure 8 could be assigned the value "archaic-period". This would indicate that, unless otherwise mentioned, a kouros belongs to the archaic period, saving the exhibit author from having to specify this information for each individual kouros.

Domain authoring in practice overlaps with the beginning of exhibit authoring. That is, domain authors typically have to insert a few exhibits of each type before they can decide about the final form of the hierarchy of entity types, their fields, and the micro-planning and document structure specifications. This is an incremental process, whereby initial information about the domain is used to generate preliminary descriptions of sample exhibits, with the generated descriptions then guiding the domain author to elaborate further the domain information. For example, new fields may be necessary to express information that is missing, new words may be needed to express new fields, and micro-planning or document structure specifications may need to be adjusted to generate texts that sound more natural. This process is repeated until the domain author is satisfied with the structure and content of the sample descriptions, at which point exhibit authors may start entering larger numbers of exhibits.

The authoring tool provides interactive facilities, designed to speed up the domain authoring process and help monitor the quality of the generated descriptions. M-PIRO expands previous ideas on interactive authoring [Paris et al. 1995; van Deemter 2000] by allowing the author to experience the effect of modifying not only the database entries, but also the database structure and the domain-specific linguistic resources. At any point, the author can ask the system to generate a description of a particular exhibit, based on the information that is currently in the database, using the current version of the domain-specific linguistic resources. This is illustrated in Figure 8, where a preview of the English description of an exhibit has been generated in the bottom right pane from the information in the database. Previews are available in all three supported languages, and it is also possible to preview phrases expressing particular fields of the database.[2] Furthermore, future versions of the authoring tool are expected to provide previews of different lengths and for different user types.

Conclusions and future work

M-PIRO's technology will allow museums and other memory institutions to enhance their electronic collections by adding textual and spoken exhibit descriptions that are generated automatically. The descriptions are produced in several languages from a single database, which reduces dramatically translation costs, and they can be tailored according to what the visitors have already seen, their backgrounds, and language skills. M-PIRO's authoring tool allows domain experts, such as museum curators, to configure the technology for new collections, to inspect and modify the system's knowledge, and to control and monitor the content and linguistic form of the resulting texts. A large-scale Web-based prototype already demonstrates the benefits of the emerging technology.

In subsequent stages of the project, M-PIRO's technology will be ported to a virtual reality environment, where visitors can approach and examine exhibits in three-dimensional space. In an immersive environment of this type, the ability to produce spoken descriptions is crucial, as it is very unnatural to require visitors to read texts. Further work is also planned to allow authors to tailor the overall structure of the generated descriptions, and to allow database records to be imported from existing museum databases. Another strand of ongoing work is developing methods to allow natural language generation to be combined with *macronodes*, a technology deriving from the HIPS project [Not and Zancanaro 2000] that allows canned texts to be customized according to the user's model, providing many of the benefits of full generation. The evaluation of M-PIRO's technology is another major component of future work. The functionality of the authoring tool will be evaluated with the help of museum experts, while the efficiency of the overall system in information provision and interaction with its visitors will be evaluated through formative and summative methods [Keene 1998]. Finally, it should be noted that apart from museum collections, M-PIRO's technology can be exploited in a variety of other contexts, including educational software, video games and on-line catalogues for electronic commerce.

Endnotes

*Acknowledgements: M-PIRO is a project of the Information Societies Programme of the European Union, running from February 2000 to January 2003. The project's consortium consists of the University of Edinburgh (UK, co-ordinator), ITC-irst (Italy), NCSR "Demokritos" (Greece), the University of Athens (Greece), the Foundation of the Hellenic World (Greece), and System Simulation Ltd (UK). More information about the project is available from: http://www.ltg.ed.ac.uk/mpiro/. The authors feel particularly indebted to Mick O'Donnell, formerly with ITC-irst, who contributed significantly to the project during its first year.

**All images of artefacts that are displayed in this paper are extracted from "The Hellenic History on the Internet" by the Foundation of the Hellenic World.

1 See http://www.cstr.ed.ac.uk/cgi-bin/ilex.cgi

2 The authoring tool is currently under development, and previews of this form can only be generated off-line. Figure 8 demonstrates the previewing facilities that are being developed.

References

Bateman, J.A., Matthiessen, C.M.I.M., Nanri, K. & Zeng., L. 1991. The Re-use of Linguistic Resources across Languages in Multilingual Generation Components, in *Proceedings of the 12th International Joint Conference on Artificial Intelligence*, Sydney, vol. 2, pp. 966–971.

Bateman, J.A. 1997. Enabling Technology for Multilingual Natural Language Generation, in *Natural Language Engineering, vol. 3, no. 1*, pp. 5–55.

Benelli, G., Bianchi, A., Marti, P., Not, E. & Sennati, D. 1999. HIPS: Hyper-Interaction within Physical Space, in *Proceedings of the IEEE International Conference on Multimedia Computing and Systems*, Florence.

Dale, R., Green, S.J., Milosavljevic, M., Paris, C., Verspoor, C. & Williams, S. 1998. Using Natural Language Generation Techniques to Produce Virtual Documents, in *Proceedings of the 3rd Australian Document Computing Symposium*, Sydney.

Dallas, J.C. 1993. Museum Practice in the Era of Information in Museums Space and Power. In *ISS ICOFOM Study Series, no. 22*, pp. 17–51.

van Deemter, K. & Power. R. 2000. Authoring Multimedia Documents using WYSIWYM Editing, in *Proceedings of the 18th International Conference on Computational Linguistics*, Saarbrucken, pp. 222–228.

Dimitromanolaki, A., Androutsopoulos, I. & Karkaletsis, V. 2001. A Large-Scale Systemic Functional Grammar of Greek. To be presented at the 5th International Conference on Greek Linguistics, Sorbonne.

Dutoit, T. 1997. *An Introduction to Text-To-Speech Synthesis*. Kluwer Academic Publishers.

Halliday, M. 1985. *An Introduction to Functional Grammar*. Edward Arnold, London.

Hooper-Greenhill, E. 1994. *Museums and their Visitors*. Routledge, London.

Hovy, E. 1993. Automated Discourse Generation Using Discourse Structure Relations. In *Artificial Intelligence, vol. 63, nos.1–2*, pp. 341–386.

Keene, S. 1998. *Digital Collections: Museums and the Information Age*. Butterworth Heinemann, London.

Kruijff, G.-J., Teich, E., Bateman, J., Kruijff-Korbayova, I. Skoumalova, H., Sharoff, S., Sokolova, L., Hartley, T., Staykova, K. & Hana, J. 2000. Multilinguality in a Text Generation System for Three Slavic Languages, in *Proceedings of the 18th International Conference on Computational Linguistics*, Saarbrucken, pp. 474–480.

Mann, W., & Thompson, S. 1988. Rhetorical Structure Theory:

Towards a Functional Theory of Text Organization. *Text, vol. 3* pp.243–281.

McDonald, D.D. 2000. Natural Language Generation", in R. Dale, H. Moisl and H. Somers (Eds.), *Handbook of Natural Language Processing*, pp. 147–179 Marcel Dekker.

McKeown, K.R. 1995. *Text Generation*. Cambridge University Press.

Mellish, C., O'Donnell, M., Oberlander, J. & Knott, A. 1998. An Architecture for Opportunistic Text Generation, in *Proceedings of the 9th International Workshop on Natural Language Generation*, Niagara-on-the-Lake, Ontario.

Mellish, C., Knott, A., Oberlander J. & O'Donnell, M. 1998. Experiments Using Stochastic Search for Text Planning. In *Proceedings of the 9th International Workshop on Natural Language Generation*, Niagara-on-the-Lake, Ontario.

Milosavljevic, M. & Dale, R. 1997. Text generation in a dynamic hypertext environment in *Proceedings of the 19th Australasian Computer Science Conference*, pp. 417–426, Melbourne.

Milosavljevic, M. & Oberlander, J. 1998. Dynamic Hypertext Catalogues: Helping Users to Help Themselves, in *Proceedings of The 9th ACM Conference on Hypertext and Hypermedia*, Pittsburgh, PA.

Milosavljevic, M. 1999. *The Automatic Generation of Comparisons in Descriptions of Entities. Thesis (PhD)*. Department of Computing, Macquarie University, Sydney.

Morales-Cueto, A. 1998. *Adaptation of the Resource Model of ILEX/WAG to Spanish. Thesis (MSc)* Division of Informatics, University of Edinburgh.

Not, E. & Zancanaro, M. 2000. The MacroNode Approach: Mediating between Adaptive and Dynamic Hypermedia, in *Proceedings of the International Conference on Adaptive Hypermedia and Adaptive Web-based Systems*, Trento.

Oberlander, J., O'Donnell, M., Knott, A. & Mellish, C. 1998. Conversation in the Museum: Experiments in Dynamic Hypermedia with the Intelligent Labelling Explorer, In *New Review of Hypermedia and Multimedia, vol. 4* pp. 11–32

O'Donnell, M. 1994. *Sentence Analysis and Generation – a Systemic Perspective. Thesis (PhD)* Linguistics Department, University of Sydney.

O'Donnell, M. 1997. Variable-Length On-Line Document Generation, in *Proceedings of the 6th European Workshop on Natural Language Generation*, Gerhard-Mercator University, Duisburg.

Paris, C., Vander Linden, K., Fischer, M., Hartley, A., Pemberton, L., Power, R. & Scott, D. 1995. A Support Tool for Writing Multilingual Instructions, in *Proceeding of the 14th International Joint Conference on Artificial Intelligence*, Montreal. pp. 1398–1404.

Reiter, E. & Dale, R. 1997. Building Applied Natural Language Generation Systems. In *Natural Language Engineering, vol. 3*, no1 pp.57–87.

Reiter, E. & Dale, R. 2000. *Building Applied Natural Language Generation Systems*. Cambridge University Press.

Teich, E. 1999. *Systemic Functional Grammar in Natural Language Generation: Linguistic Description and Computational Representation*. Cassell Academic Publishers, London.

Theune, M., Klabbers, E., De Pijper, J.R., Krahmer, E. & Odijk, J. 2001. From Data to Speech: A General Approach. In *Natural Language Engineering, vol. 7 no. 1* pp. 47–86.

Figures

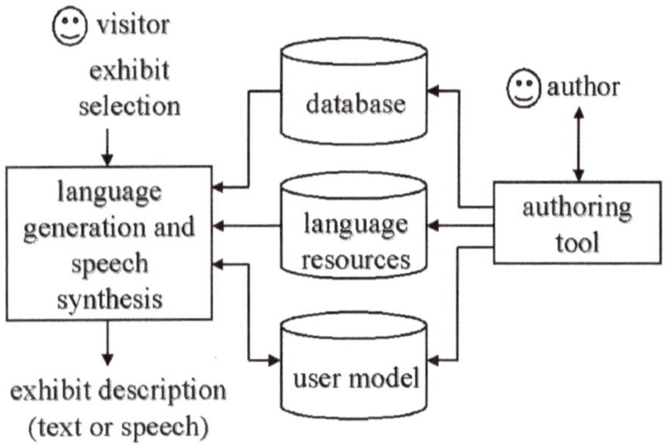

Figure 1: M-PIRO's system architecture

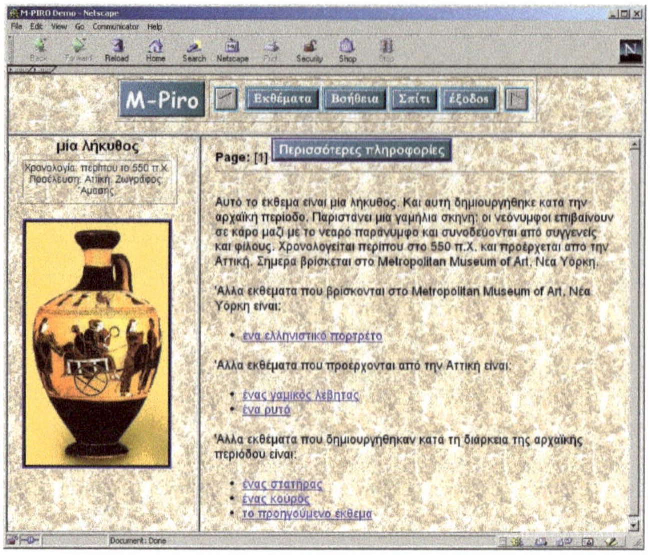

Figure 4: A dynamically generated exhibit description in Greek

Figure 2: Selecting an exhibit in the Web-based demonstrator

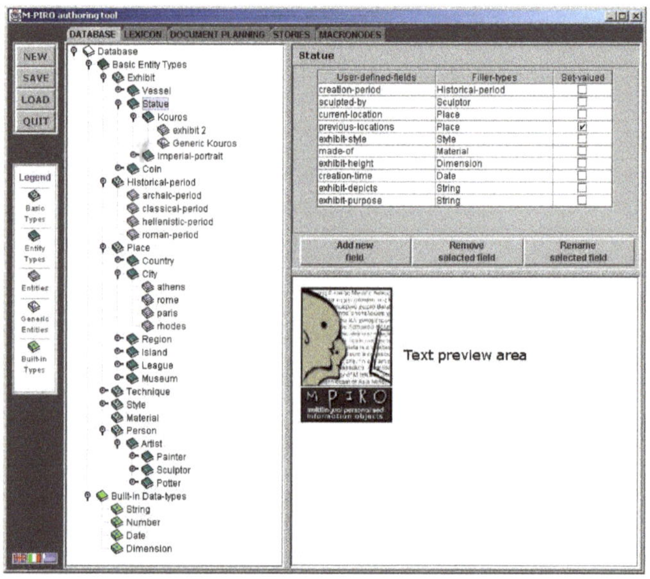

Figure 5: Editing the entity types with the authoring tool

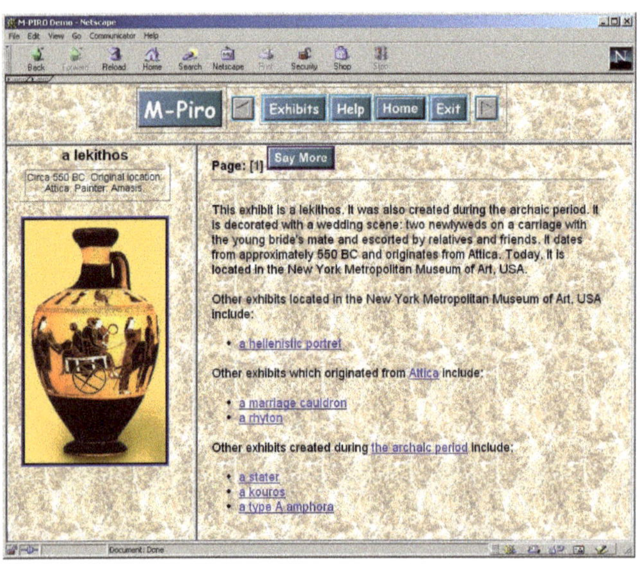

Figure 3: A dynamically generated exhibit description in English

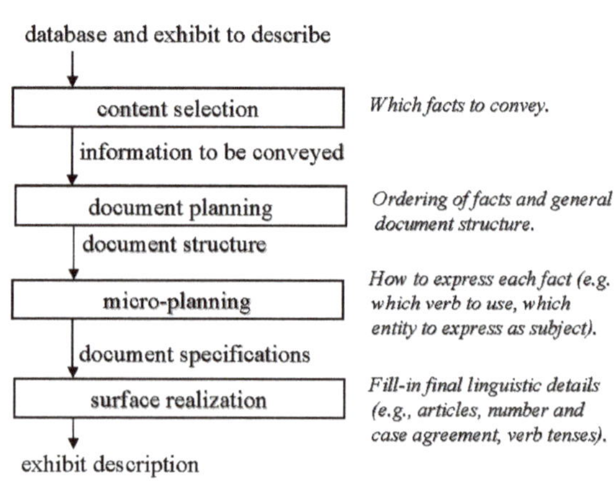

Figure 6: Stages of natural language generation in M-PIRO

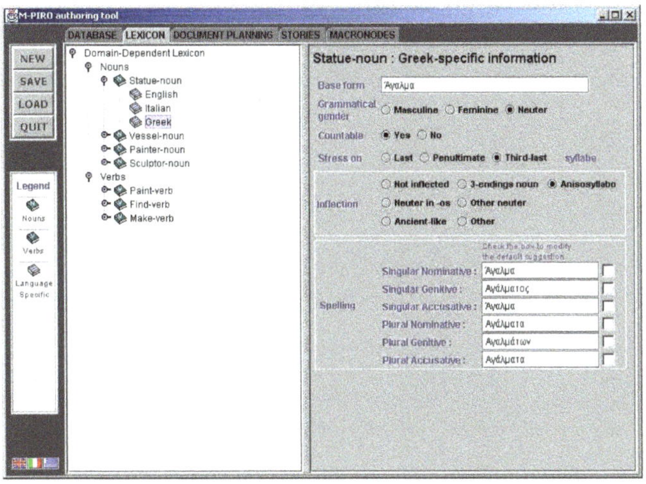

Figure 7: Updating the domain-dependent lexicon

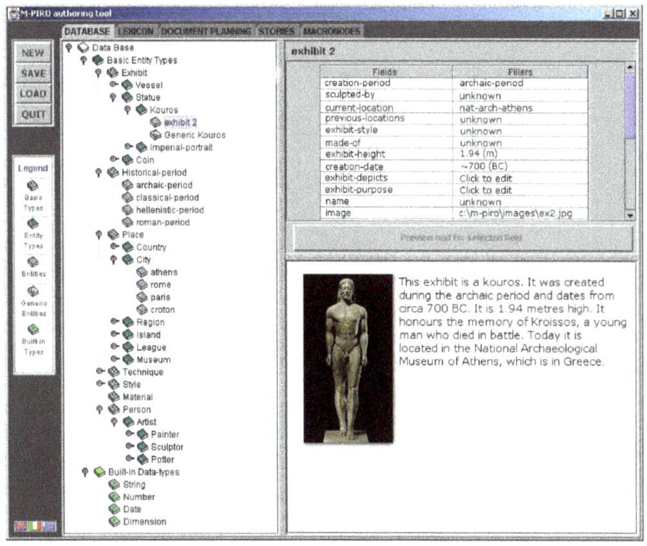

Figure 8: Entering and previewing information about an exhibit

Joined up writing: an Internet portal for research into the Historic Environment

Tony Austin
Archaeology Data Service
Department of Archaeology
The King's Manor
University of York, York, YO1 7EP, UK

Francisco Pinto
Computing Laboratory
University of Kent at Canterbury, Kent, CT2 7NF, UK

Julian Richards
Archaeology Data Service
Department of Archaeology
The King's Manor
University of York, York, YO1 7EP, UK

Nick Ryan
Computing Laboratory
University of Kent at Canterbury, Kent, CT2 7NF, UK

Abstract: Resource discovery has been a major concern of the Archaeology Data Service (ADS) since its inception in 1996. The ADS was heavily involved in the promotion and adoption of the Dublin Core (DC) metadata standard for categorising resources and how this can facilitate interoperability between distinct resources. An early example of this was the Z39.50 enabled Gateway which enables the cross searching of the metadata indexes held by the various service providers (including the ADS) that comprise the Arts and Humanities Data Service (AHDS).

Research into the Historic Environment is particularly concerned with spatial and chronological attributes of data. Dublin Core provides for this with its Coverage element. Chronological descriptors such as thesaurus derived period terms or a specific date can be referenced as can locative information such as site, place, unitary authority and country. More specific geospatial information in the form of coordinates from a grid referencing system can also be recorded. The ADS provides access to some 400,000 indexing records about the Historic Environment nearly all of which are thus referenced.

Ro:Coordinate referencing has allowed the development of various map-based search options for interfacing with ADS resources; initially relying on user input of coordinates to define a search area but more recently using click-on maps.

A recent project with the Computer Laboratory of the University of Kent at Canterbury (UKC) has successfully drawn these various strands together for the Z39.50 enabled searching of a number of geographically remote datasets. Other partners in this project include the Portable Antiquities Scheme (PAS), the Royal Commission on the Ancient and Historic Monuments of Scotland (RCAHMS) and the Scottish Cultural Resource Access Network (SCRAN) who, along with the ADS, will act as targets for a Historic Environs portal. It is expected to embrace a wider community eventually with expressions of interest from Europe and the USA.

The project allows the virtual searching of the holdings of the partner organisations as one. It has options to search on, in any combination, Who (creator), What (subject), When (coverage), Where (coverage) and coordinate defined geographic areas. Thus a user might cross search the ADS and RCAHMS (CANMORE) databases for references to Roman (when) forts (what) in the border area between England and Scotland (user defined coordinates).

The paper examines the technology, functionality, conformance and research potential of the portal.

Introduction

Governments are increasingly realising the benefits of information systems. In the United Kingdom the Joint Information Systems Committee (JISC) oversees and funds the development of such systems for Higher and Further Education and facilitates resource discovery through the Distributed National Electronic Resource (DNER). In its *Guidelines for Developing an Information Strategy* (2001) the JISC notes the potential of information systems to provide '...seamless integration between local, regional, national and international information both for research and for teaching'. They envisage 'distributed computing networks providing global, institution and local access to information for everyone' with information portals and gateways as a key to providing such access.

The Archaeology Data Service (ADS) has a remit to collect, curate and provide access to digital data resulting from archaeological research in Britain and from UK funded research elsewhere. The ADS is funded in part by the JISC and forms a part of the DNER. Other funding comes from the Arts and Humanities Research Board (AHRB) and the private sector where much archaeological research takes place within the UK. Consequently data held by the ADS needs be accessible to a wider archaeological community than the strictly academic.

The wider community also holds information independent of the ADS and the academic sector with the consequence that the provision of 'seamless integration' and the provision of 'distributed computer networks' must stretch beyond the DNER. Other senses of community suggest an even wider network with a more holistic interest in the Historic Environment both using and contributing archaeological information (e.g. Baker *et al*, 1999). Similarly the Historic Environment is unlikely to respect contemporary organisational and political boundaries. For example, researching the effect of the environment on Viking settlement might necessitate using data from several disciplines collected within the contemporary boundaries of several countries.

Resource discovery has been a major concern of the Archaeology Data Service (ADS) since its inception in 1996. The ADS was heavily involved in the promotion and adoption of the Dublin Core metadata standard for categorising resources (Miller and Greenstein, 1997) and its use in facilitating interoperability between distinct resources. An early example of this was the Z39.50 enabled Gateway which performed cross searching of metadata indexes held by the various service providers (including the ADS) that comprise the Arts and Humanities Data Service (AHDS).

Research into the Historic Environment is particularly concerned with spatial and chronological attributes of data. Dublin Core provides for this with its Coverage element. Chronological descriptors such as thesaurus derived period terms or a specific date can be referenced as can locative information such as site, place, unitary authority and country. More specific geospatial information in the form of coordinates in a geographical or projected referencing system can also be recorded. The ADS provides access to some 400,000 index records about the Historic Environment, nearly all of which are spatially and temporally referenced. Co-ordinate referencing has allowed the development of various map-based search options for interfacing with ADS resources; initially relying on user input of coordinates to define a search area but more recently using click-on maps (Austin, 2000).

A recent pilot project with the Computer Laboratory of the University of Kent at Canterbury (UKC) has successfully drawn these various strands together in a portal that provides Z39.50 enabled searching of a number of geographically remote datasets. Other partners in this project include the Portable Antiquities Scheme (PAS), the Royal Commission on the Ancient and Historic Monuments of Scotland (RCAHMS) and the Scottish Cultural Resource Access Network (SCRAN) who, along with the ADS, will act as targets for a Historic Environs Portal. The RCAHMS and the PAS are outside of the academically oriented DNER. In time, the searchable network is expected to embrace an even wider community. Expressions of interest have been received from organisations in Europe and the USA.

The project builds on the pioneering work of the Aquarelle Z39.50 enabled system (Aquarelle, 1998: Michard, 1998). It allows the virtual searching of the holdings of the partner organisations as one. It has options to search on, in any combination, Who (creator), What (subject), When (coverage), Where (coverage) and co-ordinate defined geographic areas. Thus a user might cross search the ADS and RCAHMS (CANMORE) databases for references to *Roman* (when) *forts* (what) in the border area between England and Scotland (user defined coordinates). The following sections of this paper will examine the technology, functionality, standards conformance and research potential of the portal. A technical glossary is provided at the end of the paper.

An interoperability solution

To support interoperability between diverse data sources, often with quite different internal schemas and stored data formats, a Portal needs to:

- offer a single interface to its users, irrespective of the number or type of query targets,
- translate the users' search criteria into semantically equivalent forms for each target, and
- collate and display search results when they become available.

Compliance with metadata standards is a key factor in providing interoperability between different entities as it gives well-defined semantics to create abstract levels allowing the necessary mappings.

Dublin Core (DC) is a *de facto* metadata standard. It defines a simple metadata element set consisting of 15 generic elements with well-defined semantics applicable to most, if not all, information domains. DC has been applied to several domains such as libraries, museums, geo-spatial projects, etc. The generality of elements such as creator, title, or subject illustrate its potential to be used for cross-domain searching (http://dublincore.org/ for information about DC). Some of the more specific requirements of particular domains can be met by extending DC qualifiers (DCQ) in order to provide more refined metadata elements (Dublin Core Metadata Initiative, 2000).

DC has not been free from adverse criticism having been consistently described as simplistic; however, as Weibel *et al* (1997) note it has been '*aimed at intermediate precision and high interoperability*'. More specific weaknesses have also been suggested by, for example, Godfrey Rust who see DC as fundamentally flawed as well as being simplistic especially in terms of rights management (1998). Semantically; however, it is the generality and simplicity of DC that facilitates interoperability between diverse data sets. Whilst its nature precludes the presentation of complex data structures it has seen increasing take up for resource discovery.

Z39.50 is an ISO standard defining application services and a protocol for information search and retrieval in distributed environments (NISO/ANSI/ISO, 1995). It specifies *access points* available on Z39.50 *targets* with semantics defined by information domain *profiles*. Access points specify searchable fields on virtual data sources that are managed by Z39.50 targets. Profiles are agreed pre-defined sets of rules of how to achieve interoperability by adopting well defined *attribute sets* for searching, and *abstract schemas* and *record syntaxes* for retrieving the records found in a search.

The Bath Profile is a set of rules to be adopted by the institutions providing information to obtain conformance on their Z39.50 targets to achieve interoperability (Lunau *et al*, 2001). It specifies the mandatory access points and their semantics, and the structure and syntax of returned records. Bath profile requires DC as the basic standard to provide interoperability.

Using DC, Z39.50 and the Bath profile as a basis for interoperability and the Web for accessibility, a large number of resources can be made available to users through a single portal. This Portal would be located in one institution (ADS), linking other institutions (partners) to centralise access to the partners' data sources. Users would then be able to apply high precision queries to any or all of the resources located on the connected information spaces. These access points would be accessible transparently from the Web through the Portal which connects with a target located in each entity containing access to the real fields on the databases.

This Portal aims to provide the users with a better tool to find resources either by the usual access points (e.g. Title, Subject) or by more specialised ones such as co-ordinates in a given referencing system (e.g. Referencing System, X-Coord, Y-Coord). The same results might be achieved by querying each data source using its own native query mechanism or interface, but only at a cost of considerable extra effort on the part of the user.

Implementation

The Portal Architecture provides uniform access from the Web to the data sources, launching parallel searches over the connected targets and presenting the results as soon as they arrive. It is based on a 4-tier model, where the users gain access from any Web Browser Client. The Portal is based on a Web application implemented by Java servlets and supported by a Web server and a servlet engine.

Internally, the Portal takes the form of a Portlet developed at UKC. From the Web it appears as a Java servlet which can be accessed by HTML Forms or Applets running in a browser, or by other applications such as special-purpose clients or Web robots. This servlet accepts search parameters from the Web, converts them to suit the Z39.50 world, evaluates the queries through a special purpose query processor and launches parallel searches over all or some of the connected targets according with a mediation process, as shown in Figure 1.

For Searching, HTTP requests are received by the *Portlet*, which translates them to the Z39.50 protocol using the necessary parameters (e.g. host, port, database, query-type, result set, element set, record syntax, etc). For presenting the results, HTTP requests are received by the XML/*Z39.50 TagLib*, which is supported by the *Cocoon Publishing Framework*.

Two Java packages were developed to support the Portal infrastructure, Zava and ZavaX. Zava is a Z39.50 client/server API offering basic Z39.50 features, such as Init, Search, Present and Close. Internally, Zava uses an XML parser and RDF processor to implement the profiles and to exchange the resources. ZavaX is a Web/Z39.50 API specially developed to process Z39.50 Client requests from the Web. It is supported by a Portlet for searching records according to given criteria and by a TagLib for retrieving and presenting these records. The Portlet contains a Mediator, which selects the data sources available on Targets using a query processing component, translates Web and Z39.50 transactions through a Gateway and spawns threaded searches over Z39.50 Targets using the Threaded Origin component.

The Mediator has capabilities to receive complex queries from the Web and submit them over different data sources. This mediator is an intelligent system component with crucial knowledge about the data source capabilities, which is obtained during the registration process where the data source administrator (e.g. DBA) maps the real data elements (e.g. database fields) into Z39.50 access points and record elements. Additionally, some capabilities (e.g. Spatial Referencing Systems, predefined terms belonging to thesauri) can also be registered. The query processing takes into account the mediator knowledge in order to validate, optimise and plan the queries to be submitted over the data sources.

The validation process is intended to generate a global query plan and is divided into three phases. It involves:

- checking the presence and combination of access points used in the query, which depend on the profiles implemented by the servers;
- parsing the structure of the query (e.g. boolean operators joining the sub-queries); and
- verifying the correct use of aggregated access points which only make sense if used together in order to define a given object (e.g. access points defining a spatial area within a referencing system).

The optimisation process sub-divides a given complex query into sub-queries more suitable to be dealt with by the system. It is useful, for instance, at the spatial level in order to decompose a query using the Longitude/Latitude (LL) Referencing System into sub-queries using the Ordnance Survey Great Britain (OSGB) and Ordnance Survey Northern Ireland (OSI) Referencing Systems. Therefore, data sources without LL but with OSGB and OSI capabilities will not be discarded by the mediator from a query involving the LL spatial referencing system which overlaps Great Britain and/or Ireland.

The planning process depends on the previous processes. Having a query validated and decomposed according to an optimisation scheme, an execution plan can be obtained for each data source connected to the system. Three phases are required in order to obtain the execution query plans for each data source:

- association of the query terms with *true/false* values (boolean operands) according to the data source capabilities for them;
- transformation of the query plan into a tree where the intermediary nodes are the boolean operators joining the sub-queries and the leaves are the boolean operands associated with the query terms; and
- evaluation of the query tree following a bottom-up strategy. Classic logic was found inadequate to process this tree, as some of the knowledge about the sub-queries is very weak. In fact, this knowledge is based on the assumption that an answer for a sub-query will not exist or may eventually exist. There is never strong evidence that a sub-query will be *true* which is required by the classic logic. In that way, the combination of boolean and set theory algebras proved to be the most adequate to process this query tree as it is based on sets that are empty (*false*) or may contain elements (*true*).

Having an execution query plan for each data source provides the system with decision support. This improves the overall system performance as it creates suitable queries for each data source avoiding over-complex queries or even submitting entire queries when it is known in advance that no answer will be returned.

The TagLib associates specific XML markup with associated logic in Java. It uses Z39.50 contextual XSP (XML Server Pages) tag values as parameters to establish a Z39.50 session and perform search and retrieval services. It is used both to receive Z39.50 parameters embedded in XML documents from the Web, and to convert records obtained from the Targets into XML. Subsequently, these XML 'records' can be transformed to any suitable format by using XSLT.

The targets expose one or more abstract databases providing access points for searching. Incoming search requests are mapped to the native queries (e.g. SQL) pointing to the DC metadata indexes on the underlying physical data sources. The result of each query generates *Result Sets*, which are managed locally by the Target and are used during the retrieval phase. Typically, a result set comprises a list of references to matching records, rather than copies of the records themselves.

Each target may have a different data sources and thus different access methods to the data sources (e.g. JDBC, ODBC, private API). Each data source has its own distinct schema. Having the same access points for the different schemas, one mapping is needed for each data source behind a target. The targets perform their respective mappings and native queries on their local databases, generate result sets and return the number of records found. As soon as the record counts arrive at the Portlet, they are made available to the user who may then choose to retrieve the records from the target.

Although Zava and ZavaX involve technologies usually not present on Z39.50 Servers, the Portlet is totally interoperable with any other Z39.50 Server. The only requirement for the Server is to implement the same profiles as the Portal. The records returned from the targets are always converted to XML for later processing by the publishing framework, even if the target does not implement the optional XML record syntax.

HEIRPORT, an Interoperable Portal for the Historic Environment

The ADS and RCAHMS sites have Z39.50 Servers developed by UKC, whereas SCRAN and PAS have Servers developed by Systems Simulation Ltd (SSL). Having adopted the Bath profile, all Servers expose an interface consisting of well-defined DC-based access points for searching, and abstract schemas that enable retrieval according to an agreed record syntax (e.g. SUTRS, GRS1, XML).

Typical cross-domain searching uses generic access points such as *Title, Subject, Author* or even *Any* (for searching several DC elements). As the Historic Environment emphasis is on spatial and temporal searching there are also *Where* and *When* access points. This is achieved using the DC.coverage element and feeding it with temporal and spatial descriptors based on thesauri of places and time period terms. Other access points include *Who* which allows searching by DC.creator or DC.publisher, and *What* which allows searching by DC.subject element of the resource. Additional spatial access points are based on three different spatial referencing systems, OSGB (British National Grid), OSI (Irish National Grid) and LL (Latitude and Longitude geographical coordinates). All of these can be combined into complex boolean expressions to perform precise, effective and efficient searches. These access points (4Ws and Spatial) are chosen from the CIMI Profile, a special purpose application Z39.50 profile to deal with Cultural Heritage and Museum Information (Moen, 1998).

An example query based on the generic access points on this Portal might be to use a boolean option to search by What=Fort AND When=Roman at the selected targets. A more specialised query use the spatial coordinate access points based on the British or Irish national grids, or on geographical coordinates such as What=Fort, and When=Roman, Spatial_Referencing_System=OSGB, XMin=000000.0, YMin=500000.0, XMax=400000.0 and YMin=900000.0, as shown in Figure 2.

Retrieval is based on result sets obtained during the search phase. During the search, a list of references to matching records is generated and stored in result sets on the target side. In response to a 'present' or retrieval request, the target generates and returns records in XML format. At the Portal, the Apache Cocoon Publishing Framework is then used to apply XSL Transformations to the returned XML records. In this way, the Portal can convert the XML records to a format that best suits the client application receiving the processed data. In this case, the client is normally a Web browser, so the transformation will typically generate HTML. Two examples with 'short form' and full records are shown in Figure 3.

However, as the framework has the capacity to distinguish between different browsers and other clients, it can choose to deliver different content according to the capabilities of the client. If required, it could deliver appropriate content for a small handheld computer or even for a WAP enabled mobile phone, as shown in Figure 4.

The user is given the opportunity to retrieve records from each target as soon as they are available. They can therefore begin browsing through the results as soon as the first target has completed its search, and whilst other, slower, targets are still in the search phase. Appropriate timeouts are used to ensure that the user does not wait indefinitely for a target that fails to respond to the initial search request (default 10 seconds), or fails to deliver results within a reasonable time (default 3 minutes). Other safeguards are built into the targets. To prevent large-scale data theft and to limit the load imposed on the target, the maximum number of records in any query is limited and a message is returned suggesting that the user refines their query.

Results

The main result obtained with this Portal is a solution to the interoperability problem of effective and efficient access to heterogeneous information sources.

Effective, because the access points have very precise and easily understood semantics, thus allowing users to search the Historic Environment data sets available under the targets and retrieve the records they are trying to find.

Efficient, because the searches are launched in parallel and, even if some targets are down or busy, the Portal does not wait for the last answer before presenting the results. As soon as the targets finish the local searches and send the results they are available to be retrieved from their respective sources.

The main benefit for the end-user lies in the provision of a single, simple query interface from the Web to access complex heterogeneous distributed data sources. Rather than having to locate and understand many different sources each with their own particular interface.

Conclusions

This Portal has been developed over a period of one and a half years, involving four partners as information providers and two groups of developers. Currently, ADS, RCAHMS and PAS are already connected to the Portal and SCRAN will be connected soon.

The Portal facilitates the task of finding information as it concentrates access to all of the data sources in one place with a single interface. As Dublin Core encoded in XML is used to return the records found from each target, a common presentation of results is achieved.

In summary, HEIRPORT connects the Web with Historic Environment data sources using Z39.50 and XML, thus ensuring that such sources can play a full part in the world of XML-based interoperability.

The portal can be accessed at http://ads.ahds.ac.uk/heirport/

Technical glossary

American National Standards Institute (ANSI): American body dealing with standards http://web.ansi.org/.

Application Program Interface (API): Tools and rules or protocols for building software. Blocks of code used for building programs http://webopedia.internet.com/TERM/A/API.html.

Cocoon: a 100% pure Java publishing framework that relies on new W3C technologies (such as DOM, XML, and XSL) to provide web content http://xml.apache.org/cocoon/.

Database Management System (DBMS): A collection of programs that allows to store, modify, and extract information from a database http://webopedia.internet.com/TERM/d/database_management_system_DBMS.html.

Dublin Core (DC): An interoperable metadata standard for describing resources that enable more intelligent information discovery systems. The DC schema defines elements such as dc_creator, dc_title, dc_subject and dc_coverage. The latter is for spatial and temporal and has particular relevance to Archaeology http://dublincore.org/.

Dublin Core Qualifiers (DCQ): Requirements of particular domains can be met by extending DC elements, for example, dc_coverage.period and dc_coverage.parish http://dublincore.org/documents/dcmes-qualifiers/.

Generic Record Syntax 1 (GRS1): A structured information record delivery syntax for Z39.50 technology http://lcweb.loc.gov/z3950/agency/defns/oids.html#6.

HyperText Transfer Protocol (HTTP): defines how messages are formatted and transmitted, and what actions Web servers and browsers should take in response to various commands. For example, when you enter a URL in your browser, this actually sends an HTTP command to the Web server directing it to fetch and transmit the requested Web page http://webopedia.internet.com/TERM/H/HTTP.html.

Information Standards Organization (ISO): An international organisation for standards http://www.iso.ch/iso/en/ISOOnline.frontpage.

JavaTM Servlet: This 'technology provides web developers with a simple, consistent mechanism for extending the functionality of a web server and for accessing existing business systems. A servlet can almost be thought of as an applet that runs on the server side — without a face' http://java.sun.com/products/servlet/.

Java Database Connectivity (JDBC): 'An API that lets you access virtually any tabular data source from the JavaTM programming language. It provides cross-DBMS connectivity to a wide range of SQL databases, and now, with the new JDBC API, it also provides access to other tabular data sources, such as spreadsheets or flat files.' http://java.sun.com/products/jdbc/index.html.

National Information Standards Organization NISO): 'develops and promotes technical standards used in a wide variety of information services. NISO is a non-profit association accredited as a standards developer by the ANSI' http://www.niso.org/.

Open Database Connectivity (ODBC): An API that lets you access tabular data sources.

Portal: Portal or door provides search and navigation tools on the web. Today seen as providing a user or community specific context to relevant information.

Resource Description Framework (RDF): A 'general framework for describing a Web site's metadata, or the information about the information on the site. It provides interoperability between applications that exchange machine-understandable information on the Web. RDF details information such as a site's sitemap, the dates of when updates were made, keywords that search engines look for and the Web page's intellectual property rights http://www.ukoln.ac.uk/metadata/resources/rdf/.

Structured Query Language (SQL): A 'standardized query language for requesting information from a database' http://webopedia.internet.com/TERM/S/SQL.html.

Simple Unstructured Text Record Syntax (SUTRS): A record delivery syntax for Z39.50 technology http://www.biblio-tech.com/html/z39_50_record_syntaxes.html.

Transmission Control Protocol/Internet Protocol (TCP/IP): 'Suite of communications protocols used to connect hosts on the Internet. TCP/IP uses several protocols, the two main ones being TCP and IP. TCP/IP is built into the UNIX operating system and is used by the Internet, making it the de facto standard for transmitting data over networks. Even network operating systems that have their own protocols, such as Netware, also support TCP/IP.' http://webopedia.internet.com/TERM/T/TCP_IP.html.

Wireless Application Protocol (WAP): 'Secure specification that allows users to access information instantly via handheld wireless devices such as mobile phones, pagers, two-way radios, smartphones and communicators' http://webopedia.internet.com/TERM/W/WAP.html.

eXtensible Markup Language (XML): 'A specification developed by the W3C. XML is a pared-down version of SGML, designed especially for Web documents. It allows designers to create their own customized tags, enabling the definition, transmission, validation, and interpretation of data between applications and between organizations.' http://www.w3.org/XML/.

eXtensible Style Language Transformation (XSLT): 'A language for transforming XML documents. XSL also consists of Formatting Objects: an XML vocabulary for specifying formatting semantics. An XSL style sheet specifies the presentation of a class of XML documents by describing how an instance of the class is transformed into an XML document that uses the formatting vocabulary. Transformation is the language used in XSL style sheets to transform XML documents into other XML documents. An XSL processor reads the XML document and follows the instructions in the XSL style sheet, then it outputs a new XML document or XML-document fragment http://www.w3.org/Style/XSL/.

ZAVA: Java package developed by UKC; an API offering basic Z39.50 functions such as Initialisation, Search, Present and Close. Uses XML and RDF.

ZAVAX: Provides a mapping between XML tags and Z39.50 technology for receiving XML parameters and for the conversion of records into XML.

Z39.50: An ISO standard for information search and retrieval within a distributed environment. It defines application services and a protocol for communication within the distributed system (see Miller, 1999 for a useful introduction to Z39.50 technology).

References

Aquarelle. 1998. *Final Report*, online at http://aquarelle.inria.fr/aquarelle/EN/final-report.html (downloaded 20 November 2001)

Austin, T. 2000. *An appraisal of geospatial and chronological referencing in distributed resources within the Arts and Humanities*. Unpub MSc dissertation, Staffordshire University.

Baker. D., Chitty, G., Richards, J. & Robinson, D. 1999. *HEIRNET: Mapping information resources*, on-line at http://www.britarch.ac.uk/HEIRNET/rep1.html (downloaded 27 July 2001).

Dublin Core Metadata Initiative. 2000. *Dublin Core Qualifiers*, online at http://dublincore.org/documents/dcmes-qualifiers/ (downloaded 27 July 2001).

JISC. 2001. *Guidelines for developing an information strategy*, online at http://www.jisc.ac.uk/pub/infstrat/

(downloaded 27 July 2001).

Lunau, C., Miller, P. & Moen, W. 2001. *The Bath Profile: An International Z39.50 Specification for Library Applications and Resource Discovery*, online at http://www.ukoln.ac.uk/interop-focus/bath/ (downloaded 27 July 2001).

Michard, A. 1998. Aquarelle - Sharing Cultural Heritage, *ERCIM News 33,* online edition at http://www.ercim.org/publication/Ercim_News/enw33/michard1.html (downloaded 20 November 2001)

Miller, P. and Greenstein, D. (eds). 1997. *Discovering Online Resources: A Practical Implementation of Dublin Core*, UKOLN (Bath).

Miller, P. 1999. *Z39.50 for All*, Ariadne 21 also online at http://www.ariadne.ac.uk/issue21/z3950/intro.html (downloaded 27 July 2001).

Moen, W. 1998. *The CIMI Profile: Z39.50 Application Profile for Cultural Heritage Information*, online at http://www.cimi.org/public_docs/HarmonizedProfile/HarmonProfile1.htm (downloaded 27 July 2001).

NISO/ANSI/ISO. 1995. *Z39.50-1995/ISO23950 Information Retrieval: Application Service Definition and Protocol Specification*, online at http://lcweb.loc.gov/z3950/agency/document.html (downloaded 27 July 2001).

Rust, G. 1998. Metadata: The Right Approach, *D-Lib Magazine*, July/August 1998 online at http://www.dlib.org/dlib/july98/rust/07rust.html (downloaded 20 November 2001).

Weibel, S., Iannella, R. & Cathro, W. 1997. The 4th Dublin Core Metadata Workshop Report, *D-Lib Magazine, June 1997* online at http://www.dlib.org/dlib/june97/metadata/06weibel.html (downloaded 20 November 2001)

URLs of involved organisations

Archaeology Data Service http://ads.ahds.ac.uk/
Arts and Humanities Data Service http://ahds.ac.uk/
Portable Antiquities Scheme http://www.finds.org.uk/
Royal Commission on the Ancient and Historic Monuments of Scotland http://www.rcahms.gov.uk
Scottish Cultural Resource Access Network http://www.scran.ac.uk/
University of Kent at Canterbury, Comp Lab http://www.cs.ukc.ac.uk/

Figures

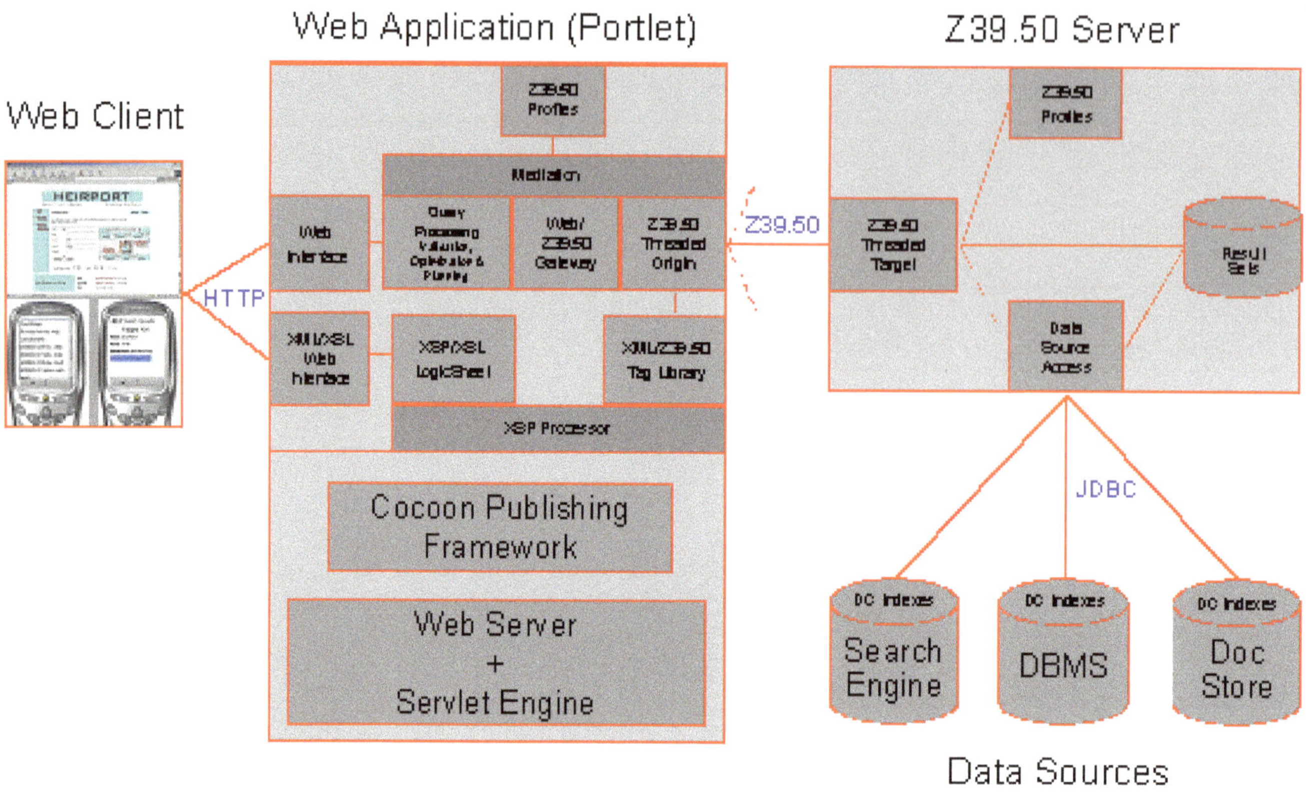

Figure 1: Portal Architecture

Figure 2: HEIRPORT Web Search Interface

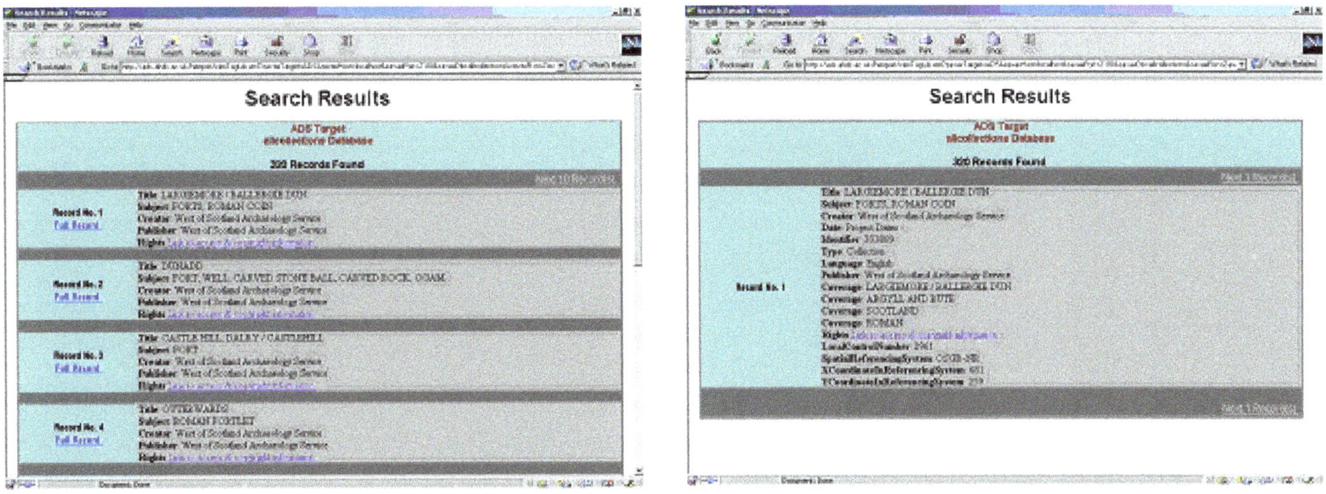

Figure 3: ADS Records converted to HTML in a Browser

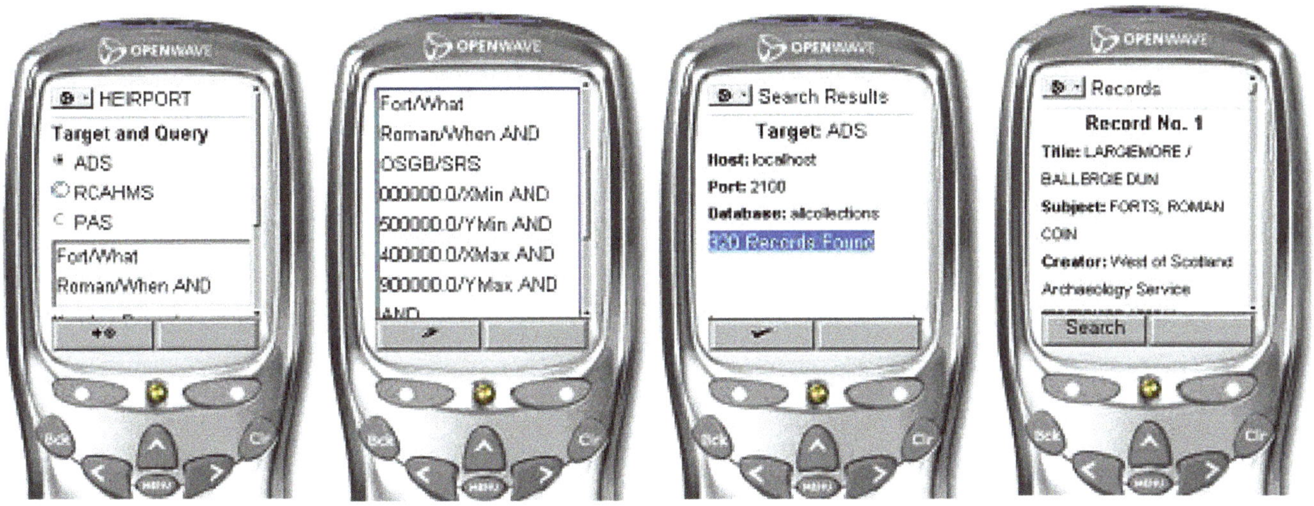

Figure 4: Portal Search Interface and ADS Record converted to WML in a WAP Client

The Archaeological Map of Egypt
(Archaeological Heritage Resource Management System)

Fathi Saleh
The National Center for Documentation of Cultural and Natural Heritage (CULTNAT)
Cairo, Egypt
E-mail: cultnat@mcit.gov.eg

Reem Bahgat
The National Center for Documentation of Cultural and Natural Heritage (CULTNAT)
Cairo, Egypt
E-mail: cultnat@mcit.gov.eg

Abstract: In this paper we will present how we employed the latest information technology to produce a powerful documentation and management tool for the Egyptian archaeological sites that are spread all over the country. We will show how we divided the information of the archaeological sites into three levels and how we implemented each of these levels, and how we used the GIS technology to add another dimension of information which is the geographical location. Also we will introduce how we used the multimedia technology in conjunction with the GIS system to document the monument details. This paper will also discuss the different types of information that are gathered and stored in our system, the user information of the system and the system outputs. The different software packages we used in the implementation are also discussed.

Introduction

Egypt, the birthplace of civilization, is pouring out a tremendous wealth of cultural artifacts, which are of world importance. It is estimated that one third of monuments and sites representing the world cultural heritage is found in Egypt. For mankind Egyptian civilization contributed in a very important manner to the making of our history and beliefs. To the Egyptians it represents a national pride and roots, it also represents a great source of revenue to the country and generates employment to a large part of the Egyptian working population. From a scientific prospective it is a source of continuous international interest. However the management and preservation of this heritage has become a serious matter. Paramount are archaeological sites considering their large number and wide distribution around the country. The wide acceleration of urbanization around the archaeological sites as well as environmental issues add another dimension to the problem. In addition to the above problems, there is currently no unified accurate documentation system for these sites.

Due to all these factors there is a demand to register this wealth and ensure its documentation, protection and preservation for future generations. This is best achieved by the creation of a geographical archaeological information system as an essential part of the cultural heritage resource management, and hence the birth of this project the archaeological map of Egypt.

The project is a joint effort between the National Center for Documentation of Cultural and Natural Heritage (CULTNAT), and the Supreme Council of Antiquities (SCA). The purpose of this project is to document all the archaeological sites of Egypt and their distribution all over the country, in addition to the documentation of all the monuments within each of these archaeological sites. The project also aims to gather all the world wide scattered information that is related to each Egyptian archaeological site and make it all accessible through a single easytouse information system, using the most up to date technology.

The Concept

The system is based on providing information at three levels: the national level, the site level and the monument level. Each of these levels is implemented using various technologies. The first level is basically implemented using Geographic Information System (GIS) technology in addition to some multimedia components. The second level is implemented using only GIS, while the third level is implemented using multimedia and virtual reality technologies.

The First Level: The National Level

The main aim of the national level is to build an inventory as well as to document and visualize the geographical distribution of the Egyptian archaeological sites. This visualization gives another dimension of information that can not be obtained from a regular information system. At this level of information, a digitized map of the entire country is displayed. Using the GIS and a vector map, the user is able to arbitrarily zoom into or out of any selected region on the map. He can also perform some geographical measurements like measuring the distance between two land marks on the map. Each archaeological site, which is the main entity at this level, is marked by a single point on the map. In addition to searching for sites from the map, the user may decide to search for sites by their names (could be Egyptian, or Greek, or Coptic name), by period, by type, by governorate, by archaeological region, or a combination of some of the above criteria. The system is bilingual. The user at any time can switch the interface and the data language from English to Arabic and vice versa. Fig. 1 shows the main screen of the system. Once a site is selected, it is marked on the map and the user can browse through different types of data that relate to that site (see fig. 2).

These data are as follows:

- Basic site information
- Recent photos of the site
- Old photos of the site
- Mission reports about that site
- Bibliography of the site
- Objects in the Egyptian Museum
- Objects in other museums
- Boundary map of the site
- Site details

The site's basic information includes the following attributes:

Site names:
This includes four fields indicating current name, ancient Egyptian name, Coptic name, and Greek name of the site. The current name is the only mandatory field in this group.

Site dominant dating:
This determines the main historical period to which the site belongs. This is a multivalue field as the site may have monuments that belong to different time periods. Each value in the site dating contains three attributes which are the pharaonic period, the dynasty and the king's name. Each dating value can contain either the three attributes or the period and dynasty attributes or just the period. For example a site can have some monuments that belong to King Ramesses II in the 19th Dynasty in the new kingdom, at the same time it can have some other monuments that date back to the old kingdom whereby the dynasty and the king are unknown.

Other periods:
In many cases the majority of the monuments at a site date back to a certain date, and a few are dated to other time periods, these non dominant periods are mentioned in this field. This field is also a multivalue field. In this case only the period is mentioned without dynasty or king.

Site type:
This determines the type of monuments at the site. This is also a multivalue data as the site can contain many different types of monuments. For example a site can have pyramids in addition to tombs and temples. The site type data is divided to two categories. A main category and sub-category. The main category has only four values which are Necropolises, Settlement, Place of Worship, and Others. Each of this categories has many values in the sub category. A site type can be determined either by a main category value or by main category and sub category value depending on the amount of information available.

Description:
This is a memo field that contains a brief description of the site. This field also contains any other information about the site that is not mentioned in any of the above fields.

PM and LÄ:
These two fields stand for:
B. Porter and R. L. B. Moss, assisted by E. W. Burney, now edited by J. Malek, *Topographical Bibliography of Ancient Egyptian Hieroglyphic Texts, Statues, Reliefs and Paintings.* Vol I-VIII; Oxford, reprinted 1994 -1999

Lexikon der Ägyptologie, edt. W. Helck, W. Westendorf; Vol. I-VII, Wiesbaden 1972-1992
Which are the most famous Egyptology encyclopedias. These two fields contain the reference of the site in the two above mentioned encyclopedias.

If the user chooses to view recent photos of the site, a set of uptodate photos of the sites are displayed.

In the old photos option a set of site photos from the beginning of the 20th century is displayed. This feature together with the recent photos feature gives a good idea about how that site changed during the last century.

The system has a link to the missions database which is currently under development by the "College de France". Using this feature enables the user to read all the excavation mission reports on work at this site.

The system is also linked to the database of the Egyptian Museum in Cairo, that enables the user to browse through the collection of artifacts that are currently in the Egyptian museum and was originally found at that site. Fig. 4 shows how the objects information is displayed.

The user can also browse through other collections of artifacts that were found at the site and are currently displayed in other museums outside Egypt. This data base will be supplied by the CHAPILLION project that has been developed by Prof. Vander Plas in University of Utrecht - Holland.

Another GIS project is currently under development in the Supreme Council of Antiquities with the help of the Finnish Government which focuses on the documentation of the boundaries of each archaeological site, and how it intersects with the surrounding urban areas, agricultural areas and so on. A direct link between our project and the boundary map project will be established at a later stage to enable the user to see the details of the site boundary and intervening parameters.

Finally, if the currently selected site contains more than one monument, the "Site Details" option will be enabled to give the user the facility to move to the second level (the site level) of data to see the details of the monuments within the site. This will be covered in the next section.

The Search methods

The user can search for any site(s) using two different methods. Either to search by site data for sites satisfying certain criteria or to select site(s) directly from the map.

The data search:
As shown in figure 1, the user has the following search options to choose from:
Search by name
Search by period
Search by type
Combined search
Search by governorate

Search by site name:
Once the user selects this search option he will then have the choice either to search by the site's current name, ancient Egyptian name, Coptic name, or Greek name. Selecting any of these options will display the corresponding list of sites names (current, ancient Egyptian, Coptic, or Greek) (see fig. 5). Once the user selects any name the corresponding site will be selected and highlighted on the map.

Search by Period:
Through this search option the user can perform three types of search. Search for sites with a specified dominant period, search for sites with a specified dynasty or search for sites with a specified non dominant period. When the user selects the first search option he will be asked to select a pharaonic time period from a list of all periods (see fig. 6). Once he selects a period, a list of all dynasties within the selected period is displayed. If the user clicks the "Search" button at this point the system will return all the sites having the selected period as a dominant period. To search for sites within a certain dynasty, he will have to select a dynasty from the list and then click the "Search" button. And finally, to search for sites with certain non dominant periods, the user will have to click the "Other Periods" button to redisplay a list of all periods to select from and then click the "Search" button.

Search by site type:
As mentioned before, the remains type data is divided into two levels: the main type and the secondary type. Using this search option the user will be able to search for sites that have a specified main type or specified secondary type. Clicking the "search by type" button will open a screen with a list of the main types (see fig 7). When the user selects a value from this list another list with all the secondary type values related to his selection will appear. To search sites with their main type just click the "search" with no secondary type selected. To search by secondary type the user will select a value from the secondary type list then click the "search" button.

Combined search:
The combined search option enables the user to perform more advanced search than those mentioned above. Using this search he will be able to form his own query combining the period and the main type fields. When the user clicks the "Combined Search" button a screen containing a query builder will appear from which he will build his query in the following way (Fig 8). First, the user will select one of the two fields "period" or "type" by clicking the corresponding button. Clicking any of the buttons will type the selected fieldname in the query box, and display a list of all possible values of the selected field. The user then should select the value he wants to search for from the list. Once a value is selected, an equal sign followed by the selected value will be typed in the query box to show the user what his query looks like. To add another condition, he has to determine first how the new condition relates to the first one. That is to be done by either clicking the "AND" button to select an AND relation or the "OR" button to select an OR relation. Selecting the relation button will also type "and" or "or" in the query box. After selecting the relation he will form the new condition in the same way he did the first one. In the same way the user can add as many conditions as he needs to build advanced queries. For example, after performing the above steps, the query box may contain a string like "([Period] = 'Old Kingdom') And ([Type] = 'Necropolis'). To prevent any confusions in sequence while using the query builder we designed it to disable any group of buttons when they should not be used.

Search by governorate:
The use of this search option enables the user to search for all sites that are located in a certain governorate. Clicking the "Search by governorate" button will display a list of all the Egyptian governorates. To search for a certain governorate, the user should select one from the list and then click the "search" button.

Search from the map:
The second search method is to search from map. Using this search method the user is able to visually select a site or a group of sites directly from the map. Clicking the "Search from map" button will open a screen with an empty list and the user will be able to either click any site on the map or draw a rectangle around a group of sites to select them. Once the user selects any sites a list of all the selected site names will be displayed in the empty list, and again the user will have the ability to make further refinement to his search by selecting only the sites of his interest from that list to display their information. This method is very useful if one is looking for a site and he knows its location but he can not remember exactly the site name. All he has to do is to select all the sites in this area and then look in their names to remember the name of the site he is looking for.

The search results:
The search results for all the previously mentioned search options are displayed in the same way. When the system displays a search result, it hides all the sites on the map except those that satisfy the search criteria and they will be displayed in a different color. At the same time, a list of all these site names is displayed for the user to select any of them to display its information. Once any site is selected it will be marked on the map to show its location. Visualizing the geographical information, especially in the search results, has a great value for researchers as it can open new fields for research. Studying the different geographical spectrums generated by different queries may result in new discoveries in Egyptian history. Fig 9 shows how the system displays a search result.

The Second Level: The Site Level

Many of the Egyptian archaeological sites, such as Luxor temple, contain only one monument. But at the same time there are some sites that contain thousands of monuments like the Giza Plateau or the Saqqara area. The main aim of the site level is to document the information on the monuments at such sites. And again one of the most important items of information is the location of each monument within the site. At this level a satellite image of the site is displayed, with each monument marked by a polygon. At this level the main entity is a single monument. For example, the Giza Plateau is represented with a single point at the first level although it contains more than 8000 monuments including the pyramids and the sphinx. When the Giza Plateau is selected, the user can click on the "Site Details" button to move to its second level of information.

Similar to level 1, the user may select a certain monument directly from the image or search for monuments in the site by their owner name, by dynasty, by monument type, or a combination of search criteria.

Once a monument has been selected, its location will be highlighted on the map and its basic information will be displayed. The monument's basic information contains the following attributes:

Owner name:
This field contains the name of the monument's owner if it is known.

The monument dating:
This indicates the period of time in which this monument was built. The structure of this attribute is similar to the dominant dating of the site at the first level, which means it also contains information about the Historic Period, Pharaonic Dynasty, and the King's name in whose reign the monument was built.

Structure:
This attribute contains information about the structure of the monument itself, for example stone built, rock cut, etc.

Type:
This indicates the type of the monument, for example Pyramid, Tomb, Mastaba, etc.

Bibliography:
This states all the references that have published any information about the monument.

PM and LÄ:
Exactly the same as at first level.

Description:
Contains a brief description of the monument.

In addition to basic information on the monument the user has the option to see more detailed information about the monument by clicking the "Monument Details" link. This will lead him to the third level of data that will be described in the next section.

The Third Level: The Monument Level

The main aim of the monument level is to document every fine detail of the monument, including its geometry, structure, the pharaonic scenes on the walls and even the translation of each hieroglyphic text that is written on these scenes. The best representation for this kind of documentation is to create a virtual tour inside the monument. This gives the user a complete feeling of what the monument really looks like, the architecture of the monument, and how the scenes are distributed on its walls, etc. Using 3D modeling and multimedia technologies, a 3D model of the monument is built and easytouse navigation tools are added to it. Moreover an exact documentation of the monument walls is implemented by employing a photogrametry system to produce high quality to scale photos of the monument walls.

Once the user clicks the "Monument Details" button he will be transferred to the virtual tour screen of the currently selected monument as in (Fig 12). The screen contains a 3D model of the monument in which the user will be able to freely walk through the monument using the arrow keys. Beside the 3D model, a plan view of the monument appears with an arrow illustrating the position and the orientation of the user at any time. In addition, the user can change his walking speed using the speed control or enlarge the 3D view to fill the screen for a better view.

At any time during the tour the user can click on any of the monument walls to zoom into the scenes engraved on it. A line drawing of the scene with multiple zoom levels will be displayed. By rolling the mouse pointer over any piece of hieroglyphic text, its English translation will be displayed. The facility of a magnifier to magnify the wall scenes is also provided. (Fig. 13)

The system's additional outputs

The system also delivers additional outputs such as printed geographical atlases of the archaeological sites of Egypt. Atlases are produced for each governorate. Each atlas includes a set of maps of the archeological sites grouped either by administrative areas, their management department or their legal status. Another output that we plan to work on is a set of multimedia CD ROMs containing virtual visits to subsets of the monuments of the system.

Current status and future plan

The project started at the beginning of year 2000, during which period we have completed the first level of data and we are now in the data revision phase. At the second level we have started with the Giza Plateau which contains more than 8000 monuments. We have already finished around 60% of the work at the site. As for the third level, which is the most timeandeffort consuminglevel, we have completed six of the Giza mastabas to date.

We are planning to complete the second level for all Egypt, within a time frame of five years. The third level is a very long task indeed. It develops at the rate of two virtual tours per month. We also plan to produce the first CD ROM including a set of the Giza Plateau mastabas by the year 2002.

Implementation Highlights

As we mentioned before the project is implemented using different technologies which are:

1) GIS
2) Data Base
3) Multimedia
4) 3D Modeling

We used Arc View V3.2 for building the GIS part of the system. Arc View was chosen because it is a very commonly used software hence the availability of developers and good technical support. The programming language of Arc View, which is called Avenue, gave us the ability to customize the user interface of the system to produce an easytouse interface for non geographic experts. Finally, Arc View has the ability to connect to any type of databases through Open Data Base Connection.

The database used is Microsoft Access. In addition to its wide popularity, Microsoft Access is a very portable and compact data base. A single file can hold all of our tables, pre-built queries, reports and also all the data entry interfaces. This file can easily be transferred from one computer to another with all of the needed data. This feature helps a lot in maintaining the data integrity especially in our case as the updating process of the data is a never-ending process. In addition to all of the above, Microsoft Access is very suitable for smaller databases which is the case in our database.

In multimedia authoring, we used Macromedia Director Shockwave Studio V8.5. During the last few years Director had proved to be one of the best multimedia authoring tools available in the market. By using Director, the developer can easily integrate all types of media files in one executable file with a very high synchronization precision. In addition to the powerful programming language the Director provides, it also gives the multimedia developer a very powerful set of commands to completely control all types of media files. Moreover the latest version of Director V8.5 has come up with a new amazing feature which is the 3D support. This feature enables the developer to seamlessly integrate a 3D world inside Director. Director 8.5 provides the developer with a large set of commands to control each element in a 3D environment at run time also receiving messages and events from any object in the 3D.

Another advantage is that the Director project can be compiled to a single readytorun executable file without any need to setup or register any extra files to the system. This feature made the process of calling this file from within Arc View very simple.

In the 3D modeling we used 3D Studio Max 4 for many reasons. The first is that 3D studio is one of the oldest and most famous 3D packages available. The second is the ease of its use compared to other 3D software. In addition the availability of modelers and animators experienced in it. Moreover 3DS Max is the best for structural non organic modeling. And finally the compatibility of 3DS Max with the new version of Macromedia Director as we can easily import any Max scene in the director project.

End Notes

Without the help and the vision of Dr. Gaballah Ali Gaballah, the secretary general of the Supreme Council of Antiquities, this project could never have come to light the way it did.

We would like to acknowledge the effort made by Eng. Mohamed Farouk for the management of the project in addition to the design and implementation of the third level ideas, Eng. Mona Tawfik for the development of all the GIS parts of the system, Mr. Maher Amer who is responsible for the geographical data of the sites, Eng. Omar El Gendi for the database administration, Eng. Motaz Mahmoud for the development of the third level, Eng. Tarek El Morry for the photogrametry work. In addition to the technical team we would also like to acknowledge the Egyptologists team for gathering, revising and refining the historical data, the team is headed by Dr. Fayza Haykal and consists of: Dr. Basma Koura, Dr. Mamdouh El Damaty, Dr. Randa Baleegh, Dr. Mohamed Hassoun, Dr. Hassan Seleem, Dr. Mostafa Attallah, Mr. Loay Saeed, and Dr. Ghihan Zaki. A very special acknowledgment goes to inspectors in every inspectorate of the Supreme Council of Antiquities who helped in the overall inventory of different sites all over the country.

References

B. Porter and R. L. B. Moss, assisted by E. W. Burney, now edited by J. Malek, *Topographical Bibliography of Ancient Egyptian Hieroglyphic Texts, Statues, Reliefs and Paintings*. Vol I-VIII; Oxford, reprinted 1994 -1999

Lexikon der Ägyptologie, edt. W. Helck, W. Westendorf; Vol. I-VII, Wiesbaden 1972-1992

W. K. Simpson. 1978. The Mastabas of Kawab, Khafkhufu I and II. G 7110-20, 7130-40 and 7150. Boston (GiMa 3), p. 9-20.

W. K. Simpson. 1976. The Mastabas of Qar and Idu. G 7101 and 7102. Boston (GiMa 2)

W. K. Simpson. 1980. Mastabas of the western Cemetery I. . Boston (GiMa 4), p. 31–32, pl LIIIa, Lva

D. Dunham & W.K. Simpson. 1974. The Mastaba of Queen Mersyankh III (in: W.K. Simpson, The Mastaba I), Boston.

G. A. Reisner. 1942. A History of the Giza Necropolis I. London.

http://www.macromedia.com/

http://www.esri.com/

http://www.descreet.com/

Figures

Figure 1. The main screen of the Archaeological Map of Egypt

Figure 4. Objects currently in the Egyptian Museum

Figure 2. Information in Level 1 of a selected site

Figure 5. Search by site name screen

Figure 3. Basic information of the site

Figure 6. Search by period screen

Figure 7. The search by type screen

Figure 8. The query builder screen

Figure 9. Fig a search result screen

Figure 10. Second level of the Giza Plateau site

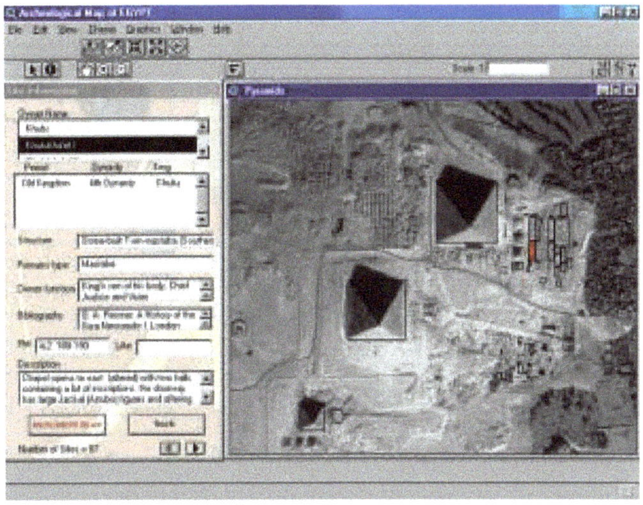

Figure 11. Monument basic information

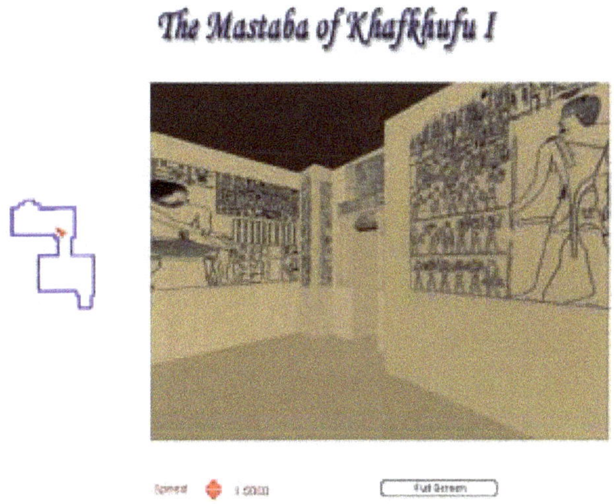

Figure 12. The virtual tour screen

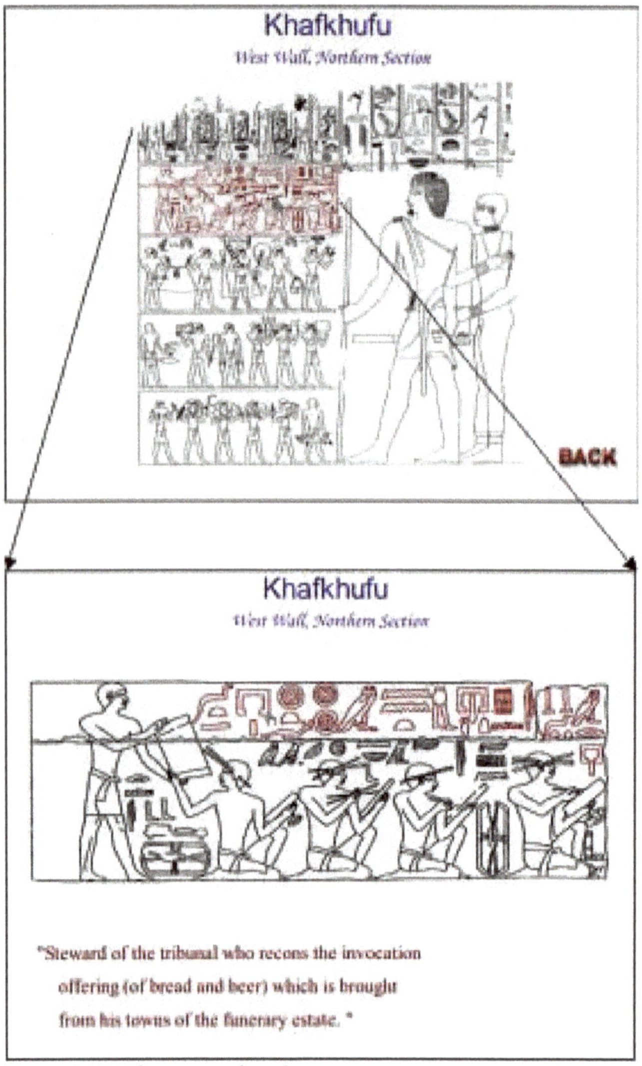

Figure 13. The scene details

Geographic Information Systems and Archaeology: the case of ancient Nora (Pula-Cagliari)

Anna Maria Colavitti
Dipartimento di Ingegneria del Territorio, Università di Cagliari
Piazza d'Armi, 16 09123, Cagliari, Italy
Tel.: +39 070 6755209 Fax:+ 39 070 6755215
E-mail: amcolavt@unica.it

Giancarlo Deplano
Dipartimento di Ingegneria del Territorio, Università di Cagliari
Piazza d'Armi, 16 09123, Cagliari, Italy
Tel.: +39 070 6755209 Fax:+ 39 070 6755215
E-mail: gdeplano@unica.it

Abstract: The Punic-Roman ancient city of Nora is located on the Sardinian south-western coast, near Cagliari, the most important sardinian city, on the south coast. Latest excavations and territorial surveys of Nora (1990-2000) have proved the continuous settlement between Phoenician-Punic-Roman and Late Roman periods (VII BC-VIII AD). The need of urgent updating of archaeological, historical, environmental data, with monitoring on ancient urban decay (ancient mosaic floorings and other structures)has led to the use of the Geographical Information System in order to solve archaeological problems and to improve research on Nora. In this prospective the study and reflections on ancient city, based on interdisciplinary research between the Dipartimento di Ingegneria del Territorio and Monuments and Fine Arts Office,have begun. The research deals with several points: Cultural Heritage, Archaeological Surveys and Field research, Intra Site Applications, Scientific studies. We evaluate problems between Urban City Planning and historical-environmental resources, respectively. The identity of present territory cannot be described as inalterable substance, but it can be considered as the continuous historical stratified result of local Community mentality. The paper aims to discuss some advancements in the Nora city planning research and to propose the description of a GIS oriented system for the archaeological assets and management of the cultural heritage.

Key words: Nora, ancient urban planning, ancient roman Sardinia, archaeological Information Systems, cultural heritage.

Introduction

The Nora peninsula stretches out into the Mediterranean Sea in an unusual shape, which, nevertheless, allows to include this site among the geomorphologically characteristic sites of a number of coastal areas in Sardinia. These include the promontories of Tharros, the lagoons of Carales and Bithia, the settlements of Bosa Vetus, Olbia, and the east coast as far as Villasimius. All these settlements are characterized by the presence of coastal lagoons, promontories, and coasts with precise harbor characteristics.

These geomorphologic characteristics partially contributed to create the privileged strategic conditions that have led to the establishment and development of the city of Nora. Research on the ancient site in the form of *excavation archaeology* started in the early 1950's. This damaged the city and caused the loss of most of the stratigraphies referable to the shallower deposits, which were irremediably compromised by the tendency to privilege the search of objects and remarkable things to preserve. In the 1990s stratigraphic excavations were at last started in Nora by a number of researchers from the Universities of Genoa, Padua, Pisa, Venice and Viterbo and from the Archaeological Superintendence for the Provinces of Cagliari and Oristano. The remaining parts of the stratigraphies destroyed in the 1950's, i.e. of the Pesce period, were excavated stratigraphically, tryng to reconstruct wherever possible the missing pieces of the history of the city of Nora. Parallel with these investigations, a collaboration between the Department of Land Engineering of the University of Cagliari and the Archaeological Superintendence for the Provinces of Cagliari and Oristano was started with the aim of studying and implementing an Information System for Nora. The Scientific Project is co-ordinated by Prof. G. Deplano.

A possible reading of the forma urbis

The reconstruction of the forma urbis of Nora is at present compromised by the actual knowledge of the city as known so far, and especially because it has not been possible to reconstruct its extension completely, though excavations have proceeded quite rapidly in the past years (MELONI 1990). Therefore the reconstructive hypothesis that will be attempted here will have to be revised when the inevitable following updates will be available as a result of further investigations. Nevertheless, considering what has emerged so far, and considering the different chronologies proposed by various scholars for the best known buildings, careful reading of the residual urban texture, shows few interesting elements that could be discussed today and in the future.

While it is not our intention to enter the debated problem of the phases of the foundation of the city of Nora (BONETTO-NOVELLO 2000), it can safely be stated that its urban phase started around the VII-VI century BC, at a crucial time in the history of this part of the Mediterranean Sea, when the balance of the forces at stake, i.e. the autochthonous population and the Phoenicians, was in favour of the latter. The original role of the city is not totally clear. It certainly must have been related both to the exploitation of the land's potentialities, whose economy depended on its rich hinterland, and to overseas relations with other Mediterranean poles of interest, such as Africa, Spain, the Balearic Islands, France, and Italy. Its relationship with Cagliari, for example, which must have been an important settlement already in the VII century BC, if not earlier, as shown by studies on the via Brenta area, is not clear (AA.VV. 1992). However, the same conformation of the small peninsula of Pula conditioned the roads which characterize its layout and which became more monumental from the Empire onwards, and therefore more clearly defined. No stratigraphic surveys have been made on road works in Nora. Nevertheless, a few data can be obtained from the supporting relationship among structures such as the theatre and the baths at sea , and road portions, considering that roads are supported and therefore are posterior to the two above structures and can be dated within the III century AD.

Even the country roads along the coast namely, the Cagliari road to the east and the *Bithia* road to the west, conditioned the city layout, since they joined the city roads at certain points that can be hypothesised with some certainty (BONETTO, forthcoming) and few observations, which should be integrated with new data from road findings, have been made. In fact two stretches of paved road in an east-west direction, parallel to the access road to the archaeological site, have been found in the north-eastern part of the city. The stretches are built with andesite slabs with a 20 cm kerb, and are 3.80 m wide and 5.70 m long. The remains of a large building tangent to the road with squared blocks of yellowish sandstone *in situ*, were also found in the same area. First of all this suggests an important fact: the city extended towards north-west covering the available space of the isthmus between the small lagoon and the sea. To what extent is difficult to tell. One thing is certain: the Sant'Efisio necropolis delimited the *pars urbana* near the sea towards north-east. On the other hand, it cannot be ruled out that the settlement used to occupy without interruption the spaces available in the hinterland and along the coastline. At a certain point this road must have joined the extraurban road system. Considering the presence of important buildings even towards north-west, i.e. at the furthest limit of the military zone, it must have joined the extraurban road system at some other place further on. It has been seen that the foundation of the city should probably be placed in the VII-VI century BC. Of this period we can only hypothesize an early organisation of spaces and roads, of which we know next to nothing, except for the banal consideration that it must have been shaped to fit the geomorphology of the site. It is obvious that, considering the strongly conditioning nature of the morphological characteristics of the small peninsula of Nora, both the Phoenicians and the Romans must have become adapted to this peculiar situation, for which an organic settlement was only during the empire, at least from the data in our possession It should be remembered that the terminology *cardus* and *decumanus*, which is often used to define the main roads in a city refers to a situation that is typical of a centuriation system, and cannot be used with another meaning (BONETTO, forthcoming).

The phases of the city

Which phases can be clearly characterised in the construction of a city layout? What is the distribution of spaces and the destination of functions? Regarding the pre-Roman phases, it can safely be stated, as already said, that the city developed in the VII-VI century BC, near the upland plain later occupied by the forum. This was an area assigned to the mixed destination of dwellings and public utility services. In the late republican age, this forum area underwent extensive town-planning restructuring (NOVELLO 2000)

In the first half of the I century BC, the area was reprogrammed to receive the forum square with its imposing paving. From a town-planning point of view, the forum is eccentric with the city core, an aspect described in Vitruvius treatise when dealing with the difference between forums in harbor cities and forums in internal cities (SOMMELLA 1988).

The location of public area par excellence is maintained in the town-planning history of the city.

As regards equipped areas, that is baths, fountains, and various works, the only clarified element so far is the link between the so-called *insula A* and the west harbor. The *insula A* is characterized by a large two-storey building complex that includes *tabernae,* actually *ergastula,* and dwelling areas on the top floor (GUALANDI-RIZZITELLI 2000). The excavation of part of the hill on which the *insula A* was later built in the middle of the III century AD (RIZZITELLI 2000) was a very significant event in the town-planning. This indicates a precise decision to refunctionalise an area related to port activity. Besides this important planned operation related to *insula A,* several other smaller works for the community life were carried out. These refer to public service areas, that were continuously on the increase during the Empire and can also be defined as the "district" level based on privileged mileage and on the inevitable attraction of crossroads: this is the case of the Small Baths that were linked to the AB block, of the Baths at Sea that were linked to the harbour, of the Central Baths that were linked to east and south oriental districts of the city, and of the Eastern Baths that were linked to the eastern districts to the city and perhaps to the supposed eastern anchorage (VERDUCHI 1995) The Small Baths building, which was originally destined to serve a block along the largest western artery of the city, was transformed perhaps in the IV century AD in successive phases from a private *balneum* to a public establishment, with the addition of new parts to evolve completely a new function of the complex (COLAVITTI, forthcoming)

The Baths at Sea are located beside an important crossroad between an east-west axis and a north-south axis. At this point the east-west road proceeded in this direction, but it can no longer be determined today because the area has been buried by bradyseism, which characterized this stretch of coast. The works were carried out in the imperial Age but before the construction of the *insula A*, therefore in the II century AD (TRONCHETTI 1984)

Sacred areas represent a separate problem in the urban context of the city. Little is known about them, because, except for the area located on the Coltellazzo promontory, they have never been investigated with the stratigraphic method. The most important buildings are located on the top of the two promontories that characterize the Nora area: i.e. the Coltellazzo sanctuary and the sacred area of Aesculapius-Eshmun. A threephase chronology, can only be proposed for the Coltellazzo area: an establishment phase in the VI century BC, a Carthaginian phase in the IV-III century BC, and a function restructuring or maintenance phase in the Roman republican period (OGGIANO, forthcoming). Other important sacred buildings must have gravitated around the forum, as the one hypothesized in the north-eastern part, where archaeological excavations are currently in progress.

The history of the settlement of Nora partly changes in the post-classical period. Few nuclei of the settlement certainly survive from the imperial Age in different forms, such as the western nucleus where a certain continuity of life planning has been observed with the continuation of building activity, the reutilisation of materials and techniques, the abandonment of structures, and the filling up with earth of a few areas (COLAVITTI.-TRONCHETTI 2000; BEJOR 2000).

Some of the main roads, such as G-H, are maintained efficiently, suggesting that the area or part of it was to receive some kind of new connotation in the general equilibrium of the site, which still escapes from our attention. Next to maintenance and the transformation of the city layout, was the phenomenon of Christianisation of the urban space, which was also felt in Nora, where it established itself and developed rapidly. The material signs of this period seem to be relegated externally with the creation of a pole around the church of Sant'Efisio, which partly re-utilizes a sepulchral pre-Christian area.

The construction of an information system for the city

First of all, why should we build an Information System for Nora? The simplest answer to this question is: to be able to manage the archaeological information on the city and its territory, by relating and comparing it to non-archaeological themes, such as the geomorphology of the site, the geopedology, the economic data of the Pula area and other themes, with the aim of revealing and enhancing an enormous heritage so intimately related to the problems of the Mediterranean space. Furthermore, with a Geographic Information System, different types of users can be brought together, such as the University, the Archaeological Superintendence, and the Local Authorities. Each of them could use can use the information according to his competence, in order to communicate such pressing themes as the infrastructures to be built in real times in touristic-cultural areas. Once this premise has been accepted, the priorities to build the entire system need to be established. The growing to update the historical, archaeological, and environmental data must proceed hand in hand with a continuous monitoring of the degradation problems on the site and of the efficiency of the established methods of protection and restoration. Therefore the decision to use the information tool, which is opened in new fields of application, comes from the prospect of solving many problems related to the complexity of the archaeological site, in order to establish a correct relationship between the processes of land use planning and its historical-environmental resources. To build an archive of the ancient city would be in itself an action of great importance in planning any future event, even in the field of activity of archaeological excavation. It has been stated that since Geographic Information Systems have not been created for archaeology, they are structured on blocks of data that are strictly related to geographic and environmental variables, which the archaeologist is incapable of adapting to his research (LOCK-STANCIC 1995).
This could be true if we don't consider that a GIS is the expression in space of an archive of data.

Therefore the **Information System** for Nora, is structured on an **Archive** of data, which shows all the necessary and useful information, aimed first of all to describe first of all the archaeological level.

This information is easily interfaced with other information levels that we consider functional in explaining the greatest possible number of elements. To this purpose, we have studied a few descriptive records of the archaeological level that make up the first phase of implementation of the entire system. The architecture of the system includes a GIS platform, a management system of relational databases, and a hypertext. Thanks to the hypertext the information can be structured without having to carry out a close examination of the vast field of spatial analysis that is the heritage of more expert figures who would still have to converse with other users. The records (1-2-3) show the main categories of spatial data and attributes that make up the system, and which can be used to carry out interrogations and analyses.

Record 1 (fig. 1)
Localisation (town–underwater land)
Conditions: Visible-non visible
Protection: Restricted-restored-not restored-not specified

Record 2 (fig.2)
Modality of recovery (from excavation -other)
Object characteristics: Immovable-free area–additional information

Record 3 (fig. 3):
Architectural complex (Fragmentary Planimetric Unit)
Functional interpretation
Chronology
Existing documentation
Literature references

The principal aim of GIS oriented system for archaeological assets is to draw up the documentary evidence from the beginning of the strathigraphic research in Nora. This fact represents a further steps to site management and perhaps to think ways for further research within urban ancient planning and contemporary environmental Nora-Pula planning. Also monitoring the dilapidated state of ancient buildings and the studing of projects and schemes where we can research on management policy can be included in the same project. Finally the whole project will provide for Nora and Pula complete thematic maps, for example on the ancient building technologies and on the preservation of buildings and environment.

References

AA.VV. 1992. *Lo scavo di via Brenta a Cagliari*, in Quaderni della Soprintendenza archeologica per le province di Cagliari e Oristano, Cagliari..

Bonetto, J., Novello, M. 2000. *Il Foro romano (Area "P")*, in AA.VV., *Ricerche su Nora I* (anni 1990-1998), by C. Tronchetti, Cagliari.

Bonetto J. I sistemi infrastrutturali di Nora romana: la viabilità e il drenaggio delle acque, in *Ricerche a Nora II* (forthcoming).

Meloni P. 1990. *La Sardegna romana*, Sassari (2°Ed.), pp. 267-271 and 497-498

Novello, M. forthcoming. *Ricerche a Nora I*, op, cit. p. 187.

Sommella, P. 1988. *Italia antica. L'urbanistica romana*, Roma, p. 240.

Gualandi, M.L., Rizzitelli, C. 2000. L'insula A, in *Ricerche a Nora I* op. cit. p. 136.

Rizzitelli, C. 2000. in *Ricerche a Nora I*, op. cit., p. 126

Verduchi, P. 1995. In AA.VV., *Atlante di Ostia antica*, Padua, pages 9-24.

Colavitti, A.M. forthcoming. *Le Piccole Terme di Nora: proposta di rilettura*, in Atti del Convegno Internaz. sull'Africa romana, Sassari December 2000, cds.

Tronchetti, C. 1984. *Nora*, Sassari , pp. 43-51.

Oggiano, I. forthcoming. *Lo spazio sacro a Nora*, in Atti Congresso Int. di studi Fenici e Punici, Palermo-Marsala 2-8 October 2000, cds.

Colavitti, A.M., Tronchetti, C. 2000. Area M. Lo scavo di un ambiente bizantino: il vano M/A; BEJOR G.. Il settore Nord-occidentale: L'area AB, in *Ricerche a Nora I*, op. cit., pp. 33-67 and 19-33

Lock, G., Stancic, Z. 1995. *Archaeology and Geographical Information Systems*, London , pp. 367-382.

Figures

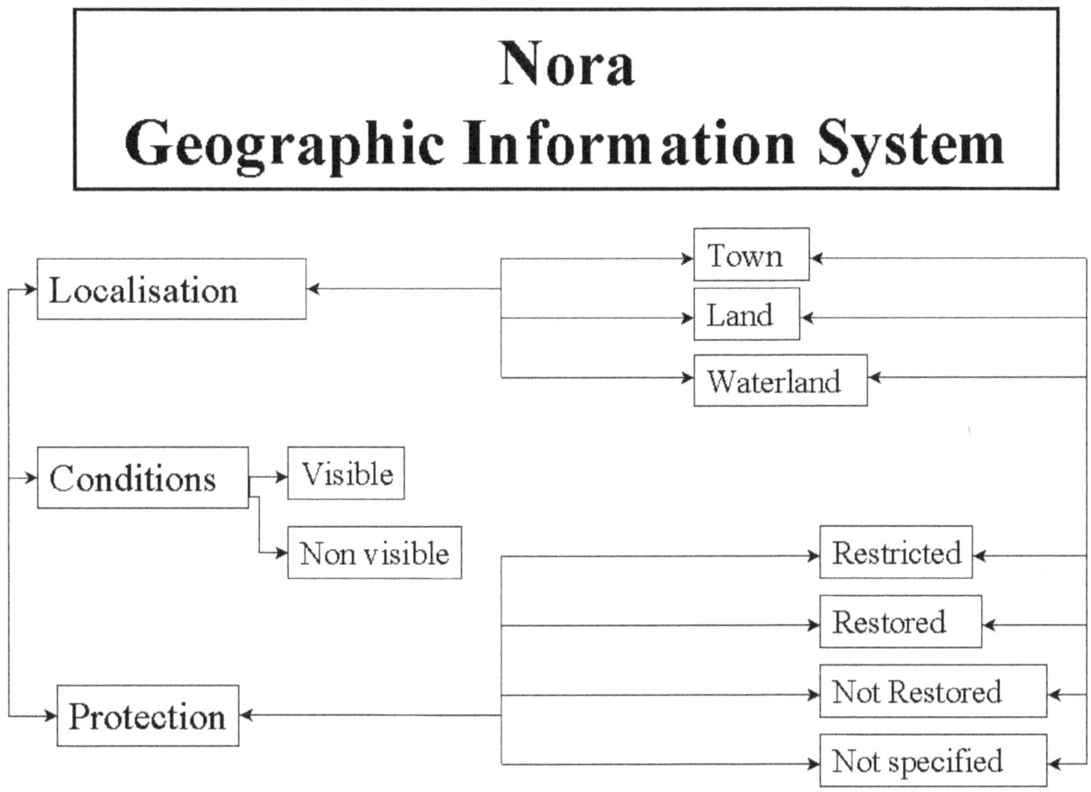

Figure 1.

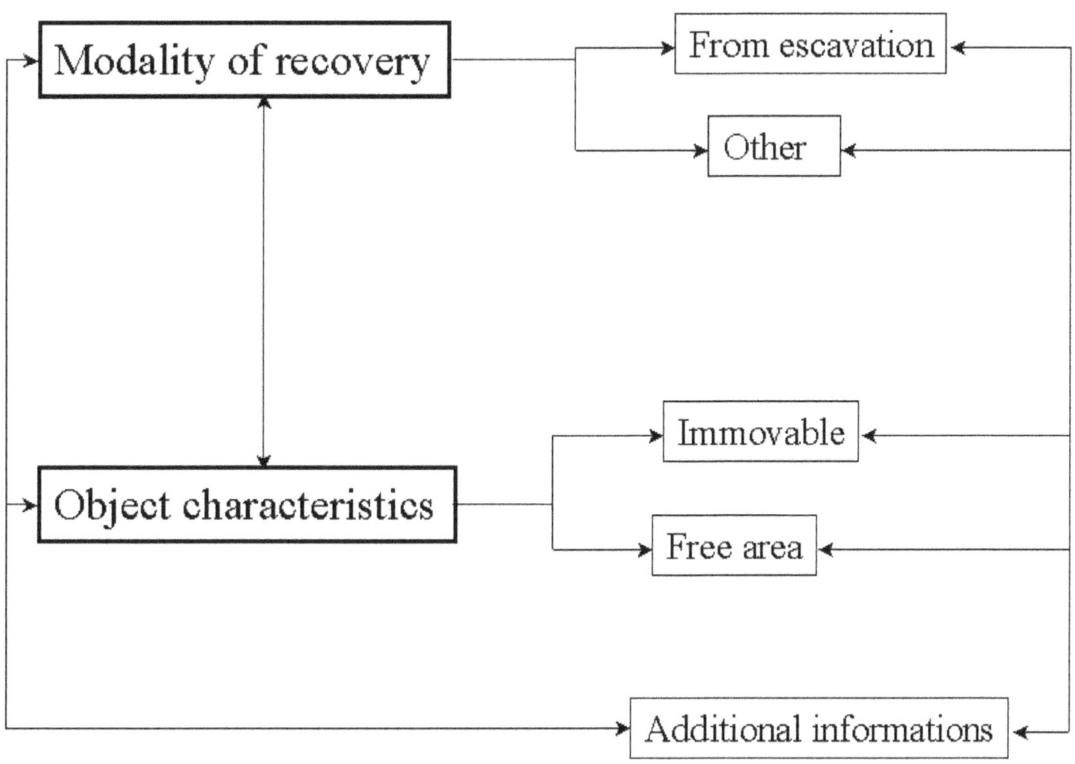

Figure 2.

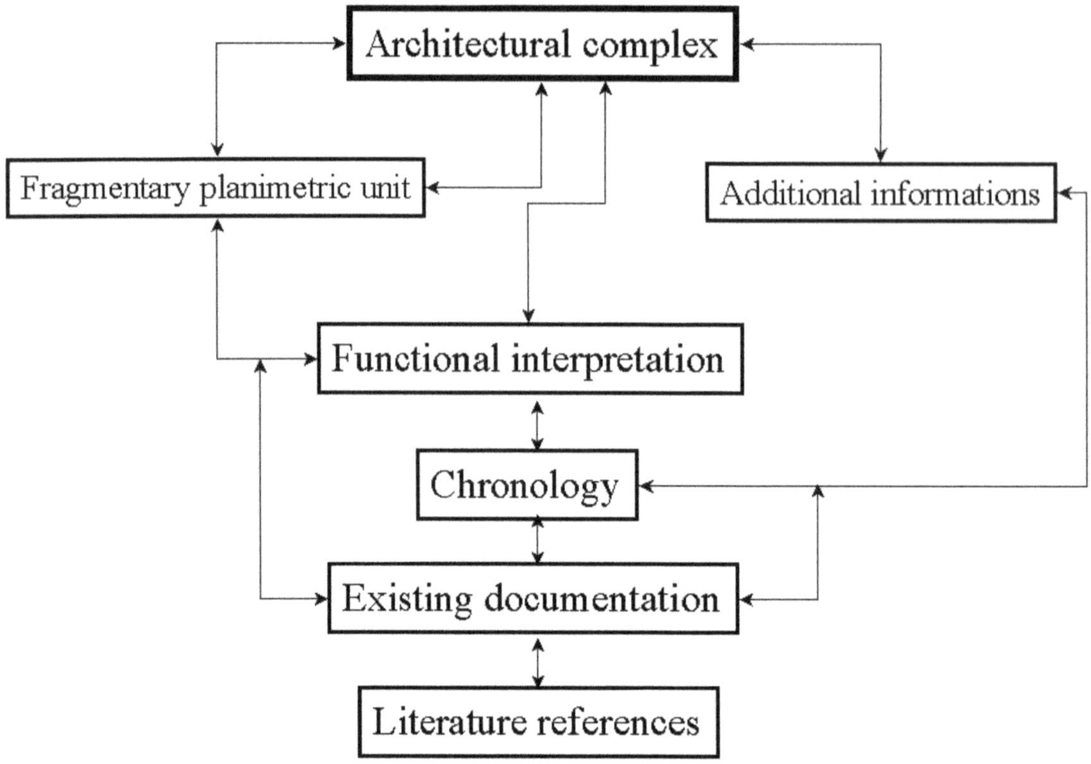

Figure 3.

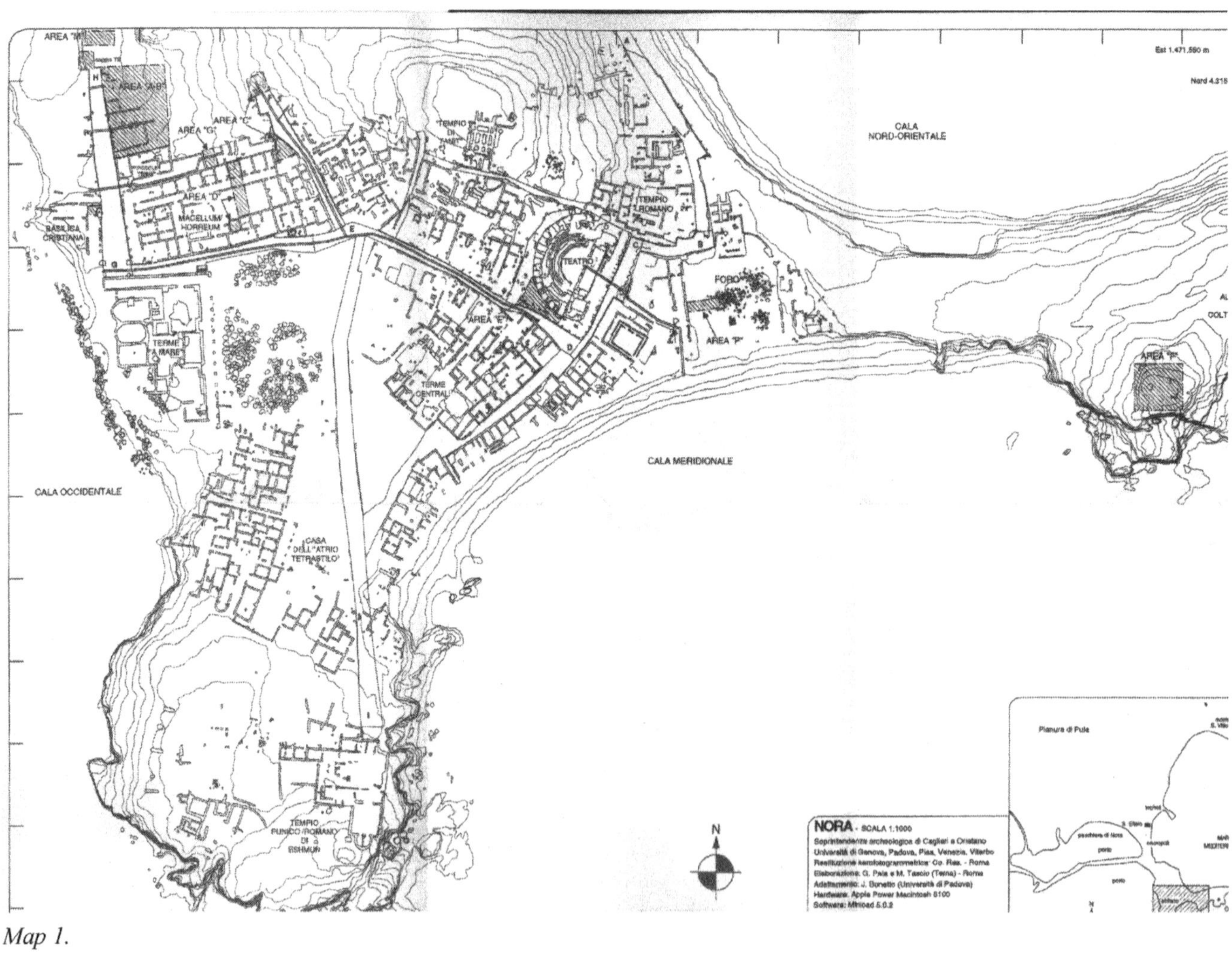

Map 1.

XML Encoding of Archaeological Unstructured Data

Marco Crescioli
Unirel srl, Firenze

Andrea D'Andrea
Istituto Universitario Orientale, Napoli

Franco Niccolucci
Università di Firenze

Abstract: The use of databases to archive archaeological records is well-established since the very beginning of the diffusion of personal computers. Using DBMS with more or less complex features is thus nowadays common practice, even in school excavations and in the most conservative and computer-unaware contexts. In some cases, however, the internal complexity of records is badly managed by the rigid structure of database management software, which requires a repetitive structure and a high level of homogeneity among data. This is the case, for example, of historical documents, which need to be linked to archaeological sources in comprehensive historical archaeology investigations. Databases are of little utility also for research that spans over many years and needs to integrate recent database archives with previous excavation diaries. In general, whenever descriptive text documents have to be analyzed together with structured archives, comparison and overall search are laborious, if ever possible, since, as it is well known, fitting text documents into a relational database may determine a substantial loss of information.

Some recent pioneering work suggests that XML may be profitably used to encode historical texts, but as search engines are not yet easily available, this may appear to be only a wishful statement.

In this paper we suggest a simple way to create and manage an archive of archaeological texts and images and a search system. This follows a research line in which we try to improve the computer tools available to archaeologists both developing extensions to manage peculiar conditions (uncertain or unstructured information) and using inexpensive software and easy-to-use, web-savvy user interfaces. The archives created in this way may then be accessed and searched using a Web interface, as stand-alone, on a Local Network or over the Internet.

Introduction

As far as excavation databases are concerned, attention has recently been drawn to an increasing proliferation of proprietary solutions, mainly oriented to a "local" data structuring and unrelated to the goal of integrating information with other solutions. It is not casual, therefore, that many applications, created to give an answer to specific investigations, have been abandoned after the publication of the research results. Notwithstanding the positive trend produced by a large process of computer literacy that invested at different levels also this sector, there is no unified perspective that may hint a final destination where all these applications may converge. Such a condition considerably limits information sharing and re-use. Thus traditional paper publication is still the main channel of knowledge diffusion, often as an interpretation synthesis and only sporadically at data level.

A short and certainly partial survey on the most recent procedures (see, for instance Arroyo-Bishop 1999 – for a comment on this paper see below – Constantinidis 2001, D'Andrea and Niccolucci 2001, Lang 2000, Madsen 2001) shows that excavation systems have been as yet characterized by solutions implemented in such a way that porting to different case studies is not easy. Even when this limit was recognized as an obstacle to data diffusion, the discussion seems to mainly concern the creation of advanced data standards from the semantic viewpoint or at the meta-structure definition level. This is, for instance, Madsen's proposal (1999: GARD project; 2001: GUARD project), who considers a difficult task the integration of different systems as far as field information description and formalisation are concerned and thus thinks to be simpler to favour an integration of different solutions by means of homogenous formal structures.

Anyway, the setup of semantic or structural standards does not appear as urgent, so invites to define common description protocols risk remaining unheard. Archaeologists differ too much in sensibility, culture, geographic areas to achieve quickly a formal homogeneity as far as description criteria and field information organization are concerned. Paradoxically, the diffusion of front-end software end-user applications and authoring systems to create personalized solutions increased this particularism by giving directly the archaeologists the responsibility of planning their data models. So archaeologists build the information representation model following their data formalization schema and aiming the system implementation to a specific archaeological investigation. This happens even with good computer solutions and with procedures that guarantee uniform digital documentation and archival criteria, without sacrificing the information management and overall processing. But the conclusion deriving from these experiences is incommunicability between different data management systems.

Between the exigency of unifying different solutions by means of standards or the creation of meta-structures, possibly it may be more useful to start reflecting and working on simplified data diffusion tools, i.e. solutions that that may use formats independent from particular software or operating systems, above all open-source. XML is such: as it is well-known, it is a meta-language based on public standards independent from the alternate fortunes or the commercial strategies of software producers. This technology uses text files and has an essential structure, so it will be able to guarantee the preservation and future use regardless new development and changes in information and technologies.

The availability of languages of the same family, aimed at the description of different data typologies already allows, moreover, levels of integration among data that would not be possible with the same flexibility using different tools. So, using different dialects of the XML family it is going to be possible not only to access on-line different archives, but also to integrate vector graphic information, for which the XML-based public standard SVG is already available, and spatial data for which GML (Geographic Markup Language) has been proposed as the XML description language of geo-referred objects.

This paper examines a first significant experiment we carried out in such a perspective. It is more than a simple hypertext-hypermedia application using the Internet interface to remotely and easily access data. Our project starts, on the contrary, from the need to convert a large database created with a proprietary application to integrate it, in a second stage, with unstructured information taken from reports published at the beginning of the twentieth century; data have then to be related to databases archiving specific object types as ceramics.

In order to avoid exporting all archives (both structured and unstructured ones) to the most recent release of a commercial package whatever, in an endless process to comply with the latest version of the same software (consider, for instance, the outcome of for distinct version of Office 95, 97, 2000 and XP in a short period, most of which show backward incompatibility for the Access DBMS format), we planned an created a meta-solution that could overcome the constraints imposed by the software market with no sacrifice on data, archives and existing information access. Thus we chose to implement the project with XML and to create a simple dynamic HTML interface to visualize and search alphanumeric and graphic data. With such a solution it is possible to use a browser, on the client side, to navigate among all these different data types and query the documentation using an appropriate search engine. It must be noticed that since XML tagging is a cumbersome job, the work is still in progress and as long as different documents are tagged there is a feedback to improve the structure of archived documents. Anyway, the principal database, created after modern excavations, is already available and may be consulted on-line and on a CD-ROM with no special software requirements.

Excavation data recording systems

Systems for archaeological excavation data recording vary from global solutions integrating alphanumeric and graphic data in the same environment to hypertext solutions addressing a simplified management of excavation data, with a remote access and easier and friendly GUI. A common tract is effectiveness as far as on-field data access is concerned. Moreover these system show a wide set of query functions and a hierarchical architecture. Most of them integrates data collected on field (stratigraphic units, drawings, photos) with laboratory-generated information (phase maps, joint maps, finds description).

In the wide diffusion of excavation systems, a specific application segment concerns the integration of both graphic and alphanumeric data within GIS systems: these are the so-called global solutions that recreate the information unity represented by the physical description of the layer-object with its spatial position but may also include data management and display or address particular inferential statistics procedures (see, for instance Djindjian 2000 with bibliography on intra-site artefact analysis) . Other solutions do not take into account the spatial component and produce exclusively alphanumeric databases. Perhaps the wide diffusion of database software – consider, for instance, the wide availability of Access, distributed as part of the Microsoft Office suite, which made it the probably most used package for data management notwithstanding his limits – and its apparent easiness of use led to a vast use of file management systems and/or (quasi) relational databases.

A negative note that accompanies most of these solutions concerns the narrow-minded attitude shown by archaeologists toward similar experiences more than software development or computer methodology. Since these solutions are extremely personalized, as noticed before, neither from the computer nor from the archaeological point of view it is relevant to compare one's experience with other solutions, in order to verify the possibility of developing common procedures or to analyze problems met while processing data. It is sufficient to give a glance to bibliographical references of different papers to notice that most teams follow their research line with little concern for similar experiences. The most noticeable case is the paper by Arroyo-Bishop (1999) on the ArcheoData project, where all the twenty-two (!!) bibliographical quotations refer to papers by the same author, with no bibliographical reference to any other solution (or to anything else). Apart from such a self-referential attitude that automatically disqualifies itself and its author and thus deserves little or no attention, the general impression is that there is a widespread superficiality and little interest, if any, for different problems and solutions (a malicious reader might perhaps say little knowledge). It seems that there is little conscience (again, knowledge?) that computer procedures have their own methods and theory and a scientific approach requires comparison and generality: for some "gro-und-scrapers" the world ends at the border of their site, and whatever is done elsewhere is neither applicable nor important, and needs to be quoted only as a courtesy.

However, in general, the diffuse opinion that one's excavation is different from any other in the universe, which by the way leads to the uncontrolled proliferation of recording forms (helped, at least in Italy, by the obsolete official ones), is the psychological basis to create slightly different records and recording systems, on paper and in the archaeologist's mind before than in the computer. Even if there is something true in this attitude, it does not favour the growth and the diffusion of excavation data and above all it commits to traditional paper publication the diffusion of scientific results. On the contrary, our solution aims at easing data access (without necessarily having special software) and is based on a development strategy/ philosophy that avoids proprietary data formats and uses instead public domain ones to integrate different data types. So, in our perspective the goal did not focus on creating a multimedia product even if the system is oriented to hypertext navigation among various archives and documents. However, the solution nucleus consists of the use a technology aimed at structuring documents lacking a rigid structure and at establishing a dialogue among different document types, thus enriching the information content of the overall database.

To make this principle assumptions concrete, and test in practice the theory enounced above, we applied this approach to a case study. It is an excavation of Cumae, a site of the greatest archaeological interest.

The excavation of Cumae

Very little is known of the organisation of urban space in Cumae, the most ancient of West Greek colonies, founded in the second half of the 8th century BC (Fig. 1).

Since 1994 the Istituto Universitario Orientale of Naples in co-operation with the Soprintendenza Archeologica di Napoli e Caserta started again an archaeological excavation focused on the topography of the ancient town. The little data as yet known about the urban settlement and the walls came from excavations carried on in early 1900 or from rescue excavations. The city extension is known according to an hypothetical reconstruction of the wall enclosure, while archaeological witnesses of the urban settlement limit to a few monumental buildings of the Sannite (5th – 4th century BC) and the Roman period (3rd century BC . 2nd century AD). The road system is witnessed by a N-S axis that crossed at least four E-W roads, discovered during a limited rescue excavation performed to build modern sewers. The choice of investigation areas was therefore finalized to the attempt of reconstructing the road network, characterizing the function of spaces with the urban perimeter, reconstructing the wall enclosure and establishing its phase chronology.

Investigations as yet mainly focused on the area between the northern walls and the forum (Fig. 2).

They allowed the discovery of the development of fortifications in a zone which probably was a hinge between the public area and the northern strip, where perhaps a lagoon landing was placed (D'Agostino and Fratta 1995).

In the same area, some sondages had been excavated between the end of the 19th century and the beginning of the 20th: there are, unfortunately, few records of these investigations, and they are kept in the Soprintendenza archives or in few concise excavation reports published in early 1900. In most cases there is no indication about the precise sondage position, while information concerning ancient buildings still visible in the 19th century may be derived from several maps compiled during the 19th century in at least three different topographic surveys on the area (Fig. 3).

Data collected during the field investigation (over 3000 stratigraphic unit records and related documents as photos, drawings, materials, samples) were recorded using the Syslat program (Lattara 10).

The SYSLAT system: archaeological data

Originally conceived as a tool to record excavation data of the proto-historic site of Lattes (Montpellier – France), since 1984, when it was first experimented, SYSLAT (Système Lattes) has undergone several releases that progressively assimilated the suggestions deriving from use. Outside France, the system has been used for instance in some excavations in Italy (d'Agostino and Fratta 1995, Gastaldi 1998, Santoriello and Scelza 2001). The experimentation so created a portable system now provided with a wide set of personalization functions, which may configured according to the needs of any excavation.

SYSLAT is not only a mere digital adaptation in digital format to the stratigraphic excavation model formalized by Harris. Even if it is based on the fundamental principles of stratigraphy, it is a tool for a "guided" organization of field data (from action recording to graphical and photographic documentation and sample collection) and for an integrated management of all data, including statistic-quantitative analysis of materials.

Created using Hypercard, one of the first authoring tools, SYSLAT represents a great container in which appropriate scripts allow to access recorded information: from stratigraphic units to "fact" and "set" records, from the photos to the graphic archive, from quantification records of materials to the typology of individual ceramic finds and so on. The global archive is structured in five different modules (terrain, objects, samples, documentation, utility) from which additional hierarchical sub-levels may be accessed.

The system includes a wide iconographic dictionary to classify ceramics (Lattara 6), and several vocabularies guide the operator in compiling forms. SYSLAT has also a module to personalize the database and to add new definitions to glossaries and dictionaries. With the latest version, 3.1, dating to 1997, the system has been exploited until reaching the limits of Hypercard: anyway, it has no multi-user version and cannot visualize spatial data. The lack of these features makes it difficult to plan further developments for the application: moreover, the fact of being limited to a specific platform does not allow porting archives to other operating system. Restructuring the system is now still more necessary since Hypercard, the software on which SYSLAT is based, will no more be supported on the current and on future versions of the Apple operating system.

Addressing the conversion of our SYSLAT archives, we had to face several problems:

a) since the original system was based on a hypertext model, descriptive fields were encouraged and these had been intensively used by the compiler; so a quantity of useful information was hidden inside the fields;
b) updating the software was unavailable, as mentioned above;
c) migrating to a different database management system would have solved the obsolescence problem only temporarily;
d) converting the archives to a RDBMS and extracting the information buried into descriptive fields would have been difficult, if not impossible; the lossy alternative would have been to leave it unchanged, renouncing to retrieve this hidden information with usual RDBMS searches.

Thus we decided to adopt a solution based on XML.

The conversion of digital archives

For the conversion, the fields included in the SYSLAT form were grouped distinguishing the ones containing information from the ones used by the system to manage the archive (a characteristic of Hypercard software). Only the latter fields were converted, associating to each one of them an XML semantic component, with identical content. Moreover, it was noticed that some elements (for instance the code of stratigraphic units) were often present inside the fields, in particular within the "Description" and "Interpretation" fields: so they were evidenced (i.e. tagged as such) in the final document, using as much as possible an automatic procedure based on the easy formal recognition of these parts inside a text (e.g. a numeric value preceded by the code US).

With such objectives, that is automatic transfer of "data" fields and automatic recognition of some references inside the fields, conversion routines written in Hypertalk, the programming language of Hypercard, created the XML files corresponding to the original records.

Having thus "saved" the database, now available as tagged text files, it may be published on Internet or consulted by the research team with no particular software or computer, but simply using an Internet browser as Internet Explorer or Netscape. Since only the latest versions of the browser are XML-compliant, to make access easier XML documents were batch converted to static HTML pages using a stylesheet and a conversion utility. The XSLT (a transformation language of the XML family) stylesheet that was created for this purpose automatically generates the link to the records that are referenced inside the fields (stratigraphic units, facts, materials, photos, drawings and maps) to allow easy navigation; different pages may also be accessed from an index page.

This solution is a temporary choice, since it has the evident drawback of requiring batch processing for data update; it has been adopted since it guarantees an immediate access to information, even with a minimal set of software resources (data can always be accessed from http://localhost) and on standalone workstations. More advanced searches require, on the contrary, the availability, at least on the network, of search engines and an XML-compliant HTTPD server as Cocoon, a freely available software which is a part of the Apache Open Source project. Moreover, as one can expect, the Microsoft browser has a tiny non-standard requirement that is easy to take into account but may make IE-compatible documents incompatible with anything else. So a more effective solution will leave the burden of XML-to-HTML conversion to the server and then send the user's browser a standard HTML document: thus pages will be dynamically created on user's demand using the XSLT stylesheet and will take into account all the variations on data. Integration with graphic vector information is already available via SVG (an XML-family description language for vector graphics) using a plug-in freely distributed by Adobe, one of the sponsors of this language; unfortunately, possibly due to Hypercard limitations original drawings were created on paper and are still such.

Other archives

In the past many archaeological investigations had been carried out in the same area as 1994 excavation. Notices about the former are, however, very scarce: they consist of a few memories and documents conserved in the archives of the Soprintendenza Archeologica and some reports concerning sondages of the beginning of the last century. Particularly interesting for the knowledge of the stratigraphy appears the information given by Maraglino (1906): He described some 1903 sondages documenting settlements in the area since the Greek period until the Roman one. The interest of this source consists not only of the rich stratigraphic information, comparable with data acquired from the most recent excavations, but also of some topographic reference, useful for an approximate placement of the sondages. They may be positioned by means of topographic references inserted into the published text that may easily be recognized in a contemporary cadaster map. Reading Maraglino's report, the stratigraphy of the area emerges clearly: it is characterized by superimposition layers under which remains of ancient buildings, overlapping preceding structures, were discovered. Of each layer the source gives a summary description, the deepness and the description of ceramics, which are relevant to date the stratigraphic unit.

Here is an example of Maraglino's text, translated into English:

> On the end of 1903 prof. Innocenzo Dall'Osso, looking after the excavations on the Correale farm-land owned by Signor Ernesto Osta, lawyer, did not omit to do some surveys around the mount of Cumae, to recognize and study the ancient features of such places.
>
> In one on these surveys, walking on the Gigante farm-land, placed some 300 m. South-West of the Acropolis, he happened to notice, in the recently moved ground, some fragments of black paste with an Italic graffiti decoration. This was a very good hint for him to suppose that there

could be settled the village, or at least the necropolis, of primitive inhabitants, occupying those place before the arrival of Greek colonists, thus preceding the colonization of Cumae.

Fortunately the owner of the Gigante farm-land is Signor Enrico Origlia, a person well educated and passionate for antiquity.

Thus prof. Dall'Osso easily persuaded Signor Origlia to pay for a sondage, which was directed by prof. Dall'Osso himself, with three workers; for example, since the Gigante farm-land was planted with a vine-yard, as almost everywhere in that place, he had to choose a point where the vines, being less dense, left a free space of about two square meters. After removing the humus layer, of the usual thickness of 30 cm., a yellowish soil followed, evidently of alluvial nature, more than one meter thick; then there appeared remains of ancient dwellings, with walls built in big blocks of yellow tuff, a local rock. Basing on numerous fragments of black painted ceramic, of Italic Greek provenance, found among the ruins, there were sufficient hints to refer the remains to the Greek-Roman period or to the Sannite domination.

Below them, however, there were remains of foundations of other more ancient buildings, among which some fragments of proto-Corinthian vase were found, undoubtedly belonging to an archaic period of pure ellenism. This first excavation shows that in that place a new population had overlapped on a preceding one and then still another until the Roman period.

As it can be seen, the text includes very useful information about early discoveries, and since the place-names appear on historical maps the settlement described in the report can be georeferenced, though imprecisely, and the information can be spatially compared with other data (Fig. 4).

To be able to perform structured searches in the text and compare results with the modern database, it is however necessary to adequately structure the source, using an element set as compatible as possible with the one used for the modern database.

The need of comparing Maraglino's information induced us to design a solution integrating all data (structured and unstructured) available for the studied area.

A further enrichment of our global solution may come from the integration within the system of another database specific to the study of ceramics coming from the latest investigations, created in order to have a precise chronology grid of recent and previous findings and to have a precise quantification of ceramic fragments for each layer, grouped by production classes and typology. This database had been implemented using Microsoft Access. It consists of two tables, linked by a one-to-many relation based on the stratigraphic unit code. The fragment table contains information on each fragment as part of the vase, number of fragments, category, production class, typology.

The multiplicity of available data (collected on field, in the laboratory, from archives) required thus a new computer solution and as already mentioned XML was the chosen one.

In particular, a great attention has been paid to the markup process of Maraglino's text since a correct definition of a translation model for similar documents was determinant for the subsequent integration of data extracted from excavation diaries, archive memories and generic texts.

It should be clear at this point that adopting a traditional technology, as relational databases, would probably imply a loss of information due to the necessity of using a common framework for rather different information structures, in particular historical documents as the above mentioned, which are usually in a discursive format and do not easy comply with database structure requirements.

Below there is a sample of marked text. It does not correspond to actual markup, which is done using the original Italian source and meaningful Italian words as tags. So this translation is given here only for the sake of clarity and to make it understandable to an international audience.

<*report*>On the end of <*year*>1903</*year*> <*responsible*>prof. Innocenzo Dall'Osso </*responsible*>, <*introduction*>looking after the excavations on the <*place_name*> Correale</*place_name*> farm-land owned by Signor <*owner*>Ernesto Osta<*owner*>, lawyer, did not omit to do some surveys around the mount of Cumae, to recognize and study the ancient features of such places.</*introduction*>

In one on these surveys, <*place_description*>walking on the <*place_name*> Gigante </*place_name*> farm-land, placed some 300 m. South-West of the Acropolis</*place_description*>, he happened to notice, <*layer level="0"*>in the recently moved ground, some <*finds type="ceramics"*>fragments of black paste with an Italic graffiti decoration</*finds*></*layer*>. This was a very good hint for him to suppose that <*interpretation*>there could be settled the village, or at least the necropolis, of primitive inhabitants, occupying those place before the arrival of Greek colonists, thus preceding the colonization of Cumae</*interpretation*>.

Fortunately the owner of the <*place_name*>Gigante</*place_name*> farm-land was Signor <*owner*>Enrico Origlia</*owner*>, a person well educated and passionate for antiquity.

Thus prof. Dall'Osso easily persuaded Signor Origlia to pay for a sondage, which was directed by <*director*>prof. Dall'Osso</*director*> himself, with <*resources*>three workers</*resources*>; for example, since <*sector*>the Gigante farm-land was planted with a vine-yard, as almost everywhere in that place, he had to choose a point where the vines, being less dense, left a free space of about <*dimensions size="2" unit="m2"*>two square meters</*dimensions*></*sector*>. After removing the <*layer order="1"*>humus layer, of the usual thickness of <*layer_depth size="0.3"*>30 cm. </*layer_depth*><*layer*/>, <*layer order="2"*>a <*color*>yellowish</*color*> soil followed, evidently of <*geology type="alluvial"*>alluvial nature</*geology*> more than <*layer_depth size="1"*>one meter thick<*layer_depth*></*layer*>; then <*layer order="3"*>there appeared <*layer_intepretation*>remains of ancient dwellings, with <*layer_description*> walls built in big blocks of yellow tuff, a local rock</*layer_description*> </*layer_interpretation*>. Basing on <*finds type="ceramics"*>numerous fragments of black painted ceramic</*finds*>, <*finds_interpretation*>of Italic Greek provenance</*finds_interpretation*>, found among the ruins, there were sufficient hints to refer the remains to the <*period*>Greek-Roman period</*period*> or to the <*period*>Sannite domination </*period*></*layer*>.

Below them, however, <*layer order="4"*>there were <*layer_intepretation*> <*layer_description*>remains of foundations of other more ancient buildings </*layer_description*>, among which <*finds type="ceramics"*>some fragments of proto-Corinthian vases</*finds*> were found, undoubtedly <*finds_interpretation*>belonging to an archaic period of pure ellenism</*finds_interpretation*></*layer*>. This first excavation shows that in that place <*site_interpretation*>a new population had overlapped over a preceding one and then still another until the Roman period</*site_interpretation*>. </*report*>

As it can be noticed, only very wide concepts have been used to tag the text. However, they allow a much more effective search when combined with the other databases.

The search engine

Since the work is currently in progress, several search engines are being tested. Moreover, while the markup of historical sources goes on, the element list (what is technically called the DTD), is often revised. So we created a very simple query interface, which can be easily modified according to variations in the DTD or different search engines.

In general, the user is presented with a HTML dynamic form where query conditions may be set. At present, the lists of searchable elements are kept separate for different document types: this is due to the above mentioned variability of the elements used to tag historical sources. Thus the list is kept in a file, which is read by the form and displayed as a combo-box. It is planned, however, to offer hints for the equivalence of elements within different element types and to propose a more limited number of search choices to a generic user.

The form, written in PHP, prepares the query strings as needed and passes them to the search engine. The result is then displayed by another page.

As far as the query involves only a database, this process is straightforward; using a native XML database makes everything a little more complicate, because such engines are still largely experimental and their interfaces are awkward.

So after the user has filled the form fields with the search parameters and the form has been submitted, the values are extracted from the fields and a command is compiled, adding the necessary codes, that is "switches" or element names. This step is transparent to the user, who needs, however, to know the data structure to choose in a correct way where the text string has to searched by the program. There is a number of options that have to be chosen: for instance one may search a specific content within an element, regardless the number of nested sub-elements occurring between the specified element and the required content, or the latter may be required to be contained directly in the indicated element. The content of attributes may be specified as a search parameter as well, thus giving a great flexibility. The equivalent of relational join operations is, on the contrary, sometimes difficult to express: a search as "construction materials used in layers containing a determined type of ceramics" is much more complicate to write than using, for instance, SQL, as the search logic is to move on the tree representing the element structure of documents, obtaining as query result a set of nodes of this tree.

The best operational results have so far been obtained using sgrep, a structured text search program developed at the University of Helsinki by Jani Jaakkola and Pekka Kilpelainen and freely available at http://www.cs.helsinki.fi/~jjaakkol/sgrep.html. While this program is very quick and effective, it uses a structure based on text "regions", which is completely different from XML nested elements philosophy, however this requires only translating from one metaphor to another and back. A program more compliant with XML specifications, in particular with XQuery, would make things simpler. Recently the beta version of an XML native database, named dbXML, has been released; it looks very promising and it is being tested, with provisional results that might be improved if a better documentation on inexplicable error messages were provided with the software.

To illustrate the procedure from the user point of view, we include a sample query form (Fig. 5).

It must be underlined that the one shown is not the real form, which is written in Italian and uses Italian names for the elements; it is just an English translation prepared for this paper, to be more intelligible to an international audience. Also the results are not the real ones, which would be in Italian as well: the real results have been translated and a specimen has been prepared. In summary, the illustration shown here is fake, but it is aimed at giving a non-Italian-speaking reader an idea of the appearance of the system to actual users. Moreover, as stated before, the form frequently changes according to variations in the DTD of historical sources, or to experiment associations among different element structures.

Conclusions and further research

As previously stated, what we present here is work in progress: it needs refinement and, above all, to prove its effectiveness it requires to turn from the present experimental status into an operational one. However, even in the present form it suggests some considerations.

Introducing XML data encoding in the humanities, as it is proposed in the present paper, might have substantial, and positive, consequences beyond the ones herein described. The Caere project (Moscati, Mariotti and Limata 1999) first introduced markup languages in archaeological data recording, with a remarkable intuition and very interesting results. It still used SGML as a reference meta-language, instead of XML, a choice which might have possible advantages but some certain drawbacks (Internet compliance first of all), and might therefore be argued. However, if incompatible SGML features are not used, as it seems from the published reports, there is practically no difference and this paper stands for its authoritative support to markup solutions.

The potential impact of XML use in human sciences and particularly in archaeology has already been evidenced in workshops (Niccolucci forthcoming) and is proving its importance as far as another related field, 3D modelling and Virtual Reality applications, is concerned (Cantone 2002; Niccolucci and Cantone forthcoming). A comprehensive and authoritative review of the importance of XML in metadata description and in general, in archaeological computing is also available (Ryan forthcoming).

Taking only archaeological databases into account, as noticed above the comparison and interchange between different investigations is still low and requires a long and fatiguing conversion among different data formats, which would be greatly facilitated by a common paradigm as XML.

Preserving digital archives is another important issue, and using proprietary data formats forces repositories (see e.g. ADS) to maintain different, and sometimes obsolete, DBMS formats to assure that data are still accessible and/or to convert them to ASCII delimited text (Richards and Robinson 2000). On the contrary, a simple and common text format, together with metadata description, would allow data access by means of current data management software and would not suffer from obsolescence.

Moreover, present trends in public administration using ICT (the so-called e-government) will lead in a near future to a vast amount of documents in electronic format, most probably XML-encoded: this seems the choice both of the Italian government [DPCM 2000] and of other European states. Thus administrative documents, for instance in Italy the administrative forms adopted by the Ministry of Cultural Heritage for monuments and archaeological excavations, could become interoperable with those produced by scientific research.
So XML might not only improve the efficiency of scientific research in the archaeological field, but also positively affect the management of cultural resources and perhaps fill the still existing gap between scientific research and administrative management, endowing both of a very powerful computer tool.

Bibliography

ADS Archaeology Data Service http://ads.ahds.ac.uk

Arroyo-Bishop D. 1999. From earth to cyberspace: the unforeseen evolution, *Archeologia e Calcolatori* 10, 7-16.

Cantone F. 2002. 3D Standards for Scientific Communication, this volume

Cantone F. and Niccolucci F., forthcoming, Legend and virtual reconstruction: Porsenna's mausoleum in X3D. Paper to be presented at the Web3D 2002 Conference.

Constantinidis D. 2001. Introspective Sitescaping with GIS, in Z. Stancic AND T. Veljanovski (eds.), *Computing Archaeology for Understanding the Past. CAA 2000. Computer Applications and Quantitative Methods in Archeology,* Proceedings of the 28th Conference, Ljubliana April 2000, BAR International Series 931, Oxford: Archaeopress, 165-172.

D'Agostino B.and Fratta F. 1995. Gli scavi dell'IUO a Cuma negli anni 10994-1995, *AION ArchStAnt* n.s. 2, 201-209.

D'Andrea A.and Niccolucci F. (eds.). 2001. *Atti del Primo Workshop nazionale di archeologia computazionale*, Napoli-Firenze 1999, Firenze: All'Insegna del Giglio.

Djindjian F. 2000. Artefact Analysis, in in Z. STANCIC AND T. VELJANOVSKI (eds.), *Computing Archaeology for Understanding the Past. CAA 2000. Computer Applications and Quantitative Methods in Archeology,* Proceedings of the 28th Conference, Ljubliana April 2000, BAR International Series 931, Oxford: Archaeopress, 41-52.

DPCM. 2000. *Decreto del Presidente del Consiglio dei Ministri 31 ottobre 2000 - Regole tecniche per il protocollo informatico di cui al decreto del Presidente della Repubblica 20 ottobre 1998, n. 428* (Decree of the President of the Council of Ministers of Italy, 31 October 2000 - Technical rules concerning the computer protocol according to the Decree of the President of the Republic, 20 October 1998, nr. 428), art. 18, available on the Official Journal of the Republic of Italy and at http://www.aipa.it/servizi[3/normativa[4/leggi[1/dpcm311000.asp.

Gastaldi P. 1998. (ed.), *Studi su Chiusi arcaica*, Annali di

Archeologia e Storia Antica, Nuova Serie n. 5, Istituto Universitario Orientale, Napoli.

Gray J. and Walford K. 1999. One good site deserves another: electronic publishing in field archaeology, in Internet Archaeology, 7.
http://intarch.ac.uk/journal/issue7/gray_toc.html.

Madsen T. 1999. Coping with complexity.Towards a formalised methodology of contextual archaeology, *Archeologia e Calcolatori* 10, 125-144.

Madsen T. 2001. Transforming Diversity into Uniformity – Experiments with Meta-structures for Database Recording, in Z. Stancic and T. Veljanovski (eds.), *Computing Archaeology for Understanding the Past. CAA 2000. Computer Applications and Quantitative Methods in Archeology,* Proceedings of the 28th Conference, Ljubliana April 2000, BAR International Series 931, Oxford: Archaeopress, 101-105.

Maraglino V. 1906. *Cuma e gli ultimi scavi*, Memoria Letta alla Reale Accademia di Archeologia, Lettere e Belle Arti di Napoli, Stabilimento Tipografico della Regia Università di Napoli, A.Tessitore e C.: Napoli.

Moscati P., Mariotti S. and Limata B. 1999. Il "progetto Caere": un esempio di informatizzazione dei diari di scavo, in *Archeologia e Calcolatori* 10, 165-188.

Lang N. 2000. *Beyond the Map: harmonising research and Cultural Resource Management*, in G. Lock (ed.), *Beyond the Map: Archaeology and Spatial Technologies*, NATO Science Series, vol. 321, IOS Press, 2000, Amsterdam,, 214-228.

Lattara 6 = Py M. (ed.). 1996. Dicocer, Lattes.

Lattara 10 = Py M. (ed.). 1997. *Syslat 3.1. Système d'Information Archéologiques. Manuel de référence*, Lattes.

Niccolucci F., forthcoming, *Dalla fonte alla rete: applicazioni di XML alla ricerca storica e archeologica* (From the source to the net: XML applications to historical and archaeological research) Proceedings of the "XML Workshop", University of Florence, March 2001.

Richards J. and Robinson D. (eds.). 2000. *Digital Archives from Excavation and Fieldwork: A guide to Good Practice*. Published for Archaeology Data Service. Oxford: Oxbow Books.

Ryan N. S., forthcoming, Documenting and Validating Virtual Archaeology, in *Archeologia e Calcolatori* 12.

Santoriello A. and Scelza A. 2001. Un sistema informativo archeologico: l'applicazione del Syslat a Fratte di Salerno, in D'Andrea and Niccolucci 2001.

All Internet references have been checked on October 10th, 2001.

Figures

Figure 1: Cumae in Campania (Italy)

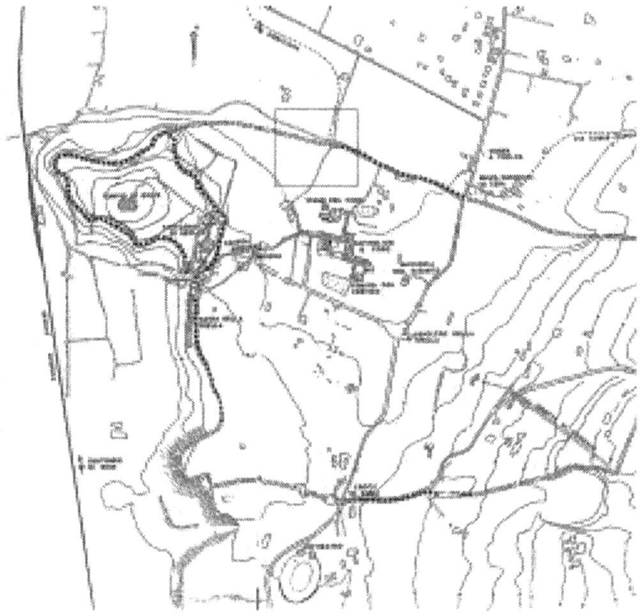

Figure 2: Archaeological remains of the ancient town of Cuma. The black square shows the localisation of the recent investigations carried out by Istituto Universitari Orientale on the area between the northern walls and the forum.

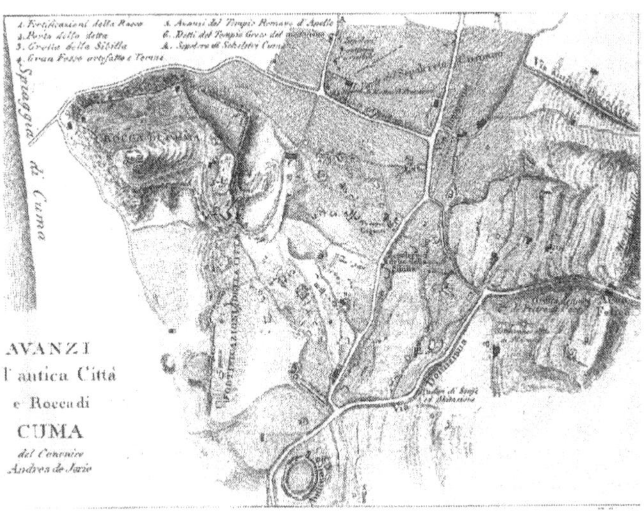

Figure 3: An example of historical map of Cumae (De Jorio 1822).

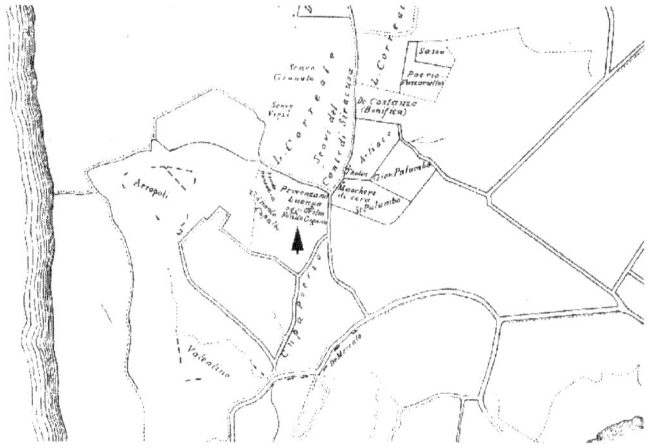

Figure 4: An historical cadaster map showing the farm-land described in the Maraglino's report.

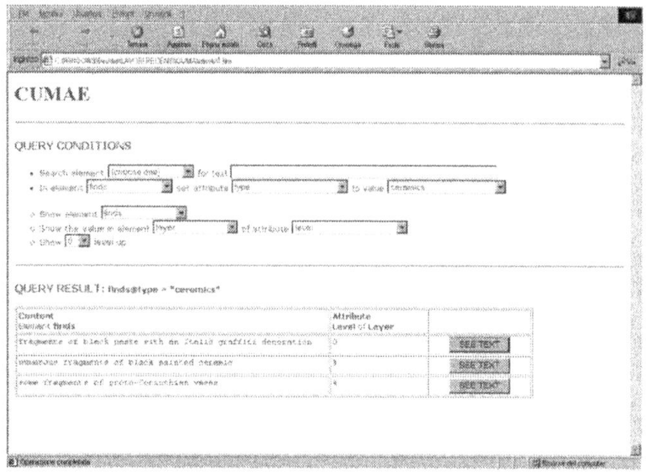

Figure 5: A sample query carried out using the search engine.

Cultural landscape, computers and characterisation:
GIS-based Historic Landscape Characterisation as a tool for archaeological resource management in England

Graham Fairclough
English Heritage
23 Savile Row,
London, W1S 2ET
Tel. + (0) 20 7973 3124 - Fax. + (0) 20 7973 3111
E-mail graham.fairclough@english-heritage.org.uk

Abstract: A programme of Historic Landscape Characterisation (HLC) using GIS is being carried out in England. HLC creates a rapid, broad-brush overview of the historic dimension – the time-depth – of the landscape for Sites and Monuments Records. The method is archaeological in its perspective, but aims to understand the archaeological and historical component of the present-day landscape. It is a method that is explicitly subjective and interpretative. This paper introduces some of the ideas behind the method, particularly its ability to deliver a wide range of practical applications, spatial planning and archaeological resource management. A summary of the method is given, with particular reference to the linked issues of defining the polygons that lie at the basis of the approach, and selecting the attributes that define their character. The likely impact of emerging new digital map-bases is also discussed.

Key Words: Cultural landscape, Historic landscape characterisation, GIS, Subjectivity, Spatial planning, Sites and Monuments Records, Archaeological resource management, Conservation

Background

Historic Landscape Characterisation ('HLC') is a national English programme of broad-based county-scale archaeological assessments of the present landscape - ie a programme of reading material culture (Fairclough et al 1999; Fairclough (ed) 1999). HLC brings together in one place, from an archaeologist's perspective, what we already know or can interpret about the historic landscape to facilitate the sustainable management of change through. It uses GIS to establish areas (polygons) defined by shared attributes based on the historic character and time-depth of the landscape. This provides a computer-based tool for sophisticated analysis and interrogation, without being wholly driven by technology or data-collation. It rests upon interpretation and perception, initially expert-led but hopefully to become more democratic. A number of examples of maps derived from HLC GIS are included in this paper, although they are inevitably highly simplified versions of what is feasible when using the HLC work on its GIS (figs 1 – 4). The programme is approaching its half-way stage and could be completed in 5-8 years (fig 5). Thirteen HLCs are both complete and have prepared method statements (Aldred 2001, Bannister 2001, Barnatt 1999, Barnatt et al 2000, Bramhill and Munby 2001, Ede 2001, Ford 1999, Herring 1998, Hoyle 1999, Lambrick and Bramhill 1999, Miller 1997, White 2001).

Each project covers a county or similar area, and is commissioned by English Heritage from local government archaeological services as part of their Sites and Monuments Records. SMRs in England provide the fundamental structure of archaeological resource management. They are maintained by every county council, sometimes in partnership with "unitary" authorities (ie mainly urban areas of single-tier local government). The functions of every SMR include spatial planning, development control, public and professional information provision, providing material for research, facilitating the active collection and curation of information about archaeological sites and their condition, landscape management, the quality control of developer-funded rescue excavations, and agri-environmental policies.

The location of the projects in SMRs is important, because HLC, although it has important research benefit, is designed first and foremost to support conservation and archaeological resource management, and to ensure that the cultural landscape is taken fully into account in decision-making. HLC is a conservation-led technique, an example of 'applied archaeology'. It is starting to fill a major gap in the archaeological resource management system in England. Similar gaps are being filled in other European countries (eg Netherlands State Government 1999 and Hallewas forthcoming; Heritage Council 2001, Danish Forest and Nature Agency 2001 and Stoumann forthcoming). As far as England is concerned, over the past three decades, the ARM system has successfully developed more or less effective ways of dealing with archaeological sites and their deposits, notably PPG16 and its related procedures (Department of the Environment 1990, Fairclough 1999(b)). It has been rather less successful, however, explaining and managing the significance of the whole landscape.

"Whole landscape" in this context means not just the ensemble of sites and monuments in an area, nor just their settings. It mainly refers to the areas between sites and most importantly of all to the fabric of the landscape itself, its fields, hedges or walls, its roads and settlement patterns, its woodlands and heath, its land cover generally. Many of these features of the landscape have not always been traditionally considered to be an archaeologist's concern, but they comprise a major part of the material remains of the landscape, and as such they invite an archaeological reading. The scale of the task of understanding the whole landscape, however, requires archaeologists to learn how to use generalised interpretations - how to work with area rather than point data, and with predominant character as well as specific detail. These different approaches have led to GIS and their related data structures, which is the main subject of this paper (one of a trilogy delivered at conferences in the spring of 2001 – Fairclough forthcoming (b), (c)).

Adopting a way of thinking that allows us to engage with the historic environment at landscape scale has also required an expansion of perception about what archaeology is and what archaeologists do, for instance:

- Recognising that significance, meaning and information about the past does not only lie within stratigraphic deposits on "sites", but throughout the landscape within all manner of structures and patterns;

- accepting that the very recent past is as much a part of landscape's historic continuum as the distant past, perhaps more so in certain ways;

- recognising that understanding the present day landscape as an archaeological artefact, unravelling its layers and components, is as much 'archaeology' as is trying to reconstruct the shape of past environments

- understanding that landscape exists for us only in the present-day, a subjective construct that uses thousands of years of environmental modification as building blocks but filtered through archaeological analysis, interpretation and perception;

- adopting new approaches to scale and generalisation using area not point data, so as to rise above unique details of places to identify significant similarities and differences;

- developing a greater acceptance that subjectivity has value and is not necessarily something to be avoided. Landscape is first and foremost a matter of perception, an emotional or intellectual construct.

HLC, finally, is not the same as landscape archaeology. The two approaches are of course related, but landscape archaeology attempts first to understand or to reconstruct past landscapes, not study the present. More often than not, despite contrary claims, it has sites rather than areas as its starting point, and normally it seeks objective rather than subjective understanding.

Non-archaeological ways of appreciating landscape character are in some ways closer in objective to HLC than is landscape archaeology. Ecologists or landscape architects for example share with HLC a concern to amalgamate landscape and its manifold attributes (historical, archaeological, biodiversity, amenity, aesthetics, association, folklore, and memory) into our own contemporary social and cultural constructions of what landscape means for people. This is notably the case for agencies like English Heritage that are primarily concerned with conservation and resource management.

The most elegant, straightforward and simple definition of landscape currently available is that of the new European Landscape Convention (Council of Europe 2000). This evolved from earlier documents (Council of Europe 1995) in parallel with the evolution of the HLC method. Its definition and objectives therefore fit HLC very closely. The Convention defines landscape as:

> *"an area, as perceived by people, whose character is the result of the action and interaction of natural and/or human factors".*

The Convention applies this definition to all parts of a country's territory. It includes urban as well as rural areas, everyday as well as exceptional landscapes, and degraded as well as well-preserved places. The definition, like HLC, emphasises change (action/interaction), people (human factors) and perception (cultural perspectives); in practice it defines 'cultural landscape', a near-synonym for historic landscape.

HLC (and the Convention) defines landscape in a way that differentiates it from "environment", partly because of its subjectivity, but mainly by an emphasis on the intervention of human consciousness and imagination in creating landscape. The creation of landscape is a human and cultural act in two ways. First, the components of landscape are created throughout the past (until yesterday), arising from the series of decisions and actions that have modified the environment, and created its material culture. Second, in today's present-time, people create from landscape from their environment as an intellectual, social or emotional construct. Archaeologists create a particular type of construct in which the issues that loom largest historical process are the remains of the past, the role of people rather than environment, the prevalence of change and the critical significance of longer-term chronology.

HLC seeks to capture all this. Almost from its inception, it has used GIS to do this. The area-based rather then record-led character of GIS has proved to be very suitable for the unusual mixture of objectivity and subjectivity that is landscape, being able to encompass concrete material things and vaguer ideas, scientific data and human perspective. It can also take into account the provisionality of any view of landscape that is caused by landscape's dynamism and the fast -changing character of perception and understanding. That is what this paper is mainly about: using computers to capture perception and understanding rather than fact and record, and doing so in ways that allow interrogation, dissemination, analysis, updating and integration with other landscape data-sets. In short, HLC is an application-led technique, and this paper therefore also tries to give an overview of the range of uses to which HLC is being put.

The paper also offers a superficial guide to the diversity of method used so far within HLC since the first, non-GIS, project in 1994 in Cornwall. One of the methods – that which is evolving for the six counties of the East of England region (East Anglia) - is described by Lynn Dyson-Bruce in some detail in a separate paper elsewhere in these proceedings. Until now, the method has been deemed to be experimental and still developing, and the invention and use of new methods has been encouraged. This has also allowed the method to develop a diversity that reflects landscape variation within England, and allows HLC to operate effectively within the many different institutional structures, spatial planning policies and computer data systems that exist across the range of England's local authorities. Local ownership and integration have so far been deemed to be more important than national consistency. At a European level, such methodological diversity may become even more important. Adoption of HLC in Scotland and Ireland has already led to significant modification of methods and aims to meet national requirements (Bruce et al 1999, Macinnes forthcoming, Cooney et al 2000, ERM/ERA 2000). A further layer of modification is being developed for its use in SW Sweden in Scania (pers comm Jenny Nord-Paulsson), as part of the EU Culture 200 project European Pathways to the Cultural Landscape, which involves ten countries in twelve linked projects. Informal assessment suggests that countries such as Portugal would require even greater adjustment of the methodology to work in different landscape types, whilst still retaining the core philosophy of HLC.

During 2001, however, English Heritage will commission a review of the English methodology. This will establish the extent of the common core of philosophy, method and structure that underlies all the separate projects, and identify the optimum techniques that have been developed.

Historic Landscape Characterisation

Ideas

Some of the precepts that underlie the idea of Historic Landscape Characterisation have already been mentioned. In summary they are that:

- "landscape" exists everywhere (and is important locally as well as nationally); but it only exists in the here and now (earlier versions of what we now call landscape did exist, and can be studied, but for contemporary purposes they are past environments not landscape) (Fairclough et al 9-16, Countryside Commission 1996).

- landscape, and landscape character, is a matter of perception more than fact (Fairclough 1998). Although the idea of landscape is often, particularly for archaeologists, made up from physical objects such as buildings hedges, walls, farms, woodland, or sites, it is only when these are viewed through perception, memory, imagination and ideas that they come together to be "landscape" rather than environment. Landscape is therefore best regarded as a matter of interpretation and perception, as an idea rather than a thing. Inevitably, therefore, there are as many landscapes as there are people to perceive them - landscape has multiple meanings and diverse values, and GIS is a useful medium for giving space to this diversity;

- landscape is defined by the interaction through time of people and their environment, and primary place in explaining and understanding historic landscape character is given to the actions in the past of people not to the ahistorical dictates of nature (Fairclough 1994). Environmental determinants exist, of course, but what interests HLC, and archaeologists generally, is the human impact on landscape, and landscape's cultural and social dimensions; these might be more nuanced, and perhaps more superficial, but they are more interesting;

- landscape is produced by process and change in the past. Its most important attribute is change, and it remains today essentially dynamic and transient. HLC data structure and GIS systems must therefore be able to deal with ephemerality, palimpsest and incompleteness. Recognition of the role of change - sometimes described by the shorthand of 'living landscape' - should be one of the foundations of sustainable conservation and sustainable development.

Character

Historic landscape character is constructed from several aspects of the archaeological and historic environment. Most central is the broad pattern of the landscape itself displayed in the diversity of attributes such as landuse, landcover, features, settlement patterns, the shapes of fields, all viewed as archaeological patterns. To understand these patterns it is necessary to look at the historic reasons for them, and at the historic processes that created the landscape's character. Relevant to HLC are attributes such as the distribution of arable, grazing, or woodlands (the patterning of woodland for example, demonstrably a manufactured not a natural pattern as it is restricted to parish edges and steep valleys), or the patterning and variety of settlement and field boundaries (the character of fields forms the backbone of the work and the main opportunity for further research). More localised features might contribute, too, such as the pattern of modified water courses, altered soils, the hedges and walls, quarries and mines, or roads, tracks, canals. Major historic processes – agriculture, quarrying, settlement - are kept at the forefront in other ways, too, as is the balance between them which has changed through time. Most GIS polygons in HLC record interpretations of previous land-uses as well as current, and thus the transition between them: ie time depth.

Archaeological sites and buildings are not mapped individually or even as distributions but can be used interactively with the area-based generalised view that HLC produces. This helps to understand site distributions and enables new types of area data to be generated from site data, a process which seems to have been largely impossible until now. A "fingerprint" of an area's archaeological character can be drawn, indicating - the type of sites that might be expected there, or which might be expected to survive given later landscape change, or the combination of different types of sites (or predicted sites) that make up the area's identity.

A less tangible aspect of HLC is territoriality, space across the landscape (Fairclough 1999(a)). Communities have always needed access to a mixture of resources, from cropland, grazing land, water resources (eg fishing) and woodland resources. This range requires many different topographical settings. Land-use territories therefore crossed the grain of the landscape, whether at parish or township scale where parishes cover several land-use zones, or at sub-regional level. There could also be temporal solutions (seasonal transhumance, for example) and solutions by trade and exchange, which also in different ways have affected the landscape's appearance. The English Heritage Rural Settlement Atlas provides an understanding of some of these cultural territories (Roberts and Wrathmell 2000), but HLC maps give us more local opportunities to see these patterns.

Finally, there is a largely unexplored area: the intangible associations of landscape and of public perception, and the personal and collective views of "real" people whether residents in an area or , visitors to it. People appreciate and value their historic landscape in so many different ways. It can be used as primary evidence for the past or to understand the current character of a place. For many people 'landscape' carries with it a host of varied associations (such as literary, mythological etc) and it is of course a source of aesthetic appreciation, meeting a basic love of 'nature' and ecology. It contains memory, both personal and social or collective; and it generates a sense of belonging (and perhaps of alienation).

Landscape in this sense varies from person to person, and from time to time, and encompasses stories, associations, memories or myths as well as history and archaeology. Capturing these ideas is a large area for future work, and HLC has so far not been able to engage much with the difficult issues it raises. Work is underway in Wales, however (eg Gwyn forthcoming), and we will experiment in England with using HLC as a baseline for public perception work, as part of the Bowland/Lune project, the English part of European Pathways to the Cultural Landscape, the EU Culture 2000 project mentioned above.

Methods

The HLC technique is not complicated. It simply assumes that the land can be divided into areas (GIS polygons) which are defined by some degree of shared character using a range of attributes or aspects. The resulting map is neither topographical nor directly reflective of physical reality, but is an interpretation of landscape character and history captured and presented as a patchwork of differentiated areas that together cover the whole map. One of the most important sources is the modern 1:10,000 or 1:25,000 map, showing field boundaries, buildings, and all land parcels. These maps are supported by up to date vertical air photos, other data-sets on habitats or woodland, and to some extent historic map.

The method works best when it is applied to large areas, by generalisation or summary, so that very broad patterns can be identified that rise above the detail of each place's unique differences. One of its characteristics is speed, to produce rapid overviews as close to a single "snapshot" as possible, by synthesising existing information. It is therefore almost wholly desk-based. Breadth – indeed comprehensiveness – rather than depth (and thus selectivity) is the aim, but of course HLC offers a new interpretation as a springboard for later, more detailed, analysis or research.

The definition of areas of land (polygons) with shared characteristics is key to HLC. This makes GIS almost essential for HLC work, although the primacy of defined areas was central to HLC even before use of GIS was adopted (Herring 1998, 1999). Intuitively defined to reflect historic landscape character (see below), the polygon is a container for all the structured data that creates the HLC.

Attributes

To produce the characterisation, all areas of the modern landscape, by interpreting maps and air photographs, are attributed to a range of 'historic landscape types'. This is sometimes a very direct process, using a prepared finite list of historic landscape types - see tables 1 and 2 (Lambrick 1999, Wills 1999, Bramhill and Munby 2001, Bannister 2001). These early types of lists, or classifications, were relatively short - 18 simplified types ('zones') in Cornwall, c80 at its most detailed in Hampshire. More recent projects have tended to record the individual attributes (indicators or "proxies") that subsequently allow high-level interpretation and analysis and the ability for different purposes for a polygon to be attributed to more than one landscape type - see table 3 (eg Aldred 2001, Ede forthcoming). The earlier approach (using a fixed number of categories or historic landscape types) has the advantage of clarity and ease, but limits interpretation to one attribute, embeds time-depth in a single view, and is difficult to simplify or re-work. The later approach is more time-consuming and difficult but it is more flexible, enabling a much wider range of analysis and levels of presentation. It also creates a very much larger, potentially infinite, number of separate types: Lancashire at its simplest (its 'Broad Types' map) has c80 types, for example, the same as Hampshire's most detailed set, but many more can be created by analysis of attributes and subdivision.

There is high degree of generalisation within HLC, so that it is the predominant aspects of the historic character of each area that is defined, rather than the detail of every distinct place. The GIS polygons against which attributes are recorded vary in size and shape in relation to the scale and grain of the landscape itself. Character is not usually defined for every land parcel or field, but for groups of land parcels. Attributes attached to a polygon, for example, might include for enclosed land, the straightness or sinuosity of boundaries, the size of fields, the existence of dog-legs indicative of removed boundaries, the morphological signs of enclosed strip fields, or the pattern of cumulative assart from woodland. In unenclosed (eg upland moors) character might be defined in terms of differences such as the existence of earlier abandoned phases of settlement or enclosure, or by the patterns of surviving prehistoric burial and ritual sites, or by industrial activity, or relationship to settlement. The likely earlier form of landscape character can also be included within polygon attributes, to allow change through time to be understood, for example whether 18th century fields were enclosed from common or common grazing, whether forestry had previously been upland moor, or what was there before urban expansion. Information is also recorded on data sources, and on the strength of confidence of interpretation.

Sites or areas of archaeology and remains are not plotted as point data, or as realistic representations. Such information exists elsewhere (eg in Sites and Monuments Records) and can be used later against the base-map that HLC provides. It is this that distinguishes HLC from other approaches to studying archaeological landscape - it does not aim to plot the components of the landscape but to map a generalised depiction of its overall historic landscape character.

TOIDs and the DNF

One technical advance that is likely to change the way HLC is carried out in England will be the issue in November 2001 of a new digital map by the Ordnance Survey, the UK's national mapping agency. This map, the Digital National Framework (DNF), converts all Ordnance Survey 'Land-line' mapping - detailed cadastral and topographic mapping captured at 1:2,500 resolution – into a seamless national dataset containing 400million individually defined and geo-referenced features, each uniquely identified (Ordnance Survey 2001).

DNF is an intelligent digital map designed especially for GIS and database use. Features defined can include points, lines and boundary lines but also areas and polygons. Each feature, including polygons, has its own unique 16-digit reference – a TOID (TOpographic IDentifier) – which in the case of area data is related to fully closed polygons. This dataset (to be issued in gml format) will therefore includes for the first time fully active pre-defined polygons for every land parcel (gardens, houses, fields, walls etc.) in the country, each capable of having additional structured data attached to it.

The full range of ways in which TOIDs will change HLC work is not yet known. The onerous, time-consuming and difficult task of digitising the boundaries of HLC polygons will largely cease, and HLC practitioners are already anticipating liberation from the exasperating business of digitising newly-defined polygons. Until now, only polygons in existing datasets (eg county datasets of, for example, mineral extraction or woodland) could be imported automatically into HLC, and even these often required modification. Every other polygon in a county's HLC map has of necessity been created anew, by 'drawing' over scanned maps or by on-screen digitising. With TOIDs, single features or groups (ie OS parcels) can be labelled as polygons to contain relevant new HLC data as described above. This will be a valuable time saving that will release more effort for the interpretation and analysis, rather than description, of historic landscape character that is the real heart of HLC. There will be a gain in precision, and much of the DNF's in-built TOID-related attribute datasets will also be useful. TOID technology will also facilitate analysis and research once HLC is complete.

The initial DNF release will not include all necessary inferred data, however. Invisible boundaries (such as that between a car-park and a road, for example, or, more significantly for HLC, the closure of fields across farm gates and other breaks in boundary) will not always be included at first. Nor at first will there be an hierarchical data structure, so that associating features into complexes – even obvious ones such as houses into a village – will not always be simple. Nor will subjective decisions be incorporated – for example, when to decide, as HLC frequently does, that detailed differences such as a single building in the corner of a field, or a small patch of woodland in fields should be submerged into a wider whole in the interests of generalisation. The availability of TOIDs will make mapped data easier to manipulate in many ways, but for HLC at least there will still need to be a large degree of further interpretation.

DNF data will be themed, however, so that, for example, features deemed to be buildings (most, not all buildings, not structures, nor their associated roads) can be displayed separately. The other eight themes, apart from Buildings, includes Water (rivers, tide lines, canals), Administrative Boundaries, and Roads/Tracks, and Land (all features that describe surface cover apart from buildings and communication routes), all useful supports for the mechanics of creating HLC.

Whilst TOIDs will streamline the technical methodology of HLC, the most interesting – and difficult – question about DNF and TOIDs is whether it should, can or will change the innate structure of HLC. Some practitioners point out that the high-level local resolution of TOIDs will allow each and every land parcel to be attributed to different HLC types or to contain a distinctive set of Historic Landscape attributes that would allow such very detailed classification. The question (putting aside the difficult practicalities, such as how TOID features can be grouped and linked to the absence of hierarchy and inferred data) is whether this is desirable or necessary. This is a theoretical not a technical issue. Detailed "recording" is not the objective of HLC, which aims for interpretation not description. The single most valuable innovation of both the first HLC projects (Cornwall without any GIS or computing; Avon with minimum, autocad-type, computer use; Peak Park with its detailed use of historic maps) and the very latest (eg Lancashire, Hertfordshire, Somerset, Herefordshire) is the ability to capture generalisation and overview. Whilst recognising that ultimately every place, every field and every land parcel has its distinctive historic landscape character when viewed closely enough and in enough detail, the real gain of HLC was to discover a way to rise above particularity. The aim is to define larger-scale differences and similarities, and to understand generic and overarching patterns, as part of the main objective mentioned earlier of creating the first ever strategic framework for decision-making that affects the wider landscape. In other words, HLC seeks to capture interpretation, meaning and subjectivity rather than to depict any real world. The ability that the DNF will give to work at a very detailed level may be an obstacle to these wider overviews.

The morphological, interpretative and stratigraphic techniques that underpin HLC require a much wider focus than individual land parcels. HLC makes assumptions about land parcels (in England, especially fields), largely on the basis of their attachment to broader patterns. It is the wider group to which a field belongs (sometimes up to several kilometres across) that allows the field's origin, purpose and history to be considered. Beyond that, it is the relationship of that group to adjacent and surrounding groups that begins to create a stratigraphy of landscape. Reduction to individual fields at too early a stage is a hindrance to this broader view (it might be desirable at later stages to isolate variations and exceptions, but the whole picture needs first to be defined). HLC polygons if they are to do their job of interpreting the landscape (and not simply recording in a different format the data from other historic maps) need to be larger than land parcels. Polygons based on a single land-parcel could admittedly be amalgamated later into larger polygons on the basis of their attributes and by other means of analysis, but this would detach analysis from observation and interpretation. The process of interpretation is so central to the philosophy of HLC that it needs to be carried out at the time of creating the GIS.

HLC is concerned with generalisation to achieve a wider view - hence the county-wide scope of each project, the insistence on the ability to produce simplified, general maps whatever more detailed data exists, the emphasis on interpretation and recording. HLC is also essentially subjective; GIS has been used to capture subjectivity and the new technique of TOIDs can contribute to this if used within that spirit. If, however, seen as a way to capture increasingly detailed objective data, it will produce merely a data-led landscape description, a database where the really interesting work of analysis, research and, most of all, understanding remains undone. The risk is also that it will describe environment, but not create landscape. HLC may therefore need to resist parts of the seductive offer that DNF and its TOIDs make of being able to capture the distinctiveness of every land parcel. When placed into a broader framework, the DNF should be capable of serving a larger purpose; but first this necessitates capturing the similarities of large groups of land parcels, defined intuitively as a polygon that may not exist in conventional terms.

Defining Polygons

How large the HLC polygons "should" be, and how they are defined, is a flexible matter, and polygon size will be variable. The basis of HLC is the presentation of a multiplicity of interpretation, ideas and theories about landscape's historic dimension as generalised areas, not as any type of real-space depiction. An HLC map is a chorophet, not a topographic map. It depicts areas characterised by some general common features, in the same way as map showing political areas (such as countries) or settlement zones (such as Roberts and Wrathmell 2000) although using a wide range of variable attributes. There are therefore two critical factors: the choice of attributes that define common character, and the choice of a level of resolution that determines the size of polygons.

The first of these critical issues - choosing the factors and attributes - is discussed elsewhere. Here the issue of polygon size is discussed, the definition of polygons within which to capture generalised, summarising data. This data (in some combination) must differentiate the polygon from its neighbours. Polygons should not therefore be devised mechanically: they should not be km-sq data collection units, nor administrative areas, even if historic), nor past settlement territories such as townships. Rather, HLC polygons need to arise out of observation and interpretation of the present-day landscape. They are defined intuitively and subjectively, but from a controlled observation of an explicit and transparent set of indicators. In this they are a little like the excavation contexts or excavation units that are defined by what makes each distinctive. In practice, HLC polygons are usually less reductive than contexts, allowing a polygons to contain a degree of heterogeneity in detail whilst maintaining a higher level, broad-brush homogeneity that allows each polygon to be described in different terms to others. Nonetheless the parallel holds, insofar as polygon boundaries are drawn around an area that at landscape scale can be seen to share, irrespective of size, key and core attributes of landscape character.

The size and shape of polygons varies because of the grain and diversity of the landscape itself, notwithstanding the obvious ability to merge polygons to create more simplified maps. In certain types of landscape each polygon might cover large areas, for instance of moorland, or 20th century intensively-farmed cereal fields ("prairie" fields). Such large polygons may include diversity of various types, such as the presence or absence of earthworks remains in disused fields, prehistoric or medieval, in moorland, and if these distinctions are required to have an area dimension the large polygon will need to be subdivided. Whether the lower level is activated however will depend on the use to which the HLC is being put.

Conversely, many landscape types will lead to much smaller polygons – functional types such as 20th century military sites (eg airfields, if they still dominate landscape character at county resolution) or industrial areas whether used or abandoned. Another trend towards smaller polygons is caused by the great diversity of field patterns, for example within the hedged landscapes of western England. These field patterns are often capable of being classified morphologically in numerous ways to represent known or assumed origins (eg, piecemeal enclosures from woodland (assarts), or long established in-field patterns, or the remnants of strip-divided fields, whether or not midland-type ("classic") open field patterns), and this tends to produce smaller polygons.

That these variations in the size of polygons reflect the grain of the landscape can be seen within individual counties, but also at regional and national level. There are areas of the country's landscape – the Hampshire chalklands, the East Anglian clay plateau - where landscape character is drawn at a very broad scale, in large homogenous tracts in significant contradiction to the diverse and detailed maps resulting from HLC in western areas such a Somerset. In part of course this is a matter of human perspectives (an inescapable part of landscape character) or the state of knowledge. For example, the field patterns of East Anglia have been little studied except through historic maps and in localised areas. In large parts of the Suffolk HLC, therefore, while polygons were drawn relatively small the distinctions between them are minor or non-existent. Given the limited extent of work that had been carried out before the HLC began (a follow-up research project is approaching completion picking up where HLC leaves off), fields were subdivided into only three types on the basis of regularity of shape, sinuosity of boundaries, size of fields. The more detailed EH-funded work now underway in East Anglia on historic fields will allow these large blocks of land to be sub-divided.

The same difficulty with knowing how to subdivide this region's landscape is visible in the Countryside Agency's national landscape character classification. This is designed to be a more detailed successor to the Countryside Character Map (Countryside Commission 1998, Countryside Agency 1999). The latter divided England into 159 Character Areas; the new emerging more detailed approach subdivides this through a set of 75 landscape types into over 2,100 (including urban areas). In East Anglia, however, the detailed map identifies only 14 types, and divides the whole region into only 40 mainly-large areas, compared to 9 original Character Areas. In the more diverse, perhaps essentially older and least changed, western parts of the country, in contrast, the effects of subdividing 159 Character Areas into 2,100 areas has been much more marked. The increase in detail is thus not spread evenly across the country, but reflects landscape character and appreciation of it. East Anglia has a relatively homogenous appearance, and to capture its subtle character requires a greater attention to detail to pick up the more minor differences in knowledge and perception. The thresholds of defining difference need to be lower in such areas. Furthermore, the indicators being used by the Countryside Agency for landscape character – the mixed pattern of agriculture, topographic indicators like slope and relief, or field boundaries – cause certain types of landscape, notably those valued by a picturesque aesthetic, to be privileged in contrast to others. Having said that, the sub-province and local regions in the Settlement Atlas are also larger in East Anglia than most other regions.

The grain of HLC – the extent to which difference is mapped, how far into detail to go, how far to summarise - needs to be established to meet user-needs. In Herefordshire (the western English midlands), the HLC polygons are drawn large because, as a county where landscape character derives from well-preserved, little changed but under-studied ancient field-patterns, the first priority is for a high level initial overview. Existing and future research at more local and detailed level can be fitted into the overview later. In Hertfordshire (Eastern England), in contrast, the landscape is more fragmented with a more diverse character, with large numbers of large urban centres, a great deal of ornamental parkland, significant post-medieval, indeed late 20th century, landscape change. This required the HLC to use a more detailed characterisation of the fewer surviving areas of earlier fields. Because of this, and because it also builds on the achievement of Suffolk, Hertfordshire HLC uses much smaller polygons, in order to capture otherwise largely invisible (and indeed until now overlooked) distinctions and survivals.

Applications

Throughout the HLC process, its very practical end-uses and applications are always borne in mind, and this has strongly influenced the way GIS is used. Being so strongly GIS-based has in itself aided HLC's adoption within various very practical activities, notably spatial planning and landscape management, because it is simple to provide large amounts of sophisticated data either on line across organisations, via CD or (in the very near future) on web pages. On-line access is particularly useful as it allows HLC to form part of much more extensive data-sets, incorporating SMR data, Urban surveys, and non-archaeological, eg land-use, data, so that integrated interrogation is feasible.

Some of the many applications that HLC is designed for, and is being used for, have been discussed elsewhere (eg Johnson 1999, Herring 1998, Fairclough (ed) 1999, pp67 -122; forthcoming (b) and (c)) but it is true to say that users of HLC are still discovering its full potential.

All in all, therefore, HLC is an application of increasingly sophisticated and sophisticated computing technology. It is aimed, not only at analysis and learning, but at broad social issues such as democratic involvement, social inclusion and participation on the one hand, and sustainable development on the other. Both of these are key themes of many current EU and Council of Europe statements of policy. The aim is nothing less than to learn to pass on a landscape to our successors that we understand and which carries in it adequate remains of its history.

Most important, because it is the foundation for almost all other applications, is that HLC enhances SMRs. It does this most simply by adding another layer of interpretation, and by bringing into the SMR new information on a wide range of semi-natural landscape components, notably historically-managed woodland, hedges, ponds and modified watercourses. These reflect the impact of people on the landscape just as much in their way as buildings or earthworks, but they have largely been overlooked. HLC also enhances the SMR more fundamentally by providing a new base map that can integrate the rest of the SMR. The type of encoded data within an HLC map is different, and for some purposes perhaps superior, to that within a standard topographic map. It does not geo-locate individual features but instead provides some explanation, through the effects of landscape change and loss through time, of the pattern of survival and condition of archaeological population. It can explain why earthwork burial mounds existed, and why they still survive in some landscape types, but not in others, for example (see Johnson 1999 for other examples). HLC provides an 'intelligent' rather than a descriptive matrix for SMR data.

HLC also offers another step towards effective site prediction. Relating HLC to recorded sites in SMRs, and thus isolating some of the reasons for survival and visibility (always remembering the effect of the vagaries of archaeological research), can identify gaps in knowledge as opposed to gaps in real distributions. It can produce "fingerprints" of an area's (or historic landscape type's) most likely predicted archaeological assemblage, based on inference and assumptions about its earlier character and later loss and change.

From such analysis it is a short step to establishing new research agendas, or conservation and management strategies. HLC for example can identify rare or typical landscape types, those that are best preserved, or most intact for a single date, or those that demonstrate a sequence of more than one period. It thus aids the very first stage of conservation, that of defining significance. Within this process, HLC can also provide a strong context for protective designation – the selection of "special" buildings that the English conservation system uses to protect a sample of monuments and buildings.

The second major group of applications concerns the use of HLC in spatial planning (Fairclough forthcoming (b)). At a generalised level, HLC can be converted straightforwardly into strategic or local forward plans, or non-land use landscape management strategies (eg Lancashire County Council 2001, New Forest District Council 2000), identifying the types of development that most fit each area's landscape character, for example. At a higher level, HLC will be able to provide a strategic contribution to wider European planning, such as the two inter-regional planning overviews that have recently been produced (Norvision 2000, NWMA Spatial Planning Group 2001). To do this effectively, HLC has also been designed to be easily laid alongside other ways of assessing landscape character that are based on non historical criteria such as scenic beauty or ecology, so as to create an integrated perspective (Fairclough 1995, Countryside Agency forthcoming).

At a more local and tactical level, HLC can be used within development control to assess the impact of individual development proposals, and to inform decisions on mitigating the effects of development. It can be used to guide the design of new buildings and settlement so that they reflect deeply embedded landscape character, thereby benefiting from the contribution that the past can make to the future.

In a similar vein, HLC can contribute to more proactive landscape management. In England, for example, the environmental strand of the EU Common Agricultural Policy consists of a series of agri-environmental grant schemes (to which c£1.6bn is dedicated over the next six years) designed to pay farmers for producing environmental and social benefits including the preservation of archaeological sites. The principal scheme for landscape is Countryside Stewardship and HLC is becoming one of the mechanisms by which the government Rural Development Agency is targetting its grants to maximise their value to the historic environment.

On the regulatory front, HLC has greatly strengthened the operation of the Hedgerow Protection regulation, designed to prevent the destruction of the most important hedges. Before HLC, the main criteria for identifying special hedges was ecological (number of species, biodiversity) and the only historic criteria were partial indicators such as parish boundaries and whether the hedge appeared on early maps. HLC enables individual hedges to be placed in context, and their date, origin and history to be evaluated as part of the wider pattern. At the same time, rates of loss over the last century can be easily measured, and issues such as rarity can be demonstrated. In Hertfordshire, for example, we now have an overall view of the differential distribution and survival of medieval or earlier fields across the whole county.

In addition to all this, the HLC maps form a benchmark for measuring future change, whether positive or negative. Change can take place in the landscape itself, and more interestingly, in our perception and understanding of it. It is the intention, exploiting the flexibility and sophistication of GIS, to keep HLCs up to date, but the first HLCs (and periodic updates) will be archived to become reference points for comparison of change across time. Such information on the rate and direction of change is extremely hard to collect others (Darvill and Fulton 1998), and this may become a major benefit of HLC. It is also possible to use HLC to measure past recent change - identifying the degree of change in area character between the picture shown by HLC and HLC derived from historic maps, for example.

Finally, HLC seems to offer an opportunity to escape a little from the expert-led interpretation of landscape towards a more democratic and wide-ranging ownership, of the kind insisted upon by the European Landscape Convention (Council of Europe 2000, Lambrick 1999, Johnson 1999, English Heritage 1997). Landscape itself is an easier and more accessible concept for most people; almost automatically it is a sort of shared common language. Maps too are relatively easy to use as vehicles of both explanation and interpretation. One of the next stages of the HLC project is to start seeking public responses to expertly-created HLC, and more importantly to ask the public for their views on what makes an area historically significant. We hope that such a process will emerge from the Culture 2000 European Pathways to Cultural Landscape project. Public criteria may well be very different to archaeologists' views, perhaps relying more on literary association, mythology, political or religious connotations, and on the more everyday details of landscape (perhaps relatively modern) that create sense of place and local identity. Adding them to more specialist archaeological interpretations will not only deepen our understanding of landscape and its social value, but will also start to create stronger support for conservation of the cultural landscape, for the sustainable management of change in the landscape, and for more practical and active management strategies.

Acknowledgements

Almost all the advances in using GIS to make HLC work better have emanated from the individuals responsible for the various projects. I am therefore very grateful to all those who have contributed to the invention of HLC. In the context of the present paper on computers, I owe particularly thanks to : Mike Chapman and Jon Hoyle for the earliest use of GIS for HLC in Avon and Gloucestershire, Paul Bramhill and George Lambrick for their more developed work in Hampshire, and Joy Ede (Lancashire), Lynn Dyson Bruce (Hertfordshire), Oscar Aldred (Somerset) and Paul White (Herefordshire) for the most recent, improved systems, the most advanced at the time of writing. Watching them work has enabled me to form a sort of overview, which lies behind much of this paper. Discussions over many years with colleagues in Scotland (Leslie Macinnes, Piers Dixon) and Ireland (Charles Mount, Gabriel Cooney) as they adapt HLC to two other very different countries, have always been enlightening for me; more recently (as part of the Culture 2000 EPCL project) discussing with Jenny Nord Paulsson how a quintessentially English approach might be transferred to the different landscape of Scania has also clarified my thoughts. Finally, as always, it is Liz Page's skill, hard work and patience that has allowed random thoughts to be drawn into a file that was worth emailing to the editors!

References

Aldred, O. 2001 *Somerset and Exmoor Historic Landscape Characterisation Project 1999-2000.* Taunton: Somerset County Council and English Heritage.
Bannister, N. 2001. *Surrey Historic Landscape Characterisation*, Kingston: Surrey County Council, English Heritage and the Countryside Agency.

Barnatt, J. 1999. *Landscape Through Time: Historic LandscapeCharacterisation in the Peak Park, draft report.* Bakewell: Peak District National Park Authority and English Heritage.

Barnatt, J., Johnson, M. and May, R. 2000. *The Derbyshire Historic Landscape Survey Character Assessment - Aims, Method and Definition of Character Types.* Bakewell: Peak District National Park Authority for English Heritage and Derbyshire County Council

Bramhill, P. and Munby, J. 2001. *Kent Historic Landscape Characterisation Report.* Maidstone: Oxford Archaeology Unit for Kent County Council and English Heritage.

Brown, I.W. and Berry, A.R. (eds). 1995. *Managing Ancient Monuments: An Integrated Approach.* Mold: Association of County Archaeology Officers and Clywd County Council.

Bruce, L. D., Dixon, P., Hingley, R. and Stevenson, J. 1999. *Historic Landuse Assessment (HLA): Development and Potential of a Technique for Assessing Historic Landuse Patterns, report of the pilot project 1996-1998.* Edinburgh: Historic Scotland and the Royal Commission on the Ancient and Historical Monuments of Scotland.

Cooney, G., Condit T. and Byrnes, E. 2001. *The Archaeological Landscapes Project: Final Report, consultation draft.* Kilkenny: University College Dublin for Heritage Council (Ireland)

Council of Europe. 1995. *Recommendation R(95)9 of the Committee of Ministers to Member States on the integrated conservation of cultural landscape areas as part of landscape policies.* Strasbourg.

Council of Europe. 2000. *European Landscape Convention.* European Treaty Series No. 176. Strasbourg.

Countryside Agency. 1999. *Countryside Character of England*: vol 4 *East Midlands* (CA10), vol 5 *West Midlands* (CA11), vol 6 *The East* (CA12), vol 7 *South East* (CA13), vol 8 *South* West (CA14). Cheltenham.

Countryside Agency, forthcoming: *Landscape Character Assessment Guidelines.* Cheltenham.

Countryside Commission. 1993. *Landscape Assessment Guidance*, CCP423, Cheltenham

Countryside Commission. 1994. *Views From the Past – historic landscape character in the English countryside* (consultation report). Cheltenham.

Countryside Commission. 1996. *Views From the Past - Historic Landscape Character in the English Countryside,* CCW4. Cheltenham.
Countryside Commission. 1998. *Countryside Character of England*: vol 1 *North East* (CCP535), vol 2 *North West* (CCP536), vol 3 *Yorkshire and the Humber* (CCP536). Cheltenham.

Danish Forest and Nature Agency. 2001. *CHIP - Cultural Heritage in Planning : Identifying valuable cultural environments through planning.* Copenhagen: Ministry of Environment and Energy.

Darvill, T. and Fulton, A.K. 1998. *The Monuments at Risk Survey of England, 1995.* Bournemouth and London: Bournemouth University and English Heritage.

Department of the Environment. 1990. *Planning Policy Guidance note 16: Archaeology and Planning.* HMSO.

Donachie, J. and Hutcheson, A. 2000. *The South West Wiltshire Historic Landscape Character Assessment.* Salisbury: Wessex Archaeology for English Heritage.

Ede, J. with Darlington, J. 2001. *The Lancashire Historic Landscape Characterisation Project Report.* Preston: Lancashire County Council and English Heritage.

English Heritage. 1997. *Sustaining the historic environment.* London.

ERM and ERA. 2000. *Pilot Study on Landscape Characterisation in County Clare.* Oxford, Dublin and Kilkenny: Environmental Resource Management and ERA-Maptec Ltd for the Heritage Council (Ireland).

Fairclough, G.J. 1994. Landscapes From the Past - Only Human Nature. In Selman (ed), 1994: 64-72.

Fairclough, G. J. 1995. The Sum of All Its Parts : An Overview of the Politics of Integrated Management in England. In Berry and Brown 1995: 17 - 28.

Fairclough, G. J. 1998. Assessment and characterisation: English Heritage's historic landscape policy. In *Landscape archaeology and ecology vol 3 : Landscape perception, recognition and management: reconciling the impossible?,* Proceedings of a conference at Sheffield, 1996). Sheffield: The Landscape Conservation Forum and Sheffield Hallam University.

Fairclough, G. J. 1999a. Protecting Time and Space: understanding historic landscape for conservation in England. In Ucko and Layton (eds) 1999: 119-134.

Fairclough, G.J. 1999b. Protecting the Cultural Landscape - national designations and local character. In Grenville 1999: 27-39.

Fairclough, G .J. 1999c. Historic Landscape Characterisation: theory, objectives and connections. In Fairclough (ed) 1999: 3-14.

Fairclough, G .J., forthcoming (a): Place and locality a non-monumental heritage? In *Proceedings of 'Interpreting Historic Places' Conference, York, 3-7 September 1997.* London: Routledge.

Fairclough, G. J. forthcoming (b). Cultural landscape and spatial planning. In *Historic Environments of the North Sea*, proceedings of an InterReg IIC conference, March 2001. South Shields.

Fairclough, G.J. forthcoming (c). Cultural landscape, sustainability and living with change. In *Proceedings of the US/ICOMOS 4th International Symposium April 5-8, 2001.* Philadelphia.

Fairclough, G. J. (ed) 1999d. *Historic Landscape Characterisation.* Papers presented at an English Heritage seminar held at the Society of Antiquaries, 11 December 1998. London: English Heritage.

Fairclough, G. J., Lambrick, G. and McNab, A. 1999. *Yesterday's World, Tomorrow's Landscape - the English Heritage Historic Landscape Project 1992-94.* London: English Heritage.

Fairclough, G. J., and Rippon S.J. forthcoming . *Europe's Cultural Landscape: archaeologists and the management of change, Europae Archaeologiae Consilium, Occasional Paper 2.* Strasbourg.

Ford, M. 1999. *Historic Landscape Characterisation in East Anglia: Part 1 - Suffolk.* Bury St Edmunds: Suffolk County Council and English Heritage.

Grenville, J. (ed). 1999. Managing the Historic Rural Environment. London: Routledge and English Heritage.

Gwyn, D. forthcoming. Associative Landscape in a Welsh Context. In Fairclough and Rippon forthcoming.

Hallewas, D. P. forthcoming. The Belvedere Project, an integrated approach, in Europe's Cultural Landscape: archaeologists and the management of change. In Fairclough and Rippon forthcoming.

Heritage Council. 2001. *Towards integrated policies for Ireland's landscape,* The Heritage Council. Kilkenny.

Herring, P. 1998. *Cornwall's Historic Landscape - presenting a method of historic landscape character assessment.* Truro: Cornwall Archaeology Unit and English Heritage, Cornwall County Council.

Herring, P. 1999. Cornwall: How the Historic Landscape Characterisation methodology was developed. In Fairclough (ed) 1999: 15-32.

Hoyle, J. 1999. *Historic Landscape Characterisation: Cotswolds AONB, draft report.* Gloucester: Gloucestershire County Council and English Heritage.

Johnson, N. 1999. Context, meaning and consequences: using the map in Cornwall. In Fairclough (ed) 1999: 117-122

Lambrick, G. 1999. Hampshire: HLC and the Community. In Fairclough (ed) 1999: 51-63.

Lambrick, G. and Bramhill, P. 1999. *Hampshire historic landscape assessment, final report.* Winchester: Oxford Archaeology Unit and Scott Wilson Resource Consultants for Hampshire County Council/English Heritage, 2 vols.

Lancashire County Council. 2001. *A Landscape Strategy for Lancashire.* Preston: Lancashire County Council.

Macinnes, L. forthcoming. *Historic Landscape Characterisation*, Historic Scotland

Miller, K.R. 1997. *The Isle of Axholme Historic Landscape Characterisation Project.* Leeds: Countryside Commission.

Netherlands State Government. 1999. *The Belvedere Memorandum - a policy document examining the relationship between cultural history and spatial planning (Belvedere – Beleidsnota over de relatie cultuurhistoire en ruimtelijke inrichting).* The Hague.

New Forest District Council. 2000. *New Forest District Landscape Character Assessment.* Lymington: New Forest District Council with Hampshire County Council, the Countryside Agency and English Heritage.

NorVision Working Group. 2000. *NorVision, A Spatial Perspective for the North Sea Region.* Essen.

NWMA Spatial Planning Group. 2001. *A Spatial Vision for North-West Europe. Rotterdam (CHECK)*: North West Metropolitan Area for Interreg IIC Secretariat.

Ordnance Survey. 2001. The Digital National Framework (What is it, Product Specification etc). Southampton: http://ordnancesurvey.co.uk/dnf

Roberts, B.K. and Wrathmell, S. 2000. *An Atlas of Rural Settlement in England.* London: English Heritage.

Selman, P. (ed). 1994. *The Ecology and Management of Cultural Landscapes - proceedings of an IALE UK conference at Cheltenham, 1993.* Landscape Issues, vol 11 No 1. Cheltenham.

Stoumann, I. forthcoming. Cultural History and Spatial Planning in Denmark. In Fairclough and Rippon forthcoming

Ucko, P.J. and Layton, R. (eds). 1999. *The Archaeology and Anthropology of Landscape.* One World Archaeology 30 (papers from the 1994 World Archaeology Congress, New Delhi), London: Routledge.

White, P. 2001. *The Herefordshire Historic Landscape Characterisation Report*, Hereford: Herefordshire County Council.

Wills, J. 1999. Cotswolds AONB: Characterisation, Classification and GIS. In Fairclough (ed) 1999 33-40.

Figures

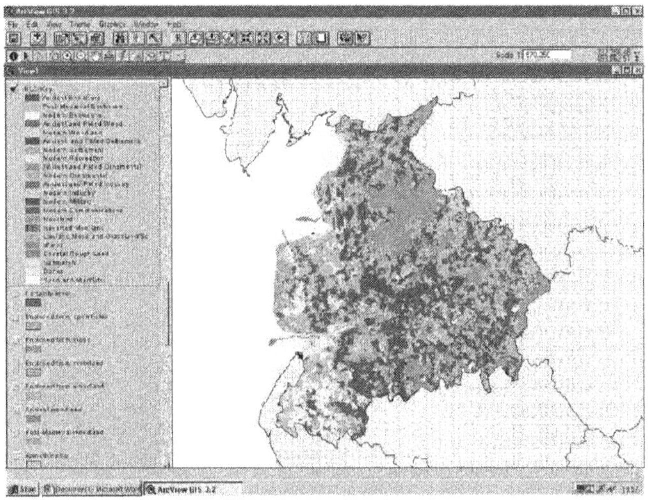

Figure 1. Lancashire (after Ede 2001): Broad HLC types

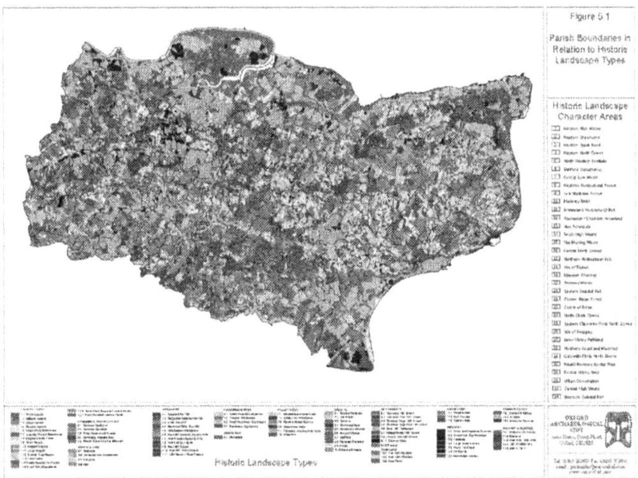

Figure 2. Kent (after Bramhill and Munby 2001): all HLC types, with parish boundaries superimposed

Figure 3A: Surrey HLC (after Bannister 2001): all HLC types (simplified)

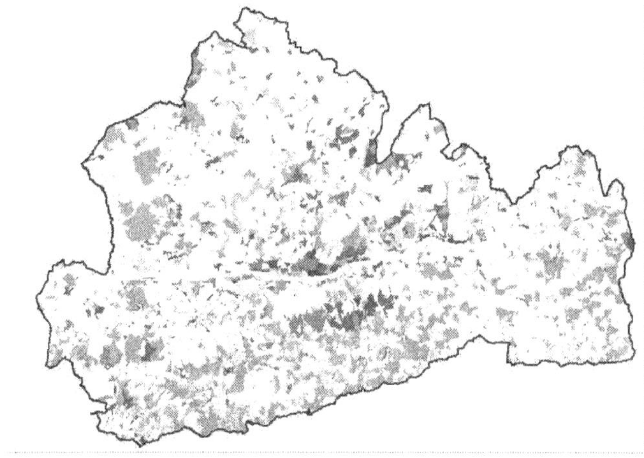

Figure 3B: Surrey HLC (after Bannister 2001): areas of predominantly pre-1810 historic landscape character

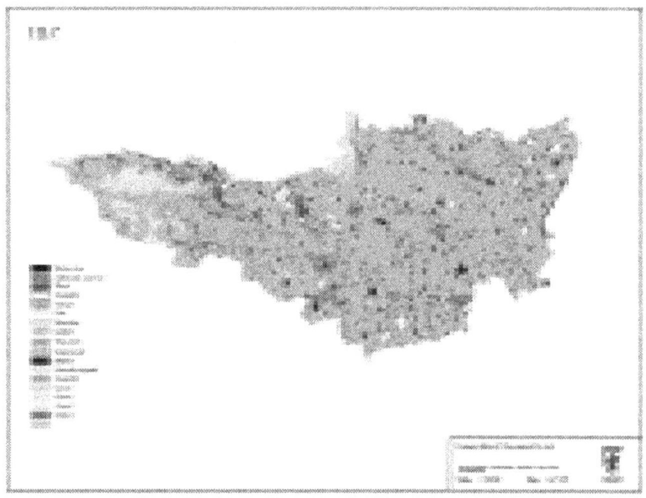

Figure 4a: Somerset HLC – all basic types, with enclosed land amalgamated as single type (after Aldred 2001)

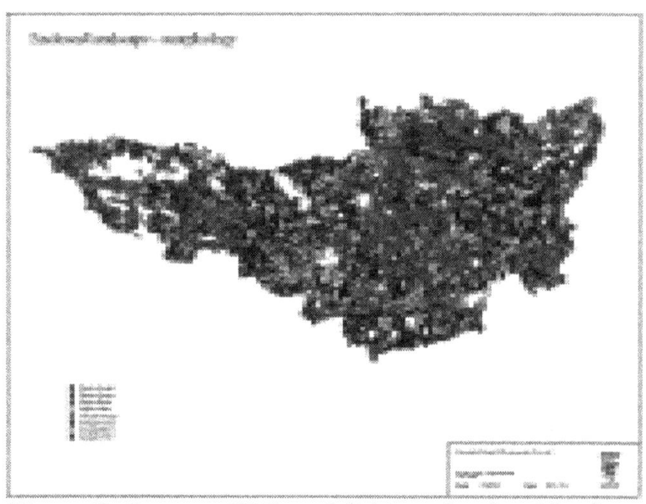

Figure 4b: Somerset HLC – subdivision of enclosed land types (after Aldred 2001)

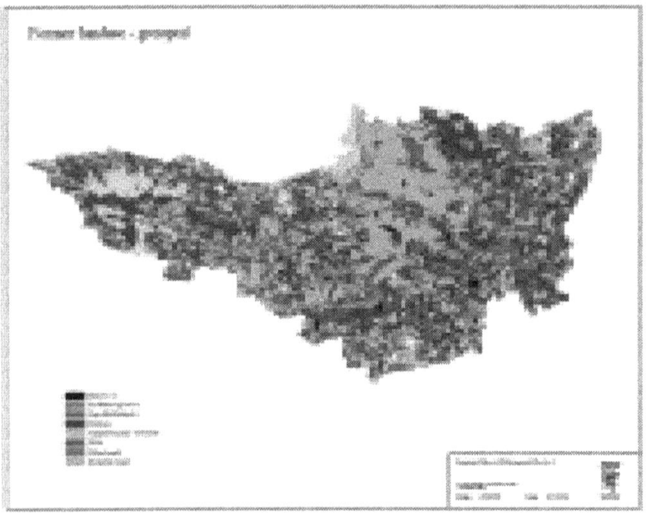

Figure 4c: Somerset HLC – model of previous landscape (time depth) (after Aldred 2001)

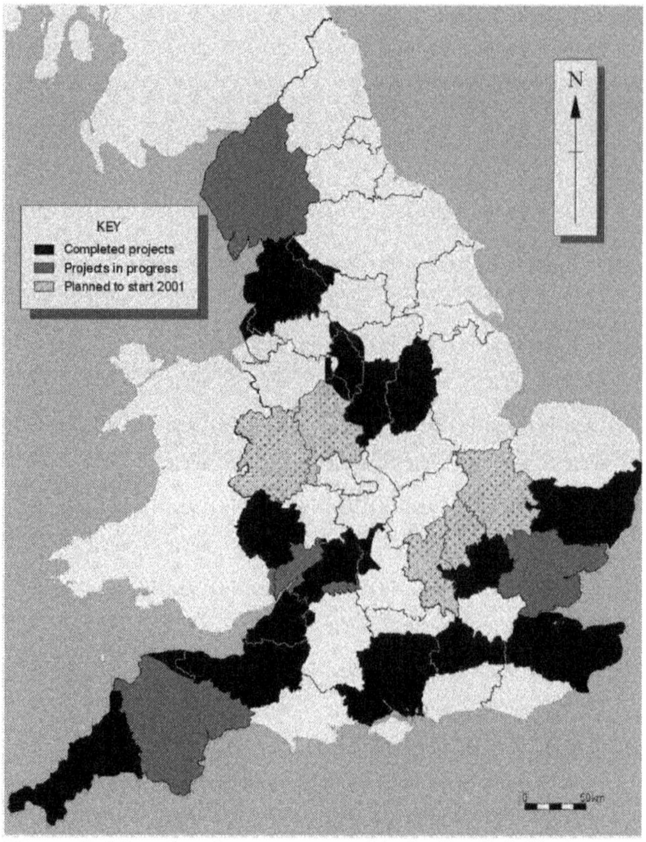

Figure 5. Map showing progress with the English Heritage programme of HLC (drawn by Vince Griffin).

Tables

Table 1: CORNWALL : Historic Landscape Character Zones

CORNWALL HISTORIC LANDSCAPE CHARACTER ZONES		
Zones (or simplified 'super-types')	**% area of Cornwall**	**Comments**
Anciently Enclosed Land	57.47	Amalgamation of the types Prehistoric Enclosures and Medieval Enclosures.
Upland Rough Ground	4.03	A subdivision of the type Rough Ground
Coastal Rough Ground	1.27	A subdivision of Rough Ground
Dunes	0.39	A topographical subdivision of Rough Ground
Recently Enclosed Land	17.44	Amalgamation of Post-medieval Enclosures and Modern Enclosures.
Anciently Enclosed Land extensively altered in the 18th and 19th centuries	0.58	Interpretation of elements of Prehistoric, Medieval and Post-medieval Enclosures.
Anciently Enclosed Land extensively altered in the 20th century	4.03	Modern Enclosures which are adaptations of Prehistoric and Medieval Enclosures.
Navigable Rivers and Creeks	1.22	Topographically derived zone using OS maps as source
Steep-sided Valleys	4.48	Topographic zone using OS maps as principal source.
Predominantly industrial	1.61	Amalgamation of Industrial (active) and Industrial (relict)
Urban development	3.83	Amalgamation of Settlements, historic and modern
Ornamental	1.32	Simple derivation from Ornamental type.
Recreation	0.34	Simple derivation from Recreational type.
Military	0.66	Subdivision of Military type, including airfields.
Upland Woods	1.13	Subdivision of Plantation and Scrub Woodland
Reservoirs	0.2	Simple derivation from Water, artificial.
Inter-tidal zone		

Table 2 : Historic Landscape Types used in the Hampshire HLC project (after Lambrick 1999)

HAMPSHIRE HISTORIC LANDSCAPE TYPES	
Categories	**Types**
1 Field Patterns	1.1 Small irregular assarts intermixed with woodland
	1.2 Medium irregular assarts and copses with wavy boundaries
	1.3 Large Irregular assarts with wavy or mixed boundaries
	1.4 Regular assarts with straight boundaries
	1.5 Enclosed strips and furlongs
	1.6 Regular form with wavy boundaries (?late med to17th/18thcentury enclosures)
	1.7 Irregular straight boundaries
	1.8 Regular "ladder" fields (long wavy boundaries subdivided by straight cross divisions)
	1.9 Small regular fields with straight boundaries (parliamentary type enclosure)
	1.10 Medium regular fields with straight boundaries (parliamentary type enclosure)
	1.11 Large regular fields with straight boundaries (parliamentary type enclosure)
	1.12 Variable size, regular fields with straight boundaries (parliamentary type enclosure)
	1.13 Not Used
	1.14 Prairie" fields (19th cent enclosure with extensive boundary loss)
	1.15 Fields predominantly bounded by tracks, roads, other rights of way
	1.16 Small rectilinear fields with wavy boundaries
2 Commons	2.1 Common heathland
	2.2 Common downland
	2.3 Other commons and greens
	2.4 Wooded over commons
3 Horticulture	3.1 Orchards
	3.2 Not Used
	3.3 Nurseries with glass houses
4 Woodland	4.1 Assarted pre-1810 woodland
	4.2 Replanted assarted pre-1810 woodland

HAMPSHIRE HISTORIC LANDSCAPE TYPES	
Categories	**Types**
	4.3 Other pre-1810 woodland
	4.4 Replanted other pre-1810 woodland
	4.5 19th century plantations (general)
	4.6 pre-1810 hangers (scarp & steep valley-side woodland)
	4.7 Post 1810 hangers
	4.8 Pre 1810 heathland enclosed woodland
	4.9 19th century heathland plantations
	4.10 Pre 1810 wood pasture
	4.11 19th century wood pasture
5 Heathland	5.1 Unenclosed heath and scrub
	5.2 Enclosed heath and scrub
	5.3 Purlieus and other enclosed heathland pastures
6 Downland	6.1 Downland
7 Valley Floor, water man't	7.1 Miscellaneous valley bottom paddocks and pastures
	7.2 Valley floor woodlands
	7.3 Marsh and rough grazing
	7.4 Water meadows
	7.5 Unimproved hay meadows or pasture
	7.6 Watercress beds
	7.7 Fishpond, hatchery complexes, natural ponds and lakes
	7.8 Watermills, mill ponds and leats
8 Coastal	8.1 Coastal wetlands
	8.2 Salt marsh
	8.3 Salterns
	8.4 Reclaimed land
	8.5 Harbours and marinas
	8.6 Shingle and dunes
	8.7 Mud flats
9 Settlements	9.1 Scattered settlement with paddocks 1800 extent

HAMPSHIRE HISTORIC LANDSCAPE TYPES	
Categories	**Types**
	9.2 Scattered settlement with paddocks (post 1800 extent)
	9.3 Common edge settlement
	9.4 Common edge settlement (post 1800 extent)
	9.5 Not Used
	9.6 Post 1810 settlement (general)
	9.7 Village/hamlet 1810 extent
	9.8 Not Used
	9.9 Town & city 1810 extent 9.10 Town & city post 1810 extent
	9.11 Caravan sites
10 Parkland & Designed	10.1 Pre-1810 parkland
	10.2 19th century and later parkland
	10.3 Deer parks
11 Recreation	11.1 Racecourses
	11.2 Golf Courses
	11.3 Major sports fields and complexes
12 Extractive & Industry	12.1 Active and disused chalk quarries
	12.2 Active and disused gravel workings
	12.3 Industrial complexes and factories
	12.4 Modern large scale industry (power stations; oil terminals etc)
	12.5 Reservoirs and water treatment
	12.6 Dockyards
13 Inland Communications	13.1 Station and sidings complexes
	13.2 Canal basin complexes
	13.3 Airfields
	13.4 Motorway service areas
14 Military and Defence	14.1 Prehistoric and Roman (eg hillforts, roman forts)
	14.2 Medieval (motte and baileys, ring works)
	14.3 Post medieval (1500-1830)
	14.4 19th century (1830-1914)
	14.5 20th century (1914-)

Table 3: Summary of data structure attached to polygons in Lancashire HLC (after Ede 2001)

1. **Pattern Recognition:**

- Habitat-based (woodland, moorland, saltmarsh etc.)
- Built (settlement, industrial, communications etc.)
- Enclosure of land

Code	**Wavy edged enclosures**	Code	**Coastal rough land**
E1	Small irregular	RL1	Dunes
E2	Other irregular	RL2	Saltmarsh
E12	Regular	RL3	Sand
	Straight edged irregular enclosures	RL4	Mudflats, shingle etc
E3	Small	RL11	Coastal rough ground
E4	Medium		**Heath/Moor incl. blanket peat etc**
E5	Large	RL5	Unenclosed
	Straight edged regular enclosures	RL6	Very large enclosures
E6	Small	RL7	Enclosed
E7	Medium	RL7A	Reverted fields since 1840s (assumed habitat change)
E8	Large		**Other unimproved land**
	Long, thin (3x long as wide) enclosures	RL9	moss - lowland, big enough areas
E9	Straight edge	RL10	grass/scrub etc, big enough areas
E10	Wavy edge		**Inactive industrial**
E13	Distinguishing features are: running at right angles to slope and paths and tracks running along edges. Some cross boundaries and loss of boundaries.	I1	Quarry
E13S	=<100m wide overall pattern	I2	Mine
E13L	=>100m wide overall pattern	I3	Other
	Clear Grid Layout		**Active industrial**
E15	Small	I5	Quarry
E16	Medium	I6	Mine
E17	Large	I7	Industrial estates
	Nurseries	I8	Other 'works'
E11	Small enclosures mainly with glasshouses	I9	Power generation
		I10	Docks
	Woodland	I11	Textile mill
WD1	Wavy edged		**Settlement**
WD2	Straight edged	S1	Settlement
	Recreation		**Ornamental**
R1	Golf course	O1	Ornamental landscape/parkland
R2	Caravan park		**Military**
R3	Playing fields/sports grounds/amenity land when marked as such on maps, and as amenity land on Phase I and large enough area or as Op on rlanduse coverage	M1	Active military
R4	Racecourse	M2	Inactive military but still present as an influence in present day landscape
R5	Holiday camp		
	Communication		**Water**
C1	Rail	W1	Inter-tidal water
C2	Canal		**Fresh water**
C3	Motorway		Man-made
C4	Other large road	W2	Reservoir
C5	Airport/Airfield	W3	mill pool
C6	Telecommunication station	W4	flooded quarry
C7	Toll roads now not large main roads only if info easily available	W7	lake/pond
	Parliamentary Enclosures (PE)		**Natural**
PE	Enclosed by Parliamentary Enclosure	W5	Lake
anc	Shown as anciently enclosed land on PE docs	W6	River

Other attributes recorded against polygons:

- Size of field or other enclosure: Small = less than about 2-3 ha
  Medium = (2-3) - about 16 ha
  Large = over 16 ha

- **Change and condition**
- **A** - altered 1840-1990 but some features remaining
- **Z** -altered a lot
- **M** - variety of patterns visible on 1st edn now altered a lot
- **W**-water boundary

- For settlement and urban:
- **MH** - mills and housing
- **R** - residential
- **MC** - municipal centre, shops

- Topography of WD coded polygons: **Y** if on steep slope (only for WD codes)

2. **Historical maps**, eg 1st Edition OS maps (mid-19th century), to sub-divide the areas into pre- and post- mid 19th century.

3. **Subjective interpretation**: additional codings for

- **dating** **1** - prehist-RB
  2 - pre 1600
  3- 1600-1st ed
  4 - post 1st ed
  5 - no date indicated
  6 - mid 18th /early 19th c
  7 - pre 1st edn.

- **identification of processes** of landscape development. interpretation of dating was applied to each polygon eg-

fossil strips	Enclosed from:
old deer park	open field
old parkland	Woodland
single farm	lowland moss
rabbit warren	Moorland
meadow grazing	the sea
Ornamental woods associated with park	by Parliamentary Enclosure

- **confidence level of interpretation,** applied to **dating** and **process**, for each polygon.

Corinth Computer Project: Internet Education

David Gilman Romano
Mediterranean Section
The University of Pennsylvania Museum of Archaeology and Anthropology
Thirty-Third and Spruce Streets
Philadelphia, Pennsylvania 19104-6324, USA
Phone: (215) 898-4437 - Fax: (215) 573-2906 - E-mail: dromano@sas.upenn.edu

Nicholas L. Stapp
Mediterranean Section
The University of Pennsylvania Museum of Archaeology and Anthropology
Thirty-Third and Spruce Streets
Philadelphia, Pennsylvania 19104-6324, USA
Phone: (215) 898-3064 - Fax: (215) 573-2906 - E-mail: nls@sas.upenn.edu

Abstract: Since 1995-1996 the Corinth Computer Project has maintained a web page that has come to serve a variety of purposes. The current site http://corinth.sas.upenn.edu has become a form of preliminary publication for the research project, as well as serving as a vehicle through which to share preliminary research results with colleagues and students. In addition to these uses, the web page has become an important teaching tool for Classical Studies and Archaeological GIS courses taught by the senior author at the University of Pennsylvania. Graduate seminars including PC virtual conferences have stimulated the interest of students and faculty alike and student related projects have been incorporated into the design and content of the web page in a number of different ways. A developing focus of the web page design has been the interest in educating the public-at-large, including elementary, junior high, high school as well as college and university students whenever possible. A number of features of the current web page are presented and discussed including the interactive archaeological state plans, virtual movies, virtual 3D fly- throughs and an example of children's educational activities.

Key Words: Elementary education, high school education, university education, GIS, HTML, AutoCAD WHIP, Apple Quicktime, virtual conference, remote sensing, Corinth, Roman cities, city planning

Introduction

Since 1996 the Corinth Computer Project has maintained a web page that has come to serve a variety of purposes. Originally a vehicle for summarizing and publicizing the research and methodology of the research project, the web page now identifies education as a primary component of its design. The Corinth Computer Project as a long-term research project, 1987- present, studying the evolution and planning of the Roman city of Corinth, has grown from a fairly straightforward computerized architectural and topographical survey of the city, looking for the contributions of the Roman surveyors to the city plan, to a broader analysis of the ways in which the Romans planned and developed the urban and rural aspects of the two successive Roman colonies at Corinth.[1]

The Corinth Computer Project is a research project of the Mediterranean Section of the University of Pennsylvania Museum of Archaeology and Anthropology, D.G. Romano, Director.[2] The summer fieldwork, 1988-1997, was carried out under the auspices of the Corinth Excavations of the American School of Classical Studies at Athens, Charles K. Williams, II, Director. The Corinth Excavations, underway since 1896, have successfully uncovered vast portions of the ancient cities of Corinth.

Archaeological, historical and literary evidence about the ancient city has been combined with elements of the current research project in the creation of the current web page to provide a fairly comprehensive introduction to the Roman city and, recently, to the Ottoman city. One of the long-term goals of the research project is to bring together a series of maps of the successive cities of Corinth to create a diachronic study of the evolution and development of the city over three millennia. Although the web page has gradually developed into an important educational tool related to the history and archaeology of the city of Corinth, there have been several evolutionary steps to its design that should be discussed.[3] Over the past five years, the web page has grown and expanded to meet not only the needs and goals of the project but also the demands of the interested public.

History of the website

To date, there have been two versions of the web page. The first version, http://ccat.sas.upenn.edu/~dromano/corinth.html was created during 1996 and released January 22, 1997, and the second, http://corinth.sas.upenn.edu created in 1999 and released March 2000. Both versions of the web page were designed, created and written by the undersigned authors with the assistance of several students, as Research Assistants, of the

Corinth Computer Project.[4] The whole idea of creating a web page for the project was one that was heavily debated between the authors before work began. The Director was concerned that the time and effort of creating and maintaining a web page would take away from the progress of the ongoing important research of the project, but he was finally convinced of the potential worth of the web page as a form of preliminary publication for the project.

After the decision was made to create the web page, time had to be dedicated to learning the basic processes for web page creation and teaching these skills to others, specifically the programming language of HTML. Creating a web page was a huge undertaking at the time for a low budget research project staffed solely by a handful of humanities-based undergraduate and graduate students typically working 5-10 hours per week. The students had to learn how to convert aspects of the published data and images to web format. This task was very time consuming and, as a result, the progress of the overall research of the project suffered. It was decided in the beginning of the web page project that the finished web page would not be posted on the internet in a piece-meal fashion, but rather that it would appear more or less as a whole in a completed form. The resulting time required to create a complete version of the web page was over one year for each of the two successive web pages. This time included not only the programming time but also the time to receive preliminary feedback about the web page from students, colleagues and interested friends.

Since the Corinth Computer Project did not own a server it was necessary to find a local server to store the data for the web page. This location was found within the School of Arts and Sciences of the University of Pennsylvania; the first generation page was located on the server at the Center for the Computer Analysis of Texts of the School of Arts and Sciences, and the second generation page is currently located on the server for the School of Arts and Sciences. We are grateful to both for their willingness to house the web page. Since its creation both successive web pages have been linked integrally to the web page of our home institution, the University of Pennsylvania Museum of Archaeology and Anthropology, http://www.upenn.edu/museum, as well as to many other web pages around the world.

The First Generation Web Page

The first generation web site was fairly simple and static. An attempt was made to combine the use of our technological tools, digital cartography, GIS, Remote sensing and 3D modeling, into the web page. This was accomplished by inserting a fairly large amount of text, combined with multiple static photographs, a 3D fly-through, and screen captures of differing elements of the research. The screen-captures included a low level air photograph, a satellite image and several landscape reconstructions based on our own DTM's and DEM's. The 3D fly through of the 35 km^2 of the central study area including Akrocorinth had been created in 1993 using AutoDESK 3D Studio with a DTM based on topographical map contours.

The result was moderately successful given the time spent and the resources available. Reflecting later on the content and format of the first web page, the color selection was poor, and the organization was coherent if not always logical. The audience had been expected to be largely academic, but it turned out to be much more varied. We attempted to design the first version of the web page with the goal of providing an easy to read summary of the project's methods and results. During the beginning of the research and design phase of the first web page, most of our energy was concerned with questions relating to copyright, publication infringement, and permissions as well as coping with the major learning curve of a programming language. It wasn't until the latter stages of our endeavor that we seriously contemplated our audience. Initially, we assumed that the web page would serve university level students and scholars who would have an interest in Corinth or the Classical world. It was not until the end of the first generation web page project that we considered the notion that the public at large might share in our enthusiasm for the research.

The organization of the original site was divided into categories, which were translated into navigation buttons in the following format: Main Page, Corinth Basics, City Plan, Methodologies, Landscape, Acknowledgments, Glossary and Bibliography (fig. 1). From the beginning of the enterprise, the sections of the web page devoted to "Methodologies" have been of great interest to the authors and a category that received full attention at the outset. The "Glossary" was a later addition that was added to help orient the public to a largely scholarly endeavor by adding definitions to words of interest to those who study ancient Roman city-planning. The "Bibliography" section was intended to list the publications that have been produced as a result of the work of the Corinth Computer Project.

During the life of this web page, 1997-1999, over 17,000 unique hits to the web page were registered with our server. We received numerous emails from around the world, highlighting the wide range of visitors to the web page. On the whole, the emails were supportive, constructive and educational for our own curiosity about who was using the web page. This information eventually influenced the creation of the next generation web page. Many of the questions that were received focused on a need for more general and background information, ranging from ancient historical questions to queries about modern methodology.

The Second Generation Web Page

Founded upon the strengths of the first web page, the second web page capitalized upon newer technical innovations of the internet and of web programming.[5] The second version of the web page included several improvements in content and presentation that enhanced the site in usability, education and overall appeal. An organizational chart was created to logically incorporate new research and technology, a broader discussion on the historical and technical background of our research was added and the authors' made a concerted effort to design the web page in a user-friendly format. This included rearranging the color scheme of the entire web page, making the text black and the background white (so users could print off a page without using excessive printer ink). Navigation through the web page was a major issue and we attempted to incorporate historical and archaeological relevance into navigation, by the use of an image of a tombstone of a Roman surveyor as the navigation bar icon.[6] Imagery content was increased, since some comments received about the first web page focused on the poor image resolution of some of the photographs as well as an interest in additional images. A more detailed discussion of the history and methodology of the project and other related scientific techniques were added, broadening the appeal of the site. Other additions included a map that clearly showed the spatial relationship of the successive Greek and Roman cities of Corinth.

A whole section was added to the first generation web page in 1998 and was subsequently improved upon in the second generation web page, being devoted to a later historical period of time, the seventeenth through the early nineteenth centuries AD, during the Ottoman occupation of Corinth (fig. 2). This portion of the web page was conceptualized and written by Ms. Leslie Kaplan, currently a PhD student in Folklore at the University of Pennsylvania, working on a related project for her dissertation. This aspect of the web page brings together literary and historical accounts as well as contemporary maps, plans, photographs and drawings of European Travelers who visited Corinth during this period of time. As such, it complements the work of the Corinth Computer Project that focuses its resources on the development of the Roman city.

Graduate Level Courses

A graduate level seminar, 'City and Landscape of Roman Corinth,' Classical Studies 625, was taught by D. G. Romano at the University of Pennsylvania during the fall semester of 1998. N.L. Stapp assisted in the course.[7] This was a seminar that focused upon the architecture of the Roman forum, and the relationship of the individual building to the Roman city plan. The students were given specific Corinth forum buildings to research, and some of them drew a ground plan of their building in AutoCAD, either representing its archaeological actual-state or its restored plan. The students were responsible for preparing a seminar paper on their building as well as delivering an oral report to the class on the same subject. A series of guest lecturers spoke about different aspects of the study of the Roman city. One of the guest lecturers for the class, Dr. Guy Sanders, Director of the Corinth Excavations of the American School of Classical Studies at Athens, participated by means of a PC video conference from Athens. This meant that the students in Philadelphia were able to directly interact with the Field Director of the ancient site from Greece and this led to some lively and interesting discussion. This seminar proved to be highly successful, with some of the results being incorporated into the second version of the web page. In each case the student's contribution, whether it was a textual building summary, an actual-state drawing or reconstruction is duly noted and credited.

Interactive Actual-State Archaeological Drawings

The study of the individual buildings of the Roman forum are a part of the consideration of the way in which the Roman city was put together. For this concept a portion of the web page has been devoted to the 'City Center, ca. 150 A.D.' This image shows the location and organization of the major public, civic and religious buildings from the center of Roman Corinth, ca. A.D. 150 (Fig. 3). The building plan is based on the drawing by C.K. Williams, II from *Hesperia* 56, 1987, fig. 1, and represents the results of fieldwork, study and publication of Williams and the Corinth Excavations over many years. The city plan has been supplemented by the results of the computerized architectural survey of the buildings and monuments of the ancient city by the Corinth Computer Project. This information relates to their location and orientation and the resulting Roman city grid (1 actus wide *insulae*).

For the purpose of the web page, the center of the Roman city is drawn and the religious buildings have been colored magenta, the markets yellow, springs and fountains blue and the places of assembly cyan. Green represents civic and political buildings and structures. Monumental arches are red and the grid of the Roman city is seen as grey. Each of the buildings in the forum area can be clicked on and the link will take the viewer to an individual page that highlights the building, providing drawings, photographs, interactive archaeological drawings and literary references.

This section of the web page was an exciting technological advance for the project. Using new web programming technologies, AutoCAD WHIP, it was possible to include in the web page a vehicle for users to access digitized actual-state drawings of specific buildings of the forum of Corinth (Fig. 4). Each of these drawings is based on the published actual-state drawings in the Corinth Excavations series. For instance, the odeum, illustrated in figure 4 is based on the drawing from Oscar Broneer, *Corinth X, Results of Excavations Conducted by the ASCSA, Cambridge, 1932, plate 11. A significant methodological advance was also incorporated into the second web page; a coding system for every drawn line contained within the individual architectural/archaeological plans of the individual buildings of the Roman forum. This system identifies material, chronology and function, all in one line. This means that the viewer can literally interact with the architectural drawings and look for information related to chronology, material or function of the building in question. Additionally, a brief historical write up of each building with modern and historical photographs and bibliographic references is included. This produces a full educational tool, not only geared towards scholars but to students of all ages.

Virtual Reality Highlights

Another new development of the second generation web page was to include more imagery highlighting the accomplishments of the research of the project (fig. 4). This section was designed to orient the user to the vast array of data on the web page. This section also provides a detailed step-by-step approach to using and viewing the data. These sections include the tutorials and downloadable internet plug-ins for interactive archaeological plans, 3D modeling, and 360-degree panoramic movies.[8] The internet plug-ins include AutoCAD WHIP and Apple Quicktime. The tutorial section includes demonstrations of each area of the web page, which requires special technical attention. For example, the panoramic movie area, not only requires a special internet plug-in but the actual use of the application needed to be explained.

Technical Explanations

The authors felt that it was important and necessary to include a detailed discussion of the technology incorporated in the second generation web page. It was recognized that all users do not have the technical ability to navigate or view many of the new innovations in the second web page; therefore, a simple description was provided for the two plug-ins, AutoCAD WHIP as well as Apple Quicktime and links directly to the vendors for the plug-in. In this way users could download the plug-in quickly and continue to use the Corinth web page.

Specialized Sections

During the year 2000, recognizing the interest in Roman gladiators and amphitheaters in general due to the movie 'Gladiator' it was decided to produce a section of the web page highlighting the Roman amphitheater in Corinth. An amphitheater was an important part of the design of any Roman city and this was also true of Corinth. During the summer of 2000 a web page dedicated to the history and function of the amphitheater in Corinth was introduced. The page consists of a physical description of the unexcavated structure and a discussion of its location within the planning of the Roman city. Also included is a brief discussion of the gladiators who would have participated in the festivals held in the facility. Historical photographs of the amphitheater were posted to the web page together with a historical drawing made of the amphitheater in 1830 by a French scholar. Together with this information was added a series of GIS images that simulate 3D views of the remains of the unexcavated amphitheater. Linked to the page is a National Public Radio Real Audio broadcast, of station WHYY, of May 6, 2000, featuring Professor Brent Shaw, Department of Classical Studies, University of Pennsylvania, speaking about Roman Gladiators. There is also a bibliography specifically geared for gladiators in general and the amphitheater at Corinth as well as suggestions for further reading.

The second web page continues to be a success, receiving a higher volume of visitors in a shorter period of time than the first web page (18,000 unique visitors during the first year or approximately three times the rate of the first generation web page). Additionally, we have received reviews from Science Magazine, Archaeology Magazine and the Philadelphia Inquirer and a second place national GIS award from Geospatial Solutions Magazine. The physical size of the web page has expanded, from approximately 12-15 MB for the first web page to a current size of ca. 25-30 MB for the second web page.

Educational Directions

The web page has been used in a number of different ways for educational purposes. As a detailed summary of the research and methodology of the project, the web page has served as a useful tool to explain the project to a scholarly audience. As a useful tool, the Project included a "links library," which is divided into categories of historical and archaeological interest. In addition to this obvious purpose there have been several other kinds of educational uses for the project. The greatest number of hits to the web page {of which the authors are aware) have come from two sources: elementary/junior high school students and religious scholars. From elementary/junior high students we have received a large number of general inquiries about the research project but also questions about the history of Corinth, ancient Greek customs and daily life in ancient Greece in general. Some of these are specific questions that have to do with the history of Corinth but many of them are more generic questions that have to do with ancient Greece. For instance, "what was the role of women in ancient Greece?" Other questions related to Corinth but were of a more general nature, for instance, "How did the lives of ancient Corinthians differ from those of ancient Athenians?" and "How was Corinthian pottery manufactured?" Other questions were focused on more specific aspects of the city of Corinth. For instance, "How often did Corinthians walk up to Akrocorinth?" Whereas none of these questions are answered by any of the elements of the web page, each of the individual questions was answered by e-mail.

The authors were not expecting the level of interest that was received by the religious community. On both national and international levels, classical scholars, religious scholars and armchair biblical tourists have communicated their interest for more information about Corinth, especially in the time of St. Paul, in the mid-first century A.D. Although the Corinth Computer Project and its web page does not specifically focus on the biblical aspects of Corinth, both the maps and plans of the ancient city as well as the building descriptions and the ancient references included within the web page have been found to be of interest to many religious groups.

The second generation web page has been made reference to by Encarta, Microsoft Corporation's digital encyclopedia, under the entry 'Corinth' which has increased the visibility of the web page.[9] Additionally, as mentioned above, the scientific and academic merits of the web page were highlighted in Science Magazine.[10] As a whole, the authors intend to offer feedback to all questions received about the web page and the research pertaining to it. Additionally, modifications continue to be made to the site, offering visitors the latest version of our research, making it a tool for the classroom, the scholar or the enthusiast and providing a resource that can be referenced for future research.

The Future of the Corinth Computer Project Web Page

As we plan for the future of the Corinth Computer Project, we continue to imagine innovative ways to develop the web page as a complete educational tool. More of the research data of the project will be made available to the public as time goes on. The public will likely play a greater role in the overall design, offering a combination of academic excellence and public interest. These plans would include expansion of existing areas, including digital cartography, GIS, 3D modeling, photography and 360-degree movies. Additionally, more educational sections for children will be added, offering exercises designed to make the student think about the Greek or Roman city while being distracted by a task or a game. Examples might include a Corinthian coloring book in which students could download a blank image of a building or monument to be colored in, or a city planning game where students would be able to design their own city using the 'component' elements of a Greek or Roman city. Incidentally both of these suggestions were made by 12-13 year old Philadelphia school children when asked what would they like to see as a future component of the Corinth Computer Project web page. Future additions might also include a complete image library, a textual archive, a larger selection of interactive plug-ins and possibly a member's section, where certified users could have access to restricted data.

Conclusion

The original worry and caution about the time commitment required to create a Corinth Computer Project web page was fully merited. In fact, the time dedicated to the web page has far exceeded even the most pessimistic estimate from 1996, partially due to the fact that the first generation web page was only the beginning of a larger, ongoing and probably never-ending web improvement project. Nevertheless the worth of the project has been demonstrated by several measures.

Probably the most valuable aspect of the web page project from the standpoint of the research is that the web page has provided an attractive and immediate form of preliminary publication. This has served a number of purposes for the scholarly community as well as for the public at-large. Furthermore, the web page has created a vehicle for the distribution of accurate, if summary, information about the Roman city of Corinth. This has proved useful in teaching graduate seminars at the University of Pennsylvania, but also it has been used by elementary, junior high and high school students around the world. Perhaps the most exciting aspect of the web page for the authors has been the ability to be in direct touch with others who find the subject of Roman Corinth both interesting and important.

Endnotes

[1] Earlier publications relating to the methodology and research of the Corinth Computer Project have been published in the following CAA publications: David Gilman Romano and Osama Tolba, "Remote Sensing, GIS and Electronic Surveying: Reconstructing the City Plan and Landscape of Roman Corinth," in Computer Applications and Quantitative Methods in Archaeology 1994, Jeremy Huggett and Nick Ryan (eds.), BAR International Series 600, 1995, pp. 163-174. David Gilman Romano and Osama Tolba, "Remote Sensing and GIS in the Study of Roman Centuriation in the Corinthia, Greece," in Interfacing the Past: Computer Applications and Quantitative Methods in Archaeology 1995, Hans Kammermans and Kelly Fennema (eds.) Leiden, 1996, pp. 457-463.

[2] The most recent article that summarizes many of the archaeological and historical results of the project is D.G. Romano, "A Tale of Two Cities: Roman Colonies at Corinth, *Colonia Laus Iulia Corinthiensis* and *Colonia Iulia Flavia Augusta Corinthiensis*," in 'Romanization and the City, Creation, Transformations and Failures,' Elizabeth Fentress, ed., *Journal of Roman Archaeology*, Supplement 38, Portsmouth, Rhode Island, 2000, pp. 83-104.

[3] A forthcoming CAA article will discuss some of the technical aspects of the second generation web page, including the layer conventions and the interactive archaeological plans. D.G. Romano and N.L. Stapp, "Stone for Stone in Roman Corinth," in Computer Applications and Quantitative Methods in Archaeology, 1999, Dublin, Ireland.

[4] The first generation web page was the result of work and design by N.L. Stapp and W. Petrofsky, under the direction of D.G. Romano. The second generation web page was designed by N.L. Stapp under the direction of D.G. Romano

[5] For the first web page, the Project solely used HTML. The second generation of the web page, the Project used HTML, Java, DHTML, Visual Basic and Macromedia Flash.

[6] Permission for use of the Roman surveyor tombstone image was received from the Ivrea Museum in Northern Italy.

[7] N.L. Stapp was responsible for teaching the students AutoCAD and assisting in other technical issues relating to the student projects.

[8] There is one Quicktime movie included in the second generation web page that illustrates a 360 degree photographic panorama of the Roman forum as it exists today.

[9] http://encarta.msn.com/find/Concise.asp?z=1&pg=2&ti=761574743

[10] http://corinth.sas.upenn.edu/awards.html#science

Figures

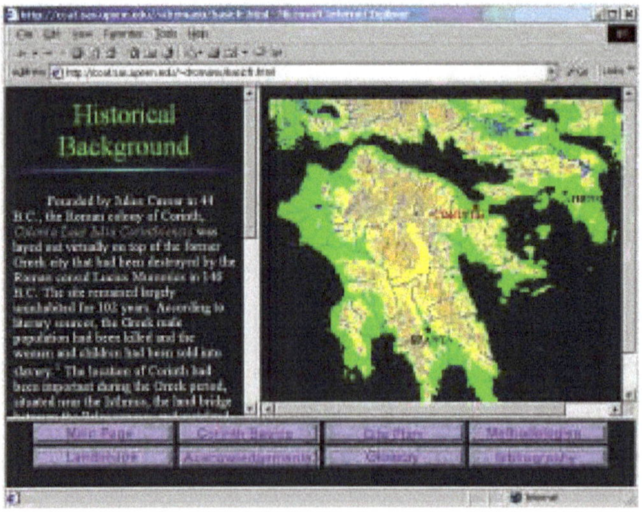

Figure 1: Example from the first generation web page, ca. 1996.

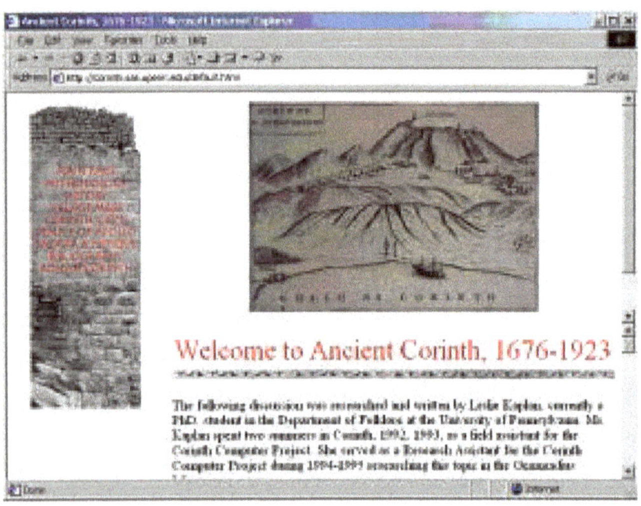

Figure 2: Introduction to Leslie Kaplan's 17th – 19th century Corinth web page.

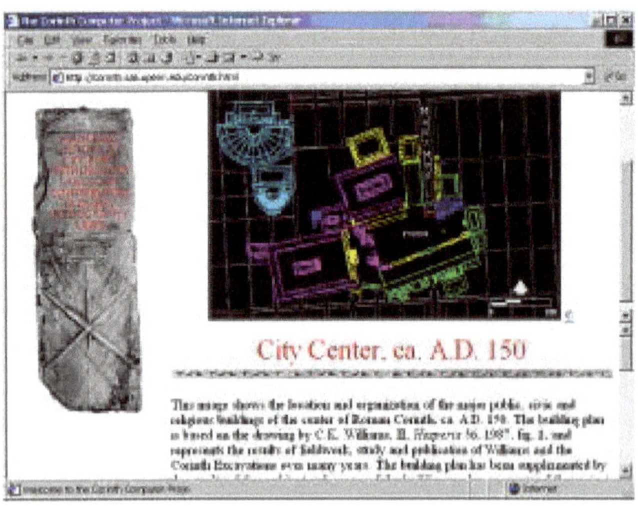

Figure 3: Clickable City Center, ca. AD 150.

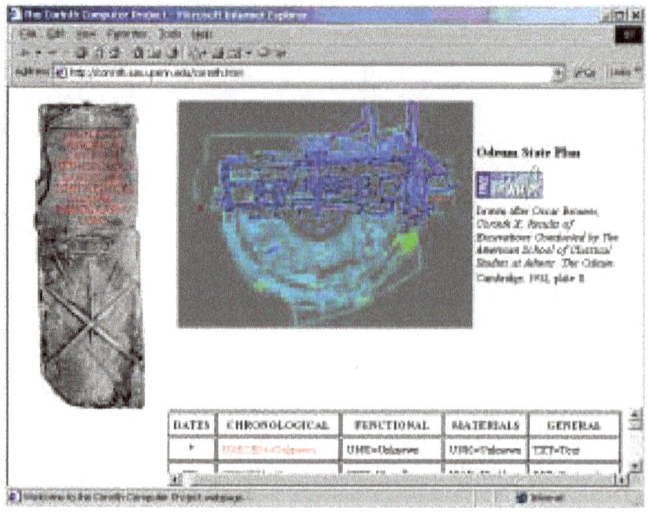

Figure 4: Example of the "Interactive State Plan" section of the web page.

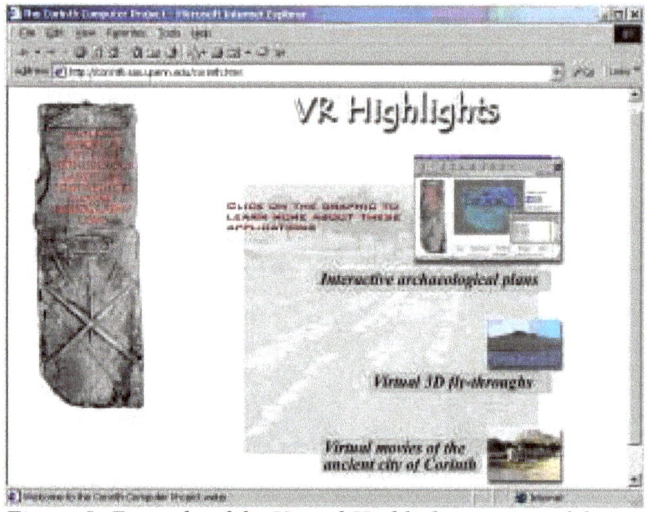

Figure 5: Example of the Virtual Highlights section of the second web page.

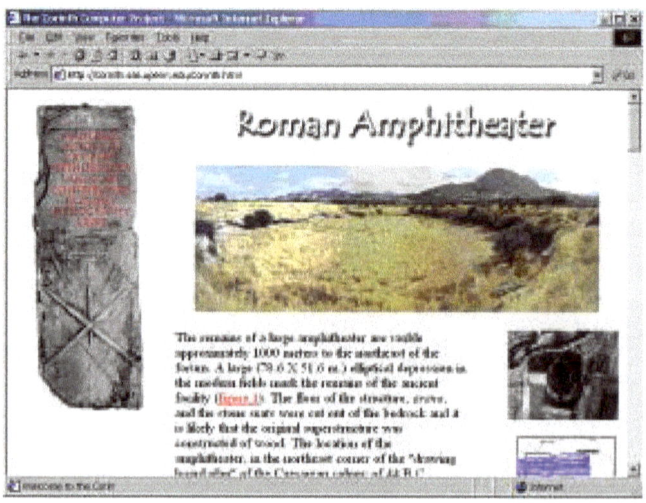

Figure 6: Detail of the section on the Roman amphitheater in Corinth.

Desktop - Photogrammetry and its Link to Web Publishing

Günter Pomaska
FH Bielefeld, University of Applied Sciences
Bielefeld, Germany
E-mail: gp@imagefact.de

Abstract: Dimensional analysis from images seems to be a field for experts involved with special knowledge and equipped with high performance tools. Since softcopy workstations replaced the analytical plotters it is not the case further on. Photogrammetry has moved to the desktop and is incorporated into CAD or image processing software. Recording is provided amongst others with digital cameras of the shelf.

Requirements to recording and documentation of archaeological sites have reached a high standard. Results have to be published not only to researchers or experts. It is important to reach a wide community for generating interest and education in history. The Internet provides this platform. Photogrammetry can be an efficient tool for on site recording as well as for exciting documentation.

This contribution focuses to some aspects of digital photogrammetry particular camera handling, recording and evaluation methods and the products one can receive from photogrammetry.

Key words: Photogrammetry, image refinement, single image evaluation, on-eye stereo, multi image photogrammetry, image objects, texture mapping, VRML description

Photogrammetric Image Acquisition

Photogrammetry is defined as recording, measurement and interpretation of 3D-objects from images with non-contact recording systems. The major application of photogrammetry is mapping from aerial photos. Non-topographic photogrammetry or close-range photogrammetry covers applications in archaeology, architecture, forensic science and industrial metrology. To produce true scale drawings from images the exposure arrangement has to be reconstructed first. Photo positions and camera parameters must be known before a detailed evaluation can be performed. Accuracy results beside other conditions from image scale and image measurement accuracy. The former analogue and analytical plotters are replaced by "softcopy workstations" today. Close-Range applications benefit from utilizing standard cameras for metric applications. A sensor supplies from a surface of 2/3 of an inch millions of pixel including colour information. The Workstation can be a notebook computer.

Figure 1 displays a "Meydenbauer camera" from 1890 designed for terrestrial applications. The image format of this camera is 30cm x 30cm, carrier for the film emulsion was glass. Another example for photogrammetric equipment is given in figure 2. The Stereoplanigraph C8 from Zeiss Germany produced 1967, connected directly to a drawing table. Driving the floating mark in the photogrammetric model was recorded directly in true scale line drawings. Today the results are presented as 3D-CAD models, orthophotos, hierarchical polygon models, high-density point clouds or models in Web-formats.

Image Refinement

Many photogrammetric evaluation methods are based upon a central projection model. A point is projected by a straight line through the projection centre (camera lens) onto the image plane. Deviations from this model can be parameterised. Without covering the details some remarks about co-ordinate systems, camera calibration and image refinement are necessary here. The location of a 3D-object is referenced in an object co-ordinate system. Image measurements, which supply after some computations the object point, are taken in an image co-ordinate system. A reference to the image co-ordinate system is given by the camera's format frame (not good), by fiducial marks (better) or by réseau (best) that is a grid plate in front of the film surface. Digital cameras include the reference in the pixel positions. The difference between the origin of the image co-ordinate system and the reference system is given by the principle point position.

Image deformations result from insufficient film flatness and radial and tangential lens distortion, which cause from variations in angular magnification and misalignment of lens components. Film flatness can be controlled and numerically corrected using a réseau-camera. Camera calibration by means of bundle adjustment provides the model correction parameters. Particular if wide-angle lenses are used a transformation to the principle point and radial lens distortion correction is necessary. The process is known as image refinement. Radial lens distortion can be simply detected by observing straight lines of the object shown as curves in the image. A stereoscopic image pair of a plane surface gives an example. Without image refinement, the plane is presented as a curved surface. Image processing programs enable subtraction of two images. The difference between a distorted image and the according refined image illustrates the influence of distortion as well.

Strategies for camera calibrations should not discuss here. Figure

3 is an image of a plain calibration field, taken with a 2-megapixel camera. The radial lens distortion curve of this camera is shown in figure 4. The amount of distortion in the image corner is about 40 microns. Please note that this value in object space increases with the factor of the image scale.

One mega pixel camera, which corresponds to the metric requirements, is the Rollei d7 metric[2]. The calibration parameters are stored in the digital image. The original image data is protected in the raw data format that cannot be manipulated. Image refinement is processed in advance of the photogrammetric evaluation.

Object resolution

The discussion of image resolution using digital cameras and in case of film scanning as well should be directed to object resolution. What object area should be covered by one pixel? The image scale is given by comparison of object size with the according image size. Full format photographing an object of 50-metre extension with a medium format camera supplies an image scale of 1:830. For a 35mm camera this would be 1:1400, a 2/3 inch CCD displays the object with a scale of 1:2950.

Here is an example from the point of view of accuracy calculated only by the thumb rule that the object accuracy is derived from the image measurement accuracy multiplied by image scale. Measurements from images should deliver an object accuracy of 1 cm. The sensor used is sized 2/3 of an inch with a resolution of 1800 pixel. Pixel size is approx. 10 microns, equal to the estimated measurement accuracy. Image scale should be 1:1000 that means an object size of approx. 16 metres could be imaged. But do not forget the scanning theorem that the scanning frequency has to be at least twice as the local frequency. Desktop Scanners with film adapter units are equipped with optical resolutions of 1200 ppi and therefore usable for formats from 35 mm to medium format and larger. 35mm film scanner supply resolutions up to 2700 ppi optical.

To summarize: A medium format camera in combination with a desktop scanner supplies the same image resolution as a digital camera of the shelf. A disadvantage of the digital camera is the small format. To fulfil high accuracy requirements, a réseau camera provides best results. Every photogrammetric evaluation software package needs a camera description file including, focal length of lens, principle point position, lens distortion parameters, image format width and height, shape and position of reference for image co-ordinate system.

Photogrammetric recording and evaluation strategies

The number of photos taken and the method used for object reconstruction can classify photogrammetric methods. Single image methods require some object information. Two photos taken as a stereo pair represent a model for evaluation based on parallax measurements. Several models are combined by adjustment. Multiple images enable point computation by intersection and over determination.

Single image rectification

Single image methods need additional information about the object shape. If the object shape is a plane surface, using the relationship between object plane and image plane can perform rectification. The perspective projection parameters can be calculated by four given control points in the object plane. Relative positions of these points in the image must be measured. Architectural applications can benefit from parallel lines in the object. Two orthogonal located pairs of parallel lines have to be marked in the image. The geometric condition will be restituted after rectification. Results of image rectifications are high-resolution image maps. Image rectification in this way works without metric camera modifications and unknown focal length.

Multi Image Potogrammetry

Multi image photogrammetry works with convergent exposure arrangements. Object co-ordinates are computed by intersection methods. An object is completely covered by an image bundle. The photo positions have to be calculated by best-fit algorithms. Multi image photogrammetry supplies geometric object information with high accuracy and reliability. Interactive detailed evaluation enables production of high quality large-scale line drawings. There are no limitations to camera positions. Multi image photogrammetry needs at minimum 2 photos; applications with several hundreds of images are known. Figure 6 shows the principle of multi image photogrammetry.

Stereoscopic Photogrammetry

A stereoscopic image pair corresponds to the human viewing system. Reconstruction of the object is given via disparity or parallax measurements. Using a stereoscoping viewing systems enables visual evaluation by driving a floating mark in the stereoscopic model. Automatic object reconstruction is possible by means of correlation. In this case monoscopic viewing is sufficient, systems are called "one-eye stereo systems" and represent a major advantage of digital photogrammetry.

Orthophoto

Particular the digital orthophoto is an example for automatic image evaluation. A digital surface model is calculated automatically. Photos are resampled from perspective view to orthogonal projection taking into consideration of the predefined surface.

Numerous software packages for PC-workstations like RolleiMetric CDW, RolleiMetric MSR, Elcovision 10, Photomodeler, Photo3D and others are available for object reconstruction from photos. Conventional photogrammetric products are line drawings, CAD-Models, point-clouds, image maps and orthophotos. With the ongoing development of Web technologies the demand to new products like image objects, panoramas, digital 3D-models for interactive navigation increases. Hierarchical polygon models replace model descriptions by contour lines. Texture mapping became more popular. In the following chapter some methods of image based documentation and the according Web publishing techniques are introduced.

Web Publishing

3D-Models and Texture Mapping

With the definition of VRML virtual modeling language a standard for interactive inspection of 3D worlds is designed. For increasing the reality imagination, the geometric model is covered with texture. A texture is a 2D bitmap. Every 3D corner point of a surface of the geometric model is connected with a 2D point in one of the texture files. Current photogrammetric software packages include geometric modeling and texture extraction from photos. Since not all applications require the high accuracy low end products like Photo3D provide solutions to get a VRML model from standard 35 mm photos or digital cameras. Figure 8 displays a sample. On the left is the original photo, in the middle a scaled image map of the facades. On the right a screenshot of a VRML view.

Orthophoto and DTM

Monoplotting from stereo pairs by means of correlation seems to be one of the major advantages of digital photogrammetry. Under this circumstance a digital orthophoto is the appropriate solution for automated mapping. The geometric shape of the object surface is automatically detected. The orthophoto in plane view provides a true scale image map. An example is given in figure 9. A DTM and the orthophoto mosaic resulted from a low altitude image flight. Some buildings of particular interest are presented in a 3D structure. All the data was converted in a VRML format. The buildings are connected by a link to other URLs. Detailed Information will display after a mouse click.

Panoramas

There are several methods existing to shot a panorama. The conventional method crops the image to a long format. When the negative is printed it looks like a long narrow panorama but the field of view is still unchanged. The APS cameras today provide such an imagination.

Field of views up to 180 degree can be photographed with a swing-lens camera. The lens is mounted in a rotating cylinder while the film plane is curved. During exposure the lens rotates and paints the image through a slit onto the film.

The rotational method needs a camera that can rotate about 360 degree while the film turns synchronously opposite to the camera rotation. This technique is still very popular. An advantage is, that moving objects can be recorded without any problems. While taking street scenes with the segmented method, people or cars can occur several times in different positions.

A series of images taken with a rotating exposure direction and displaying the photos next to another is the oldest method to show a wide field of view. The advantage of digital image technology makes the segmented method very popular today. Taking the photos with a digital camera enables an automated process to display scenes for interactive viewing. Stitching software processes the single pictures to a cylinder, sphere or cube. The most precise panorama consists of evenly spaced photos using a tripod. A special panoramic head assures that the camera rotates on a level plane around the centre of the lens. A number of 12 photos using a wide-angle lens of 35 mm in portrait format can be sufficient for outdoor scenes.

Publication via the Internet enables free selection of the viewing direction and the distance to the object by zooming in and out. Plug-ins or Java applets are required. Panoramas on Web pages can be linked to other sources by putting "hot spots" into the images.

Image Objects

Taking a series of images from a fixed camera position while the object is rotating on a turntable provides photos for an image object. An image object can interactively inspected by selecting the rotation and the distance to the object. Processing those images can be done with QuickTime or MGI Picture Object Modeler. The software enables image selection, image adjustment and storage in different data formats like .avi or *.fpx. The latter is a hierarchical image format, designed for distribution of high-resolution images. The Web server must be equipped with the according software to support the format. Figure 11 demonstrates the production of spin photo objects, here in stereo mode. Two digital cameras are computer controlled. Photos are taken in a time interval; the object rotates on the turntable in steps of 15 degrees for example. The software mentioned above supports masking as well, so that the environment can be corrected digital.

Stereoscopic Images

The usage of stereoscopic images covers a wide range of applications of computer visualization. Those can be entertainment, medical and forensic applications, topographic mapping or architecture. In all cases stereoscopic displayed data is far more effective for visualization then a flat presentation. Monoplotting from stereo pairs by means of correlation seems to be one of the major advantages of digital photogrammetry. Displaying the results in stereo with standard equipment on a PC-Workstation is one step more for a better acceptance of photogrammetry. Image separation is realized by spatial separation (cyber head), temporal separation (shutter glasses) or filtering (red/green anaglyphs).

Stereoscopic image pairs can be stored in one image file. Side by side is the basic arrangement of a stereo pair according to the position of the human eyes. From the two free viewing modes parallel and cross-eyed, parallel is limited to very small images. Therefore cross-eyed is the default mode of a JPS stereo pair. JPS is in fact a JPEG format, which allows specific stereo software to handle those images. For example a Web browser plug-in from Vrex Corp. interprets that format and converts it to the viewing mode available.

According to BMP or PNG (portable network graphics) the stereo formats BMS and PNS exist. While BMS has no direct form of compression, PNS is a no-loss form of compression. Using the JPS format, a moderate amount of compression is recommended, since a major loss of stereo quality will occur while reducing image information.

Figure 12 demonstrates the use of the Java applet stereoscope. The stereo image will be calculated upon request client sided. The server provides the left and right image. Alternatives for image separation can be selected from the applet menu.

Image pairs can be photographed by using a stereo bar as shown in figure 11. For artefacts a desktop scanner provides high quality images. Computer models must be rendered from different viewpoints.

Conclusion

In addition to the conventional photogrammetric products like line drawings or contour line plotting new techniques and data formats for recording and documentation of 3d objects from images are developed. Image objects, VRML-models and hierarchical polygon models with texture mapping are distributed via the Internet. New data formats and plug-ins for Web browsers are introduced for sufficient transfer speed, image quality and interactive presentations.

The paper is available in pdf and HTML-format at www.imagefact.de, images and examples are given there in colour and with the above-mentioned interactive facilities.

References

Pomaska, Günter. 1999. Stereoscopic Images for Visualization of Cultural Heritage, Contribution to ISPRS International Workshop, in *Photogrammetric Measurement, Object Modelling and Documentation in Architecture and Industry*, Thessalonica, Greece.

Pomaska, Günter. 1999. First Experiences with the new Digital Camera Rollei d7 metric, in *Contribution to XVII CIPA International Symposium, Recife*, Brazil, October 3-6.

Pomaska, Günter. 1999. Documentation and Internet Presentation of Cultural Heritage Using Panoramic Image Technology, in *Contribution to XVII CIPA International Symposium, Recife*, Brazil, October 3-6.

Pomaska, Günter. 2000. Präsentation von Architekturdaten im Internet, in *Deutsche Gesellschaft für Photogrammetrie und Fernerkundung e.V., 20. Wissenschaftliche Jahrestagung im Rahmen der InterGEO 2000 Berlin, 11. -13. Oktober, AK Nahbereichsphotogrammetrie.*

References are published under http://www.imagefact.de/ as well. More literature for further reading is announced there.

Figures

Figure 1: Meydenbauer camera built1890 for terrestrial photogrammetry

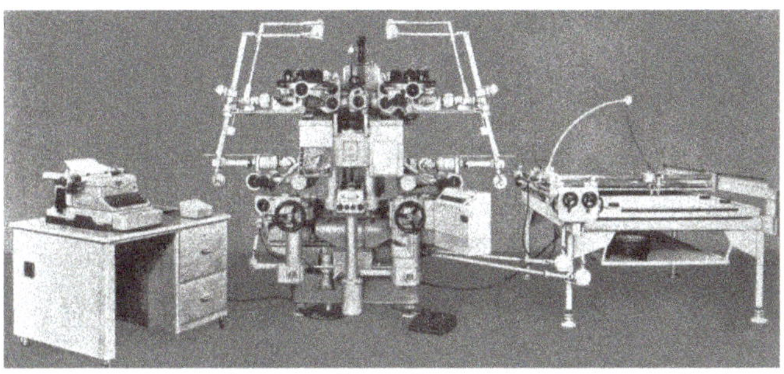

Figure 2: Stereoplanigraph C8 from Zeiss Germany, built 1967

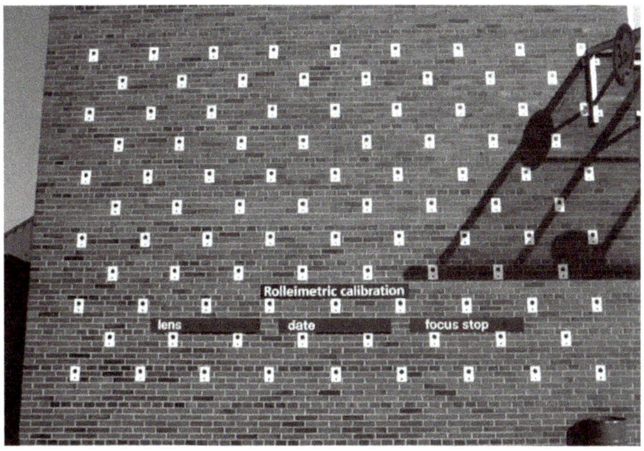

Figure 3: Calibration wall for camera field calibration, Rollei Fototechnic, Braunschweig

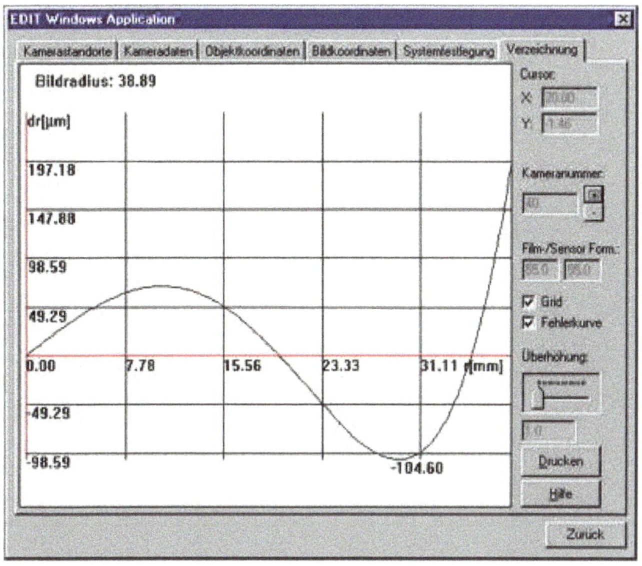

Figure 4: Lens distortion of a CCD camera with an 8 mm lens

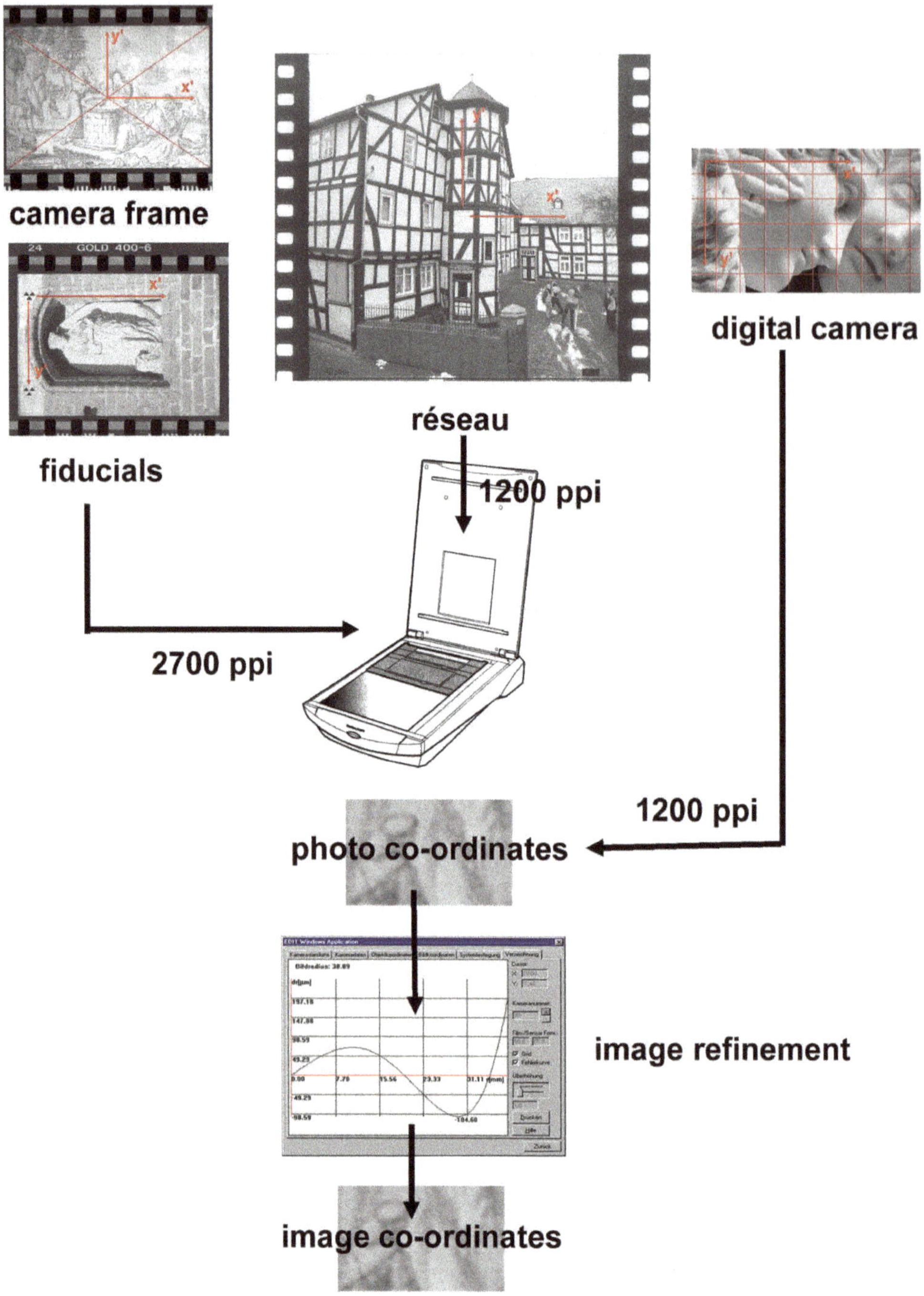

Figure 5: References to the image co-ordinate system. Format frame, fiducials as recommended by EOS Systems, réseau image and digital image

Figure 6: Exposure arrangement for multi image photogrammetry

Figure 7: Stereoscopic recording simultaneously with two cameras

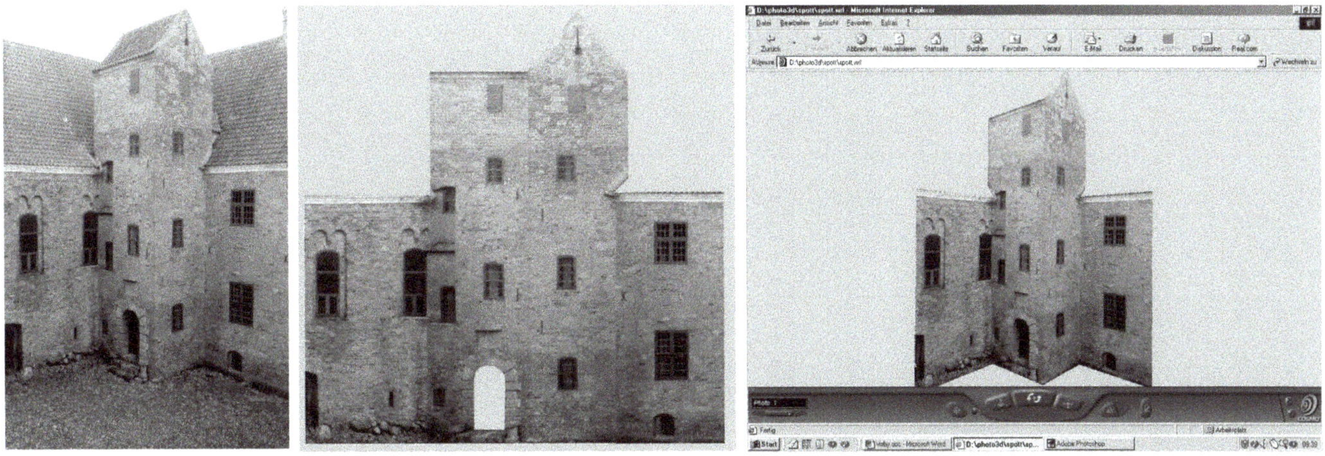

Figure 8: From a single 35mm photo, to a scaled rectified facade map and a VRML-model with PHOTO3D software for the PC

Figure 9: Orthophoto, DTM and buildings of particular interest in a VRML-File with links to detailed description

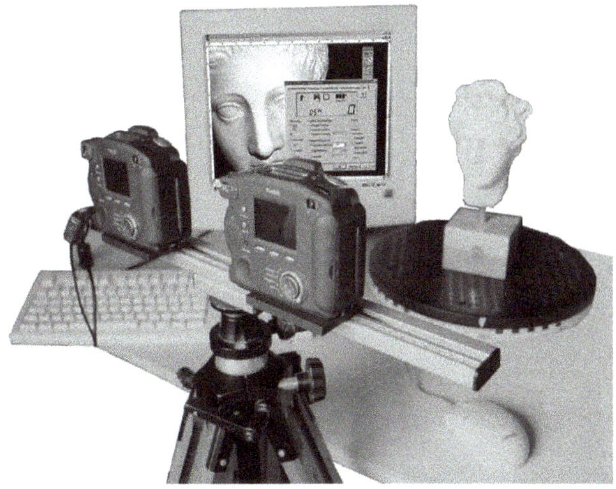

Figure 11: Computer controlled recording of image objects. Here a stereo arrangement with the object on a turntable.

Figure 10: Panorama of the inner court of the Convento Sao Francisco in Recife / Brazil. The curved lines will be straightened while viewing in a Web browser using a plug-in or applet. This panorama was taken free hand. Therefore it is not so precise as it should be.

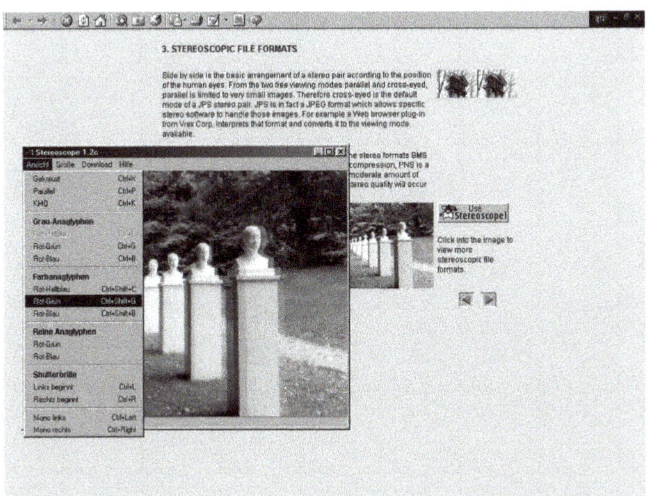

Figure 12: Web page including stereoscopic images to be viewed with stereoscope, a Java applet from Andreas Petersik, see [1], [4].

A Web-based Digital Archaeological Map of Lasithi, E. Crete.

Apostolos Sarris
Katerina Bichta
Marina Giasta
Anthi Giourou
Evaggelia Karimali
Vaggelis Kevgas
Kostas Margetousakis
Eleni Peraki
Steven Soetens
Katerina Tzaneteas
SofiaTopouzi
AchilleasTripolitsiotis

Laboratory of Geophysical - Satellite Remote Sensing & Archaeo-environment
Institute of Mediterranean Studies – Foundation of Research & Technology, Hellas (FORTH)
Melissinou & Nikiphorou Foka 130, PO. Box 119, Rethymnon 74100, Crete, Greece
Tel. +30-831-56627, 25146 - Fax. +30-831-25810 - E-mail: asaris@ret.forthnet.gr

Abstract: Digital archaeological spatial databases can contribute significantly to the management and protection of archaeological resources, as well as to the effective use of the environment. The diffusion of these data through the Web has also major advantages in terms of modelling the past usage of the landscape as well as in the future preservation of sites within the general framework of large-scale construction works.

*In the absence of a detailed archaeological registry, the Laboratory of Geophysical-Satellite Remote Sensing & Archaeo-environment of the Institute for Mediterranean Studies/Foundation of Research & Technology, Hellas (FORTH) undertook the creation of a Web-based GIS platform mapping of the known archaeological sites (more than 900) of the wider region of Lasithi, located in E. Crete (*http://www1.ims.forth.gr/maps/website/imslasithi2001*). The research project was carried out under the framework of Applied Research program of the Prefectorial Functional Program of Crete (1994-1999) and was funded by the Prefecture of Crete, through the European framework of Development-II.*

The final product consists of an integrated archaeological geographical database of the cultural resources of the Lasithi district, based on the principles of a Geographic Information System. The background layers of the digital archaeological map of the region consist of a complete map mosaic (topography, geology, landuse, land capability maps, DEM), SPOT and Landsat satellite images, which are used for a better presentation of the distribution of archaeological sites. The geographic database contains a catalogue of the known archaeological sites of the Lasithi district, environmental information, chronology, general information, references and photographic material from the archaeological sites. The Geographical Information System offers the capability of synthesis of different information layers and the creation of thematic maps of the archaeological sites in the region. In this way, this project can be considered as an initiative for the development of a digital archaeological cadastral registry of Crete. It has the dynamic potential of a tool in the decision making process of the development of the region and constitutes a detailed research guide of the cultural resources of the region, with direct consequences in the protection, conservation and management of the ancient environment.

Through the mapping of the geomorphological and archaeological variables of the region, the system can be used for the development of settlement pattern models, the enhancement and management of cultural resources (protection zones, archaeological parks, conservation risk areas), the protection of monuments and the parallel economic and tourist development of the region.

Key words: Web, GIS, Lasithi, Crete, digital archaeological map, satellite, GPS, CRM.

Introduction

In the last years, the profile of cultural resources management (CRM) has changed due to the large construction works, the increase of tourist industry and the requirements for a different city planning and a sustainable development (Sarris 2000). This development has been intensified through the expansion of the communication networks and the new advances of information technologies.

The turning of the century and the millennium is mostly characterized by the exponential creation of digital data that have a major impact on all sectors of society. Culture is one of them. Cultural resources need a special approach as they belong to the boundary that distinguishes technology from art, development from preservation, digital from archaeological information. Although the above fields may entail certain contradictions, it is possible to interact in favor of each other (Sarris 2001:2).

Geophysical prospection, satellite remote sensing and GIS are considered important modules for acquiring digital information regarding the cultural topography and landscape. Geophysical prospection techniques constitute a valuable tool in the mapping and management of archaeological sites during the course of large development works. Aerial and satellite images can contribute to the detection of new archaeological sites and the outline of protection zones of known archaeological regions. The recent developments in satellite remote sensing and Geographical Information Systems (GIS) have offered new standards to both historical and archaeological research and the management and protection of cultural and natural resources (fig. 1).

Innovative satellite sensors of high spatial and spectral resolution, along with the use of high accuracy Global Positioning (GPS) systems and enhanced image processing systems offer great possibilities in the mapping of archaeological sites. In addition, the combination of the above technologies with other databases, which contain archaeological and environmental information and with socio-economic models has direct consequences on our knowledge of use of cultural space in antiquity as well as on management policies of archaeological sites today.

Moving one step further, Geographical Information Systems could be used for integrating archaeological data, together with satellite and aerial images, topographic and geological maps and other digital environmental and cultural information. One of the most important applications of the Geographic Information Systems is the development of predictive models for archaeological site assessment based on image and statistical processing techniques of satellite imagery and environmental information. These methods are aiming to the management of cultural resources and the decision making process in large development works. Avoiding high-risk areas, namely areas that have a large probability to contain archaeological sites, it is possible to insure the protection of monuments, a better planning of the development works and the proper accommodation of large amounts of funds.

The need to develop of a Geographic Information System of cultural resources, with capabilities of processing and modelling digital images, is actually imposed by the effort of accommodating funds, due to the high cost of surface surveying and archaeological site registration and assessment during or prior the course of large scale construction works (e.g. highway or railway construction, expansion of rural estates, exploitation of coastal areas, construction of waste dump areas, a.o.). The adoption of such a system has direct consequences in enhancing the current inventory systems and electronic databases and in upgrading the current models of protection and the general strategy of management of cultural resources. A further advantage of GIS lies in their ability of updating their geographical information index in a continuous and interactive mode, processing and storing large volume of diverse origin data and creating thematic maps based on specific inquiries. The above can be used in archaeological research for modelling the settlement patterns of a region, locating and outlining high probability archaeological candidate sites, studying the communication or defence networks, specifying cost surface regions used for the exploitation of natural resources, etc. The creation of electronic thematic maps that present various cultural and environmental information simultaneously, could be extremely useful in solving problems resulted by the environmental and development plans, suggesting specific solutions for the protection, preservation and management of ancient monuments.

A priori condition for any project dealing with the protection, promotion and management of archaeological sites, as well as with archaeological landscape and settlement modelling, is the accurate mapping of sites and their correlation with the corresponding environmental information. Such an endeavor was undertaken by the Laboratory of Geophysical-Satellite Remote Sensing & Archaeo-environment of the Institute of Mediterranean Studies – FORTH aiming towards the creation of a Web-based digital archaeological map of Lasithi district in east Crete, operating on a GIS platform (Sarris *et al* 2000). The following actions and activities took place in the course of the project (fig. 2):

- Retrieval of published archaeological information, maps and field surveys. Emphasis was given on Prehistoric to Roman times.
- Construction of an archaeological database, which was implemented in the GIS. The database was also installed on a different platform, having the potential of updating the current information and querying through its data, without being necessarily within the GIS environment. Within the GIS environment, thematic maps can be produced based on simple or combined queries (eg. type, chronology, etc.).
- Construction of the Digital elevation model (DEM) from SPOT orthophotos and thematic maps modelling slope, aspect, inclination and shading.
- Digitization of 7 geological maps of the Institute of Geological and Mineral Exploration (IGME), 8 landuse and 8 forestry capability maps of the Ministry of Agriculture and other topographic data (coastline, villages, main and secondary road network, caves, quarries, etc) covering the same area of interest.

- Processing of satellite imagery taken by Landsat Thematic Mapper (30m resolution) and SPOT satellite (10 & 20m resolution). Real color and pseudocolor images were also produced by the synthesis of the different spectral bands.
- Construction of archaeological thematic maps representing sites as dots, lines and polygons (depending their extent and outline of them in cases where GPS readings were taken). All sites were connected to the corresponding archaeological database.
- DGPS mapping of more than 200 archaeological sites. Reduction of all GPS readings, satellite and aerial imagery and digitized maps to EGSA'87 coordinate system. In this way it is possible to overlay the data of the estate registry on the digital archaeological map or vice versa.
- Design and construction of a web site, which hosts the GIS-based digital archaeological map along with the rest of the results of the project. The web site is serving both the digital map within a GIS environment and the database. Thus, the user, accessing an HTML viewer, is able to define the map layers he wants to view. At the same time he has full access to the archaeological database for searching and printing the registered information. The web site includes additional information regarding the research project, such as general information, research team, coding information, technical reports regarding digitization of maps, GPS mapping and retrieval of archaeological information, together with other useful links that can be found in the net. The database and most of the technical reports are bilingual (greek and english).

The following sections analyze in detail the above activities.

Archaeological Database

A relational database was constructed for the registration of the archaeological information that was retrieved from the published references. The format and general design of the database was based on the principle of compatibility with the rest modules of the Web-based GIS platform and the possibility of expanding the project within the wider region of Crete, or even in other areas of Greece. The database was initially constructed in Access and then it was transformed to DBF and SQL formats. The DBF format of the database included the coordinates of the sites and it was connected to the rest geo-referenced maps. The SQL format of the database was used for querying the archaeological information and producing reports of the sites of interest.

The database consisted of 17 basic tables (fig. 3) which included various types of information, such as name of the sites, accessibility, nearest village, type of site, chronological periods, information regarding the excavation or surface survey, excavator/surveyor, bibliography, geophysical prospection techniques, archaeo-environmental data, present environment, bibliography, photographs, plans, GPS coordinates, general comments, a.o. The tables were related to each other through a common coding system, so that they guaranteed the successful retrieval of information through possible queries. The main tables of the database referred to the sites and the corresponding bibliography or references. Other tables contained the full path name of the files that referred to the images or topographic plans of the sites.

Every table or form of the database required the completion of different types of information (fig. 4). Macros were included to make easier the movement from one form to the other or the change of the language (greek or english) used for filling the fields that exist on the forms. In a few cases, look-up tables, numerical elements and predefined tick boxes were used to minimize the time and errors of the data entry process, automate the updating of both the Greek and English versions of the database and avoid possible errors that could be caused due to the discrepancies of terminology, such as in the case of chronological periods. In other fields, the user was able to enter additional comments in text format. The above design guaranteed the uniformity of the database that was necessary in establishing a connection to the GIS environment.

Different tables were created to distinguish the sites that were topographically mapped through a GPS survey from the sites that were mapped through digitization of maps or diagrams. A similar table was created containing the centroid of the sites instead of the total number of coordinates that was taken in the specific location and which was related to the extent of the sites or the outline of the architectural remnants.

The database was implemented in the GIS, which was used for the digital map. The database was also installed on a different platform (SQL), having the potential of updating the current information and querying through its data, without being necessarily within the GIS environment. The Web user is able to search the database by specifying either the full name or part of the letters that form the name of the site. If more than 1 match is encountered, all of the site names that satisfy the query parameters are listed and the user can decide which one is the proper site that he is interested in. Querying can be also carried out in the GIS platform through a number of tools or Boolean operators. Based on the query results, thematic maps can be produced based on simple or combined queries (eg. type, chronology, etc.). The user has also the ability to print specially designed reports for each site, even through the WEB. The reports of the sites were used for the better presentation of the information that is included in the database (fig. 5). All the site reports pull information through the database and one is able to print them either in Greek or in English.

Topographic mapping of archaeological sites through the use of Global Positioning Systems (G.P.S.)

Accurate mapping of the most important archaeological sites of the Lasithi district was carried out using 2 sub-cm geodetic GPS receivers (Ashtech Z-12 double frequency receivers) with a static differential mapping procedure (choke ring and 700718'_Geo.III L1/L2 antennas were used for the particular purpose).

Prior to the topographic mapping, a catalogue containing all the sites and important relevant information about their location and type was created. The approximate location of the sites was marked on the 1:50,000 topographic maps of the Geographical Service of the Hellenic Army, according to the descriptions and maps found in the relevant bibliography. The fieldwork (fig. 6) was mainly contacted in 6 periods, covering a total period of about 80 days, within which more than 200 sites were mapped of a total of 900 that have been registered in the database. Data collected prior to the initiation of the project were also incorporated to the database.

Static differential positioning technique (Leick 1990, Kaplan 1996) was systematically employed in all stages of the research project. The base station receiver, connected to a Choke Ring antenna, backed up by both a 12V battery and electric power, so as to prevent loss of data in case of power failure, was recording every 10-30sec. The base stations were different for each survey period and always within a distance of 30km from the surveyed sites. The coordinates of the base stations were established in the beginning of each survey period, by taking measurements for a period of over one hour at 2-3 close-by army datum points and comparing them with their known coordinates. During fieldwork, the rover receiver was transported at the sites of interest, where it recorded for a period of approximately 10-20 minutes, with a mask angle of 15°, so as to be able to locate from 4 to 9 satellites. Each measurement was coded with a distinct 4-digit code name, while a small draft plan of the site marked the locations that were measured.

The selection of points recorded each time varied according to the case and type of the archaeological site. In the case of surface distribution of sherds or visible architectural remnants, measurements were taken to define the extent of the site or the outline of the architectural features. Generally, more than one measurement were taken in each site (in a few instances a few dozens of readings were taken), aiming towards the definition of the limits of the settlements. If a topographic plan of the site was available, specific points of the features were recorded and assigned on the plans. In the case of caves, one point close to the entrance of it was measured. Similar strategies have been followed for similar features such as isolated tombs, houses, peak sanctuaries, etc. Besides the archaeological sites, GPS measurements were also taken in Ground Control Points (GCPs), in locations visible in airphotos and satellite images, which were used for the georeference of the images and their registration to the national coordinate system (EGSA'87).

Processing of the data was conducted daily, so as to evaluate the results and the accuracy of the measurements. Every measurement was processed and corrected (using Ashtech Office Suite software) according to the data recorded by the base. GPS measurements were recorded in WGS'84 and they were later transformed to the EGSA'87 coordinate reference system. A comparison of the measurements taken at specific datum points with those provided by the Geographical Service of the Hellenic Army resulted to an average percentage error of less than 10cm.

The use of accurately defined coordinates is of critical importance in a Geographic Information System. Measurements can be used for the necessary geometric corrections and transformation of all data in a common reference system. Furthermore, the accurate mapping of sites can refine the subsequent spatial analysis of the data, resolving problems related to the viewshed analysis, the precise definition of buffer zones (such as protection areas) and the accurate correlation of sites with the underlying environmental layers, especially critical for the derivation of the spectral signatures of sites (Topouzi *et al* 2001).

References & Bibliography used for Archaeological Data Retrieval

The information that accompanies the digital archaeological map of Lasithi was retrieved by a number of Greek and foreign journals and books. Emphasis was given to sites spanning from Prehistoric era to Roman period. The more recent monuments and sites (Byzantine and later period) were selectively mapped. The following information was retrieved for each archaeological site (depending on the availability of data):

- Name of the site/settlement
- Municipality – province in which it belongs to, nearby town or modern settlement
- Information concerning excavation (systematic, rescue, test, illegal)/ survey method
- Name of the excavators/researchers
- Year/years of survey or excavation
- Extent or area coverage of the archaeological site/ settlement
- Chronological periods of occupation or usage
- Type or category of the site at the particular chronological period or at other chronological periods (settlements, industrial site, religious site, burial site, cave, etc.)
- General remarks, based on bibliography, describing the general characteristics of the archaeological site.
- Important references and bibliography

The above information regarding the sites was registered, provided that systematic research had been carried out and results of this research had been made public (fig. 7). In specific cases, such as a) when surface survey was taken place without excavation, b) when excavation was not yet completed and c) when the study of the ceramics, finds or other structural features was not published, it was not possible to provide a full record of the above information.

When evidence was not sufficient to support the definition of a specific chronology or category of site, a more general date (e.g. LM instead of LMIIIC) and type (e.g. archaeological site instead of settlement, town, village, metallurgical site, shrine, sanctuary, cemetery, isolated building, farmstead, villa etc.) was provided. Since there are also arguments, of archaeological ground, concerning the type and chronology of sites, especially when sites have not been excavated yet or systematically studied, the adoption of different terms or contrasting suggestions (e.g. *farmstead* instead of *country house* or *villa*, or *LMIA* instead of *LMIB*) were mentioned in the comments.

Digitization of Maps and Photographic material

Topographic & Archaeological Maps

AutoCAD14/2000, connected with a CalComp DrawingBoard III digitizer, was employed in the digitization process of maps. A number of archaeological and topographic features were digitized to be included in the thematic map list that can be activated in the final product. Topographical data included the coastline, the main and secondary road network and about 344 caves, 90 quarries and 338 village polygons (outline of village limits). The digitized data were treated similarly, together with the rest GPS measurements, while a number of geometrical symbols were used for mapping the archaeological sites and features. More specifically, point symbols were used for mapping the archaeological sites, based either on the GPS measurements or the calculation of the centroid of a group of measurements that defined the approximate limits of the archaeological site. In other cases, polygons, cirles or lines were used for describing some particular details (table 1) (Paraschakis, *et al* 1991, Hoffman 1992). In the case of polygons, the database included site names, estimated area and perimeter of the polygon. The following convention (table 1) includes the cases that were taken in account in the symbolism definition within the digitization process.

Table 1: Digitization convention.

Digitization of Topographic Elements	*Digital Mapping of Archaeological Sites*
Lines: Coastline (Lasithi district)	Polygons: Outline of buildings Estimated extent of sites Quarry pits (eg. Kalamaki quarry)
Lines: Main Road network	Circles: Tomb Cave entrance Estimated extent of sites, defined by 2 points
Lines: Secondary road network / dust roads	One Point: Approximate centroids of sites Sites not found-coordinates taken in the approximate vicinity of the site Sites discovered through surface survey
One Point: Caves (Total = 344)	Two Points: Approximate site limits (ex. Agia Fotia – necropolis)
One Point: Quarries (Total = 90).	Lines: Walls and roads
Polygons: Village polygons/Modern towns & villages (Total = 338)	

Different layers were created in order to provide the capability of separation of the cartographic products. Different colours were assigned to each layer. The topographic features that were digitized from the 1:50,000 scale maps are presented in fig. 8. For the wider Palaikastro area (north of the village of Palaikastro), 4m contour lines were digitized from 1:5,000 scale maps.

Geological, land use and land capability for forestry maps

A similar procedure was followed in the digitization of the geological and landuse maps of the region. The different formations of the corresponding maps were represented by closed polygons. Thus, digitization of double lines was prevented, avoiding problems that would have arisen in assigning the formation attributes. Each one of the above environmental maps was placed in the same layer (fig. 9).

Following the digitization of the maps, georeferencing of them was performed with TNT Mips software, by choosing specific GCPs in the corresponding maps and the digitized costline. Later on, the relevant attributes were filled in the corresponding database.

The geological attributes included a polygon consecutive number, the geological formation code similar to the one assigned by the Institute of Geology and Mineral Exploration (e.g. "al" for alluvium), a general type and description of the geological formation, the corresponding geological chronological period (e.g. upper cretaceous/lower Eocene) and the coding of the digitized map sheet. The accompanied database, consisting of 99 distinct geological formations, included also information related to the area (in square meters and acres) and the perimeter of each formation polygon. Discrepancies in the coding due to the different year of map design and production were also taken into account. Formations with different codes between new and old maps were registered as being the same when description and geological period were identified as the same. In the case of an overlapping region between new and old maps, the most recent and more detailed description was recorded, coded appropriately to refer to the corresponding map. The following table (Table 2) summarizes the maps that were digitized and were implemented in the Geographical Information System database.

Table 2: List of topographic, landuse and geological maps that were digitized for the purposes of the project.

Topographic Maps Geographic Service of the Army (scale 1:1,5000 – 50,000)	*Soil Maps (Land use & Land capability for forestry) Ministry of Agriculture Scale 1:50,000*	*Geological Maps Institute of Geological & Mineral Exploration (IGME) Scale 1:50,000*
DIONYSADES, 1994	DIONYSADES, 1996	
PALAIKASTRO, 1994	PALAIKASTRO, 1996	ZIROS, 1959
SITEIA, 1994	SITEIA, 1996	SITEIA, 1959
ATO CHORIO, 1994	ATO CHORIO, 1996	RAPETRA, 1993
ERAPETRA, 1994	RAPETRA, 1996	RAPETRA, 1959
AGIOS NIKOLAOS, 1993	GIOS NIKOLAOS, 1996	GIOS NIKOLAOS, 1987
CHOS, 1993	CHOS, 1996	CHOS, 1989
ANO VIANNOS, 1994	ANO VIANNOS, 1996	ANO VIANNOS, 2000

The registered attributes of the landuse and land capability for forestry are listed in Tables 3-4, defining approximately 365 different landuse categories.

Table 3: Recorded attributes of the Landuse maps

Landuse Attributes
Surficial Geology (hard limestones, tertiary deposits and conglomerates, schists, alluvium, mixed flysch, limestone colluviums, dolines, deposition cones)
Physiography (flat areas, steep summits, escarpments, cliffs, rounded summits, upper slopes, middle slopes, terraces, lower slopes, open valleys, closed valleys)
Soil Depth (deep, deep and shallow, shallow and deep, shallow, shallow and bare, bare and deep, bare and shallow, bare)
Erosion (none, none and moderate, none and severe, moderate and none, moderate, moderate and severe, severe and none, severe and moderate, severe)
Slope (gentle, gentle and moderate, gentle and steep, moderate and gentle, moderate, moderate and steep, steep and gentle, steep and moderate, steep)
Ecological land region (broadleaved evergreen zone, pinus brutia, oak zone, pseudoalpine zone)
Degree of artificilization of vegetation/ Human provoked vegetation (nil, moderate, severe, no vegetation, grassland, phrygana, agriculture)
Aspect (north, north and south, variable, south and north, south, nil, nil and north, nil and south, north and nil, south and nil, variable and nil, nil and variable)
Code of the map sheet.

Table 4: Recorded attributes of the Land capability for forestry maps

Land Capability for Forestry	
Field	Name Description
Category 1:	Lands without limitations for timber growth
Category 2:	Lands with slight limitations for timber growth
Category 3:	Lands with moderate limitations for timber growth
Category 4:	Lands with severe limitations for timber growth
Category 5:	Lands with extremely severe limitations for timber growth.

Photographic material

Photographic images were taken from all the archaeological sites that were mapped with GPS units. The photos were scanned with a Hewlett-Packard ScanJet 4c scanner, with 150dpi resolution, and they were saved in jpeg format, to be easily retrieved by the database by specifying the path of the directory of the photographic registry. In a few cases, such as peak sanctuaries, Photopaint 9 was employed for creating panoramic views of the wider area surrounding the archaeological sites (fig. 10).

Satellite Remote Sensing – Digital Elevation Model.

SPOT and Landsat TM images have been georeferenced using the coastline and the SPOT derived DEM (resolution of 50m). All images were georeferenced to the Greek Geodetic reference system (EGSA' 87), with a Transverse Mercator projection and a WGS84 datum. All images were compressed with MrSid encoder down to a factor of 75-90%, resulting in images of total size less than 800KB, which were easily served in the Web. All images were processed through the application of contrast enhancement techniques and high pass filtering. Due to the limited surface information, classification of the images produced poor results.

An Ortho Pan Spot image was used as the basic background layer of information, having a resolution of 10x10m and a georeferencing accuracy of +/-12m. Three more panchromatic Spot images were used in the map registry, having a less degree of georeferencing accuracy (+/-13-170m). Similarly, a Spot multispectral image (resolution of 20m) was used for creating a pseudocolor image, emphasizing mainly the vegetation of the area of interest (fig. 11).

Finally, six of the seven bands of the Landsat Thematic Mapper (30x30m) were used for constructing false color composites (combinations 1-2-3, 3-4-5, 4-5-7). TM images were initially resampled to 10m cell size, achieving a georeferencing accuracy of less than 5 meters.

The construction of the digital elevation model (DEM) was based on 14 level 1A Spot images. The resolution of the DEM was 50m and its accuracy after georeferencing was of the order of 35m. A lookup table was formed for associating the grey scale of the original image (0-255) to the actual altitude values (0-2084). Different colors were assigned to each altitude class (every 10m for altitudes 0-100m, 50m for altitudes 101-1000m and 100m for altitudes 1001-1084m).

Surface modeling was carried out through processing of the DEM, including the construction of the 50m-elevation contour map of the region, together with a slope map (0-90°), an aspect map (0°:N, 90°:E, 180°:S, 275°:W) and a hillshade map (external light source with azimuth: 315°, elevation: 45°). These images (fig. 12) can be useful in the further study of the archaeological sites and the settlement pattern analysis.

Dissemination of results. Creation of the Web-based digital archaeological map of Lasithi.

For the better dissemination of the results of the project, it was decided to serve all the information layers through the Web. In this way, the user was provided access not only to colourful static maps, but instead he was allowed to interact with the system, activating any number of information layers at any time at his own will.

Still, transferring data of different type, format and size to the cyberspace needed to take into account the size of files to be served (and thus the speed of data presentation), the Web publishing support, the client software support, software and hardware compatibility and the reliability (and security) of data serving within a user-friendly environment. ESRI's ArcIMS (version 3) proved satisfactory for the above requirements, providing handy tools within an Arcview-like environment.

ArcIMS was installed on a Windows NT 4.0 (SP4, 5, 6A) server. IIS was used as the Web server environment, together with a servelet engine, which was employed for the connection to ArcIMS map server. The manager facility of ArcIMS provided a Web-based interface supporting three main operations: author mapping, web page design and administrator services. The server side component of ArcIMS contains modules for processing the client's request and serving the spatial data (image or feature files), monitoring the requests' traffic and the automatic initialization of map services. On the other hand, from the client side, Java viewers (standard and custom) or HTML viewers are both capable on presenting the results of a request that has been processed by the server. Taking in account the advantages and disadvantages of the two viewers, it was decided to employ the HTML viewer as the client viewer. Although the HTML viewer does not exploit all the functions of ArcIMS, it covers all web browsers, it has a relatively simple interface, and it does not require any additional modules to be downloaded or installed. On the other hand, the Java viewer (~5MB) needs to be downloaded by the client and it requires the installation of JRE, without having full support by the Netscape browser.

Web publishing of large size (>100MB) satellite images was also made possible through the use of MrSid Image Server (http://www.lizardtech.com/), which was supported by ArcIMS. Prior to publishing, satellite images were compressed down to approximately 700-500KB using the corresponding compressor (MrSid Compressor). In this way, it became possible to incorporate a number of SPOT and Landsat images, together with the digital elevation model and its secondary products, as different information layers of the GIS platform, upon which the user is capable to overlay any other raster or vector format layer to obtain a visual comparison or correlation of the spatial data. A specific scale-dependent renderer was set to establish a minimum display scale for the satellite images (1:3,000), above which images rasterize the screen of the client.

The initial web page of the digital map is presented in a very simplistic form, consisting just of the coastline and the main and secondary road network. The user can activate any layer and query information included in the menu. The query results that satisfy the criteria of the client are presented at the lower portion of the screen, consisting of all the descriptive and spatial information that the administrator has allowed to be accessible by the client (fig. 13).

A number of tools are also available for the navigation of the map. They include zoom in and zoom out buttons, zoom to full or active extent, pan movement (top, bottom, left, right), overview map window, distance measure tool, unit definition, selection by line/polygon or rectangle, a query builder and identify tool. The user can print our any portion of the map resulting by a combination of layers or query result.

Future prospects

The next step of the project will be devoted to the analysis of the spatial information and the study of the diachronic settlement pattern analysis of Lasithi. Having accurate topographic and environmental information, the archaeological data will be processed through different GIS analytical techniques, such as viewshed analysis, spatial patterning and distribution, least cost surface analysis and site prediction. The incorporation of the satellite images could also lead to the spectral signature recognition of archaeological sites of the area.

A joint research and technology programme between the Institute for Mediterranean Studies - FORTH and the Institute of Archaeology - Chinese Academy of Social Sciences has been also initiated, aiming towards the study of archaeological sites and their management in semi-arid and desertification affected regions. The project will make use of the digital archaeological map of Lasithi and it will compare the environmental conditions of the location of the sites and their preservation status with those in the Xinjiang district in west China. A similar collaboration with the Université Catholique de Louvain, aiming towards the study of the Minoan cultural landscape resulting from the spatial characteristics of the peak sanctuaries, is under progress and will produce similar results for the whole island of Crete (Soetens *et al* 2001).

The digital archaeological map of Lasithi constitutes the first step towards a more systematic plan for the management and preservation of the cultural resources of Greece. The expansion of the project can be easily achieved, following similar steps to those that were used for the Lasithi district. The final product of this approach could be accessible to the interested parties through the WEB, assisting the planning and state authorities, supporting the local and regional growth, enhancing the tourist and cultural industry and offering a multidimensional and integrated solution to a number of problems encountered in the management of cultural heritage. As all archaeological artifacts, from ceramics to megalithic monuments and from surface concentrations of sherds to natural landscapes, play an integral role in the reconstruction of cultural heritage, the need of technological means to record and disseminate the corresponding information to all sectors of society is more urgent than ever, in order to provide the necessary knowledge framework that will enhance the decision-making and policy-making process that concerns the protection and preservation of cultural resources. Within the current era of digitization, the construction of a digital cultural world will broaden cultural policy frameworks & will strengthen the design of cultural policies (Sarris 2001:2).

References

Hoffman K.R. 1992. *Application note: X-Y Digitizing Techniques and Applications*, Reference Manual of the TNT-mips Software of Microimages Inc, Lincoln

Kaplan, D. E. 1996. *Understanding GPS Principles and Applications*, Boston.

Leick, A. 1990. *GPS Satellite Surveying*, U.S.A.

Paraschakis, G., Papadopoulou, M. & Patias, P. 1991. *Automated Chartography*, Thessaloniki.

Sarris, A. 2000. Cultural Resources Management Through the Application of Ground-Based and Satellite Remote Sensing and Geographical Information Systems: A New Digital Cultural World, ECPR Conference on *Culture and Regional Economic Development – Cultural, Political and Social Perspectives,* organized by the Department of Planning and Regional Development (DPRD), Centre for European Studies, University of West of England (UWE), European Consortium for Political Research (ECPR) Standing Group on Regionalism, School of European Studies, Gardiff University in Wales, University of Thessaly, Volos, 28-30 September 2000.

Sarris, A., Giasta, A., Gkiourou, A., Karimali, E., Kevgas, V., Margetousakis, K., Bichta, K., Peraki, E., Soetens, S., Tzaneteas, K., Topouzi, S. & Tripolitsiotis, A. 2000. Digital Archaeological Map of Lasithi, Crete. A New Approach in the Management of Cultural Resources, *10th Greek ArcInfo-Arcview Users meeting*, Athens, 9-10 November 2000.

Sarris, A. 2001. Moving Cultural Resources Management in Digital and Cyber Space, *Archaeo-Telepiskopika Nea*, Sarris, A. (ed.), no 4, January 2001, Rethymno, Crete.

Soetens, S., Sarris, A., Topouzi, S. & Tripolitsiotis, A. 2001. GIS Modeling of the Minoan Peak Sanctuaries of East Crete, *Computer Applications and Quantitative Methods in Archaeology Conference (CAA 2001)*, April 25-29, 2001, Visby, Gotland, Sweden.

Topouzi, S., Tripolitsiotis, A., Sarris, A., Mertikas, S. & Soetens, S. 2001. Errors & Inaccuracies in Repositioning of Archaeological Sites, *Computer Applications and Quantitative Methods in Archaeology Conference (CAA 2001)*, April 25-29, 2001, Visby, Gotland, Sweden.

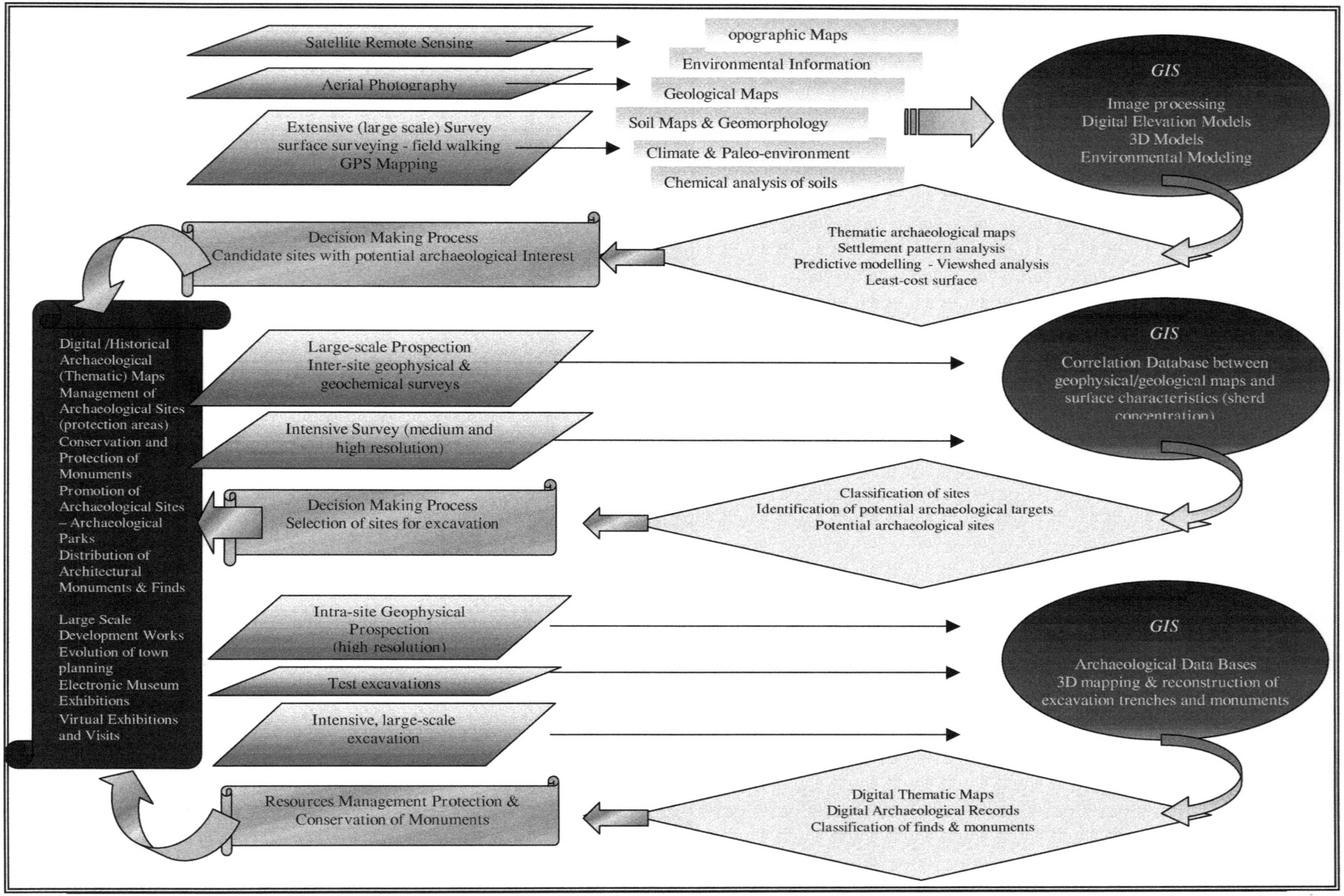

Figure 1: Macro, medium and micro scale applications of Geographical Information Systems (G.I.S.) in archaeological research and CRM.

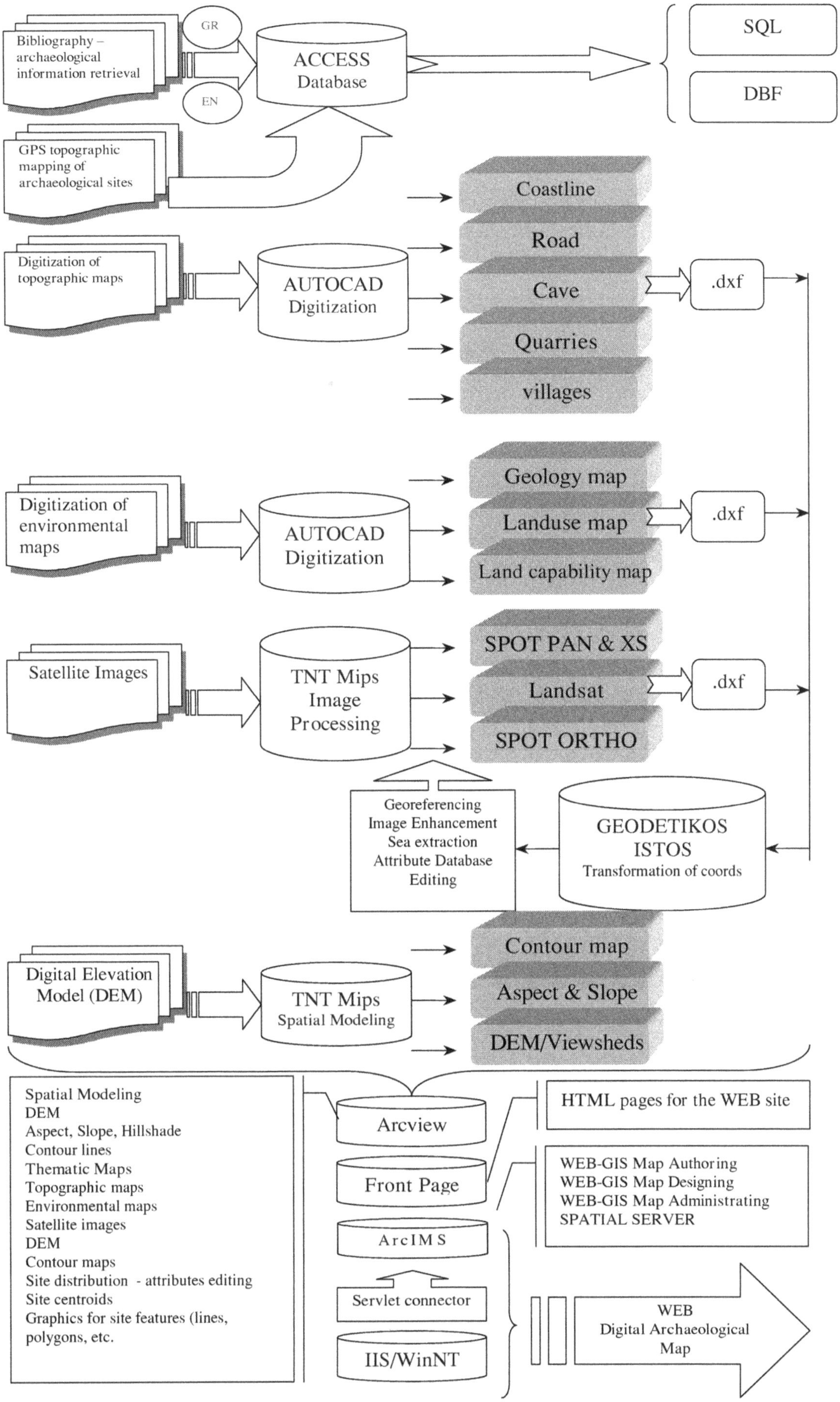

Figure 2: Flowchart of the actions and the activities that took place for the construction of the digital archaeological map of Lasithi district, together with the software modules that were employed.

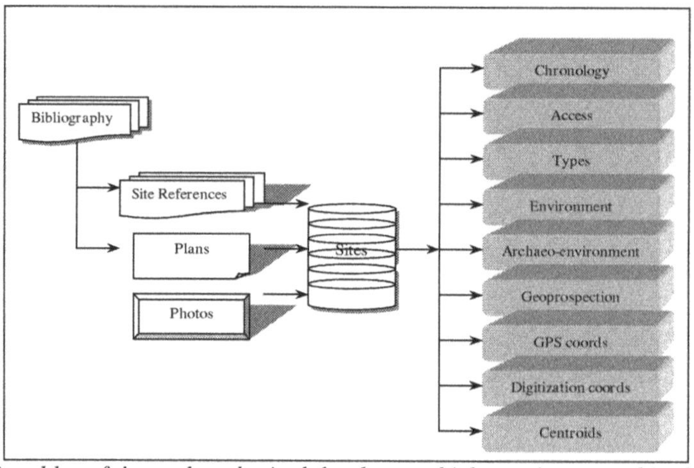

Figure 3: Analysis of the basic tables of the archaeological database, which was integrated in the GIS of the digital archaeological map.

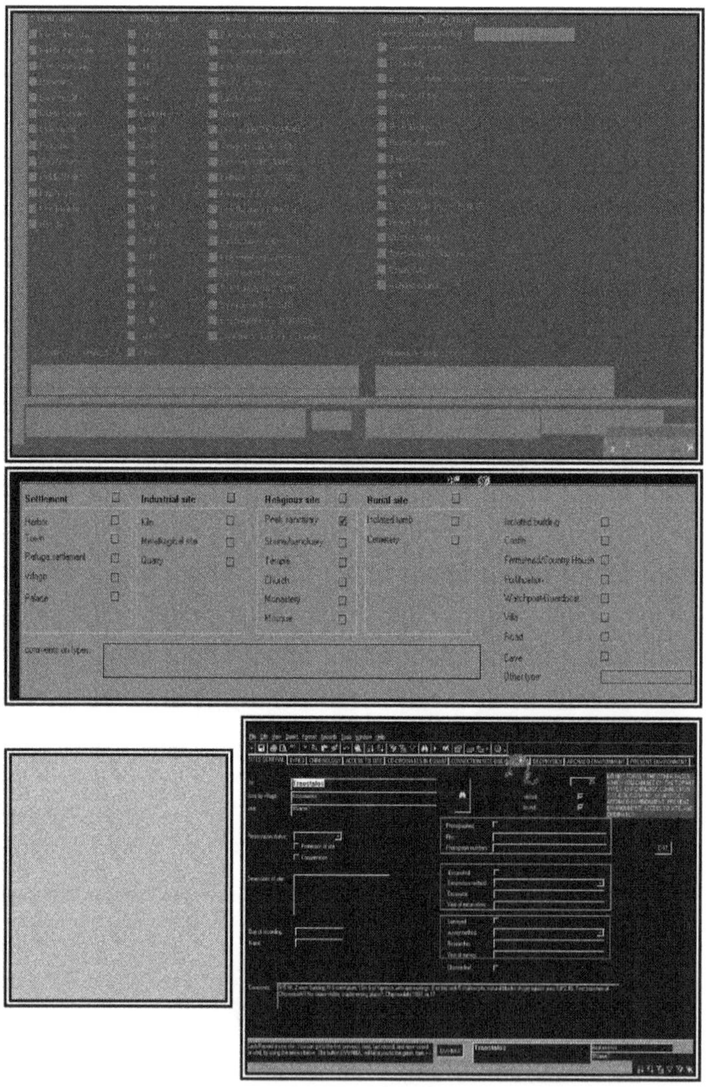

Figure 4: Samples of the Access database forms that were designed for the registration of the archaeological information.

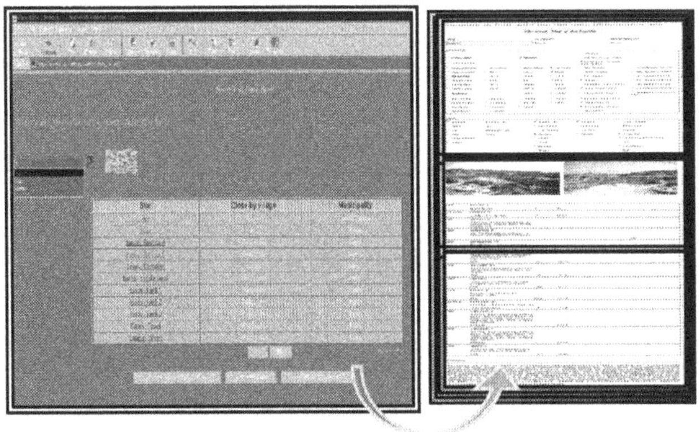

Figure 5: Site querry and the corresponding archaeological site report that the user can print for his own use.

Figure 6: Details of the GPS fieldwork activities. Recording at the site of Monastiraki – Katalimata (left) and Modi (right).

Figure 7: Archaeological site of Gournia

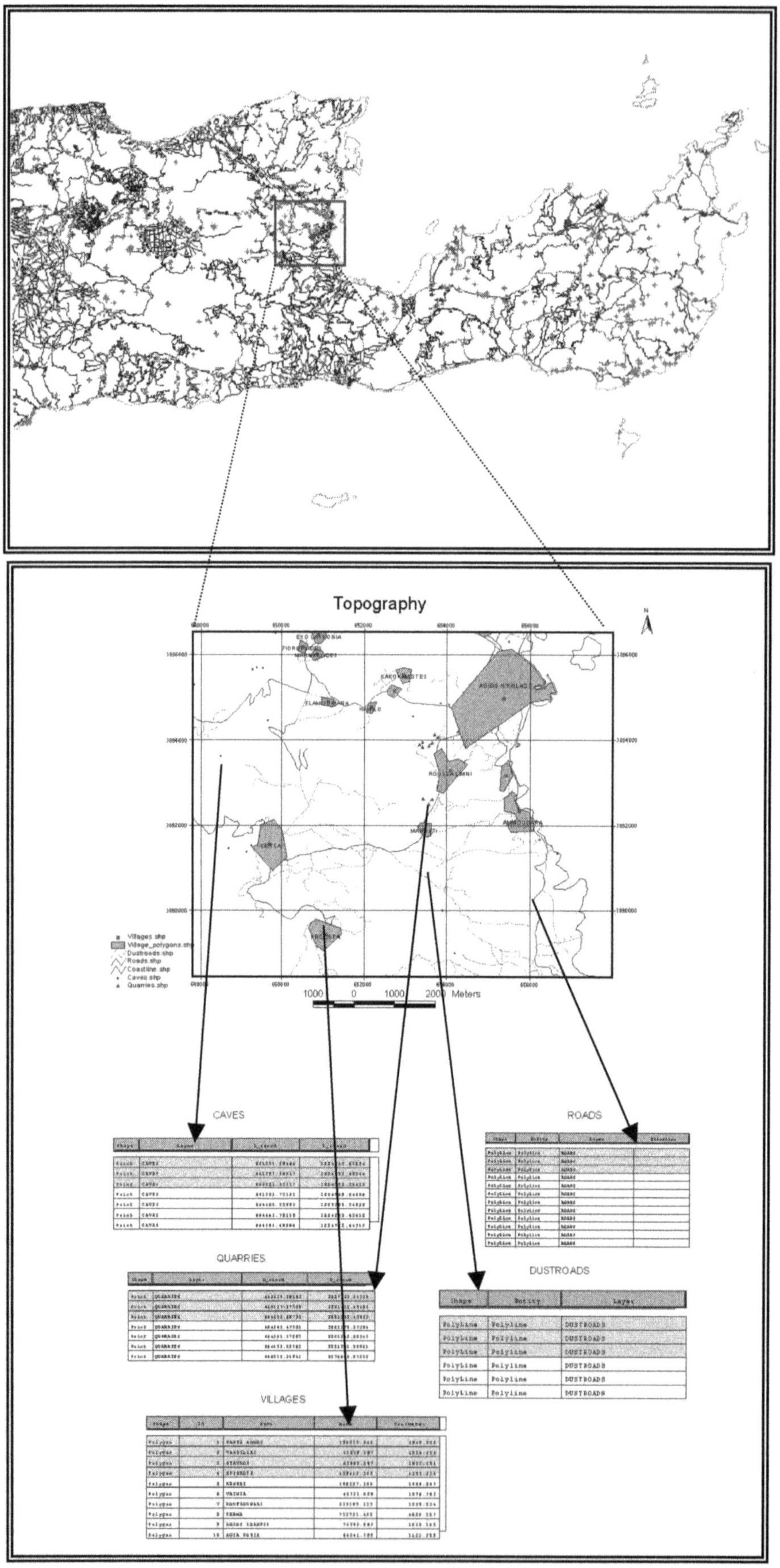

Figure 8: Digitizing and mapping of topographic features of the Lasithi district.

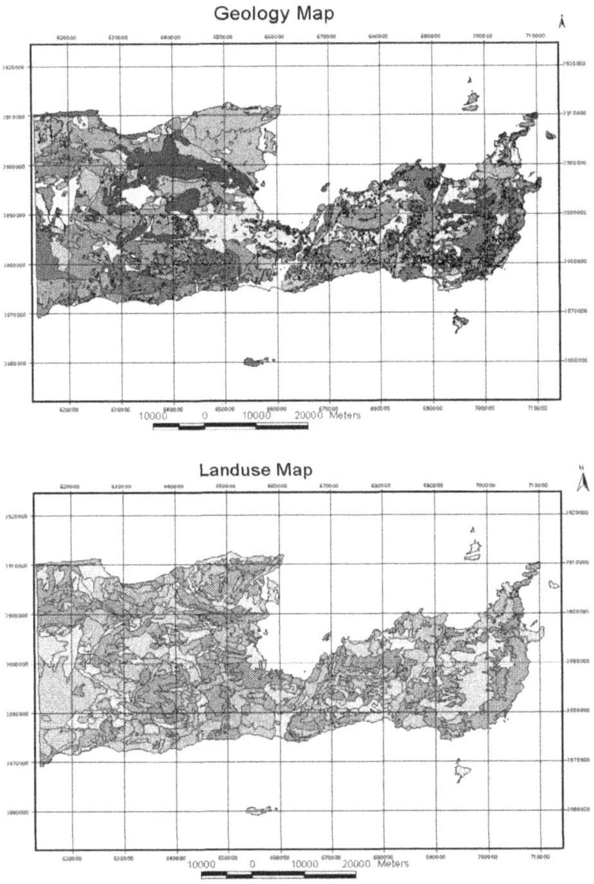

Figure 9: Digital thematic maps. Geological map (above) and landuse map (below) of the Lasithi District. Different colors represent distinct formations.

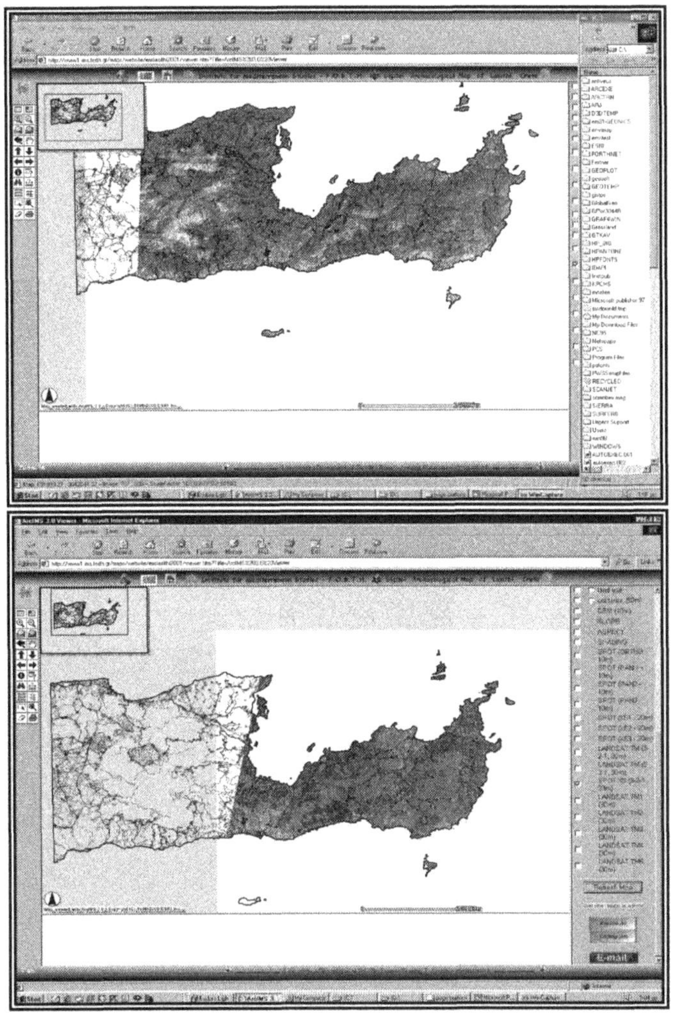

Figure 11: Real color (RGB: 3-2-1) Landsat satellite image (above) and pseudocolor SPOT XS image (below).

Figure 10: Panoramic view from Monastiraki Katalimata.

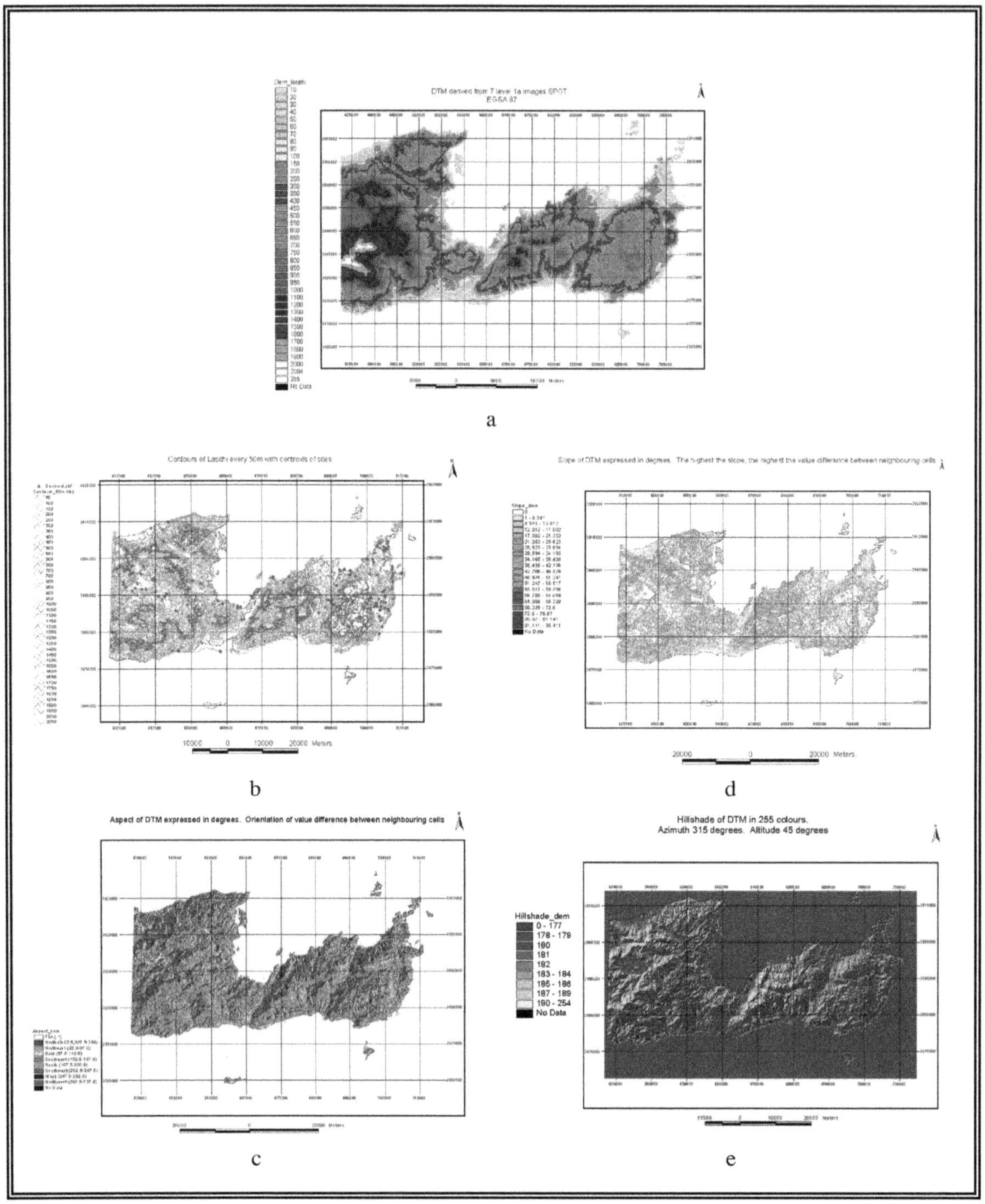

Figure 12. *a. Digital elevation model (50m)*
b. Elevation map (contours of 50m)
c. Slope map
d. Aspect map
e. Hillshade map

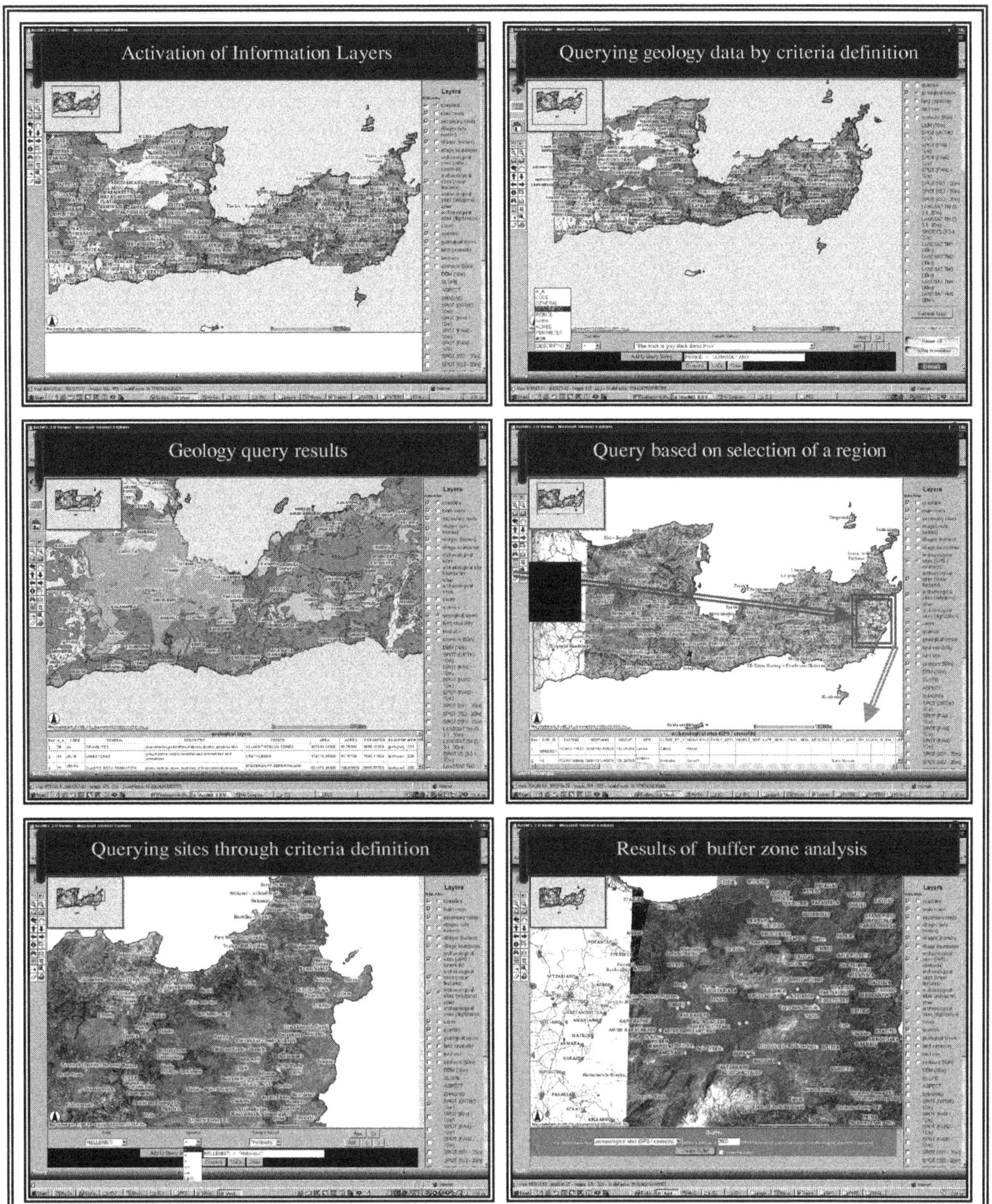

Figure 13: Screens showing different functions of the Web-based digital archaeological map of Lasithi. The user is able through the GIS Web server to activate any number of information layers, perform queries based on criteria definition or selection of a region and even perform simple spatial analysis, such as the construction of buffer zones around archaeological sites. The results of the queries are displayed on the lower part of the screen, consisting of all the fields that the site administrator has allowed to be visible.

The Archaeologist Files: An approach to the digital contextualization of archaeological finds in user adaptive information systems

Isto Vatanen
Aboa Vetus & Ars Nova Museums, Finland
Itäinen Rantakatu 4-6
FIN-20700 TURKU
Phone: +358-(0)2-2794922 - Mob: +358-(0)43-822794922 - Fax: +358-(0)2-2794909 - E-mail: isto.vatanen@aboavetusarsnova.fi

Abstract: While surveying the opinions of the general public it becomes evident that archaeological finds tend to taken as hard to understand without expert knowledge on the subject. The consecutive problem is that this makes the finds easily rather uninteresting to the amateur audience. The presentation of context and giving meaning for archaeological finds in wider historical perspective is an apparent problem, and a considerable challenge for educational institutions and especially for museums. One plausible approach to this challenge is the use of digital presentation techniques and user adaptive information systems in giving context to archaeological collections. The Archaeologist Files project has explored different methods for providing museum visitors a more holistic approach, comparing to the traditional, to the archaeological material in historical perspective. The results do suggest that dynamic and adaptive hypermedia interfaces combining intelligent search facilities with narratives could be a feasible method for achieving these goals.

Key words: The Archaeologist Files, context, archaeology, human computer interfaces, narratives, hypermedia, adaptation, information systems

Introduction

While surveying the use of computer multimedia workstations in museum use, it has become increasingly evident that a museum is a rather demanding environment to set up functional information system for public use in an exhibition. This article discusses the general problematic of archaeological finds in a public display by exploring a possible solution to increase meaningful communication between the archaeological exhibition objects and the visitors by using digital information media applications.

The Archaeologist Files project has been a logical step in the development process of archaeological computing applications in Aboa Vetus Museum. Aboa Vetus is a rather small museum of archaeology and history located in Turku, in south western Finland. Despite the size of the institution, it has been a major player in the field of archaeological computing and hypermedia presentation in Finland since its founding in the mid 1990's.

The multimedia information systems applications of Aboa Vetus before the Archaeologist Files include the Lost Town –project which was launched to produce a complete 3D model of late medieval Turku based on exact archaeological data (Sartes & Uotila 2000). The other major project has been "Cheers! Fragments from the Middle Ages" which was to exhibit medieval North European glass vessels as a multimedia application containing video clips of digitally reconstructed objects. The museum has also launched a number experiments in smaller scale aiming to explore various possibilities of computer technology in presenting archaeological data.

The first computer multimedia applications of Aboa Vetus were developed in co-operation with a professional multimedia design company. In spite of the fruitful co-operation the following projects were began as in-house operations because of the demands to streamline the production processes and the vision which saw the computer applications as an integral part of publication work and research done in the museum. Experiences of this renewal of processes have been encouraging and no reasons for outsourcing the CA projects has been encountered since then.

The project

Because the archaeological collection of Aboa Vetus is unique regarding its size and comprehensiveness in Nordic context, it was clear from the beginning, that this material should be forged into digital format sooner or later. The Aboa Vetus collection consists of the all the archaeological material excavated from the site where the museum is located. The collection is very comprehensive consisting of nearly 40.000 individual catalogued finds dating from the early middle ages to the beginning of the 20th century. This kind of material is however far more demanding for a museum than a traditional collection of antiquities and art objects. Because of its origins there is a rather small amount of completely preserved items and fine arts objects comparing to the huge amount of sherds and other small remains. The somewhat limited visual attractiveness of the objects does not decrease their scholarly importance or the usability value in museum displays, but do increase the demand for supplying a comprehensive context to make them understandable. This is a common problem with all archaeological material. Only a minimal part of any finds can be considered interesting to amateurs as such. This leads to a controversy of having tens of thousands of finds in a collection without having but a few dozen which are of apparent interest to a common visitor.

The Archaeologist Files is based on the response gathered from the museum visitors and surveys done among users, which all do suggest that a new comprehensive method for presenting archaeological data to the wider audience was needed to satisfy the information needs in all user categories excluding the professional archaeologists. The lack of context and special expertise needed to interpret the basic archaeological data were the most critical factors hindering the communication. That is the reason why they were chosen to be the primary goals for the new project.

The Archaeologist Files tries to explain what a tiny piece of wood or ceramics can really contribute to research, what it once was, where it was used and for what purpose, and what are the finds lying behind the closed doors of the museum storerooms. The project was titled The Archaeologist Files with purpose to emphasize presentative approach and the need to increase the attractiveness of finds especially among the general public.

The idea was further developed to a sketch of an information system capable of providing multiple media content to mass of different kinds of users giving them much demanded context and explanatory material.
Because of the promising earlier experiences the results were decided to be published on a number of platforms. The rather flexible portability of the raw media has enabled increasingly streamlined production of content to be published in museum based multimedia stations, on a CD and a DVD-ROM formats, and on interactive web pages. These all formats were included in project definition. They were prioritized to be completed in order to prepare first the multimedia station version for prototyping and gathering experiences, and then to proceed to further developed versions on CD-ROM and on the web platform.

The actual project was started in co-operation with a private enterprise Muuritutkimus ky. Aboa Vetus Museum has been providing the project framework, management services, project design and co-ordination and research work while the partner has been concentrating on the visualization of objects and scholarly consultation on some specific finds. The funding for the project has been provided by Matti Koivurinta Foundation and the Finnish Ministry of Education.

Theoretical background

Apart from being a commercial project with directly applicable results for the museum, the Archaeologist Files is also a research project for developing the next generation of digital media systems for museum environment. The fact is that a simple computer multimedia programme placed in a museum is not anymore a valid answer to the integration of digital media use to the contemporary visions of the museums today or in the future.

The traditional view of delivering information in digital form with traditional paper oriented layout and systems design was considered inadequate for the goals of The Archaeologist Files project. A picture laid out with static text on the screen is not a plausible way to present information on a computer system. Digital information processing allows use of inter linked hypermedia content, dynamic processing and presentation and tools for human computer interaction.

The other view to the traditional versus more up to date way of presenting information on a digital media platform concerns the media content itself. Supplying an increasing amount of visual data in form of figures, photographs, computer modeled and reconstructed objects is plausible in a way because its apparent value in educational terms. In spite of that while implementing more visual data, the question arises, whether it does actually satisfy a need for specific information or answer a concrete question concerning some subject. While becoming more visual the information systems are actually facing a danger of loosing the vital interdependence between different kinds of content and becoming blind to the increasing amount of available data in comparison to the actual information it contains.

Visual images are perhaps the best example of this danger. They are self-explaining to some degree but rarely produced with such care that they would fulfill their full potential. The problem is tied especially to the virtual realities and 3d reconstructed finds and structures which are being produced nowadays in all increasing numbers. Reconstructed objects tend to be visually attractive. High resolution graphics, sound, perhaps even possibility to smell the item with so called natural user interface do increase the immersion. It is self evident that using advanced technology is plausible, and does apparently give added value as the level of virtual experience and its authenticity enchances. But the limit of such a solution is that a reconstruction do still quite rarely provide more than only a limited amount of actual information. A picture of an object, even reconstructed one, is still rather obscure if spectator is without any special knowledge on the given kind of objects.

During the earlier Aboa Vetus CA projects and according to the experience gathered during their evaluation it has become clear, that a museum exhibition is highly demanding environment for working IS applications. The most critical factors to meet were according to the experiences and surveys the variable background of potential users and extremely short time each individual actually uses a computer station placed in the museum environment. Of the different background factor two of the most difficult issues to comply were apparently the relatively large potential user group with only basic computer literacy, and the widely different levels of interest and previous knowledge on the subject among the users. The problem of extremely short time usage frame was decided to be included in the specification, but it was apparent that it was plausible rather to try to increase the usage time than to try to adapt to it.

While concentrating more on the user issues it was noticed during the research, that the user needs were still formed in a rather common structure, similar than for instance in the AVANTI project described in Fink, Kobsa & Schrenck 1997. The resemblance was apparent even despite the fact that networking applications and some special groups were decided to be omitted during the first development phases in The Archaeologist Files. It was apparent that The Archaeologist Files should be implemented with fast but still reliably adaptive information retrieval procedures and a user interface that would use common easy-to-learn metaphors.

As the key principles for creating the environment were chosen the previously evaluated criteria for well behaving user adapting information systems. According to the criteria the environment should be 1) appealing as the user should be feeling home using it, despite the premises. Appeal should mean also attractiveness, intuitiveness, understandable metaphors and consistent interaction. Secondly the system should be 2) communicative as being able to interact effectively with the user. The third criteria is the 3) knowledge or information content meaning that the system should be able to deliver relevant information in such quantities that will satisfy the user needs (Vatanen 2000).

As during the Archaeologist Files project the main concern was to counter the potential lack of meaning and absence of relevant information by applying digital technology to place individual finds and pieces of information into a common context, it resembles rather well common information distribution systems. The principle behind The Archaeologist Files was therefore decided to be to weave a network of chronology, of applicable written data, of historical background and of scientific interpretation done by professionals, and to tie all the finds in this structure of information.

The contextualization process was decided to be approached from two main directions. The interactive and adaptive user interface to the knowledge base was the first method and the second was the rebuilding of the information itself into narratives requiring only basic previous experience with computers and a limited amount of interactivity with the interface. The power of narrative in human computer interface had been pointed out by Abbe Don (Don 1990) but as the experiences gathered in the Aboa Vetus also suggested that most of the users were extremely interested in what was visible on the screen by the time they passed by the computer but hesitated to touch it and were literally afraid to interact with the programmes, the narrative seemed to be a correct approach to the problem of limited computer illiteracy.

A narrative was definitely an answer but it required quite extensive research to fulfill all the expectations. The main factor that determines the success of narratives was that they had to be relatively short to capture and keep the attention of an occasional visitor because of the average interest span of only a couple minutes that was determined in the computer use behaviour surveys. Another point was that the narratives had to be relatively easy to interrupt to cruise between them. It was also determined that the programme has to propose a new narrative after one had ended and be able to proceed without direct or definite user interaction.

It was noticed that The Archaeologist Files project had to deal with a variety of design issues regarding the data itself, interface design and implementation and fundamental database management issues. The focus is however on the development of plausible information interrelation mechanisms and their presentation and implementation on the user interface.

Technical considerations

The Archaeologist Files platform was decided to be built on a potentially scalable but still standard office database platform. After some rather extensive testing and evaluation the platform was chosen to be an ordinary Microsoft Access database. The choice was made mostly because of readily available integration tools for interfacing in the existing production environment, the fact that Access is today a de facto standard in small scale database management also in a number of museums, and because it seemed to offer potential scalability to more powerful SQL Server environment if required in the future evolution versions and in the forthcoming web applications. Another point of interest was the announced coming support for XML which was specified as an imperative requirement for the database tool, because it was decided that the project had to be able to follow the international standardization process. If the specifications would have been required more demand on extensive scalability, performance with huge amounts of data or uptime specific requirements, the choice could have been some other database but in the current situation Access was evaluated to be the most feasible alternative.

The process of defining the database structure to ensure its coming interoperability on the XML based data transfer was apparently rather time consuming. A number of propositions and projects have been established to develop DTD's for archaeological XML (e.g. Gray and Walford 1999), but until now, no definite and widely accepted results have been achieved. As the global standards for representation of archaeological data in XML-format will develop the chosen implementation will be further developed to fit these specifications.

As being a hybrid of a pure technology project and a project with more strictly defined business plan, it was also vital to define the requirements for the actual data processing tools and for the front end design. If there would not have been any previous experiences on developing information systems with hypermedia interfaces this would have been a rather demanding phase in the specification cycle. The experiences gathered during the previous projects did however encourage to do only slight modifications to the existing workflow.

The key media content was produced in a number of applications. The textual material is basically the most straight forward type of media but the developed type of hyper textuality employed in the system required developing a specific notation for nested linking which could be translated into XML in the future.

The aim was to use digital-only workflow as far as possible. The most notable exceptions have been the bulk of artistic graphics in the narratives and approximately one third of all photographic material. The finds have been digitized using digital cameras, digital video cameras and scanned drawings which have been imported into AutoCAD and to 3D MAX for visualizations.

The user front end was decided to be designed in two phases. The main front end was decided to be a Windows and possibly a Macintosh hypermedia application authored in Macromedia Director environment. It was decided to be be linked dynamically to the databases and appropriate media contents. An adapted version of this interface application was decided to be produced for web use and various desktop applications to maximize its portability.

The idea behind this concept of designing different versions of application interface is to be able to adapt the most costly part of this structure – the information – into as many uses as possible. The major drawback of multiple interfacing is potential for limited data access with some secondary interface and the need to design more than one front end. The first drawback is possible to avoid by strict specifications and well preplanned concepts. The second one is a major drawback, but with the evolutionary approach The Archaeologist Files is being developed, it is far less a drawback than in the traditional cyclic models of information systems development (for instance STRADIS methodology).

Present evolution versions: the concepts of the Hyper Objects and of the Hyper Doc

The present evolution version of The Archaeologist Files application interface is strongly based on the theoretical background of developing a dynamic interface which allows both narrative and interactive approaches to the information. The interface is intelligent into limited degree and allows seamless integration between different elements and analysis of the user behaviour.

Technically the application is divided into two main interfaces and one secondary desktop. The main view, called the Hyper Objects, allows user to interact with basically all kinds of media content available (Fig. 1). It is a dynamic rearranging interface of individual, in limited manner interdependent, data objects. The key features include the possibility to browse hyperlinked information, to create associations between different kinds of information. An integral part of the design are the features which allow user to choose between active, semi-active and passive interaction with the system. The principle is that more passive the user is, more narrative the system becomes and vice versa.

The working logic of Hyper Objects approach lies in the analysis of the user behaviour. Hyper Objects interface offers first time user a basic interface layout and begins to rearrange it slowly depending on user actions. The rearrangements made are very gradual and the interface never changes more than slightly. The basic methods for adaptation are rearrangement of interface objects into an order which follows the expected user preference, enlargement and diminishing of more and less relevant items, and propositions made by front end of potentially interesting alternatives.

The technique is based on limited goal composition (Nielsen 1994) and on-fly user modelling. Hyper Objects does not at the present implementation preserve user models from one session to another as the user is not generally expected to return to the museum exhibition within reasonable time period again. This approach do pose disadvantages because the expected session is relatively short and the programme has to adapt itself to the user very fast while still trying to be as accurate as possible. The main advantages in the approach are that the system does not require space to preserve the user profiles, and security and user identification issues may be omitted (compare Fink, Kobsa, Schreck 1997).

The other main interface, a hypertext environment, called Hyper Doc (Fig 2), may be seen in some sense as a more traditional interface to the information. The traditional features are however mostly limited to the approach which expects user to have a slightly more formal question than when using the Hyper Objects. Hyper Doc was mainly designed for a limited audience using the system as a search tool to the available information resources and having more exact idea what they are looking for. While the main interface is designed around the concept of in a limited manner interdependent data objects, the hypermedia environment is based on a concept of dynamic self reorganizing hypermedia document. It is not traditional in the sense that a hyper media environment should consist of inter linked documents. The key idea is to dynamically create and maintain an intelligent desktop document containing required information according to the expectations and information needs of different users in different media formats.

The third interface, a map browser, was implemented on the front end application to allow more flexible access to cartographic information. A separate map browser was designed for basically two reasons. While the integrated Hyper Objects desktop allows user to browse the cartographic content, the possibility to inspect this kind of content in three dimensional graphic and temporal space required more specific browsing facilities. The second and more restrictive reason was the limited space available on standard graphic displays. An object on the integrated environment is always, despite the possibility to zoom, limited by the size of the desktop space available. By designing a separate cartographic browser enabled to use the available screen more efficiently and implement all the features expected by the specification.

In the current evolution version the Archaeologist Files front end design is based on a networked approach of different interfaces including the two primary interfaces and secondary browsers. The main difference between the two primary interfaces, the Hyper Doc and the Hyper Objects is in the approach. Hyper Objects is designed for browsing while it allows also more formal requests for information. The Hyper Doc is a logical opposite designed for formal queries, but with browsing functions implemented for refinement and evolvement of the retrieved results. The workflow diagrams of both interface designs (Figs 3-4) give a graphic view of the working logic of both approaches.

Conclusion and future prospects

During the summer 2001 The Archaeologist Files project is well under way heading to the next development phase which will result in actual production applications. The present state of the project is promising and the lately closed last definition phase did produce even more far reaching results it was expected beforehand. The coming tests will provide the project with more actual data on user behaviour for developing further the two interface concepts which will apparently push forward the research.

The main results from The Archaeologist Files have been so far in the development of new working concepts for presenting, storing and managing archaeological data for an audience with extremely varying background knowledge, interests and concepts to approach the data retrieved in archaeological excavations in form of objects combined with other information and data recoverable of the given place and period. The two concepts: Hyper Objects and Hyper Doc will be in the centre of development in the future applications as well as the current production versions. Within the Archaeologist Files results these two interfacing concepts are a major starting point for beginning be to able to provide more meaning to the finds, to increase something that could be called "factual immersion", and to make the finds more understandable especially to the public audience. The principal theme that no one can overemphasize is that the archaeological data have to be presented, and deserves to be presented, in a way which is the best possible also according to the standards of information storing and retrieval, and of communication.

Apart from continuing with the original goals, the research has so far opened a number of new interesting leads. The two interfacing concepts will be no doubt pushed forward and implemented also on more scientifically oriented information systems. Even in the beginning it was apparent that the problem of context with archaeological data did concern also professional archaeologists and historians doing research. Managing the data during the process of trying to understand the past is extremely time consuming when trying to get a somewhat holistic picture of the field of study. The plausible way of organizing data and storing information is vital also for experts who do all have the basic knowledge, but who would still gain advantage by being able to manage their research data more efficiently. The possibilities of implementing the information system including the interfacing engines created during the Archaeologist Files project, and expanding it to wider use, would be a very promising prospect for future work. The theoretical background of creating computer aided archaeological workspace applied in the Archaeologist Files project could contribute to the development of. a common interdisciplinary tool for researchers studying the past.

Acknowledgements

I would like to express my gratitude to following persons and institutions: Aboa Vetus & Ars Nova –museums including the individual members of its personnel, Matti Koivurinta Foundation, The Ministry of Education, museum director Minna Sartes, PhD Kari Uotila and MA Carita Tulkki.

References

Don, Abbe. 1990. Narrative and The Interface. In *The Art of Human-Computer Interface*. Edited by Brenda Laurel. Apple Computer.

Fink, Josef, Kobsa, Alfred, Schreck, Jörg. 1997. Personalized Hypermedia Information Provision through Adaptive and Adaptable System Features: User Modeling, Privacy and Security Issues. In *Proceedings of the workshop "Adaptive Systems and User Modeling on the World Wide Web", Sixth International Conference on User Modeling*, Chia Laguna; Sardinia, 2-5 June 1997.

Gray, John, Walford, Karen. 1999. One Good Site Deserves Another: Electronic Publishing in Field Archaeology. In *Internet Archaeology Issue 7.*

Nielsen, Jakob. 1994. *Goal Composition: Extending Task Analysis to Predict Things people May Want to Do.* http://www.useit.com/papers/goalcomposition.html (downloaded Jan 19th 2001).

Uotila, Kari and Sartes, Minna. 2000. Medieval Turku: The Lost City. A Project trying to reconstruct a medieval town in Finland. In *Virtual Reality in Archaeology. Edited by Juan A. Barceló, Maurizio Forte and Donald H. Sanders. BAR International Series 843*. pp. 219-223.

Vatanen, Isto. 2000. *Effective User Adaptation in Information Systems*. An unpublished manuscript.

Figures

Figure 1A: screen shot of the Hyper Objects interface

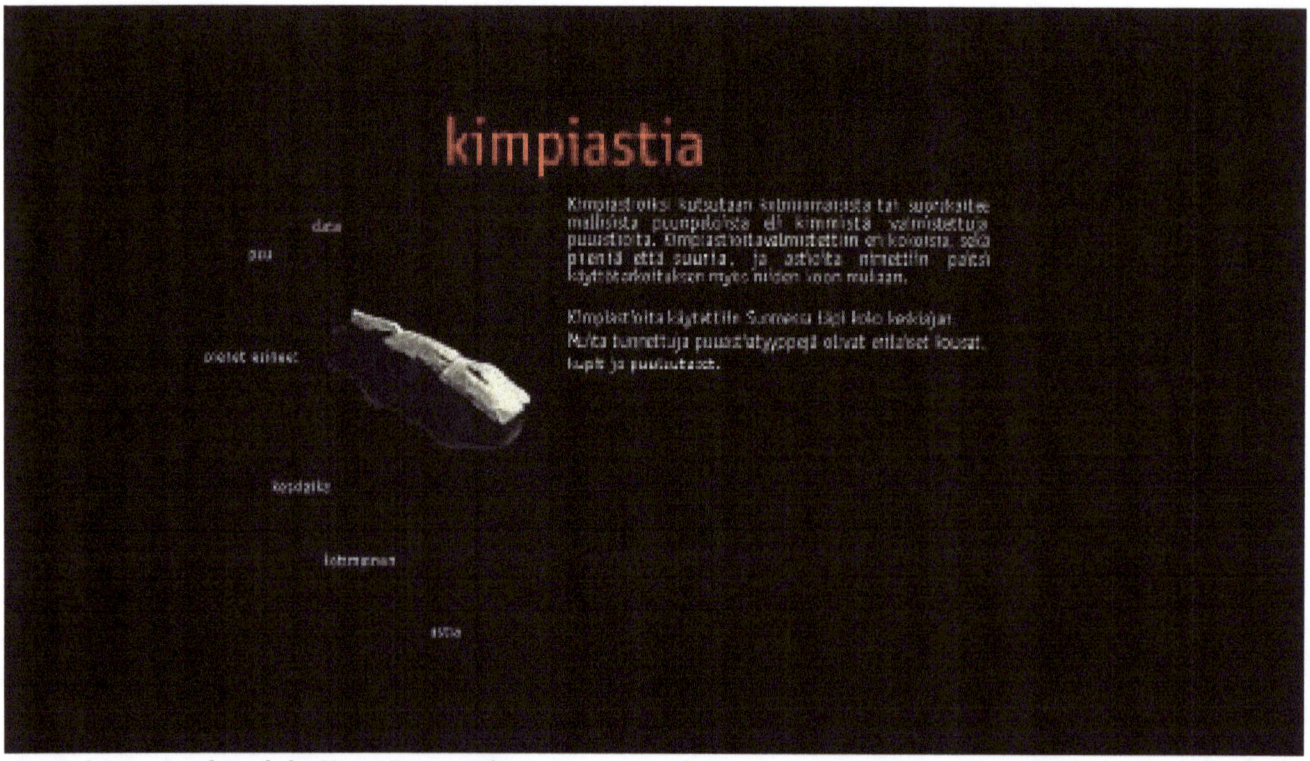

Figure 2A: screen shot of the Hyper Doc interface

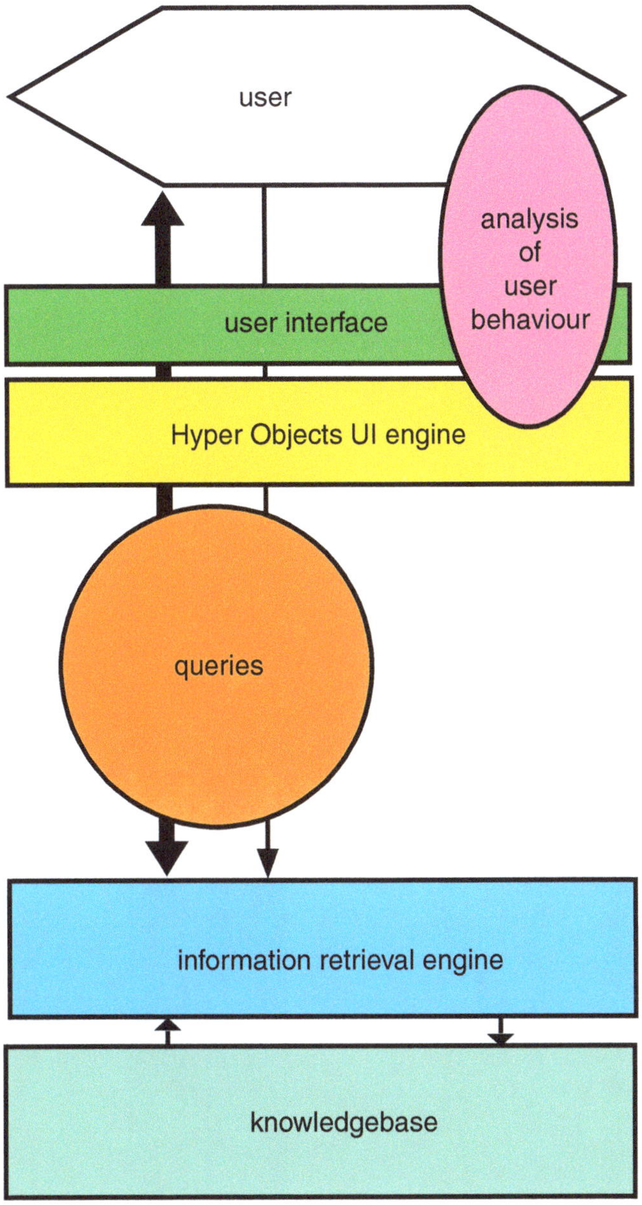

Figure 3: Workflow diagram of the Hyper Objects interface

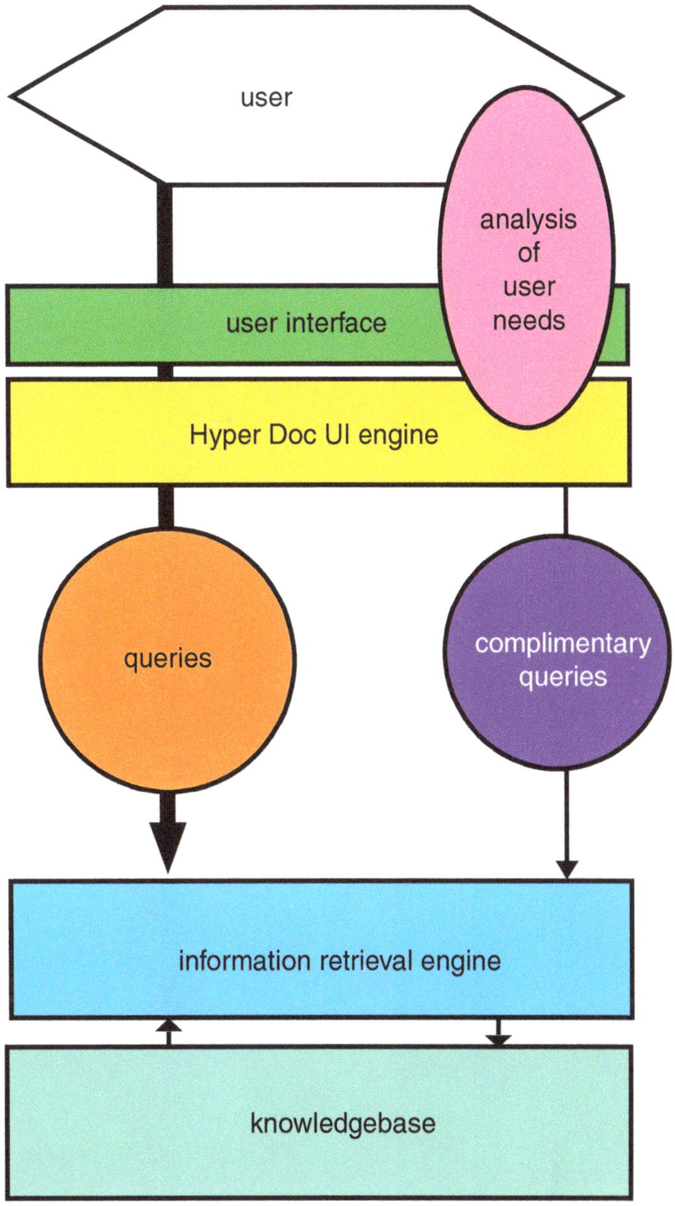

Figure 4: Workflow diagram of the Hyper Doc interface

Papers in the Survey and Mapping, Archaeometry, GPS and CAD categories

New Tools for Understanding Plains Indian Sites in Grasslands National Park, Canada

Gary Adams
Western Canada Service Centre
Parks Canada
145 McDermot Avenue
Winnipeg, Manitoba, R3B 0R9
E-mail: gary_adams@pch.gc.ca

Abstract: With over 3000 Native sites containing more than 20,000 features, site survey in Grasslands National Park has an ongoing issue of trying to get good research information while conducting rapid surface surveys and still providing cultural resource management tools. All of the original survey was done by hand recording techniques. Recently, we have been using GPS to record sites at the feature level and taking some basic feature information. This information has been analyzed within SPANS GIS to provide far more accurate and detailed site maps, at times raising the whole concept of what is a site. It has also been used to analyze tipi rings in large sites to look for patterns of settlement, as it is highly unlikely that large sites with 20 to 200 rings resulted from a single occupation. This technique has successfully broken some of these large sites into patterns that can be hypothesized as individual camps within the larger area. This talk will demonstrate how we have improved on initial survey techniques and how the intrasite analysis has been used with some positive results.

Introduction

This paper presents some initial findings on survey field methods that take advantage of Geographic Positioning and Geographic Information technologies to produce better results than previous methods, saving time, money, and promoting better research in an applied situation. The paper has three basic objectives:

1. To identify issues of survey mapping
2. To describe field recording techniques used in Grasslands National Park, and
3. To present results of an analysis of feature mapping.

Grasslands National Park was established in 1981 to represent Canada's mixed grass prairie. With its establishment, a small portion of North American Great Plains has been protected and made available for public use. The Great Plains are characterized by relatively little topographic relief, wide expanses of grassland, and a semiarid climate. The northern portion, where this park sits, is further characterized by short, hot summers, long, cold winters, and periodic, intense droughts. The former habitat of the bison, it is still the home of pronghorn antelope, elk, and such endangered species as the swift fox, burrowing owl, and prairie dog. The park is accented by a massive glacial spillway through which the Frenchman River flows.

The park is established in two separate blocks. The West Block features the Frenchman River valley and its surrounding terrain. The East Block is oriented to open prairies, running from the Wood Mountains, prominent hills that mark the north-south continental divide, south to a large erosional feature called the Kildeer badlands. The entire park is actively being created from the purchase of private lands on a willing buyer - willing seller basis. To date, approximately 50% of each block has been transferred to the park.

The Evolving Methodology of Area Surveys

Simply because it is so open and accessible, the park initiated an intensive archaeological survey of the surface remains on all owned properties. At that stage, the primary objective was simply to find the sites (Adams and Filopoulos 1995:3-6) and describe their approximate nature and extent. Two field workers spent a month or two each summer over six years between 1985 and 1995 hiking the park and recording its cultural features. The basic methodology was to establish a daily traverse over a given set of terrain features, using air photographs as guides. Each encountered site was recorded on a survey form that included basic site information and a sketch map. Even the most significant of sites with 50 or more features seldom took a half hour. Locations were then transferred from the air photos to a 1:50,000 topographic map to obtain coordinate information and to a 1:20,000 air photo mosaic to identify terrain features and the site boundaries. To determine site entities in a highly concentrated area of cultural features, the practice was to create a new site if 100m or more distance occurred between cultural entities, or, if the terrain contained a major break, like a cliff wall. The surveyors ignored small, isolated flakes of chipped quartzite, otherwise the whole park would have been designated as a half dozen sites.

From the onset, it was realized that following this form of survey, the surveyors could miss major features and links between sites. In one practical test of the methodology, an instrument survey of a single site in 1991 revealed just how much error there might be. The site was originally described as 35 tent rings spread over a piece of knob and kettle terrain around a small slough. Three people working for two days with an alidade and plane table, broadened that to 65 tent rings that came dangerously close to merging with three other sites.

At the conclusion of the survey, 3100 archaeological sites were inventoried and entered into a database. The major features that were encountered included:

- Tipi Rings: 13,000 circular configurations of cobbles left behind from a tipi. They were originally used to hold down the skin covering. These are critical archaeological resources as they outline households, define living space, and establish both community and landscape settlement patterns
- Cairns: 2,700 large piles of rocks with many purposes including burials, caches, beacons, ceremonial functions
- Bison Kills: 11 major communal hunting sites where bands congregated to herd bison over cliffs or into pounds where they could be dispatched in large numbers for feasts, immediate needs, and preserves of meat. Identified by linear rock lanes that extend up to 10 km and funnel toward the kill site,
- Vision quests: 14 small cobble nests where a young man would retreat for a period of fasting and meditation to obtain a vision that would set his name and his destiny
- Effigies: three cobble configurations laid out in human or animal forms
- Medicine wheels: 11 wheel-like rings with spokes and a central hub that were probably used in ceremonial activities
- Cobble configurations: 62 undefined alignments of cobbles.

The original methodology suited the purposes of rapid recording and creating an overall inventory, but had severe limitations on quality of data. Periodic quality checks demonstrated that small sites were occasionally misplaced due to the problems of map placement. The survey test identified that large, but nonlinear sites could have as many as half the features unrecorded. There was very little data being generated for further analysis and, finally, there was little confidence that the original site border descriptions were accurate. However, this was offset by the benefit of a complete initial mapping of the park. The quality checks indicate that approximately 95% of all surface sites larger than 10 m^2 have been recorded. It gave both park managers and researchers a base line inventory of what to expect when called back to evaluate land for environmental impact assessments, trail development, prairie fire damage, threatened resources, or new research.

The known data gaps and quality problems began to be addressed in 1998 (Adams 1999). Some of the initial sites are now being reinvestigated as part of an ongoing project to monitor and evaluate resources. Though this program is multifaceted and conducted with limited resources, one aspect is to examine the whole survey methodology. The current survey component is focused on intrasite mapping of selected sites where significant research or interpretation potential was evaluated in the initial visit. The methodology for the re-survey has changed substantially.

First, two surveyors traverse a defined geographic area with a single site or several closely spaced sites, locating as many features as possible and marking each with a flag. Once an area is prepared, the team returns with the G.P.S., a tape measure, and a compass. While the coordinate readings are being taken, a specially prepared sheet is filled out to identify and number the feature, measure its diameter in two directions, and record incidental notes. If it is a tipi ring, it is divided into eight pie-shaped pieces from magnetic north, and the number of perimeter cobbles for each segment is counted. Additional features like internal hearths, double lines, and cobble platforms were also noted. In some sites, a minimum of 40 individual readings are taken with a Trimble Navigator Basic Plus at the centre of each feature. Readings were differentially corrected with Pathfinder Office version 1.10 software using base station data from the U.S.F.W. office in Lewiston, Montana and averaged to provide a point location with about +/- 3m of accuracy. When the selected availability is turned off, Garmin G.P.S. 12 XL units are used without correction to obtain averaged readings with similar error magnitudes. That done, the surveyors recheck the area for any additional features. The flags are all removed at the end of the survey. Two people, using a form for consistent observation and recording, can map and describe between 50 and 150 features in a day, depending on the number of attributes recorded.

Feature mapping has one very practical benefit. It allows the archaeologists to create reasonably precise site maps with a minimum of effort by superimposing the archaeological data points over a geographic base map. One major problem is that the scale of feature mapping incorporates accuracy factors that are not yet available in the core data sets. Most of the background data for the park was generated from the 1:50,000 National Topographic System maps. When attempting to overlay individual features with coordinates that are often less than 10 m apart onto this background, numerous small errors have very visible implications.

The Problem of Large, Complex Sites

In many parts of the park, the initial survey had identified clusters of large tipi ring sites, converging on a favourable location or accompanying a communal bison drive site. After the experience of hand mapping one such area and doubling the number of rings in the site, the archaeologists were concerned about validity of the previously identified site boundaries and the accuracy of the original feature counts. In these compacted habitation areas, there was also some doubt as to the applicability of the original definitions of a site. The working definition used to characterize a site has created a situation where the Archaeologists have defined their very nature is based upon a few spatial constructs that make good management sense but may not have related to the original settlement pattern.

The first part of this problem, a more accurate count and description of the features has been adequately addressed by the shift in methods. Though there undoubtedly a few features within a given site that are still unrecorded, the vast majority are now identified, correctly located in relation to each other and described with enough attribute data to permit further analysis of settlement patterning.

An experiment to address the aspect of border definition was conducted at a locale called the "Buttress" (a term given to the hills because of their fortress-like shape). The Buttress contains a particularly dense collection of sites that incorporates almost every kind of feature within a small geographic area. The original, manual survey identified 18 sites in a small land holding containing less than two sections (518 ha). In this area, the northern sites are all habitations of four to 50 tipi rings and associated features. The southeastern sites are processing sites with few features. Bison drive lanes run right through the complex. This configuration has always implied some interesting research problems related to settlement patterns. Was the Buttress heavily used as the gathering spot for small groups to camp together and direct a communal bison hunt? It appears to have all of the constituent parts to make that a possibility; ring sites, processing areas, drive lanes, large and small cairns, diverse terrain, access to water, and even a bison gathering basin. Alternatively, was it a favorite campsite for small groups that reused the area repeatedly for centuries, creating the impression of a large site complex?

In a specific portion of the Buttress, measuring 1400 by 600 m, the initial survey had identified ten sites and half of an eleventh site that continued outside the test area. This entire area was re-surveyed using the feature methodology, with no reference to the known sites and the data from the two surveys were compared by superimposing them on a GIS map. It became immediately apparent that the original survey data bore little resemblance to the spatial distribution of the features. While some of the smaller sites retained their original distinctions, three of the largest coalesced into a single "supersite" even though the total number of features within the test area remained relatively similar. Experiments in two other site concentrations have not yielded the same magnitude of error. It is now believed that the degree of error is directly related to the surrounding terrain. The problem is considerable where there are no natural boundaries like river channels, valley rims, or large hills that the surveyors could use to establish and reaffirm their bearings in the initial survey.

Intrasite Feature Patterning to Define Occupations

The shift in tipi ring patterning noted in the test area led to a more detailed analysis to determine whether culturally defined settlements might be hypothesized within the test area and its 89 habitation features. The test for cultural patterning presupposed that there was some organization to how camps were arranged. Laubin and Laubin (1957:231-242) go into some detail about camp circles and the different ways large gatherings establish circular camp patterns in summer and linear patterns in winter. Other historic accounts refer to camp circles being preferred in homogenous environments while linear arrangements were common along watercourses or other linear features. Small camps rarely conformed to these patterns. Instead, the primary determinants of camp patterns in small sites were simply distance and family relationships (reviewed in Quigg and Brumley 1984). Across the Northwestern Plains, archaeological surveys mostly encounter the remains of small camps with no apparent pattern. However, even larger archaeological sites seldom exhibit any distinctive camp pattern. One explanation might be that superimposed camping obscures the patterns of the individual camps. If a large site was a preferred camp location, several small encampments would leave the impression of one large one on the ground. One 38-ring site in Alberta is known to have at least six cultural groups represented over a range of 4000 years. Yet every ring was visible on the surface (Stuart 1990).

Within the test area, a map of the features surveyed revealed two obvious patterns of rings. Each of these concentrations has four or five rings in a north-south alignment, and a single tent to the east. The north to south alignment is a good initial indicator as it is an ideal arrangement to provide an opening to the rising sun. Using these two subsets as guides, a series of assumptions was used to test whether or not other patterns could be discerned.

The first assumption was that if families lived in a single group at one point in time, and shared a very similar technology, then it would follow that their tents would be very similar and their footprint on the ground would be similarly sized. The validity of this assumption is very questionable as no researcher into Northwestern Plains tipi rings has yet to demonstrate any relationship between ring diameter and age or cultural affiliation. Quigg and Brumley (1984:30), in fact, offer a completely different perspective. However, for the purpose of this study, a contour analysis was run to create contour intervals for each size of ring in metres. In that analysis, the two guiding areas appeared to have similar sized rings though the ring to the east was different in both cases.

A long-standing argument in archaeological literature debates the value of "rock loading," a process by which more rocks are found in the octants (one-eighth) of a stone circle that face the prevailing winds. Brumley and Dau (1988:127-130) found a general correlation between mean octant rock loading and mean yearly wind velocity in Forty Mile Coulee but were less successful in identifying the month of year or contemporaneity with a set of rings (see also Krozser and Hjermstad 1995 for a discussion). The numbers of rocks for each octant, starting from magnetic north, were recorded. The initial mapping of these contained too many tied numbers and spurious concentrations so the data was reorganized in two segment units. The two adjacent segments with the highest numbers of stones were plotted. This analysis demonstrated that there was no precise uniformity in the two guiding areas but there was general agreement. For instance, in one control, two rings had rock loading to octants 1&2, one to 2&3, and one to 3&4. This led to a categorization of rings with identical rock loading and adjacent rock loading characteristics.

Many tipi sites have central hearths within them for cooking, heat, and light but many do not. There are also several styles of hearths, some with rock liners or rock bases and others that lie flat on the prairie with no liners. Absence of internal fires can only be a summer phenomenon when there are long light periods, and cooking fires would actually create more heat than is desirable. This resulted in the third assumption that rings in close proximity with or without surface evidence of fires may indicate contemporaneous seasonal occupation. Since none of the sites are excavated, evidence of fire was taken in a hierarchical approach. No evidence of fire; evidence of fire broken rocks in or near the ring; small clusters of cobbles within rings that could be hearths; and both forms of evidence; increasingly indicate the presence of an internal fire.

The final assumption made was that tipis within a single occupation would be set up in relatively close proximity to each other. Therefore, tents with very similar attributes but great distances apart would not be considered as part of the same set.

Even at this simplified level of analysis, the variables were difficult to handle. Eventually, 13 clusters (B to N) were isolated, each with consistent attributes of proximity, size, rock loading, and fireplaces. Category M was unusual in that all of the rings were of large diameters, a metre larger than the other rings but they did not cluster together. Instead they seemed to be almost regularly spaced across the large hill that served as the focal point of the study area. The residual rings were then reexamined to find close, but not perfect, fits with the existing categories. In this way, eight of the categories developed variants (C1 and C2 or D1 and D2, for example). These variants were interpreted together as a single category since the differences between the rings were minor. See fig. 1.

Several of the clusters (B, C, D, H, and L) did organize into straight lines, suggesting linear arrangements of three to four tipis. Only one group (M) appeared to form an arc of five rings. As noted above, the largest rings (J) seemed to be spread out widely over the entire site area. If these are contemporaneous, the form a pattern as yet undefined in the literature. One other group (E) seems to be too widespread to be considered a single occupation though all three examples are within a small portion of the study area. Group G was discounted as it has only two rings. The remaining groups (F, I, K, and N) all appeared to be in tight, relatively formless clusters.

The existence of the groupings also delineated another pattern, formerly unrecognized. If the patterned groups of M and B are considered to be separate components and are removed from the landscape, the remaining rings in the northwest corned of the study area form an almost perfect camp circle, 120 m in diameter, incorporating group L and a number of ungrouped rings.

Concluding Comments

The observed patterns are all extremely tentative and of little direct applicability for interpretation. However, they have established a testable hypothesis for further investigation. This same technique can be applied in other large sites to determine which rings should be excavated in situations where developmental pressures are requiring environmental impact mitigation or where excavation research funding is available.

This paper was developed to demonstrate how Geographic Information Systems and Global Positioning Systems have improved or will benefit a field archaeologist, working under cultural resource management frameworks. Within that context, two points appear to stand out.

1. Without the GPS, none of this would have been started. In a non-research format, financial support does not exist to allow traditional intra site mapping on this kind of scale.

2. Without GIS, the analysis of the data would probably not have provided the same observations. The ability to read the attribute data spatially and make interpretations on the nature of that data was a crucial step in the analysis.

References

Adams, Gary. 1999. *Grasslands Monitoring and Field Investigations, 1998 Field Season.* Manuscript on file at Grasslands National Park, Val Marie, Saskatchewan.

Adams, Gary and Filopoulos, Peter. 1995. *Grasslands National Park Archaeological Inventory, A Place for All Seasons* (2nd Edition). Manuscript on file at Grasslands National Park, Val Marie, Saskatchewan.

Brumley, J. H. and Dau, B.J. 1988. Historical Resource Investigations Within Forty Mile Coulee Reservoir. In *Archaeological Survey of Alberta Occasional Papers No. 13.* Alberta Culture, Edmonton.

Krozser, Kit and Hjermstad, Ben. 1995. *Stone Circle Treatment Review and Workshop Final Report. Prepared for Resource Management - Archaeology*, Heritage Branch, Saskatchewan Municipal Government. Regina.

Laubin, Reginald and Laubin, Gladys. 1971. *The Indian Tipi: Its History, Construction and Use.* Ballantine Books, New York.

Quigg, J.M. and Brumley, J.H. 1984. *Stone Circles: A Review, Appraisal and Future Directions.* State Historical Society of North Dakota, North Dakota Heritage Center, Bismark.

Stuart, Glenn S.L. 1990. The Cranford Site (DiPb-2) A Multicomponent Stone Circle Site on the Oldman River. In *Archaeological Survey of Alberta Manuscript Series No. 17.* Alberta Culture and Multiculturalism, Edmonton.

Figures

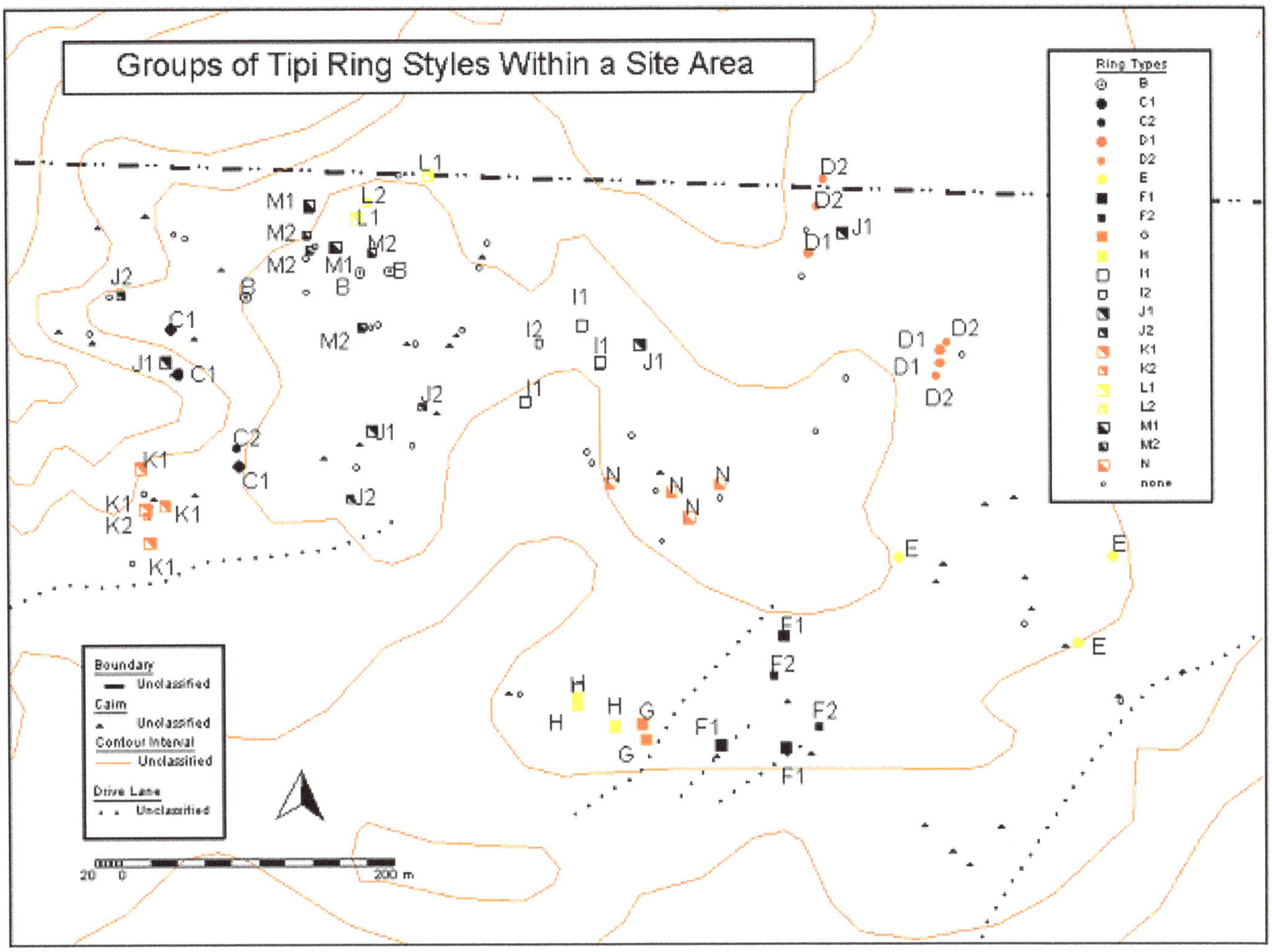

Figure 1.

Archaeological Survey as Optimal Search

E. B. Banning
Department of Anthropology
University of Toronto, Toronto, Canada M5S 3G3
Phone: +1-416-978-2315 - Fax: +1-416-978-3217 - E-mail: banning@chass.utoronto.ca

Abstract: The research designs of many archaeological surveys suffer from a failure to distinguish between the goals of prospection, estimation, and pattern recognition. Often surveyors expect sampling methods, which are designed for estimation, to act as detection tools, or use predictive models without regard for their purpose. This paper shows how many problems of archaeological detection are suited to the principles of optimal search in Operations Research, and illustrates this with an example of optimal allocation of search effort.

Key words: archaeological survey, Bayesian probability, detectability, Game Theory, optimal search, Operations Research, prospection, purposive survey, spatial sampling

Introduction

It is easy to find examples of archaeological surveys that failed to detect a Teotihuacán (Cowgill 1975: 260, Flannery 1976: 133-134), that provided biased estimates of site density, or that yielded none of the economic or contextual information needed to investigate a particular research problem. Yet surveys are also uniquely able to provide some kinds of important data, and especially ones at the regional scale. The key to more useful surveys is to tailor surveys' designs to their objectives. The following paper will introduce a set of tools that can improve the effectiveness and reliability of one type of survey in particular, while also having more general implications.

Types and goals of archaeological survey

I classify surveys broadly into three categories, on the basis of their principal goals, although some surveys certainly combine aspects of more than one.

Today, most surveys are of the sampling variety, designed to estimate parameters, such as site density, mean edge angles, population size, or the proportion of sites that are Saxon settlements. These surveys can also be used to test statistical hypotheses or to devise predictive models for the location of archaeological materials. A common design for such a survey is to divide some region of space into regular, geometric spatial units, which serve as the sampling elements. In that case, sampled artifacts or sites typically constitute a cluster sample of the population of sites or artifacts in the region. Note that here I use the term "sample" in its narrower, statistical sense, and not merely to describe some subset of a larger whole.

Other surveys are meant to detect and understand spatial structure, such as Christalleran settlement lattices, rank-size distributions, the scale of artifact clustering over continous space, or the organization of road or canal networks. Usually such goals would be ill-served by a spatial sample because it would not provide sufficiently continuous coverage of the region for the patterns of interest to be recognizable. Some archaeologists, consequently, have argued for so-called "total survey" in these situations (e.g., Ebert 1992, Fish and Kowalewslei 1990).

Finally, what I call here "prospection" is survey meant to find particular targets or kinds of targets, to test hypotheses that predict the specific locations of targets in space, or to ensure that the other two types of survey are sufficiently thorough to achieve their goals. Effective or successful prospection, or "purposive survey," as it is sometimes called, finds sites, features, or artifacts. In other words, its purpose is detection. Here you will note that I use "prospection" quite broadly, and not only to refer to geophysical remote sensing.

Many archaeologists have shunned purposive survey and used sampling as a discovery method, apparently out of fear of seeming unscientific. Statistical or sampling surveys are, however, a poor way to find rare or specific targets, as archaeologists have known for a long time, because sampling is meant to find typical or common ones, or to represent central tendency in a population (cf. Cowgill 1975: 260).

Prospection, by contrast, takes advantage of information that may improve our chances of finding the archaeological materials of interest, even when they are rare, but may be a poor basis for statistical generalizations. Prospection also allows us to balance the costs of survey and risks of missing these materials, which for convenience I will call "sites" or "targets," although they can be "non-site" evidence or individual artifacts (e.g., Dunnell and Dancey 1983, Ebert 1992, Foley 1981). Among the goals that prospection may face are minimizing costs or search time, operating within resource constraints, or minimizing the risk of missing targets. Where prior information is available, purposive survey is more productive than a sample (Cowgill 1975: 260, 27, Read 1975: 45, Schiffer et al. 1978: 18).

Prospection can involve searching for specific, historically documented targets or, more generally, can help us find examples of some class of sites, especially rare or unusual ones, which random sampling and its variations tend to miss unless sample size is unusually large.

Prospection of the former type is perhaps most familiar in maritime applications, when we know the date and sometimes even the cargo of a shipwreck, but not its exact location. For example, historical documents explicitly describe the orientation and approximate location of ships of the Chesapeake Fleet, which the Americans scuttled in the Patuxent River in 1814 to keep the ships from falling into the hands of British forces (Shomette and Eshelman 1981). Prospection builds such information into a search strategy to improve our chances of finding the ships with limited resources. Such a strategy helped Tommy Thompson, a non-archaeologist, locate the SS Central America in 1986, which was known to have sunk in the Atlantic in 1857 (Kinder 1998).

Prospection of this type is also applicable, however, to some examples of terrestrial survey, such as cases where historical documentation refers to some fort or settlement, the precise location of which has been forgotten. For many years archaeologists have used historical information to narrow such a search, as in the search for Columbus's settlement at La Navidad (Deagan 1989), the search for Mission Santa Catalina de Guale, or even Schliemann's search for Troy. Yet, although recognizing the need for "a systematic approach to prospection, reducing the area of search and thereby incrementally increasing the probability of success" (Garrison et al. 1985: 302), searchers have rarely selected near-optimal strategies for this purpose, and sometimes even begin with a random sample that does not take advantage of prior information. In a more productive vein, Lolley (1996) used historic maps to predict the locations of Upper Creek Muskogee settlements in Alabama. He combined this prior information to make a more accurate map on which he marked the predicted locations of these sites with circles whose areas represented the uncertainty of location, and used these to recommend the distribution of survey effort.

As an example of the second type, searching for some class of sites rather than a particular site, some researchers (e.g., Storck 1978, 1982, 1984) have long used the tendency of previously known Paleoindian sites to occur on particular landforms, such as fossil beach ridges, to narrow the search for undocumented Paleoindian sites. Today the use of GIS predictive modelling (discussed below) has refined this kind of approach.

In spite of such applications, archaeologists have been largely unaware of a large body of theory explicitly devoted to searches of these kinds, found in Operations Research, that could help them improve the effectiveness of their surveys. Sometimes they have also reinvented tools that had already been available for several decades.

Operations Research

Operations Research began with efforts by Koopman (1980) and other mathematicians to optimize the US Navy's searches for German U-Boats almost 60 years ago. These researchers also worked out near-optimal ways to conduct search-and-rescue operations (Richardson and Discenza 1980), including transect spacing and orientation and distribution of search effort, and to track multiple moving targets (Stone et al. 1999). The principles were later extended to fields such as petroleum exploration and mineral exploration (e.g., Cozzolino 1972, Drew 1967, 1979, Harbaugh et al. 1977, McCammon 1977, Savinskii 1965, Singer 1975, Singer and Drew 1976, Singer and Wickman 1969), which share many of the same problems that survey archaeologists face, and especially the need to detect subsurface targets.

Archaeologists have rarely referred to any of this literature, in spite of previous archaeological interest in Optimal Foraging Theory, which uses some of the same principles and concepts, including encounter rate, patchiness, and cost functions (see Winterhalder and Smith 2000 for a recent review).

The one area in which Operations Research has had some archaeological impact is in the optimal arrangement of augers, cores, and test pits and the evaluation of their detection probabilities, including work by Kintigh (1988), Krakker et al. (1983), and Zeidler (1995), all of whom cite geological applications in this area. Other research of this kind (e.g., Nance 1979, 1983, Nance and Ball 1989; Shott 1985) and work on the edge effects of transects and other geometric survey units by Plog et al. (1978) progressed independently and apparently without benefit of the earlier publications in Operations Research or earth sciences. Even archaeologists working for the US Department of Defense seem unaware of the US Navy's substantial and long-standing contributions to search theory.

Approaches to optimal search in Operations Research

Broadly speaking, "classic" optimal survey strategies have emphasized one of two approaches. Koopman (1980), Stone (1989), and others who pioneered Operations Research emphasized Bayesian approaches (Dobbie 1968, Washburn 1981). Corwin (1981), Gal (1980, 1989) and others explored the Game Theory approach. Recent advances in search theory have brought novel methods, including application of genetic algorithms (see Reynoso and Jezierski, this volume) to search problems, and some of these may conceivably have potential in archaeological searches.

The Bayesian approach uses prior probabilities based on existing information, analysis of expert opinion (Morris 1977, Savage 1971), or even educated guesses, to try to optimize the allocation of search effort over space (Koopman 1980, Stone 1989, see also Buck et al. 1996). It can also adapt to new information as survey progresses to refine the survey strategy. Typically it assumes some probability distribution for the target's location, such as a bivariate normal distribution over a potential shipwreck. In the worst-case scenario, it might assume a uniform distribution, with all parts of the survey space equally likely to contain the target or targets. In such instances we end up with a strategy resembling the long-familiar sampling surveys.

Game Theory uses some methods already familiar from

Optimal Foraging Theory, such as the Minimax approach, and treats the search as a two-person, zero-sum game between the searcher (the archaeologist) and the hider (the target or site). This is like Optimal Foraging Theory's focus on foragers trying to optimize their encounters with prey (e.g., Paloheimo 1971). Often the goal is to find the trajectory or sequence of search spaces that will probably find the target most quickly. Sometimes, alternatively, it is to ensure against the risk of not finding the target at all. As in Optimal Foraging Theory, it uses a cost function to evaluate the risks and benefits of various strategies. Generally it assumes that search consists of movement through a sequence of spaces or a continuous sweep of space, as in the transects that archaeologists typically use for surface survey by visual inspection.

Factors affecting discovery (detectability)

In both Bayesian and Game-Theory approaches to search, we also need to assure ourselves that interception of the target will probably lead to its detection. However, as archaeologists are well aware, many factors can confound detection, even when we are standing right on a site. Archaeologists (McMannamon 1984, Plog et al. 1978, Schiffer et al. 1978) have identified some of these factors as visibility, obtrusiveness, intensity, and so on. Others have stressed the distinction between the probability of intersecting a site and the probability of detecting it, given that one is present (e.g., Krakker et al. 1983, Lightfoot 1986, Zeidler 1995). Operations Research addresses these same issues, although usually under different names, and provides a framework to minimize the risk that we will fail to detect a site we intersect, and the converse risk that we will mistake a "false target" for a site.

Among the factors that affect discovery are ones concerning the properties of the target. Archaeologists have summarized these as "obtrusiveness" (Schiffer et al. 1978: 6). In fact obtrusiveness is a function of both the target's properties and those of the target's immediate environment. Successful detection depends on a contrast between these. For example, during a magnetic survey, an iron cannonball is highly obtrusive if it is buried in typical soil or sediment, but it would escape detection if it were buried in iron filings because there would be no magnetic contrast between it and its environment. For artifact scatters detected by visual inspection, test-pitting, or augers, obtrusiveness is largely a function of artifact density (Krakker et al., 1983, Stone 1981) and artifact clustering (Nance and Ball 1986, 1989), although other factors come into play.

Another factor that affects detectability is the type of signal that communicates information about these properties from the target to a sensor. For example, the signal could be seismic, magnetic, or electromagnetic, and includes visible light.

A third factor is the medium of signal propagation, which archaeologists have summarized as "visibility" (Schiffer et al. 1978: 6). Visibility through a medium such as sediment or air may be poor or even zero for some types of signal while the same medium is perfectly transparent to some other type of signal.

Another factor is the kind of sensor or method of inspection. For example, we might use our unaided eyes to conduct survey for surface artifacts, but a magnetometer to detect anomalies in the magnetic field due to buried iron objects or sediments whose magnetic susceptibilities contrast with that of their environment.

Yet another major factor is our ability to recognize the signal and correctly classify or identify it. In practical situations, there are often "false targets." These are phenomena that send signals we find difficult to distinguish from those of targets of interest. For example, some buried natural irregularity in the bedrock could produce a magnetic or electromagnetic signal that we mistake for that of a buried pithouse, or the discovery of a single artifact in a test pit may be ambiguous evidence for the presence of a site (or cluster of artifacts). The incidence of false targets can cause us to waste valuable time digging test trenches to investigate ambiguous remote-sensing anomalies or putting additional test pits in the area around isolated artifact discoveries to see if more evidence for a site is forthcoming (e.g., Lightfoot 1986).

Some of the many other factors we need to consider are resource constraints, crew training and motivation, accessibility, density of survey effort, and even weather and time of day.

Factors in optimization

Some of the factors that Operations Research considers in order to optimize searches include cost (in search time or effort), resource constraints, the area or number of spaces to be searched, the expected number of targets, the likelihood of false targets, and the relationship between point or transect spacing (or sweep width), the range of our eye or other sensor, and the probability of detection. Archaeologists have usually used the term "intensity" to refer to the cost or effort involved in searching a space, but the term tends to be confusing because it can refer to total effort, to amount of effort per unit area (density of effort), or to the spacing (resolution) of transects, auger holes, or magnetometer measurements. Operations Research treats these separately, and cost refers to the total effort devoted to a search space, usually measured in either time or person-hours. The range is often conceptualized as the lateral (or perpendicular) distance between a transect and a target but, in remote-sensing applications, can also refer simply to the distance between the target and a sensor.

Unlike U-boats and antelopes, and fortunately for archaeologists, archaeological targets are stationary, which makes the search problem much easier. However, we still need to consider whether it makes more sense to conceive of the search space as a set of discrete spaces, such as topographic features of the landscape, arbitrary geometric spaces, or the pixels of a GIS Digital Terrain Model, or as continuous space, such as part of the open sea over a shipwreck. Our way of sensing the archaeological remains can also be discrete or continuous. Resistivity measurements, crew visits, auger holes and test pits involve discrete "peeks" at what lies below, while surface survey and Ground-Penetrating Radar may involve virtually continuous scans or sweeps of the ground.

Discrete peeks or glimpses run the risk of missing remains that lie between observation points. Some of the literature concerns ways to minimize this risk, and several authors have published on the optimum spacing and pattern of points to detect circular (e.g., Kintigh 1988) and elliptical targets (e.g., Drew 1979, McCammon 1977, Singer 1969).

Sweeping the ground, however, does not always assure detection either. Sometimes it is convenient to assume that detection is assured as long as the target comes within range R of a transect, sometimes called the discovery radius. In that case, a transect spacing of 2R makes a "clean sweep" of an area or region. So-called total survey depends on this assumption, which Operations Research describes as the "Law of Definite Detection" (fig. 1, Koopman 1980: 82-83). Of course, different kinds of sites have different discovery radii, and it may only be possible to make a clean sweep of the more obtrusive sites.

In many instances, furthermore, we know that the probability of detection falls off rapidly with distance. Often, the Inverse-Cube Law pertains, with detectability of targets inversely proportional to the cube of the range (fig. 1b, Koopman 1980: 57-67). Consequently, in these cases, we could fail to detect sites even when they are quite close to a transect. This has led to surface surveys with narrow transect intervals, or even crawling on the ground, to minimize the maximum distance between observer and target.

Archaeologists have also sometimes tried to address this problem with a variety of search geometries, or by surveying each space with two complementary geometries (fig. 2). For example, in surface survey, "wavy" transects are intended to decrease the systematic omission of between-transect observations (e.g., Mortensen 1974), but make it more difficult to evaluate detectability. In surface survey or distributional archaeology (Ebert 1992), a second set of transects is often at right angles to the first, the rationale being that repeating the survey from a different angle (and with different lighting conditions) will reveal some artifacts missed in the first pass. Operations Research has also investigated this problem and found that, for double sets of transects, it is optimal for the second set to be diagonal to the first (fig. 2d), rather than at right angles, because it maximizes the area of new ground covered .

In the Bayesian approach, meanwhile, some situations (namely when the prior probabilities make a circular normal density function) call for a spiral trajectory expanding from the point of highest prior probability, as in the case of a shipwreck or historic settlement whose approximate coordinates can be estimated. For practical reasons, it is usual to approximate this with an Expanding Square centered on the Point of Fix (mode of the probability distribution; fig. 3, Koopmans 1980: 214-227), and with a spacing interval (S) that depends on the distribution's standard deviation (σ) and the sweep width (W) or "effective visibility" (E) of the detector. E is half the sweep spacing that results in a detection probability of 0.5. For definite detection, W = E, but for inverse-cube detection, W = 1.076E (Koopman 1980: 76-77, 217). We can then define the spacing interval as

$$S = 0.75\sqrt{E\sigma}$$

Note that, if definite detection applies, posterior probabilities fall rapidly to zero as search progresses without finding anything, so that the posterior probability distribution can quickly become doughnut-shaped (Koopman 1980: 216), rather than bivariate-normal.

Target detection

Not surprisingly, the conditional probability of detecting a target, given that the target is present, increases with the amount of time we devote to searching for it. For continuous searches with definite detection and accurate transect trajectories, this relationship is linear. For discrete searches with uncertain detection, however, the quantity, g, is the instantaneous probability of detection by one glimpse. Thus g summarizes the factors, including visibility and obtrusiveness, that contribute to detectability. If each glimpse is independent of the others, the probability of each failed glimpse is (1 – g) and we multiply this n times to find the probability of not detecting the target in n glimpses. The converse of this, that is, the probability of detection in n glimpses, is then

$$P_n = 1 - (1 - g)^n$$

More generally, the probability of detection in the time interval dt is

$$p(t) = 1 - e^{\gamma t}$$

This is one way to express the exponential detection function, where g and gð depend on physical conditions of visibility and obtrusiveness, and t is the amount of time devoted to search. Although the exponential detection function it is somewhat unrealistic, in that it assumes that glimpses are randomly distributed and independent of one another, it actually is a conservative detection function (fig. 4 and Stone 1983: 218) because it predicts a lower detectability than either definite detection or inverse-cube detection.

The important thing, for our purposes here, is that the exponential detection function describes a curve that levels off. This means that we can expect the payoff for increasing search effort to show diminishing returns (fig. 4). At some point, it may be more profitable to search elsewhere, rather than continue searching in the same space, under these circumstances. As it turns out, a wide variety of circumstances can lead to an exponential detection function, including inaccurate placement of survey transects and exponential falloff in the probability of "seeing" the target as range increases. The inverse-cube detection function also has diminishing returns, but is more optimistic than the exponential function (fig. 4).

Steps in planning prospection

There is not space here to discuss optimal search plans in detail. Once searchers have identified the goal of search, however, the search involves a number of tasks (modifed after Stone 1983: 208):

1. Estimating the prior distribution of the target location ,
2. Characterizing the target or targets and selecting appropriate sensing methods,
3. Estimating the capabilities of sensors to detect targets,
4. Estimating the impacts of visibility,
5. Estimating the impact of false targets,
6. Determining a detection function, including estimated detectability under various circumstances,
7. Developing a search plan and estimating its cost and probability of success,
8. Updating the plan as the search yields new information, and
9. Evaluating the search's effectiveness.

Note that many of these steps would apply equally to surveys of the sampling variety. In fact, one could well question whether we can justify extrapolating from a sample to a population without estimating some of the impacts included in this list.

Examples of prospection problems

This section will briefly present some examples of search problems that could benefit from the approaches of Operations Research. In the interest of brevity, I will concentrate on allocation of effort, as this is a problem integral to all kinds of surveys.

As an example of discrete search, but in cells that are small enough to approximate continuous search, we could use Bayesian methods to find the optimal allocation of effort among the pixels in a GIS predictive model, such as Minnesota's Archaeological Predictive Model.[1] The pixels, each 30 m x 30 m in size, have been classified to eight levels of prior probability. Commonly, archaeologists make little use of this information except to warn planners of probable CRM impacts, yet more specific models of this type can also guide survey for rare targets or what Altschul (1990) calls "red flags": surprising results rather than the typical and highly predictable ones that sampling surveys normally report.

Another example that could benefit from optimal allocation of effort is Florida State University's survey for Paleoindian sites on Florida's continental shelf, flooded by rising Holocene sea levels (Faught and Donoghue 1997). The FSU survey has focussed on geological outcrops along drowned river valleys that are likely to contain chert sources, which in turn have relatively high prior probabilities of having prehistoric sites in their vicinities. This is an intuitive application of the principles of optimal search, yet does not take advantage of the theory available to optimize search effort explicitly (cf. Mangel 1983).

In both these cases, and ones like them, where we can reduce space to a set of discrete, smaller spaces, each with different areas and probabilities of containing targets, the optimum densities of search effort are equal to the logarithms of quantities proportional to the probability densities of the spaces (Koopman 1980: 149). When we are looking for a single target, and it can only be in one of the spaces, all the probabilities sum to 1.0. In cases where we are only looking for a class of targets, and do not even know how many targets there are, the probabilities are independent. The total density of search effort (fð) in the i-th space of n spaces in a region of area A, is:

$$\varphi_i = \log \rho_i - \frac{1}{A}(A_1 \log \rho_1 + A_2 \log \rho_2 + \ldots + A_n \log \rho_n) + \frac{\phi}{A}$$

Here ρ_i is the probability density of space i (the prior probability that space i contains the target divided by the area of space i), n is the number of spaces in the search area, A_i is the area of space i, ϕ = total search effort, expressed as area covered (for transects ϕ = WL), and ϕ_i is the "coverage" or "sampling fraction" of area i:

$$\varphi_i = \frac{WL_i}{A_i}$$

where W and L are respectively the width and total length of transects. ϕ/A, then, is what archaeologist's have usually called "coverage." This can be expressed just as easily with the total area of test pits, augers or cores.

It turns out that, when the total amount of search effort is very small relative to the region surveyed, it is optimal to devote the entire search effort to the space or spaces with the highest probability density. This is much like what happens in typical use of predictive models in Cultural Resource Management today. With middling amounts of search effort available, we distribute it among the spaces according to the expression above. We can test each space to see whether it meets the threshold for allocation of space. If, for any two spaces,

$$\rho_1 \leq \rho_2 e^{-\phi/A_1}$$

then no search effort is allocated to space 1 (Koopman 1980: 149). When search effort is very large, however, the amounts of effort per unit area for all spaces become equal; this amounts to a simple, proportionally stratified sample, a type of survey strategy with which archaeologists are very familiar. Unfortunately, we usually do not have the luxury of nearly unlimited survey effort.

Simplified examples of optimal allocation

To provide a more concrete example, let us assume a simple case with a survey region stratified into three spaces with areas of 3 ha, 2 ha and 1 ha, respectively. Let us furthermore assume that we are searching for a historic fort, but have no prior information about it other than that it should occur somewhere in the region we are surveying. A common archaeological plan is to divide search effort in proportion to the three areas, which amounts to assigning the three spaces prior probabilities of 0.50, 0.3333, and 0.1667. In that case the value of ρ_i for every space will be the same: 0.1667, and most of the terms in the equation cancel out (Table 1A). Thus, the result is simply a proportionally stratified sample with equal coverage of each stratum. Let us assume, furthermore, that poor surface visibility and low expected artifact density leads us to survey with test pits, each 1 m^2 in size, and the amount of survey effort available (ϕ) is 3000 m^2 (3000 test pits, or roughly 1500 person-days of effort [Zeidler 1995:5.5]). Dividing this admittedly huge resource by the total survey area A, then gives us ϕ/A of 0.05 (5% "coverage"), so that a total of 3000 test pits would be allocated as 1500, 1000, and 500 test pits, respectively.

An alternative, if there is no reason to think it more likely that the fort would occur in any one of these spaces is to assign them all equal prior probabilities. This may seem strange, but the practice in the last paragraph actually makes the untested assumption that large spaces are more likely to contain the fort even though we know that strategic locations for forts, such as hilltops or fords, often occupy very small proportions of space. Since the fort can only be in one place, the probabilities have to sum to 1.0, so the priors are all 0.3333. In that case, however, the probability density of the smallest space is so great relative to the others that we would allocate all search effort to it (Table 1B). Neither space 1 nor space 2 meets the threshold given in (6). Needless to say, assuming equal probabilities in cases where we know nothing except that the spaces vary considerably in size would usually be risky, so it makes sense to try to estimate probabilities more carefully.

Sometimes we have statistics from previous surveys and background research that gives us a basis for estimating the fort's probability of being in various kinds of location. For example, a sample of a dozen similar forts in the broader region or information from historic maps might allow us to estimate prior probabilities of 0.45, 0.35, and 0.20, respectively (Table 1D). In that case, dividing by their surface areas, the probability densities (ρ_i), are 0.15, 0.175 and 0.2 respectively. Note that the probability *densities* are ordered quite differently than the ordinary probabilities on which they are based. This results in a ϕ_1 of 0.007 (206 test pits spread over 3 ha), ϕ_2 of 0.074 (1476 test pits over 2 ha), and ϕ_3 of 0.132 (1318 test pits). A more generous, if unrealistic, total effort allocates a much greater number of test pits to areas 1 and 2 (Table 1C). Note, however, that if our resources only allow us to dig 1200 test pits (ca. 600 person-days), then the "coverage" becomes 2% and space 1 no longer meets the threshold of probability density. Consequently, no effort is allocated to space 1 in this scenario because its probability density of 0.15 is simply too low. We now include only spaces 2 and 3 in the calculation, so that ϕ_2 becomes 0.021 (413 test pits), and ϕ_3 becomes 0.0.079 (787 test pits)(Table 1E). Note how the proportions of effort change when the resources decrease, and how the greatest amount of effort, somewhat counter-intuitively, is concentrated in space 3, even though it had the lowest prior probability (but highest probability density).

It is important to keep in mind that adopting a strategy like this does not mean that the target is most likely to occur in space 3 - after all, we have estimated that there is a 45% chance that it is in space 1 - but only that looking in space 1 is more likely to be productive, given our resource constraints. Allocating a small amount of effort to such a large area as space 1, by contrast, is quite unlikely to find the fort even if it is there. We must remember that in this case the goal is to find the fort as quickly as possible, when resources are limited. Had the goal been to characterize the three spaces, obviously the allocation would have been very different.

More realistic allocations are made incrementally, using information from a previous, unsuccessful search with effort ϕ to plan the next increment with effort ϕ' (Koopman 1980:150). Stone (1975:108-109) has shown that incremental plans that always put the next search in the space with the highest ratio of detection probability to search cost will be optimal in this situation.

Conclusions

It is important for the designs of surveys to be suited to their goals. Where the goal is to find particular targets or kinds of targets, rather than to make parameter estimates, build predictive models, or detect spatial structure, we should use prospection, by which we try to optimize the recovery or detection of those targets.

> There is little question that purposive selection is preferable to sampling whenever selection is feasible, sufficient for one's research objectives, and not wasteful. One of the reasons why improved techniques of data detection ... are so important is that they offer new possibilities for feasible and effective data selection. (Cowgill 1975: 260).

Here I have tried to provide a taste of the tools that Operations Research, and particularly optimal search methods, can offer to help archaeologists accomplish this. With its emphasis on measuring and evaluating detectability, this kind of research also has implications for the other kinds of survey. Sampling surveys, for example, will provide biased estimates of population parameters if their practitioners employ unrealistic assumptions about the detectability of targets that are supposed to be included in the sample.

Operations Research is underexploited as a source of well-developed tools for survey archaeologists and can help us with all three kinds of surveys, including more efficient "total survey" and "non-site survey."

To conclude, we should take prospection, or "purposive survey," seriously where it is warranted, and not treat it as a haphazard "poor cousin" to sampling space. Prospection provides some kinds of information sampling cannot, such as documenting the presence of rare sites. It allows us to test hypotheses that make specific predictions about the locations of archaeological materials much more efficiently than a sampling survey (Cowgill 1975: 260-261). Its tools also help ensure that sampling surveys do not omit parts of the intended sample, creating bias, or at least allow us to evaluate the seriousness of such bias.

Prospection succeeds through the combination of basic mathematical principles and archaeologists' experience and prior knowledge. Background research should not be "a mechanical task, a token interlude between the signing of a contract and the beginning of fieldwork..." (Wobst 1983: 58). We can focus background research on the goals of a survey by using it to model the expected distribution of archaeological materials (Wobst 1983: 62) or to estimate the prior probabilities that particular kinds of targets will occur in particular locations. Operations Research provides the framework with which to make a fruitful partnership of background research and survey design.

References

Altschul, J. H. 1990. Red flag models: The use of modelling in management contexts. In Allen, K. M. S., Green, S. W., and Zubrow, E. B. W. (eds.), *Interpreting Space: GIS and Archaeology*. Taylor and Francis, London: 226-238.

Buck, C. E., Cavanagh, W. G., and Litton, C. D., 1996. *Bayesian Approach to Interpreting Archaeological Data.* John Wiley & Sons, New York.

Corwin, T. L., 1981. Search for a moving target under Minimax estimation. *Journal of Applied Probability* 18: 167-180.

Cowgill, G. L., 1975. A selection of samplers: Comments on archaeo-statistics. In Mueller, J. W. (ed.), *Sampling in Archaeology*. University of Arizona, Tucson: 258-274.

Cozzolino, J. M., 1972. Sequential search for an unknown number of bjects of nonuniform size. *Operation Research* 20: 293-308.

Deagan, K. A., 1989. The search for La Navidad, Columbus' 1492 settlement. In Milanich, J. T., and Milbrath, S. (eds.), *First Encounters: Spanish Explorations in the Caribbean and the United States, 1492-1570.* University of Florida Press, Gainesville: 41-54.

Discenza, J. H., 1980. A solution for the optimal multiple rectangle search. In Haley, K. B., and Stone, L. D. (eds.), *Search Theory and Applications.* Plenum Press, New York.

Dobbie, J. M., 1968. A survey of search theory. *Operations Research* 16: 525-537.

Drew, L., 1967. Grid-drilling exploration and its application to the search for petroleum. *Economic Geology* 62: 698-710.

Drew, L., 1979. Pattern drilling exploration: Optimum pattern types and hole spacings when searching for elliptical shaped targets. *Mathematical Geology* 11: 223-254.

Dunnell, R. C., and Dancey, W. S., 1983. The siteless survey: A regional scale data collection strategy. *Advances in Archaeological Method and Theory* 6: 267-287.

Ebert, J. I., 1992. *Distributional Archaeology.* University of New Mexico Press, Albuquerque.

Faught, M. K., and Donoghue, J. F., 1997. Marine-inundated archaeological sites and paleofluvial systems: Examples from a karst-controlled continental shelf setting in the Apalachee Bay, northeastern Gulf of Mexico. *Geoarchaeology* 12: 417-458.

Fish, S., and Kowalewslei, S. (eds.), 1990. *The Archaeology of Regions: A Case for Full-coverage Survey.* Smithsonian Institution Press, Washington, DC.

Flannery, K. V. (ed.), 1976. *The Early Mesoamerican Village.* Academic Press, New York.

Foley, R., 1981. Off-site archaeology: An alternative approach for the short-sited. In Hodder, I., Isaac, G., and Hammond, N. (eds.), *Patterns of the Past: Studies in Honour of David Clarke.* Cambridge University Press, Cambridge: 157-183.

Gal, S., 1980. *Search Games.* Academic Press, New York.

Gal, S., 1989. Continuous search games. In Chudnovsky, D. V., and Chudnovsky, G. V. (eds.), *Search Theory: Some Recent Developments.* Marcel Dekker, Inc., New York: 33-53.

Garrison, E. G., Baker, J. G., and Thomas, D. H., 1985. Magnetic prospection and the discovery of Mission Santa Catalina de Guale, Georgia. *Journal of Field Archaeology* 12: 299-313.

Harbaugh, J. W., Doveton, J. H., and Davis, J. C., 1977. *Probability Methods in Oil Exploration.* John Wiley & Sons, New York.

Ives, J., 1982. Evaluating the effectiveness of site discovery techniques in boreal forest environments. In Francis, P., and Poplin, E. (eds.), *Directions in Archaeology: A Question of Goals.* University of Calgary, Alberta: 95-114.

Kinder, G., 1998. *Ship of Gold in the Deep Blue Sea.* Random House, Toronto.

Kintigh, K. W., 1988. The effectiveness of subsurface testing; A simulation approach. *American Antiquity* 53: 686-707.

Koopman, B. O., 1980. *Search and Screening: General Principles with Historical Applications.* Pergamon Press, New York.

Krakker, J. J., Shott, M. J., and Welch, P. D., 1983. Design and evaluation of shovel-test sampling in regional archaeological survey. *Journal of Field Archaeology* 10: 469-480.

Lightfoot, K. G., 1986. Regional surveys in the eastern United States: The strengths and weakness of implementing subsurface testing programs. *American Antiquity* 51: 484-504.

Lightfoot, K. G., 1989. A defense of shovel test sampling: A reply to Shott. *American Antiquity* 54: 413-416.

Lolley, T. L., 1996. Ethnohistory and archaeology: A map method for locating historic Upper Creek Indian towns and villages. *Journal of Alabama Archaeology* 42: 1-81.

Lynch, M., 1980. Site artifact density and the effectiveness of shovel probes. *Current Anthropology* 21: 516-517.

Mangel, M., 1983. Optimal search for and mining of underwater resources. *SIAM Journal of Applied Mathematics* 43: 99-106.

McCammon, R. B., 1977. Target interception probabilities for parallel-line and continuous-grid types of search. *Mathematical Geology*: 9: 369-382.

McManamon, F. P., 1984. Discovering sites unseen. *Advances in Archaeological Method and Theory* 8: 223-292.

Morris, P. A., 1977. Combining expert judgments: A Bayesian approach. *Management Science* 23: 679-693.

Mortensen, P., 1974. A survey of prehistoric sites in northern Luristan. *Acta Archaeologica* 45: 1-47.

Nance, J. D., 1979. Regional subsampling and statistical inference in forested habitats. *American Antiquity* 44: 172-176.

Nance, J. D., 1983. Regional sampling in archaeological survey: The statistical perspective. *Advances in Archaeological Method and Theory* 6: 289-356.

Nance, J. D., and Ball, B. F., 1986. No surprises? The reliability and validity of test pit sampling. *American Antiquity* 51: 457-483.

Nance, J. D., and Ball, B. F., 1989. A shot in the dark: Shott's comments on Nance and Ball. *American Antiquity* 54: 405-412.

Paloheimo, J. E., 1971. On a theory of search. *Biometrika* 58: 61-75.

Plog, S., Plog, F., and Wait, W., 1978. Decision making in modern surveys. *Advances in Archaeological Method and Theory* 1: 384-421.

Read, D. W., 1975. Regional sampling. In Mueller, J. W. (ed.), *Sampling in Archaeology*. University of Arizona, Tucson: 45-60.

Richardson, H. R., and Discenza, J. H., 1980. The United States Coast Guard computer-assisted search planning system (CASP). *Naval Research Logistics Quarterly* 27:659-680.

Reynoso, C., and Jezierski, E., this volume. A genetic algorithm problem solver for archaeology. In xxx (ed.), *Archaeological Informatics — Pushing the Envelope.* Computer Applications and Quantitative Methods in Archaeology. BAR International Series. Hadrian Books, Oxford: xxx-xxx.

Savage, L. J., 1971. Elicitations of personal probabilities and expectations. *Journal of the American Statistical Association* 66: 783-801.

Savinskii, I. D., 1965. *Probability Tables for Locating Elliptical Underground Masses with a Rectangular Grid.* Consultants Bureau, New York.

Schiffer, M. B., Sullivan, A. P., and Klinger, T. C., 1978. The design of archaeological surveys. *World Archaeology* 10: 1-28.

Shomette, D. G., and Eshelman, R. E., 1981. A developmental model for survey and inventory of submerged archaeological resources in a riverine system: The Patuxent River, Maryland. In Watts, G. P. (ed.), *Underwater Archaeology: The Challenge Before Us.* Fathom Eight, San Marino, CA: 159-171.

Shott, M. J., 1985. Shovel-test sampling as a site discovery technique: A case study from Michigan. *Journal of Field Archaeology* 12: 458-469.

Shott, M. J., 1989. Shovel test sampling in archaeological survey: Comments on Nance and Ball, and Lightfoot. *American Antiquity* 54: 396-404.

Singer, D. A., 1975. Relative efficiencies of square and triangular grids in the search for elliptically shaped resource targets. *United States Geological Survey Journal of Research* 3: 163-167.

Singer, D. A., and Drew, L. J., 1976. The area of influence of an exploratory hole. *Economic Geology* 71: 642-647.

Singer, D. A., and Wickman, F. E., 1969. *Probability Tables for Locating Elliptical Targets with Square, Rectangular and Hexagonal Pointnets.* Pennsylvania State University Mineral Science Experiment Station Special Publication 1-69.

Stone, G. D., 1981. Artifact density and shovel probes. *Current Anthropology* 22: 182-183.

Stone, L. D., 1975. *Theory of Optimal Search.* Academic Press, New York.

Stone, L. D., 1983. The process of search planning: Current approaches and continuing problems.*Operations Research* 31: 207-233.

Stone, L. D., Barlow, C. A., and Corwin, T. L., 1999. *Bayesian Multiple Target Tracking.* Artech House.

Storck, P. L., 1978. Some recent developments in the search for early man in Ontario. *Ontario Archaeology* 29: 3-16.

Storck, P. L., 1982. Palaeo-indian settlement patterns associated with the strandline of glacial Lake Algonquin in southcentral Ontario. *Canadian Journal of Archaeology* 6: 1-31.

Storck, P. L., 1984. Glacial Lake Algonquin and Paleo-indian settlement patterns in southcentral Ontario. *Archaeology of Eastern North America* 12: 286-298..

Washburn, A. R., 1981. *Search and Detection.* Operations Research Society of America, Arlington, VA.

Winterhalder, B., and Smith, E. A., 2000. Analyzing adaptive stragegies: Human behavioral ecology at twenty-five. *Evolutionary Anthropology* 9: 51-

Zeidler, J. A., 1995. *Archaeological Inventory Survey Standards and Cost-estimation Guidelines for the Department of Defense.* US Army Corps of Engineers Research Laboratories Special Report 96/40.

Endnotes

[1] The use of the Minnesota Archaeological Predictive Model is courtesy of Minnesota's Department of Transport and the US Federal Roadway Authority. I thank Michael Faught and Joe Donoghue for the use of information from the FSU survey of Florida's continental shelf. The author's research has been supported by the Social Sciences and Humanities Research Council of Canada.

Tables

Space	A_i (ha)	p_i	ρ_i	Φ/A	Φ (ha)	ϕ_i	No. Test Pits	Total
A								
1	3	0.500	0.167	0.05	0.3	0.05	1500	
2	2	0.333	0.167	0.05	0.3	0.05	1000	
3	1	0.167	0.167	0.05	0.3	0.05	500	3000
B								
1	3	0.33	0.111	0.05	0.3	0	0	
2	2	0.33	0.167	0.05	0.3	0	0	
3	1	0.33	0.333	0.05	0.3	1.0	3000	3000
C								
1	3	0.45	0.15	0.10	0.6	0.057	1706	
2	2	0.35	0.175	0.10	0.6	0.124	2476	
3	1	0.20	0.20	0.10	0.6	0.182	1818	6000
D								
1	3	0.45	0.15	0.05	0.3	0.007	206	
2	2	0.35	0.175	0.05	0.3	0.074	1476	
3	1	0.20	0.20	0.05	0.3	0.132	1318	3000
E								
2	2	0.35	0.175	0.02	0.12	0.021	413	
3	1	0.20	0.20	0.02	0.12	0.079	787	1200

Table 1. Examples of allocations to three spaces with unequal areas, proportional (A), equal (B) and unequal probabilities (C-E), and different amounts of total survey effort (F). Note that scenario A is equivalent to a proportionally stratified sample, and that in scenario B spaces 1 and 2 do not qualify for any survey at all.

Figures

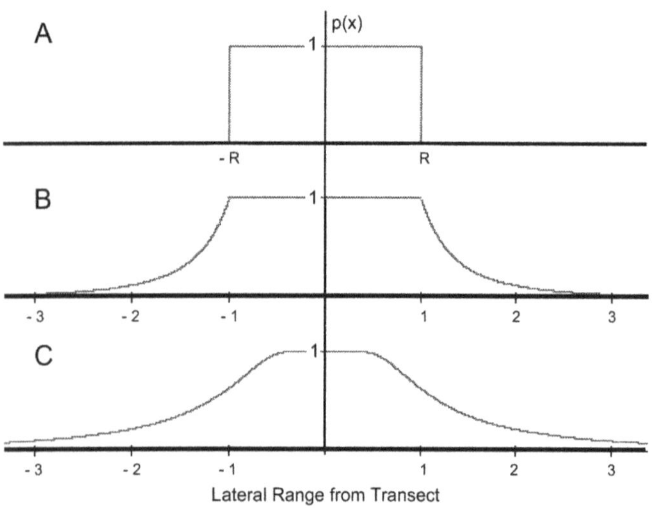

Figure 1. Cross-sectional plots of the effect of range away from a transect on detection probability: (a) definite detection, (b) inverse-cube detection, and (c) exponential detection.

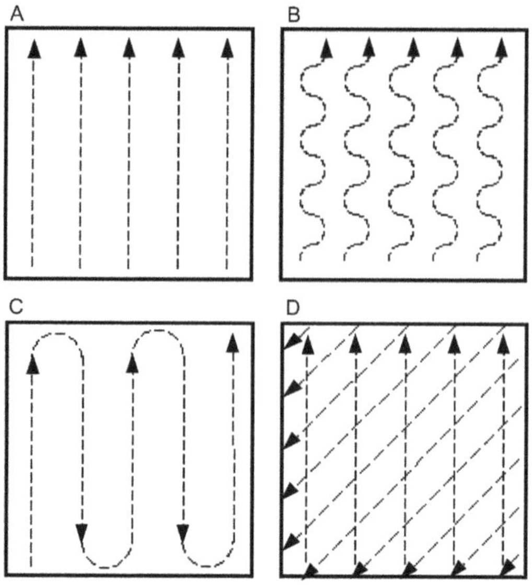

Figure 2. A selection of transect geometries (a-c) and two passes by transects oriented 45° (d).

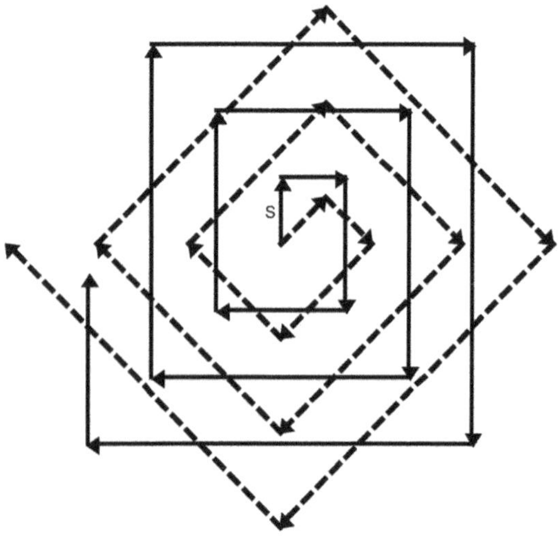

Figure 3. "Expanding square" search over a target with a circular normal prior probability distribution. Note that a repeat search (dashed lines) is optimally oriented 45° to the first (after Koopman 1980:214-221).

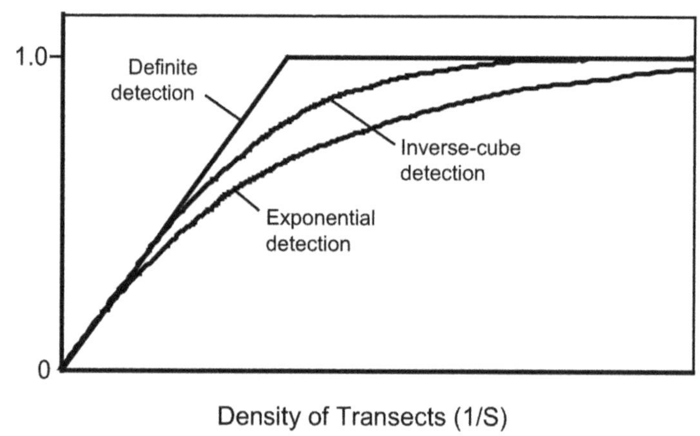

Figure 4. Detection functions for parallel transects with a regular spacing of S, assuming definite detection, inverse-cube detection, and exponential detection (after Koopman 1980:78).

Archaeological spatial modelling
A case study from Beagle Channel (Argentina).

Juan A. Barcelo
Divisió de Prehistoria. Facultat de Lletres. Edifici B
Universitat Autonoma de Barcelona
E-08193 Bellaterra, Spain
E-mail: ilphd@blues.uab.es

Ernesto L. Piana
Centro Austral de Investigaciones Científicas
Av. Malvinas Argentinas s/n
V9410BFD Ushuaia, Tierra del Fuego (Argentina)
E-mail: arqueologia@tierradelfuego.org.ar

Daniel R. Martinioni
Subsecretaría de Planeamiento
Gobierno de Tierra del Fuego
Antártida e Islas del Atlántico Sur
E-mail: danymart@satlink.com

Abstract: Social actions are always performed in locations intrinsically better or worse for some purpose, because of their position relative to another location used for any other action or for the reproduction of the same action. The objective is to analyse how a social action "varies from one location to another". The analysis then pretends to examine if the characteristics in one location have any relationship with those of a neighbouring location. This goal can be made possible through the definition of a general model of spatial dependence. In other words, the main objective of spatial analysis should be the spatial correlation of different social actions: how the spatial distribution of an action has an influence over the spatial distribution of other action(s).

This paper discusses the probabilistic nature of spatial causality. Once it is known whether social actions at neighbouring locations are similar or not, it can be explained why the location of social actions are homogeneous or heterogeneous in the area defined by the performance of that actions. That means that spatial causality can be studied in terms of the "influence" landscape features have over the location of social action or the "influence" the location of social action has on landscape features.

The discussion is presented with the help of an intensive and extended archaeological survey of shell-middens, ranging from 6000 ^{14}C years B.P. from the northern coast of Beagle Channel, Tierra del Fuego (Argentina): "the uttermost part of the world".

Key words: Spatial Archaeology, Settlement Patterns, Hunter-Gatherer, American archaeology, Correspondence Analysis

Hunting, fishing and gathering at the Beagle Channel

The southernmost extremity of the Americas was the last major landmass to be settled by human beings. This area has always been dominated by hunting-fishing-gathering economies. In Tierra de Fuego neither agriculture nor fully-fledged pastoralism ever emerged. However, it would be a mistake to characterise those populations as an "archaic" adaptation to hard environmental conditions (Orquera & Piana 1996; 1999a). The idea of this paper is to show how simple environmental determinism and easy generalisation are misleading.

When the first Europeans arrived, the South American territory southward from 34°S latitudes was inhabited by nomadic hunter-gatherer groups with a subsistence economy whose environmental adaptation patterns differed. Europeans found two fundamentally contrasting socio-economic systems, which can be characterised as the "terrestrial" and "canoe -or- sea nomad" cultures. The terrestrial people focused on the exploitation of terrestrial big mammals -mainly guanaco-, whereas the subsistence of "sea nomad" people was heavily dependent on the consumption of sea mammals, mussels, marine birds, and fish (Estévez & Vila 1995; Orquera 1999a; Orquera & Piana 1999a; Piana *et al.* 2000). When the opportunity presented, however, "sea nomad" populations hunted otters or pursued terrestrial mammals, and terrestrial hunters had access to coastal resources.

Isotopic dates suggest that human populations have been exploiting coastal resources for the last 6200 uncorrected radiocarbon years (Orquera & Piana 1999b). The beginning of such life style was not much older (Orquera & Piana 1988), with the oldest population covering at least down to the southern coast of Navarino island (Legoupil 1995). With an unknown sort of navigation device, they were capable of sea travelling from the very beginning of their colonisation (Piana 1984; Legoupil 1993; Piana & Orquera 1998), but known data of their presence as far south as Cape Horn is restricted to later moments (Legoupil 1993). Steep relief and tangled unending woodland appear unattractive for human settlement, since they provide few vegetable or animal food-resources. Abundant littoral resources largely counteract that fact: pinnipeds, molluscs, occasional stranded cetaceans, sea birds, fishes, crustaceans, etc. There, abundance, concentration, predictability and the diversification of resources provide enough food to support relatively dense populations with high mobility within small ranges (Orquera & Piana 1999b, Piana et al. 1992).

This paper deals only with the exploitation of coastal resources and spatial properties of human settlement along the northern shore of Beagle Channel. Massive accumulations of shellfish constitute the main archaeological evidence in the area. Within them, remains of many other activities such as other resources processing, fires either for cooking or heat, lithic chipping, bone instrument making, and so on, have always been found (Clemente 1997, Piqué 1999). Fires for cooking and warmth were tended inside the huts and the refuse, primarily the quickly accumulated shells, were tossed outside huts. In this manner the depressions were maintained and served as partial shelter from the wind (Orquera & Piana 1992). Archaeologically, these structures may appear as domes, annular structures, thinner lenses, or a combination of them, containing an inordinate amount of charcoal and humus in comparison with intervening midden consisting mainly of shells and giving way to a hummocky topography (Orquera & Piana 1991).

The spatial nature of social action

Social actions are performed in specific locations because of their position relative to another location used for any other action or a reproduction of the same action. Consequently, actions performed at one location may have some relationship with actions performed at a neighbour location. For instance, when an action is performed *here*, it increases the probability of some posterior action and decreases the probability of some others. Therefore, if the location of action influences, conditions or determines the location of other actions, we can calculate a general model of spatial dependence.

In this case, the action we are studying is *settlement*. Humans decide the location of their settlements according to many reasons, but more specifically we make "placement" decisions based on the social strategies for resource management in the area around the settlement. What means a sophisticated effort balance between social needs, available techniques and resources?

Since the human population under study concentrated on the consumption of marine resources, it is to be imagined that settlements were mostly concentrated along the shoreline. The question to be solved is how and in which way did the natural configuration of the coast and coastal landscape, and local availability of resources along the northern shore of Beagle Channel determine human settlement during the last six millennia.

Traditional conceptions about hunting-gathering argue for the axiom that space determines settlement. That is, that settlement is a consequence of the environmental features or spatial properties of resources. In this sense, the distribution of sea nomads occupation of this region (*sensu* Orquera & Piana 1999b) -when no other factors like exposition to stormy open sea limited the constant peopling- coincides with the distribution of austral beech forest. Other than the marine resources, the sea nomad's life style relied heavily on the availability of extensive amounts of firewood and large trees. This explains why the territory of those populations dependent more on maritime resources extended only as far as the distribution of forest permitted (Orquera & Piana 1988). This fact underlines the fundamental role of wood and tree bark for these cultures, of which they used to make items ranging from canoes to buckets (Borrero 1997). Nevertheless, this explanation is only to be put forward when analysing the whole sea nomad regional limits, though it is not useful to analyse causal selection of spots for settlements *within* that region.

"Space" is not a cause, but a material product of social processes acting back on social structures by limiting, constraining, and, in some cases, determining future actions. Human settlements may be tied together by proximity, but the causal mechanism is not physical proximity. The real cause should be explained in terms of the "influence" an action performed at a location has over all locations in the proximity (Barceló & Pallarés 1998) and the future use of the same spot. Thus, social actions are not adapted to the environment, but productive actions (hunting, fishing, gathering) determine the location of residential actions (settlement). The suggestion is that a social action can generate the reproduction of similar actions around it, or it can prevent any other similar action in the same vicinity. In this case, the "visible" characteristics of shell middens increase the probability that new human sites appear in the vicinity of older sites (*cf.* Orquera & Piana 1991; 1992).

Settlement (*stricto sensu*) is not adapted to environmental conditions or resources, but the place where social agents perform actions like shellfish gathering, sea lions or sea birds hunting, and fishing. Human action increases the probability of settlement in the vicinity or at the same spot, while it diminishes the probability of performance of other social actions, for instance, of non-residential nature.

Distance is a much better candidate for a causal mechanism, because it is usually assumed that "everything is related to everything else, but near things are more related than distant things" (Tobler's law). This assumption is based on an intuitive Neighbourhood Principle, which relates the intensity of influences converging to a single location from the spatially neighbouring locations. Therefore, it is no surprise to discover that all shellfish gatherers choose to locate their camps near collecting spots or, at least, process shellfish at the shore and carry only the meat back to the camp. Thus, be it transitory shellfish preparation sites or base camps of shellfish gatherers, both are predicted to be located near the collection point (*cf.*

Waselkov 1982). Shellfish, from the viewpoint of a human predator, are small packages of meat sealed in heavy inedible shells. In terms of energy efficiency, this fact sets definite limits in the distance live molluscs can be transported with simple technologies, beyond which, energy expended in transport will exceed that gained from the food (Orquera 1999b).

The approach here relies on a prior hypothesis of spatial smoothness (see also Barceló & Pallarés 1998), which considers that two neighbouring observations of the similar kind are supposed to have been more likely originated from the same cause than two of such observations lying far apart. Therefore, if spatial variation among settlements were not wholly erratic, it can be said that there is a certain degree of spatial dependence between spatial units. This is the reason why most settlements are concentrated on the shoreline: social work took place where it was most efficient. Herein it is not described how settlement is adapted to the environment, but how some specific social actions are related with other actions, because of their neighbourhood relations.

Instead of "adaptation", the term *attraction* is more suited to the study of social space from a dynamic standpoint. Let's consider all possible locations of human settlement along the shores of Beagle Channel. There are a vast number of possible settlement locations, depending on the places where social action may be performed. If the range of possible locations is so great, why do spatial effects of social actions appear at first glance always regular? Why is so usual to find spatial dependency, if spatial consequences may be so diverse? In this way, Social Activity areas should be considered as spatial attractors. Any social action is performed in space, and as a consequence it produces a social activity area acting as spatial attractor and conditioning the performance of the action and successive actions performed in the neighbourhood.

The observation and statistical measurement of "regularity" in human settlement can be misleading. In archaeological spatial modelling, spatial correlation is commonly used as an evidence of regularity. This paper intends to discuss some usual mistakes connected with the trivial inductive explanation of spatial regularity. Observed regularity is meaningless, if a model of social actions as spatial attractors is not proposed. Anything can be related with anything, and if the proper method is used, it will always produce a measure of regularity. But not every regularity can be explained as a result of social action.

Variability of settlement along the Beagle Channel

Survey at the Beagle Channel and adjacent areas

The *Proyecto Arquelógico Canal Beagle* development lead to the observation and location of more than 500 archaeological shell-middens since 1975, because of the use of different scopes and methodologies along time though, the information gathered is not uniform. Since 1994 European Union Project "Beagle Channel Marine Resources Prior to the Industrial Exploitation (CI1*-CT93-0015)" enabled the systematic record of sites, which made 318 of them to be homogeneously described and compiled in a geo-referenced data base (Martinioni 1998; Piana *et al.* 1998). Data involved here 318 shell-middens-result from an intensive and long term surveying of archaeological sites along a fringe of circa 1 km depth along the northern shore of Beagle Channel, i.e. the southern hill slope of the Andes Fueguinos and the coastal area of the Isla Grande de Tierra del Fuego between roughly 68° 36' W & 66° 45' W latitudes. Under proper conditions, i.e. a valley transverse to the coast, the survey was carried uphill and inland, up to the vegetation limit at circa 700 m above sea level. In special locations, where topographic characteristics were unsuitable for human settlement, e.g. a cliff, the survey was restricted to less than 1 km from modern shore.

Landscape correlations

The goal is to test if all archaeological sites present in the database are in similar landscapes, or if some physical features within those landscapes are spatial attractors for human settlement. The database of the *Proyecto Arqueológico Canal Beagle* includes subsidiary and particular observations and descriptions not relevant for statistical analysis. The information useful for landscape correlations involves the following features:

1. Geographic location, archaeological structures (morphology of the shell-middens accumulations),
2. quantity of observable structures (discriminated by morphology),
3. size of the sites (range of surface units),
4. substratum (geomorphologic features of the site location),
5. back-of-the-site (geomorphologic and topographical characteristics of those parts of the site opposite to the area open to the coast),
6. relationship with moraine deposits,
7. distance to the (modern) shoreline,
8. kind of shore (slope and composition),
9. kind of coast (geomorphologic features),
10. distance to (modern) woods,
11. distance to (modern/ancient) water sources,
12. kind of available (modern/ancient) water sources,
13. wind protection (percentage of total protected area),
14. wind protection (orientation of protected area),
15. kind of protection (only topographic features),
16. height above sea level, chronology (^{14}C available dates).

Since dating was done under a different scope and looking forward recent sites, these data are biased. Nevertheless, the total number of dated sites is low, and presence of layers included in the known range for sea nomads occupation is to be expected. Archaeological sites are mostly grouped along the coastline and appear in large quantities. The distance between two neighbouring sites is usually less than a few hundred meters, with the exception of few non-habitable very steep or wall like coastal segments. There seems to be no difference in density, size or aspect. A preliminary inspection of data would allow concluding:

- the majority of shell-middens are near or very near the actual shoreline (81% of all sites closer than 100 meters),
- related to modern beaches with smooth slope (67%), n near a modern and/or ancient water source (76% of all sites closer than 100 meters), and

- near moraine deposits (69% of all sites closer than 100 meters), which serve as a back-of-the site (56%).
- archaeological sites overlay ancient pebble/cobble beaches (63%).

According to these first results, the main *spatial attractors* for human settlement activities seem to be:
- the location and slope of the modern shoreline,
- a substratum composed of materials from ancient pebble/ cobble beaches,
- modern and/or ancient water sources,
- and moraine deposits, seem to be determining human choices.

In contrast, the location of modern woods and the existence or absence of wind protection do not seem to determine human settlement. It is easy to see that those results are contradictory: in which way modern landscape features can be spatial attractors for social actions performed in the past?

It is true that isostatic movements uplifted the landmass resulting in drop of the relative sea level of a maximum of 10 m (*Rabassa* et al. 2000). Consequently, since the general slope is relatively steep, the retreat by coastal elevation, in general, did not affect more than 600 horizontal meters. So that there are several locations within 200 m off the modern shoreline that were always available for settlement. Nevertheless, to pay attention to this fact does not solve the previously stated contradictions.

The contradictory nature of preliminary results is enhanced by the fact that less than 4% of all archaeological sites are attracted *simultaneously* by the most common landscape features. For instance, only 49% of all sites are near (less than 100 meter) the modern shoreline *and* near a water source. Many coastal sites are far from water, and even worst, those sites near the modern shoreline are those that are more distant to modern water sources.

There is however some evidence of two-dimensional correlation between the best candidates for spatial attractors –notably between distance to the modern shoreline, and the location of the site on pebble/cobble ancient beaches. If the shoreline has been since the beginning of human occupation a spatial attractor for human settlement along the Beagle Channel, then we would understand why archaeological sites are simultaneously related with past evidence of shoreline, preserved as the substratum of archaeological sites, and modern ones. The correlation between ancient beaches and the modern shoreline suggests that changes in sea level have not affected the general orientation along the Channel. The majority (82%) of the archaeological sites with pebble/cobble *substrata* (evidence of ancient beaches) are near or very near the modern shoreline. As the sites are less related to the modern coast, they are less related to pebble/ cobble ancient beaches.

We have calculated a Multiple Correspondence Analysis to test those hypotheses. We have done the calculations for Multiple Correspondence Analysis using SPSS 10, optimal scaling series of commands. A preliminary solution, using all non-missing data (183 sites), shows that 45% of total variance in the data set is produced by the existence of three outliers. Rock-shelters seem totally different from the rest of the sites, probably because their eccentric relationship with modern shoreline, which is the most dominant variable.

Deleting the 3 most notorious outliers, the results are slightly different, showing a greater variability between sites. A three dimensional Correspondence Analysis solution only accounts for the 32% of total variance explained (eigen values: 0.365, 0.332 and 0.275). This gives strong support to the absence of explainable spatial regularity.

Regularity accounts only for a third of total variation:
- sites far from the modern shoreline are inside the modern forest, very near to moraine deposits, and located on organic substratum areas, and very much protected from winds.
- sites far from modern woods, and far from moraine deposits, and located on ancient pebble/cobble beaches.

It is very interesting to show that those sites, which are well protected from the wind, seem to be in similar geomorphologic areas, that is, wind protection is only possible in some specific conditions, and not everywhere along the coast.

Those sites, which are not *on* the beach, not *inside* the woods or which are not immediate to moraine deposits seem also very similar between them. This observed regularity suggests that there is only one cause of exceptional location.

The Second Dimension discriminates between three types of back-of-the site features: depression, smooth or plain and abrupt, and associates sites on smooth or plain modern beaches with sites on substratum defined by the presence of materials from pebble/cobble ancient beaches.

Finally, the Third Dimension differentiates sites on abrupt shores and half protected to the wind from sites on sandy beaches and mostly open to the winds. That means that wind protection is a cause of settlement location only in a few cases. In general, discrimination measures suggest that the variables SUBSTRATUM, DISTANCE TO THE SHORELINE, RELATION TO WOODS, and WIND PROTECTION are the most influent to explain spatial variability. If we calculate the Correspondence Analysis using only these variables, we obtain a much more adjusted solution, which accounts for 54% of the total variance explained (eigenvalues = 0.622 and 0.458).

Figure 4 (*a* and *b*) graph shows an increasing difference from sites on pebble/cobble ancient beaches (E) or on stony ancient beaches (D). Sites on those locations are not in the woods (O: very far, N: far, M near), and are mostly open to the winds (P). Sites on ancient sandy beaches (C) are associated to half protected from wind sites (Q) and sites very near the shoreline (G). Sites a bit farther from the shoreline (H) are predominantly on moraine deposits, very near the woods, but not inside them, and are well protected from wind (R). Finally, the most opposite category to the first one are sites on organic substratum (A), in the woods (K), completely protected from wind (S), and far from the shoreline (I: far, J: very far). The fact that there is a non negligible correlation between ancient pebble/cobble beaches and proximity to modern shoreline suggests that measured variation cannot be explained in chronological terms.

The results suggest that the main spatial attractors are the modern shoreline and the modern wood. The first one *attracts* settlement, the second one prevents settlement. The relationship between both attractors is, however, characteristically non-linear (Figure 5).

These results are misleading, because human groups are not avoiding woods. Statistical results show that woods are also settled (28% of total sample). Along the northern coast of Beagle Channel, forests arrive to the shoreline only in few occasions. Austral beeches need organic soils to grow, a certain pH and stable substratum for their superficial roots. These characteristics are not fulfilled by soils on ancient beaches pebble/cobble substratum. Elsewhere the woods can arrive to the beach, *e.g.* where moraine deposits are near the shore. In those cases, coastal lines become rocky and steep, which makes social action performance much more difficult, and human settlement avoids them. Consequently, if human action is produced inside the woods, then it should be mostly performed in woods which are relatively far from the shore, with organic substratum, and related to moraines.

51% of archaeological sites inside woods are *also* near the modern shoreline. Only a 9% of sites are inside woods *and* far from the modern shoreline. Those far sites seem very different from the rest of the sample.

The possibilities for wind protection do not appear as a spatial attractor. Only 25% of the total sample is well protected from wind. The strong positive correlation between wind protection and proximity to the shoreline, suggests that modern shoreline is the most influential spatial attractor, and only within its attraction basin, the most protected locations attract human settlement.

Smooth modern beaches act as a spatial attractor, however, we cannot say that, within the beaches, abrupt ones appear as negative attractors, because 26% of sites appear with this feature. Correspondence Analysis has not discovered any significative correlation with this fact.

Archaeological correlations

The most characteristic archaeological site along the northern shore of the Beagle Channel is the shell-midden, constituted by the accumulation of a big quantity of mussels and other molluscs shells, discarded tools, charcoal, stone sherds, mammal and bird bones, etc. (Orquera & Piana 1991; 1999b: 23, Estévez & Vila 1995). Given the results of the intensive survey, the most frequent morphology is the ring-shaped mound (more than 50%). It is also interesting to remark the existence of sites composed of "isolated structures" (33,2%) and sites characterised by the presence of "grouped structures" (66.8%). Correlating the morphology and topology of observable archaeological sites, the majority of cases can be divided in two kinds:

- Ring-shaped grouped mounds (40%)
- Bell-shaped isolated mounds (11 %)

The quantity of isolated ring shaped mounds is very low (9%), as it is the quantity of grouped bell-shaped ones (4.5%). Size differences between isolated and grouped structures confirm this point: largest sites are always composed of many grouped ring-shaped or bell-shaped mounds, whereas, the smallest sites are usually non-ring isolated structures.

This relationship is statistically significant using the chi-squared test, and ordinal correlation tests –Sommers *d* and Kendall *t*- give figures around 0.60 for the relationship between isolation and size. Given that size is the result of accumulation of discarded material, it is easy to understand that the more complex the site (more mounds), the more surface it occupies. Highly probable, the largest sites are also those were material has been accumulated for a longer period of time as a result of a high amount of resettlements at the same place independently of the settlers number. Very few sites with thin shell-midden accumulations covering a relatively large area, as Lanashuaia (cf. Piana *et al.* 2000), may have been produced by a large number of occupants along a short period. It is important to realise that isolated mounds show clear evidences of reoccupation, as it is the case for the more complex structures made of a great number of middens. The possibilities of reoccupation do not explain the size or complexity of an archaeological site.

The shape of the site structures seems to be irrelevant for a spatial model of landscape use. Using chi-square, Cramer V and ordinal correlation measures, we have determined only a significative relationship with DISTANCE TO THE WOOD: bell-shaped mounds are mostly *in* the woods, while ring-shaped are *near*, not always *in*. We have calculated a Categorical Regression or Regression with Optimal Scaling (Meulmann & Heiser 1999) to analyse the properties of this relationship. The idea is to consider the predictive ability of WOOD on the shape of sites. R^2 is quite low (0.090) because the intensity of relationship is also very low. As distance to wood increases, predicted shape increases, and that means, that the location of ring-shaped structures is more probable far from the wood, than inside or near it. We should remember that, although present, this non-linear relationship accounts for a very low quantity of total variance. That means that not all sites inside woods can be described as isolated bell-shaped mounds.

We have also studied the relationship between isolated and grouped structures and landscape characteristics. Again, we have used chi-square tests, Cramer V and ordinal correlations. We have obtained significant values in the case of SUBSTRATUM (Cramer V= 0.413) and DISTANCE TO MORAINE DEPOSITS (Cramer V= 0.331). In the first case, isolated structures may appear in any kind of substratum (organic, pebble/cobble, stony, sandy), but grouped structures tend to be concentrated on pebble/cobble ancient beaches. In the second case, grouped structures tend to appear far from moraine deposits, while it is difficult to discover isolated structures far from them.

The intensity of relationship is relatively high (R^2 = 0.301), explaining 30% of total variance.

The number of ring-shaped structures increases as the prediction of substratum (from organic to pebble/cobble ancient beaches) decreases. The same relationship is also possible for bell-shaped mounds, although the relationship is smaller; it explains only 8% of accumulated variance.

Given that distance to moraine deposits seems to be related with substratum (organic substratum is positively correlated with moraine deposits and therefore those sites are near moraines), we obtain good relationship also with this variable (10% of total variance). Since some sites composed of isolated mounds tend to appear with organic substratum, there is a higher probability that isolated structures appear near the moraine deposits, than grouped structures.

The linear correlation between morphology/topology and distance to the coast is negative. This fact can be explained in terms that only isolated mounds appear far from the modern shoreline. Grouped mounds are *attracted* significatively by the modern shore-line.

If we consider now the surface of surveyed sites as dependent variable, there is a significant relationship with:

- SUBSTRATUM: small sites tend to overlay organic soils, while large sites overlay substratum defined by the presence of pebble/cobble as evidence of ancient beaches.
- DISTANCE TO THE COAST: 18.8% of small sites are far from the modern shoreline –200 m or more- while only 7.2 % of large sites are that far from the shoreline.
- DISTANCE TO THE WOODS: small sites may appear *inside* the woods, while large sites are usually not in the woods.
- AVAILABLE WATER SOURCES: However, largest sites are not always concentrated where the most significant (or permanent) water sources are

Using Categorical Regression, we observe that SUBSTRATUM explains 23% of total variance, AVAILABLE WATER SOURCES explains 16 %, DISTANCE TO THE SHORELINE explains 11% and DISTANCE TO THE WOOD 10%. Pratt's measure of relative importance (Meulmann & Heiser 1999) gives the same results.

Again, WIND PROTECTION is neither a spatial attractor, nor in the positive, nor in the negative sense. ABRUPT MODERN BEACHES may be considered as a negative spatial attractor in the case of the largest and the most complex mounds, but 23% of grouped structures appear in those areas. The WOOD is also a negative spatial attractor for grouped structures. We have seen why the formation of organic substratum is exclusive of wooded areas. The recurrence of organic substratum with small sites inside woods is then a spurious correlation of the geomorphologic characteristic of wood areas.

Chronological variability

Several ^{14}C dates indicate that a life-style based on a high dependence on littoral resources took place in the region of the Magellan-Fueguian channels and islands from some 6500 uncorrected radiocarbon years to the 19th century AD (Orquera & Piana 1988; Piana *et al.* 1992). From the data available, it can be assumed that the occupation of the northern border of Beagle Channel throughout the period was rather stable (Orquera & Piana 1999b, and references therein; Estévez *et al.* 2001). Support for this hypothesis comes from the fact that ancient pebble/cobble beaches act as spatial attractor in the same way as the modern shoreline. That means that the shoreline has been the main spatial attractor all along the period.

Research done provided 54 ^{14}C dates from the surveyed sites. Nevertheless, the data cannot describe precisely the temporal evolution of human settlement in the area, because in most cases, only the most recent layers have been dated. Sites with modern dates hide the existence of older layers at the same location. Results of this study can thus be confusing, because the number of old sites is biased.

In general, it is hard to see any temporal variation on spatial effects of human actions along the Beagle Channel. Modern sites have characteristics similar to those known of older times, although it seems possible to suggest some differences between the oldest and the most modern sites. Correspondence Analysis shows that the most modern sites are similar among them, what is not true for the rest of the sample. We have segmented the results of preliminary Multiple Correspondence Analysis using two chronological filters. Figure 7a shows in green colour the placement of old sites (older than 2000 ^{14}C years BP) in a graph of factorial scores, and the Figure 7b shows in green colour the placement of the most modern sites (after 500 ^{14}C years BP). In Figure 7b, modern sites appear much more concentrated in the margins of the distribution. It is important to remember that this particular factorial representation accounts for 50% of total variance.

Very old sites tend to be smaller and more isolated than the most modern. Spatial attractors as Wind Protection and Location on Smooth Slopes determine the location of human settlement as soon as the sites begin to grow in size and complexity, and better areas (beaches with smooth slopes) are being settled.

Abrupt beaches were occupied only at the end of the period. This might be related with the formation of new coastal features -related to abrupt beaches- that became available for settlement as a result of the previously mentioned land uplift with the consequent shoreline retreat. Meanwhile old by-shore locations were progressively farther away from the shoreline. This leads us to consider a very close relationship between coeval settlement and shoreline position. Some of the old sites were on that moment abrupt beaches as it is the case of Mischiuen I, presently under study (Piana unpublished data). Of course, it could also be related to a supposed higher demography in the later times.

It is to be expected that, when foreign colonisation began to interfere with the indigenous social and economic organisation, *i.e.* greeting sedentarism and teaching incipient agriculture and cattle breeding, variables for settlement site selection changed because of changes in social activities. This expectation is based in ethnohistorical information (*cf.* Orquera & Piana 1999a), though, it was not carried out extensively until the last decades of the 19th century, when the social activities of the first Anglican Mission at Ushuaia (1867-1886), the city of Ushuaia (1884), the two first ranches of the Beagle Channel, namely Harberton (1886) and Puerto Remolino (1898), were social attractors for the remaining aborigines. Up to that moment, the majority of the Yamana maintained the basis of their settlement pattern as shown in the only two archaeological sites studied from this period (Estévez & Vila 1995; Orquera 1999a; Orquera *et al.* 1993; Piana et al. 2000).

Conclusions

Along the Beagle Channel, the shoreline has been the main spatial attractor for human settlement for the last 6000 ^{14}C years. Nevertheless, it should be remembered that *space* is not a cause of spatial variability. If geomorphologic features are correlated with the location of human settlement, it is because of spurious correlation: they are features of coastal areas. It is human work made on the coast, that attracts human settlement. The production unit and the residence unit are indistinguishable among the hunter-fisher-gatherer groups studied. Consequently, it can be concluded that people lived where they worked. For instance, a compact resource such as a large mussel bed at a shoal may circumscribe the choice of suitable nearby campsite locations. On the other hand, a dispersed resource such as an extensive rocky shoreline with gastropods can be profitably exploited from a number of points. In the first case, the few preferred camp sites are liable to be frequently reoccupied as long as shellfish may be gathered. As for the latter, campsite selection is arbitrary, not related to the mollusc resource while the other factors remain constant, so the same spot will seldom be reoccupied (*cf.* Waselkov 1982).

From a regional standpoint, staple survival resources for the sea nomads at the Beagle Channel were evenly-spaced distributed within a range of half a day mobility, be it in canoe or walking. This favours human dispersion in the smallest viable social units, evenly spaced in tight proximity to each other and with high mobility within small ranges. Of course, the coastal environment is not homogenous; local differences in landscape morphology are related to differential costs of access to the diverse resources (Piana *et al.* 1992, Orquera & Piana 1999b). This fact explains local diversity of settlement locations among archaeological sites.

Archaeological correlations show also the ambivalent nature of the modern shoreline as spatial attractor. All kinds of sites are near the modern shoreline, but not all. Some human actions were performed far from it, and those locations seem archaeologically different from the others: they are smaller and topologically less complex. Only isolated mounds appear far from the modern shoreline. Grouped and largest mounds are *attracted* significatively by the modern shoreline or by the evidences of the ancient shoreline (pebble/cobble substratum). Settlement inside woods is not avoided, but forest resources do not attract human settlement in the same way as coastal resources. It should be remembered that although most of the sites are near or very near the shoreline, the same sites are also *near* the woods (although most of them not *inside* them). In general, it can be said that the location of ring-shaped structures is more probable far from the wood, than inside or near it.

Hypothetically, it can be concluded that the areas far from the coast were occupied with less intensity than the beaches.

Social space forms an integrated part of social strategies for subsistence and/or social processes, because such strategies and processes inherently involve a spatial (and temporal) dimension. We think that the place of *human work* and *settlement* is the cause of the reproduction of both generic actions *at the same place*. Consequently, the spatial structure of social activities is not constant, neither static. It is dynamic because it is socially *caused,* and simultaneously it *determines* society. Social space is not absolute, but relational. It depends on the underlying network of social actions, which are the interrelationships between objects, objects and individuals, individuals and individuals, individuals and activities.

The spatial model presented in this paper is obviously too simplistic to be considered a model of spatial causality. Nevertheless, our application shows some very important aspects of archaeological spatial modelling:

- The use of only landscape features to define the model parameters is misleading, because most of those features show different degrees of latent correlation, and prevent the detection of the real spatial attractors.
- Archaeological surveying tends to use modern landscape features (observable) instead of past evidences (non observable). This fact leads to affirm the causal nature of modern features on ancient elements.

References

Barcelo, J.A., Pallares, M. 1998. Beyond GIS. The Archaeology of social spaces. *Archeologia e Calcolatori* vol. 9, pp. 47-80.

Borrero, L.A. 1997. The Origins of Ethnographic Subsistence patterns in Fuego-Patagonia. In: *Patagonia. Natural History, Prehistory and Ethnography at the Uttermost End of the Earth.* Edited by C. McEwan, L.A. Borrero & A. Prieto. London, British Museum: 60-81.

Clemente, I. 1998. *Los instrumentos líticos de Túnel VII: una aproximación etnoarqueológica.* Serie Treballs d'Etnoarqueologia No. 2. Edición Conjunta: Publicacions de la UAB- Publicacions del CSIC

Estevez, J., E. Piana, A. Schiavini, N. Juan-Muns. 2001. Archaeological analysis of shell middens in the Beagle Channel, Tierra del Fuego Island. In: *International Journal of Osteoarchaeology 11:24-33.* London.

Estevez, J., A. Vila (Editors). 1995. *Encuentros en los conchales fueguinos. Treballs d'Etnoarqueologia, vol. 1.* CSIC-UAB, Barcelona.

Legoupil, D. 1993 –94. El archipiélago del Cabo de Hornos y la costa sur de la isla Navarino: poblamiento y modelos económicos. *Anales del Instituto de Punta Arenas 22:101-121.* Punta Arenas

Legoupil, D. 1995. Des indigènes au Cap Horn: conquêted'un territoire eet modèle de peupplement aux xonfins du continent sud-americain. *Journal de la Société des Americanistes 81: 9-45.* Pris

Martinioni, D.R. 1998. *Proyecto Informatización y Georreferenciamiento de Sitios Arqueológicos.* E.L. Piana & R. Agüero (Supervisors). Informe Final (Primera Etapa): 250 pp.

Cadic-Conicet & Secretaría de Desarrollo y Planeamiento del

Gobierno de la Provincia de Tierra del Fuego. Ushuaia. MS.

Meulmann, J.J., W.J. Heiser. 1999. SPSS Categories (©) 10.0. Chicago, SPSS Inc.

Orquera, L.A. 1999a. Análisis de los conchales fueguinos y de la distribución espacial interna del Sitio Túnel VII. En: *Actas del XII Congreso Nacional de Arqueología Argentina: 66-72*. Buenos Aires.

Orquera, L.A., 1999b. El consumo de moluscos por los canoeros del Extremo Sur. Relaciones de la Sociedad Argentina de Antropología XXIV: 307-328. Buenos Aires.

Orquera, L.A., E.L. Piana. 1988. Human littoral adaptation in the Beagle Channel region: maximum possible age. *Quaternary of South America and Antarctic Peninsula, vol. 5* (1987): 133-162. Rotterdam.

Orquera, L.A., E.L. Piana.1991. La formación de los montículos arqueológicos de la región del canal Beagle. *Runa, vol. XIX* (1989-1990): 59-82. Buenos Aires.

Orquera, L.A., E.L. Piana.1992. Un paso hacia la resolución del palimpsesto. En: *Análisis espacial en la arqueología patagónica*. L.A. Borrero & J.L. Lanata (Coord.). Ed. Ayllu S.R.L.: 21-52. La Plata.

Orquera, L.A., E.L. Piana.1996. La imagen de los canoeros magallánico-fueguinos: conceptos y tendencias. *Runa, vol. XXII*: 187-245. Buenos Aires.

Orquera, L.A., E.L. Piana.1999a. *La vida social y material de los Yámana. Eudeba-IFIC*, 567 pp. Buenos Aires.

Orquera, L.A., E.L. Piana.1999b, *Arqueología de la región del canal Beagle (Tierra del Fuego, República Argentina)*. Sociedad Argentina de Antropología, 146 pp. Buenos Aires.

Orquera, L.A., E.L. Piana., J. Estevez Escalera, A. Vila Mitja. 1993. El fin de un sistema: un sitio de canoeros fueguinos del siglo XIX. En: *Actas del XII° Congreso Nacional de Arqueología Chilena (1991). Boletín del Museo Regional de la Araucanía, n° 4, tomo I: 89-94*. Temuco.

Piana, E. L, 1984. *Arrinconamiento o adaptación en Tierra del Fuego. Ensayos de Antropología Argentina (año 1984)*, Editorial de la Universidad de Belgrano: 7-110. Buenos Aires.

Piana, E., J. Estevez, A. Vila. 2000. Lanashuaia: un sitio de canoeros del siglo pasado en la costa norte del canal Beagle. En: *Desde el País de los Gigantes. Perspectivas arqueológicas en Patagonia*. J.B. Belardi, F. Carballo M. & S. Espinosa (Eds.): 455-469. Río Gallegos.

Piana, E. L., D. R. Martinioni, R. Agüero. 1998. Planificación territorial y protección de sitios arqueológicos en Tierra del Fuego. En: *Cuartas Jornadas de Arqueología de la Patagonia. Resúmenes de ponencias: 3*. Río Gallegos.

Piana, E. L., L. A. Orquera. 1999. Canoe fuegine: etnografia storica e archeologia. *L'esemplare del Museo L. Pigorini. Bulletino di Paletnologia di Roma. N° 89*, 397-445. Roma.

Piana, E., A. Vila, L.A. Orquera, J. Estevez. 1992. Chronicles of 'Ona-Ashaga': archaeology in the Beagle Channel (Tierra del Fuego-Argentina). *Antiquity 66 (252)*: 771-83 . London.

Pique, R. 1999. *Producción y uso del combustible: una evaluación arqueológica.* Serie Treballs d'Etnoarqueologia No. 2. Edición Conjunta: Publicacions de la UAB- Publicacions del CSIC

Rabassa, J; A. Coronato; G. Bujalesky; M. Salemme; C. Roig; A. Meglioli; C. Heusser; S. Gordillo; F. Roig; A. Borromei, M. Quattrochio. 2000. Quaternary of Tierra del Fuego, Southernmost South America: an updated review. *Quaternary International 60-71*: 217-240. London.

Waswlkov, G.A. 1982. *Shellfish gathering and shell midden archaeology*. PhD. dissertation. The University of North Carolina at Chapel Hill. University Microfilms International. Ann Arbor.

Figures

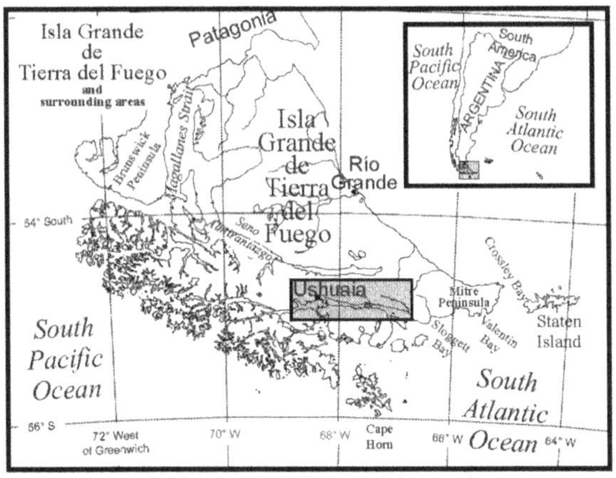

Figure 1. Tierra del Fuego: The World's End

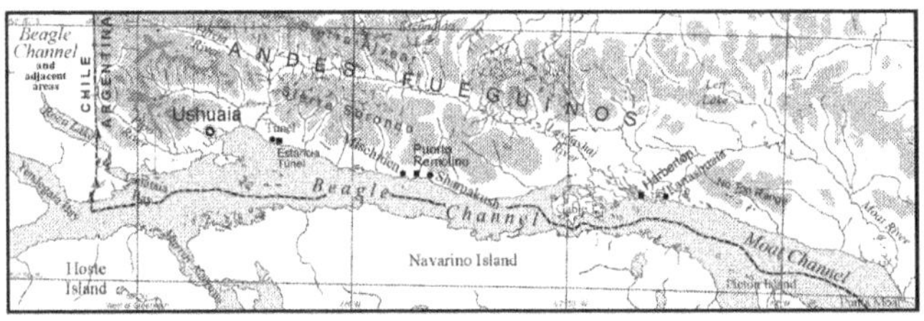

Figure 2. Area of Study

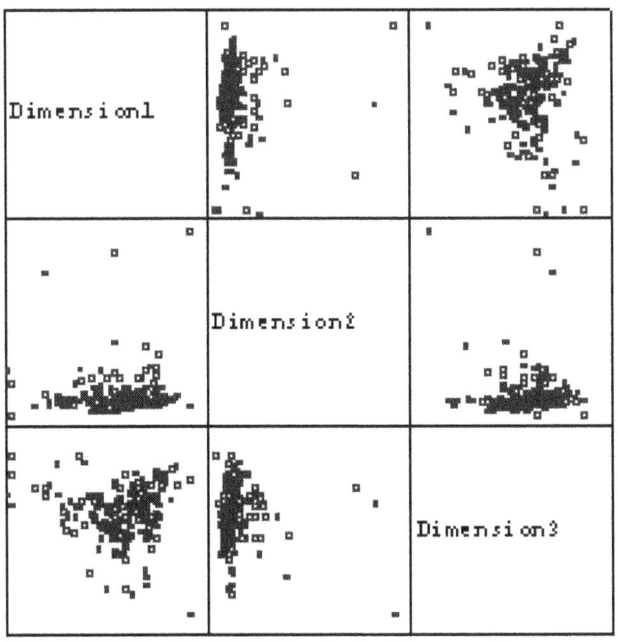

Figure 3. Correspondence Analysis without outliers. Individual Scores

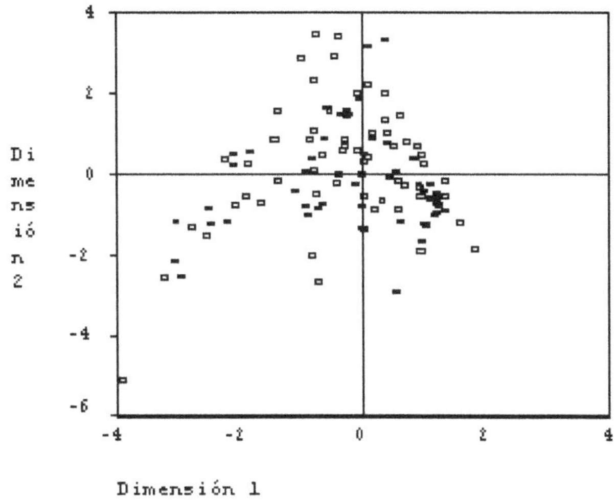

Figure 4a. Correspondence Analysis of selected variables. Individual Scores

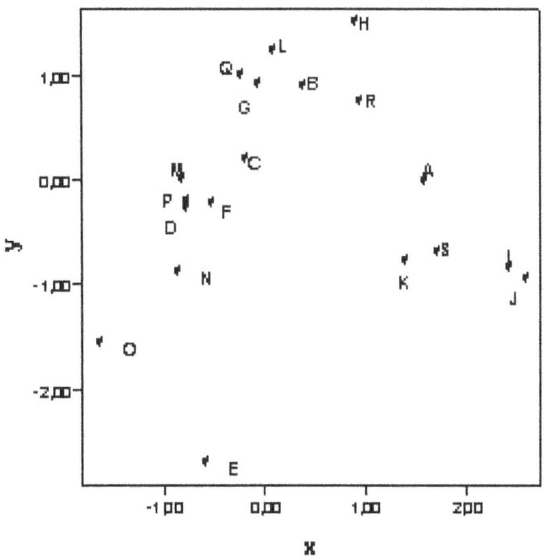

Figure 4b. Correspondence Analysis of selected variables. Variable Plotting

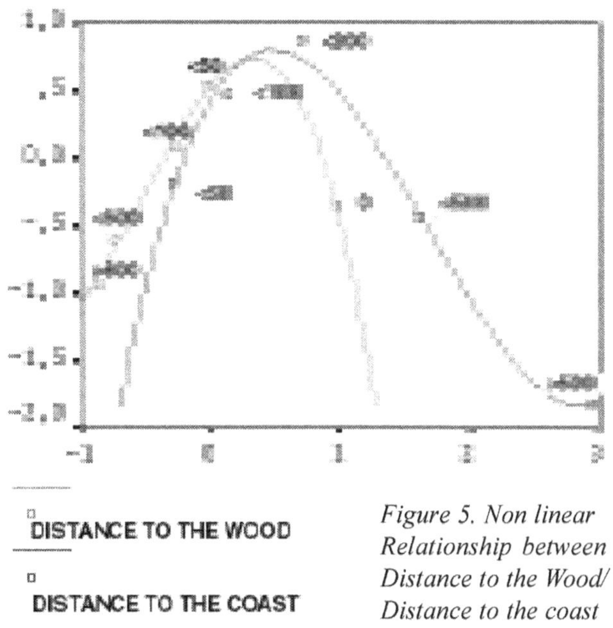

Figure 5. Non linear Relationship between Distance to the Wood/ Distance to the coast

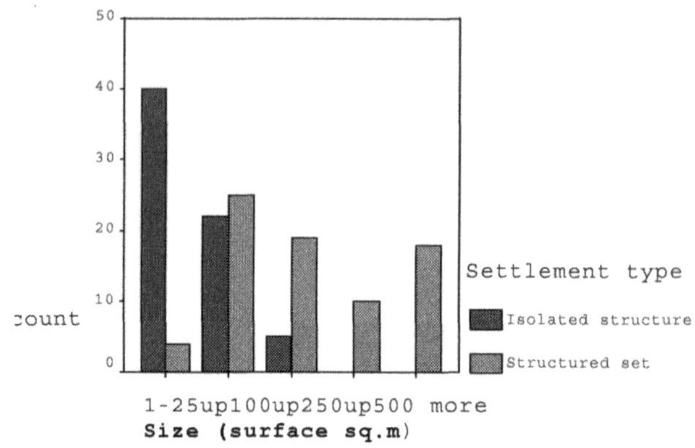

Figure 6. Relationship between shape, topology and size.

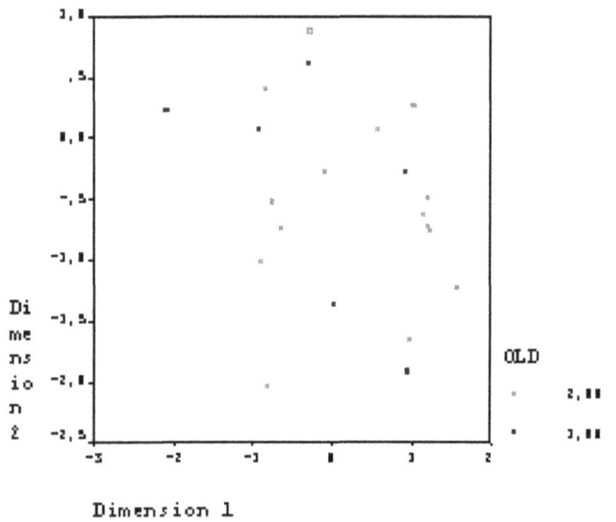

Figure 7a

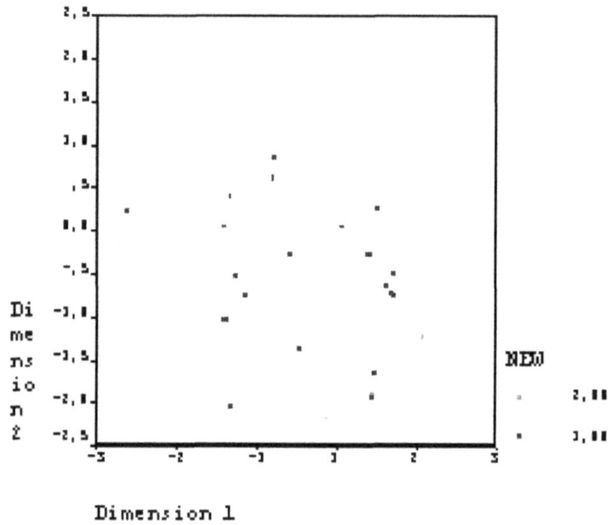

Figure 7b

Figure 7. Correspondence Analysis with chronological information

Views on "castrum cuscin"
Archeology and archaeometry under water

Ralf Bleile
Ernst-Moritz-Arndt-Universität Greifswald
Historisches Institut
Lehrstuhl für Osteuropäische Geschichte
Domstraße 9a
17489 Greifswald
Phone: 03834-863311; 03834-509857 - E-mail: he980141@uni-greifswald.de

Cornelius Meyer
Eastern Atlas Leipzig
Spinnereistraße 7
04179 Leipzig
Phone: 0341-4980500 - Fax: 0341-4980509 - E-mail: info@eastern-atlas.de

Burkhardt Ullrich
Eastern Atlas Leipzig
Spinnereistraße 7
04179 Leipzig
Phone: 0341-4980500 - Fax: 0341-4980509 - E-mail: info@eastern-atlas.de

Abstract: The landscape of Mecklenburg-Vorpommern in northern Germany is characterized by more than 700 lakes. Remains of wooden bridges verify a run to the sheltered islands in the late-slavonian, indicated by dendrochronology to the 11th and 12th century AD. A research project on the exploitation and variability in the district of the upper lakes (obere Seen) from the 10th to the 13th century was started by the department of Eastern European History at Greifswald University in cooperation with the National Agency for Archaeological Research of Mecklenburg-Vorpommern.

ḤCastrum cuscinḤ belonging to a late š lavic castle is the main area of investigation. This island is situated in the Plauer See. Different on-shore and off-shore techniques, like under water survey, geophysical prospection, sediment sonar and drilling are used to get information about the changes of water-level as well as the visible and hidden remains around the island. All obtained data, including aerial and water photography are provided in a data base. At the presentation different views on Ḥcastrum cuscinḤare given. The application of GIS allows to compare details and find archaeological features of the š lavic fortification.

The used technologies, mostly based on non-destructive techniques, will be applied on other islands of the Germania-Slavica project. By the connected GIS we get comprehensive answers as well as new questions to documented projects. May it support description and protection of archaeological places under water.

Introduction

Mecklenburg Vorpommern is situated in the north-east of Germany. It borders to Schleswig Holstein in the west, Poland in the east and Berlin-Brandenburg in the south. Mecklenburg Vorpommern's landscape is shaped by more than 750 freshwater lakes. They are aligned from south-west to north-east. This area is called "Mecklenburger Seenplatte". Among others the largest lakes are the Müritz, the Plauer See, the Kummerower See, and the Schweriner See. This area was inhabited by Slavic people at least since the 8th century AD. They settled on islands and peninsulas; their existence is proved by numerous archaeological sites that were discovered on islands and along the shore.

In the 19th century, the first lacustriane dwellings were explored in southern Germany and Switzerland. But also in the north of Germany freshwater lakes became subject of prehistoric studies. In search of the legendary temple "Rethra", scientists investigated bridge posts. Using so called "glass boxes", the archaeologists crossed the lakes in boats in order to examine wooden posts between the former settlements and the mainland.[1] More than 60 sites, dating back to the late Slavic period, were recorded up to the present.[2] They are located in freshwater lakes, rivers and lowland areas. Since the 1960ies divers, formerly lead by the "Akademie der Wissenschaften" (Academy of Sciences) of the GDR, have contributed to the study of bridge constructions.[3] They took part in excavations of Slavic castle mounds. Such mounds are named, for example, Teterow in the Teterower See and Fergitz in the Oberuecker See.[4] Since 1990 the "Landesamt für Bodendenkmalpflege Mecklenburg-Vorpommern" (National Agency for Archaeological

Research) has been responsible for archaeological research under water. Surveys are conducted together with the "Landesverband für Unterwasserarchäologie Mecklenburg-Vorpommern" (National Association for Nautical Archaeology Mecklenburg-Vorpommern), an association for archaeologists, professional and sports divers who share an interest in nautical archaeology. The major task is protection, preservation and documentation of submerged sites.[5] Since 1996, the archaeologists have also been responsible for the study of sites in freshwater lakes, dating back to the Slavic period.[6] During the systematic documentation of Slavic bridge remains, the scientists discovered a remarkable site in the Plauer See, where features promised extensive information on late Slavic water bodies. Together with the Flesensee, the Kölpinsee and the Müritz, the Plauer See belongs to the so called "upper lakes". The average water level of 62 m MSL is relatively high. All lakes are connected by the river "Elde" and form a closed system of water bodies. Numerous archaeological sites as well as historical and geological research work concerning the water level fluctuation of lakes provide a solid base for this interdisciplinary study of the regional development.[7]

Since 2000, the Institution for East European History at the University of Greifswald and the Landesamt für Bodendenkmalpflege Mecklenburg-Vorpommern have investigated the use of and the man made influences on lakes and rivers during the late Slavic and early German period. The research work covers the time from the 10th to the 13th century AD.[8]

This interdisciplinary study is supposed to extent the knowledge of the Germania-Slavica-research work. The registration of archaeological sites in freshwater lakes is important for the protection of the national heritage. In addition the studies increase methodical skills that are needed for underwater prospections. Moreover, this work is concentrated on the cultural structure and means of transportation in the late Slavic and early German period. Furthermore, a certain relevance is put on the man made influence on the changes of the water body. The main interest is the variation of the water level and its consequences for settlements. To answer these questions, the interdisciplinary co-operation of different scientific branches, as archaeology, geology, geophysics, botany, and history is necessary. Finally, a GIS (Geographical Information System) helped to combine all present results. By means of this system, the mapped lakes and sites were connected to a two- and three dimensional projection of the documented archaeological features.

This report shall give an overview of the results won since last year. The research work was concentrated on the Kohlinsel and its surroundings.

First surveys in the Plauer See

The Kohlinsel is situated in the Plauer See (fig. 1), 120 m off the north-western shore. The island is 100 m long and 50 m wide. It rises only 1 m above the surface (fig. 2). To a high extent, the island's shore is affected by strong waves from southern and south-eastern directions (fig. 3). They cause a continuous erosion; this process is mirrored by the island's present form. There are bays along the shore in the east and in the west, as well as numerous trees that fell into the water.

Right in front of the shore wooden remains were continuously scoured (fig. 4). The discovered wooden objects and the high amount of other late Slavic finds, as for example pottery, iron knives, parts of jewellery, and weapons proofed the existence of the late Slavic nobility castle "cuscin" on the island.[9] About 1900, scientists already suspected Slavic remains on the Kohlinsel. First excavations took place in the 1950ies. A relation between the castle and a rampart, dating back to the 11th or 12th century, was suggested.[10]

The present work aims at mapping a part of the island shore in order to ascertain the functional affliction of the wooden remains and taking samples for dendrochronological determination. The results are supposed to explain the relation between single features wich were found on the island.

To gain the necessary information, measurements had to be taken. A local grid was established in up to 1 m deep water, the grid extended to 10 x 10 m squares. The measuring points under water facilitated the drawing of the features. Every single square was marked with measuring tape and galvanized poles. This arrangement enabled the drawing of all wooden remains on a scale of 1:20. 839 wooden remains were recorded on the bottom of the lake (fig. 5). 444 of them could be identified as former posts. 90 were found in their original constructional assemblage, as for example an anchor beam and an eyebeam (fig. 6). They are still fastened to the bottom by posts driven through eyes. In addition, there are 305 round timbers, boards and planks. Their position might have been changed over the years.

After the digitalisation, the master plan of the features shows two functional units (fig. 5):

> In up to 70 cm deep water the mentioned remains of a rampart were discovered. They are situated in front of the shore. This construction consists of wooden boxes wich were put up one behind the other. Posts were used to fix long cut timbers and sharpened anchor beams to the outer castle mound (fig. 6). Additionally, long cut timbers and crossbeams were put one upon the other. The high amount of wooden posts and the sharpened anchor beams in situ suggest the situation of rampart's outside in this area. It cannot be said yet whether the wooden remains lying in front of this construction eventually belong to it or were part of an earlier rampart (fig. 5). They might also have been part of a run around corduroy road. To avoid an interference into the ecological sensitive surface along the shore, we mapped the mentioned area with the help of a magnetometer (fig. 7). This work was conducted by Eastern-Atlas, Leipzig.[11] The raster graphic shows semicircular constructional interferences into the lake bottom, pointing to a fragment which consists of two eyebeams and with wisely lying round timbers (fig. 5). In addition, those structures are

orientated according to course of the outer castle mound. A continuation of the magnetometer tests this year will hopefully shed light on the multiple phases of the rampart.

Up to now 30 samples were taken from this site in order to date the rampart. Sevem of them led to the following results (fig. 6): Four oak samples wich were taken from the constructional assemblage dated back to AD 1095 (Waldkante), AD 1040 (without sapwood), AD 1089 (without sapwood) and AD 1123 (without sapwood). Only feature no. 2685 was completely preserved. The other three samples might be much younger than it is suggested here. Taking into account the average of all samples, the relatively young date of feature no. 2438 (AD 1123) is noteworthy. In this case a younger construction or repairing phase can be assumed. Three other oak samples which were probably not found in situ complete the picture. Number 2357 and 2371 are lying next to each other, pointing to the long cut timber no. 2384. Especially remarkable is the result of feature no 2357 which dates back to AD 1095 (Waldkante). This allows the conclusion that it may belong to an older construction- or repairing phase.

After all wooden remains along the south-eastern shore were measured, the course of the rampart can be imagined. During the survey on the island, a fragment of a wooden box was discovered exactly an the place where it was suspected. This is proof of an oval rampart, running around the island.

Apart from the wood found in front of the Kohlinsel, divers discovered a row of posts (fig. 5). This wooden palisade varies in its breadth. It was found in 2 m deep water along a former embankment. This embankment had already earlier been recognized in the relief.

The row of posts consists of very thin oak and birch poles of less than 10 cm in diameter. The row continues around the south eastern shore of the Kohlinsel but it is not preserved around the whole island.

One of the oak sticks dated back to AD 1089 (Waldkante). This result proves the simultaneousness of the palisade and older construction or repairing phases of the rampart.

As the row of thin poles faces the lake and due to ist situation close to the former embankment, it allows the conclusion that it served as a shore protection against wave erosion.

Archaeological excavations on the island

During a second survey numerous late Slavic occupation layers were discovered. On the bottom of a trench a thick wooden grating was found. It partly consists of carbonised oak trunks and is situated in a height of 60,7 m MSL. This grating can be expected on the whole island and give an explanation for the fact that it was impossible to take drilling cores wich would have been needed for geological research and pollen analysis. However the core drill was not able to penetrate the grating.

Samples taken from above the grating dated back to AD 945 (Waldkante) and AD 972 (Waldkante). They are distinctively older than the earliest rampart remains.

Dendro dates taken in 1996/97 seem to confirm our results. Those samples were taken from a bridge that leads to the mainland (AD 956 Waldkante and AD 976 without sapwood).
To ascertain the water level of the lake Plauer See in the late Slavic period, a chart combining the results of the geological drillings and the excavation was drafted.

It becomes obvious that the remains of the rampart are to be found between 61,51 m MSL and 61,11 m MSL on the bottom of the lake (fig. 8). On the island, it was discovered at 61,64 m MSL. The upper edge of the wooden grating lies at 60,74 m MSL in test trench I. The palisade posts were discovered between 60,38 m MSL and 59,91 m MSL.

Assuming a thickness of 30 cm for the wooden grating, the water level of the Plauer See did not rise above 60,44 m MSL during the second half of the 10th century.

The question of the palisade's location during the late Slavic period seems to be very important: In case it served as protection against the water, the palisade was probably situated very close to the shore in low waters. On the other hand, if it was built as a defensive measure, the palisade might have been located on the island.

Thus, the water depth measured in the area close to the posts can not be compared to the surface in late Slavic times without further proof. Only future geological studies can give final certainty.

The consequences that water level alterations of only a few centimetre can have on water bodies are visible in a chart made with the help of a high and low frequency sonar by K. Storch, SonderSonar Jena (fig. 9).

The islands former extension can clearly be recognized. Furthermore, a dike-like low elevation rises from the sea bottom between the island and the mainland. This elevation does not correspond to the bridge that dates back to the 10th and 12th century AD.

The poles belonging to the palisade are located on the former embankment. By laying a depth contour line across the highest points of the palisade, the island character of the Kohlinsel is confirmed. It may be concluded that the nobility castle was built on an island.

Picturing a depth contour line that is put across the lowest points of the posts, a connection between island and mainland is to be noticed.

It has to be mentioned that for the end of the 11th century AD no dendrochronological dates for the bridge were recorded although the rampart and the palisade already existed in those days. Only two samples dating back to the 10th century and two samples of the 12th century were documented.[12] The breadth of the bridge and the location of the posts in relation to the length of the crossbeams suggest a construction that was built in at least two phases. The dam might be an explanation for the missing proof of a bridge in the 11th century.

With the help of wood samples and pollen analysis the regional development of climate and vegetation will be investigated. There are indications for a supraregional climate change.

We might not be able to reconstruct a single, uninterrupted cultural settlement landscape during the late Slavic period but have to take changes of the water bodies into account.

These changes might have had a deep influence on the settlements. Whether humans were causally connected to the changes and how people adjusted to new living conditions will be subject of future research work.[13]

Endnotes

[1] Virchow 1881:267.
[2] Bleile 1999:127-169.
[3] Herrmann 1975:138-141.
[4] Herrmann 1966, Unverzagt/Schuldt 1963.
[5] Lübke 2000:17-35, Förster 2000:46-49.
[6] Bleile 1998.
[7] Kaiser 1998, Ruchhöft 1999.
[8] These studies are sponsored by the Deutsche Forschungsgemeinschaft DFG (German Research Association), the Kultusministerium Mecklenburg Vorpommern (Ministry of Culture), the Arbeitsgemeinschaft limnische Archäologie e.V. (Limnic Archaeology Working Group) and by the Mecklenburger Brewery Lübz GmbH.
[9] Helmold:92, 310, 328, 344.
[10] Bleile 1999a:33.
[11] Ettel/Meyer 2000:143-149.
[12] Bleile 1999:132-139.
[13] Translation: K. Auer and O. Uhlmann.

References

Bleile, R., 1998. Unterwasserarchäologische Untersuchungen im Dümmer See, Landkreis Ludwigslust. *Archäologische Berichte aus Mecklenburg-Vorpommern 5*, 54-61.

Bleile, R., 1999. *Slawische Brücken in Mecklenburg-Vorpommern. Bodendenkmalpflege in Mecklenburg-Vorpommern. Jahrbuch 1998-46*, Lübstorf 1999, 127-169.

Bleile, R., 1999a. Vorbericht zu unterwasserarchäologischen Untersuchungen an einer slawischen Brückenanlage im Plauer See bei Quetzin, Landkreis Parchim (Mecklenburg-Vorpommern). *Nachrichtenblatt Arbeitskreis Unterwasserarchäologie 5*, 32-35.

Ettel, P., Meyer, C., 2000- Die Burg von Dobin – Vorbericht zur geomagnetischen Prospektion und Sondagegrabung 1999/2000. *Archäologische Berichte aus Mecklenburg-Vorpommern 7*, 139-151.

Förster, Th., 2000. Stand der schiffsarchäologischen Untersuchungen in der Wismarbucht. *Nachrichtenblatt Arbeitskreis Unterwasserarchäologie 7*, 46-49.

Helmold von Bosau. 1963. Slawenchronik. Übertragen und erläutert von Heinz stoob. *Ausgewählte Quellen zur Deutschen Geschichte des Mittelalters* (Freiherr vom Stein-Gedächtnisausgabe), hrsg. von Rudolf Buchner, Band 19, Berlin.

Herrmann, J., 1966. Die slawischen Brücken aus dem 12. Jahrhundert im Ober-Ückersee bei Prenzlau. Ergebnisse der archäologischen Unterwasserforschung. *Ausgrabungen und Funde 11, Heft 4*, 215-230.

Herrmann, J., 1975. Underwater archaeological resaerch in the German Democratic Republik, *International Journal of Nautical Archaeology and Underwater Exploration. 4/1975*, 138-141.

Kaiser, K., 1998. Die hydrologische Entwicklung der Müritz im jüngeren Quartär – Befunde und ihre Interpretation. *Zeitschrift für Geomorphologie, Neue Folge, Supplement-Band 112*, 143-176.

Lübke, H., 2000. Timmendorf-Nordmole und Jäckelberg-Nord. Erste Untersuchungsergebnisse zu submarinen Siedlungsplätzen der endmesolithischen Ertebølle-Kultur in der Wismar-Bucht, Mecklenburg-Vorpommern. *Nachrichtenblatt Arbeitskreis Unterwasserarchäologie 7*, 17-35.

Ruchhöft, F., 1999. Der Wasserstand der „Oberen Seen" in Mecklenburg in historischer Zeit. *Nachrichtenblatt Arbeitskreis Unterwasserarchäologie 5*, 36-39.

Unverzagt, W., Schuldt, E., 1963. *Teterow. Ein slawischer Burgwall in Mecklenburg*. Deutsche Akademie der Wissenschaften zu Berlin. Schriften der Sektion für Vor- und Frühgeschichte 13. Berlin.

Virchow, R., 1881. *Verhandlungen der Berliner Gesellschaft für Anthropologie, Ethnologie und Urgeschichte, Jahrgang 1881*. 267.

Figures

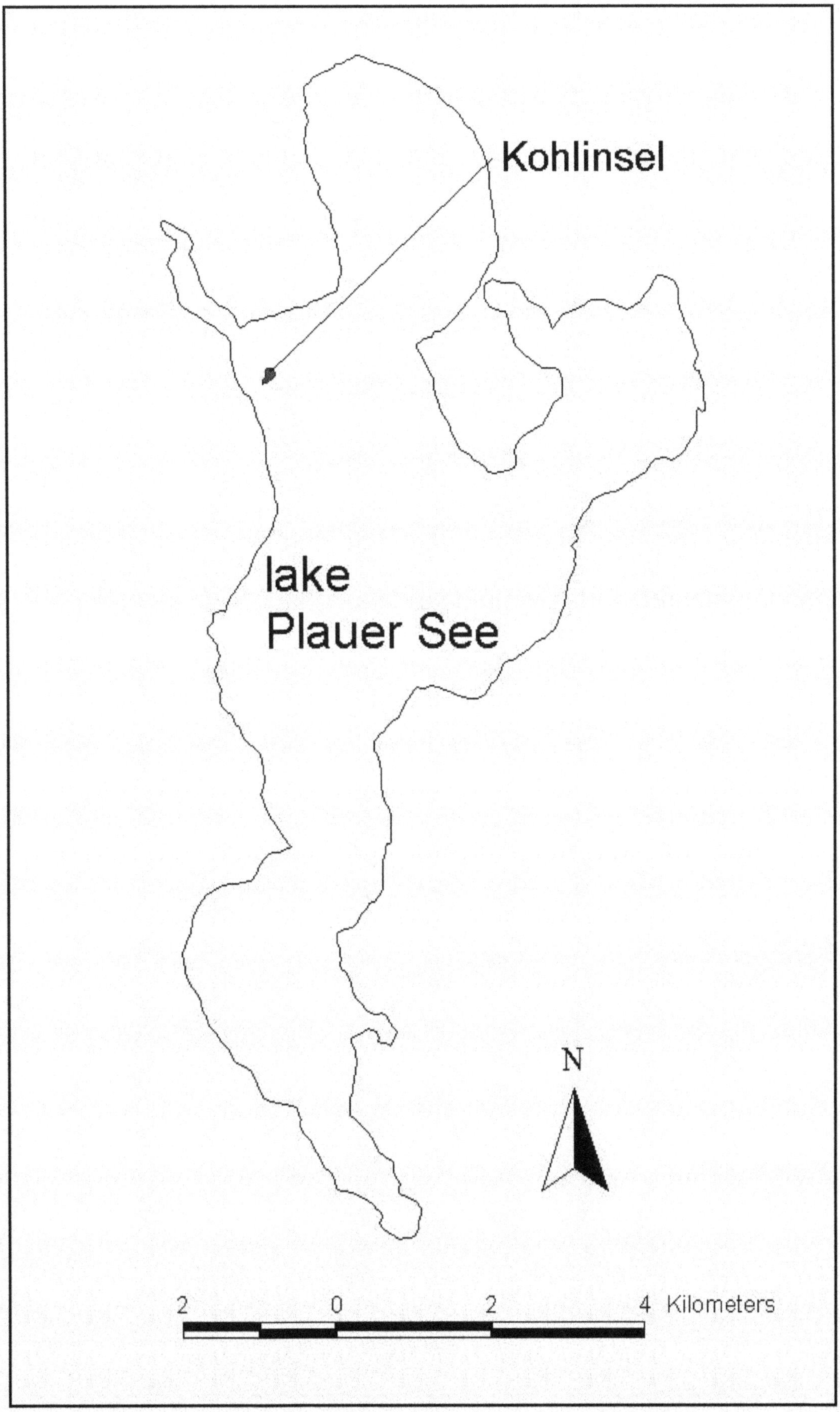

Figure 1. Germany, Mecklenburg-Vorpommern: lake Plauer See and Kohlinsel.

Figure 2. Plauer See, Kohlinsel rises only 1 m above the surface.

Figure 3. Plauer See, Kohlinsel: The island's shore is affected by strong waves from southern and south-eastern directions.

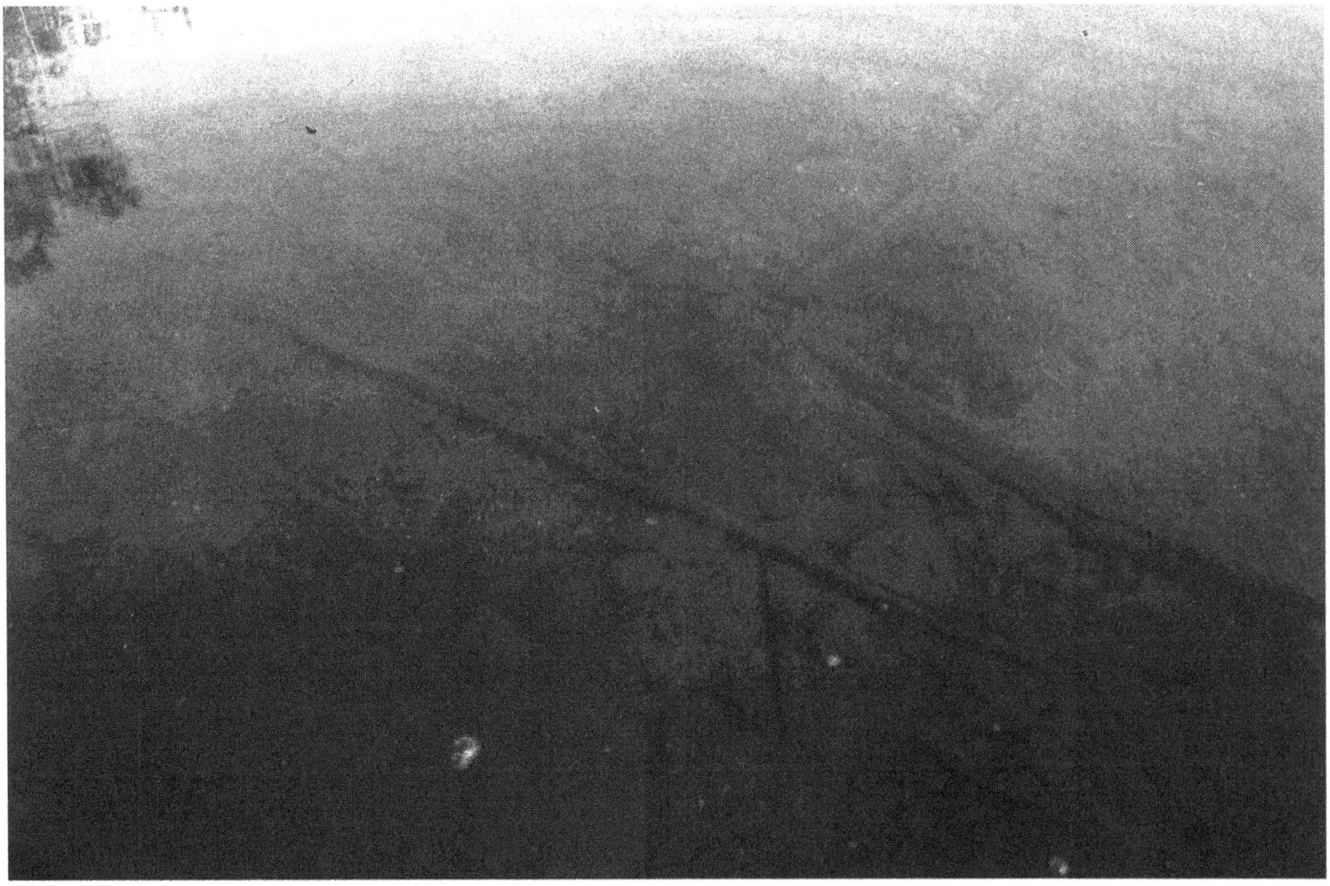

Figure 4. Plauer See, Kohlinsel: Wooden remains on the bottom.

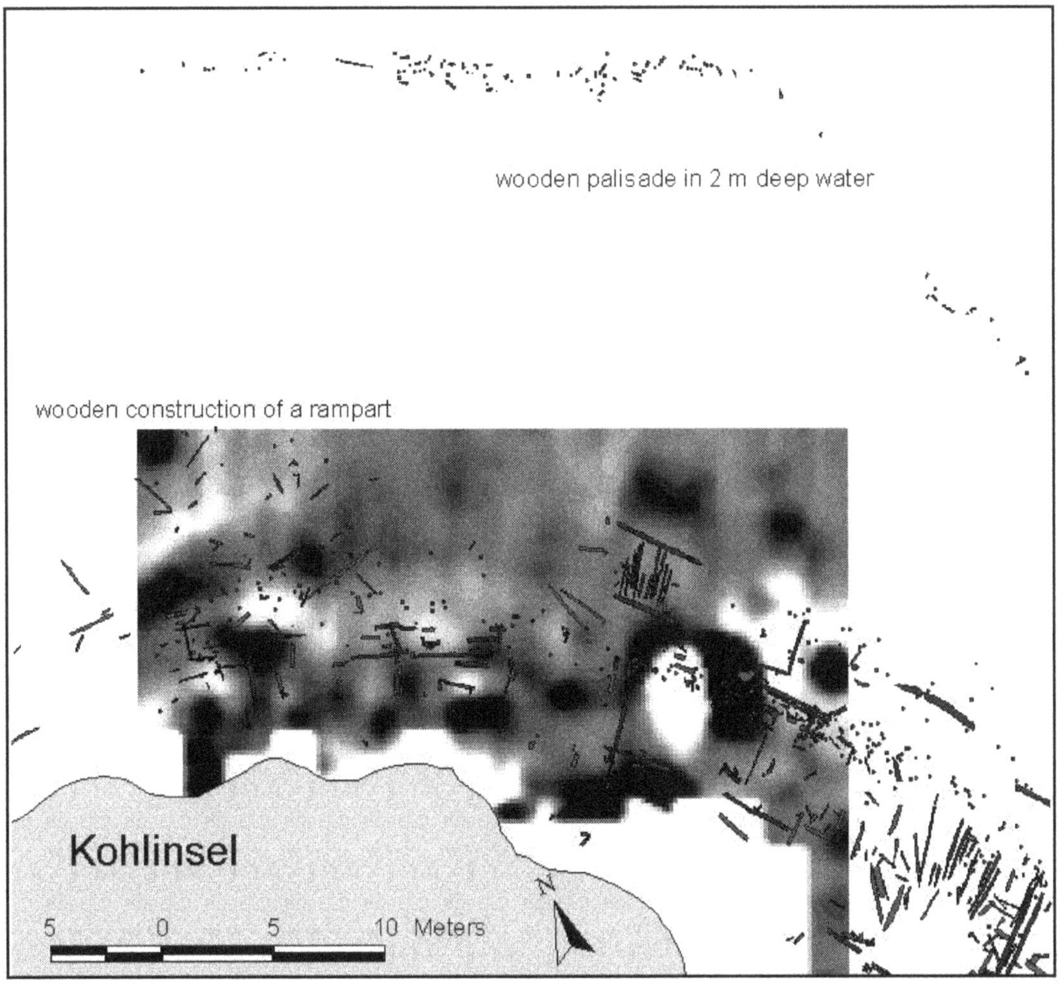

Figure 5. Plauer See, Kohlinsel: Master plan of the features and raster graphic of the magnetometer mapping.

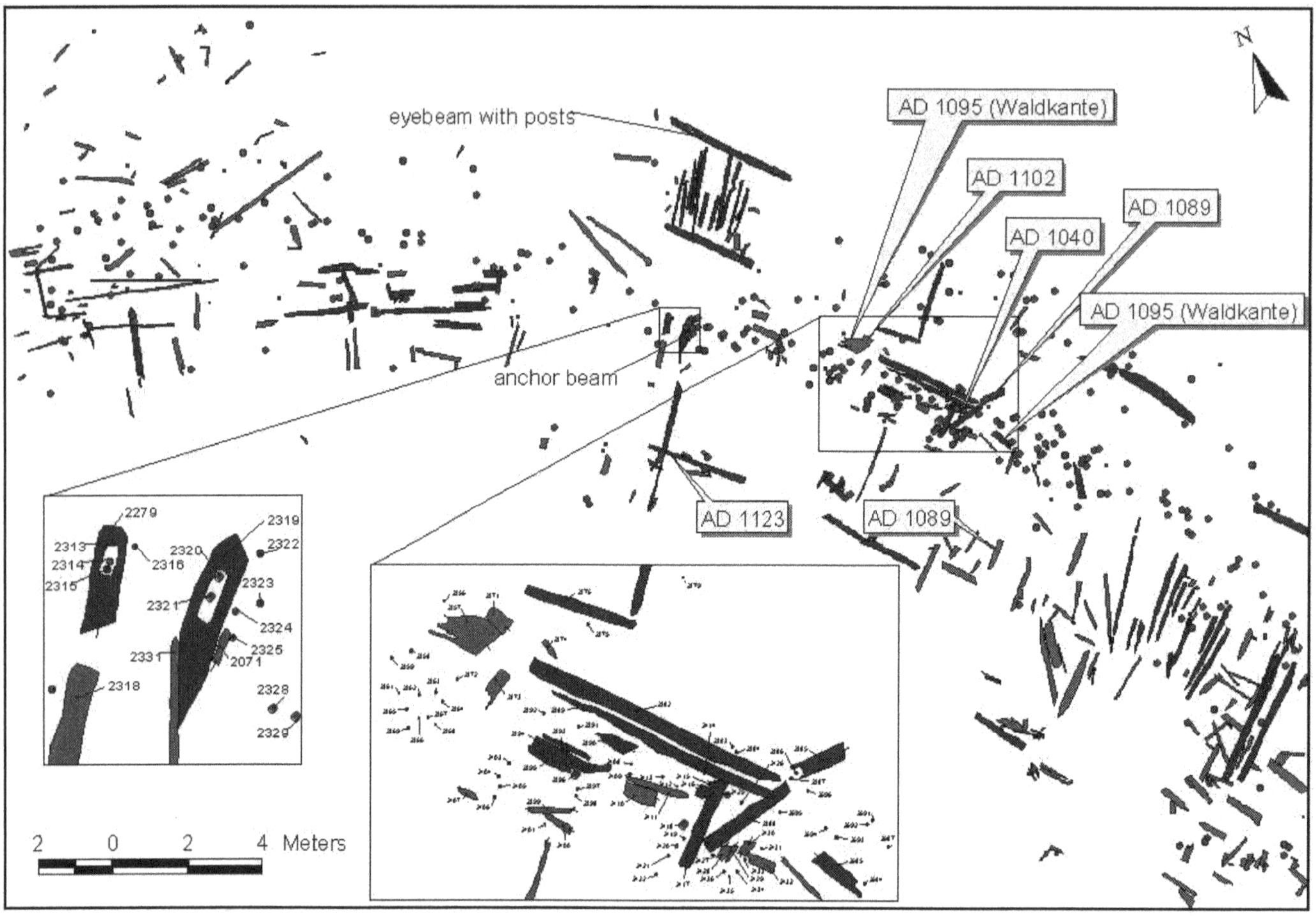

Figure 6. Plauer See, Kohlinsel: Wooden construction of the rampart and dendrochronological dates.

Figure 7. Plauer See, Kohlinsel: Magnetometer on the water.

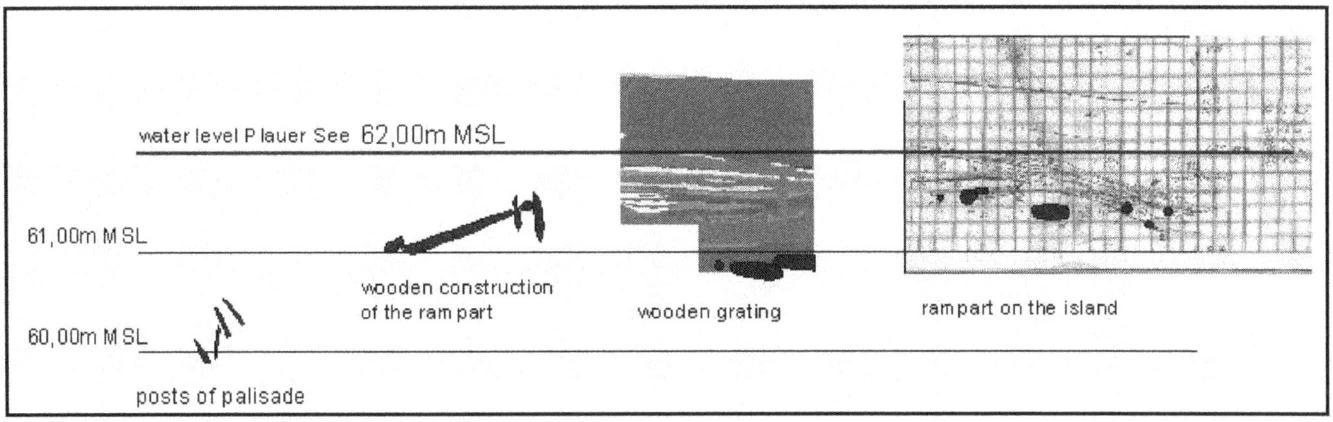

Figure 8. Plauer See, Kohlinsel: Wooden construction under the water level.

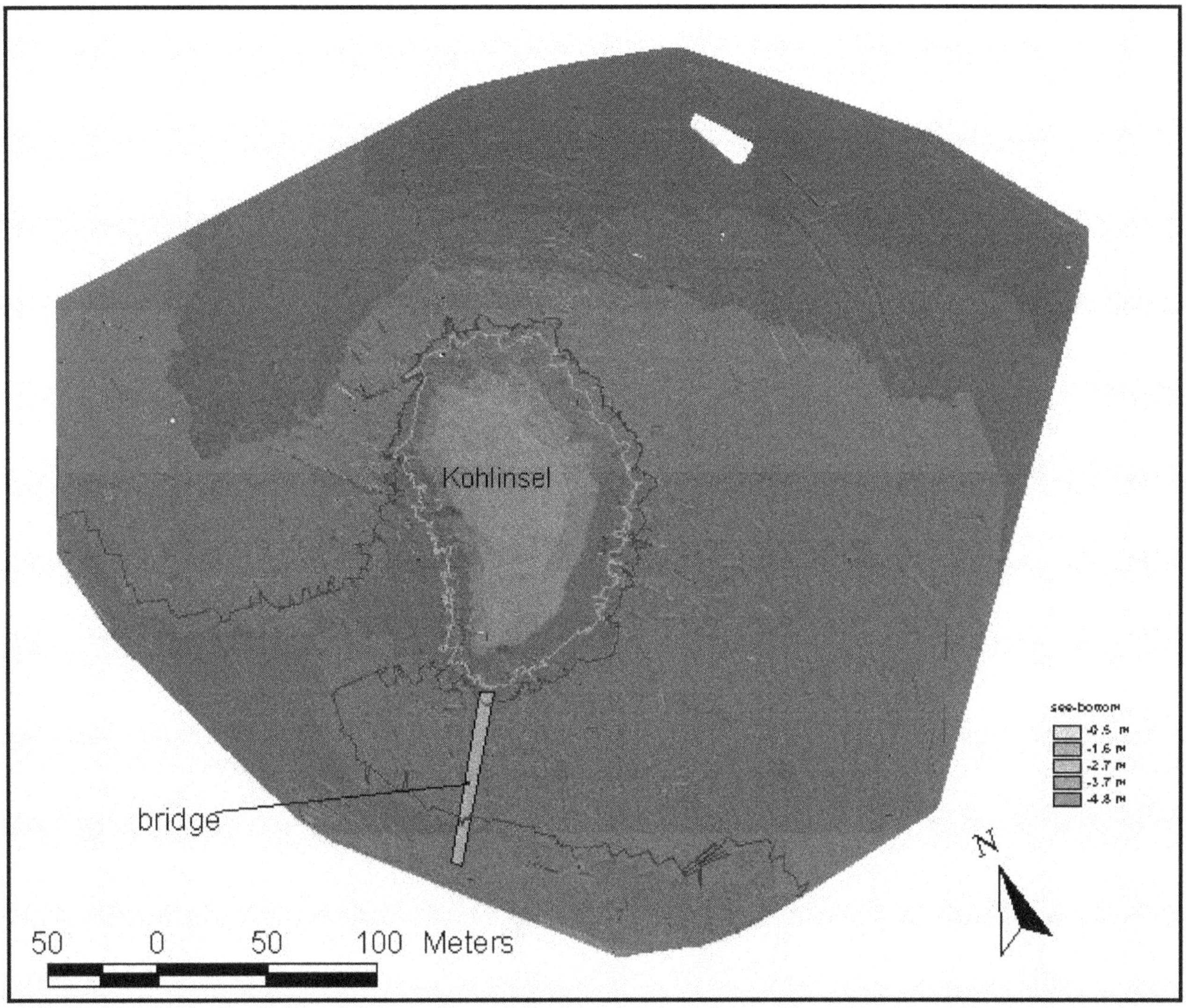

Figure 9. Plauer See, Kohlinsel: See-bottom around the island.

A PC-based stereoscopic measurement system for the generation of digital object models

Frank Boochs
FH Mainz, University of Applied Sciences, Germany
i3mainz, Institute for Spatial Information and Surveying Technology
boochs@geoinform.fh-mainz.de

Stephan Eckhardt
FH Mainz, University of Applied Sciences, Germany
i3mainz, Institute for Spatial Information and Surveying Technology
boochs@geoinform.fh-mainz.de

Ben Fischer
FH Mainz, University of Applied Sciences, Germany
i3mainz, Institute for Spatial Information and Surveying Technology
boochs@geoinform.fh-mainz.de

Abstract: A stereo system designed and used for the measurement of 3D-coordinates within metric stereo image pairs will be presented. First, the motivation for the development allowing to evaluate stereo images is shown. As the use and availability of digital metric images rapidly increases, corresponding equipment for the evaluation and measuring process is needed. Systems which have been developed up to now are either very special ones, founded on high end graphics workstations priced accordingly, or simple ones with restricted measuring functionality. A new conception will be shown, avoiding special high end graphics hardware but providing the measuring functionality required. The presented stereo system is based on PC-hardware equipped with a graphic board, uses an object oriented programming technique and allows a connection to CAD packages like AUTOCAD. The specific conception of the system is shown. The key role of OpenGL is described, which supplies some elementary graphic functions, being directly supported by graphic boards and thus provides the performance needed. Further, important aspects as modularity and hardware independence and their value for the solution are shown. Finally, the functionality of the software is presented in more detail.

Key words: Digital Photogrammetry, Photogrammetric workstation, Digital object model, Stereoscopy, 3D-Measurement, CAD, Object-orientation, OpenGL.

Introduction

Digital images are increasingly playing an important role in the photogrammetric business. Their primary advantage is to permit a higher degree of automation within evaluation processes. They of course allow additionally to apply conventional actions onto images like manual stereoscopic measurements, for example. However, the size of the images to be displayed during such interactive evaluation steps makes it necessary to have powerful computer equipment allowing to effectively handle all the data. Therefore conventional photogrammetric workstations are based on specialized graphic hardware with corresponding special software solutions. This concept is accompanied with high costs, dependency on proprietary hardware and low flexibility.

These disadvantages might be avoided by use of powerful standard hardware together with an open and modular software structure. This allows to reduce hardware costs, to simplify and accelerate the process of development and to adjust the functionality to different applications. For the domain of close range photogrammetry, for example, it might be of value to integrate existing CAD packages into an evaluation environment.

Based on the aspects of simplicity, modularity and flexibility a digital workstation has been designed and is under development. Further more, motivation and background for the proposed solution are outlined, the design of the system is sketched and some technical details are presented.

Photogrammetric tools

General

Since the end of the 70ies, photogrammetric instruments work under computer control and allow to generate digital object data. They are controlled by a human operator and meet all requirements for an efficient data collection system.

Actually, the technological step towards purely digital systems is ongoing. Digital systems have the advantage to contribute to the collection process by some automatic evaluation tools exploiting the content of the digital data. However, their use for a complete 3D-data collection is restricted to objects being represented by a certain number of directly mounted targets. Here, adopted search algorithms can be applied allowing to detect the targets and to determine their positions in space. Under certain conditions even real-time applications are possible using integrated camera systems capturing the image data and automatically exploiting them by means of processes running on the integrated computer system (Imetric, 1999; Leica, 1998).

However, in general, purely digital systems need human control, at least for the collection of the object 3D-data. Then, one demand for these systems is to handle the image data during the 3D evaluation process. Typically, images may have sizes ranging from 10.000 by 10.000 pixel up to 30.000 by 30.000 pixel each. Accordingly, the work load for the computer system is high within all those steps dynamically accessing the image data.

Digital workstations

Conventional digital photogrammetric workstations support such huge digital images. They are generally based on powerful graphic workstations equipped with special viewer components needed for the stereoscopic perception process. Moreover the manufacturers often have adopted there software solutions to the special characteristics of these workstations in order to maximize the performance.

This leads to some disadvantages. The nature of these instruments causes

- high costs
- dependence on the specialized hardware and on the providing company
- low flexibility to technological changes

Most of these items are caused by the proprietary characteristic of the instruments. The manufacturer has to invest great effort into an own solution which has to be amortized. The customer has to rely on the provider and his capability to keep track with the technological process. Furthermore, third party solutions being available are hardly to integrate, because of the hardware dependence of most proprietary equipment.

Alternative

It seems promising to avoid hardware dependence. With respect to flexibility it is necessary to have a modular and open software structure allowing to substitute individual components as there are better ones available. Furthermore, one should try to found an approach on available standards and other open resources in order to reduce the development effort.

Nevertheless, the demanding characteristic of photogrammetric evaluation processes remains and has to be considered. In order to sketch just some of the needs, the following list gives an impression of the features to be met:

- allowing images of any size and depth (up to 900 Mpixel per image, B/W and color)
- fixed floating mark with moving image
- continuous, endless image move with different speeds
- single, double image and 3D-viewing measuring mode with appropriate interactive control features
- registration of all kinds of coordinates together with semantic object information
- ergonomic environment and an agreeable, not tiring stereoscopic viewing system

These requirements need considerable computer power and an adopted control and application software. On the other hand, the technological progress on the computer market is extremely high allowing to realize solutions based on standard PC systems.

Design of the PC-system

General

As conclusion from the deficiencies of specialized solutions we have tried to find a modular concept based on available standards and the corresponding resources. Fortunately, we can profit from the rapid development of the market of PC hardware and applications, which meanwhile even has produced components for a stereoscopic environment. Although this is driven by the game and virtual reality market it allows to choose a solution founded on standard PC components, which should be powerful enough to realize a 3D-measurement application. In order to keep a maximum of portability, device dependent control components have been avoided. This might result in a slight loss of capacity, but that should be acceptable in comparison to the cost of device dependency coming with specialized solutions. Stereoscopy and graphic control therefore are based on the VESA and OpenGL standards.

The aspect of portability is extended to the use of other operating systems than MIRCOSOFT™ based ones resulting in the need to have system specific software elements controlling the graphic devices. As it should not be subject to this project to develop system specialities it has been decided to incorporate the wxWindows Graphical User Interface (Smart, 1999). This provides all necessary routines for the installation of graphic solutions within different environments and additionally gives the corresponding look and feel.

Look and feeling is of supplementary value for the acceptance of the graphical interface by the user community. Once used to work on a PC one is in general familiar with MIRCOSOFT™ and the window technique utilized there. The acceptance of products coherently increases, if the interface seems to be well-known. The design of the graphical user interface takes these aspects into account and tries to provide simple and easy to use tools.

Hardware

Central part of the hardware is a Pentium III CPU with 500 MHz equipped with 512 MB memory. The CPU provides standard processing speed with respect to the actual PC generation. Of importance is the equipment with sufficient main memory, because the main work load during the measuring process comes from the transportation of image data. Here a large intermediate buffer is necessary, to avoid waiting periods induced by the slow access to the mass storage. The selected amount of memory seems to be a minimum quantity required. Correspondingly the amount of mass storage should not fall beyond 5GB, as standard projects may require up to 3 GB storage capacity (cf. Fig. 1).

Stereoscopy is based on a powerful graphic device (ELSA Gloria XL) a stereo monitor and stereo glasses (Stereographics® CrystalEyes®) with appropriate control. The graphic device shows up some features, which are important for a 3D-measuremnt environment:

- high resolution (up to 1280*1024 true color mode)
- VESA interface for direct connection to stereo glasses
- hardware acceleration for several OpenGL-functions
- stereo-in-window technique
- stereo refresh rate up to 120Hz

This graphic card is one representative of a new generation of graphic boards widely supporting OpenGL functionality onboard and being designed for stereoscopic applications. Their appearance will increase in the future allowing to choose between several alternatives, if a software is founded on a graphic standard like OpenGL.

Of interest for our solution are the direct support of stereo glasses, the high refresh rate guaranteeing flicker free stereo viewing and the stereo-in-window technique. The latter has some advantages compared to the standard above-below stereo solutions, which are splitting the screen in a top and bottom part each containing one of the stereo images. By this the whole screen is affected and can only be operated in stereo or non stereo mode avoiding to simply mix different windows on the screen. Furthermore, there is a loss of 50% resolution. Both drawbacks are avoided with the stereo-in-window principle, which is possible due to the stereo mode available in OpenGL. This means to have a quad buffered window, where front and back buffer are split into two pieces each referenced to one image of the stereo pair. Finally, we have the hardware acceleration provided by the board, what of course is very useful, although software solutions would do their task likewise.

Software

Design aspects

Concluding from the deficiencies of specialized solutions we use a modular concept based on available standards and corresponding resources. With respect to the display of images and handling of graphic data, for example, we use the most common standard: OpenGL. All available system platforms have at least a software solution for OpenGL and many graphic boards provide on board intelligence giving the acceleration needed for an acceptable measuring environment.

As OpenGL is supported by all common computer systems, this offers the opportunity to build a solution being independent from the manufacturer. However, the graphical user interface then has to be developed independent from the system environment too. In general, this causes some additional efforts, because UNIX and MICROSOFT™ use different graphic software. But fortunately developments from other sides handling the problem of graphic interfaces in different computer environments do exist. One very powerful solution is the wxWindows Graphical User Interface[1]. It is an object oriented software package providing all necessary components to build user interfaces and is founded on system dependent libraries. This assures the look and feel the user expects and allows to install the same package on differing computer environments simply by exchanging the underlying graphic library. The user community of this package permanently grows and ranges from scientific users to ones[2].

The use of OpenGL and wxWindows assures the requirements coming from the wish to avoid hardware dependency, to be flexible to the technological progress and to exchange components, if necessary, or to incorporate new ones. Based on these standards now an application software has to be built, which has to fulfil very different functional needs (cf. Fig. 2), should reduce the effort for maintenance and modifications and should allow to compose packages being optimally structured to serve different type of applications. As a consequence, there is only one way to support all these wishes, that is to use an object oriented programming approach within C++. This language combines the advantages of object orientation with the possibility to incorporate time effective functions written in C.

Photogrammetry

Several tasks have to be performed when a photogrammetric project has to be handled:

- Project set up
- Interior orientation
- Exterior orientation
- Generation of stereo image pairs
- 3D data evaluation

This list looks very conventional for a photogrammetric processing chain, but it might need much more effort than just handling two images forming a stereo model, which will be used for the data evaluation. Therefore, some aspects will be mentioned documenting the flexible nature of the software developed.

The set up allows to store all necessary external information (point coordinates, calibration values,...) available and has to introduce the control of the following processing steps. Here flexibility is expressed by

- the number of images, which might be processed simultaneously. This number has to be only restricted by the available storage capacity
- the type of external information usable. Types have to be extended to the use of any kind of geodetic data, allowing to process distances, angles, geometric constraints and so on.

The step of the interior orientation comprises the measurement of the fiducial marks and the calculation of the geometric relation between camera system and digital system. Here additional performance is introduced, when the system allows any kind of camera type. This makes it necessary to process reseau camera data or image geometries differing from the normal central perspective model.

The step of the exterior orientation is very simple, if there exist simply one image pair. This is not necessarily the standard situation, when surveying objects of any type. Conventionally, objects are observed by several images capturing all the parts having to be geometrically evaluated (cf. Fig.3). It's then the job of this orientation step to combine all images into a stable and fixed image block. The only way to do this is the measurement of tie and pass points expressing image rays which have to intersect and to hit the object at the same position. In principle, this might not appear as a demanding step, but if one considers, that

- images have different perspectives
- images might be exposed under differing illuminations
- most object points are signalized
- objects show occlusions and discontinuities

  the process of measuring tie and pass points obviously is difficult and possibly needs some iterations until good points are found.

The generation of stereo image pairs is indispensable, because, except of standard aerial image blocks, images forming a model allowing to perceive the data stereoscopically do not exist. Therefore one has to introduce a processing step, which generates epipolar image pairs serving as visual basis for the 3D-evaluation procedure (cf. Fig. 4).

Finally, there is the 3D-evaluation process. Here, challenges come from the fact, that the measuring environment has to fulfil the needs of a photogrammetric workplace. For example, photogrammetric operators expect to have certain control elements allowing to move through the 3D-space. They await to find the floating mark pointing to the object details resting fixed in the middle of the viewer. This makes it necessary to permanently move through the image space, which means to shift the image data continuously through the viewer – a challenging aspect, having the amount of image data in mind, which has to be handled.

The amount of image data, in general, introduces hard challenges. The storage amount of aerial images starts with moderate 21 MB (50 mðm Pixel, B/W image) and ends with 1587 MB (10 mðm Pixel, color image), when the image is prepared for an interactive evaluation, needing to provide all necessary information details. This data has to be processed during the 3D-evaluation without any narrowing of the movement during the measuring procedure. The step of continuous measuring therefore need an intelligent solution for the image display.

Solution aspects

Image Loading

First, a decision had to be found, how the data can be accessed with maximum speed. Several other photogrammetric packages use proprietary formats designed to easily get access to the image data. This doesn't seem to be necessary, because there exist an efficient format with tools available for data transfer: TIFF. We use TIFF, as it allows to hold all formats one can imagine. It's even possible to create an own format. In its latest version, TIFF does already provide everything needed to implement the loading component. Furthermore, TIFF is supported by most conventional image processing tools, what simplifies the data conversion from or to other sources and avoids the need to develop own conversion tools. TIFF is even flexible enough to load the images in tiles, which is inevitable with the large images sizes to be handled. With tiles one is free to load as much of the image as necessary, but not more, thus reducing the memory load of the application. Sam Leffler's libtiff library[7] already comes with a small tool to tile existing TIFF images. This step is sometimes necessary in order to convert external TIFF data into tiled one, because this is not necessarily supported by all image processing tools.

With respect to the size of the tiles, experience has shown, that 256 x 256 pixels for each tile are an adequate compromise between storage load and transfer speed. Other sizes are possible, but up to now there is no need to select a different one. Within the measure component the one class (AerialImage) is responsible to handle these tiled TIFF images. Fortunately, the libtiff library already offers all functionality to load single tiles from a tiled image so one just has to map the corresponding libtiff functions to methods of class AerialImage. Additionally, the AerialImage class holds a tile table to handle the memory addresses of already loaded tiles.

Image Display

Once the image file is tiled it's possible to load just those tiles needed to fill the viewport within the display window. (cf. Fig. 5). The TileManager class is responsible to manage the tiles in memory. After creation of the display window and the OpenGL viewport within this window the TileManager class calculates the number of tiles necessary to fill the viewport. It will recalculate the number every time the user resizes the display window. Next, it selects the tiles to be displayed and loads them. Then, this object waits for user interaction. When the image is moved by the operator, this will be registered by the TileManager class. Dependent on the actual viewer position, it decides whether it is necessary to load new tiles and to destroy old tiles or if the already displayed area of tiles has still enough data to fill the viewport. All image movement directions are handled separately. That means positive X, negative X, positive Y and negative Y movement is handled separately. As a result, the steps necessary to move the image just have to be done for one direction.

This buffering procedure assures that the main memory always holds enough data, when further movement in the negative X direction occurs and new image information has to be loaded into the viewport. Obviously, there are only a few instructions necessary to handle image movement. To actually display the tiles within the viewport there're just two loops over the tile area needed (cf. Figure 5), one to display the rows and one to display the columns. OpenGL can handle raw pixel data directly so there is no need for further calculations or conversation. Experience has shown, that drawing raw pixel data works better than using textures. These two concepts, few instructions for image movement together with fast raw pixel display using OpenGL results in very high display speed even on standard PC hardware.

Stereoscopic Viewing

The OpenGL standard has built -in support for stereoscopic view of rendered content. Using this build-in functionality it's quite easy to get a stereoscopic view of stereo images. As all functionality to display images is encapsulated in the TileManager class it's relative straightforward to extend this class to handle stereo images. There're four things one has to take care of, when rendering stereo images.

- a stereo hardware environment is needed. That means a graphics card that supports stereoscopic view and some equipment to separate the left image for the left eye and the right image for the right eye (We use shutter glasses for image separation)
- a stereoscopic window is needed. Some operating systems and graphic card drivers support the stereo-in-window technique, where the stereoscopic view is restricted to just one window and not the whole screen (We use this technique under Windows NT)
- one has to render the content twice. Once for the left image and once for the right image (The TileManager class of the measure component handles this)
- one has to apply an offset between the left and the right image to actually create the stereo effect (The TileManager class of the measure component handles this)

So, the modular structure of the software supported by the graphics standard selected allows easily to extend the whole image management to stereoscopic images and therefore assures an effective handling of large data sets even in stereo mode and provides the functionality needed for a powerful measurement environment.

Simultaneous viewing of multiple images

For the majority of terrestrial applications images are taken in blocks surrounding the object in question. Prior to any three dimensional measurements the images need to be oriented and then transformed into epipolar image pairs allowing to view stereoscopically. The step of orientation simply needs some measurements of tie and -if available- pass points. However, this produces some problems, if the object is neither equipped with point signals nor showing enough details allowing to clearly and unambiguously identify points.

A simple solution for such difficulties is possible, if the selection, identification and measurement of the points will be done in all images simultaneously. Then a visual comparison of the object as it is represented in those images covering the same part allows to carefully select and identify identical points, avoiding errors due to wrong interpretations. In order to achieve this, all images have to be managed at the same time. This produces a lot more effort from the side of programming, because one has to handle multiple images, with multiple object points, each having multiple image measurements at one time. However, such a conception is very lucrative, because it increases the flexibility of the program and allows to handle any constellation of images and measurements therein.

Fig. 6 shows the environment for the measurement of tie and control points as arising from 4 images. Within the image frame all images available in the project are listed. This allows to control which images are displayed and gives the possibility to interactively change the active images. All images active have an overview frame. By means of these frames the navigation through all images of interest will be done, when looking for an appropriate object point. For the measurement, the image to be treated is loaded into one of two measure frames. This frame has the resolution needed for an accurate measurement and is used for the collection of the coordinates. The results of the data collection are displayed in the control frame. Here, the user finds the point, which has been selected for the measurement , the name of the active image and the coordinate values. Finally, there exists the point frame. Here available object points are listed. They may come from previous definitions of new tie points or may be introduced as pass points.

Conclusion

A modular conception based on graphic standards has been presented allowing to handle digital photogrammetric image blocks, to produce stereoscopic image pairs and to compile them performing 3D-measurements in the virtual 3D space spanned by these models. The use of the OpenGL standard even permits to use hardware acceleration providing high processing speed. Furthermore, standardization allows to incorporate from third party most of the graphic functionality necessary for user dialogs and image viewing processes, considerably reducing the effort of development. An intelligent image handling drops the amount of data transfer to the inevitable minimum and permits to handle very large images as they can only conventionally be handled within specific photogrammetric workstations.
This shows, that the use of standards together with a modular object oriented approach provides a performance up to now being reserved to high end photogrammetric workstations.

References

Boochs F., Gehrhoff A., Neifer M. 2000. *An object-oriented stereo system for 3D-measurements*, SPIE, Vol. 3957.

Davis T., Neider J., Woo M., 1993. *OpenGL Programming Guide*, Addison-Wesley, New York

Imetric SA, 1999. *ICAM6 und ICAM 28 – Die Imetric Messkamera. Product description*, Porrentruy.

Kilgard M. J., 1999. *Avoiding 16 Common OpenGL*. Pitfalls.

Leica Geosystems AG. 1998. *INCA – an intelligent camera system. Product description*, Unterentfelden.

Smart J. 1999. *wxWindows Documentation.*

http://www.i3mainz.fh-mainz.de/projekte/dista.html

http://www.wxwindows.org

Figures

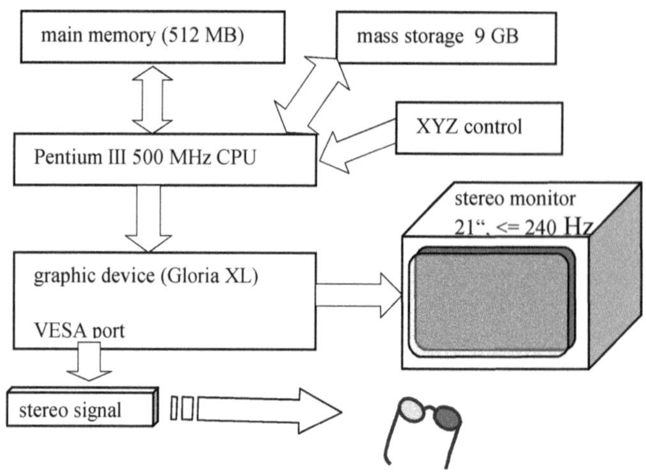

Figure 1: system components

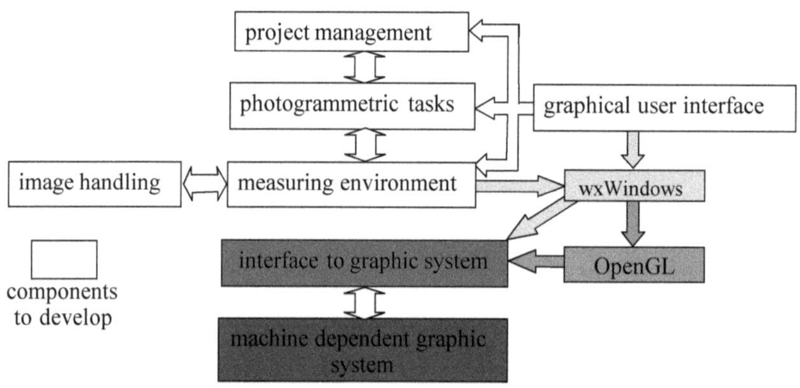

Figure 2: software components

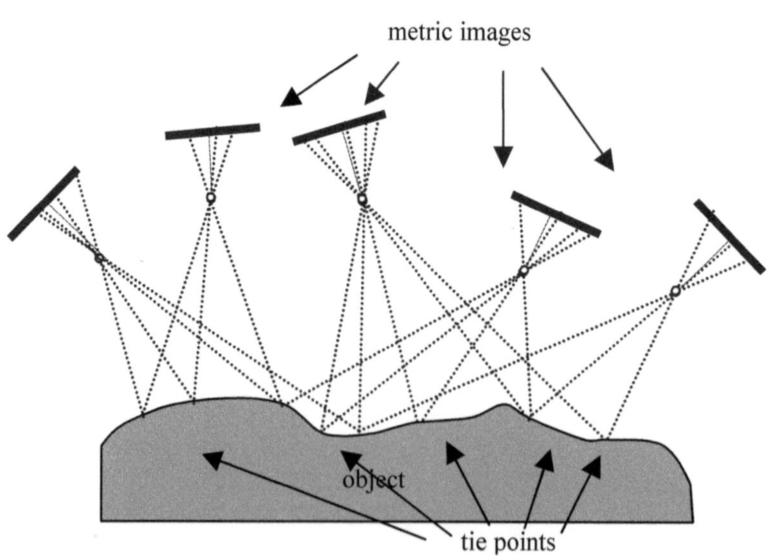

Figure 3: photogrammetric images forming a small block

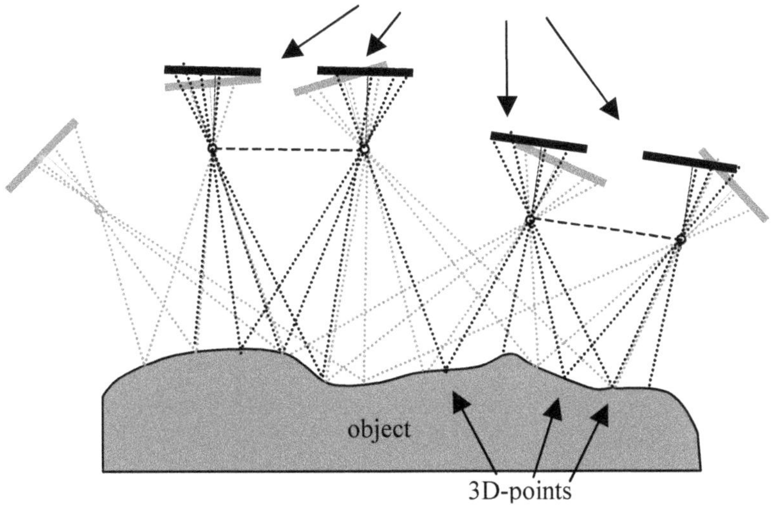

Figure 4: Generated stereoscopic image pairs used for 3D-

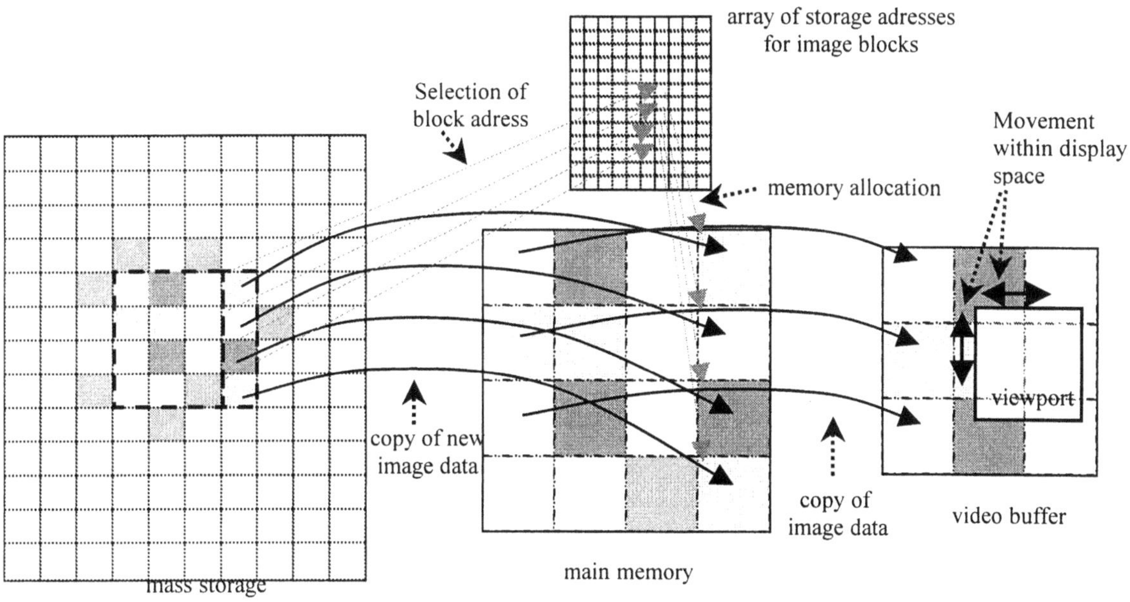

Figure 5. Dynamic handling of image data

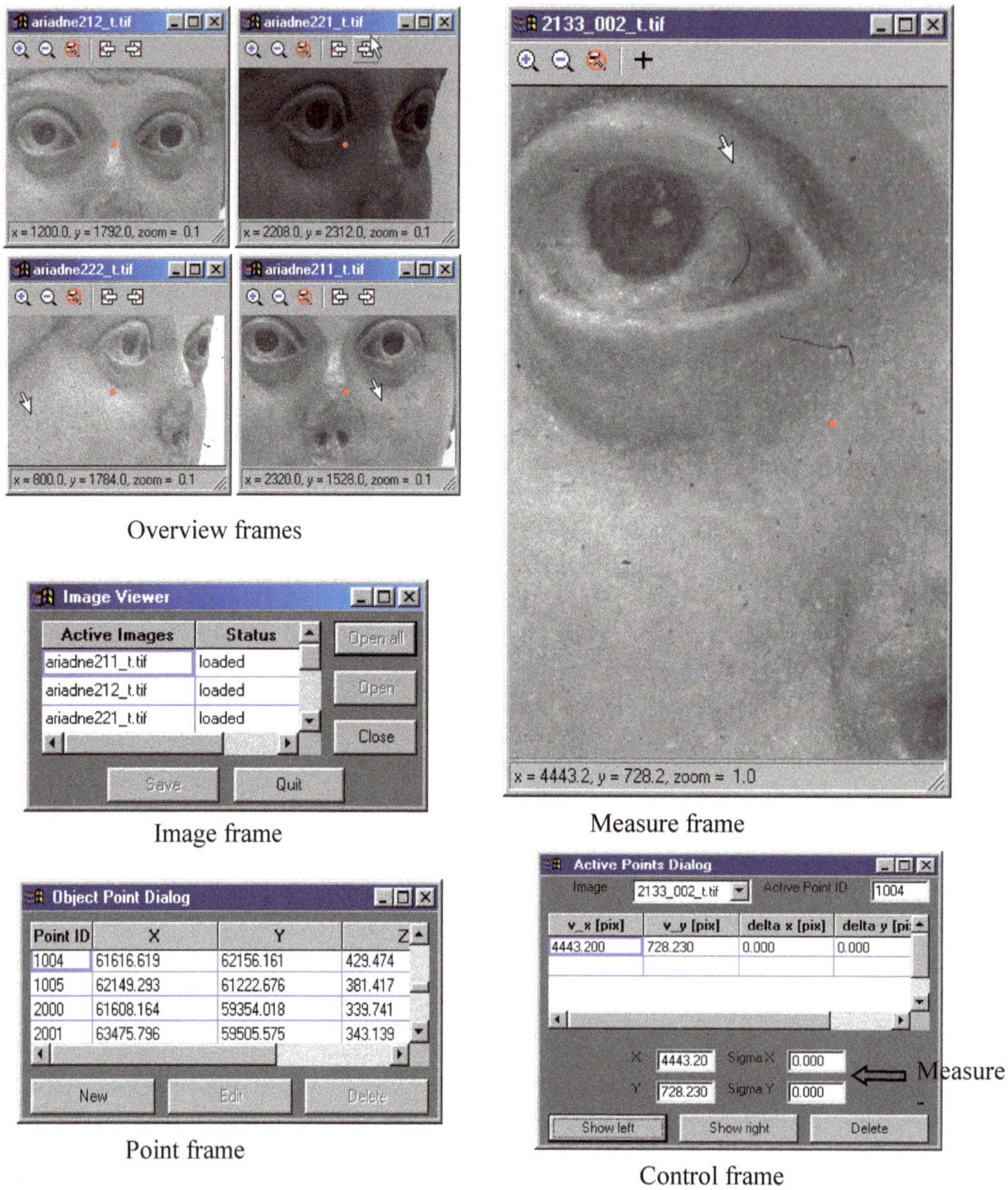

Figure 6: Environment for tie and cointrol point measurement

Infra-red (Reflectance) Spectroscopy of Ceramics from Tell Halawa, Syria

Murray Lee Eiland
Institut für Mineralogie
Postfach 11 19 32
Senckenberganlage 28
D-60054 Frankfurt aM, Germany
E-mail: meyelandes@hotmail.com

Quentin Williams
Department of Earth Sciences
University of California
Santa Cruz, CA, 95064, USA
E-mail: quentw@es.ucsc.edu

Abstract: Many scientific studies characterizing archaeological ceramics rely upon techniques that are becoming obsolete. Infra-red microspectroscopy (IR), long in use by the materials science community, is here applied to archaeological ceramics. This fast, non-destructive technique provides constraints on the mineralogic make-up of samples. Examining a group of ceramics from Tell Halawa, Syria, from the early Bronze Age (before c. 2800 BC) we note variability in firing technology.

Key words: Infra-red spectroscopy, archaeological ceramics, Tell Halawa, Syria, firing technology

Introduction: Determination of Firing Conditions: Previous Studies

While there are a number of methods that can be used to characterize ancient firing environments, few have become standard practice. The technique for examining the colour (using the Munsell colour system) of ancient ceramic materials to propose firing temperature has been used for a number of years (Matson, 1971). Many excavations as a matter of course categorize all diagnostic shards using this technique. Ideally, test-briquettes are made from local clay and fired in controlled conditions. These samples are then graded according to colour, and compared with archaeological samples. This method requires the investigator to analyze separately variations in value, chroma, and hue, as each aspect records a facet of source materials and firing environment. This method has a number of serious drawbacks:

A. Because assessment of colour differs between individuals, one person must be entrusted with recording colour, which can be a time-consuming task, prone to error.

B. In cases where re-firing experiments are performed, there are several important variables. Clay preparation can reduce the larger fraction of particles, remove organic matter, altering the colour of the clay.

C. Without petrographic or chemical tests one cannot be certain that particular wares were not imported. There is the possibility that a fraction of the sample group will be from a distinct clay source that will not match the parameters of the local clay.

D. This complicated visual method of investigation has never gained universal acceptance, and the reproducibility of the results is doubtful. Methods of recording colour not using the Munsell system are also used, particularly in many European countries (Orton et al., 1993).

In contrast, IR offers several advantages over this technique:

A. The majority of samples require essentially no preparation.

B. Analysis time for the majority of samples is several minutes.

C. Information about the firing environment, as well as the raw materials, is presented. Spectra may be quickly matched with similar spectra, making a rapid assessment of many samples possible.

Calcite is particularly reflective, and can serve as a useful diagnostic mineral to estimate maximum firing temperature. There is also an increasing demand for non-destructive methods of analysis. Museum curators are becoming less enthusiastic about releasing shards for destructive tests, while complete vessels have almost never been sampled. Infra-red spectroscopy when used in reflectance mode is totally non-destructive, and if a sample of powder is required, it can often be obtained from a broken surface of a sample. With further work to define the parameters of the method, this technique can be applied to a wide range of materials from a number of sources.

Previous Archaeological Studies

IR has been used to determine the nature of resinous deposits adhering to pottery (Sandermann, 1965; Sauter, 1967). The technique was also used to characterize deposits on the inside of a jar from Godin Tepe, Iran, dating to the mid-fourth millennium B.C. The presence of tartaric acid in the samples demonstrates that this vessel was used to store wine (Badler, 1996). Amber is the archaeological material that is most closely associated with IR, and it is here that an archaeological methodology is effectively used.

Early chemical investigations into the amber trade confirmed assumptions - based largely upon textual sources - that amber was traded from the Baltic region. Due to the time required to prepare amber for wet chemical analysis, this method was used to analyze a small number of samples. Using new information from IR spectroscopy (Beck, 1986; Beck and Shennan, 1991; Beck et. al., 1964), it was found that the amber trade was much more complicated. It also demonstrated that IR spectroscopy could be successfully applied to archaeological materials. There are also some isolated applications of the technique to firing history. Maniatis et al. (1982) characterized the hydration state of low fired terra-cotta statuettes from Greece, and Blasius et al. (1983) examined Roman tile fragments using IR. Duraiswamy et al. (1992) used the technique to examine Indian clays from an archaeological site.

The Technique

Infrared spectroscopy has been extensively utilized to characterize modern industrial ceramic materials, as well as a broad range of inorganic, organic, and mineralogic specimens. It is also gaining ground among conservation scientists (Derrick et al. 1999). A large number of reviews of the applications and theoretical aspects of infrared spectroscopy exist (eg McMillan and Hofmeister, 1988 Mitra, 1989; Williams, 1995; McMillan et al., 1996), and we provide only an abbreviated description of the technique here. The usefulness of infrared spectroscopy in characterizing archaeological ceramics lies primarily in its ability to yield a 'fingerprint' spectral pattern that can be readily associated with the mineralogic constituents of samples. Essentially every mineral of importance to ceramics has had its infrared spectra characterized (eg Farmer, 1974, Salisbury et al., 1991), and these spectra fall into different groupings depending on the type of bonding within the different minerals of the sample. For example, carbonates are readily distinguishable from framework silicates (such as quartz and feldspars) which in turn are easily distinguished from chain silicates (amphiboles and pyroxenes). The overall mineralogic make-up of samples can be used to provide constraints on the firing temperature of the ceramics, an effect noted by Maggetti (1982).

The simplest underlying principle of infrared spectroscopy has long been used in the archeological sciences, that is, characterizing the colour of a sample as a classification tool. Infrared spectroscopy involves characterizing the colour of samples in the wavelength range of 1-200 microns, or equivalently 10000-50 cm^{-1} in units of frequency. For ceramic characterization, we focus primarily on the frequency range between 500-2000 cm^{-1} (20 to 5 microns), as this spectral region proves to be the most diagnostic of different ceramic types. This range of infrared light is at comparable energies to those of interatomic vibrations of solids, liquids, and gasses. The energies of these vibrations are determined by the strengths of bonds within a sample, and the masses of the atoms. Stronger bonds produce higher frequency vibrational bands, as do less massive atoms. The local symmetry surrounding the atoms (particularly in crystal structures) also plays a major role in determining which types of vibrations are infrared-active within a crystal. For example, calcite ($CaCO_3$) has eight symmetry-allowed infrared-active vibrations (Williams, 1995), with three of these lying above 600 cm^{-1} (typically, spectra of mixed phase materials such as ceramics are not notably diagnostic below 600 cm^{-1}). Of these three, two (near 1400 cm^{-1} and 880 cm^{-1}) are strong and completely diagnostic of the presence of calcite within the sample.

We found that reasonably high quality and completely reproducible data can be produced from flat regions of shards (or shards oriented so that their curvature is small in the region being sampled) in areas as small as 50 microns in diameter. It is of considerable importance that the characteristic grain sizes in the bulk (inclusion-free portions) of archeological ceramics are generally smaller than the wavelength of the incident infrared light. Large grains of material can bias the observed reflectance spectra, and we occasionally observe such effects when, for example, reflectance spectra are taken near macroscopic inclusions. In many instances, we found that light scraping of the surface were necessary to remove dirt or surface alteration products from the sample. In the majority of cases the uppermost 100 - 500 microns were removed. Because of the spatial selectivity of the infrared reflectance technique, such scraped areas could be confined to spots less than 1 mm^2 in area.

The increased utility of IR can largely be traced to technologic advancements in over the last two decades. Archaeological ceramic samples are not particularly reflective in the infrared (only about 1% that of a metallic mirror). Accordingly, to conduct spatially selective measurements, in which reflected light is collected from areas with lateral dimensions as small as 50 microns, requires that infrared light be extremely efficiently collected and detected. We utilize a Fourier Transform Infrared (FTIR) Spectrometer coupled with an infrared microscope to conduct all of our measurements. The infrared microscope is equipped with Cassegrain all-reflecting optics (reflecting optics are required for IR microscopes, as conventional lenses absorb infrared light). Specifically, we utilize a Bruker Industries IFS-66v evacuated FTIR spectrometer, equipped with a KBr beamsplitter and glowbar source to conduct all of our measurements. The reflectance measurements are conducted on an infrared microscope attachment (also Bruker Industries) equipped with a liquid-nitrogen cooled intermediate range mercury cadmium telluride (MCT) detector. All spectra were run with 4 cm^{-1} resolution, with typical collection times of approximately 1-3 minutes for each reflectance spectrum. The spectra were referenced against a front-silvered aluminum mirror (which has a flat spectral response across the frequency range of interest) to remove instrument response from the spectra. In most cases, multiple spectra were collected from each sherd to both ensure reproducibility and to verify that the spectra were not aliased by a shallow sub-surface inclusion.

IR Applied to Archaeological Ceramics

The detailed mineralogy of ceramic samples is intimately connected to both their firing temperatures and the composition of the original source materials of the clay. As a simple example, pure kaolinite ($Al_2Si_2O_5(OH)_4$), a major component of many ceramics, undergoes a sequence of mineralogic transformations as it is treated at progressively higher temperatures. Specifically, kaolinite dehydrates near 550EC to an anhydrous phase with a chemical composition of $Al_2Si_2O_7$ (metakaolin). This phase alters to a mixture of SiO_2 (initially amorphous, but which can ultimately crystallize to quartz or cristobalite) and a structurally complex and defect-rich spinel phase with a stoichiometry such as $Al_{21.33}O_{32}$ at 950-1000EC (Carty and Senapti, 1998). This spinel phase reacts in turn to form mullite ($Al_6Si_2O_{13}$) near 1075EC. The critical point raised by the transformation sequence undergone by this simple clay is that even though a sample containing a kaolinite-rich starting material undergoes no changes in its chemistry (other than the loss of water near 550EC) during firing, the resultant mineralogy of the sample shifts markedly as a function of the temperature to which the sample has been exposed. Thus, kaolinite starting materials converge at high temperatures to SiO_2-bearing assemblages with aluminum-bearing accessory phases, and this progression can be monitored (and firing temperature bounded) through mineralogic characterization. Clearly, when more chemically complex clays such as smectites and illites are present, the range of minerals that can result during firing is markedly expanded (Kingery, 1976). Mineralogic characterization can not only yield bounds on the firing temperature of a ceramic sample, but also the starting materials mixed into clays (as well as constraining the identities of the clays).

Determination of the presence of calcite within ceramics is not straightforward using a number of analytic techniques. For example, bulk chemical analyses of ceramic samples do not typically record the contents of elements such as carbon and hydrogen, so the presence or absence of calcite cannot be readily derived from such measurements. Furthermore, as with the mineralogic changes undergone by kaolinite as a function of temperature, the presence of calcite also provides a first-order bound on the temperature of firing: calcite breaks down to lime (CaO) and carbon dioxide at a temperature of 894EC, which is thus a firm upper bound on the temperature of firing of calcite-bearing ceramics (eg. Southard and Royster 1936). The firmness of this upper bound is accentuated by the effect that free SiO_2 can have on lowering of the decarbonation temperature of calcite: the reaction of $CaCO_3$ + SiO_2 -> $CaSiO_3$-wollastonite + CO_2 is well known to proceed at lower temperatures than the simple decarbonation of calcite alone (Wyllie and Huang, 1976).

Samples

Tell Halawa , in northern Syria about 100 km east of Aleppo, is a multi period site. There are many graves and in general the ceramic sequence is well stratified, yielding remains that are characteristic of the region (Orthmann, 1981;1989). While details of the site will not be covered here, the ceramics fit into an ongoing research programme into Mesopotamian ceramic technology (Eiland in press a and b). Archaeological ceramics from Tell Brak has received detailed consideration using IR as well (Eiland and Williams 2000). Here we present several samples - housed at the Institut für Mineralogie, Frankfurt - all taken from Layer 3, early Bronze IV period (before c. 2800 BC) from Tell Halawa. Colour is noted but as this is such a small sample no Munsell numbers will be used.

HL 13 Thin dark grey 'Euphrates ware' shard. Very hard texture. This would have been a 'luxury' vessel.
HL 8 Thick grey (typically Mesopotamian) fabric. Coarse and friable ceramic body.
HL 7 Red fabric similar texture to HLW 8 barring colour.
HL 3 Buff fabric with large number of calcite inclusions visible by eye. Thick fabric is visually much like HLW 8 and HL 7, barring colour.

Interestingly, chemical analysis (X-ray fluorescence) showed little difference between the Euphrates ware and the common ware (HL13 and 8 respectively), though there is a clear difference by eye. This is particularly the case for bulk elements, here rounded to the nearest percent, and is also the case for trace elements (not considered here). Silica, roughly representing sand, and aluminium, roughly the feldspar-clay fraction, are essentially identical in the two samples, as is iron. Calcite would be expected to vary both according to initial mixture as well as firing temperature.

	SiO_2	Al_2O_3	Fe_2O_3	CaO
HLW 13	49	12	8	14
HLW 8	49	12	8	15

While this may suggest that these two samples were made with similar raw materials (at the same locality) using similar techniques, the IR spectra present a different picture. (Fig. 1) The roughness (or lack thereof) in the region of the spectra from 1300-1400 cm^{-1} shows variations in the particle sizes of the various samples. HL 13 has little carbonate, and the major peaks at 1250 cm^{-1} and 1080 cm^{-1} are quartz, which is about the only phase clearly shown in the spectrum, implying that the sample is crystallized.

HL 3 and HL 8 show clear evidence of carbonate, with the amount in HL3 being much greater (the peak at 860 cm^{-1} is diagnostic, and a portion of the 1200 1400 cm^{-1} slope is also due to carbonate). HL 8 may have a small amount as well, with a weak 860 cm^{-1} feature and a smaller feature (than HL 3 and HL 8) near 1380 cm^{-1}. Otherwise, the silicate fraction of the bottom three spectra are similar, with a major peak at 1050 cm^{-1} for all three. This feature is probably feldspar related. The most parsimonious interpretation is that firing temperature increases from top to bottom. Alternatively, it could be that HL 3 had a large amount of carbonate, and was fired to about the same temperature as HL 7. We suggest that HL 8 has been fired to a slightly higher temperature. From the differences in the spectra, it appears much of the initial carbonate has been broken down. The latter observation has been supported by thin section studies that show the carbonate as voids of a characteristic shape with a reaction rim with the ceramic matrix. All these observations are in keeping with what may be deduced from the ceramic assemblage logically. While a 'fineware' vessel may have been fired to a temperature where it was vitrified and could hold water, more common vessels were fired in a range of environments. Most likely, as long as the vessel was 'baked' into a ceramic, it was deemed acceptable, as it is doubtful that thousands of everyday wares would have warranted careful attention during firing.

While the latter observation may cast doubt upon the necessity of recording the colour of potsherds, it cannot be doubted that given a wide temporal range, changes in firing technology - from oxidising to reducing atmosphere - can be detected. At the same time it should be realized that there can be considerable variation in ceramics from a single segment of archaeological time.

Further Research

Perhaps the most limiting aspect of the technique is that methods of presentation have largely been limited to 'squiggly lines', which do not lend themselves to numerical presentation. The interpretation of the spectra depend upon the relationships with other mineral phases, and frequently the presence of a mineral is indicated by more than one peak in different spectral regions. The result is that few spectra can be effectively presented. The majority - particularly at archaeological conferences - will not be able to easily assimilate the differences (though gross morphology of the spectra will be obvious). Several suggestions for presentation were made during the conference, and the authors would welcome others.

Particularly regarding the well stratified sequence of ceramics from Tell Halawa, there exists the opportunity to contrast the Munsell method with an independent test. While much more work needs to be done, it is clear that while major differences in firing temperatures are reflected in colour, there are a number of cases where this criteria alone can be misleading. While a difference in colour can indicate a change from one sample to another, different IR spectra can yield a temperature range. With the gradual reduction in cost, and weight, of IR equipment, it should be possible to take a small apparatus into the field. After a range of thousands of shards had been examined there would both be an extensive database, and the opportunity to retain material of interest for further research after a large scale programme of analysis.

Perhaps the applications that lie further in the future are of even greater interest. It is clear that infrared equipped satellites are increasingly being used to characterize earth materials. For many areas of the Near East, with a minimum of plant cover to obscure signal, it should be possible to correlate the IR signal obtained from a satellite with spectra obtained from archaeological ceramics. The major question is if it is possible, even if the region offers mineralogic diversity, to spatially discern what may be slight differences in outcrops. Bearing in mind that raw clays can be heterogeneous and that one can expect changes in mineralogy after firing, any correlation would assume several factors. At the same time, given a wide enough range, it may be possible to propose, at least in a tentative way, if a chemically anomalous sample is from a nearby region, or from a distant source.

Acknowledgments

The Mineral Physics Laboratory at UCSC is supported by the NSF and the W.M. Keck Foundation. One of the authors (Eiland) would also like to thank the organizing committee for providing funds to travel to the conference. The wide range of presentations was much appreciated. Thanks are due to Professor Meyer (Near Eastern Archaeology) for providing the samples and offering comment, as well as the Institut für Mineralogie in Frankfurt for support.

References

Badler, V.R.,1996. Drink and be merry! Infrared spectroscopy and ancient Near Eastern wine. *MASCA Research Papers in Science and Archaeology,7*: 25-36.

Beck, C.W.,1986. Spectroscopic studies of amber. *Applied Spectroscopy Reviews, 22*: 57-110.

Beck, C.W., and Shennan, S.,1991. *Amber in Prehistoric Britain*. Oxbow, Oxford.

Beck C.W., Wilbur, E., and Meret, S.,1964. Infrared spectra and the origins of amber. *Nature, 201*: 256-257.

Blasius, E., Wagner, H., Braun, H., Krumbholz, R., and Schwartz, B. 1983. Tile fragments as characteristic evidence for ancient Roman settlements: Scientific investigations. *Archaeometry 25(2)*: 165-178.

Carty, W.M., and Senapti, U. 1998. Porcelain - raw materials, processing, phase evolution, and mechanical behavior. *Journal of the American Ceramic Society, 81*: 3-20.

Derrick, M.R., Stulik, D., and Landry, J.M., 1999. *Infrared Spectroscopy in Conservation Science*. The Getty Conservation Institute, Los Angeles.

Duraiswamy, D., Ramaswamy, K., and Venkatachalapathy, R., 1992. Infrared spectroscopic studies of archaeological materials: Determination of firing temperature. In Ganorkar, M.C. and Rama Rao, N. (eds.), *Role of Chemistry in Archaeology: 1st International Colloquium Held in Hyderabad (India), 15-18 Nov. 1991*: 47-54. Birla Institute of Scientific Research, Hyderabad, India.

Eiland, M.L. in press a. Pottery technology. In Matthews, R.J. (ed), *Sampling an Urban Centre: Excavations at TellBrak 1994-1996*. British School of Archaeology in Iraq/McDonald Institute for Archaeological Research, Cambridge.

Eiland, M.L. in press b. Ceramic replication at Tell Brak, Syria. *Newsletter: Department of Pottery Technology* (Leiden University).

Eiland, M.L. and Williams, Q., 2000. Infra-red spectroscopy of Ceramics from Tell Brak, Syria. *Journal of Archaeological Science, 27*: 993-1006.

Farmer, V.C. (Ed.),1974. *The Infrared Spectra of Minerals.* Mineralogical Society,London.

Kingery, W.D.,1976. *Introduction to Ceramics*. Wiley, New York.

Maggetti, M., 1982. Phase analysis and its significance for technology and origin. In Olin, J.S., and Franklin, A.D. (eds.), *Archaeological Ceramics:121-133*. Smithsonian Institution Press, Washington.

Maniatis, Y., Katsanos, A., and Caskey, M.E., 1982. Technological examination of low-fired terracotta statues from Ayia Irini, Kea. *Archaeometry 24(2)*: 191-198.

Matson, F.R., 1971. A study of temperatures used in firing ancient Mesopotamian pottery. In Brill, R. (ed.), *Science and Archaeology: 65-80*. MIT Press, Cambridge.

McMillan, P.F. and Hoffmeister, A.M.,1988. Infrared and Raman spectroscopy. *Review of Mineralogy, 18*: 99-159.

McMillan, P., Hemley, R.J., and Gillet, P.,1996. Vibrational spectroscopy of mantle minerals. *Geochemical Society Special Publications, 5*: 175-213.

Mitra, S., 1989. *Fundamentals of Optical, Spectroscopic and X-ray Mineralogy*. John Wiley & Sons, New Delhi.

Orthmann, W., 1981. *Halawa 1977 bis 1979*. Rudolf Habelt Verlag, Bonn.

Orthmann, W., 1989. *Halawa 1980 bis 1986*. Rudolf Habelt Verlag, Bonn.

Orton, C., Tyres, P., and Vince, A., 1993. *Pottery in archaeology*, Cambridge University Press, Cambridge.

Salisbury, J.W., Walter, L.S., Vergo, N. and D'Aria, D.M., 1991. *Infrared (2.1-25 microns) spectra of Minerals*. Johns Hopkins University Press, Baltimore, MO, USA.

Sandermann, W., 1965. Untersuchung vorgeschichichtlicher 'Gräber-harze' und Kitte. *Technische Beiträge zur Archäeologie, 2*: 58-73.

Sauter, F.,1967. Chemische untersuchung von 'Harzüberzügen' auf hallstattzeitlicher keramik. *Archaeologia Austriaca, 41*: 25-36.

Williams, Q. 1995. Infrared, Raman and optical spectroscopy of earth materials. In Ahrens, T.J. (ed.), *Mineral physics and crystallography: A Handbook of Physical Constants*: 291-302. American Geophysical Union, Washington, D.C.

Wyllie, P.J. and Huang, W.L.,1976. Carbonation and melting reactions in the system $CaO-MgO-SiO_2-CO_2$ at mantle pressures with geophysical and petrological applications. *Contributions to Mineralogy and Petrology, 54*: 79-107.

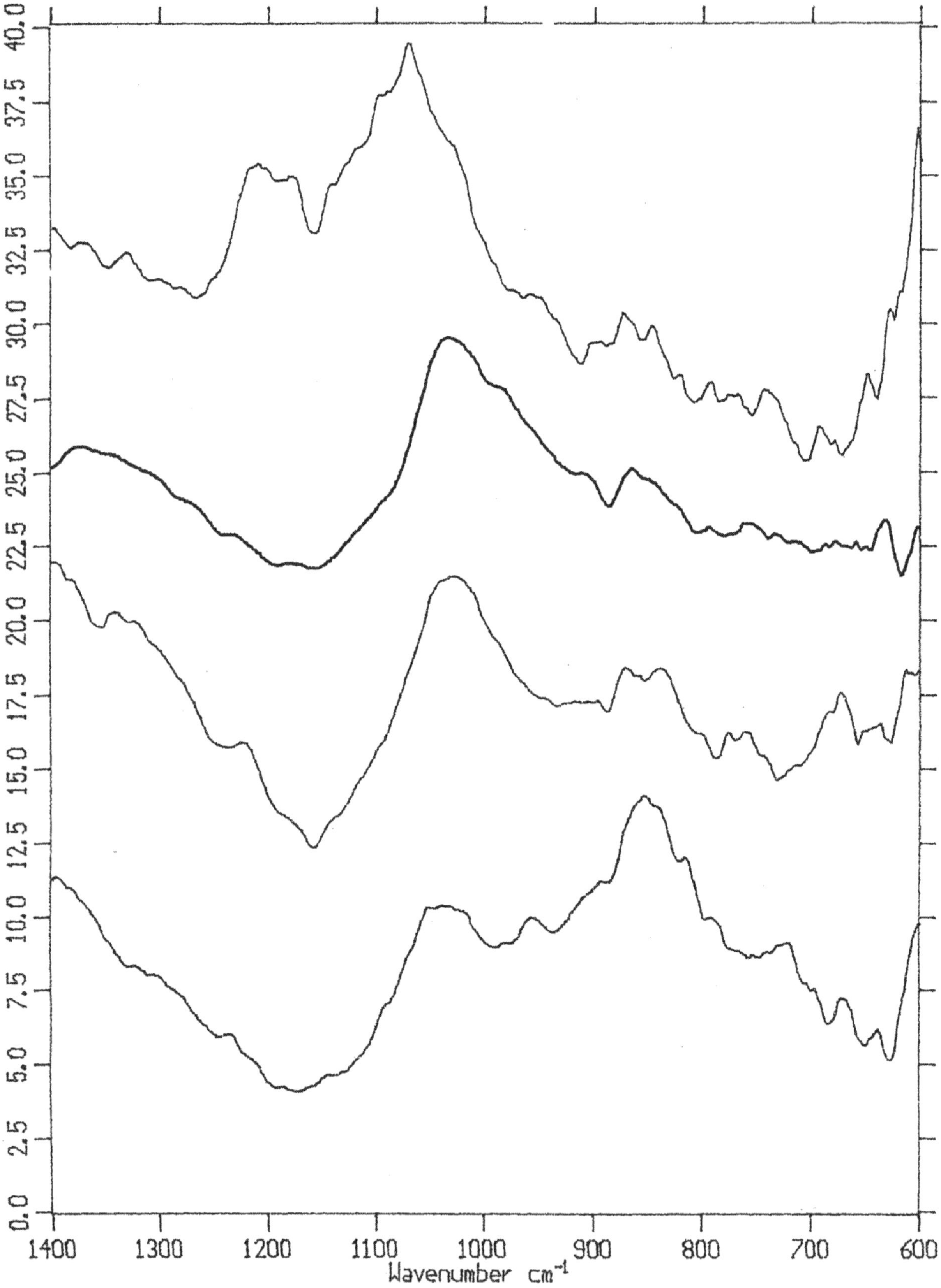
40.0
37.5
35.0
32.5
30.0
27.5
25.0
22.5
20.0
17.5
15.0
12.5
10.0
7.5
5.0
2.5
0.0
1400
1300
1200
1100
1000
900
800
700
600
Wavenumber cm⁻¹

Figure 1.

Detailed Topography and Surface Survey. What is the point? Tanagra City Survey 2000

Emeri Farinetti
Faculteit der Archeologie
Reuvensplaats 4,
Postbus 9515,
2300 RA Leiden, The Netherlands
Phone: +31 71 5272931 - Fax: +31 71 5272429 - E-mail. emerif@tin.it

Lefteris Sigalos
Faculteit der Archeologie
Reuvensplaats 4,
Postbus 9515,
2300 RA Leiden, The Netherlands
Phone: +31 71 5272425 - Fax: +31 71 5272429 - E-mail. e.sigalos@arch.leidenuniv.nl

Abstract: Topographical surveys and the creation of digital elevation models have become quite common in the archaeological projects. In the excavations they are used as a frame to locate the excavated area in a wider context. On the other hand, in artifact surface surveys, detailed topographical surveys produce a base layer over which the results of the collection can be lain.

The paper aims to illustrate the topographical survey and the mapping of the architectural features visible on the surface carried out at the ancient city of Tanagra in Boeotia, Central Greece, as a part of a wider city survey project. The methodology used will be illustrated and examined, and the results will be analyzed to evaluate their effectiveness for the final goals of the artifact surface survey project. Technical aspects will be illustrated and discussed together with the validity of the methods employed.

Furthermore, from this analysis crucial methodological questions can be raised: is a topographical detailed survey useful only to map data resulting from the artifact surface collection, and to give us a general picture of the area we are working on? Or rather, by inserting the results as a layer into a Geographical Information System and linking it to a diversified database, could we gain new information and squeeze fresh juice out of something that would not be a mere Digital Elevation Model?

Key words: survey, topography, digital elevation models, geographical information systems, cityscape

Archaeologists of the Classical world, traditionally, regard computer applications and especially GIS as a good means of visualization of the results they have strenuously studied and processed by hand. The immense number of dots covering maps of publications is the greatest proof. Presentation of data is of course a basic function, but computer applications can be used for the analysis of data, the combination of information and the creation of new data that have not been collected from the field.

The study of the city of Tanagra, under the direction of Professor John Bintliff from Leiden University, having as a primary aim to extract as much information about the city as possible with non-destructive means, such as excavation, has been utilizing technological means so as to manage and analyze the vast in number and variety data collected on a day-to-day basis. Our target is to reconstruct the cityscape and activity zones of Tanagra at its peaks, that is the late Classical and late Roman periods, and, additionally, investigate the trends of occupation and reoccupation of the site from prehistoric times to nowadays. For this purpose three teams have been created for the collection and study of surface remains, mainly ceramic, the architectural and topographical survey, and the geophysical prospection.

The ancient site of Tanagra, locally known as Grimada, lies in southeast Boiotia in Central Greece, close to the borders with Attica (fig.1). The settlement is situated on the precipitous northern slope of Mt Soros, identified as the ancient Mt Kerykion. The ridge of Mt Soros separates the two Boiotian plains of Thebes and the coastal areas to the east, respectively. At a distance of less than one kilometre to the south of the city runs the river Asopos that used to be the ancient border between Boiotia and Attica. To the west of Tanagra is a little stream, nowadays called Lari that joins the river Asopos SW of the city. The city itself is marked by its late Classical fortification walls covering a total area of 30 hectares. Being situated on sloping ground, the city has been built on different levels and terraces. The theatre, dominating the site, is built against a steep slope at the SW corner of the city and the agora with the public buildings on a flat high terrace, or acropolis, raised above the rest of the city (fig.2).

The city had been identified already in the 19th century by travellers in the process of the rediscovery of Greece. The first attempt, though, to study it systematically was made by Professor Roller with a team from the university of Wilfrid Laurier in Canada. The aim of the team was the survey and mapping of the architectural features of the site and locating the quarries of the city. The interpretation of the plotted architectural remains allowed the team to locate various aspects of the plan of the city. Thus, they identified a series of roads running parallel to each other with a SW-NE direction at approximately 52m intervals, and avenues crossing them at right angles every 102m. The residential blocks would, therefore, be rectangular with dimensions of 50x100, if we allow 4m for the width of the roads (Roller 1987).

In this paper we will present some preliminary results of the 2000 season, focusing our attention on the micro-topographical survey and its relevance to the general surface survey project. Our aim is to stress the importance of such a method for the reconstruction of the topography of the city. Not only are we interested in locating architectural features, ceramic assemblages and scatters, geophysical data on a detailed Digital Elevation Model (DEM), but also our main intention is to attempt to combine the various levels of information to produce data that is not readily available in the field. The paper is divided into two sections. Firstly, we will discuss the methods applied by the groups for the retrieval of the data as well as their nature. Consequently, we will not only assess the value of the topographical data as a mere displaying background, but also as an active layer in a complex Geographical Information System (GIS) environment, examining processes, such as erosion patterns and their affect on the data retrieved.

Thus, having Roller's interpretations at hand, we decided to divide the walled area of the city in units of 50 by 50 m following the directions of the proposed *insulae*. These units were approximately half the size of the occupational blocks (*see* Roller 1987). Once a number of units were laid out both survey and topographical teams could get into action. After evaluating the visibility of the unit, the teams would count the ceramic material from the surface. Due to the large number of specimens within each unit, the most diagnostic and peculiar sherds would be collected ranging from 150 to 250 pieces per unit. Simultaneously, two people of each team would collect all the ceramics from a lane of 10 by 30 m at the south eastern corner of the unit. Finally, a third group, the so-called specialist group, would cover the unit once again conducting a targeted search for particular ceramics and other artifacts, not easily detected otherwise. All samples would be placed in different bags with different recording sheets related to them and sent for study and processing at the lab. There, every find would be given a unique number and its characteristics, function and date would be inserted in a database for further analysis. During the first season, the area covered by the ceramic survey teams was approximately 1/3 of the total city, or a little less than 12 hectares.

The units used for the city survey were also the physical framework of the topography team. Each unit was surveyed in detail taking measurements at 4m intervals. Ideally, approximately 145 measurements would be taken per unit capturing as many details as possible. Special attention was paid to topographical features such as terraces and depressions that could easily be missed by the grid of points. Furthermore, all visible architectural features whether *in situ* (walls, terrace walls, cornerstones and large masoned fragments) or scattered in the cityscape (column drums, capitals, friezes, triglyphs and rubble heaps) were located and recorded both manually and electronically. For this purpose we used a SOKKISHA Total Station provided by the University of Leiden. This method was a development of the one first introduced to the preceding Boiotia project by Mark Gillings, who very successfully managed to capture invaluable information about the residential area of the Frankish and Early Ottoman site VM4 or Palaio-Panagia, in the vicinity of ancient Askra, in the Valley of the Muses (Gillings 2000). Due to the number of measurements and the variety of architectural data collected, the team only managed to cover a considerably smaller region than that of the ceramic collection (approximately 6 hectares).

The question that may be raised is: Why should one get in so much trouble? Beyond the self-evident reason for capturing the topography and testing Roller's hypotheses, there is an additional one and it can be summarized in the following question: To what extent is it possible to reconstruct a cityscape with non-destructive means? Or, otherwise, can we study a city without excavation? And that was exactly the aim of the Tanagra city project as a whole.

Before looking at some preliminary interpretations two aspects of the city should be mentioned. In the first place, it has been known that Tanagra during the 4th century BC got a new fortification wall and possibly a regular street plan. A characteristic example is Olynthos, in Northern Greece. Secondly, Tanagra being situated at an irregular slope would have required extensive terracing so as to allow a regular street plan to be introduced. Based on these two assumptions the detailed topographical survey would permit the plotting of such terracing to be identified.

Since the area covered topographically did not have slope differences exceeding 2-3m within each unit it was regarded acceptable to produce contour lines at 0.20 metres intervals (fig. 3). This was also permitted by the frequency of our measurements resulting a pattern of approximately 2 contour lines between two measurements. From the site plan produced it seemed that the direction of the walls were coinciding with the majority of the scattered architectural features, represented by points, the latter following a directional pattern too. In addition, the plan that our team produced seems to have many similarities with that proposed by Roller. Concerning the roads, that is the streets running from SW-NE direction, both Roller's and our plans seem to coincide. On the contrary, the avenues, or the streets parallel to the roads, seem to be differently located (see below).

As for the architectural features themselves, marked in bold (fig. 4), a series of walls were discovered and recorded that in most cases, especially in the SW corner of the city, seem to have belonged to road terracing or continuous house walls bounding the streets. Some more elaborate structures are evident in the NW side of the area covered during the first season. These particular structures are at most 20 cm above the ground surface, but their location on the top of a ridge and the nature of the masonry, imply that either a public building or a temple was located on the site. Furthermore, a capital located towards the south of the city may indicate the presence of a temple or a civic building in the vicinity (Roller 1987). At the S and SW side of the surveyed area a section of the fortification wall was measured which in large was concealed by soil and rubble.

Moving from the methodologies used in the field and the preliminary interpretations of the architectural features, we will examine the DEM interpolation methods applied and some GIS applications based on preliminary results. The area covered by the topographical survey consists of a small part of the most rugged part of the city, to the South, the eastern part of the ridge prominent over the lower city to the North, a small part of the slope gently descending from the ridge towards the North, and a depression to the South-East, terminating against the eastern side of the city wall. The elevation difference is 29 m and topographical features are characterized by different directions (fig.5).

The point elevation data (a quasi-regular grid of three-dimensional point coordinates) collected during the survey have been interpolated in order to build a DEM, a virtual continuous representation of the surface topography. The points, as we said, have been taken on a 4 x 4 m grid and the accuracy of z values is 1 cm. The raw data matrix has then been enhanced by different kind of interpolation methods so as to produce a simulated surface with an interval of 0.2 m between contour lines. We tested various interpolation methods, changing anisotropic rates and search radius, in the attempt of detecting significant anomalies and structural evidence, taking into account topographical discontinuities, which characterize the surveyed area.

The application of the *Shepard's* method, useful to maintain inhalterated line distances, created the so-called Bull's Eye effects and the resulting picture was very different from the actual topography. The *Radial Basis Functions* method, which takes into account the exact location of control points and does not smooth the data, was not of any help, as it created strange effects (trying to maintain the exact location of the points without calculating an average). The *TIN* method though, despite not giving average results either, creates triangles having as vertices the measured points (fig.3). It is usually suitable for perfect grid (in our case it was regular but not perfect) and very steep areas (which we do have in some locations). We tried also to apply an anisotropy ratio (1.5 and 2) along the direction of the slope towards the theatre, but no new information seemed to appear. The results obtained applying the TIN method, however, seemed to be the best for visualization purposes. Although this procedure does not create any smoothing effect itself, the resulting image shows a smoothed surface, without sharp interruptions but being very steep in certain areas. The main macro-features characterizing the topography are the ridge that crosses the city along the central axis, from the West to the East, and the hollow close to the city wall to the East.

The micro-features clearly discernible are quite wide terraces in the area to the South of the ridge, going up towards the West, and narrower and steeper terraces going up towards the South, to reach the theatre. These terraces were meant to sustain the roads and the buildings along them. The correlation between terrace features and roads can be clearly seen in fig.6, displaying the road system of the ancient city as suggested by Roller (1987). As for the roads running SW-NE we have plentiful evidence. On the other hand, as for the avenues, or the streets running perpendicularly to the previous, we have topographical proof in only one case. Here terraces have been built perpendicularly to support a crossroad. Another feature clearly visible is the steep slope leading to the easternmost edge of the ridge, starting few meters from the city wall.

We tried then to apply *Kriging's* interpolation method. It smooths data too much and in our case it did not give the possibility to detect anomalies, even when considering the direction of the terraces. The general resulting image is nice to see though and, if we produce an interpolated surface (fig.7), consisting of continuous data instead of contours, it shows better the detectable features - terraces in our case. Using the *Inverse Distance to a Power* method, we could notice some discontinuities on the simulated surface, but nothing new. The same happened with the *Minimum Curvature* method (fig.8). In both resulting DEMs some small terraces in the slope gently descending from the ridge towards the North seem to appear. Applying the Minimum Curvature method though, which smoothens the data but in the meanwhile tends to be as exact as possible, the results can be distorted sometimes and they need to be checked.

Generally speaking, topographical surveys and the creation of DEMs have become quite common in the archaeological projects. In the excavations they are used as a frame to locate the excavated area in a wider context. On the other hand, in artefact surface surveys, detailed topographical surveys produce a base layer over which the results of the collection can be laid.
In the framework of the Tanagra Survey Project, the purpose of a detailed DEM was, as mentioned above, twofold: surveying topographically the area covered by the artifact collection in order to produce a detailed base map over which to overlay and 'over-display' collected data, and processing and analysing topographical data as an autonomous data set. According to this approach, present topography is interpreted and studied in a research framework, in order to capture any surviving trace concerning past use and spatial organisation of the city.

Examining the mere DEM, only few micro-features and the topographic anomalies noticed earlier can be clearly detected. However, new information can be added to the 1:5000 map of the area (the only one available before our survey), where only the general sloping nature of the hillside and macro-features such as the central ridge are represented. For instance, the exact location of the terraces was crucial for the interpretation of the plan of the city.

Furthermore, when the architectural features recorded during our survey are overlain, adding a new information layer, the picture becomes clearer. Some terrace walls that follow topography can be detected. The remains of other terrace walls are probably under the surface soil, but the terraces are clearly visible through topography. Looking closely at the architectural features recorded (fig.4), they represent, as stated above, more or less what Roller detected some years ago (Roller 1987). In recording them, we distinguished features (stone blocks) *in situ* in evident connection with each other and belonging to walls (walls *in situ*), from what we have called *disiecta membra* (the architectural blocks not in their original place or if *in situ* not in evident connection with others). We attempted to compare the density of the two categories, according to their presence in each collection unit, but no clear pattern appeared. On the other hand, considering and plotting them individually on the general topography we can easily see that they follow directions and present clear patterns in some cases. In the S part of the surveyed area, not only the few preserved walls but also the scattered architectural features, represented by points, follow the main direction of terraces and proposed roads. On the contrary, as pointed out earlier, some of the avenues (perpendicular to the roads) seem to be differently located (fig.9). They are, at least in this area, running in the same direction with the slope resulting them to be more vulnerable to erosion. On the top of the ridge, where we have the 'plateau' gently ascending towards the East, the pattern is less clear, but it is easy to connect the position of the features with the presence of larger and possibly public buildings, characterized by different sizes as well as different orientation. On the slope to the North we cannot detect any pattern, and the majority of the present blocks cannot be considered *in situ*. They have probably rolled down from the structures on the top of the ridge.

Overlaying the collection grid upon filtered DEM, we can directly compare the results of the artifact surface survey, such as the density of pottery per collection unit corrected by visibility, with topography and topographical features (fig.10). As we can see, inside the city wall, the largest concentration of pottery is on the slope coming down from the main ridge. This can be due to erosion factors, if we also consider that we have the highest evidence of prehistoric pottery at this location. Moreover, looking at the relationship between density of pottery and the presence of architectural features on the surface, we may notice a presence of the latter where the former is smaller. Considering the architectural remains certainly *in situ* together with the topography, we can see that they are visible on the surface especially on the bare area on the top of the ridge. The surface soil, rich in ceramic material, has moved and rolled down the slope. Erosion can, therefore, explain pottery concentrations (on the slope) and architectural features concentrations (on the flattened bare area), as well as the apparent absence of some probable avenues at the lower areas.

Later in the year, in November, jointly with a Slovenian team from the University of Ljubljana, we carried out a campaign of geological investigation and geophysical prospection, in a rather flat area close to the southern city wall, probably residential in the urban plan of the 4th century BC. The aim of the geophysical prospection were in many respects similar to those of the topographical survey, ranging from the detection of structural features to that of functional areas and organization of space inside the urban area (Bintliff *et al.* 2000). For our purposes it is of interest noting that it confirmed our hypothesis of the urban street plan and supported the interpretation of the topographical features clearly emerging from our survey.

To store all data coming from the field and derived information, both from the artifact surface collection and the architectural features survey, geophysics and geology, a database was structured and implemented in Microsoft Access 2000. All the information recorded in the field on the recording sheets have been inserted into the database, providing also a quantity of further insights concerning the way of collecting data. The set of basic units of the information collection has been carefully constructed as topologically consistent. Once spatially plotted, they have been linked directly to the attribute information relating to the finds made within each of the grid units and all the features have been assigned to different information layers.

Within the GIS system then, the resulting survey maps have been geo-referenced and integrated with the available paper maps at smaller scale (1:5000 – 1:50000) converted into a digital format, using control points taken in the field.
Moving beyond this relatively simple task of spatially distributed data management, the GIS has been used to process and explore the data set resulting from the topographical survey, integrating it with the architectural features recorded and the artifact collection data.

To sum up, from the described case-study crucial methodological questions can be risen: is a topographical detailed survey useful only to map data resulting from the artifact surface collection, and to give us a general picture of the area we are working on? Or rather, by inserting the results as a layer into a Geographical Information System and linking it to a diversified database, could we gain new information and squeeze fresh juice out of something that would not be a mere Digital Elevation Model?

As we have shown above, certainly the topographical survey has enabled us to plot and overlay the artifact surface survey results on a more detailed terrain model of the surveyed area. Additionally, though, a GIS based recording procedure has provided us with a uniform framework, based on multilayer and multiscale procedures, within which to bring together and curate data derived from a varied range of sources. Furthermore, the relationship between Database and GIS has been shown as complementary and one of synergy. Thanks to a well-constructed Database, GIS plays a crucial role in the analysis of a complex site, as Tanagra has proved to be, since it works as a single, integrated, highly flexible analytical environment. Within it, every single layer contributes to the analyses concerning archaeological problematics, and also Topography becomes a meaningful, high information potential data set.

Acknowledgements

We would like to express our gratitude for the encouragement and support of Professor Dr John L. Bintliff, Leiden University, who provided us with the opportunity to work for the Tanagra City Survey Project. We would also like to thank the University of Leiden for its technical support, as well as the students that assisted us with practical aspects of the project. It would be an omission not to thank Professor Bozidar Slapšak and Dr Branko Mušic for having provided us with their expertise on Geophysical prospection.

Bibliography

Bintliff, J.L., *et al.* 2000. The Tanagra Survey. Report on the 2000 season, *Pharos* 8, pp. 95-129

Gillings, M., 2000. The Utility of the GIS Approach in the Collection Management, Storage and Analysis of Surface Survey Data, *The Future of Surface Artefact Survey in Europe*, Bintliff, J.L., Kuna, M., Venclova, N. (eds), pp.105-120

Mušic, B., 200. An evaluation of the potential of geophysical prospection in difficult environments. The silent presence of GIS, *On the good use of Geographic Information Systems in archaeological landscape studies*, COST Action G2, Paysages Antiques et Structures Rurales, Slapšak, B. (ed), European Commission, pp.127-144

Roller, D., 1987. Tanagra Survey Project 1985, The Site of Grimadha, *Annual of the British School at Athens* vol. 82, pp.213-232

Figures

Figure 1.

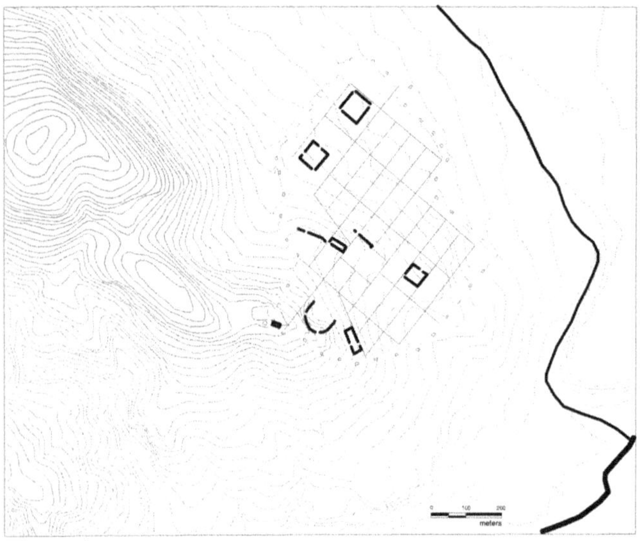

Figure 2.

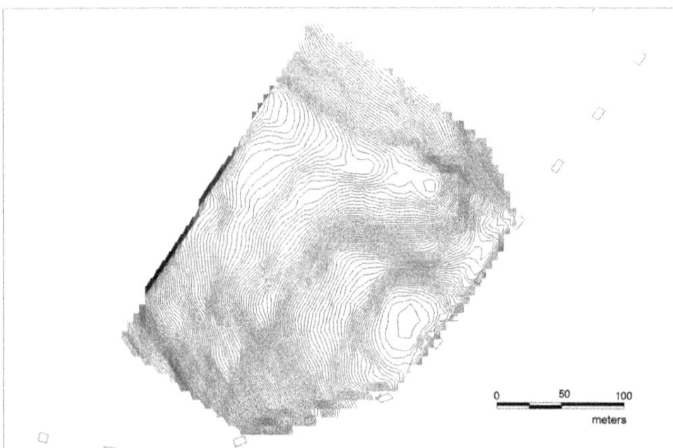

Figure 3.

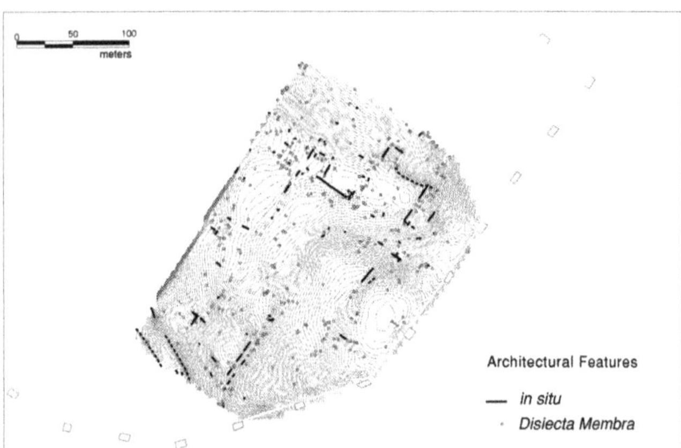

Figure 4.

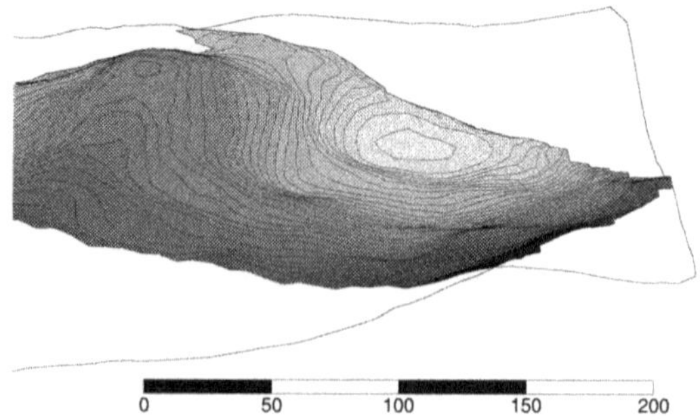

Figure 5.

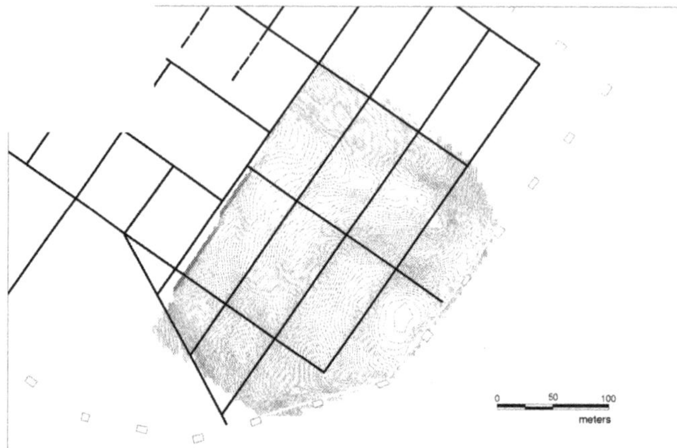

Figure 6.

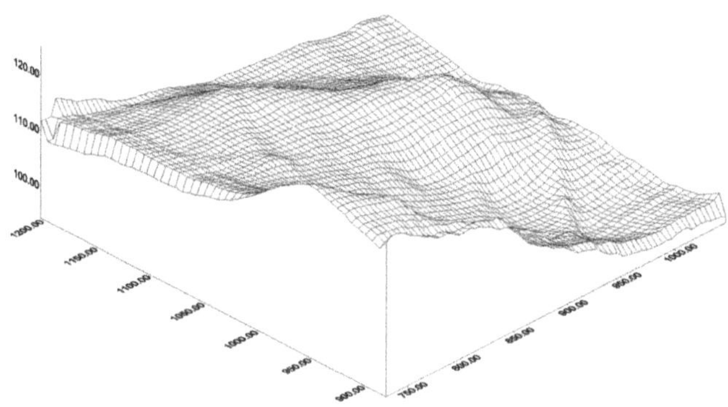

Figure 7.

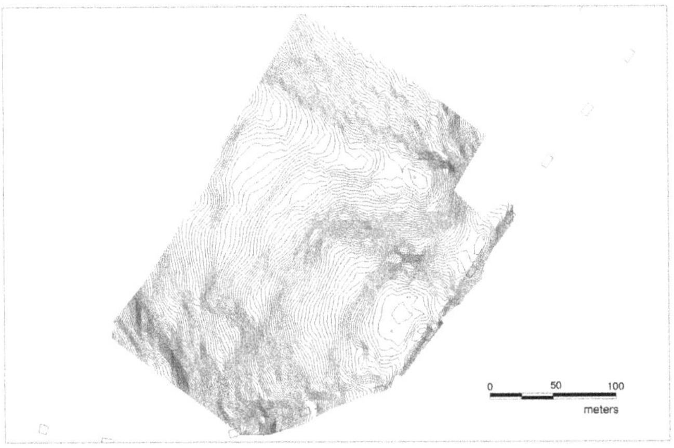

Figure 8.

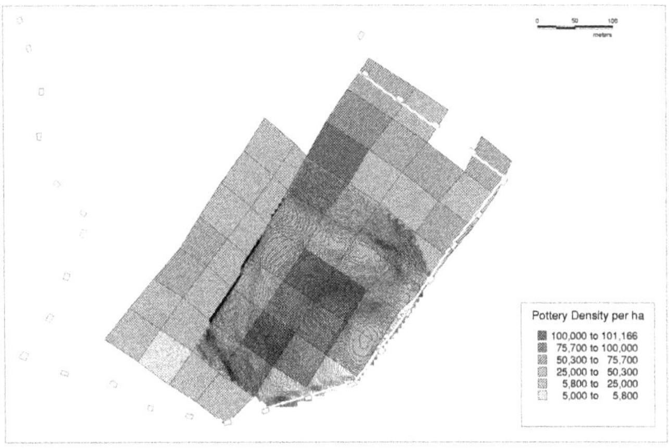

Figure 10.

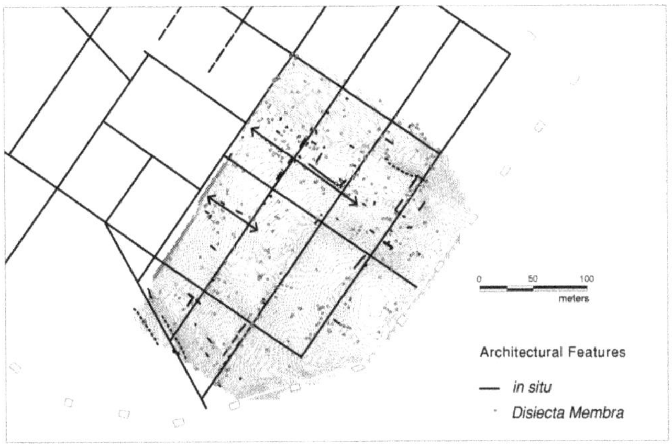

Figure 9.

The application of high-resolution satellite imagery for the detection of ancient Minoan features on Crete

Lefki Pavlidis
Department of Geomatics
Fourth floor, Civil Engineering Building, Grattan Street
The University of Melbourne, VIC 3010, Australia
Phone: +61 3 8344 9901 - Fax: +61 3 9347 2916 - E-mail: lefki@sunrise.sli.unimelb.edu.au

Clive S Fraser
Department of Geomatics
Fourth floor, Civil Engineering Building, Grattan Street
The University of Melbourne, VIC 3010
Phone: +61 3 8344 6806 - Fax: +61 3 9347 2916 - E-mail: c.fraser@eng.unimelb.edu.au

Cliff Ogleby
Department of Geomatics
Fourth floor, Civil Engineering Building, Grattan Street
The University of Melbourne, VIC 3010
Phone: +61 3 8344 6806 - Fax: +61 3 9347 2916 - E-mail: c.ogleby@eng.unimelb.edu.au

Abstract: Archaeological features, such as ancient settlements, roads or other indications of human activity on the ground during ancient times can sometimes be detected with remote sensing techniques. Two modes of above-ground remote detection, aerial photography and satellite sensed imagery, have been used in the field of archaeology. Aerial photography has to date been employed more often due to both the superior ground resolution traditionally available from photography, and the fact that aerial imaging is a well established technology. The use of satellite imagery in archaeology has been used in the recent past with varying degrees of success. Multi-spectral satellite imaging has the benefit of allowing the archaeologist to view parts of the electromagnetic spectrum not visible to the naked eye and also not detectable with photographic techniques. The development of commercial high-resolution satellite imaging, as exemplified by the deployment of the satellite Ikonos-2, *offers new possibilities for imaging archaeological features to a ground resolution of 1-metre in panchromatic mode and 4-metres in four multi-spectral bands.*

This paper describes an investigation in which high-resolution Ikonos-2 *satellite imagery is employed as an archaeological tool to discover, locate and map ancient Minoan roads and other archaeological features, using the Greek Island of Crete as a study area.*

Key words: remote sensing, high-resolution satellite imagery, visual interpretation, field verification, multi-spectral analyses, image classification.

Introduction

Archaeological features whether they are situated sub-surface or on the surface of the earth can be potentially detected with remote sensing techniques. Airborne or spaceborne remote sensing offers a manner of discovering remnants from our human past on the ground from a high viewpoint. Viewing archaeological remains from ground level does not clearly identify the spatial significance of those remains or their relationship to surrounding archaeological sites. In some instances archaeological sites are not apparent from ground level but become obvious once viewed from above.

Two types of overhead remote sensing used for archaeological feature detection are aerial photography and satellite imaging. Aerial photography has the advantage of offering high spatial resolution enabling detailed visualisation of archaeological structures. Satellites produce imagery with a lower resolution but cover both the visible and non-visible parts of the electromagnetic (EM) spectrum, which is useful for multispectral analyses.

With the deployment of the *Ikonos-2* satellite in September 1999, the combination of high-resolution and multispectral imagery is available from one sensor. *Ikonos-2* is capable of producing high-resolution satellite imagery (HRSI) with a 1-metre spatial resolution in panchromatic mode and 4-metre resolution in the multispectral.

Archaeological investigations utilising the recent technological advancements in satellite imagery are in their infancy. The aim of this investigation is to explore the use of HRSI for archaeological feature detection, mapping and classification using *Ikonos-2* imagery of eastern Crete.

This paper is divided into two parts. Part I deals with the necessary background information regarding remote sensing issues and the satellite imaging system of *Ikonos-2*. While Part II focuses on the actual investigation encompassing the project characteristics, the methodology to achieve the aim, the results and conclusions.

Part I – Background

Remote Sensing in Archaeology

From about the turn of the twentieth century aerial photographs have been used to aid archaeological research. One of the first archaeological aerial photos was taken of Stone Henge, England from a hot-air balloon [Doneus, 1999 #28]. In the 1920s, OGS Crawford, one of the pioneers of aerial archaeology formulated a set of site classifications for determining archaeological features using aerial photographs. Table 1 lists the different types of classifications along with their descriptions.

These classifications are quite useful for both aerial photographs and satellite images providing that the spatial resolution of the satellite image is high enough for the interpreter to discern what is revealed within the image. Although not all archaeological features that are large enough to be detected from the air are immediately apparent in imagery. The site classifications in Table 1 are reliant on the visible patterns created by the archaeological remains. In some cases it is beneficial to investigate imagery within the non-visible part of the spectrum for archaeological features. The use of the infrared (IR) region of the electromagnetic (EM) spectrum provides additional information about the growth patterns of vegetation. Buried remains can cause changes in vegetation growth, as in the crop marked sites of Table 1, but sometimes these changes may be too subtle to be seen with the naked eye. With IR imaging these growth variations are enhanced allowing the interpreter to investigate vegetational changes within the non-visible spectrum.

Compared to aerial photography not as many archaeological investigations have been conducted using satellite imagery, mainly due to the ground resolution constraints and cost and availability. However the spectral range of satellite imagery is greater than that of aerial photography due to the capabilities of multi-spectral sensors, allowing for a more comprehensive analysis. Satellite images exhibit greater spectral sensitivity and contrast in ground reflectance, in comparison to aerial photographs [Fowler, 1996 #36]. Most satellite sensors have the ability to capture data within the visible and non-visible parts of the EM spectrum encompassing a portion of the ultraviolet region, the visible and the near IR (NIR). Other remote sensors are capable of capturing the thermal IR region, such as the sensor in the satellite LANDSAT TM [Kruckman, 1987 #10].

IKONOS-2 Capabilities

The satellite *Ikonos-2* orbits the earth at an altitude of 680 kilometres and travels at a ground speed of 7km per second. It captures 1-metre panchromatic digital images and 4-metre multi-spectral images covering the blue, green, red and NIR bands. The two data sets are captured simultaneously over an 11km swath [SpaceImaging, 1999 #32]. The satellite has the ability to acquire stereo images for the creation of 3-D data. It has a nominal 26° angle of convergence for stereo scenes and can collect data of the same scene three times for stereo applications by first tilting the sensor forward by 26°, then at nadir and again when the sensor is tilted back 26°. The size of a single *Ikonos* scene is normally 11km by 11 km.

Space Imaging, the company that owns the *Ikonos-2* satellite, produces a number of image products for commercial use. For instance the *Geo* product is geometrically corrected and georeferenced to a user-specified ellipsoid and map projection. The absolute positional accuracy of the image is ±25m (RMS, 1-sigma) [SIE, 2000 #88]. The user has the option of ordering the image as a 1-metre pan-sharpened product, which is effectively a colour image with a 1-metre resolution. The image of eastern Crete used in this investigation is from the *Geo* range, being georeferenced to WGS84 with a UTM projection and ordered as 1-metre panchromatic and the 4-band multispectral.

When thinking about satellite imagery for archaeological application two main features are necessary: 1) high spatial resolution, and 2) multi-spectral bands for quanitative pixel analyses. *Ikonos-2* possesses both of these important features offering high-resolution capabilities and basic multispectral content.

Part II – The investigation

Project Characteristics

The Area of Study

The area of study is situated in eastern Crete, Greece. Crete is a topographically diverse island where the ancient people, known as the Minoans (as they were named at the turn of the 20th century) flourished during the Bronze Age from approximately 3000 to 1100 BC [Hood, 1971 #17].

The *Ikonos-2* image of the study area was captured on March 9, 2000 at the requested time of 8:30 in the morning to highlight any shadow marked sites. The image covers an area of 9km by 11km and includes the ancient Minoan palace site at Kato Zakros. Figure 1 shows a true-colour composite image of the area of study.

Like most of the island, eastern Crete is a rocky and barren region with very few natural springs. The mountains are mostly composed of grey limestone and earlier deposits of brown, grey and greenish schists, quartzites and shales. Around the coastal areas are homogeneous marls – a soft marly limestone also known as *kouskouras* [Gifford, 1992 #15]. The schists were used by the Minoans for paving while limestone materials and mud-bricks were used for buildings.

Ancient Minoan Archaeological Features
The Minoans had developed external trade relations with the Aegean islands, Egypt and the Near East. Their harbours were important for commercial reasons so major settlements were built at these locations. There is evidence in the form of traded goods to suggest that there was contact between these major sites [Tzedakis, 1990 #40]. The topography of the island played a major role in moulding Minoan society with respect to the location of settlements and the routes of internal interaction between settlements.

From the work carried out in eastern Crete by a team of archaeologists led by Yiannis Tzedakis and Stella Chryssoulaki we know the location of some Minoan sites and roads. Investigations led the researchers to the discovery of ancient Minoan roads and other sites that were verified through excavations and were subsequently mapped [Tzedakis, 1989 #3; Tzedakis, 1990 #40; Tzedakis, 1990 #41]. Table 2 sets out Minoan archaeological features and their characteristics compiled from the extensive research of the area by Tzedakis and colleagues.

It is important to have the background knowledge of the geological composition of the island to assist with the multi-spectral analysis and classification of the image. The summarized characteristics of Minoan features in Table 2 were compiled to assist with the spatial interpretation of the imagery.

Research Methodology

The methodology was developed by taking into consideration the existing remote sensing techniques, Crawford's classifications and the background knowledge regarding the island and the Minoan people. The different phases of the methodology are as follows:

A. Pan-sharpening
B. Ancient Feature Overlay
C. Visual Interpretation
D. Field Verification Survey
E. Remote Sensing Analyses

A. Pan-sharpening
The pan-sharpening of the 4 multispectral bands using the 1m panchromatic image was carried out to improve the spatial resolution of the multi-spectral images so as to assist in the visual interpretation. The pan-sharpening process consisted of converting the multispectral band combination from the three colour guns red, green and blue (RGB) into hue, saturation and intensity (HSI). The intensity layer was replaced by the panchromatic band and the HSI was then converted back into RGB [Carper, 1990 #73], resulting in the pan-sharpened multispectral image with a spatial resolution of 1 metre.

B. Ancient Feature Overlay
This process involved the tentative location of ancient Minoan features using the archaeological map created by Tzedakis and colleagues. This was done so that the interpreter could gain a better idea as to where to look for the ancient roads and sites in the image.

C. Visual Interpretation
This phase involved a visual search of the image for linear or geometric features with possible archaeological significance. It was accomplished by taking into account the spatial characteristics of ancient features as set out in Table 2, and using the ancient feature overlay as a guide. The whole image was systematically visually scanned to manually detect any of the characteristics. Once detected the features were digitized on the image into a vector layer with various symbols denoting different features, such as roads, palace and forts.

D. Field Verification Survey
The verification process involved a field trip to the area of study in eastern Crete to determine whether what was visible in the image corresponded with what is on the ground. The field survey incorporated a GPS survey to locate ground control points (GCPs) and the verified features.

E. Remote Sensing Analyses
A number of steps were implemented for the remote sensing analysis, once the image had undergone enhancement and re-georeferencing using the GCPs from the field survey.

- Band ratios and combinations to highlight certain ground cover types.
- Edge detection filters for enhancing linear and geometric features.
- Classification of the image to emphasize sites with the same spectral signature.

Software
The software packages used to carry out the various phases of the methodology were ERMapper v:6.1 and ERDAS Imagine 8.4, both being used on a Pentium platform under Windows.

ERMapper and ERDAS Imagine are both suites of software for remote sensing analysis. ERMapper was primarily employed for the pan-sharpening and the ancient road overlay, whereas ERDAS Imagine was preferred because it supported more rapid processing of large data volumes. Also, it is equipped to perform any type of GIS and image processing application, and it has the facility to execute photogrammetric processes.

Results

In Phase C of the methodology the visually detected features were mapped for verification in the study area. The field verification survey was conducted in conjunction with the Institute of Mediterranean Studies (IMS) at Rethymno, Crete, who supplied the high-accuracy GPS (Global Positioning System) equipment, namely 2 Ashtech Z-12 receivers. The 4 GCPs located in the GPS control survey were used to transform the image coordinates to accord with the coordinates of the GCPs, thus improving the absolute positional accuracy of the image to ±2m.

Not all detected features were searched for on the ground due to time constraints. A total of six days were allocated to perform both the field verification and the GPS control survey. In some cases the process of verifying at least one feature took a full day due to the rugged terrain and vehicle inaccessibility at most areas visited. Some of the detected features were verified whereas others were not. Figure 2 is a portion of the image showing a detected linear feature. The pink line is from the ancient feature overlay and represents a Minoan road, the distinctive curving white line is a dry creek bed and the faint linear feature that forks out is a detected linear feature, corresponding to the Minoan road overlay. When the survey team visited the area to verify the linear feature we could not find anything on the ground that resembled a road, path or track. A few placed stones were found parallel to each other at a distance of about 2 metres wide running for approximately 6 metres, but these were not representative of what is visible in the image. Possible explanations for this are the different seasons of image capture and field survey causing the feature to appear more emphasized in the image; and, the linear feature is better seen from an aerial view rather than from the ground.

Remote Sensing Analyses

ERDAS Imagine was used to perform the remote sensing analysis, which included supervised classification, band ratioing and edge detection. The supervised classification involved the classification of the image into different ground cover types according to the interpreter's *a priori* knowledge of the area. Band ratios were implemented since they can highlight a particular surface type, while edge detection filters were used to enhance linear and geometric features through mathematical operators.

A Supervised classification is dependent on the analyst's familiarity with the various ground cover types within the image and relies on knowledge from field surveys, maps, aerial photographs or personal experience. In the software, the analyst defines a sample area on the image that is representative of a specific ground cover or information class. This is called a *training area*. Once the analyst trains all areas of interest to represent the known different ground cover types in the image the computer then classifies the remaining pixels according to the training areas. It does this by using the spectral signatures of the trained pixels to identify and categorize every remaining pixel in the image with similar spectral properties [ERDAS, 1999 #80].

Twenty ground cover types were 'trained' in the *Ikonos* image. They ranged from olive groves, bitumen roads, bare soil, gravel roads, the sea and crop fields. The image was then processed to classify all the pixels according to the 20 ground cover types. One of the chosen training samples was the archaeological site of the Minoan palace at Kato Zakros. The colour magenta was used to represent the archaeological class. As a result of the supervised classification the Minoan forts near Karoumes bay were classified in the archaeological group along with some other areas, like the linear feature running in an east to west direction in the upper part of figure 3. The reason why the Minoan palace at Zakros and forts at Karoumes have a common spectral signature is because they are excavated sites, revealing a common sub-surface layer with ruins made from similar building materials. The other areas that were also highlighted under the 'archaeological' class are areas that have experienced some erosion exposing a similar ground surface, such as the bottom of the gorge in figure 3, which undergoes weathering processes during the wet months.

The classification of ancient roads is rather difficult since the roads were constructed from local materials, which have the same spectral response as the surrounding environment. But in the regions where the natural surface has been disturbed there may be a different response. The other factor limiting the classification of the ancient roads or paths is that some roads are only 1 to 2 pixels wide, which is not enough for the software to reasonably select those pixels for classification into a class of their own. Instead these pixels are grouped into a class that resembles their spectral response the most [Gorte, 1999 #91].

Band Ratioing involves the manipulation of multi-spectral bands by dividing one band with the other. This is useful for enhancing subtle differences between rocks and soils [Jensen, 1996 #76]. As mentioned, limestone was used by the Minoans as a building material. The ratio of NIR to Blue was selected because the reflectance level of limestone is lower in the blue than it is in the NIR and by dividing the blue band values into the NIR, the limestone values increase causing higher reflectance and appearing as a bright colour. This ratio also highlights vegetation while bodies of water are shown as black as is evident in figure 4. The portion of the image in figure 4 shows the area of the Minoan palace at Zakros Bay. The result of the NIR/Blue ratio image was not as expected. The palace area is shown as dark but was expected to appear as a light grey since it is known that it was constructed with limestone blocks, mudbricks and sandstone for paving. The most likely explanation for this is a high soil-moisture content resulting from the water table being close to the surface.

Edge detection enables the spatial enhancement of geometric or linear features in an image. General filtering techniques or convolution operators are employed to detect edges or lines by changing the spatial frequency characteristics of the image [Richards, 1999 #74]. Convolution operators are designed to highlight or conceal specific features in an image, based on their spatial frequencies. A low-frequency area in an image is where there are minimal changes in brightness values of pixels. For high-frequency areas the opposite is true. Areas of high-frequency are those where maximum changes in brightness values occur over short distances [Jensen, 1996 #76].

A 9x9 high-pass filter was passed over the whole image to enhance edges. Figure 5 shows a portion of the colour infrared image with a soil marked site that underwent the 9x9 high-pass filter. The portion is part of the false colour image where areas of red represent vegetation and the greenish areas are bare soil. The high-pass filter enhanced the olive trees and the linear feature running south-west and forks out in the bare field, is more apparent.

Discussion of Results

The visual interpretation of the *Ikonos* image was moderately successful as confirmed by the field verification survey. Some of the detected features were verified on the ground demonstrating that HRSI is useful for visual interpretations. The supervised classification showed that it is suitable for detecting the known archaeological sites although it is not a totally automated approach, since other areas with similar spectral responses are highlighted also. Therefore, it is a matter of verifying on the ground all the areas classified with the archaeological site training region, to check whether they are archaeologically significant or not. For the classification of ancient roads, it is recommended that imagery with even higher spatial resolution be used to pick up the spectral signature of the road pixels.

The band ratio of NIR/Blue did not produce the expected outcome but it still managed to emphasize the palace and surrounding excavated settlement as a dark colour. However, this is not as useful as the supervised classification where the analyst has more command over the selection of sites to be highlighted or grouped. A greater range of the EM spectrum needs to be covered for the differences in rock types to be discernible with band ratioing. As mentioned, the NIR/Blue ratio also emphasized the vegetation within the image. If the area of study were more vegetated, the use of the ratioed NIR band would be useful for highlighting any crop marked site.

The edge detection filters are effective for enhancing existing lines or edges which may not be greatly apparent in the original image. In this way features can be 'detected' more easily after the image has undergone a high-pass filtering, as compared to their original state. It must be noted that edge detection cannot detect any lines or features that are not present in the original image.

Conclusion

Archaeological remote sensing is best served by technology which produces high-resolution, multispectral images since it is a large scale application and depends upon visual detection. The investigation for the detection of ancient ground features using *Ikonos-2* imagery over eastern Crete has shown that HRSI has practical potential for archaeological applications mainly in the areas of visual interpretation and the creation of archaeological maps.

HRSI is particularly useful when combined with a field verification survey to confirm the results of the visual interpretation of the image. This usefulness is further enhanced via a GPS control survey to satisfy accuracy needs for the creation of archaeological maps.

It is evident from the investigation that *Ikonos* imagery has further archaeological potential in areas of study where there is more vegetation, in order to take full advantage of the NIR channel.

Overall, the investigation so far has shown that HRSI is useful for qualitative archaeological image analyses, whereas the remote sensing analyses have not been as successful, although not all possible avenues of remote sensing techniques have been considered. Thus, the research is continuing with emphasis on the quantitative analysis of the image, by further examining its multispectral properties with the use of various techniques such as principal component analyses.

Tables

Table 1: Crawford's Site Classifications (Scollar et al., 1990)	
Shadow Marked Sites	Are those sites where the remains are partly above the ground's surface and are visible from the air in early morning or late afternoon sun, through shadows cast by the above ground irregularities. Visibility of features depends on the contrast they create in the photo.
Moisture or Soil Marked Sites	From past constructions the physical order of the soil is disturbed, due to the introduction of foreign materials into the local soil. Past features become visible from colour variations or tonal differences in the soil.
Crop Marked Sites	Plants grow differently depending on the contents of the soil. Sub-surface archaeological features become visible from differences in height and colour of the crop.
Frost Marked Sites	Timing and the appropriate season need to be chosen for this type of marking to appear. When air temperatures are around freezing point, thermal gradients become visible as differences in thawing and freezing of frost or light snow occur.

Table 2: Minoan archaeological features and characteristics

Archaeological Feature	Geometric Characteristics	Additional Characteristics
Ancient road	Linear, 1.5m to 3m wide	Paved with limestone slabs (closer to settlements) Follow natural course of terrain Enlargement of naturally narrow curves
Guard post	Approx. 10m x 10m	Situated at road junctions Constructed from limestone megaliths
Palatial complex	Approx. 120m x 100m	Contains central courtyard 30m x 12m, treasury, royal rooms, other adjoining rooms Surrounded by Minoan urban settlement Positioned near harbour
Peak sanctuary	No structure or dimensions	Located on top of mountain peaks Situated close to settlements

Figures

Figure 1. Ikanos-2 image of area of study (eastern Crete).

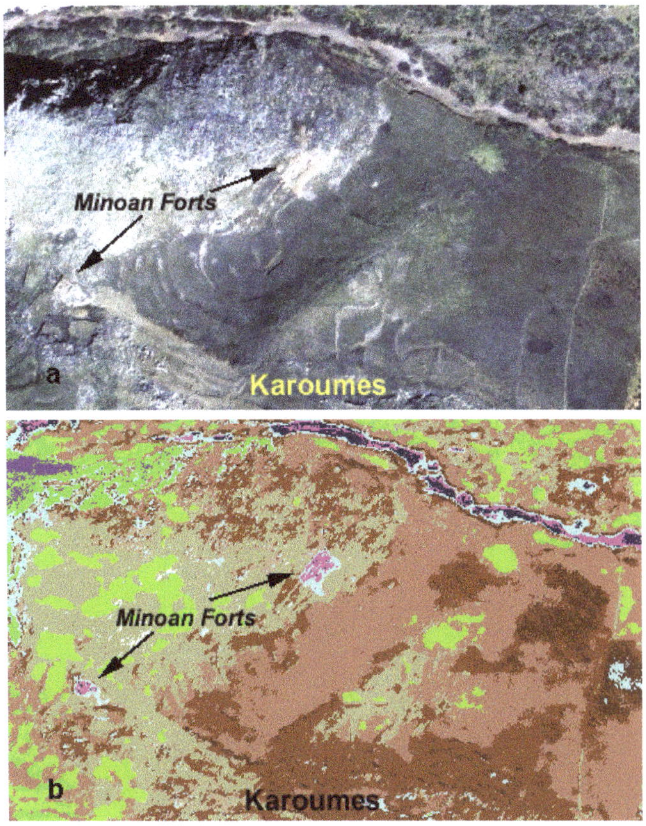

Figure 3. The supervised classification showing the two Minoan forts in magenta and the linear feature to the north.

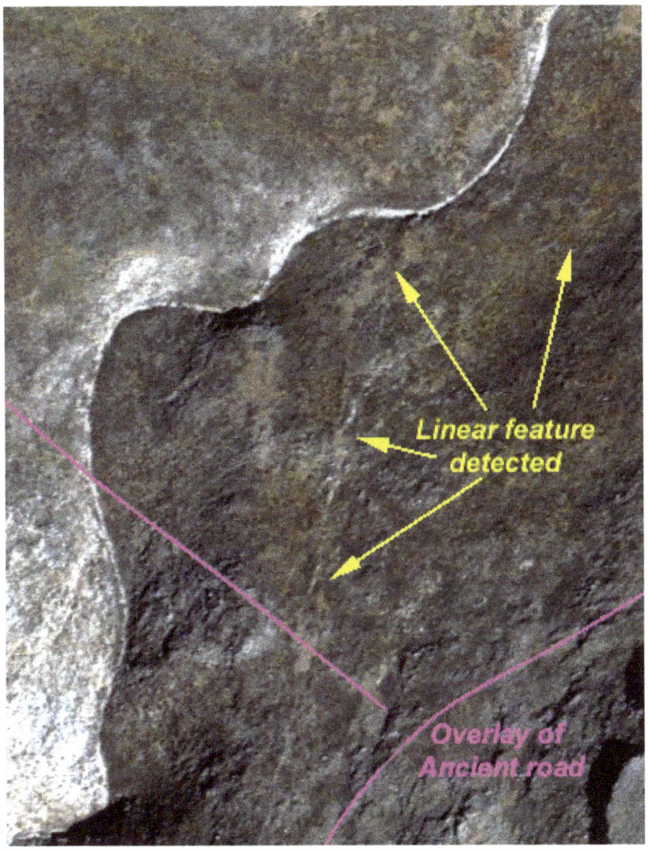

Figure 2. detected linear feature in Ikonos image.

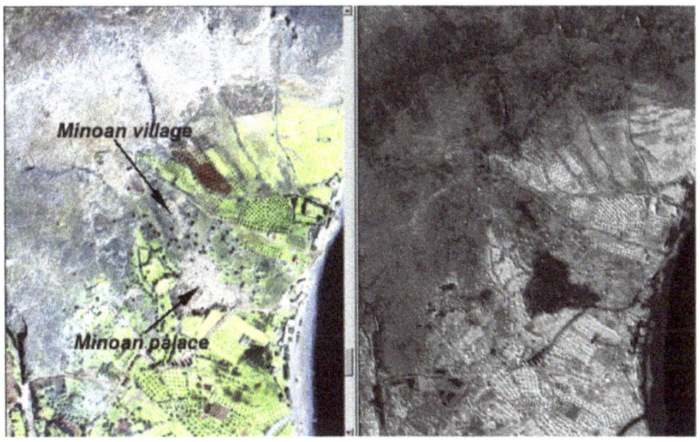

Figure 4. Ratio of NIR:Blue on the right, showing the Zakros palace complex and the true-colour image on the left

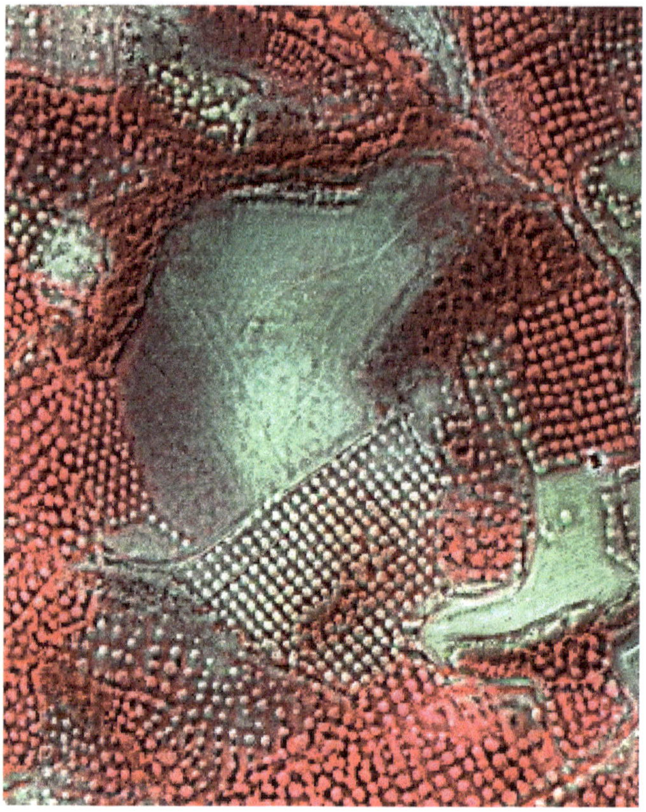

Figure 5. CIR portion of the image after high-pass filtering

Educating the Digital Fieldwork Assistant

Nick Ryan
Computing Laboratory
University of Kent at Canterbury
Canterbury
Kent, CT2 7NF, UK
Phone: +44 1227 827699 - Fax: +44 1227 762811 - E-mail: N.S.Ryan@ukc.ac.uk

Martijn van Leusen
Groningen Institute of Archaeology
Poststraat 6, 9712 ER Groningen, The Netherlands
Phone: +31 50 3636717 - Fax: +31 50 3636992

Abstract: This paper describes ongoing development of the FieldNote system at the Department of Computer Science, University of Kent at Canterbury (UK). A new version of this portable system for self-location, mapping, and note-taking during archaeological fieldwork was field tested in October 2000 during a systematic survey by teams from the archaeological departments of Groningen University and the Free University of Amsterdam (NL) in the Sibaritide region of northern Calabria (Italy).

In these tests, the system was for the first time used for wide-area mapping tasks, and proved to be very useful in mapping field boundaries, highland transhumance routes, and archaeological sites to a specified accuracy and in the absence of detailed up-to-date topographic maps. It also proved useful in navigation and in re-locating archaeological sites mapped in the 1960's.

The results of the SIBA2000 field tests are presented here, followed by a discussion of their significance for future field survey design and methodology, with an emphasis on the trade-off between speed of operation and accuracy. Lines for further development of the system, including improvements to both interface and functionality, are set out as well.

Key words: Field survey methods, mobile computing, context awareness, spatial accuracy, GPS

1. Introduction

1.1 Improving the efficiency and accuracy of field work procedures

Modern field walking surveys have become increasingly labour-intensive both because a higher coverage rate (often with total collection of artefacts) is now thought to be essential, and because more stringent demands are now put on the precision and accuracy of field recording methods. The management and analysis of modern survey data within GIS requires the accurate mapping of large numbers of small collection units so that minor variations in the densities of all material categories can be detected. The processing of large numbers of, often undiagnostic, finds puts a strain on the ceramic specialists. In some recent surveys, the overall speed has slowed to as little as 1 hectare per person/day spent in the field; there is therefore a need to improve efficiency by any means available.

Modern survey practise also requires the accurate mapping of collection units, and the detailed recording of environmental variables which may influence the finds circumstances ('visibility conditions'). Collection units are usually mapped directly onto a large-scale topographic or cadastral map of the survey area, using existing landmarks for orientation, and a series of forms is normally used to record information about the collection unit and the finds made in it. Forms are also used to track finds through the various pre-processing steps and the specialist classification stage into storage. The information is later transferred from the forms into a DBMS. In current practise, a certain number of errors, omissions, and illegal entries is unavoidable because maps may be incorrectly interpreted or outdated and hardcopy forms cannot prevent erroneous or illegible entries; further errors may (and will) arise where the procedure requires a transcription step. Procedures based on independent self-location and direct digital data entry should prevent most types of errors from being made.

Finally, modern field surveys are multidisciplinary. Typically, geopedological research and a study of historical maps and records will take place in conjunction with the field walking. In many cases it would be helpful if the information collected by each of the participants were available to the others in the field. For example, archaeologists might want to steer away from areas where the geopedologist has mapped severe erosion or deposition, and might want to have 19th century maps available in order to correlate finds distributions and landscape features to the pre-industrial landscape. In practise, this is hardly ever possible because the additional hardcopy maps, if they are available in time, are too cumbersome to carry about in the field.

1.2 Development of a digital Field-Work Assistant

A potential solution to this problem presented itself in the form of digital Fieldwork Assistants (dFA's) based on small handheld computers linked to GPS receivers. One such system that has been developed to address the requirements of a range of field sciences is *FieldNote*. This originated in a project that set out to examine the utility of "context-aware" systems as fieldwork tools in a range of disciplines, including archaeology and ecology (Ryan *et al.* 1998). A context-aware system is one that actively monitors its working environment, or context, and adjusts its behaviour according to changes in that context. Contextual input may be supplied by the user, or derived automatically from internal or external sensors.

FieldNote includes several modules that work together to monitor context and trigger events under user-defined conditions. Essentially, they provide context-aware and communications services to support a range of applications designed to meet specific fieldwork requirements. The most frequently used modules are the *LocationTracker* and *ContextClient*. *LocationTrackers* exist in various specialised forms designed to work with different proprietary GPS receiver protocols as well as the widely used, but limited, NMEA protocol. The *ContextClient* handles network data exchange between handhelds and other machines.

Note taking and mapping modules *FieldNote* and *FieldMap* build upon this infrastructure to directly support the fieldworker's recording and navigation needs. *FieldNote* uses the context services in two ways. Firstly, when recording information, it automatically tags all records with additional context, including the current location, derived from a GPS receiver, the date and time, and the user's identity. Secondly, whenever the context of a recorded note matches the user's current context, that note will be 'triggered' and its contents displayed on-screen. Although many criteria might be used in matching recorded notes to the user's context, proximity of location is most commonly used. When the user approaches the location of some previously recorded information, it will be brought to their attention.

The notes, together with background maps and other data, may be displayed using the *FieldMap* program. This uses information from the context service to keep its display centred on the user's current location and to select which map layers (note symbols or cartography) are displayed.

This system had originally been developed for the now defunct Apple Newton, and a similar, but conceptually simpler, set of programs, the Stick-e-note suite, had been developed and tested on early Palm Pilot devices. These had been evaluated in several field trials with archaeologists from Southampton University (Ryan *et al.* 1999a) and ecologists from Manchester Metropolitan University (Pascoe *et al.* 1998). These early trials confirmed the essential utility and potential of the system and, in 1999, work began on a new version of the system based on the experience gained in these trials.

Previous versions had all been implemented using what was, at the time, the most appropriate language for the target device: NewtonScript for the Newton and C/C++ for the Palm. To ease portability between different hardware platforms, this new version was written in Java and has been successfully implemented on a variety of handheld devices including those based on Windows CE and EPOC operating systems, as well as laptop and desktop Windows, Macintosh and Unix machines[1].

The note taking software had previously used either plain text or HTML formats for storage and exchange between handheld and desktop. Although some experiments in using XML based formats for data exchange had been carried out, this aspect of the system was not extensively developed (Ryan *et al.* 1999b). The new version provided an opportunity to fully implement these areas with the aim of simplifying the process of exchanging data between handheld field computers and either a remote server, or a laptop or desktop machine at the survey team's base.

One of the main strengths of *FieldNote* is that the user can carry their own and other researchers' georeferenced data around in the field. Their current location is always available, either as raw coordinates or as a moving cursor on a map display. This facility becomes particularly useful when there is a need to confirm earlier work or to re-examine an area of interest located by another member of the team. Potentially, this might be exploited to provide continuity between successive seasons of work in the same area.

An opportunity to test the new version of *FieldNote* came when, in the spring of 2000, a survey team composed of staff and students from two Dutch universities (University of Groningen and the Free University of Amsterdam) prepared to conduct a three week survey campaign in October of that year, near the village of Francavilla Marittima in the Sibaritide region, province of Calabria, in Italy.

1.3 The SIBA2000 campaign

The Sibaritide is a small alluvial plain on the Ionian seacoast, named after the Archaic and Classical Greek colonial town of Sybaris. It is surrounded by the limestone massifs of the Pollino and Sila ranges, and its margins are formed by a series of terraces of marine and fluvial origin mostly composed of pebbly sands and conglomerates, dissected by the wide pebble beds of several seasonal rivers. Early archaeological research in the area concentrated on the rediscovery of the Greek colony of Sybaris and its Pan-Hellenic and Roman successors (Thurioi and Copia), which are now covered by up to 6 meters of alluvium. Later, archaeological interest expanded to include the larger indigenous settlements and sanctuaries surrounding the plain, and the University of Groningen became involved in research at the indigenous hilltop sanctuary of Timpone della Motta and its nearby necropolis (Kleibrink 1993, 2000). Next to the small number of sites being excavated over the decades, the wider regional archaeological record was mainly created by in a large-scale survey conducted by Lorenzo Quilici and his teams in the late 1960s. These surveys, mapped at scale 1:10,000 but available only at the published scale of 1:200,000 (De Rossi et al 1969), were intended to provide the context for the ongoing excavations at Sybaris and its successors. Largely targeted at the Hellenistic-Roman landscape, these data were more recently complemented by a regional compilation of pre- and protohistoric sites supervised by Renato Peroni (eg, Peroni & Trucco 1994).

The SIBA2000 fieldwork campaign forms part of the wider Regional Pathways to Complexity (RPC) project running at the University of Groningen and the Free University of Amsterdam (for an overview, see Attema et al. 1998). Its main objective was to assess the quality of the existing archaeological record in preparation for an analysis of the regional settlement history in the light of processes of centralisation, urbanisation and colonisation occurring from the Iron Age onwards. Specifically, it was unknown which fields and areas had been visited by the Quilici teams, and whether the large-scale clustering of sites visible in the Quilici maps might be related to differences in the accessibility and surface visibility of agricultural fields in the 1960s. It was decided to approach this objective through a systematic survey of a representative section of the transitional landscape unit (see figure 1) using both intensive and extensive methods. This would allow us, on the one hand, to check the recorded size, location, and interpretation of the sites mapped by the Quilici teams (plus one site mapped by Peroni) and, on the other, to collect the distributional data needed to establish the presence and nature of any spatial patterning in the extant archaeological record. A preliminary report on the campaign is in preparation (Van Leusen in prep).

As in previous RPC project surveys, an important objective of the fieldwork was to develop appropriate and efficient survey methodology for the local circumstances. The Sibaritide was known to be relatively poorly mapped in a series of 1:10,000 scale map sheets produced in the 1950s, and parts of the survey zone were expected to contain few if any topographical reference points by which collection units might be located and sites relocated. Hence the decision to experiment with the *FieldNote* system developed by Ryan and his team for efficient and accurate relocating and mapping of fields and features. The system was expected to be particularly useful in a subsidiary project being carried out at the same time as the main survey. This aimed to relocate and accurately map protohistoric settlements and caves in the hinterland and to map subrecent transhumance routes from Francavilla Marittima into the mountains of the Pollino massif (see figure 2). This project was carried out by a small team under the guidance of local speleologist/archaeologist Nino Larocca.

Prior to the SIBA2000 campaign, the new version of the *FieldNote* system had received only limited testing in familiar environments and by users who were experienced with its predecessors. The XML-based mechanisms for exchange of data between the handheld devices and base system had yet to be tested with large volumes of data. Experiments with the archaeological use of earlier versions had all taken place in relatively well mapped areas and within clearly constrained areas, typically within, or very close to, the boundaries of large urban sites. The SIBA2000 project provided an opportunity to evaluate the new system on a significantly larger geographical scale under poorer map control, and with a group of users who were not familiar with the system or its predecessors.

This campaign also provided an opportunity for the first significant testing of the system following the decision by the US military to remove the deliberate degradation of GPS accuracy known as Selective Availability (SA). This change took place on May 1st 2000 with the effect that even simple handheld GPS receivers improved their apparent accuracy from around +/-80m to better than +/-10m. Previously we had needed either to use an additional receiver for live differential correction, or to record raw satellite range data and to post-process all GPS measurements against data from fixed base stations in order to remove the effects of SA. There was a need to evaluate whether the recently improved accuracy obtainable with simple equipment was sufficient for a range of fieldwork needs.

2. The Field Tests

Field tests of the dFA during the SIBA2000 campaign were conducted in a range of conditions, from extremely mobile recording of movements and observations during a survey of highland transhumance routes, to the detailed recording of grids during intensive survey. The results of these tests are described and evaluated in the following sections.

2.1 Highland Survey: the Recording of Transhumance Routes

Since the indigenous societies of southern Italy are assumed to have had largely pastoral lifestyles into the early Iron Age, the presence of late prehistoric and early protohistoric remains was to be expected both in the foothills and in the mountainous hinterland of the Sibaritide, and possibly correlating with sub-recent transhumance corridors. In preparation for the design of an appropriate research proposal, the team decided to test the kit by walking along some of these routes (see figure 2), making digital notes of archaeological material found along the way while at the same time recording accurate locations for a number of known highland prehistoric settlements and cave sites. The equipment functioned as expected, and a large number of observations, including georeferenced photographs, were recorded along routes running from the foothills near Francavilla Marittima toward the highland villages of Alessandria Carreta and San Lorenzo Bellizi. These included observations of Hellenistic farmsteads (up to 1000m asl), junctions of transhumance routes, subrecent structures relating to pastoralism, natural springs, potential locations for pollen cores, cave sites, and even individual sherds. The GPS trails of these surveys are depicted in figure 3; figure 4 gives examples of notes taken at such observation points.

2.2 (Re-) location and Recording of Sites

The capabilities of the dFA system for wide-area mapping tasks were tested during the SIBA2000 campaign by using it to relocate sites mapped in the 1960s by the Quilici teams, and by tracing agricultural field boundaries and centroids and circumferences of archaeological sites to a specified accuracy. In the absence of detailed up-to-date topographic maps for the area, the latter test was of direct practical utility to the survey team, as it turned out that the infrastructure of roads and tracks had changed considerably over the years and changes in land use and ownership had resulted in the removal of microrelief and old field boundaries.

Thus, in many cases it was no longer possible to relocate sites mapped more than 30 years earlier by reference to mapped landmarks only. In contrast, the use of the dFA as a navigation instrument allowed existing sites (the positions and identities of which had been pre-loaded onto the kit) to be relocated in a straightforward manner. Both the map layer containing the Quilici sites and the operator's current position were marked on the display, and with the kit set to respond to the operator's current geographical position, nearby sites were brought to our attention by a beep followed by a display of the site's database record.

Only a few field and site boundaries were digitally recorded, but this was enough to show that the system functioned well. However, it was noted that the current procedure for tracing field boundaries on foot is inefficient, and alternative methods should be explored (see section 3). The digital recording of the centroids of ceramic scatters using a single GPS reading with an attached note, which only involved previously tested functionality of the dFA, again was a trivial exercise.

It should be noted parenthetically that the criterion used to draw site boundaries during the SIBA2000 survey was a simple finds density drop-off, and the reason for recording these boundaries was not analytical but practical, enabling easy and reliable relocation at a later stage. It is recognised that the concept of 'site' has come under attack from many quarters in recent years, and that future surveys may choose to ignore it altogether, preferring instead to record collection units at higher resolutions and accuracies.

2.3 Recording of topographic reference points and collection units

Doubts about the quality of the available topographic maps and about the possibility of accurately mapping gridded collection units were the main reason for including measurements of topographic reference points in the SIBA2000 fieldwork. Separate measurement grids, consisting of square units approximately 50 by 50 m (0.25 hectare) in size, were set up in preparation to surveying (groups of) agricultural fields, and cardinal points in each grid were located in reference to these topographic landmarks. As the survey grids were established using a combination of sighting and pacing methods in sometimes difficult terrain, we felt it would be a good idea to obtain additional GPS measurements of the grids; accordingly, in some cases the locations of the canes used to mark the corners of collection units were also recorded using the dFA. All grids, landmarks, and GPS positions were mapped on field maps at a scale of 1:2000. Each of these field maps was later digitised along with its GPS points, and the latter were used to calculate a simple 1st order georeferencing transformation[2].

Although having a large number of GPS points, with attached notes, available during GIS processing did allow us to resolve some mapping problems, it was found that the procedures described above were insufficient for obtaining the desired mapping accuracy for many of the collection units. Where a large number of corners of collection units had been measured by GPS, the collection units could be mapped directly, without having recourse to transformations of the field maps. But for most field maps only a small number (from 3 to 8) of GPS points had been taken, and it proved impossible to 'rubber sheet' some of these onto the relevant GPS points in a satisfactory manner – apparently because there were too many internal distortions to the survey grids. Figure 6 gives an example of the problems encountered.

The georeferencing of the field maps brought to light other problems as well; in a few cases GPS points were so poorly placed for georeferencing transformations that additional points had to be constructed using plane geometry. In others, the inherent (standard) locational error of the GPS points confounded our attempts at georeferencing. Clearly, we have to rethink our approach in this regard (see discussion in sections 2.4, 2.5 and 3.1).

2.4 GPS Accuracy

During the SIBA2000 survey, we experimented with three types of GPS measurements at different levels of accuracy:

1. For most readings (eg, field boundaries) speed is more important than accuracy, so we took the readings on the move or even set the equipment to continuous logging, which typically yields positions with a standard error better than 7m.
2. For field reference points (topographical landmarks, cardinal survey grid points) our experience suggests that the accuracy should be increased to 2 to 3m by taking several minutes for each reading.
3. For base reference points the accuracy can be increased even further to about 1 to 2m by leaving the equipment to record a position for several hours.

Figure 6 shows a plot of GPS measured points collected over a period of nearly six hours at the Francavilla Marittima museum on the morning of 26th October 2000. The antenna was fixed to the top of a ranging rod which was then attached to the south-west corner of the railing at the front of the museum. Measurements were collected at approximately 10 second intervals for a period of about three hours. The antenna was then moved to the south-east corner of the railing and a further three hour sequence of measurements was recorded.

The observations at the western point were made at a time when the geometry of the visible satellite constellation was near-optimal, and are probably indicative of the best that can be achieved with a stand-alone single-frequency receiver. Those at the eastern point include a period of poor geometry and the re-establishment of position following a failure of the GPS receiver battery. They are probably more typical of average reception conditions. However, the introduction over the next few years of additional satellites to provide Wide Area Augmentation Systems (WAAS), primarily to support improved aircraft navigation, can be expected to bring typical performance into line with the better results seen here from the western point. Although a single receiver can be now used to obtain the level of spatial accuracy required for archaeological surveys, it comes at the price of reduced measurement speed. It is also clear that "mission planning" software, which predicts the number of visible satellites and the quality of their geometry, still has an important role to play. However, the combination of extended point occupation times and the need to avoid periods of poor geometry, which may amount to one or two hours of the working day, hardly contribute to streamlining the fieldwork process.

For our second and third types of measurements there remains a case for using differential techniques to improve accuracy without sacrificing speed.

The accuracy of GPS measurements made by a single receiver can be significantly improved using differential methods (DGPS). In its simplest form, this involves using satellite-receiver distance measurements taken by another receiver at a known fixed location. Details of the orbital position of a satellite can be obtained as part of the satellite signal, so it is possible to calculate the difference between the measured and "correct" satellite-receiver distances. These differences can then be used to correct measurements made by a roving receiver. This simple technique effectively removes the measurement errors due to atmospheric effects on the propagation of satellite signals. With two receivers separated by only a few kilometres, accuracy can be improved to +/-1m or better. More complex techniques involving signal carrier-phase measurements may be used to improve this to sub-metre levels. All this can be achieved using inexpensive single-frequency GPS receivers; greater accuracy requires the use of purpose-built and much more expensive dual-frequency 'survey-quality' equipment.

Real-time differential correction services are available in many areas. These include freely available signals from coastal beacon stations, intended for maritime use but often available at a considerable distance inland. Their main benefit is as a source of reliable correction data for inexpensive GPS receivers intended primarily for navigational use. Typically, the error in individual position fixes can be reduced to around +/-3m, increasing with distance from the beacon station. Various commercial services are also available, but these usually broadcast encrypted signals for which payment of license fees and a special receiver and decoder are required.

If live correction is not required, much better results may be obtained by post-processing recorded measurements against full measurement data from a second receiver. In many parts of the world, there are stations operated as part of the International GPS Service (IGS, igscb.jpl.nasa.gov) which make records available on a daily basis. For many areas, this is an invaluable source of highly accurate data. Unfortunately, the nearest observatory to the SIBA2000 survey area is at Matera, some 100 kilometres to the north. This distance is towards the upper limit of baseline distances for reliable corrections.

In view of the limited correction quality obtainable by post-processing against data from distant observatories, a further option would be to operate our own base station at a fixed location throughout the survey. This would require a dedicated computer and GPS receiver at the project base, recording continuously whilst survey teams were in the field. Given a maximum range to the edges of the extensive survey area of no more than 5 km and to the furthest point of the highland survey of about 15 km, position accuracy relative to the base station of between one and two metres could be expected by this method. In addition, long duration observations would help to significantly improve the absolute position of the base station by comparison with data from an observatory such as Matera. Clearly, for those applications where the accuracy requirements are higher than what can be achieved with a single receiver, the last option would give the most satisfactory results. It is, of course, still a post-processing option so would not offer any improvement in the real-time positioning in the field but, as already mentioned, this does not appear to be a high priority because *FieldMap* provides sufficient accuracy for the field-workers' location needs. Should it become necessary, corrections could be broadcast from the base station and received at the rovers by using conventional wireless-modem transceivers.

2.1 Spatial accuracy and the problem of identity

With the advent of accurate GPS location (and even before that with the increasing use of accurate field equipment) a peculiar problem has begun to haunt archaeologists: conflicts between field measurements and existing cartography. The position of topographical features as measured by GPS may not agree with their position as mapped on the most accurate available maps[3]. In the case of the SIBA2000 survey, this problem expressed itself in many conflicts between the GPS positions and the 1:10,000 scale topographic map; since the latter already had a bad reputation we 'resolved' the conflict by believing the GPS data to be the more accurate. On the other hand, some disagreements between paced distances on our field maps and GPS-measured distances could not be resolved at all because they were larger than could be explained by the standard GPS error. Using differential GPS or taking redundant readings suggest themselves as potential ways out of such conflicts. From this, and our problems in attempting to georeference the field maps, we learnt that field mapping methods based on estimates of distances and bearings are insufficiently precise in the kind of rolling terrain encountered in the Sibaritide foothills.

With respect to the sites mapped by the Quilici teams in the 1960s, the same problem was expressed in a more archaeologically relevant set of decisions. How much disagreement should we allow between the notional and measured locations of sites, before deciding that the two are *not* identical? Given the small scale of the published map data from which we had to work, it will not come as a surprise to learn that many Quilici sites were relocated up to 50m from their notional location. With larger disagreements it is no longer clear whether we have relocated an existing Quilici site, or have found a new one. While efforts continue to locate the Quilici's original 1:10,000 scaled field maps we may still hope that some of the remaining conflicts will be resolved.

3 Further work

Development of the dFA is ongoing; further work is needed to improve the functionality of both hardware and software components, and the user interfaces. The system should be able to deal with a range of fieldwork tasks under various field conditions, and should be to a large extent configurable by the user. In the following section we discuss plans and potential for further work on the functionality of the current dFA kit, on customising it with additional hardware and communication features for survey work, and on improvements to user and networking interfaces.

3.1 Enhancing current functionality

The functionality of the system as tested during the SIBA2000 fieldwork should be improved by some fairly straightforward changes to both hardware and software components, as outlined below.

Firstly, current systems are not well adapted to typical Mediterranean field conditions. Although monochrome screens work well, colour displays are preferred for situations where multiple map layers are displayed. Unfortunately, most colour displays are dependent on backlighting and are difficult to read in direct sunlight. Manufacturers have been slow to realise that a mobile device might be used outdoors, but a solution may be found in some newer PDA models with partially reflective screens intended for outdoor use.

Data input is typically by an on-screen virtual keyboard or character or handwriting recognition software. Although the recognition systems are much improved when compared with those available on early handheld devices, these are still relatively slow processes, particularly for users who have not had considerable practice. The main interfaces of the *FieldMap* and *FieldNote* programs have been designed to minimise the amount of written input and, with large buttons and other controls, most interaction can be performed using a fingertip, making the pen almost redundant. This facility needs to be extended to the options screens, and we anticipate further development based on the Minimal Attention User Interface (MAUI) devised in earlier work (Pascoe *et al.* 1999). This employed large input controls that could be driven by the user's thumb, thus enabling one-handed control of the software.

Simple use of voice actuation for selections from a constrained list of options is worth investigating as well. Recognition and automated transcription of voice input is not yet possible on these devices but, with rapid increases in computational power, may become available in the near future. In the meantime, we will investigate the use of the built-in voice recording capabilities of the PDA for adding voice notes as an alternative to text input.

Recording of geometric elements such as lines and polygons was a feature of earlier versions that was omitted from the new version because of development time constraints. Although the necessary data is recorded as part of the track log, restoring explicit user control of the geometry associated with a note is now a high priority.

Software should also be developed to facilitate conducting various types of 'gridded' surveys. The simplest option would be to record the locations of unit corners as they are being set out by the survey team, but more helpful alternatives should also be made available. For example, the system could be set to indicate the locations of unit corners at specified intervals and bearings from a given origin, or it could simply track the ground covered by teams as they work and warn them at specified intervals that a new unit is required. A major advantage of locating collection units in any of these ways, at least in rough and poorly controlled terrain, is that the spatial error is non-cumulative. Overall error can never be larger than that of an individual GPS point – about 5 to 10 m. A second advantage of the latter option is that the time consuming stage of setting up survey grids can be largely skipped, leading to greater efficiency.

A third area in which the functionality of the system may be improved is in the ease with which data can be exchanged between kits and with the project's base computer systems. Properly defined XML-coded data formats will ensure that information can be exchanged and downloaded, stored and edited easily, and can be accessed by web servers and browsers (see section 3.3).

Lastly, a number of minor potential improvements to the functionality of the system were identified during the SIBA2000 experiments:

- Addition of several simple utility functions, to allow, among other things, on-screen measurement of distances and bearings using the pen
- The readability of the display may be further increased by the addition of line and area symbols for monochrome display (eg, dotted lines for grid edges, thick line for paths; dithering for images such as air photos), and by the use of transparency when displaying icons.

3.2 Hardware and Communications

If the paper recording trail is to be obviated altogether by the digital recording equipment proposed here, then one further step should be taken – the use of a barcode reader to link bags of finds to collection units. Handheld computers with attached barcode readers have been in use in archaeological excavation and survey since the mid-1980s, and the addition to the system of a barcode reader on a CompactFlash card is trivial.

A downside of the current system is that it can only record GPS locations of the kit itself, forcing the operator to walk along the features that are to be mapped. The mapping capabilities of the system would be clearly enhanced if it were possible to record the locations of distant features as well (eg, suitable areas for further work, caves seen in cliff faces). The time consuming and often strenuous task of following field boundaries could also be replaced by either of the following methods:

- by manual on-screen mapping as proposed in section 3.1, using a georeferenced large-scale topographic map as a reference; in this case the position of the features is estimated; or
- by attaching laser range-finder binoculars, which allow the measurement of distance and bearing. Commercially available models can be as accurate as +- 2m at distances up to 2000 m.

As the functionality of the digital fieldwork assistant increases, more hardware components are added, leading to a shortage of ports. Many components require a serial connection so not all of them can be connected to the PDA at once without some form of intermediate switch; and even if they can, the additional cables and connectors make the system increasingly cumbersome. Even with a well-designed harness to keep these under control, there is an ever-present danger of catching cables on trees and other obstacles. We feel that the most practical solution to these problems lies in the adoption of wireless (radio) communications between system components. Ideally, each device would contain its own radio and would collaborate with the others to form a 'Piconet' or 'Body Area Network'. The long-awaited arrival of Bluetooth devices (www.bluetooth.com) which are intended to support short range (<10m) wireless networking may provide a solution.

An ever-present concern with mobile equipment is battery life. Battery technologies are improving but this is at least partly offset by increasing power demands as handheld computers become more powerful. The major limitation here is that most manufacturers design their systems for occasional and brief usage, whereas field computers are typically used more frequently and for longer periods. Many devices that are aimed at a consumer PDA market do not have adequate battery life for a full day's work in the field, particularly when heavy use is made of their serial interface to receive GPS data. Other similar machines intended for industrial/commercial use are, however, equipped with higher capacity batteries that have proved equal to our demands.

The power requirements of GPS receivers have reduced considerably in the last few years as manufacturers strive to develop receivers that can be embedded in other equipment, such as PDAs and mobile phones. At present, a few receivers are available in PC card format, and smaller CompactFlash devices may appear in the near future. Whilst these have the advantage of reducing the number of cables used in the system, they rely on the PDA as a power source and therefore put an additional strain on its batteries. Whilst we intend to experiment with integrated components such as these, it will probably remain necessary to carry spare or external battery packs to support the combined load.

Communication for data exchange between handheld and desktop machines typically uses serial or wired Ethernet connections. For those devices capable of using PC cards, local communication over a range of about 150m is possible using conventional 'Wireless Ethernet' cards. As yet, however, no such card is available in the CompactFlash format more commonly supported by PDAs. Away from the team's base, mobile phones provide a suitable communication medium, provided that there is adequate network coverage in the survey area. In uncovered areas, other devices such as wireless modems might be used.

The main limitation of current mobile phone technology as a data transfer medium is cost. At the time of the SIBA2000 campaign, data calls on GSM digital network still required the pretence of analogue transmission and hence the use of a modem. As a result, the exchange of any amount of data, no matter how small, requires a lengthy negotiation phase as the modem attempts to establish a connection with an ISP. It is this rather than the low data rates (typically 9600bps) that constitutes the main limitation. *FieldNote* transfers often involve sending and receiving only a few hundred bytes, but long connection times mean that any transfer takes a minimum of about 70 seconds.

Since the SIBA2000 campaign, several telecom networks have begun to roll-out a GPRS service. Whilst still using the basic GSM technology, this provides a fully digital connection, similar to that of ISDN systems, albeit at a much lower data rate. The mobile phone can maintain an 'always-on' IP connection to a network server and charging is by data volume rather than usage time. Initial experiments with such a system in the UK suggest that this may become a viable communications medium for future field campaigns. Over the next few years, further developments are scheduled. The next major advance will be the so-called 'third generation' networks which will offer far higher data rates (up to 2Mbps) and the possibility of live multimedia and video links.

In parallel with infrastructure developments, a convergence of mobile phone and PDA technologies is under way. Eventually, the question of whether to add mobile communications to our field tools may become irrelevant as these will be part of the normal function of a PDA.

3.3 User and network interfaces

Extending the functionality of the PDA in all the directions suggested above will require some rethinking of its user interfaces, which will have to allow full configuration of task and display options and efficient ad hoc switching between tasks and displays. As more intelligent use of the system will require the simultaneous display of more different types of information and the availability of more options on a limited screen size, the design of intuitive and effective user interfaces will become essential.

A complex multi-component system will also require extensive configuration. Here we envisage that the design phase of a typical fieldwork project would include the preparation of configuration scripts which are uploaded onto each individual kit. An on-screen menu will then give access, firstly, to the list of available configurations (tasks such as 'create new grid' or 'relocate features'), and secondly, to a list of the configurable variables for that task (e.g., 'set grid size' or 'set alert distance'). The configuration data can also specify which options will be available as on-screen buttons and/or as voice actuated options. Given that the system already contains components for handling data exchange in XML format, and that XML techniques for managing software configuration are becoming widespread elsewhere (see, for example, Austin *et al.* this volume), it is likely that this approach will be used to provide these configuration scripts.

At present, we see on-line communications between field crews as something of a luxury, but it has the potential to significantly improve the reliability of the system. If newly recorded data is rapidly mirrored on a remote server, we can overcome the ever-present fear of losses arising from device failures in the field. There may also be benefits in making preliminary environmental and archaeological maps available to crews as they are produced. As the costs of use and features offered by the mobile networks become more favourable, we anticipate exploring ways in which such connectivity might be exploited, particularly as the field teams are already carrying mobile phones for voice communications.

Work is also under way to provide a mechanism for automatically generating a web interface to notes, maps and photographs recorded in the field. This has the obvious benefit of making the field data accessible in a widely used form but, with on-line field communications, it would enable all participants in the field, at the campaign base, and at their home institutions to have direct, near real-time, access to the progress of the campaign. This opens up the possibility of 'remote' specialists taking part in the field work, for example providing determinations of enigmatic finds.

Further development of the desktop data management tools developed for the current and earlier versions of the *FieldNote* system is required. At present, these tools support conversion to and from appropriate GIS data formats and simple display and editing of data collected in the field. A revised version of the desktop component should include facilities for managing the configuration scripts discussed above. We envisage that responsibility for the dFA will become part of the Data Manager's task, and that procedures for the acquisition and distribution of data on dFA's will need to be fully integrated with the broader survey design.

4 Conclusions

The experiments described here have confirmed the potential of the dFA system for both speeding up field recording procedures and reducing the number and size of errors made during the recording process. The system's potential for easing navigation and the sharing of information during surveys has not been fully explored, but our experience in (re-)locating archaeological sites mapped in the 1960's indicates that it will also prove useful in that area.

With the limited enhancements to functionality discussed in section 3.1, and the further improvements to the standard accuracy discussed in section 2.4, the system can profitably be used in any type of archaeological survey. With full technical and procedural integration of a professional version of the kit into fieldwork methodology, along the lines suggested in sections 3.2 and 3.3, dFAs will begin to transform fieldwork practise. This will require further extensive testing of system components, software, and field procedures.

In recent years, the use of professional GPS surveying equipment in archaeological fieldwork has become much more popular (cf. De Wulf *et al.* 2000a,b), and some teams are adapting commercially available products in order to obtain GIS-like capabilities in combination with GPS (Johnson & Wilson n.d.). These high-powered approaches, while providing very high accuracy and versatility, require considerable expense and highly skilled personnel, and cannot provide a true field information system4. The advantages of the digital Field Assistant system described here over such alternative approaches can be summarised as follows:

- Inexpensive - it is possible to fully equip a fifteen to twenty person team for the price of one typical professional survey kit;
- Immediate feedback - the data collected in the field are made available for use straightaway in a process contributing to and enhancing the available pre-loaded data, rather than being taken away for later processing and use. They can also be made context-aware, presenting themselves actively rather than passively as in field GIS;
- Portability and unobtrusiveness - the equipment weighs very little, will fit in your trouser or vest pockets, and requires very little training. It is designed to avoid distracting the user from the tasks at hand;
- Utility - the equipment, as envisaged, performs typical and frequently occurring archaeological fieldwork tasks such as setting out grids, mapping collection units and points of interest, and recording finds information.

With respect to the GPS component of the system, the availability of a good location device is a crucial feature in the recent shift in emphasis of archaeological survey work away from the well-mapped and well-controlled coastal zones of Italy, to the more rugged and less well-mapped inland zones. For archaeological applications where the accuracy requirements are higher than what can be achieved with a single receiver, the addition of a GPS base station for differential correction would give the most satisfactory results. Post-processing, of course, would not offer any improvement in real-time positioning in the field but so far we have not identified any reason why this should be a high priority. Should it become necessary, corrections could be broadcast from the base station and received at the rovers by using conventional wireless-modems.

References

Attema, P. A. J., G.-J. Burgers, M. Kleibrink & D. G. Yntema, 1998. Case studies in indigenous developments in early Italian centralization and urbanization: a Dutch perspective, *European Journal of Archaeology* 1(3):326-381.

De Wulf, A. et al., 2000a. GPS Surveying Techniques in Archaeology: Topographical Survey of the Thorikos Region (Greece), in Vermeulen, F. & M. de Dapper (eds), *Geoarchaeology of the landscapes of classical antiquity* (Babesch Suppl. 5):189-196.

De Wulf, A. et al., 2000b. Analysis of the Efficiency in Archaeology of GPS Satellite Surveying versus Classical Surveying using Totalstations: Applications in the Thorikos Region and on the Pyrgari (Greece), in Vermeulen, F. & M. de Dapper (eds), *Geoarchaeology of the landscapes of classical antiquity* (Babesch Suppl. 5):197-208.

Johnson, I. & A. Wilson, n.d. *Making The Most Of Maps: Field Survey On The Island Of Kythera.*

Kleibrink, M., 1993. Religious activities on the "Timpone della Motta", Francavilla Marittima, and the identification of Lagaria, *BABesch* 68:1-49.

Kleibrink, M., 2000. Early cults in the Athenaion at Francavilla Marittima as Evidence for a pre-Colonial Circulation of nostoi Stories, in Gassner, V. et al. (eds), *Die Ägäis und das westliche Mittelmeer; Beziehungen und Wechselwirkungen 8. Bis 5. Jh. V. Chr.* (Archäologische Forschungen 4): 165-192. Wien: Österreichische Akademie der Wissenschaften.

De Rossi, G. M., L. Pala, L. Quilici & S. Quilici-Gigli, 1969. *Carta Archeologica della piana di Sibari*. Roma: Società Magna Grecia.

Pascoe, J., D. R. Morse, & N. S. Ryan, 1998. Developing personal technology for the field, *Personal Technologies* 2:28-36, August 1998.

Peroni, R. & F. Trucco, 1994. *Enotri e Micenei nella Sibaritide*. Tarento.

Ryan, N. S., J. Pascoe & D. R. Morse, 1998. Enhanced Reality Fieldwork: the Context-aware Archaeological Assisstant, in Dingwall, L. et al. (eds), *Computer Applications in Archaeology 1997* (BAR Int Ser 751):269-274. Archaeopress, Oxford.

Ryan, N. S., J. Pascoe & D. R. Morse, 1999a. FieldNote: extending a GIS into the field, in Barcelo, J. A., I. Briz & A. Vila (eds), *New Techniques for Old Times: Computer Applications in Archaeology, 1998, Proceedings of the Barcelona Conference, March 1998* (BAR Int Ser 755):. *Oxford:* Archaeopress.

Ryan, N. S., J Pascoe & D. R. Morse, 1999b. FieldNote: a Handheld Information System for the Field, in Laurini, R. (ed), *Proc. TeleGeo'99, 1st International Workshop on TeleGeoProcessing*, Lyon.

Van Leusen, P. M., in prep. *The SIBA2000 survey campaign: preliminary report.*

Van Leusen, P. M., 1998. The Viroconium Cornoviorum Atlas: high resolution, high precision non-invasive mapping of a Roman civitas capital in Britain, *European Journal of Archaeology* 2(3): 313-325.

End notes

1 In practice, we have not achieved the full platform independence potentially offered by Java. The WindowsCE version of FieldNote used in this campaign was in fact written in *waba* (see www.wabasoft.com), a Java-like language with its own libraries and independently developed virtual machine. Despite these differences, a high proportion of the code is common to this and the pure Java versions.

2 For these operations, we used PC Arc/Info; all other processing took place under GRASS 4.1.

3 This was noted, for example, when geophysical survey grids were very accurately positioned with the help of differential GPS at the buried Roman town of Wroxeter (Van Leusen 1998).

4 ESRI offer the field mapping system ArcPad (which runs under Windows CE); Trimble offer the Pro XRS system with Asset Surveyor and Pathfinder Office software for use on notebook computers. This will view and edit pre-loaded maps and interface to ArcView for subsequent query, analysis and presentation. Total cost for this system, used by the Australian Paliochora-Kythera Project APKAS to map survey units and landscape/archaeological features and to upload and correct the data in ArcView at the end of each day, is $12K. De Wulf et al. (2000a,b) investigated the use of survey quality GPS in an archaeological topographical survey of the Thorikos region in Greece, concluding that it is more than twice as efficient as a total station (EDM) survey but also twice as expensive at the highest accuracy class; moreover, the processing, quality control, and interpretation of GPS measurements required a higher degree of skill than was needed with a total station.

Figures

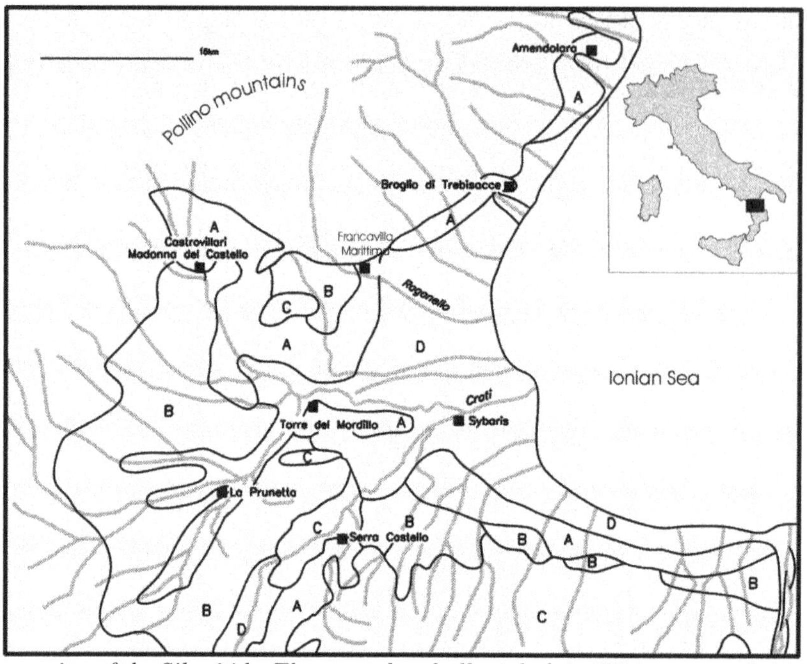

Figure 1: Major landscape units of the Sibaritide. The coastal and alluvial plain (D) is surrounded by a series of marine and fluvial terraces (A) which merge into the lower slopes of the Pollino and Sila ranges (B).

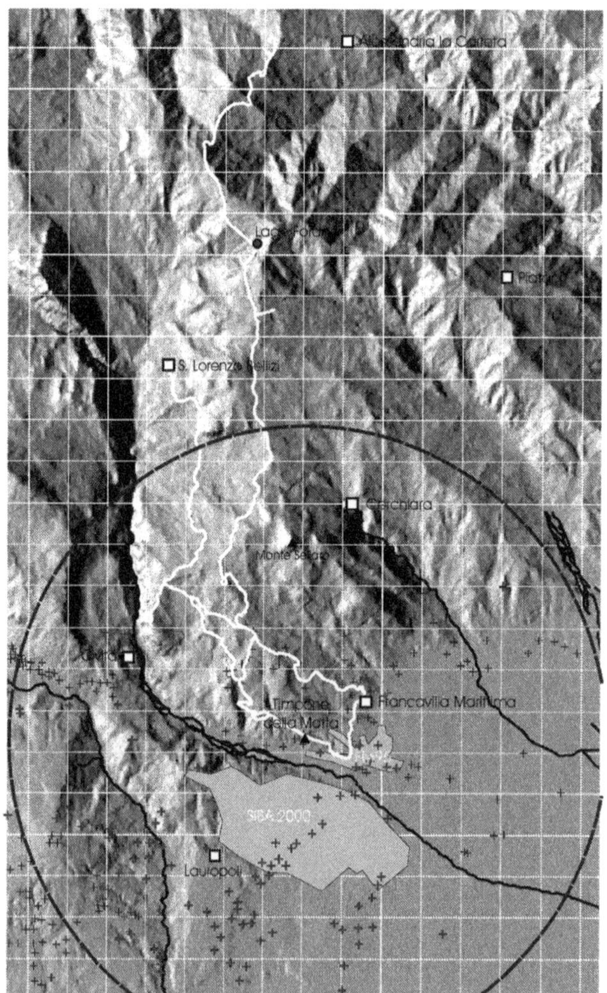

Figure 2: The SIBA2000 survey zone centres on the Late Iron Age to Hellenistic hilltop sanctuary at the Timpone della Motta de Francavilla Marittima, under excavation by the Groningen Institute of Archaeology since 1991. The locations of the survey areas, GPS tracks of highland survey, and Quilici sites are indicated. Grid size: 1 km.

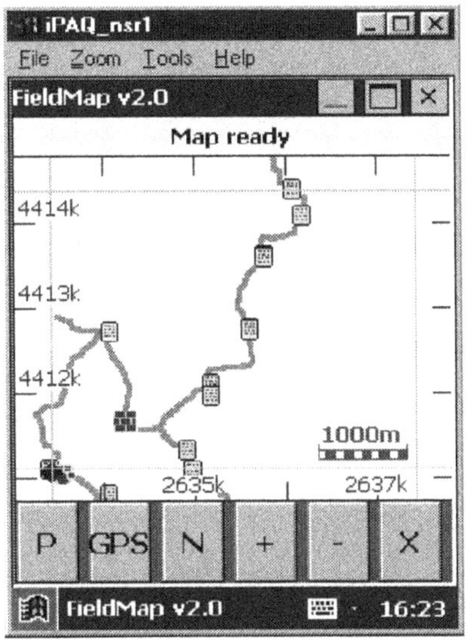

Figure 3: PDA window with GPS trace and observation points for exploratory highland survey along a transhumance route.

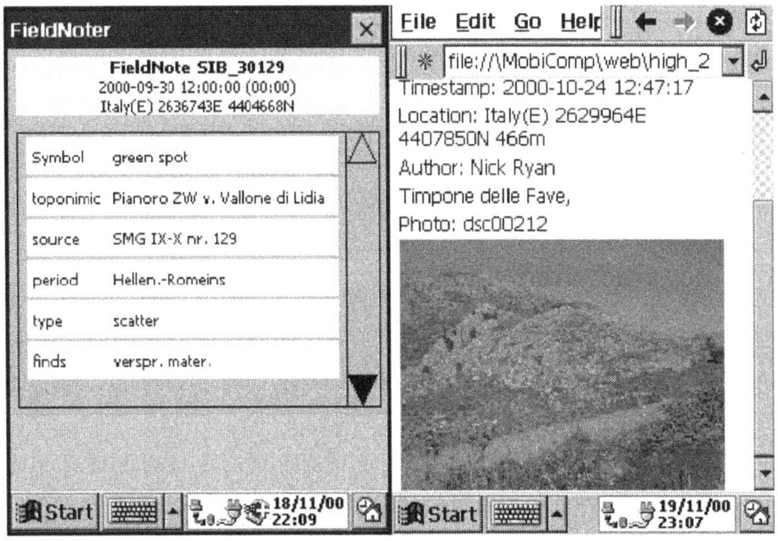

Figure 4: Examples of archaeological FieldNotes:
(a) popup note for Quilici record no. 129; (b) a georeferenced photographic note.

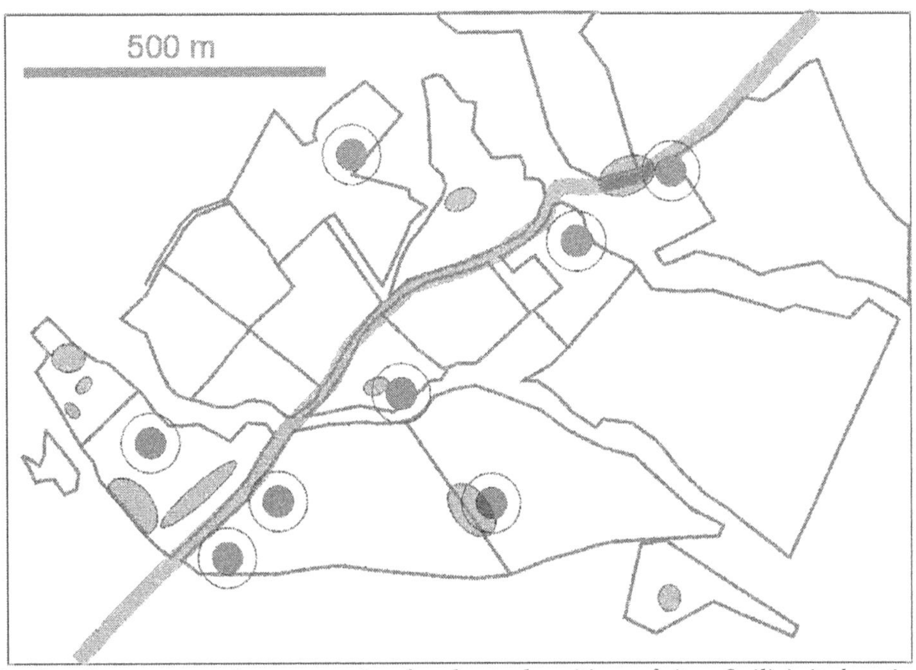

Figure 5: Detail, showing disagreements between notional and actual positions of sites. Quilici site locations as derived from De Rossi et al. 1969 are indicated with 25m and 50m radius; approximate shape and location of RPC surface scatters is indicated by grey ellipses.

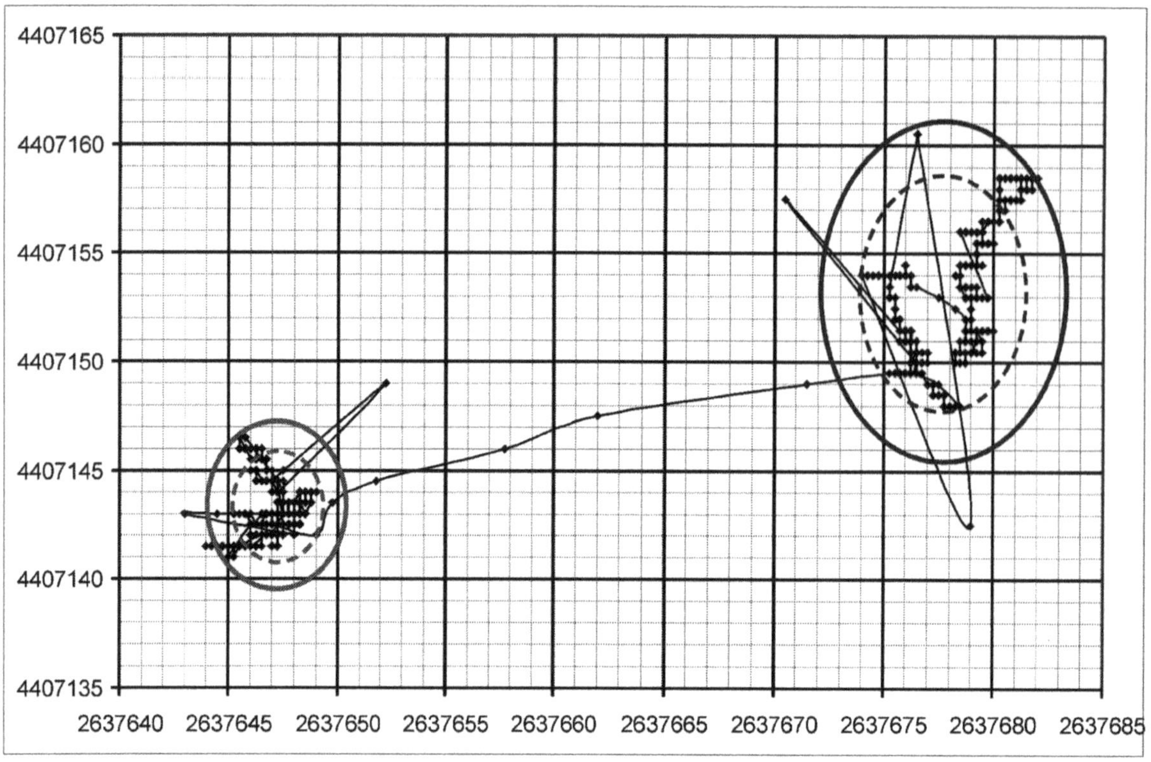

Figure 6: GPS measurements of the two ends of the Timpone della Motta site museum balcony, illustrating the accuracy obtainable with a single receiver (Trimble SK8). Ellipses indicate 1 and 2 SD of location. Interval: 10 seconds, duration: 6 hours.

From Field Books to Powerbook
Computer Applications and the Promontory Palace of Herod the Great

James G. Schryver
Cyprus American Archaeological Research Institute
11 Andreas Demetriou Street
Nicosia 1066, CYPRUS
Tel: +357 2 67-0832 - Fax: +357 2 67-1147 - E-mail: mdvlarchaeologist@hotmail.com

Abstract: Since 1976, the Hebrew University of Jerusalem and University of Pennsylvania Museum's joint excavation of over fifty trenches has revealed the promontory palace built by Herod the Great at Caesarea Maritima, Israel. Foremost among the many tasks that now face the excavators is a site-wide analysis of the stratigraphy. During the author's study of the destruction of the palace, the analytical and visualization capabilities available through traditional stratigraphic analysis incorporating Harris Matrices and CAD plans of excavated loci proved limiting because of the size of the data set. To address these limitations, a GIS combining the most desirable quantities from both approaches was created. This has allowed for the creation of an intra-site database, as well as the examination and visualization of stratigraphical relationships in three dimensions. The project is still underway, but this paper discusses the process that brought the excavation team to the present point and the difficulties that were encountered along the way.

Keywords: Caesarea, Promontory Palace, Herod, GIS, intra-site analysis, 3D, Harris Matrix, stratigraphy

Introduction

The accurate recreation of excavated stratigraphy in three dimensions continues to be an important goal for archaeologists. The possibilities for utilizing a GIS to accomplish this goal have been previously discussed by Trevor Harris and Gary Lock (Harris and Lock, 1996). Among the different solutions proposed and attempted by others since then are those incorporating various 3-D modeling programs, such as the project currently being undertaken as part of the Brown University excavations at the Great Temple, Petra, Jordan (Vote, et al., 2001). Recent work on the creation of an analytical tool for the University of Pennsylvania's excavations at Herod the Great's Promontory Palace in Caesarea Maritima has shown that another possible solution is to combine a site-wide Harris Matrix with an intra-site GIS using the 3-D capabilities of ArcView's 3D Analyst extensions.

The promontory palace at Caesarea Maritima, Israel, was built in two stages in the first century BC. The first, the Lower Palace at sea level, was excavated in 1976 by Ehud Netzer and Lee I. Levine of Hebrew University, Jerusalem (Levine and Netzer, 1986). They revealed the remains of a palatial complex surrounding a rock-cut pool (fig. 1). In 1990, Kathryn Gleason and Barbara Burrell of the University of Pennsylvania Museum joined Netzer to resume excavations on the top of the promontory, where they discovered an attached Upper Palace (Gleason and Burrell, 1998) (fig. 2). In 1992, Burrell and Gleason assumed the excavation license and carried out annual excavations until 1996. In 1997, they began a series of annual study seasons that have continued until the present.

By 2000, work done on the promontory palace had produced a very large data set. Analysis of this data set was complicated by the fact that the 1993-1996 excavations had been carried out under "rescue" conditions in advance of the development of major new tourism facilities within this portion of the Caesarea National Park. There were also records that had been made under less than ideal conditions, such as a series of photographs and sectional drawings of the long baulks left standing at the end of the 1994 season just before they were bulldozed for development.

In most respects, however, the excavation continued as it had before the 1993-1996 seasons. For example, one procedure that forms a basis for this study is the "overlay" system, in which each member of the staff recorded in plan view each locus or stratigraphic unit excavated in their trench. In many cases these plan views are complete with elevations. As part of the plans for publication, these drawings, photographs, and overlays needed to be correlated with the trench stratigraphy.

Objectives

One of the main objectives of this project has been an analysis of the relationships both among the individual trenches and between these trenches and the site as a whole using the sources discussed above. The analysis of each of these pieces of data has presented unique challenges requiring specific solutions. The photographs of the long baulks taken at the end of the 1994 season show continuous stratigraphy across the majority of the site. However, no one excavator had dug all of the trenches along a full length of baulk and not all of the baulks had been drawn along any particular transect. To prepare the excavation for publication, the author was asked to develop a site-wide Harris Matrix. Along with defining the exact limits of the different strata found on the site, such as the destruction layers being studied by the author, the Harris Matrix was to provide the excavators with the means to confirm or disprove ideas that had been forming throughout the years concerning the phasing of the architecture. For example, was the destruction of the palace complex was wholesale, or piece by piece?

Process

To produce this Harris Matrix, a windows-based Harris Matrix program, ArchEd, was employed. Unlike many flow-chart design programs, ArchEd allows the user to add as much information as needed for each excavation unit or locus, such as information concerning shape, elevation, contents and date, without necessarily affecting the initial appearance of the rectangle or oval representing that stratum. Instead, by clicking on the appropriate locus, information about elevation, soil type, content, and date can all be added in a separate window that appears on the screen (fig. 3). This feature allowed the incorporation of all relevant data into each particular locus. This data, organized by locus and in its stratigraphic context, could then be shared with all of the other excavators working on the analysis of different aspects of the excavations.

The first step taken was the creation of separate matrices for each of the individual trenches excavated over the six field seasons of the Penn Project (1990-1996). Information from the excavators' field books and locus sheets such as that described above was added to each locus using the inherent capabilities of the ArchEd software. At this scale, the program proved completely adequate.

The next step was to combine these individual matrices with neighboring trenches linking similar loci into continuous strata across the site. This proved especially feasible in the upper strata of the excavated area that were consistent across the whole site. Once all of the neighboring trenches had been combined, the goal was to create one large, site-wide Harris Matrix. This method was at first attempted when all of the loci from the Lower Palace excavations had been entered into the computer. Unfortunately, this was the point at which the software proved somewhat limiting. In order to combine matrices, for example two separate matrices from neighboring trenches sharing the same stratigraphy, first one file and then the next had to be exported from ArchEd to a text-based program such as Notepad. Once this was done, the information from one Harris Matrix in text form could be copied and pasted into that of the other file. The new combined text file was then re-imported into ArchEd where the two matrices appeared side by side. They could then be joined manually at the appropriate loci. Unfortunately, when combining matrices, not all of the information attached to a particular strata or locus concerning date or contents was exported with all of the other information concerning the placement of that particular locus within the actual Harris Matrix. In addition, exporting and re-importing the files erased any changes made to the dimensions of the shape representing any particular locus. As a result, any data associated with individual loci that the user wanted to have within the larger matrix had to be recopied each time that the matrix was enlarged in this manner.

This problem may prove to be a stumbling block for some projects, but is in no way insurmountable if one has enough patience and manpower to cut and paste the lost information into the new diagram and to double-check the relationships established between these loci. If one is willing to work with a skeleton Harris Matrix, serving only as a visual aid, these issues likewise cause no problems. However, since part of the excavators' intention was to use the Harris Matrix as a repository and source for information about each locus, the information was needed. It is for this same reason that simply re-drawing an increasingly larger Harris Matrix was not a viable option.

Another difficulty encountered was that of the correlation of the loci within the Harris Matrix with their physical location on the site. As more and more loci were entered into the computer, the number of components in the Harris Matrix naturally increased. As the site-wide matrix grew, it became increasingly difficult for users to relate different loci with their actual in-situ location through the Harris Matrix alone. This proved true even though one can include information such as location, soil type, and/or date into the label of the locus to avoid the need for continual clicking back and forth within ArchEd. Still, the team eventually reached the point where we thought it would be really useful to be able to see the location of these loci on-site as we examined the information being provided by the Harris Matrix, especially since there were already so many different loci to account for, and this number would only continue to grow as more loci were added.

To resituate the loci onto the site, the plan views or overlays of the different loci that had been drawn during excavation were added onto the existing AutoCAD site map. To handle the large number of layers within AutoCAD that this process required, the site was divided into a number of areas and each of the plan views was drawn separately onto the appropriate one. Separate files could then be associated with one another to get a view of the stratigraphy over specific areas of the palace. In addition to allowing users to view the loci in their original locations within the site, this regrouping facilitated interpretation of the architecture.

Analysis

Initially, the AutoCAD program proved sufficient to run the sorts of analyses that we were interested in for the lower palace. Once all of the loci for the lower portion of the palace had been entered, the plans could be cross-referenced with the information on the Harris Matrix. The analysis we were most interested in at the time involved the *terminus post quem* for each of the loci covering various walls and therefore indicating the destruction or cessation of different rooms of the Lower Palace complex. To do this, information concerning the dates of individual loci from the Harris Matrix covering the lower palace was used to select which layers to leave on and which to turn off. After all of the loci with the selected date had been turned on and all others turned off, the resulting image could then be exported from CAD and another date could be selected. This was done for loci dating to the 1-2 centuries AD, the 2-3 centuries AD, the 3-4 centuries AD, and so on, based on the pottery readings.

To work with the information stored in the Harris Matrix, it was necessary to print hard copies of the Harris Matrices. One subsidiary issue then became the display of the date of each locus on these hard copies. The dates were indicated in one of two ways: 1) printing out the matrices as they were and hand-writing the dates next to them; or 2) adding the date of each locus to the locus' label and then printing out the matrices.

In the end, all of the CAD images that had been exported were arranged in chronological order in a simple slide-show presentation to show the chronological progression of the strata across the Lower Palace. Viewing the strata in this way allowed the excavators to confirm some of their earlier suspicions. As mentioned, because of the nature of the dig, the excavators had assumed that something interesting was going on here, but until we had performed this analysis, the patterns were somewhat speculative.

At this point, the next difficulty was recognized. Once the other five years worth of excavation data representing the loci covering the Upper Palace (eastern area of the site) was entered, this method of analysis, that is, selectively turning different CAD layers on and off depending on the criteria selected for would become more complicated, not to mention time consuming, considering the thousand or so loci excavated. At the same time, it became clear that by turning these different AutoCAD layers on and off in the way that we had done, we were essentially mocking a GIS analysis. So, we decided to move ahead with a site-based GIS.

The GIS and its Results

To begin, the drawings and the data for the loci covering the lower palace were imported into ArcView. The data from the Harris Matrix for each locus was then added to the appropriate field in a theme table. The result is an intra-site database in which each locus is considered an individual entity with specific identifiable characteristics. Analyses can be performed by querying these characteristics. This type of analysis provides much quicker results than the previously mentioned CAD method. In addition, by utilizing ArcView's 3D Analyst extension, we not only have the option of seeing these results in plan view, but in 3-dimensional space as well. This can be done by creating a 3D-shape file from the 2D-shape file resulting from any particular query, and assigning the elevation of each strata as the attribute from which the z value is selected (figs 4-5). By doing the same to the architectural elements and then extruding them by subtracting their elevation from the maximum z value, we can create solid architectural blocks and see their relationship to the loci in 3-dimensional space (figs 6-7). Should it be desired, the depth of individual loci can also be seen in 3D by entering the depth into the appropriate theme table as the loci's thickness and then extruding the loci by subtracting the thickness from the maximum z value. In cases of complex stratigraphy, relationships can also be exaggerated vertically to create a clearer picture. These results can in turn be compared with specific Harris Matrices to clear up stratigraphical relationships on a site-wide level. This is especially important for our goal of examining architectural phases and determining when walls/rooms were covered and out of use. Once more data is entered, we will be able to recreate the surfaces of the different strata and see how they relate to the architecture across the site.

Conclusions

Although not part of the original research design, an intra-site GIS is proving to be an important analytical tool for stratigraphical analysis at Caesarea Maritima. The results are preliminary, but already suggest a pattern of inhabitation and abandonment in the Lower Palace that is somewhat different from the original interpretation of the excavators (Burrell, Gleason, Netzer 1993). The distribution of loci seen in the GIS images suggests that the northern area of the building had a longer, more complex history than the southern portion. This pattern needs further evaluation before a final interpretation can be made, but clearly the process is giving us the kind of complexity and specificity needed at this stage of the project. GIS has proven to be an exciting and effective tool for intra-site and stratigraphical analysis.

Acknowledgements

The author would like to thank Drs Kathryn L Gleason and Barbara Burrell for permission to work on this material, as well as their advice on this paper; and Dr Bruce Ralston, Joseph D Nigro, and Daniel L Costura for their continued help along the way. This paper was also made possible by generous travel grants from both the Cornell University Graduate School and the Conference Organizers.

References

Burrell, Barbara, K.L. Gleason, and E. Netzer, 1993. Uncovering Herod's Seaside Palace. *British Archaeological Reports.* 19:50-57, 76.

Gleason, Kathryn L. and Barbara Burrell, 1998. The promontory palace at Caesarea Maritima:preliminary evidence for Herod's Praetorium. *Journal of Roman Archaeology.* 11: 23-52.

Harris, Trevor M. and Gary R. Lock, 1996. Multi-dimensional GIS: exploratory approaches to spatial and temporal relationships within archaeological stratigraphy. In H. Kamermans and K. Fennema (eds.), *Interfacing the Past. Computer Applications and Quantitative Methods in Archaeology, 1995:* vol. II, 307-316. University of Leiden (Analecta Praehistorica Leidensia 28).

Levine, Lee I. and Ehud Netzer, 1986. Excavations at Caesarea Maritima, 1975, 1976, 1979: Final Report. *Qedem* 21. Jerusalem (Institute of Archaeology).

Vote, Eileen L., et al., 2001. ARCHAVE: A Virtual Environment for Archaeological Research. In Z. Stancic and T. Vejanovski (eds.), *Computing Archaeology for Understanding the Past. Computer Applications and Quantitative Methods in Archaeology, 2000*: 313-316. British Archaeological Reports (International Series 931).

Figures

Figure 1. Lower Palace, Caesarea Maritima, Israel

Figure 2. Upper Palace, Caesarea Maritima, Israel

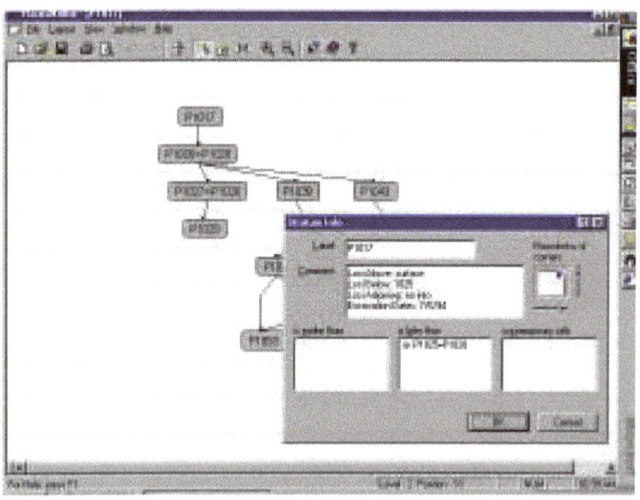

Figure 3. Harris Matrix software

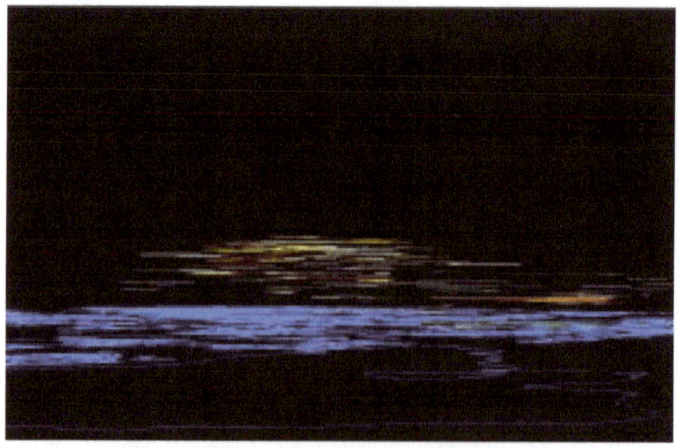

Figure 4. 1990-1992 excavated loci in citu

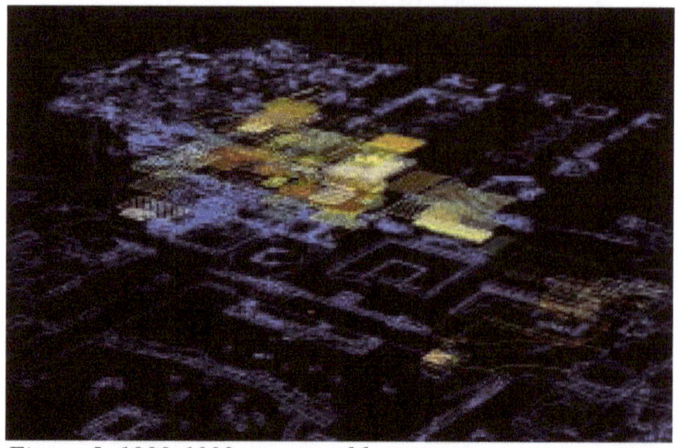

Figure 5. 1990-1992 excavated loci in situ

Figure 6. 1990-1992 excavated loci in situ with

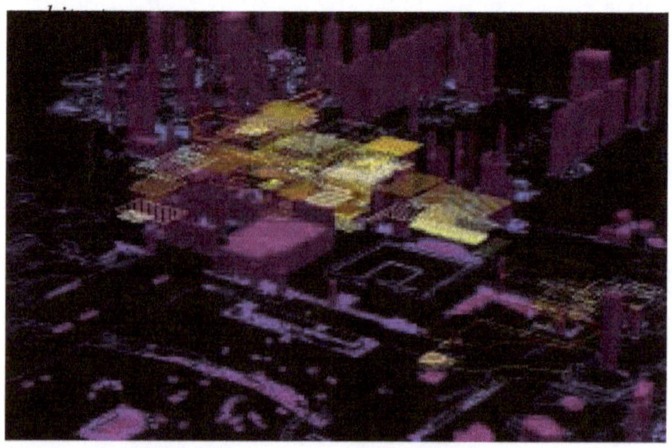

Figure 7. 1990-1992 excavated loci in situ with architecture

Errors & Inaccuracies in Repositioning of Archaeological Sites

S. Topouzi
Laboratory of Geophysical - Satellite Remote Sensing & Archaeo-environment
Institute of Mediterranean Studies – Foundation of Research & Technology (F.O.R.T.H.)
Melissinou & Nik. Foka 130, PO. Box 119
GR-74100 Rethymnon, Crete, Greece
Phone: +30831-56627 - Fax: +30831-25810 - E-mail: topouzi@students.phl.uoc.gr or topouzi@yahoo.gr

A. Tripolitsiotis
Laboratory of Geodesy and Geomatics Engineering
Department of Mineral Resources Engineering
Technical University of Crete
GR-73100 Chania, Crete, Greece
Phone: +30821-37429 - Fax: +30821-37476 - E-mail: atripol@mred.tuc.gr or atripol@yahoo.com

A. Sarris
Laboratory of Geophysical - Satellite Remote Sensing & Archaeo-environment
Institute of Mediterranean Studies – Foundation of Research & Technology (F.O.R.T.H.)
Melissinou & Nik. Foka 130, PO. Box 119
GR- 74100 Rethymnon, Crete, Greece
Phone: +30831-56627 - Fax: +30831-25810 - E-mail: asaris@ret.forthnet.gr

S. Mertikas
Laboratory of Geodesy and Geomatics
Department of Mineral Resources Engineering
Technical University of Crete
GR-73100 Chania, Crete, greece
Phone: +30821-37429 - Fax: +30821-37476 - E-mail: mertikas@mred.tuc.gr

S. Soetens
Laboratory of Geophysical - Satellite Remote Sensing & Archaeo-environment
Institute of Mediterranean Studies – Foundation of Research & Technology (F.O.R.T.H.)
Melissinou & Nik. Foka 130, PO. Box 119
GR-74100 Rethymnon, Crete, Greece
Phone: +30831-56627 - Fax: +30831-25810 - E-mail: ssoetens@yahoo.com.

Abstract: In archaeological surveys and excavations, positioning and mapping of sites and monuments is essential to studies related to settlement pattern, viewshed analysis as well as cultural resource management. Various methods and techniques have been used in the past for the mapping of sites according to the technological means that were available at each time. These methods had a variable degree of accuracy, creating a number of problems in the repositioning of archaeological sites today.

The errors and inaccuracies of the past mapping techniques are discussed and they are compared with the latest developments of the GPS technology. The various GPS geodetic surveying techniques employed today are here addressed, along with a statistical analysis of the position data, as a function of time and distance.

Problems related to the transformation of geodetic coordinates from one reference system to another and the geometric registration of aerial and satellite imagery are addressed, as well as the effects of error propagation in the analysis of viewshed or topographic slope.

Key words : GPS, differential processing, GIS, viewshed analysis.

Introduction

Even before the beginning of the first archaeological excavations, travellers designed and recorded on maps the remnants of the visible ancient monuments, which they met in their route. The purpose was to provide the prospective researchers a way to identify and reposition the archaeological sites. The methods employed in the past were able to record the approximate location of the sites in a very relative way. Problems in accuracy and consistency, both in terms of coordinates and map scale, are obvious today.

Though positioning and mapping of sites was always regarded as a necessity, there are not many things that have changed in the corresponding approach of the archaeologists until the very recent years. The methods used were more or less similar to those of the travellers : a compass, a pencil and a paper map.

Later on the use of theodolite upgraded archaeological mapping and cartography. A theodolite provides coordinates of a site by measuring horizontal and vertical angles from a base station. Of similar use is the electromagnetic distance measurement (EDM), invented in the mid-20th century. The latter made it possible to measure distances as accurately and easily as theodolite, by electronically timing the passage of radiation over the distance to be measured. Although the above techniques provided advantages in local scale mapping of sites and monuments, problems arose in large-scale extensive mapping of sites that require a visual contact between the observation stations and the recording of coordinates in local reference datums.

Traditional surveying techniques commonly use a total geodetic station with a theodolite, an Electronic Distance Measuring (EDM) device and several reflectors to determine monument coordinates. The reflectors are mounted on rods while intervisibility between the surveying instrument and the reflectors should always be guaranteed. Errors of determined coordinates are in that case proportional to the separation distance and hence the larger the spacing, the larger the surveying error.

Although the use of theodolites and EDMs is rather easy and inexpensive, these instruments have only occasionally been used to record and map survey sites (MacDonald – Rapp 1972, Whitley 1999). The more common method of recording sites in the course of a surface survey is by sketching the surveyed areas on 1:5000 scale maps or on aerial photos and then adding spots to represent the sites that have been found. Information about the size, the location and the surrounding area is usually recorded in the survey notebooks (Macready – Tompson 1985; Wright et al. 1985, 1990; Fish – Kowalewski 1990; Alcock 1991; Hayden et al. 1992; Watrous 1993; Jameson – Runnels – van Andel 1994; Haggis 1996; Branigan 1998).

Due to the subjectivity of the descriptions of various researchers and the recent changes of the environmental conditions (both by natural or anthropogenic causes), repositioning of sites based on a survey published report has been proven a difficult, if not impossible, task. On the other hand the accurate relocation and mapping of sites has become an essential module in archaeological research, whether it is related to topographic studies or settlement pattern analysis and cultural resource management.

An alternative methodology, which includes satellite-based positioning technology is offered by the Global Positioning System (GPS) (Federal Radionavigation Plan, 1999).

The Global Positioning System

The most up-to-date way of specifying the horizontal coordinates of a site, so as its elevation, is by using the Global Positioning System (GPS).

GPS is a satellite-based technology developed by the U.S. Department of Defense. It provides accurate, continuous, three-dimensional position and velocity information to users with appropriate receivers. GPS also disseminates a form of Coordinated Universal Time (UTC).

The satellite constellation consists of 24 satellites in 6 orbital planes, with 4 satellites in each plane. Each satellite transmits ranging codes and navigation data on two frequencies (L1 = 1575.42 MHz and L2 = 1227.6 MHz). The basic concept of the GPS estimation of coordinates is the measurement of the distance of a point on the surface of the Earth from four or more satellites, using a receiver.

In any case, the constellation of the satellites is planned in a way to provide at least four satellites in good geometric position 24 hours per day at any position on earth. The available satellites over a location can even reach the number of ten.

Using a network of five worldwide monitoring stations, navigation data are uploaded to satellites. Thus, a user can accurately determine the position, velocity and time of his receiver, by measuring the one-way time of arrival (TOA) to the satellites. On July 17, 1995, the US Department of Defense announced the full operational capability of the GPS system.

Observation Techniques

Until some time ago, recording with a single GPS receiver provided an accuracy of the order of plus or minus 100m, because it only had the ability to use one frequency. After the recent release of the second frequency, it is estimated that measurements will have an even better accuracy, even with only one receiver. Still, some applications demand even a sub-centimetre accuracy. In that case, the satellite positioning techniques require at least two receivers to measure the GPS satellite signals simultaneously. A reference station is located at a fixed point, within the area of interest, while the second receiver is set on another point (roving station), whose coordinates are to be established. For post processing applications, the actual observations from the reference station are provided to the roving station. The positions of the roving station are then computed through special data reduction software. These GPS surveying techniques are briefly described below. Figure 1 presents the concept of modern surveying techniques (Hofmann-Wellenhof, et al., 1994).

Relative Static Positioning is the classical GPS surveying technique, where the reference, as well as the secondary, stations are set up over static points, to measure satellite signals for long observation sessions (ranging from 30 min to several hours to days). The length of each observation session is a function of the measured length of the baseline between the two stations, the number of the satellites being tracked and the satellite geometry. Static positioning determines the three-dimensional vector from the base to the secondary station and provides the most accurate and reliable results, in the order of millimeters and sub-millimeters (fig. 2).

To measure as many baselines as possible, in the briefest observation sessions (i.e., minutes or even seconds) and to obtain results while the secondary station is in motion, several modern GPS surveying techniques have been developed recently. These are the Rapid Static, the Reoccupation, the Stop&Go and the Kinematic positioning techniques. All these techniques require special hardware and software, as well as specific field procedures.

The Rapid Static technique requires shorter observation periods than the classical static GPS surveying, typically in the order of 5 min to 20 min, depending on the baseline length and the number of satellites being tracked. For example, for a baseline of less than 5 km and tracking 6 or more satellites, the required observation time is about 5 min with accuracies in the order of one centimeter.

The Reoccupation positioning technique is faster than the classical static one and requires observation sessions of a few minutes at each secondary station, but collected about one or more hours apart. That means that, while the base station remains stationary, the roving receiver must reoccupy the same secondary point one or more hours later to allow for changes in satellite geometry, in order to achieve centimeter accuracies. It should be noted that the roving receiver could be switched off between revisiting sessions. In this case, no special hardware is needed but only appropriate software for data post-processing.

The Stop&Go surveying is a semi-kinematic technique. The roving receiver has to stop over stationary points to collect a few minutes of data in order to determine their coordinates. In addition, it continues to move while tracking GPS satellites from one site to the next. The trajectory of the roving antenna is not of interest here but only the stationary points. Special initialization field procedures are necessary before a Stop&Go survey begins and the technique may not allow signal disruptions, while moving from site to site (i.e., going under a bridge or thick foliage). Accuracies of 1 to 2 centimeters can be obtained with a Stop&Go procedure.

True kinematic GPS procedures attaining accuracies of centimeters for the moving antenna have been under development recently. To avoid re-initialization of field procedures, as required in the Stop&Go method, several sophisticated software programs (e.g., on-the-fly ambiguity resolution techniques) are being developed to secure and account for GPS signal interruptions that may occur during motion.

Real-time kinematic (RTK) is another GPS procedure, where positioning accuracies of the order of centimeters can be achieved in real time. It requires, however, a radio communication link (such as VHF, UHF etc.) to transmit GPS correction data, from the reference receiver to the roving receivers (fig. 3).

Advantages and disadvantages in the use of GPS for archaeological sites positioning

Total geodetic stations composed of mechanical, optical and electronic components have not undergone the dramatic price reduction that the computer market is experiencing these days. The average cost of a total geodetic station never fell below the range of US $10,000 to $20,000. On the other hand, GPS receivers, composed mainly of electronic parts, have experienced a tremendous price reduction similar to that of the computer market; starting with a price of $100,000 in 1980, and falling down to the $10,000-$15,000 range in 2001, for basic models measuring phase. It should be noted, however, that for accurate relative position determination, a pair of GPS receivers is needed, thus doubling the cost of the GPS instrumentation.

GPS technology, however, has several advantages over conventional terrestrial techniques in terms of instrument cost, ease of operation and productivity. It can also be used easily as a data capture tool for Geographic Information Systems (GIS) because the data provided are digital.

Further, this GPS technology has the following advantages over conventional terrestrial surveying methods:

- It requires no intervisibility between GPS stations,
- The coordinates are recorded in a global reference system (the World Geodetic System '84 – WGS'84), which is independent of local reference datums,
- All GPS operations are weather independent and can work around the clock,
- The positioning is carried out accurately and quickly,
- GPS has the ability to measure simultaneously the three components of coordinate vector (i.e., horizontal position and height) of any point on the surface of the earth.

Nevertheless, GPS surveying has disadvantages over the conventional terrestrial techniques:

- The receivers can not be used in a subterranean survey (such as tombs, caves, tunnels etc.) or under trees, because there is not visual contact with the satellites,
- Frequently the results, being in a global reference system (WGS '84), have to be transformed in the local datums, an act that can often cause transformation inaccuracies,
- The GPS elevations are not orthometric, which means that they are not measured from the surface of the sea,
- The cost of the receivers is high, for a moderate user,
- Training and experience is required in order to use the receivers.

GPS surveying – Methodology

During the last four years, the Laboratory of Geophysical – Satellite Remote Sensing and Archaeo-environment of the Institute of Mediterranean Studies – F.O.R.T.H. has been involved in a number of GIS projects, involving GPS mapping of archaeological sites. In one of the projects, dealing with the construction of a digital archaeological registry of Lasithi, in East Crete, more than 200 sites were positioned and mapped through the use of GPS techniques. A number of ground control points (GCPs) were also recorded to be used for the geometric rectification and registration of aerial and satellite images.

After retrieving information (from the published bibliography related to the cultural resources of the region), a total of 972 sites were registered in a database, which contained, among other, information related to the general location of the sites, their type and chronology, environmental data etc. The next step was to position the sites with accurate coordinates in order to map and correlate them with the rest cartographic material, such as digitized maps and satellite images. Two state-of-the-art geodetic receivers (Ashtech Z-12 double frequency receivers) were deployed to relocate the archaeological sites.

Still finding the sites was not always an easy task. Relaying on the published plans of the survey reports, some sites were difficult o impossible to be found again. That fact made once again clear the need of accurate positioning of the sites.

For those sites found, data were collected using the static differential positioning, in order to achieve a higher degree of accuracy. The base station was connected to a Choke-ring antenna and to both a battery and the electric power, so as to prevent loss of data in case of power failure, and took measurements every 30 seconds for the most part of the day, while the second receiver was moving at the sites of interest.

The second receiver (Rover) was used in the field, connected to a geodetic antenna (usually placed on a tripod) and to a battery. The receiver recorded for a period of approximately 15 minutes, locating 4 to 10 satellites each time. For each point recorded, the proper parameters were inserted in the device, while the signal was stored with a different distinctive 4-digit code name.

The selection of points to be recorded, varied according to the case. In areas where the archaeological remains were visible, an effort was made to describe the perimeter of the site, by recording several points of it (fig.4). In cases where topographic plans of the site existed, distinctive points of the plans were recorded and assigned on them. In sites with undefined boundaries, a couple of points were recorded in order to define the size of them. One measurement was recorded for tombs, in the entrance of caves or at sites with indefinite evidence of archeological remains (but with confirmed testimonies of existence).

Besides the archaeological sites, a number of ground control points (GCPs) were established for georeference and registration of the images to the national coordinate system. Thus, the points were chosen in such places so they would be visible in both air photos and satellite images.

Processing of the data was conducted daily, in order to keep track of the accuracy and the measurement quality. Every measurement was processed and corrected according to the data recorded by the base, so as to obtain the minimum percentage of error. The coordinates of the base were corrected in the beginning of each survey, taking 1 hour measurements with the static positioning technique and by using the known coordinates of a close-by army datum point.

The coordinates of the sites were recorded in both the WGS'84 and the EGSA'87 reference systems. While WGS'84 is the reference system used from the GPS, EGSA'87 is the system used from the National Cadastre of Greece. Processing was followed by exporting the coordinates in spreadsheets (excel format) in order to import them in the database of the Geographic Information System.

On average the accuracy of position determination was found to be less than 10 centimetres. The accuracy of the GPS data was tested by checking misclosures of datum points of known coordinates, established by the Greek military.

Statistical processing – Errors and suggestions

The recorded GPS points (over 350) were subjected to statistical processing in order to determine the relation among the distance between the rover receiver and the base station, the recording time and the deviation in the accuracy of estimating the coordinates (x, y, z) of a site.

After sorting the deviation values for x (Easting), y (Northing) and z (Altitude), the following accuracy and distance classes were created :

Accuracy in mm	0-5	5,1-10	10,1-15	15,1-20	20,1 <	
Distance in m	0-2000	2000,1-5000	5000,1-10000	10000,1-15000	15000,1-20000	20000,1 <

The percentage of points assigned to each accuracy class was appointed to the corresponding distance class. The results follow in figures 5, 6 and 7.

Obviously the accuracy, when recording within a distance of 0 to 2000m from the base station, is quite impressive in all axes, reaching the percentage of 95 to 100. Furthermore it remains in good levels as to the Easting and Northing until the distance of 20000m, a fact that proves the precision of the GPS equipment. The relatively higher error in altitude is due to the known defect of the GPS in positioning heights within sort periods of observation, a fact that can be outweighed when assigning to the measurements the correct geoid.

When the factor of time was considered together with distance, accuracy was divided in smaller classes : 0 – 2, 2.1 – 4, 4.1 – 6, 6.1 – 10, 10.1 – 15, 15 – 40 (60 for Easting and 100 for Altitude). Figure 8 represents the error in accuracy as a function of the distance in two ways : by graduation of shape size and by different colors.

The large concentration of GPS points is within the distance of

5000 to 15000m from the base, with an observation time of 1000 to 2000 sec. The accuracies are kept in good levels. The few exceptions are due to the small number of satellites during the recording time or their poor satellite-receiver geometry, meaning that they are not equally spaced in the horizon, which causes difficulty in calculating the pseudoranges. When plotting the deviation as function of time (fig.9) and as function of distance (fig.10), the accuracy levels remain unchanged. Still there is a sufficient good accuracy when recording in a close distance, no matter how sort the observation time is. When one records in a large distance, the error in measurements is substantially decreased by increasing the observation time.

The use of GPS measurements in the geometric registration of spatial data

GPS coordinates of sites can be used in a number of applications, one of which is the geometric registration of aerial and satellite imagery. The procedure is simple : First a number of visible on the images points are selected in order to be used as ground control points – GCPs. Images are georeferenced by assigning the actual geographic coordinates to the ground control points (on the images).

However, before deciding to use the GPS technology for georeferencing satellite imagery, the image resolution must be examined first. In order to register a Landsat image with a pixel size of 30´30m, it is not necessary to contact a GPS survey with an accuracy of 1cm. A good result can be obtained by using digitised map data at a scale of 1:5000. On the other hand, the geometric registration of a 1´1m ICONOS image implies the need of high accuracy GPS coordinates of a better than 0.5m order.

GPS coordinates can also be used in a number of GIS analyses, such as viewshed or slope, in a way that they are accurately positioned on a digital elevation model. What seems to be the problem is the error that appears in the GPS positioning of Altitude. However, for measurements in a small distance from the base station, such an error is systematic. This can be somewhat levelled by allowing a specific shift of all elevation measurements towards the actual elevation values, after comparing them with the known elevation values of the army datum points.

Conclusions

Within the era of cultural resource management and heritage protection, positioning a site with accurate coordinates is essential. The use of the Global Positioning System has made it possible in international level, since it records in a common reference system, the WGS '84.

Being a relatively accessible technology, it enables the researcher to choose from a variety of equipment, according to the available financial resources and a number of observation techniques, one of which will provide the desired accuracy in the available time.

Judging from the statistical results it is suggested that recording within a distance of 15000m from the base station, when using the differential positioning, provides more accurate coordinates. The observation period must be decided upon the number of available satellites and their geometry, the distance from the base and the interval of recording. Thus the archeological sites can be mapped and repositioned from the later generations.

References

Alcock S., 1991. Urban survey and the polis of Phlius, *Hesperia 60 (1991)* : 421 – 463.

Branigan K., 1998. Prehistoric and early historic settlement in the Ziros Region, Eastern Crete, *BSA 93 (1998)* : 23 – 90.

Peter H. Dana, 1999. *Global Positioning System Overview, The Geographer's Craft Project*, in http://www.colorado.edu/geography/gcraft/notes/gps/gps_f.html
Department of Geography, The University of Texas at Austin.

Federal Radionavigation Plan, 1999. Federal Radionavigation Plan, Final Report, US Department of Transportation and Department of Defense, *NTIS No.: DOD-4650.5/DOT-VNTSC-RSPA-98-1.*

Fish S.K. and Kowalewski A. (eds), 1990. *The Archeology of Regions, A case for full-coverage survey.*

Haggis D.C., 1996. Archeological Survey at Kavousi : A preliminary report, *Hesperia 65 (1996)*: 373 – 433.

Hayden B., Moody J. and Rackham O., 1992. The Vrokastro Survey Project, 1986 – 1989 Research design and preliminary results, *Hesperia 61 (1992)*: 294 – 349.

Hoffmann-Wellenhof B., Lichtenegger H. and Collins J., 1992. *Global Positioning System, Theory and Practice*, second edition, Wien and New York.

Jameson M.H., Runnels C.N. and van Andel T.H., 1994. *A Greek Countryside : the southern Argolid from prehistory to the present day*, Stanford.

Macready S. and Thompson F.H. (eds), 1985. *Archaeological field survey in Britain and abroad*, London.

McDonald W.A. and Rapp G.R. Jr. (eds), 1972. *The Minnesota Messenia Expedition, Reconstructing a bronze age regional environment*, Minnesota.

Sarris A., 2001. Digital Archaeological Map of Lasithi, E.Crete, Greece, *Archaeo-Telepiskopika Nea 4, Rethymno.*

Watrous L.V. et al., 1993. A survey of the western Mesara plain in Crete, *Hesperia 62 (1993)*: 192 – 198.

Wright J. et al., 1985. *To ereunitiko archaiologiko programma stin koilada tis Nemeas kata ta eti 1984 – 1985*, "' 18 (1985) : 86 – 104.

Wright et al., 1990. The Nemea Valley Archaeological Project : A preliminary report, *Hesperia 59 (1990)*: 579 – 645.

Whitley J. et al., 1999. Praisos IV: A preliminary report on the 1993 and 1994 survey seasons, *BSA 94 (1999)*: 205 – 247.

Figures

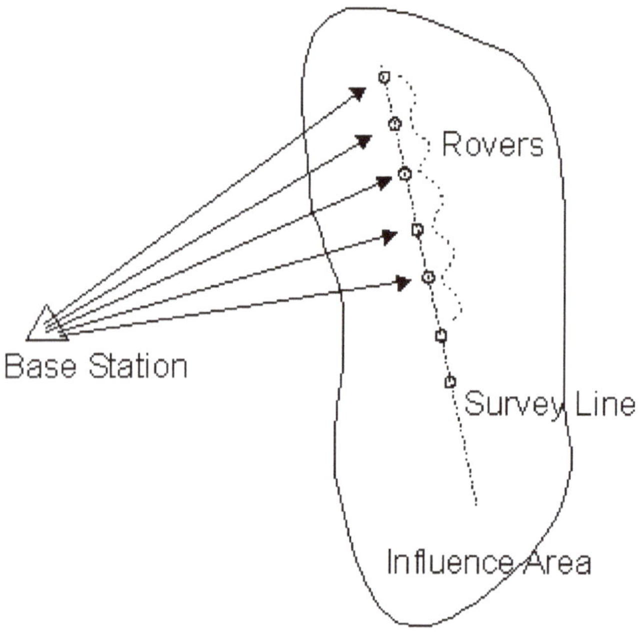

Figure 1. Modern GPS surveying techniques

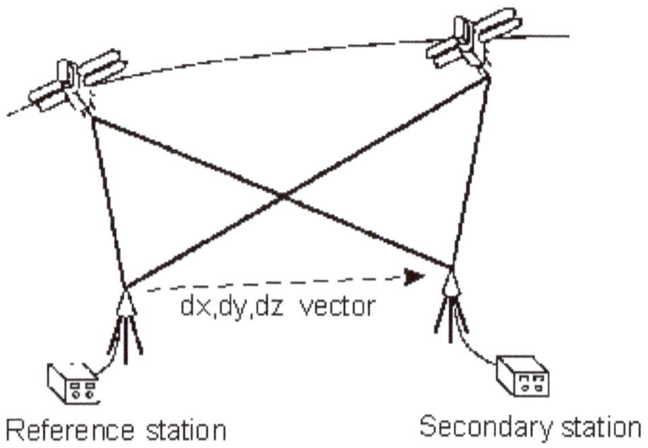

Figure 2. Relative static positioning

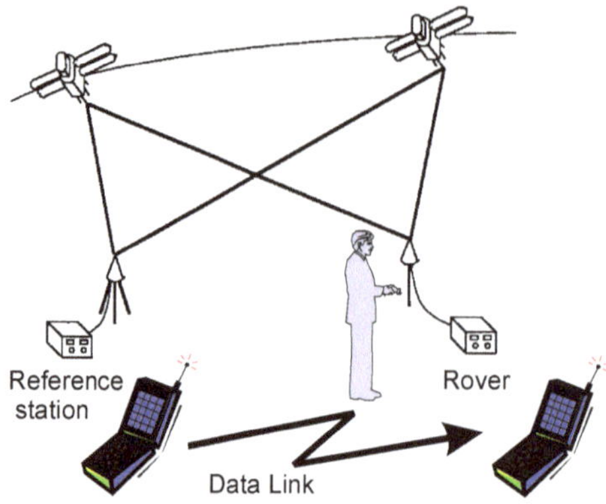

Figure 3. Real Time Kinematic GPS

Figure 4. Simultaneous mapping with GPS and EDM at Chalinomouri, E.Crete

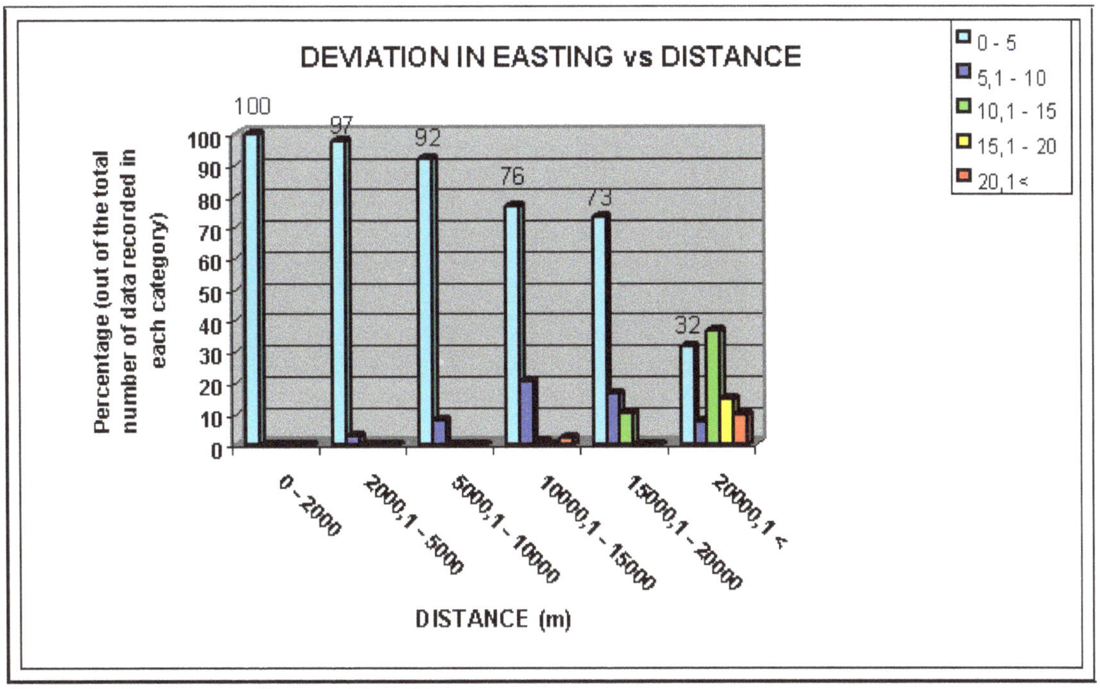

Figure 5. Deviation in Easting (x) vs Distance

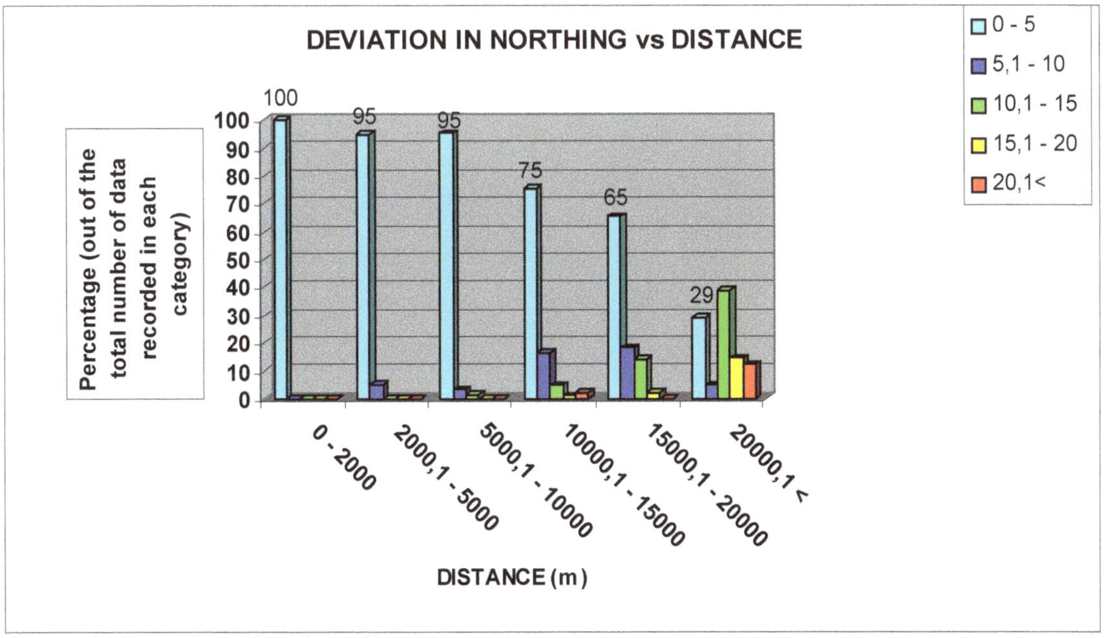

Figure 6. Deviation in Northing (y) vs Distance

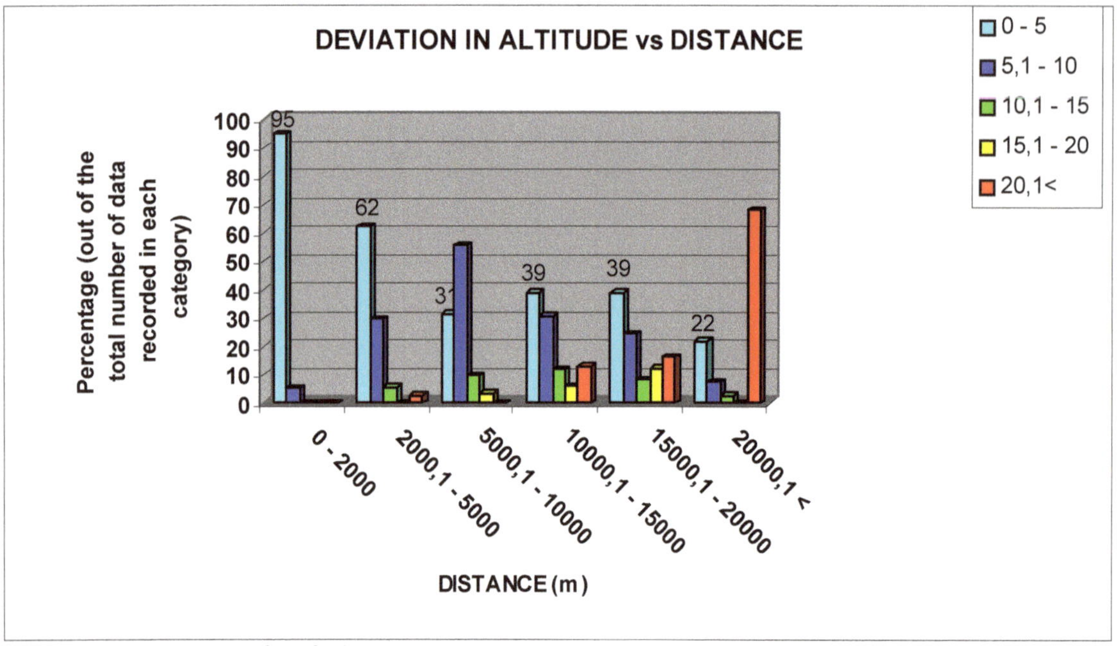

Figure 7. Deviation in Altitude (z) vs Distance

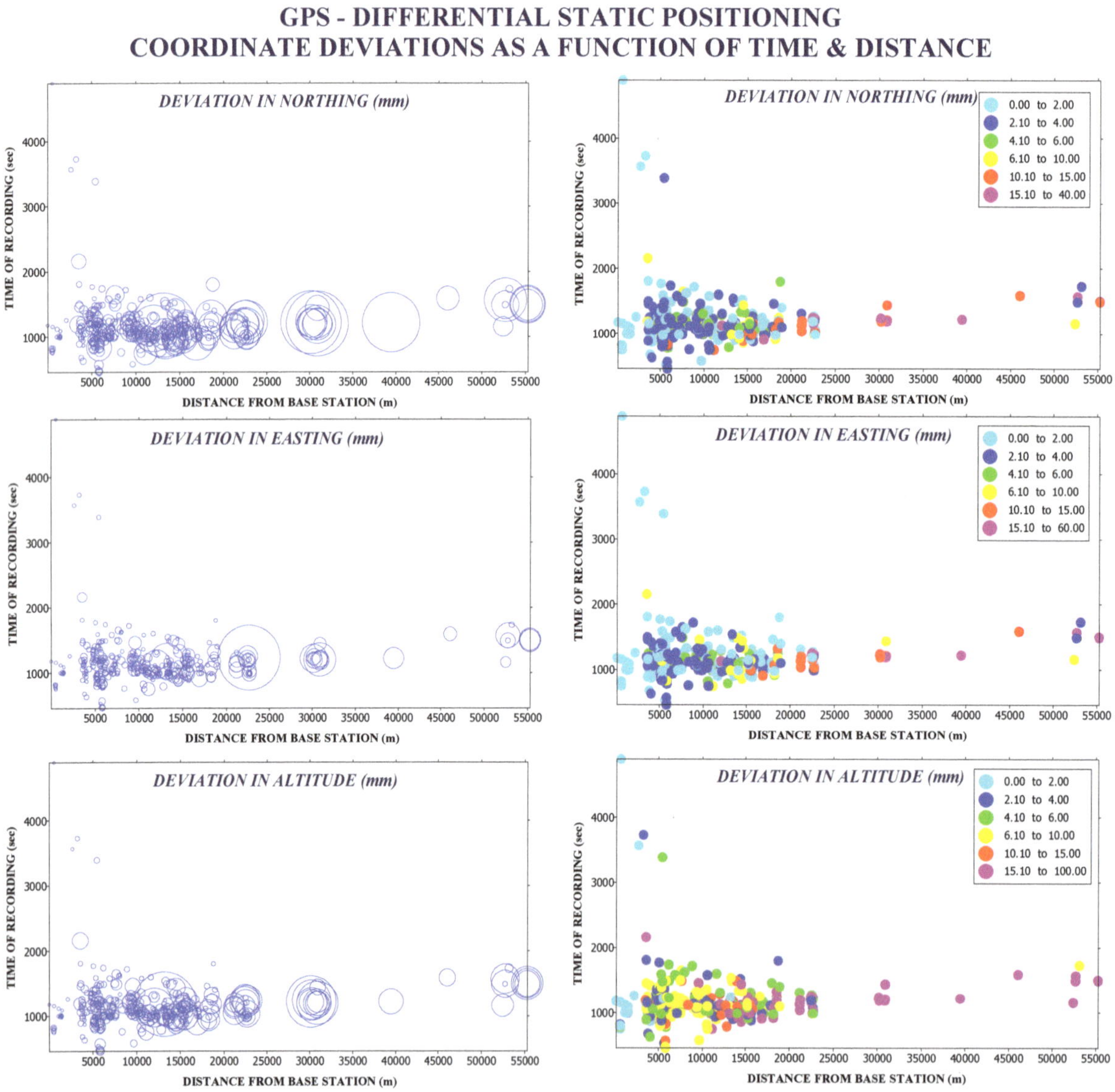

Figure 8. Correlation of observation time, distance and deviation in the three dimensions. Representation of error in graduation of shape size (left) and color (right).

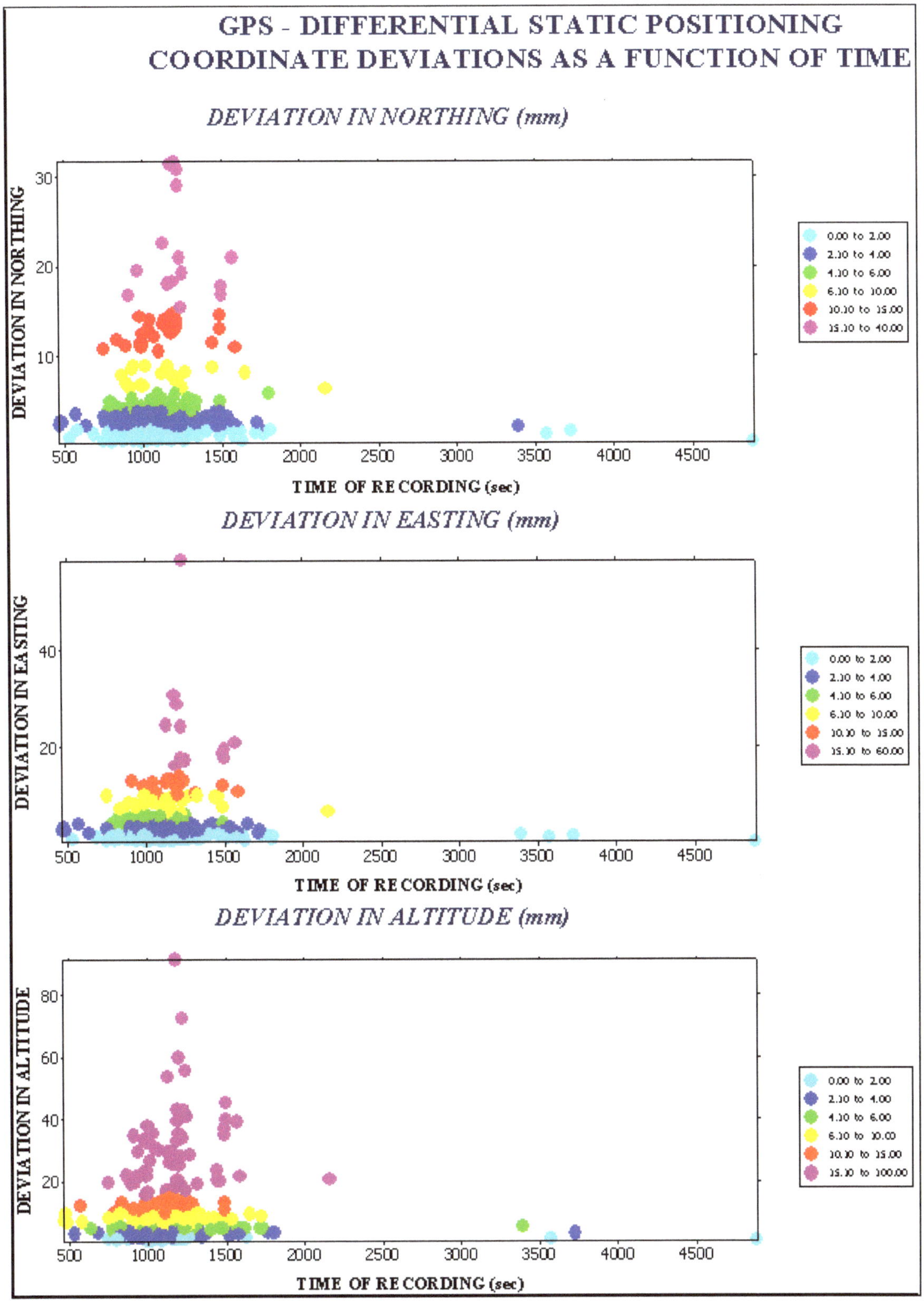

Figure 9 : Representation of the deviations in the three dimensions as a function of time

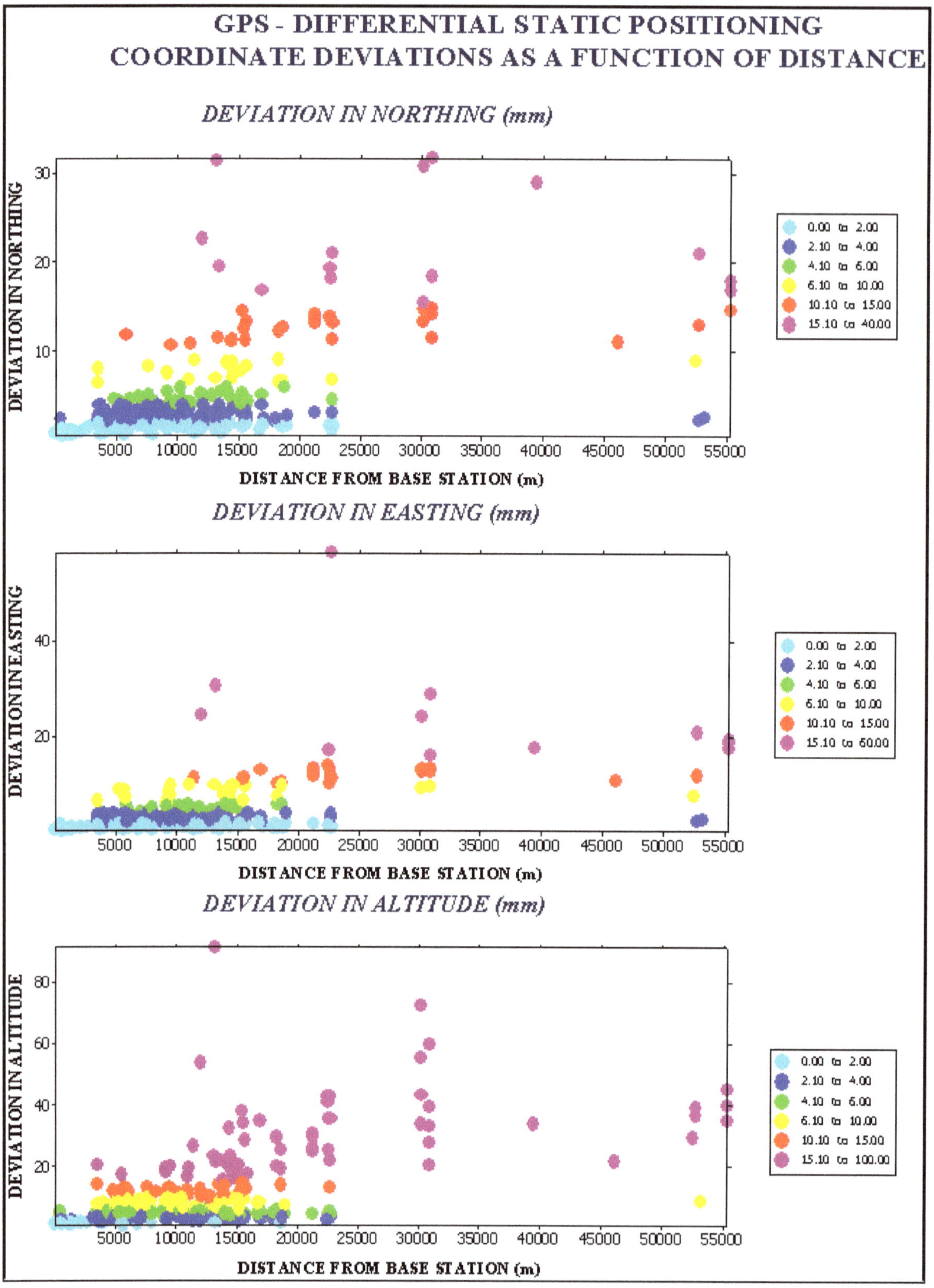

Figure 10. Representation of the deviations in the three dimensions as a function of distance

Three-dimensional excavation plans and 3D Studio Max

Experiences from the excavations of the medieval town of Naantali, Finland

Kari Uotila
University of Turku
20014 Turku. Finland

Carita Tulkki
Muuritutkimus ky.
Suovillankatu 3
20780 Kaarina. Finland

Abstract: During fieldwork all structures and soil layers were documented according to the matrix and a drawing was made straight from a total station to a laptop. The archaeologists used a Finnish land survey program called 3Dwin.

The data is imported in DXF format from the land survey program into AutoCAD, where the code of the third field moves straight into different layers. This means that you can treat the measured data by units and combine, for example, wooden and stone structures or different groups of finds. The basic work on the research material is carried out in AutoCAD and the finished excavation plans are printed from there. Additionally, the chosen objects to be modeled can be imported, for example, in DXF format into 3D Studio MAX.

A quick model of the example case has been created in 3D Studio MAX, in which you can adjust the visibility of different structures and soil layers. This means that you can easily examine the different units and the relations between them.

The stratigraphy of different layers can also be easily modeled and presented with the help of 3D Studio MAX. In this picture, the soil layers and structures of the example case are seen from several different directions.

Introduction

In Finland archaeologists began collecting digitized research material during excavations in the 1990s. On prehistoric excavation sites – such as Stone Age excavations – finds such as flakes and potsherds were recorded as points. On historical sites, however, finds are mainly recovered in large quantities and documented in bulk, and therefore measuring different features as lines or points became the focal point. One of the first wider research surveys in Finland was the measuring of a palisade in the Episcopal castle of Kuusisto, where over 3 000 pine posts in large palisade in one-metre-deep water were mapped during 1992-93. Mapping the most important features as 3D polyline was also used in, for instance, the research of the Aboa Vetus -museum (In Turku) in 1993-94 and in the excavations of the Bridgettine Convent Church in Naantali in 1996-97.

Excavations in Naantali in late autumn 2000

In October 2000, archaeologists unexpectedly had to document a trench dug in the street for sewerage work in the small medieval town of Naantali, located near Turku on the southwestern coast of Finland. The town of Naantali grew next to a Bridgittine monastery in the 1440s and remained a small coastal trading town throughout the centuries. Today Naantali is one of the most popular travel destinations in the country and tourists know the town especially from its Moominworld. (fig 1.)

Naantali's well-preserved wooden town quarter and its plan date from the 19th century, and there is very little information about the earlier town. There had previously only been one larger excavation in the area of the old town in the 1980s, so the sewerage works offered a new opportunity to study the history of Naantali.

In Finland, the light and temperature conditions in October and November are not the most favourable for archaeological excavations. Documentation had to be carried out so that archaeologists supervised the digging. Monitoring the soil layers removed by machine, the digging was only stopped in order to record clear features. It was a salvage excavation, where archaeologists tried to recover the most important archaeological data.

Right from the beginning of the excavations, the chosen method for documentation was digital measuring carried out with total station, which the archaeological excavation company Muuritutkimus ky has specialized in already for a few years. New to Finland at least, all new significant features (both stone and wooden) were measured digitally and that all observed soil layers were measured as stratigraphic units. Each unit was given its own context number according to the principles of the Harris matrix and the bounding edges of each unit were measured three-dimensionally. Additionally, during excavation the positions of all observed artefact finds and different kinds of soil samples were measured with the total station prior to being collected. In this way, a completely three-dimensional digitized research material was created already in the field.

Documentation was carried out during the excavations in such a manner that all significant features and natural soil layers were numbered according to the matrix and a drawing was made straight from the total station to a computer, and thus it could be viewed instantly on a laptop. The archaeologists also had an A3-sized printer that was used for printing site plans and sections as needed and small details were added to them – such as the details of wooden structures, small stones and fragments of brick. When making the final drawings, these hand-drawn additions were added to the AutoCAD drawings as raster images by setting them at the right level in the measured layers and features.

Handling the measured data

The measured data is transferred from the total station to a computer with the help of a Finnish land survey program (3Dwin), which is used in Finland both by private companies as well as the state and the municipalities. A common program was chosen in order to have functioning relations between all parties carrying out detailed surveys. The idea is that using a common program offers good opportunities for archaeologists to cooperate with, for instance, the town of Naantali. The town officials can deliver the needed measured data in the right format for fieldwork and the archaeological measurements can be transferred directly to the use of the town planning organization even during the same day. Additionally, having as many users of the program as possible guarantees that there will be technical help and cooperation partners available around the country. (fig. 2.)

The digitized data is transferred from the survey program either to a database program (such as MS Access) or to an actual map drawing program, which in our case is AutoCAD. Using an accurate engineering program for drawing has its origin at the beginning of the 1990s, when the cooperation contacts of archaeologists were directed towards the CAD world and contacts with universities and their GIS programs were few. It is difficult to take a stand on the everlasting question of which one is the better program, but at least in our experience AutoCAD is fast and accurate in composing a functioning set of archaeological drawings – at least in 2D form. We have had experience of this from different research sites for almost ten years. (fig 3.)

Documenting the measurements functions so that we normally use four different code fields in the program during fieldwork, to which you can input different types of data about the context being measured. When needed, the program can use altogether six code fields. The most essential factor of the code fields is that the code of each field is transferred into its own layer in AutoCAD. This has the benefit of being able to immediately separate the units measured with the same code from the other layers – in other words, the measured data is arranged in exactly the same way as the person who was drawing planned it in the field. Each unit stands out as its own and each artefact find can be positioned accurately.(figs 2-3)

Making a plan three-dimensional

The measured data is transferred in 3D form from the land survey program to AutoCAD, where the data can, in principle, be viewed three-dimensionally, but distinguishing the different lines and features is very difficult and certainly not very visual. With developing technology, 3D visualization of an excavation area has become a new opportunity, which in my opinion opens up completely new ways to interpret the sometimes complex combinations of units and wide excavation areas. However, modeling illustrative excavation sites has been restricted mainly by the fact that 3D mesh models use a great deal of computer resources.

In our case, the tool used for visualization in different kinds of multimedia and virtual reality projects has for several years already been the 3D Studio MAX program, which now runs in its fourth version. The "Terrain" command, which was a new feature introduced in version 3, has made it easier to model simple and sometimes more complex 3D meshes. With the help of the program you can also model the bottom surface of a pit-like layer, as long as it has been measured during excavation.(figs 4-6)

The carefully documented data from the excavations at Naantali can, thus, first be transferred to the land survey program, and from there to AutoCAD for the actual drawing, after which the ready units and features can be imported into 3D Studio MAX where a 3D model of the excavation area can be created. It is true that you also have to modify and model the data at each stage of this process, but in principle the excavation site is as accurate in 3D Studio MAX as it was during the measuring stage in the field. (figs 4-6)

In 3D Studio MAX, the colour and material of each stratigraphic unit can be specified, as well as, for example, their transparency. It is relatively easy to visualize relationships between layers and features, and stratigraphy can also be easily presented for instance with the help of animation, as all of those who were present at the CAA seminar in Visby could see.

The time when any archaeologist can create a 3D model of an excavation area in a simple, functional and inexpensive manner is not quite yet at hand, because the price of the required computer equipment and programs is perhaps some 10 000 euros, but the equipment clearly has developed and new opportunities are on the horizon.

As regards the excavations at Naantali, 3D modeling of the data has so far been mainly used in visualization and in testing and developing measuring equipment, but as the excavations continue in 2002, a collection of all the data from the researched area will be completed. We can use this information to make observations that are important for research and will perhaps be able to outline the nature of the whole medieval town and the village settlement that preceded it.

Figures

Figure 1. In the Middle Ages there were altogether six towns in Finland, the most important one being the Cathedral city of Turku. Naantali was the last town to be founded during the Middle Ages and it grew next to a Bridgettine Convent founded in 1443. The population of the town in the Middle Ages was perhaps some 300-400 people, so it was a fairly small town in Finland's circumstances. Naantali has retained its old town plan and wooden buildings, which mainly date already from the beginning of the 19th century. (Drawing K. Uotila)

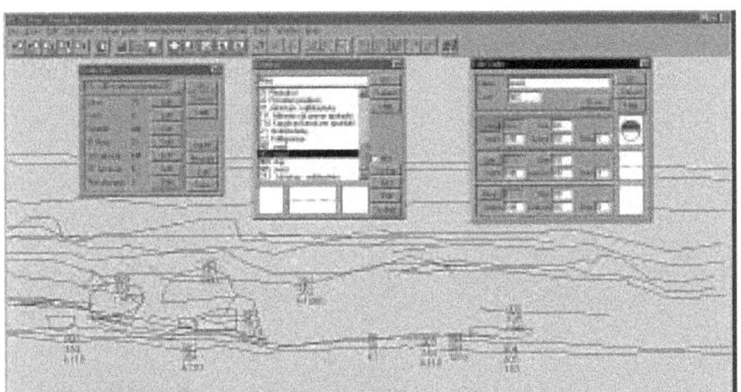

Figure 2. During fieldwork, all structures and soil layers were documented according to the matrix and a drawing was made straight from a total station to a laptop. The archaeologists used a Finnish land survey program called 3Dwin. In this example, soil layers from the eastern section of the excavation can be seen. (Drawing P. Hakanpää & C. Tulkki)

Point info – in this example there are four fields in use in the code section of the window. The first of them (904 in the picture) documents the type of soil, the second gives the number of the measured line (305 in the picture), the third has the context number (110) and the fourth gives the number of the measured point (273 in the picture). Additionally, it is possible to collect different kinds of data into two open fields in the program.

In this example, the program is showing the soil type (in the middle of the picture e.g. 901), the measured line (352) and the context number of the points in the section. In the program you can choose which of the measured data you wish to display for study.

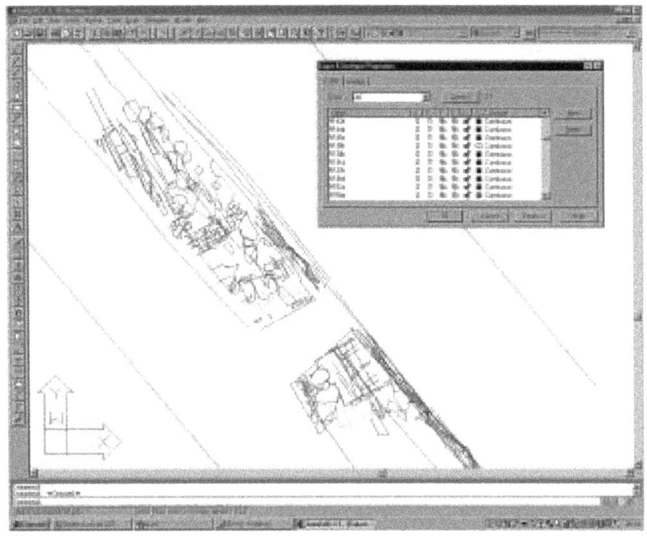

Figure 3. The data is imported in DXF format from the land survey program into AutoCAD, where the code of the third field moves straight into different layers. This means that you can treat the measured data by units and combine, for example, wooden and stone structures or different groups of finds. The basic work on the research material is carried out in AutoCAD and the finished excavation plans are printed from there. Additionally, the chosen objects to be modeled can be imported, for example, in DXF format into 3D Studio MAX. (Drawing P. Hakanpää)

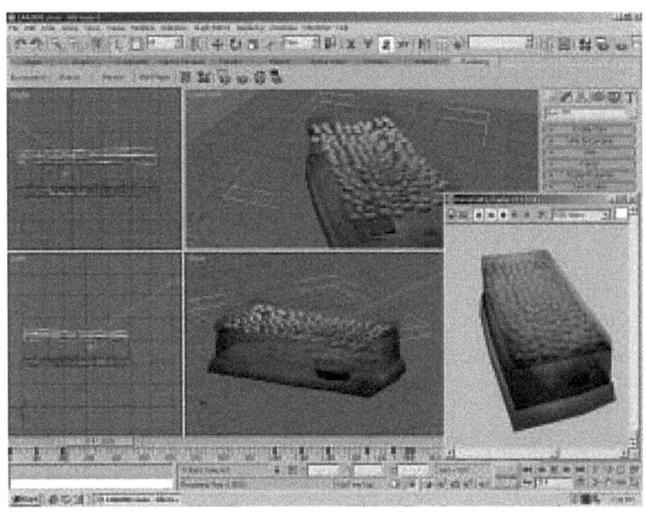

Figure 4. A quick model of the example case has been created in 3D Studio MAX, in which you can adjust the visibility of different structures and soil layers. This means that you can easily examine the different units and the relations between them. In this example picture, one can see the higher street paving and the layers related to it. (Drawing K. Uotila)

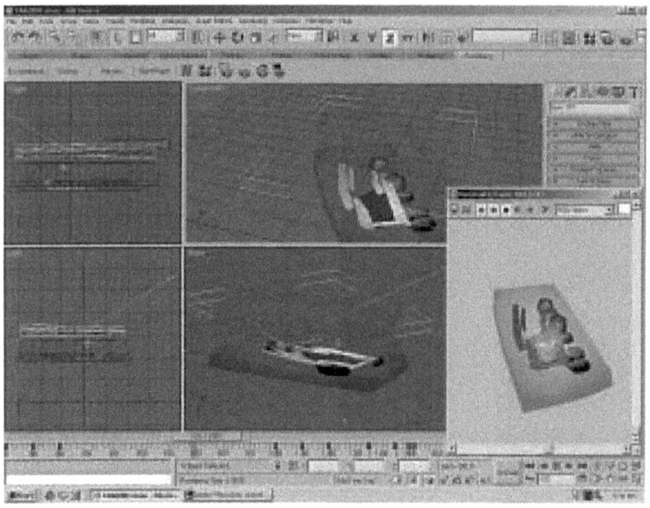

Figure 5. In this picture one can see the lower wooden structure and its stone foundation. (Drawing K.Uotila)

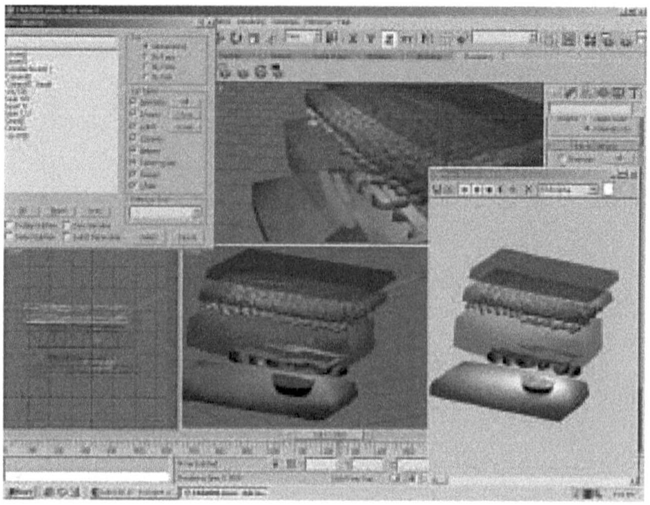

Figure 6. The stratigraphy of different layers can also be easily modeled and presented with the help of 3D Studio MAX. In this picture, the soil layers and structures of the example case are seen from several different directions. Similarly, the stratification of soil layers can easily be presented in an animation, which shows the chronological sequence of the stratified layers in the area. It is obvious that these kinds of animations are impressive to look at, but they also have scientific significance when analysing wide and complex excavation areas. (Drawing K. Uotila)

Handling digital 3-D record of archaeological excavation data

Mikhail Zhukovsky
Moscow State University
Department of archaeology
Russia

Abstract: Inspired by the evident methodological stagnation of the archaeological science in Russia the workgroup of Smolensk archaeological expedition launched a fundamental research project in summer year 2000. The expedition has been conducting a research work on the largest medieval complex of archaeological sites - Gnezdovo, belonging to the Viking period, for more then 50 years.

The mainstream of efforts of our workgroup was concentrated on the development of excavation data recording techniques. In fact, excavation process is a mean of analytical transformation of a raw data "buried" in soil into an array of archaeological information, followed by inevitable data losses and distortions. The draft survey showed a very dull picture of 50-60 % of data, that could be traced and recorded, neglected in the process of excavations on Russian archaeological sites. One of the main reasons affecting such a dramatic loss is the very principle of traditional paper-form recording of stratification, when originally 3-D spatial archaeological data (deposition lithography, artificial and natural remains, distribution of finds, etc) is projected or "compressed" onto consequent 2-D sketch plans drawn with fixed elevation interval. Therefore, the sketches contain the discrete data from the layer borders, while the evidence concealed within the layer is generally roughly recorded or even neglected. All finds and other artificial evidence are registered within logical squares with uneven precision in each individual case.

This paper is focused on an alternative strategy of handling excavation record: direct copying of evidence data from a true 3-D space of an excavation ground to a modeled 3-D virtual space. AutoCAD 2000 was used as a core modeling environment for the project. A notion of all-embracing digital environment for handling archaeological data is introduced; and the crucial problems of recording stratification, artificial and natural remains, and finds are successively discussed in the presented paper. An overview of major innovative capabilities and its immerse potential for development is given. Finally, an issue of applying the digital excavation record technique to previously collected archive excavation data is discussed.

Key words: methodology, excavation, stratification, 3-Dimensional, record, virtual data, database, hyperlinks

Introduction

Research is brought to life by information. Archaeological information is extracted from a great variety of sources. In the last years, their number is being multiplied, while the information, as a whole is getting increasingly diverse and heterogeneous. Archaeology seems to shift to ever widening use of non-destructive methods of survey, such as remote sensing or geophysical survey. Still, excavations remain major providers of archaeological data. The results of non-destructive surveying are commonly tested against sample excavations. Another prominent tendency is research concerned with re-utilization of previously obtained excavation data. This awareness is marked by a growing number of publications, presenting re-evaluated, re-processed, re-analyzed archive data from excavations and surveys, conducted for the last two centuries. This can be regarded as a manifest of the major significance of excavation record in a mass of archaeological information.

The excavation techniques have been rapidly developing and improving in the last decades, but still very unevenly in different countries, especially those with limited international scientific links. Methodology, forms of record, storage and representation of data collected in a course of archaeological excavations are very far from international or even regional level unification.

Russian archaeology today

Generally, the present state of Russian archaeological science can be described as conservative methodological stagnation. The main efforts of research are concentrated on extensive continuous excavations. This major preference to wide-scale fieldwork results in lack of attention and enthusiasm to the collected data analysis, interpretation, and publication. It is being piled year after year, waiting for the researchers for decades. From the other hand, the excavation technique, which is currently widely used in Russia show no significant development since 1950-70-s.

In fact, excavation process is a mean of analytical transformation of a raw data, "buried" in soil into a systemized array of archaeological information, followed by inevitable data loses and distortions. A draft survey showed a very dull picture of 40-50 % of data, that could be traced and recorded, neglected in the process of excavations on Russian archaeological sites.

To some extend it is provoked by the notion of wide-scale excavations, when quality is sacrificed to speed. Although the main reason, effecting in such a dramatic loss, is within the technique of excavation record itself.

Methodological dead-end

The dominating excavation strategy used in Russia is planar/section method (planum system). The future excavation ground is regularly divided into 1x1 or 2x2 m logical squares. After that, a site is excavated by consequent strictly horizontal layers with predefined elevation interval set most commonly to 10/20 cm – or spits. The revealed archaeological evidence is manually recorded at sketch plans made after each spit is removed (fig. 1). Therefore, the sketches contain the discrete data from the "logical layer" borders, while the evidence concealed within the spit is generally roughly recorded or even neglected. All finds and other artificial evidence are registered within logical squares with uneven precision in each individual case, and are drawn in index books (fig. 2).

Much more attention is paid to vertical sections, which are thoroughly drawn with great detail once the excavation is finished. Thus, vertical stratification is regarded as the major source for chronological/periodic reconstructions and archaeological interpretation of the revealed evidence (fig. 3). For more detailed description of the planum system refer to the recently published article by Darvill (Darvill 2000:30-36).

The disadvantages of the described excavation technique are evident:

1. The archaeological evidence concealed within cultural layers is recorded with insufficient detail and recision.
2. Vertical sections are used as a primary source of stratification sequence record and archaeological interpretation of excavated evidence only once the excavation is finished. This makes a flexible and efficient direction of excavation process and on-the-fly interpretation of data impossible. From the other hand, vertical sections disintegrate the whole archaeological context of a site into regular yet incoherent mosaic.
3. Archaeological evidence revealed in the course of excavations in true 3-D space is projected or "compressed" onto consequent planar sketches with considerable data loses and distortions
4. Finally, excavation record is stored in a paper-form in a sole copy, dramatically limiting the access and interpretation of data by multiple researchers, especially those not involved in particular excavations, as well as possibilities for its safe storage, publication and wide-scale distribution.

Further refinement of the described excavation and recording technique is dead-ended, providing no noticeable data surplus and overburdening the excavation process.

Seeking for solutions – a 3-D approach

Inspired by the evident methodological stagnation of the archaeological science in Russia the workgroup of Smolensk archaeological expedition launched a fundamental research project in summer year 2000. The expedition has been conducting a research work on the largest medieval complex of archaeological sites in Russian inland - Gnezdovo, belonging to the Viking period for more then 50 years.

The mainstream of our efforts was concentrated on the development of digital excavation data recording techniques. In recent years, there have appeared a number of publications proposing different strategies for computerization of excavation record (e.g. see Huggett 2000:117-122; Ryan, Pascoe, Morse 1999: 269-274; Norbach 1999:275-278 and other), handling particular blocks of excavation record; though no versatile digital technique covering all stages of excavation process and all types of excavation data was suggested. Our workgroup has presented an initiative of an all-embracing recording strategy rejecting the principle of data projection. Instead, the notion of direct copying of evidence data from a true 3-D space of an excavation ground to a modeled 3-D virtual space was introduced. The completeness and integrity of a recorded data, which is neither "compressed" nor projected, thus, is limited only by the capability to trace it. Thereby, the complete change of a recording technique demands a change in an excavation methodology. The single context method seems to suit the best the contemporary requirements for precision and integrity of recovered archaeological data. At the same time, the excavation sequence implied by the method, when each archaeological context is excavated, examined, and recorded individually in a successive order, fits the demands of a 3-D digital data record in a computer. While the combination of the introduced recording technique and the single context method is the most efficient for the future excavations in Russia, there is a vital need for utilization and re-interpretation of previously collected excavation data. Therefore, the capabilities of digitized data for 3-D excavation record reconstruction are also to be examined.

In order to fulfill both tasks it was proposed to test direct 3-D recording technique against traditional excavation strategy in a sample excavation. Thus, along with the in-the-field elaboration of the computer recording methods the potential of the traditionally recorded evidence was evaluated.

Digital excavation record. Theory and practice

Software environment

The excavation record is a complex of spatial (geometrical), text, and photo and video data. Digital data procession unleashes rich potential for its integration, unification, and analysis. Thereby, evidently the most powerful solution for handling digital excavation record is a GIS-based approach. At the same time, while GIS software systems have proved their efficiency with 2D spatial data archives, their 3-D modeling and analysis capabilities are insufficient.

Spatial data is either the most important excavation record or the most difficult to obtain and process digitally. Taking all these points into consideration, AutoCAD 2000 was used as a core modeling environment for the project. Its powerful

COGO (coordinate geometry) engine satisfies strict requirements for accurate and fast direct recording of excavation record on-the-fly in situ, and paper-form data digitizing and procession with minimum of conjectures and approximations. Both vector and raster data support along with integrated database engine for all common formats and internet/intranet support make AutoCAD 2000 the preferable solution for mixed-type and mixed-format data integration.

Experimental excavations: extending theory in the field

The huge complex of archaeological sites, situated on the upper of the Dnepr river, – Gnezdovo – consists of the central fortified settlement, the hill-fort or "gorodishe", surrounding rural settlement with area totalling c. 16 hectares, and four groups of mounds, which number little less then 4000 burials (figs 4-5). Belonging to the Viking period, Gnezdovo is considered the major principal proto-urban trade center in the Russian inland on the way from Scandinavia to Byzantium. It has been strongly believed, that the archaeological complex is limited by the first riverbed terrace until mid-1990-s, when the traces of the settlement in the flood-lands of Dnepr was discovered. A series of prospecting excavations has revealed wet cultural layers rich of organic material of an excellent preservation.

The experimental excavation ground was started on a slope on a terrace edge covering both terrace and a flood plain on an area of 40 m². Thus, the stratification link between flood plain and terrace was intended to be examined. The excavation record has been kept either traditionally, with paper-form sketch plans and drawings, or directly modeled in AutoCAD; for the potential of both recording strategies could be explored and compared, and the integrity of record mutually checked one against another. At the same time the excavation technique, based on the principle of the described above planar/square system, was modified, where it was possible, so that each major stratification unit was excavated individually in a strict consequence of their deposition (fig. 6).

The digital excavation record was held with a Notebook computer in three main parallel directions:

Stratification record

Stratification record was the most difficult to handle digitally in 3-D modeled space. Geometrically each stratification unit is a solid of a complex irregular shape. Two main techniques were used to measure its surfaces. In case the "upper" surface of a stratification unit was completely and clearly revealed, the measurements were taken by a 40x40 cm elevation grid, while the horizontal contour of a SU was limited by the vertical sections going along the borders of the excavation ground, or by a contour line/multiple lines with a fixed elevation mark (fig. 7). The "lower" surface of a SU coincides with the "upper" surface of the consequent stratification unit/units.

The rather different strategy was used when stratification units were sliced by horizontal spits with an elevation interval set to 10 cm. In this case, the SU limits were traced on a planar surface of a spit and marked by one (SU completely inside the excavation ground) or two (SU extends the excavation ground) contour lines (fig. 8). In both cases, if a SU border exceeded the borders of the excavation ground, vertical sections, which stratification was thoroughly recorded once the excavation was finished, provided the wireframe 3-D "upper" and "lower" contours of each stratification unit (fig. 9).

Due to budget limitations, the data collection was limited to manual and optical theodolite measurements. When it was possible, the stratification record, elevation grid points, contour lines, and all other graphical data were directly input or copied in AutoCAD-based virtual 3-D excavation ground environment. Though, weather conditions, battery life limitations, and other unpredictable circumstances made direct on-the-fly computer record impossible. The voids in digital record were compensated with traditionally taken paper-form plans, drawings and sketches, which were scanned and then manually vectorized in AutoCAD module – CAD Overlay (fig. 10).

Thus, raw data of stratification units' measurements were collected. The next step was to create the most precise 3-D virtual solids of SU from the obtained elevation points and contour lines data array. The extrapolation principle of digital relief reconstruction was assumed to be the most appropriate. Stratification unit's surface is very similar, not to say - exact, to a relief surface. We strongly believe, that elevation points extrapolation in this case is not less precise and accurate then those used in digital cartography. The most common technique, which was chosen, is geometrical extrapolation. The initial points and contour lines for each SU were integrated in a triangular irregulated network (TIN) (fig. 11), which was followed by creation of a regular extrapolation grid with 20x20 cm cell density (fig. 12). Merged together, the extrapolated surfaces formed the final shape of stratification units' virtual 3-D solids (fig. 13-14).

Artificial and natural remains record

Natural stone scattered chaotically or organized geometrically in the cultural layers; chopped, cut or broken wood of both natural and artificial origin; bones were recovered in hundreds in the process of excavation. Some of them appeared in the cultural layers accidentally, being a natural deposition, the rest were intended by inhabitants of Gnezdovo as parts of different structures and hearths. Large amount of these remains along with very slow speed of manual and optical theodolite measurements made it impossible to transmit their shape to the 3-D modeled AutoCAD environment identically. A series of approximations had to be accepted.

All unearthed stone was of irregular natural shape; its dimensions did not exceed 20x20x20 cm. Therefore, after its' precise position and dimensions were marked in the virtual excavation ground, the sample stone library, consisting of about two dozens of common stone shapes, was used to find the most similar shape to fit the original (fig. 15).

This kind of approximation strategy, appropriate for stone, could not satisfy the wooden remains and bones. No sample would fit their individual unique shape and represent the significance of small details of appearance. 3-D solids of wooden and bone remains were substituted by their 2D contour outlines taken from the top view and several thickness measurements as it is done in a traditional paper-form excavation record. In some cases, multiple contour outlines could be recorded to give a rather rough idea of its 3-D shape (fig. 16).

Finds record

All artificial finds were registered in a common format Access database with a traditional "logical square" reference. A range of standard fields contain find's id numbers; year, id and author of an excavation ground; position within cultural layer, "logical layer", "logical square", context and a description of each find. A description record is split between "functional group", "name", "shape", "material", "colour", and "state of preservation" fields. Precise location of a find was marked in a virtual 3-D excavation ground space. AutoCAD database engine allowed to integrate an Access database in the modeled excavation record environment and establish links between spatial location points of finds and database records (fig. 17). Thus, common GIS spatial databases analysis features were added to the CAD environment, enabling visual estimation of standard database queries along with "visual queries" to a database. In addition, finds' records are referenced to 3-D stratification unit solids, what definitely extend the capabilities for data analysis.

Results

The final integrated digital excavation record encourages us to consider this ambitious initiative to be generally successful. The experiment showed that digital and traditional record could be used for virtual excavation record reconstructive modeling, for two stratification-recording techniques have been tested for both types of record. Paper-form find indexes could be easily transformed to a database format, though this technically evident process is very time consuming. The potential of digitally obtained, processed, modeled, analyzed and represented excavation record is immense and is summarized below:

- Digital format ensures not only safety of data storage and back-up copying, and significantly reduces the risk of its damage or loss, but also provides multi-user access to the data and unlimited potential for its distribution, publication and representation.
- Direct copying of excavation record into a 3-D virtual modeled environment permits to avoid data projection or "compression", resulting in its inevitable loses and distortions, and makes it more precise, comprehensive and integrated.
- Archaeology has always been a destructive science. Digital record grants a researcher a unique chance to re-excavate a site virtually in any sequence from any point, and to re-estimate and re-analyze excavation data.
- Virtual modeling environment immensely extends the potential of stratification analysis. A stratification record may be filtered to show only individual stratification units, or be examined as a whole system. Virtual environment enables a researcher to navigate freely in 3-D excavation record modeled space and to estimate it visually from any viewpoint, distance, or angle, inside or outside an excavation ground. Virtual profiles and vertical sections may be cut through cultural deposits in any direction and plane, or true vertical sections may be easily hidden, so a site could be visually estimated as an integrated whole. Virtual modeling environment provides tools to analyze slope/elevation of stratification units, to calculate area and volume of any natural or artificial evidence (stratification units, structures, hearths and any type of other remains), and to measure any distances in 3-D space.
- AutoCAD provides full common format database support. Besides standard set of features, it enables to link database records with 3-D objects and to construct both common and "visual" queries to a database, thus adding GIS system capabilities to an AutoCAD environment.
- AutoCAD allows to hyperlink each object in virtual 3-D environment with Internet resources, containing video, photo, text, and any other data. Thereby, powered with barely unlimited Internet potential for data integration, CAD integrates the whole complex of data collected in the process of excavations. At the same time, Internet technology provides multi-user remote access to data.

The potential for development

Though these are the major features of digital excavation record, their range is much wider. Some of them become prominent only after a certain period of practical work. One of the noticeable capabilities, not mentioned above, is the potential for development. Inspired by an evident success of the initiative, we have planned to proceed with further improvement and extension of digital recording technique in two parallel directions: continuation of sample excavations in years 2001-2002 and digital re-procession of data obtained in previous excavations.

The fieldwork is intended to introduce some new features:

- One of the most crucial points of digital excavation data record is speed and precision of measurements. It makes the use of laser Totalstation indispensable for recovered data integrity and completeness. Totalstation permits not only to get much more accurate stratification record, but to provide direct 3-D record of artificial and natural remains, which was impossible with manual measurements. As a result, the whole excavation record becomes 3-dimentional, thus minimizing distortions of initial data.
- Digital photography and video shooting of stratification and artifacts are to be used more extensively, which will particularly allow creating photogrammetric coverages for surfaces of stratification units.
- Database support for all types of data/objects in digital excavation record. In its present state, solely a descriptive database for finds is integrated in the 3-D environment. Each object is to be linked with a

database record, containing its extensive photo, video and text description.
- All data is integrated in AutoCAD environment and accessed through Internet interface. Data exchange and interlink with GIS systems is provided by ArcCAD – the GIS extension to AutoCAD.

Conclusion

The digital excavation record technique is designed to overcome the methodological stagnation in Russian archaeology. We strongly believe that our efforts will stimulate the development of both the methodology of excavations and the technique of excavation data record on Russian archaeological sites. However, the initiative that we have introduced is directed either in the future or in the past. The Gnezdovo complex of archaeological sites has been excavated for more then 125 years, which makes the re-procession of all of the collected data indispensable for the research. Digital integration of these data is our main goal, where the presented technique is only one small step.

Acknowledgments

I would like to express my gratitude to all the team of the Smolensk archaeological expedition, which contribution made this work possible, especially Veronica Murasheva – for providing excavation documentation, and the members of my workgroup – Anastasia Krasinskaya, Boris Meledin and Vasiliy Novikov – for their enthusiastic efforts. I would like also to direct my cordial gratitude to my bride, Julia Simonova, for her tenderness and support in all my projects.

References

Darvill, T., 2000. *Recording systems in archaeological excavation: introducing ARTHUR, Anglo-Russian Archaeology Seminar: Recording Systems for Archaeological Projects*. Bournemouth and Moscow.

Huggett, J., 2000. Looking at intra-site GIS, in *Computer Applications and Quantitative methods in Archaeology*, BAR International Series 845.

Norbach, L.C., 1999. IT-based Documentation of Large Scale Excavations, in Drengsted: *A Case Study, Computer Applications and Quantitative methods in Archaeology*, BAR International Series 750.

Ryan, N., Pascoe, J., Morse, D., 1999. Enhanced Reality Fieldwork: the Context Aware Archaeological Assistant, in *Computer Applications and Quantitative methods in Archaeology*, BAR International Series 750.

Figures

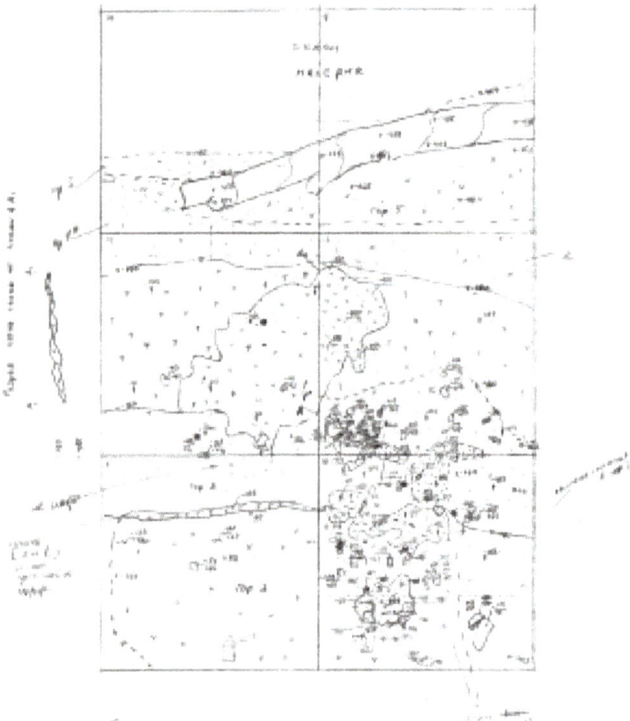

Figure 1. A sketch plan of a spit ("logical layer")

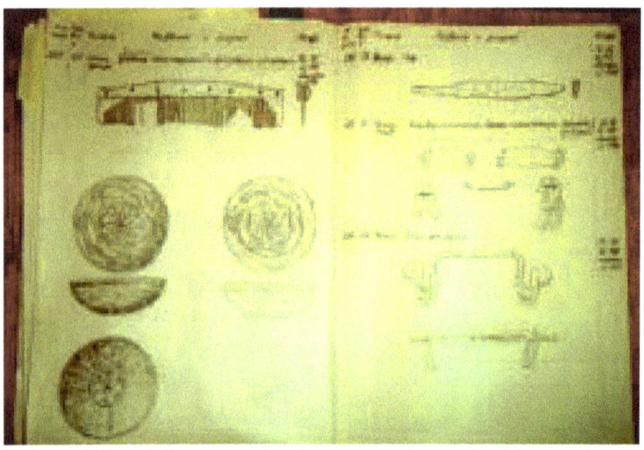

Figure 2. A finds' index book

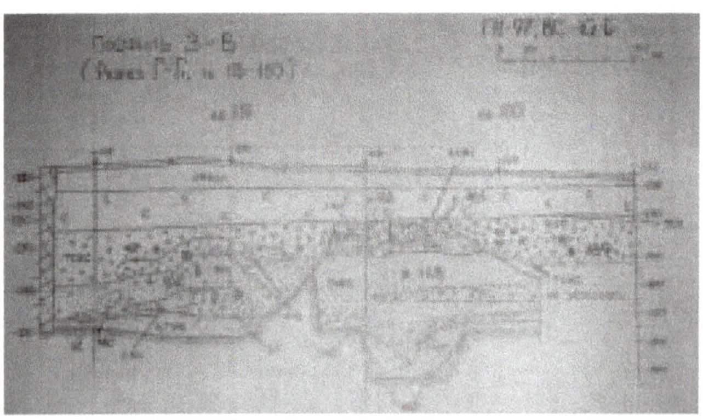

Figure 3. A vertical section sketch plan

Figure 4. Gnezdovo complex of archaeological sites

Figure 5. Gnezdovo complex of archaeological sites

Figure 6. The experimental excavation ground, year 2000

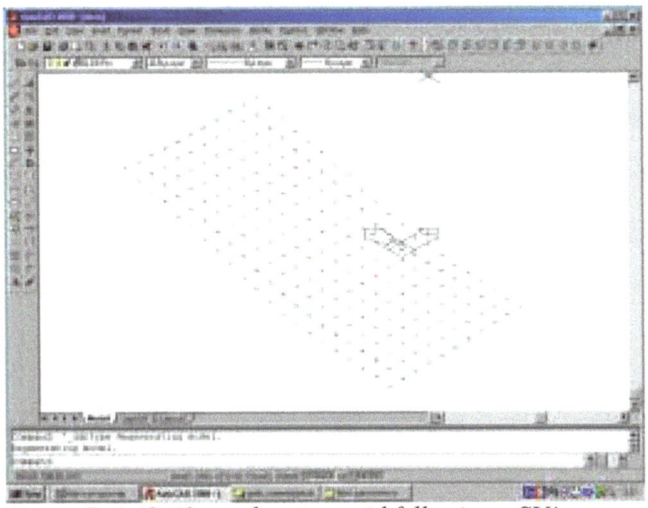

Figure 7. A 40x40 cm elevation grid following a SU's surface

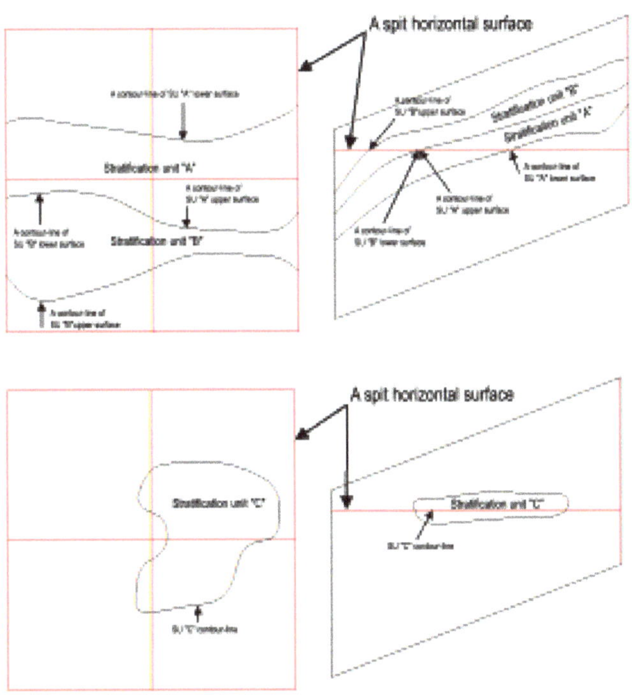

Figure 8. Spit sliced SU recording strategy

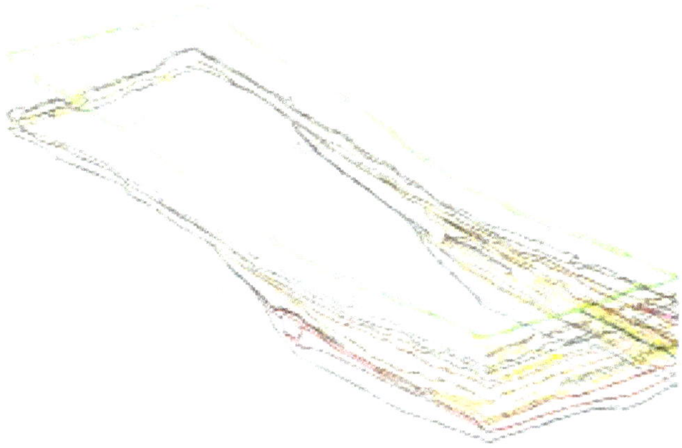

Figure 9. Vertical stratification wire-frame 3-D record

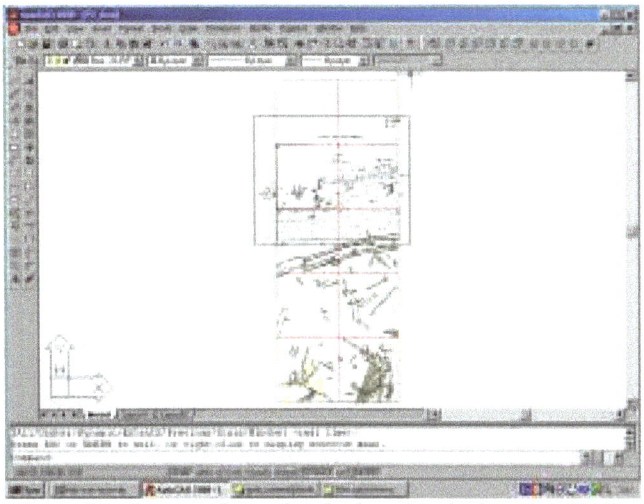

Figure 10. A paper-form sketch plan attached to AutoCAD

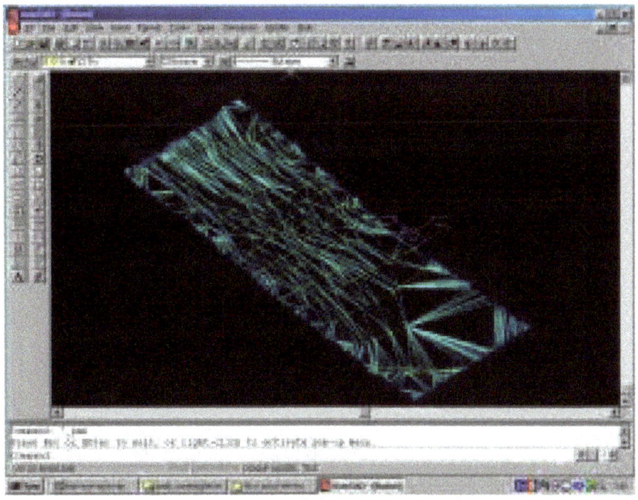

Figure 11. A TIN constructed from contour lines

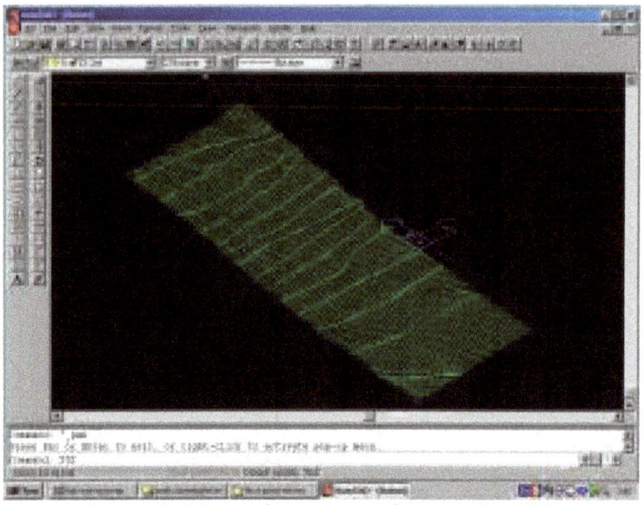

Figure 12. A 10x10 cm elevation grid extrapolated from TIN

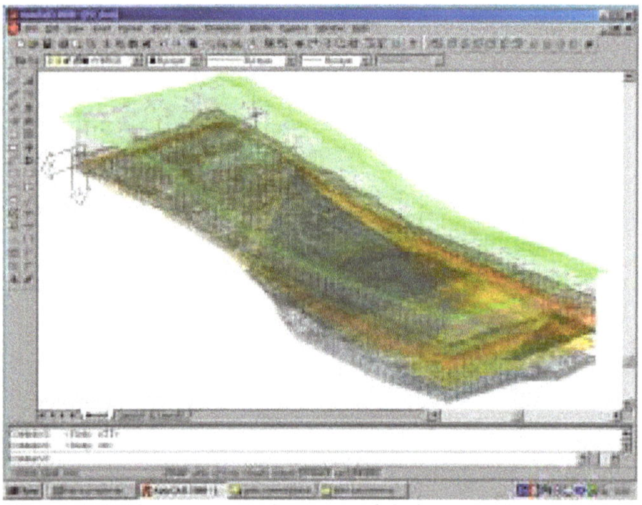

Figure 13. The 3-D wire-frame model showing the whole stratification

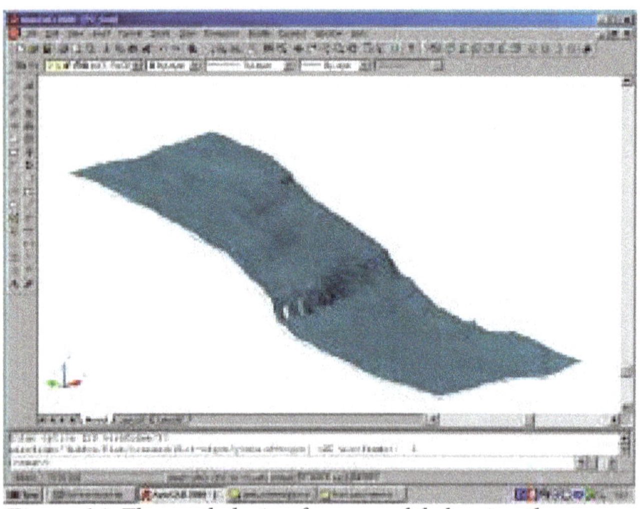

Figure 14. The rended wire-frame model showing the sequence of SU's deposition

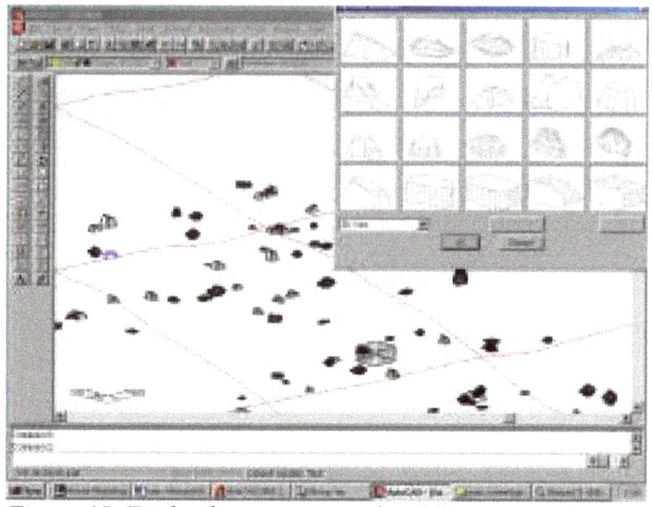

Figure 15. Finds of stone presented in 3-D environment

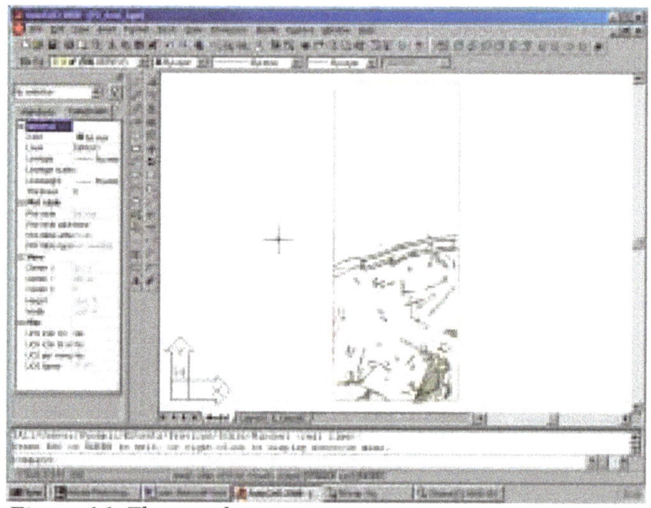

Figure 16. The wooden remains

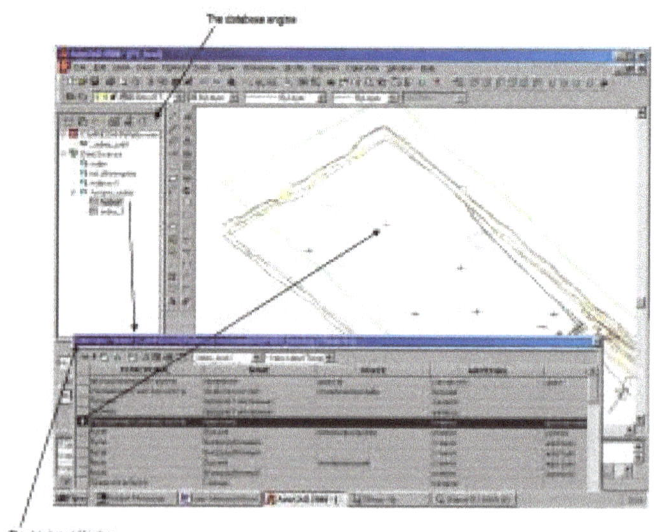

Figure 17. The finds record database in AutoCAD environment

Papers in the Database Applications and Statistics and Quantitative Methods categories

Approaches to petrographic data analysis using S-Plus

C.C. Beardah, M.J. Baxter and I. Papageorgiou
Department of Mathematics
Statistics and Operational Research
The Nottingham Trent University
Clifton, Nottingham NG11 8NS, UK

M.A. Cau
Department of Archaeology and Prehistory
The University of Sheffield
Northgate House, West Street
Sheffield S1 4ET, UK

Abstract: In this paper we shall show how S-Plus can be used to address issues involved in clustering artefacts on the basis of petrographic thin-section data. If such data are coded as a set of categorical variables or as presence/absence data a variety of analytical options are open. These involve choosing how to measure similarity between cases and how, subsequently, to group these using clustering or scaling techniques. Using S-Plus functions that come with the package, or which are freely available in libraries developed by S-Plus users, options include multiple correspondence analysis and metric and non-metric scaling methods including Sammon mapping and isotonic multidimensional scaling. Some of these will be illustrated, and it will be shown how different methods can produce results that reveal different aspects of a data set. Our interest in pursuing this is that we would like to investigate the possibility of including both petrographic and geochemical data in a statistical analysis of artefact compositional data, and on an equal footing. We conclude by indicating how this might be achieved within the methodological framework discussed in the paper.

Key words: ceramics; petrographic thin-sections; multivariate analysis; cluster analysis; S-Plus.

Introduction

This paper presents the development of previously published work (Baxter et al. in press; Beardah and Baxter 2001) by members of the EU funded GEOPRO TMR Network. The general aims of the project are to integrate geochemical and mineralogical techniques in the study of raw materials and archaeological ceramic provenance. From a statistical point of view, a specific interest is the possibility of incorporating both geochemical and mineralogical data into statistical analyses. Using these two types of data in combination (so-called *mixed-mode* data), or by independently applying appropriate techniques to each, we hope to identify groups within a set of data. Such groups could be assumed to indicate, for example, distinct origins of the artefacts or raw materials used in their manufacture. Our ultimate aim is to produce user-friendly and accessible software, developed using the S-Plus package (Mathsoft 1999; Venables and Ripley 1999, 2000) and allowing "state of the art" analysis and presentation of geochemical, mineralogical and mixed-mode data. The reasons for choosing S-Plus as the basis for the development and distribution of our software are reported in Beardah and Baxter (2001).

In the same paper we showed how a clustering methodology for grouping chemical compositional data from artefacts, developed by Beier and Mommsen (1994), could be implemented in the S-Plus package. Geochemical data is typically continuous in nature. By contrast, as discussed in section 2 below, mineralogical data in its raw form can consist of variables of a mixture of types, including continuous, categorical and presence/absence. In this paper we concentrate solely on the implementation of statistical methods for mineralogical data.

More background on data arising from the analysis of petrographic thin-sections is given in the next section. Section 3 of the paper presents a brief discussion on techniques for grouping petrographic data. Such methods include multiple correspondence analysis and metric and non-metric scaling methods including Sammon mapping and isotonic multidimensional scaling. In section 4 we present an archaeometric case study. This is followed in section 5 by our conclusions.

Analysis and coding of ceramic thin-sections

Examples of ceramic thin-sections are shown in Figure 1. Usually a single analyst, on the basis of examination and comparison, carries out the classification of a collection of thin-sections into groups based upon provenance. This time-consuming approach is well known to rely upon the experience of the observer and the availability of reference groups and collections and furthermore it can be "heavily biased by the personal approach of the analyst involved" (Whitbread 1991) and lack reproducibility.

In attempting to develop an alternative, less "subjective" method of classification, the first challenge we face is how to describe a collection of thin-sections in a manner that is amenable to statistical analysis. In order to do so, one approach is to represent each thin-section in terms of qualitative or categorical variables. Such variables can then be coded in a manner that can be input into a quantitative statistical analysis. Cau et al. (in preparation) review some reasons for, and attempts at, quantifying thin-section data, and describe the coding system used in our examples.

In order to describe the thin-sections, a system of 19 categorical variables covering the most common features normally recorded in ceramic petrology has been developed (Cau et al., in preparation). Variables 1-5 record aspects of the thin-section that can be used to reveal "technological information", for example firing temperature or forming techniques. The main rock types, organized by general families, and the main rock forming minerals that are found in Mediterranean ceramics are recorded in variables 6-18. Finally, variable 19 records packing, an estimate of the general amount of particles in a given sample.

It is important to note that in the system reported by Cau et al., a combination of categories within a specimen will result in the formation of a new category. As a result, the number of categories for each variable can be as many as 30, or as few as two (presence/absence). Furthermore, the system is open, that is new variables and categories can be added with relative ease.

Statistical methodologies

After description using the system described in section 2, a collection of n thin-sections can be represented by a table with n rows and 19 columns. Each row contains the description of a single thin-section; each column represents a variable; and each cell contains a number indicating the category. An important point is that *any* system similar to that of Cau et al., which results in an n by p table of categorical data, where p is the number of descriptive variables, can be treated using the statistical methods described in this section.

Converting categorical data into 0/1 data

In order to apply statistical methods, one approach is to convert the table of categorical data into a data matrix consisting of 0/1 entries. There are two ways of doing this, illustrated here by means of a simple example. Table 1 shows an example of a 4 by 1 table of categorical data. Here the single variable is divided into four categories, the last of which consists of the presence of both category 1 *and* 3. (Recall that in the system reported by Cau et al., a combination of categories within a specimen will result in the formation of a new category). For this particular data set, category 2 is not represented.

Coding method 1

Here we introduce one dummy variable for each category appearing within the data set, with a case being coded 1 for the variable whose category it belongs to and zero otherwise. Thus applying coding method 1 to the categorical data of Table 1 results in the 4 by 3 data matrix, denoted G, shown in the bottom-right of Table 2. Here the number of categories present in the variable is three (1, 3 and 4); no column appears for category 2 as this does not appear with these data. (Any zero columns would need to be omitted prior to the subsequent statistical analysis.)

Also, note that since our example has only one categorical variable, this coding method results in a single non-zero entry for each row of the matrix G. More generally, for an n by p table of categorical data, where variable i has L_i levels present in the table, we introduce L_i dummy variables corresponding to the levels of variable i. This gives rise to an n by L data matrix, G, where

$$L = \sum_{i=1}^{p} L_i$$

and the sum of the entries in each row is p.

Coding method 2

Using the descriptive system of Cau et al. we note that many categories are in fact combinations of those previously defined. For example, category 4 of our example consists of the combination of categories 1 and 3. An alternative method of coding the data is to only record the presence or absence of the "core categories" (1, 2 and 3 for our example). Thus applying coding method 2 to the categorical data of Table 1 results in the 4 by 2 data matrix shown in the bottom-right of Table 3. Here the number of core categories present in the variable is two (categories 1 and 3). Again, zero columns would need to be omitted prior to the subsequent statistical analysis and are therefore not included.

Although coding method 2 is conceptually simpler than method 1, it results in a more unwieldy data matrix, G. In particular, for our example the sum of the entries in each row is no longer 1, as was the case for coding method 1. More generally, for an n by p table of categorical data coding method 2 leads to an n by L data matrix, G, where the sum of the entries in each row varies and in each row is $>=p$.

Statistical methodologies

A variety of statistical methodologies can be used to investigate the data resulting, in each case, in a 'map' showing how similar cases are to each other.

- Multiple Correspondence Analysis (MCA) can be used if the data are coded using method one, and is simply correspondence analysis applied to the matrix G. This will have the effect of giving greater weight to rare categories.
- Principal component analysis (PCA) can be applied directly to G and, using coding method one, is equivalent to classical multidimensional scaling (MDS) applied to a dissimilarity matrix whose elements are the square-roots of the elements of the matrix $(P - GG^T/p)$ where P is an $(n \times n)$ matrix of 1s and ^{T} indicates a matrix transpose.
- Non-metric MDS methods, including Sammon's

mapping and Kruskal's isotonic MDS (Venables and Ripley 1999: 333-335) can be applied to the dissimilarity matrix described above.

- Correspondence Analysis of the 0/1 matrix *G* can be applied to data derived from coding method two, and any of the MDS methods can be applied to a suitable dissimilarity matrix derived from *G*.

All of these methods result in graphical output similar in nature to that of a PCA. The number of components used to display the results can be varied.

Case study

The methodologies introduced in the previous section are now illustrated using mineralogical data arising from a collection of Late Roman Cooking Ware (LRCW) from the Balearic Islands and the eastern Iberian peninsula. In examining these data our aims included the determination of origin, in other words, were the wares local or imported? The full data-set contains 115 samples of LRCW from different sites in the region. (For these data we have both chemical and petrographic information.) A subset of these data contains 25 samples of LRCW from Can Sora (Eivissa). These samples come from a cistern, used as a rubbish dump during Late Antiquity. Samples came from two different layers (from the 5th and 6th centuries A.D.)

Data format and conversion

Figure 2(a) shows a part of the 25 by 19 table of categorical data for these data as it appears in S-Plus. The second column, whose entry in row number 1 is "CS-2 (pl)", contains labels for each case. Such labels can be extremely useful when analysing (in S-Plus) graphical output of the type generated by the methods of section 3. Symbols in brackets denote the result of a preliminary classification carried out by one of us (MAC) and based solely upon a petrographic analysis of the kind discussed in section 2. This analysis led to the identification of five types, labelled *pl*, *v*, *m*, *p* and *f*, and five outliers labelled *o* (Cau et al., in preparation).

Conversion between categorical representation and 0/1 representation can be achieved via our **catdist.mat** routine. This routine can be called from the GEOPRO menu (figure 3(a)) and results in the dialog box shown in figure 3(b).

This returns a list, **cansora.GX**, containing the matrices *G* and *X* (see section 3). To extract *G* and store it in a separate, appropriately named variable we can use the S-Plus command

cansora.G <- cansora.GX[["G"]]

Figure 2(b) shows part of the matrix (*G*) of 0/1 entries so obtained, again as it appears in S-Plus. Coding method 1 has been used (see section 3.1.1). For illustration, note that variable 1 (optical activity) has three categories (1. active, 2. inactive and 3. intermediate). Only the final case in the data set, labelled CS-27, falls into category 2 for this variable. As a result, the matrix *G* has three dummy variables, labelled **opt.act.1**, **opt.act.2** and **opt.act.3**, corresponding to the categorical variable describing optical activity. Those cases, for example CS-2 and CS-3, falling into category 1 for optical activity thus have a 1 in the column of G labelled **opt.act.1** and zero entries in those columns of *G* labelled **opt.act.2** and **opt.act.3**.

Of course, the matrix *G* may have already been coded (using method 1 or method 2) and entered directly. This possibility is allowed for in our implementation of the various statistical methods for the analysis of petrographic data. In either case we must be wary of columns whose sum is zero. Such columns are automatically detected and removed upon submission to the routines for performing MCA, MDS etc.

Analysis of petrographic data

Using a combination of existing S-Plus library routines (Venables and Ripley 1999) and new implementations of existing techniques, adapted to make the most of the graphical capabilities of the package, we have made the following methods available.

1. Multiple Correspondence Analysis (MCA).
2. Classical Multi-Dimensional Scaling (MDS).
3. Non-metric MDS methods: Sammon's mapping and Kruskal's isotonic MDS.
4. Correspondence Analysis of the 0/1 matrix *G*.

We initially use the Can Sora data set for illustration (recall that this is a subset of the full LRCW data set).

In order to make the most of the interactivity available within S-plus, the application of these methods is a two-stage process:

1. Application of the method (**MCA/MDS Scores** menu item, figure 4(a));
2. Plotting the results graphically (**MCA/MDS Plots** menu item, figure 4(b)).

The calculation stage results in the creation of a variable named according to the method used, for example, if we use Multiple Correspondence Analysis, then the variable is called **last.MCA**. It is important to note that the output from MCA *must* be stored in this variable.

Different methods result in different numbers of component scores that can be plotted. Table 4 shows the maximum number of these currently available for each method.

Highlighting subgroups

Figure 5 shows the graphical output upon application of MCA to the Can Sora data set. Graphical output based upon three components has been displayed. The window on the left shows a display based upon the first three components, while that on the right shows all possible plots based upon two of the first three components. The subgroup classified petrographically and labelled *v* has been highlighted. This is achieved by clicking rows in the **last.MCA** data window. (The centre of the three windows shown in Figure 5; the **ctrl** and **shift** keys can be used to highlight individual rows or sets of consecutive rows.) Corresponding points in the component plots are automatically highlighted. It is clear from the left window of Figure 5 that this subgroup also separates on the basis of a plot of the first three components of the MCA. If this were not immediately apparent, the three-dimensional plot could be rotated and examined from different viewpoints. Furthermore, plots based upon components 1 and 2, and 2 and 3 also exhibit clear separation of this subgroup. Based upon the plot of components 1 and 3, separation is less clear. Despite this, there is overwhelming evidence that this subgroup would be identified by this statistical technique (MCA) independently of any knowledge of the prior petrographic analysis.

Of course, we may not have a petrographic classification available. In these circumstances we will be attempting to group the data by using purely statistical techniques, possibly including methods (such as PCA) for the analysis of geochemical data, if available. S-plus can be used to identify and highlight subgroups in many different ways. Figure 4(b) shows a simple example. Here we have used the "Highlight rows" box to identify objects where the second component of the MCA output is negative. However, more complicated expressions can be used to highlight rows. In fact we can enter any valid S-plus expression in the "Highlight rows" box. For example, the subgroup shown in figure 5 could be defined by noting that these are the cases where the first and second MCA components are negative. Hence we could use the S-plus expression

last.MCA[,1]<0 & last.MCA[,2]<0

in the "Highlight rows" box to get the same effect as that shown in figure 5. Furthermore, the expression used could depend upon the analysis using a different method entirely. So we might choose to highlight (in the MCA output) the rows where the second *MDS* component is positive. This might be useful way of comparing the subgroups obtained using different methods, especially if the dataset is large. Simple exploratory techniques, for example histograms, can be used to look at the distribution of individual components in an effort to identify criteria for separating groups in this way. A slight twist on the traditional histogram, a labelled histogram, is provided via the GEOPRO menu. An example is shown in figure 6. This shows that the Can Sora data set separates clearly into two groups on the basis of whether the second MCA component is positive or negative.

More usually, and perhaps especially in analyses involving smaller datasets, subgroups may be identified visually, directly from the graphical output. By positioning the mouse pointer over a plotted point, the row number (or label) of the point and its component values are shown (see the left window in figure 5; here the mouse was positioned over the rightmost point of the subgroup, representing the case labelled CS-17). For example, as previously discussed, it is likely that the subgroup highlighted in figure 5 would have been identified in the absence of a prior petrographical classification, by visual observation of the MCA output alone. (Since it separates nicely in plots of both the first two, and first three, components.) By using the mouse, we can easily identify that this subgroup consists of rows 8, 9, 10, 13, 14, and 15 of the dataset (see the centre window of figure 5).

Finally, and perhaps most conveniently, subgroups can be highlighted directly in windows displaying graphical output by means of the **Select Data Points** feature of the **Graph Tools** menu within S-Plus. Clicking and dragging the mouse can be used to highlight sets of points in the graphics window. (The **ctrl** key can be used to highlight subgroups consisting of several individual clusters.) Selected points are automatically highlighted in the data window also (as illustrated in figure 5).

It would be useful to investigate whether the subgroup previously identified separates just as easily when we use other methods. Returning to the **MCA/MDS Plots** dialog we can now generate the graphical output resulting from, for example, Kruskal's non-metric MDS (**IsoMDS**). The dialog box is shown in figure 7. Note that the entry in the "Highlight rows" box is generated automatically, since these rows were previously highlighted in a data window. The graphical output so generated reveals that the aforementioned subgroup separates just as easily with this method. In fact, all five methods discussed here identify this subgroup with little difficulty.

Outlier removal

It will often be the case that examination of the graphical output from one or more methods will reveal the presence of outliers. Removal of outliers can help to further separate other genuine subgroups within the data set. For example, figure 8 reveals that, based upon the output from MCA, it is fairly clear that rows 6 and 11 (cases CS-7 and CS-13) could be considered to be outliers. This observation is supported, to a greater or lesser extent, by the other methods. At this point it is worth recalling that, for the Can Sora dataset, five cases were identified petrographically as outliers. Rows 6 and 11 were two of these. The others were rows 7, 22 and 23 (cases CS-8, CS-24 and CS-25). There is fairly strong evidence, from methods other than MCA, that row 23 is an outlier, however the situation is less clear, for rows 7 and 22, which seem to associate quite well with the subgroup labelled *pl* on the basis of petrography. This is an example of where the application of statistical techniques possibly results in a different conclusion to the application of traditional petrographic analysis.

Outliers can be removed at the first (calculation) stage of the application of our methods. Figure 9 shows the application of MCA to the Can Sora dataset minus rows 6, 11 and 23.

Combining petrographic and geochemical analyses

For the full LRCW data set (115 cases) we have both geochemical and petrographic information. The former consists of 25 concentration values and the latter has already been coded as a matrix (*G*) of 0/1 entries using coding method 1. We can calculate and plot PCA scores in much the same way as previously discussed. Figure 10 shows the dialog box called from the GEOPRO menu entry **PCA Scores**.

The initial plots of the component scores are quite cluttered and would possibly benefit from the removal of some outliers. The most obvious outliers are cases MC-16 and MC-19 (rows 33 and 34 respectively; these cases were also identified as outliers on the basis of the petrographic analysis). Re-calculating the PCA scores with these cases omitted gives a slightly clearer plot. To do this we would use the same dialog as in figure 8, but with **c(33,34)** entered in the **Omit these cases** box. There is now evidence of an outlying group consisting of cases U-1 and U-10 (rows 60 and 69 of this reduced dataset). Care needs to be taken here, as cases U-1 and U-10 correspond to rows 62 and 71 of the original dataset (with no outliers removed). This shows the value of assigning labels to cases! Re-calculating the PCA scores with rows 33, 34, 62 and 71 omitted gives a plot that is rather more spread out and that could be used to identify more subgroups on the basis of chemistry.

Of course we can also apply our various methods for grouping petrographic data to the LRCW dataset with rows 33, 34, 62 and 71 omitted. Analysing the two-dimensional output from MCA reveals a potential subgroup of four cases (CS-18, 19, 20 and 25) in the bottom-centre of the plot of components 2 and 3. Interestingly, on the basis of the three-dimensional output, we could probably discard CS-25 from this potential subgroup. CS-18, 19 and 20 has also been identified as a subgroup on the basis of petrography. (CS-25 was identified as an outlier on the basis of the petrographic analysis.) However, highlighting these cases in the plot of the PCA scores (with rows 33, 34, 62 and 71 omitted) reveals that based upon the analysis so far, these cases do not separate on the basis of chemistry.

Mixed-mode data

In the previous section we have seen that one way of incorporating both petrographic and geochemical information in our statistical analysis is to analyse the chemistry and petrography separately, but possibly concurrently, using methods appropriate for each. Using this approach, we can investigate whether subgroups identified with the various methods for analyzing petrographic data are also identified using exploratory methods for analyzing geochemical data. Such methods available within S-Plus include, for example, PCA, cluster analysis, and other, less traditional methods such as that proposed by Beier and Mommsen (1994) and discussed in Beardah and Baxter (2001).

However, within the framework outlined so far, we can incorporate both chemical and petrographic data. This could be achieved, for example, by first converting the elemental concentrations of the chemical data into categorical data before applying the techniques outlined here. Alternatively, some methods can deal directly with data of mixed type (e.g. continuous chemical data and 0/1 petrographic data). To do so, we need to measure dissimilarity between objects of mixed type (Gower and Hand 1996; Kaufman and Rousseeuw 1990). Such approaches will be the focus of future work.

Summary and conclusions

In this paper we have seen that

- Otherwise complex methodology for the analysis of petrographic data can be implemented in S-Plus with relative ease. In addition, many techniques (e.g. the non-metric methods) have been implemented in S-Plus and made freely available, for example as part of the MASS library (Venables and Ripley 1999).
- The ability to manipulate the user interface makes it possible to design user-friendly routines.
- The nature of the S-plus interface makes it possible to (a) easily identify sub-groups within the data and (b) compare the results when using different methods independently, including those appropriate for the analysis of geochemical data.
- Some of the methods reported here can be used to analyse data of mixed type.

Our implementation has been illustrated using data arising from the petrographic and chemical analysis of 115 specimens of Late Roman Cooking Ware (LRCW) from the Balearic Islands and the eastern Iberian peninsula. Whilst not fully illustrated in this paper, the general approach to the data analysis is an iterative process consisting of identifying obvious groups and outliers, "peeling" these away from the data set, and proceeding with an examination of what remains. See Cau et al. (in preparation) and Papageorgiou et al. (2001) for more detail on this approach.

Finally, as stated at the outset, our involvement in the GEOPRO project has the aim of developing an S-plus library of user-friendly routines for grouping ceramics using chemical and/or mineralogical data. In support of the software, materials such as documentation and tutorials, both paper and web based, will be provided. The final collection of routines will be made freely available, via the Internet, to the archaeometric community.

Acknowledgements

This work forms part of the GEOPRO Research Network funded by the DGXII of the European Commission, under the TMR Network Programme (Contract Number ERBFMRX-CT98-0165).

References

Baxter, M.J., I. Papageorgiou, M.A. Cau, P.M. Day and C.M. Jackson. In press. Integrating geochemical and mineralogical data in studies of ceramic provenance: some statistical issues (the GEOPRO project). Proceedings of CAA99, Dublin. Oxford, BAR.

Beardah, C.C. and M.J. Baxter. 2001. Grouping ceramic compositional data: an S-plus implementation. Proceedings of CAA2000, Ljubljana. Oxford, BAR.

Beier, T. and H. Mommsen. 1994. Modified Mahalanobis filters for grouping pottery by chemical composition, *Archaeometry* 36, 287-306.

Cau, M.A., I. Papageorgiou and M.J. Baxter. In preparation. Exploring automatic grouping procedures in ceramic petrology.

Gower, J.C. and D.J. Hand. 1996. *Biplots*. London: Chapman and Hall.

Kaufman, L. and P.J. Rousseeuw. 1990. *Finding Groups in Data*. New York: John Wiley.

Mathsoft. 1999. *S-Plus 2000 Programmer's Guide*. Seattle: MathSoft, Inc.

Papageorgiou, I., M.J. Baxter and M.A. Cau. 2001. Model-based cluster analysis of artefact compositional data. (Submitted for publication.)

Venables, W.N. and B.D. Ripley. 1999. *Modern Applied Statistics using S-Plus*. New York: Springer-Verlag.

Venables, W.N. and B.D. Ripley. 2000. *S Programming*. New York: Springer-Verlag.

Whitbread, I.K. 1991. Image and data processing in ceramic petrology, in A. Middleton & I. Freestone (Eds.), *Recent Developments in Ceramic Petrology*, British Museum, Occasional Paper no. 81, pp. 369-388.

Figures

Figure 1. Examples of ceramic thin-sections

		1	2	3	4	5	6	7	8
		opt.act	min.orient	Voids.ori.	texture	pecial.comp	plut.rocks	volc.rocks	metam.rocks
1	CS-2 (pl)	1	1	1	4	11	3	1	1
2	CS-3 (pl)	1	1	1	4	11	3	1	1
3	CS-4 (pl)	3	1	1	5	9	3	1	1
4	CS-5 (pl)	1	2	1	4	11	3	1	10
5	CS-6 (pl)	3	2	1	4	11	3	1	1
6	CS-7 (o)	3	1	1	3	9	2	1	4
7	CS-8 (o)	3	2	1	5	9	3	1	1
8	CS-9 (v)	3	1	1	4	11	1	2	1
9	CS-10 (v)	1	1	1	4	11	1	2	1
10	CS-11 (v)	1	1	1	4	11	1	2	1

		1	2	3	4	5	6	7	8
		opt.act.1	opt.act.2	opt.act.3	min.orient.1	min.orient.2	1	texture.3	texture.4
1	CS-2 (pl)	1.00	0.00	0.00	1.00	0.00	1.00	0.00	1.00
2	CS-3 (pl)	1.00	0.00	0.00	1.00	0.00	1.00	0.00	1.00
3	CS-4 (pl)	0.00	0.00	1.00	1.00	0.00	1.00	0.00	0.00
4	CS-5 (pl)	1.00	0.00	0.00	0.00	1.00	1.00	0.00	1.00
5	CS-6 (pl)	0.00	0.00	1.00	0.00	1.00	1.00	0.00	1.00
6	CS-7 (o)	0.00	0.00	1.00	1.00	0.00	1.00	1.00	0.00
7	CS-8 (o)	0.00	0.00	1.00	0.00	1.00	1.00	0.00	0.00
8	CS-9 (v)	0.00	0.00	1.00	1.00	0.00	1.00	0.00	1.00
9	CS-10 (v)	1.00	0.00	0.00	1.00	0.00	1.00	0.00	1.00
10	CS-11 (v)	1.00	0.00	0.00	1.00	0.00	1.00	0.00	1.00

Figure 2. The Can Sora data-set expressed as (a) a table of categorical data, (b) a matrix (G) of 0/1 entries obtained using coding method 1. Symbols in brackets denote the result of a preliminary petrographic classification

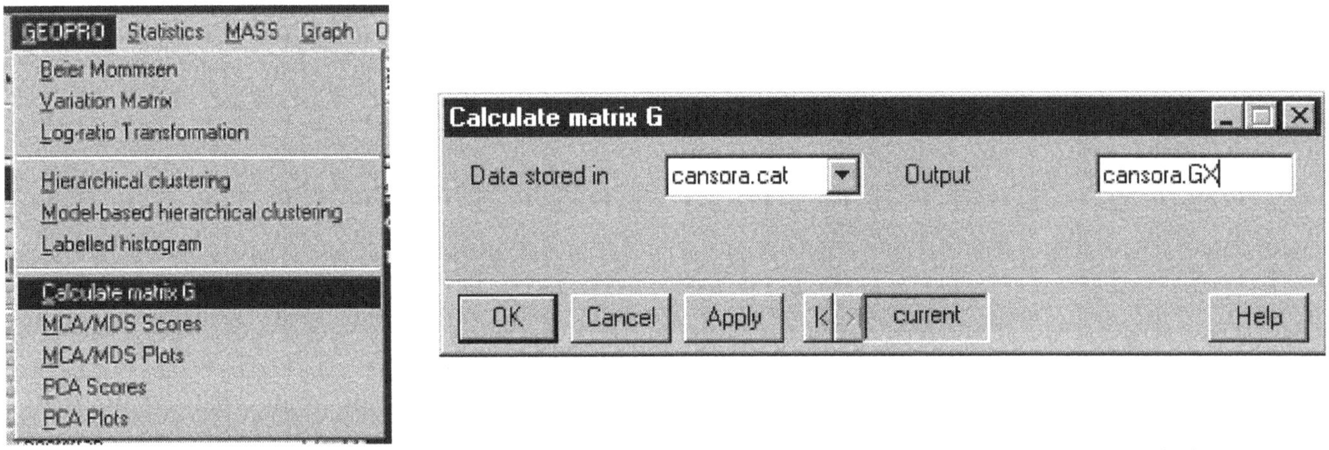

*Figure 3. (a) the GEOPRO menu, (b) dialog box enabling conversion between categorical representation and 0/1 representation (**catdist.mat** routine).*

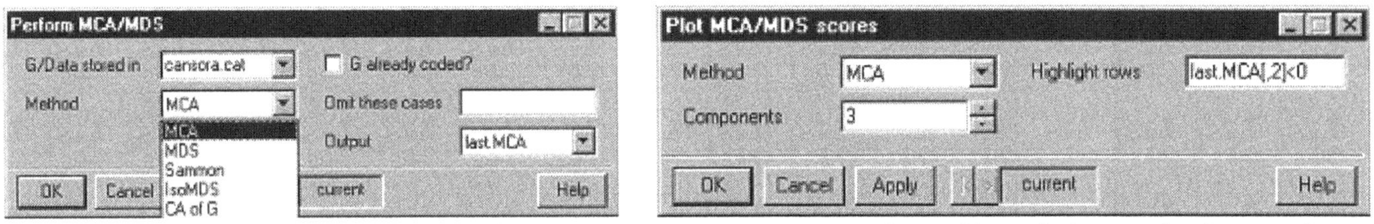

Figure 4. Dialog boxes for (a) application of various automatic grouping procedures, (b) producing graphical output from the calculation in (a).

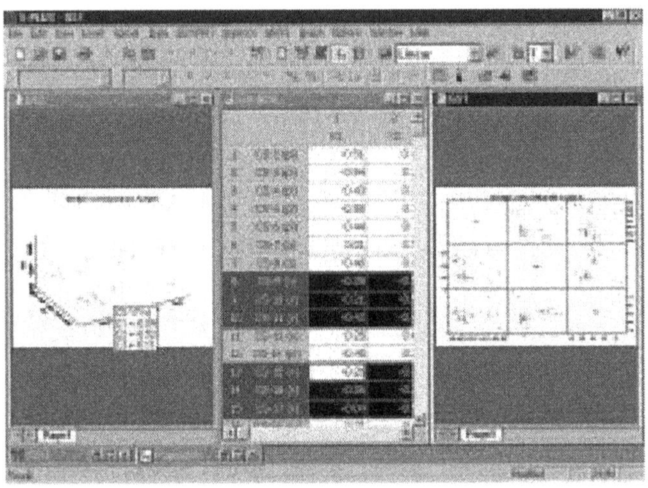

Figure 5. Application of MCA to the Can Sora data set.

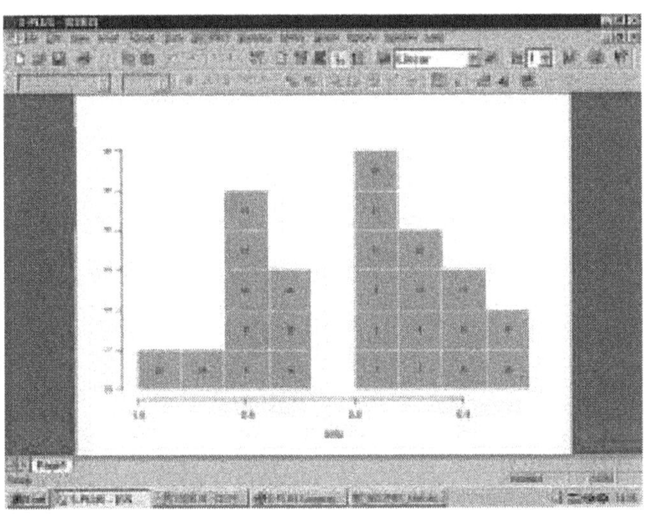

Figure 6. A labelled histogram (second component of MCA output for Can Sora dataset).

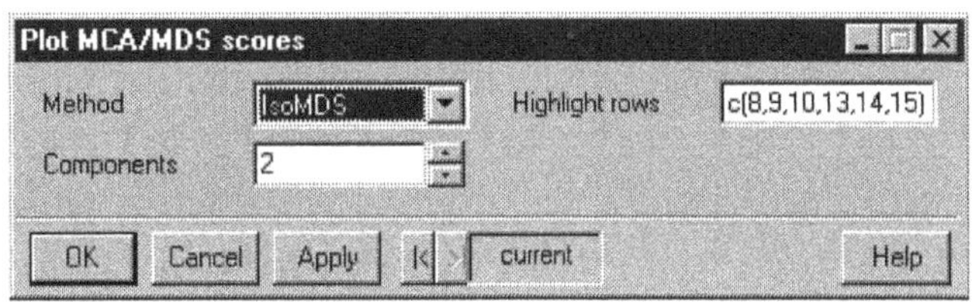

Figure 7. Generating the graphical output resulting from Kruskal's non-metric MDS. Rows 8, 9, 10, 13, 14 and 15 will be highlighted (this was the subgroup identified earlier using MCA, see figure 5).

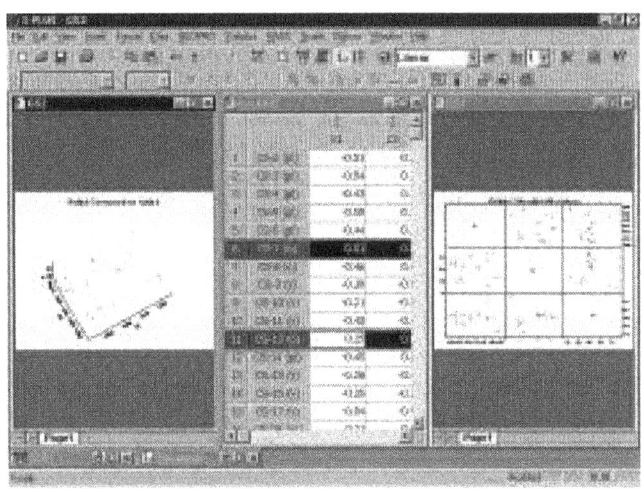

Figure 8. MCA output revealing that rows 6 and 11 (cases CS-7 and CS-13) of the Can Sora data set are fairly clear outliers.

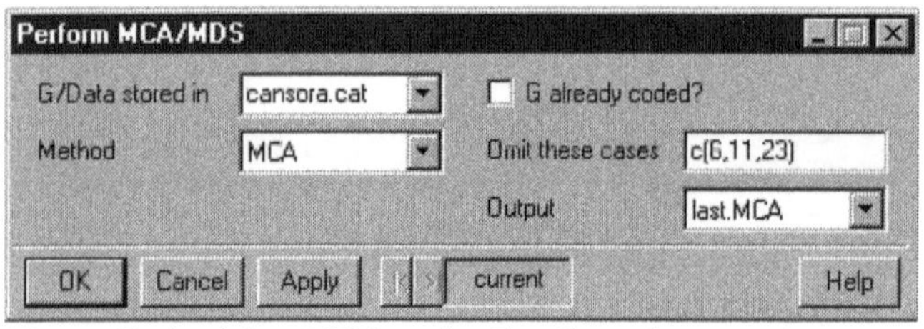

Figure 9. Applying MCA to the Can Sora dataset minus rows 6, 11 and 23 (cases CS-8, CS-24 and CS-25).

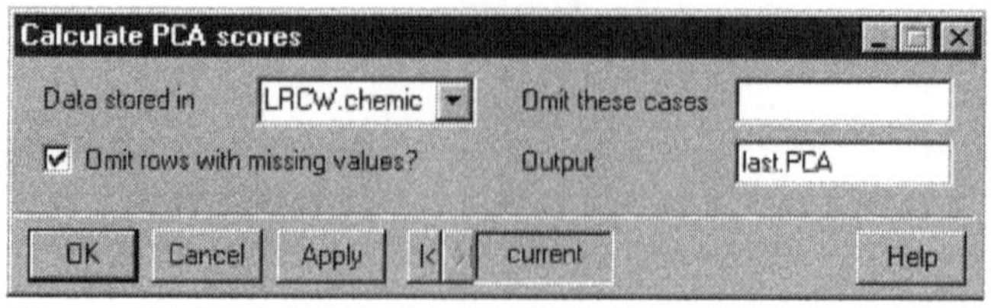

Figure 10. Applying PCA to the LRCW dataset.

Tables

Object number	Category
1	1
2	4 (= 1 + 3)
3	3
4	1

Table 1. A simple example of a 4 by 1 table of categorical data. The single variable is divided into categories 1, 2, 3 and 4, the last of which consists of the presence of both category 1 and 3. Also note that category 2 is missing in this particular data set.

Object number	**1**	**3**	**4 = 1 + 3**
1	1	0	0
2	0	0	1
3	0	1	0
4	1	0	0

Table 2. Coding method 1.

Object number	**1**	**3**
1	1	0
2	1	1
3	0	1
4	1	0

Table 3. Coding method 2.

Method	Acronym	Variable name	Components (maximum)
Multiple Correspondence Analysis	MCA	last.MCA	4
Multi-Dimensional Scaling	MDS	last.MDS	4
Sammon's Non-linear Mapping	Sammon	last.SAM	2
Kruskal's Non-metric MDS	IsoMDS	last.ISO	2
Correspondence Analysis of G	CA of G	last.CAG	4

Table 4. Summary of available methods.

Developing an information system for archaeological sites and monuments – administration and maintenance model

Malin Blomqvist
National Heritage Board
Box 5405
114 84 Stockholm
Sweden
Phone: +46 8 5191 8588 - Fax: +46 8 667 89 43 - E-mail: malin.blomqvist@raa.se

Cissela Génetay-Lindholm
National Heritage Board
Box 5405
114 84 Stockholm
Sweden
Phone: +46 8 5191 8551 - Mobile: +46 708 83 80 21 - Fax: +46 8 667 89 43 - E-mail: cissela.genetay-lindholm@raa.se

Abstract: The Swedish National Heritage Board is currently developing an information system for archaeological sites and monuments. The system is primarily intended as a support for cultural heritage management, forest protection management, planning of physical environment and scientific research.

The project is now at the second stage, which is composed of construction and test of the information system and preparation of a model for the administration and maintenance of the system.

Details are given about the background to the development project and about the work done on an administration and maintenance model. Further topics are the expectations of more efficient information handling and increased information quality, induced by the chosen model.

Key words: information system, GIS, archaeological sites and monuments, system administration, system management

Introduction

The topic of this paper is a development project where the aim is an information system for archaeological sites and monuments. The paper will elaborate on its use, administration and management. It also includes a short background to the project and a brief description of Swedish cultural heritage management. The development project is carried out at the National Heritage Board which is the central authority in Sweden for matters concerning the cultural environment and the cultural heritage.

Background

In 1938 a nationwide survey of archaeological sites and monuments was started by the National Heritage Board. The reason for this survey was a government decision on mapping the land use of the country. Archaeological sites and monuments protected by law were to be included on the Land use Map published by the National Land Survey. Information on the recorded sites, in the form of geographic positions on maps and descriptions in notebooks was gathered in the Register of Archaeological Sites and Monuments, in Swedish ḤFornminnesregistretḤ Information from subsequent surveys, from archaeological investigations as well as reports on finds made by private citizens have since been added to the register. The register is maintained by the National Heritage Board.

Today the parish-based register contains information on roughly 1 million objects at close to 400 000 sites. Not all of the sites and monuments in the register are covered by the Protection of the Cultural Monuments Act, the register includes for instance some early industrial sites which are not protected by law at present. The information in the Register has widespread use in Swedish society. It is needed in physical planning and decisions concerning land-use, in research and in the accumulation and communication of knowledge about the cultural heritage. Organisations using the information include the National Heritage Board, the County Administrations, the county museums, forestry organisations, large infrastructure developers, municipalities and private landowners.

The County Administration Boards are responsible for supervising cultural heritage management at regional level. Decisions on whether an archaeological site or monument is to be excavated and removed is made by the County Administration. In their decision they are obliged to consider whether the importance of preserving the site is greater than the importance of carrying out the development project in question. Information from the register is vital in this process.

The County museums are responsible for the accumulation of knowledge and research on the cultural heritage in their county. They are also responsible for communicating information about the cultural heritage to the public. Information on archaeological sites is part of these responsibilities.

The municipalities are responsible for physical planning on a local level and consequently they too need to have access to information from the Register.

Organisations involved in contract archaeology use information from the Register as a basis for planning, interpretation and for making prognoses on the presence of archaeological sites in areas where development is planned. This information is of course also of great interest to entrepreneurs planning construction work.

And finally, the public also have a need of and a wish for this information, not least because landowners are required by law to show consideration to archaeological sites on their land.

Today the need for this information involves large amounts of paper copies being sent between different organisations. Since the major part of other basic information used in planning is digital, it is becoming increasingly important that cultural information is also accessible in digital form.

In the middle of the 1980s the National Heritage Board started to digitize the descriptions in the register. The geographical information had been partly digitized by the National Land Survey.

The efforts of the National Heritage Board was never brought to a satisfactory conclusion and in 1998 the Board was given a commission by the government to solve the problem of digital access to this kind of information. In 1999 a development project was started.

The development project

The result of the development project is to be threefold:

- an information system, presenting information about archaeological sites and monuments together with geographical data from the databases of the National Land Survey
- a model for system administration and management
- support for the deployment of the system, for instance instruction in its use

The information system is to be in operation by the beginning of 2002. The information in the system will be more or less the same as that in the analogue register.
One of the tasks of the development project has been to make an evaluation of the quality of the existing digital information, both from a technical and from an archaeological point of view. The ambition is to re-use as much as possible of the existing digital information but it is already clear that some of it is of too poor a quality and will have to be digitized anew.

One of the basic aims of the future system is that information should be entered into the system as close to the source of the information as possible. Hopefully this will increase the quality of the information and minimize the timelapse between the discovery of a site and the time when information about it is entered into the national databases.

As a first step only the County Adminstrations will be able to enter data into the system. The change in responsibility for the entering of data from the National Heritage Board and to the County Administrations will however probably take some time.

A so-called Ḫunbroken digital chain of productionḪis a vision for the future. Whichever organisation produces data enters the relevant information into the system, for instance different organisations doing archaeological excavations. One example could be an archaeological unit using a field documentation system on a dig. When the excavation is finished someone from the unit transfers relevant parts of the digital information directly into the information system.

From a decentralised data entry follows a shared responsibility for the information in the system. The organisation responsible for entering information is also responsible for the information having the right quality. Decentralised entering of data, however, needs to be combined with different types of support from the National Heritage Board and an effective quality assurance system.

Initially the information system will be accessible to professionals working in cultural heritage management. Later on the information will be accessible to the public by way of the Internet. One of the reasons for the delay with going public is that the information contains many archaeological terms and may be difficult to interpret and understand for the layman. Time is needed to make the information more accessible to everyone.

System administration and management model

In addition to the system, the development project has the responsibility of creating a model for system administration and management. This includes management of the information in the system as well as management of a more technical nature. This is work in progress and the following is more a presentation of method and visions rather than a finished product.

Present situation at the National Heritage Board

A large number of information systems have been acquired by the National Heritage Board over the years. Some of them have been especially developed for specific purposes, others are standard products and have been used more or less without customization. The non-standard systems are primarily used to manage information on archaeological sites and monuments, buildings and also archival material, suchs as documents and photos.

Since the information systems have mostly been acquired to fulfil a specific purpose they are rather varied as regards development tools, software, accessibility, data storage etc. This makes administration and management difficult and expensive. The fact that much of the information concerned has a geographical dimension has unfortunately not always influenced the construction of the different systems.

Information on archaeological sites and monuments as well as buildings is to a great extent of common interest for the entire cultural heritage sector and this information is collected and used at several levels. However, the construction, administration and management of these systems have not always been sufficiently influenced by this fact. The development of the systems have seldom been made in co-operation with interested parties. This leads to problems with data capture, updating, accessibility etc.

Strategy for the use of information technology

At present the National Heritage Board is working on a strategy for the use of information technology. The aim is to make the use more uniform and efficient. It is also meant to elucidate on how this technology is to be used in order to fulfill the assignment given by the government.

The long term vision of the National Heritage Board is to have a common technical platform for several information systems. This means that information on archaeological sites and monuments, buildings and historical land use is stored in an information system for common use, a kind of common cultural heritage database.

This vision naturally demands a common administration and management organisation where synergy effects of joint system developments arise and is made use of, where different systems can use the same components and where there are established rules and routines for the administration and management of information.

In this way the emphasis is moved from the system to the information. Applications for working with the information is independent of the information itself and can be changed over time. In conclusion the information can be regarded as lasting and independent of technology.

No fixed timetable exists for the common technical platform and the common cultural heritage database but the information system for archaeological sites and monuments will be the first system built with an eye to the future.

Administration and management

An information system comprises manual routines, use of information as well as computer hardware and software.

To provide a system with information is often quite expensive. It is therefore very important to have an administration and management model which really works and which keeps the information up-to-date and with a guarantee of quality. Often the demands for changes regarding how to work with and compile information arise with those who work with it. The administration and management model must therefore be able to respond to these demands and allow certain changes. A system should never be so fixed that changes are impossible.

The administration and management model is an issue for those who work within the field of cultural heritage management. They own the system and make necessary decisions. It is not a technical issue.

When defining a model like this one some of the more important tasks are to identify all necessary functions, to clarify and describe different responsibilities and, finally, to make sure that everyone has understood and that they all agree. This entire process must be fully documented. This has been an extensive part of the work so far.

The next step has been to put the different functions in to comprehensive groups.

- Information provision
- Administration and management of the information
- Extending information
- Administration and management of the system
- Technical maintenance and support
- Helpdesk

Information provision describes the different ways of entering information into the system. It deals with the actual registration process and the data capture that is made during field work.

Administration and management of the information includes, among other things, all the manual routines that will be necessary both for today's analogue register as well as for the future system. It also includes education activities, information support, the execution of information compilations and analysis.

A reference organisation for *extending the information* involves the creation of formalised ways of contact, for example groups of users who discuss different aspects of the information and the information system. These groups include:

- A user forum which is formed by representatives of different user organisations where you can discuss defects in the system, proposed changes and suggestions for development.
- A forum for discussion on definitions and denominations of different types of archaeological sites and monuments. This group also needs a broad representation of users.

A vital process is the one concerning which types of cultural remains should be entered into the system. The discussion also includes whether they should be protected by law. The definition of what constitutes a protected site or monument is constantly developing and changing. The Cultural Monuments Act gives a framework which has to be discussed and reinterpreted. The National Heritage Board is by tradition the interpreter of the legislation, and the interpretation made by the

Board has become practice. Other institutions and organisations have previously only been marginally involved in this process. In the future the discussion should involve more people and institutions in the sector.

Administration and management of the system includes for example economic management, co-ordination of functions within the organisation, system development etc.

Technical maintenance and support includes among other things being responsible for the performance of the technology, for the accessibility of the system, for making back-ups, for the maintenance of hardware and software, for technical systems administration and for questions on security.

Helpdesk involves one thing only, to distribute, within the National Heritage Board, incoming questions to the right place in the organisation. Helpdesk is a vital function as it is the first contact with the National Heritage Board for the users. A user is normally not so much interested in how the National Heritage Board is organised, as getting correct and quick answers to questions. Helpdesk will be the receiving point of all questions concerning support on information and technique for most of the areas of activity of the National Heritage Board. Helpdesk is not responsible for solving problems, that is the responsibility of the support organisation.

Two models for administration and maintenance

The work has so far resulted in two models:

The first model is more of a temporary solution. The administration and management is organised as a project for a limited time, a year or two. In this way we avoid forcing the administration and management into an existing structure. The information system will result in a number of new tasks and responsibilities. In order to make those as clear as possible it is better to make a break with the organisation of today.

The administration and management could function in this way until there is a reason for having a common organisation for this purpose, that is when several of the information systems managed by the National Heritage Board are based on a common technical platform.

The second model deals with exactly this situation. Several information systems are based on a common technical platform. Another prerequisite is efficient co-ordination within the National Heritage Board in order to make use of synergy effects in, for instance, systems development. This model is a realization of the National Heritage Board's strategy regarding information technology and a common technical platform.

It is essential that the staff working with administration and management do not lose contact with daily issues and the information in the system. One way of assuring this could be that these individuals divide their time between day to day cultural heritage management and systems administration and management. Another way could be to circulate staff between different tasks, thus spreading competence in the organisation.

It is of great importance that both general and system specific manuals regarding system administration and management are produced at the National Heritage Board. The organisation, tasks and activities are constantly changing, people change jobs etc. Therefore, up to date and adequate documents describing the administration, management and maintenance, are vital.

In conclusion, we believe that no matter how the system administration and management is going to be organised, a greater extent of co-operation than previously will be required by the sector when this system is in operation.

Conclusion

Finally some thoughts on the use of an information system like the one under development. There is a great need today for analyses and compilation of the existing information, but as long as they are only available in analogue form these analyses are difficult to carry out.

Examples of possible use of the information system are:

- to investigate the degree of preservation and destruction of different types of archaeological sites and monuments;
- to study sites and monuments and current land use in order to take preventive measures, for instance, as regards to the damage caused by forestry;
- to compare how monuments of the same type have been classified in different parts of the country.

Hopefully, by the beginning of next year mission will be accomplished and the Swedish cultural heritage management will have access to an information system on archaeological sites and monuments. In a couple of years the information will also be accessible to the general public.

You are welcome to contact us at: fornminnesinfo@raa.se
More information can be found at:
http://www.raa.se/fornminnesinformation/ (in Swedish only)

How can a database full of Bugs help reconstruct the climate?

Phil Buckland
Environmental Archaeology Lab.
Department of Archaeology & Sami Studies
University of Umeå,
901 87 Umeå
Sweden
Phone: + 46 90 786 9792 - Fax: + 46 90 786 7663 - E-mail: phil.buckland@arke.umu.se

Paul Buckland
Department of Archaeology & Prehistory
University of Sheffield
Northgate House, West Street
Sheffield, S1 4ET, UK
Phone: +44 114 22 22913 - Fax: +44 114 27 22 563 - E-mail p.buckland@sheffield.ac.uk

Abstract: The BUGS Insect Ecology Package was originally constructed (using Dbase and Clipper) to compile Coleoptera (beetle) habitat and distribution data from a myriad of sources into one, easy to use, and publicly available database. Its primary users were researchers and teachers within the palaeoentomology field. The present system, five versions and many revisions later, is built in MS Access 2000, and covers some 5300 species, 2000 references, and 240 sites (archaeological and Quaternary), and is of value to archaeologists, ecologists, and conservationists alike.

BUGS is essentially a relational database management system constructed around three components:

- *the species data (modern ecology and distribution)*
- *the bibliography*
- *the site data with species lists*

Its implementation in several institutions has greatly accelerated the efficiency with which palaeoentomological investigations can be carried out, and greatly improved the teaching of the subject.

Palaeoenvironmental reconstructions are performed by the superimposition of the ecology and distribution of modern insect populations over fossil assemblages. At the moment, this is essentially performed semi-quantitatively by cross-reference of the data (which BUGS collates for a species list and then exports as an RTF file to any word processing package). BUGS contains a wealth of ecological data which can be employed in the interpretation of archaeological sites and contexts. In natural deposits, away from the artificial heat islands created by human activity, insect distributions are essentially constrained by climatic parameters. Tim Atkinson (UEA) and Dave Perry (formerly at Birmingham University) digitally encoded the temperature range data for over 400 species into a program for the calculation of palaeoclimates through the MCR (Mutual Climatic Range) method, and this has been extensively used in the modelling of Quaternary climates from beetle remains. The aim of our present phase of BUGS development is to implement MCR functionality into the BUGS database system. From this point it should be possible to move on to other ecological variables such as habitat and vegetation types, and increase the precision of modern climatic data, thus enhancing the value of insects in archaeological interpretation and the modelling of past climates.

Key words: Bugs, beetles, Coleoptera, database, ecology, palaeoecology, climate change, environmental archaeology, palaeoentomology

Introduction

The phylum Insects (= Hexapoda) makes up approximately 98% of the planet's animal biodiversity. Some 75% of these species are beetles, most of which have very specific habitat requirements, within which temperature is a very important variable. In addition, they are perhaps the most common –and useful- macrofossil the archaeologist is likely to encounter both on site and in suitable adjacent localities from which local palaeoenvironments may be reconstructed (cf. Buckland 2000; Ashworth *et al.* 1997; Buckland & Coope 1991; Buckland & Sadler 2000; Elias 1994). Insect remains are abundant in all anaerobic, permanently dry or permanently frozen contexts, from Eskimo middens (Böcher & Fredskild 1993) through Norse Greenland farms (Buckland *et al.* 1996), urban York (Kenward & Hall 1995) to the deserts of Egypt (Panagiotakopulu 2000). The reconstruction of these past environments relies upon the ability to project modern ecological data over fossil assemblages, allowing models to be built, often of very intimate places (cf. Kenward 1999; Panagiotakopulu & Buckland 1999). The close climatic constraints on the distribution of many species also means that insects are sensitive indicators of climate change, and it was study of fossil beetle faunas which first indicated the precipitate nature of the end of the last glaciation (Ashworth 1972; Osborne 1972; 1980; Bishop & Coope 1977; Coope & Brophy 1973), long before its confirmation in the Greenland ice core data (Mayewski *et al.* 1996; White 1993). The large sample size (> 5 litres preferred) and the slow process of identification by comparison with modern reference collections, however, has made this a field in which there are few practitioners. Interpretation also has had problems (cf. Kenward 1976), and habitat data on individual taxa are not only widely dispersed through an often obscure literature but also sometimes of doubtful quality – readers of Donisthorpe's (1939) *Coleoptera of Windsor Forest* for example would be convinced that most beetles just sat around on flowers! The BUGS database collates habitat (including environmental dependency variables) and distribution (both modern and fossil) for over 5000 species of beetle – the complete UK fauna, and a significant part of the North European fauna. With its twin on the Egyptian beetle fauna (Buckland *et al.* in press), they contain in excess of 100 mb of information, of use not only to palaeoecologists but also to applied entomologists and those involved in species' conservation. As part of a Leverhulme Trust funded project developing the use of flies in archaeological interpretation, a similar database is being constructed for the Palaearctic Diptera, and a molluscan database is also in process of development. The basic engine developed in Microsoft Access © can be applied to any biological group.

A detailed description of the Bugs Coleopteran Ecology Package, albeit an older version, has been presented elsewhere (Buckland *et al.* 1997), and so only a brief outline will be provided here. Instead this paper concentrate on the enhancements planned in the current expansion phase of the project, centring on the development of a semi-automated system for producing proxy climate data from fossil insect remains.

From a palaeoclimatological point of view, many groups of insects have the useful trait of being specific to particular habitats (stenotopic) or ranges of environmental variables. Despite earlier ideas that insects represent rapidly evolving lineages, it has become evident though work on the Quaternary fossil record that, during periods of climate change, the more stenothermal species of insect tend to migrate as a population, tracking the changing environment, rather than evolving (Coope 1990; 1995). Potentially therefore, they may provide at least two million years of proxy climatic data. Taken with their propensity to occupy very specific microhabitats, it allows us to reconstruct fossil environments, both human and natural, to a very detailed level, confined only by the limits of sample resolution (cf. Buckland & Perry 1989). In the case of climate reconstruction, the most significant recent work has involved the larvae of non-biting midges, the Chironomidae (cf. Brooks & Birks 2000), where sample resolution has allowed significant refinement on earlier models of Lateglacial Holocene transition, but the group is essentially restricted to lacustrine sediments and do not allow the parallel reconstruction of regional environmental changes. In this the most useful group is the Coleoptera (beetles), which include taxa, which are dependent on plant foods, as well as highly mobile predators, such as many of the species of carabid (ground beetles). These latter species, independent of specific plant or animal hosts, have the ability to migrate quickly as populations to take advantage of climate changes. They may, for example, be among the first species to arrive at newly exposed glacial tills during a retreat phase after rapid climatic amelioration. By collating the temperature tolerance data for species, essentially via the proxy of modern distributional data, found fossil in a sample it is possible to begin to construct snapshots of past environments and the changes that occurred in them, constrained only by sample resolution. This technique, called Mutual Climatic Range Analysis (MCR) (Atkinson *et al.* 1986; 1987) relies upon the definition of individual climatic envelopes for each species, and then overlying these to define the range in which they may mutually occur. It is the intention to expand the BUGS program to include a facility for calculating palaeo-temperatures.

BUGS in Brief

BUGS, the Coleopteran Ecology Package, was designed as a research aid for both palaeoecologists and entomologists. Its roots lie in a DOS database (DBase IV & Clipper) database created by Jon Sadler and Mike Rains (Sadler *et al.* 1992), and the present form, built in Microsoft Access ©, BUGS 2000v5, consists of a primary database, of ecology, distribution and fossil record data for over 5,300 species, continually updated (with images and key identification characters being added slowly); the associated bibliography lists over 2500 references (Fig 1). A separate listing Qbib includes all known papers on fossil insects from the Quaternary, including the archaeological record. In addition, to this, a database of about 240 sites, archaeological and Quaternary, contains species spread sheets in taxonomic order, which can be accessed through BUGS or Microsoft Excel © (Fig.2). In the process of data entry, BUGS automatically orders any entered species list into taxonomic order. Only basic site data is included, such as latitude and longitude, elevation, interpretation, references, and dating evidence, to avoid the duplication of parallel site databases.

Basic query functions allow the user to retrieve a subset of the species database, which is restricted to their search criteria. This could be used, for example, to find all the species, which are attracted to fire, and found in Britain at the present day (Bugs names 54 species). This feature is particularly useful when a sample produces species that are no longer found in the existing environment of the sample, or even the country under research. Similarly, the Red Data Book (RDB) field allows a researcher to see whether a species is present/rare at the present day.

Built in reports provide simple, but very useful, printable outputs for: species data; search results (just the names, or full details); site species lists; ecology and distribution for species at a site; and references/bibliography. These reports, which can be printed directly, or saved as rtf files, effectively reduce the time necessary to produce meaningful results from an insect assemblage. Habitat codes, based upon those used by Koch (1989a & b; 1992) and incorporating the ideas of both Kenward (1978) and Robinson (1983; 1991), are being developed to help in the process of dimensioning a faunal assemblage across taxonomic boundaries to refine interpretation.

Where beetles like to be – the Mutual Climatic Range

Any species of beetle (or animal/plant for that matter) can only survive within a certain temperature range. So, when finding a fossil population of a certain species in a sample it is possible to say that the temperature of the environment represented by the sample must have been within the range of tolerance for that species. This may sound perfectly logical, but there always questions of interpretation to consider – such as what does the sample *actually represent*. Depositional conditions, as well as post depositional disturbances, redeposition or bioturbation, have to be worked out with the archaeologists/geologists during both primary sampling and interpretation; the interdisciplinary nature of both Quaternary studies and archaeoentomology cannot be over stressed. It must always be considered whether the individuals found actually lived in the sample environment – possibly as a breeding population, or were just passing through, so to speak. This is nowhere more problematic than in the archaeological context, where the artificially warmed habitats created by both human and domestic animal waste products, as well as house, barn and byre interiors, provide heat islands well north of natural distribution areas for many species (Barlow *et al.* 1998). A wide range of species, termed synanthropic, largely accidentally transported by Man, occur in these contexts, from the house fly, *Musca domestica*, probably of south Mediterranean or Old World subtropical origin (Skidmore 1985) to the grain weevil, *Sitophilus granarius*, an early fellow traveller (Büchner & Wolf 1997; Buckland 1990; Panagiotakopulu 2000) and not now known from a natural habitat. In addition, the accidental import of insects in dunnage and ballast on ships may lead to either temporary establishment of local populations or an enigmatic fossil find in an archaeological context. The single find of the rather southern scarabaeid *Oxyomus sylvestris* at medieval Bessastaðir in Iceland for example can easily be dismissed as an accidental traveller (Amorosi *et al.* 1992), and the ants from post-medieval Reykholt, Iceland (Buckland *et al.* 1992), probably arrived in straw around the priest's wine bottles, but the several specimens of the small rove beetle *Metopsia retusa/clypeata*, now only found in the very south of Norway and Sweden (Lundberg 1995), in the midden of the medieval fishing station at Langenes in the Norwegian Arctic is more problematic. Was the medieval thermal optimum this much warmer (Buckland & Wagner 2001), or is this again a ballast and dunnage accidental traveller? Only further refinement of the climate model, and more fossil faunas from the site can provide the answers.

Setting aside these essentially taphonomic problems, and taking as given that a sample accurately represents the environment in question, then it is possible to go on to refine the climatic interpretation by overlaying the individual climatic envelopes derived from modern distributional data to find the range of temperatures in which all species present could survive (Fig.3). This is where choice of species is fundamentally important. When considering transient environments, or where a chronology is required, it is necessary to employ species that are mobile (but not necessarily winged), omnivorous, and not plant host specific. In this way, it should be possible to eliminate the effects of any other variables from the equation (although these variables will of no doubt be important for the overall site interpretation). Where only a snapshot reconstruction is required, then the migration abilities of the species used in the calculations is of less importance, but it is apparent that succession is important in terms of what arrives first. Those species which form part of what Crowson (1981) termed the aerial plankton appear first onto any *tabula rasa* created for example by deglaciation. Thus chironomids and caddis flies (Trichoptera) may arrive before any beetles, and incidentally their food amongst the diatoms and algae must also be the first colonisers. In most contexts, however, sample resolution is insufficient to define this process.

Existing Mutual Climatic Range (MCR) Software

BUGS runs in the MS Windows© environment, and is built around the Access 2000© database engine. Having been developed in DOS© during the 1980s, MCR has yet to undergo an updating process, and at present two pieces of software exist for calculating MCR envelopes from species lists. Both run in MSDOS and are referred to as MCR Birmingham, and MCR UEA (University of East Anglia) after the institutions where they were created (Atkinson *et al.* 1986). Although both systems were excellent tools at their time of creation, and Sinka and Atkinson (1999) have recently developed a version which utilises plant distribution, their interfaces are difficult to use and their data storage system cumbersome in terms of the present day Windows environment. Both systems used a binary temperature matrix (Fig.4) to store species thermal envelopes, which allows a resolution of only 1° C, although it should be noted that further refinement is ultimately constrained by the coarse nature of modern distribution data, often conflating records of over a hundred years of insect collecting. The use of the present form of MCR to reconstruct climate from archaeological assemblages from medieval sites is therefore questionable (cf. Hellqvist & Lemdahl 1996; Buckland & Wagner 2001).

Implementing MCR in BUGS (Figs 5-6)

As a first step, the existing digital MCR data has been ported over into BUGS as a working base. However, it is generally agreed that these datasets, initially obtained by digitising small scale distribution maps of species' Palaearctic range (Atkinson *et al.* 1986; Perry 1986), are no longer considered to be of high enough resolution, and will eventually be superseded by new datasets. The old form of MCR was very much restricted by the technology of the 1980's and the availability of computers. The new system is intended as an upgradeable expert system, which can be asked to recalculate thermal envelopes and MCR overlaps in the light of new data. As with the previous version, however, data entry remains the main problem, since each record requires location, elevation, and year of capture to be collated against the relevant interpolated climate data, something which would have been impossible with the equipment, and datasets available twenty years ago.

Data storage (fig.5)

Base distribution data for species comes from a variety of

sources, both maps and text, mainly printed material, which will need to be digitised (fig.4a). In Scandinavia, the published distributional checklists (e.g. Lindroth 1961; Lundberg 1995) are based upon political units, which cut completely across geographical units, from sea level to mountain peaks. For some relevant groups, however, such as the Carabidae (Lindroth 1945) more closely sourced data are readily available, and in Britain a series of distribution maps using 10 km^2 is in process of publication by the Biological Records Centre (e.g. Luff 1998). ArcView's © digitising capabilities are used to convert mapped locations directly into latitude and longitude points which are transferred into BUGS. (Note that these values are not stored in ArcView since it is intended that Bugs (Access) will be the repository for as much data as possible, and will perform reconstruction calculations without the use of ArcView). Text sources are entered directly into Access tables, and figure 5b shows the structure. Weather station/temperature data and topography data will be stored within an ArcView project which will link to the appropriate tables in BUGS on demand (Fig.5c).

ArcView's spatial analysis extensions will be used to convert the lat/long values into TMAX and TRANGE (see fig.3) values for each species entered. This will feed back into Access to fill the empty columns of the MCR tables, rather than producing a binary temperature matrix as the previous MCR programs did, thereby increasing the resolution of the data, and making additions easier.

Retrieval – Reconstructing the Climate (Fig.6)

Currently there is a backlog of available data waiting entry, but once a substantial dataset, initially of the Carabidae, has been constructed, a list of which is maintained in BUGS, it will be possible to begin to look at sample assemblages. Each sample from the selected site can be compared with the MCR species list and the common species extracted into an MCR subset (Fig.6a). This can then be used to calculate the MCR – the Mutual Climatic Range overlap diagram being the visual graphical representation (Fig.6b). This provides a frequency diagram, indicating the number of species utilised in the reconstruction, their degree of overlap, and frequency. The option of constructing time series diagrams is also be available for sites, which have dates and/or originate from profiles or cores, such as from lake or bog sediments (Fig.6c).

Asking Questions

Correlation coefficients have frequently been used to compare sample assemblages, and Kenward (1978) has consistently championed Fisher's α in the comparison of archaeological assemblages. BUGS is being upgraded to include a series of common biological comparative methods (e.g. Jaccard, modified Sorenson (Southwood & Henderson 2000), so that similarities between samples and sites can be sought. The querying form- is also customised to allow comparison within thermal groups, or to take thermal envelopes into consideration when comparing assemblages. This should give opportunities to think of new questions to ask the datasets, which were previously difficult to handle.

Issues yet to be Resolved

The final nature of calculation of the TMAX/TRANGE values from the temperature/lat/long data has yet to be resolved, for example, whether we will use a basic 6°C per 1000 m altitude lapse rate, or cubic splining of the temperature trend surfaces. It is possible that a better approach would be to use remotely sensed data rather than interpolations from weather stations, although the constraints here are likely to be financial, and the data is only available for the last couple of decades. The availability of weather station data varies from country to country, whereas satellite data is more uniformly available, albeit more expensive. On the other hand binary raster data take more processing power/time and could prove impractical for the purposes of interpolation, although one can always limit the processed area within ArcView. The mathematics involved in the calculation of MCR overlaps are significantly complicated by the switch from binary temperature matrices (fig.4) to linear data tables (Fig.5b), even if it does have great advantages in terms of expandability and accuracy. Further consideration needs to go into the statistical implications, and associated error sources, of the various methods available.

ArcView does have some database manipulation capabilities, a query engine, and the Avenue scripting language, and it is possible that some calculations will be more easily performed in ArcView than Access where spatial data are concerned. The practical desire to keep as much calculating as possible within Access (for reasons of portability) may have to be compromised for the sake of simplicity. However, with the incorporation of the Visual-BASIC language into the latest version of ArcView, this may become less of a problem, with both engines using different programming languages. The desire to keep BUGS as a freeware product, which can be used by those who do not have expensive software such as ArcView (or even Access, with the runtime installation), may lead to Access as the motor for most of the calculations.

In a project of this type, data entry is always a very time consuming task, and makes up a considerable part of the workload. It has therefore been decided to limit initially the study area to Scandinavia and the UK. The primary dataset will also only include the Carabidae. The geographic areas have good histories of insect recording, and have easily available distribution and weather data, whilst the carabids have been extensively collected in both regions (e.g. Lindroth 1985; 1986; Luff 1998). Even though this is a compromise, and excludes species which occur in Quaternary fossil deposits with either Mediterranean or Siberian distributions (Angus 1973; Coope 1990), it still allows a 3000 km north-south transect, and the system is designed with expandability firmly in mind. It goes without saying, the more detailed and extensive the base data are, the better the accuracy of the system. The Icelandic and Greenlandic fossil insect data would suggest that the coldest part of the Holocene was the Little Ice Age of the post-medieval period (Buckland & Wagner 2001). Throughout the Holocene, however, the fossil insect evidence, as Blytt and Sernander (1908) long ago realised with fossil plant distribution, implies oscillation between relative oceanicity and continentality, and the east-west transect provided by the dataset is at least as important as the north-south; the difficulty, as always, is taking Man, and human impact on available habitats, out of the equation (Wagner 1997).

Future Applications of MCR Bugs

As Perry showed nearly twenty years ago in terms of the Icelandic beetle fauna (Buckland *et al.* 1983), MCR can be used for -predictions, in both the past and the future. Given a particular climatic envelope, it is possible to provide a list of what species should be present given unimpeded access. Simulation is an extremely important area within meteorology, geography, and ecology, and BUGS is designed with this in mind. Useful questions might be, for example, if the temperature were to decrease by 1°C what would happen to the distribution of species X, which is a predator on crop pest Y? In the archaeological context, there are opportunities for re-examining the Elm decline here (Girling & Greig 1985; Robinson 2000), utilising fine climate data derived from contemporary insect faunas – does the expansion of the elm bark beetle, *Scolytus scolytus* coincide with a period of climatic stress? The fine resolution of beetle chronologies has been instrumental in illustrating the rapid nature of climatic change, and has complemented other research, which stresses the often precipitate nature of glacial/interglacial switches (e.g. Taylor *et al.* 1993), and there is much to be gained from further study.

Modern insect distribution maps are often full of gaps where collecting, for one reason or another, has been scarce. This may leave holes in distributions where the species quite probably could be found, if there was anybody interested enough to look for it there. MCR BUGS provides a database, which can be used as a source to direct collectors to particular areas. Andersen (1993; 1996) has queried whether fossil beetles can be used to retrodict climate, and although his arguments were effectively countered by Coope and Lemdahl (1996), there is no doubt that MCR is currently a very coarse though effective tool. Its development within the BUGS framework should help to dispel any remaining doubts.

Conclusion

Temperature is just one many environmental variables that influence insect distribution. Rainfall and number of daylight hoursday length are perhaps two of the more obvious others that could be considered in palaeoenvironmental reconstructions, both being essential when crop yields are considered, and even more so in the potentially extreme limits of farming systems (cf. Barlow *et al.* 1998). BUGS is designed to be expandable, and extra columns fields can be added to the MCR tables to store as many variables as required. In the long term, its various facets should add much to archaeological and palaeoecological interpretation, although it should perhaps be added finally that all palaeoecological reconstruction is only as good as the initial field sampling, and no amount of computer application can replace good field techniques.

BUGS is distributed for free over the net from a website (http:www.umu.se/envarchlab), or by post to those who request it.

Acknowledgements

The two elements currently being combined into BUGS began with Jon Sadler, at the University of Birmingham, and Tim Atkinson at University of East Anglia in Norwich. Both owe much to the primary stimulus of Russell Coope, then at Birmingham, later at Royal Holloway, University of London. All are gratefully acknowledged. Development of the initial Access version of BUGS was by Yuan Zhuo Don, whilst an MSc student at the University of Sheffield, and later revisions came from the Genisys development group there. The authors are grateful to Roger Engelmark, University of Umeå for continued interest in the project, and for funding to the Northern Crossroads project devised by Noel Broadbent, also at Umeå. Various parts of the work have been funded by an NSF grant to Tom McGovern, City University of New York, as part of NABO, the North Atlantic Biocultural Organisation, and by the Leverhulme Trust, whom are hereby thanked.

References

Amorosi, T., Buckland, P.C., Ólafsson, G., Sadler, J. P., & Skidmore, P., 1992. Site Status and the Palaeoecological Record: A Discussion of the Results from Bessastaðir, Iceland. In C. D. Morris & D. J. Rackham (eds.) *Norse and Later Settlement and Subsistence in the North Atlantic*, 169-192. Dept. of Archaeology, University of Glasgow.

Andersen, J., 1993. Beetle remains as indicators of the climate in the Quaternary. *Journal of Biogeography*, **20**, 557-562.

Andersen, J., 1996. Do beetle remains reliably reflect the macroclimate in the past? - a reply to Coope & Lemdahl. *Journal of Biogeography*, **23**, 120-121.

Angus, R. B., 1973. Pleistocene *Helophorus* (Coleoptera, Hydrophilidae) from Borislav and Starunia in the Western Ukraine, with a reinterpretation of M. Lomnicki's species, description of a new Siberian species, and a comparison with British Weichselian faunas. *Philosophical Transactions of the Royal Society of London*, **B265**, 299-326.

Ashworth, A. C., 1972. A Late-glacial Insect Fauna from Red Moss, Lancashire, England. *Entomologica Scandinavica*, **3**, 211-224.

Ashworth, A. C., Buckland, P. C. & Sadler, J. P. (eds.., 1997. *Studies in Quaternary Entomology - An Inordinate Fondness for Insects. Quaternary Proceedings* **5**.

Atkinson, T. C., Briffa, K. R., Coope, G. R., Joachim, J. M. & Perry, D. W., 1986. Climatic calibration of coleopteran data. In B. E. Berglund (ed.) *Handbook of Holocene Palaeoecology and Palaeohydrology*, 851-858. J. Wiley & Son, Chichester.

Atkinson, T. C., Briffa, K. R. & Coope, G. R., 1987. Seasonal temperatures in Britain during the past 22,000 years, reconstructed using beetle remains. *Nature* (London), **325**, 587-592.

Barlow, L. K., Sadler, J. P., Ogilvie, A., Buckland, P. C., Amorosi, T., Ingimundarsson, J. H., Skidmore, P., Dugmore, A. J. & McGovern, T. H., 1998.. Ice core and environmental evidence for the end of Norse Greenland. *The Holocene* **7**: 489-499.

Bishop, W. W. & Coope, G. R., 1977. Stratigraphical and faunal evidence for Lateglacial and early Flandrian environments in South-West Scotland. In J. M. Gray & J. J. Lowe (eds.) *Studies in the Scottish Lateglacial Environment*, 61-88. Pergamon, Oxford.

Brooks, S. J. & Birks, H. J. B., 2000. Chironomid-inferred Lateglacial and early-Holocene mean July air temperatures for Kråkenes lake, western Norway. *Journal of Paleolimnology*, **23**, 77-89.

Buckland, P. C., 1990. Granaries Stores and Insects. The Archaeology of Insect Synanthropy. In D. Fournier & F. Sigaut (eds.) *La préparation alimentaire des céréales,* 69-81. Rapports présentés à la Table ronde, Ravello au Centre Universitaire pour les Biens culturels, avril 1988. PACT, Rixensart, Belgium.

Buckland, P. C., Amorosi, T., Barlow, L. K., Dugmore, A. J., Mayewski, P. A., McGovern, T. H., Ogilvie, A. E. J., Sadler, J. P. & Skidmore, P., 1996. Bioarchaeological and climatological evidence for the fate of Norse farmers in medieval Greenland. *Antiquity*, **70**, 88-96.

Buckland, P. C. & Coope, G. R., 1991. *A Bibliography and Literature Review of Quaternary Entomology*. J. Collis Publications, University of Sheffield.

Buckland, P. C. & Perry, D. W., 1989. Ectoparasites of Sheep from Storaborg, Iceland and their interpretation. Piss, parasites and people, a palaeoecological perspective. *hikuin*, **15**, 37-46.

Buckland, P. C., Perry, D. & Sveinbjarnardóttir, G., 1983. *Hydraena britteni* Joy (Coleoptera, Hydraenidae) fundin á Íslandi í setlögum frá því seint á nútima. *Náttúrufræðingurinn*, **52**, 37-44.

Buckland, P. C. & Sadler, J. P., 2000. Animal remains, identification and analysis: Insects. In, L. Ellis (ed.) *Archaeological Method and Theory: An Encyclopaedia*, 21-26. Garland, New York & London.

Buckland, P.C., Sadler, J. P. & Sveinbjarnardóttir, G., 1992. Palaeoecological Investigations at Reykholt, Western Iceland. In, C. J. Morris & D. J. Rackman (eds.) *Norse and Later Settlement and Subsistence in the North Atlantic*, 149-168. Dept. of Archaeology, University of Glasgow.

Buckland, P. C. & Wagner, P., 2001. Is there an insect signal for the Little Ice Age? *Climatic Change,* **48**, 137-149.

Buckland, P. I., 2000. *An introduction to palaeoentomology in archaeology and the BUGS database management system*. Institute for Archaeology and Sami Studies, University of Umeå.

Buckland, P. I., Buckland, P. C., Panagiotakopulu, E. & Sadler, J. P. (in press) EGBUGS a database for Egyptian entomology. *Egyptian Journal of Entomology*.

Buckland, P. I., Yuan Zhuo, D. & Buckland, P. C., 1997. Towards an expert system in Palaeoentomology. In A.C.Ashworth, P.C.Buckland & J.P.Sadler (eds.) *Studies in Quaternary Entomology - An Inordinate Fondness for Insects. Quaternary Proceedings* **5**, 67-78.

Böcher, J. & Fredskild, B., 1993. Plant and arthropod remains from the palaeo-Eskimo site on Qeqertasussuk, West Greenland. *Meddelelser om Grønland, Geoscience*, **30**.

Büchner, S. & Wolf, G., 1997. Der Kornkäfer - *Sitophilus granarius* (Linné) - aus einer bandkeramischen Grube bei Göttingen. *Archäologisches Korrespondenzblatt*, **27**, 211-220.

Coope, G. R., 1990. The invasion of Northern Europe during the Pleistocene by Mediterranean species of Coleoptera. In F. di Castri, A. J. Hansen & M. Debussche (eds.) *Biological Invasions in Europe and the Mediterranean Basin*, 203-215. Kluwer, Dordrecht.

Coope, G. R., 1995. The Effects of Quaternary Climatic Changes on Insect Populations: Lessons from the Past. In R. Harrington & N. E. Stork (eds.) *Insects in a changing environment*, 29-48. Academic Press, London.

Coope, G. R. & Brophy, J. A. , 1972. Late Glacial environmental changes indicated by a coleopteran succession from North Wales. *Boreas*, **1**, 97-142.

Coope, G. R. & Lemdahl, G., 1996. Validations for the use of beetle remains as reliable indicators of Quaternary climates: a reply to the criticisms by Johan Andersen. *Journal of Biogeography*, **23**, 115-120.

Crowson, R. A., 1981.. *The biology of the Coleoptera*. London, Academic Press.

Donisthorpe, H. St. J. K., 1939. *The Coleoptera of Windsor Forest.* Publ. Privately., London.

Elias, S. A., 1994. *Quaternary Insects and Their Environments*. Smithsonian Institution Press, Washington.

Girling, M. A. & Greig, J. R. A. , 1985. A first fossil record for *Scolytus scolytus* (F.) (Elm Bark Beetle) : its occurrence in Elm Decline deposits from London and the implications for Neolithic Elm Disease. *Journal of Archaeological Science*, **12**, 347-352.

Hellqvist, M. & Lemdahl, G., 1996. Insect Assemblages and Local Environment in the Mediaeval Town of Uppsala, Sweden. *Journal of Archaeological Science* **23**, 873-881.

Kenward, H. K. , 1976. Reconstructing ancient ecological conditions from insects remains : some problems and an experimental approach. *Ecological Entomology*, **1**, 7-17.

Kenward, H. K. , 1978. *The Analysis of Archaeological Insect Assemblages : a New Approach.* Archaeology of York, **19/1**. Council for British Archaeology for York Archaeological Trust.
Kenward, H. K., 1999. Pubic lice (*Pthirus pubis*) were present in Roman and medieval Britain. *Antiquity*, **73**, 911-915.

Kenward, H. K. & Hall, A. R., 1995. *Biological Evidence from 16-22 Coppergate*. Archaeology of York, **14/7**. Council for British Archaeology for York Archaeological Trust, York.

Koch, K., 1989a & b; 1992. *Die Käfer Mitteleuropas. Ökologie*, **1-3**. Goecke & Evers, Krefeld.

Lindroth, C. H., 1945.. Die Fennoskandischen Carabidae I-III. *Goteborgs K. Vetensk. o VitterhSamh. Handl.*(6) B, 4. Goteborg.

Lindroth, C. H., 1961.. *Catalogus Coleopterorum Fennoscandiae et Daniae.*

Lindroth, C. H., 1985;1986. The Carabidae (Coleoptera) of Fennoscandia and Denmark. *Fauna Entomologica Scandinavica*, **15**. E.J.Brill, Leiden.

Luff, M. L., 1998. *Provisional atlas of the ground beetles (Coleoptera, Carabidae) of Britain.* Centre for Ecology & Hydrology, Biological Records Centre, Abbots Ripton.

Lundberg, S., 1995.. *Catalogus Coleopterorum Sueciae*. Stockholm, Naturhistoriska Riksmuseet.

Mayewski, P. A., Buckland, P. C., Edwards, K. J., Meeker, L. D. & O'Brien, S., 1996.. Climate change events as seen in the Greenland ice core (GISP2). *The early prehistory of Scotland.* T. Pollard and A. Morrison. Edinburgh, Edinburgh University Press: 74-86.

Osborne, P. J. , 1972. Insect faunas of Late Devensian and Flandrian age from Church Stretton, Shropshire. *Philosophical Transactions of the Royal Society of London*, **B263**, 327-367.

Osborne, P. J. , 1980. The Late Devensian-Flandrian transition depicted by serial insect faunas from West Bromwich, Staffordshire, England. *Boreas*, **9**, 139-147.

Panagiotakopulu, E., 2000. *Archaeology and entomology in the eastern Mediterranean. Research into the history of insect synanthropy in Greece and Egypt.* British Archaeological Reports International Series **836**. Oxford.

Panagiotakopulu, E. & Buckland, P. C., 1999. The bed bug, *Cimex lectularius* L. from Pharaonic Egypt. *Antiquity*, **73**, 908-911.

Perry, D. W., 1986. *The Analysis of sub-fossil insect assemblages : a Numerical Approach.* Unpubl. Ph.D. thesis, University of Birmingham.

Robinson, M. A. , 1983. Arable/Pastoral Ratios from Insects. In, M. Jones (ed.) *Integrating the Subsistence Economy*, 19-47. British Archaeological Reports **S181**, Oxford.

Robinson, M. A., 1991. The Neolithic and Late Bronze Age insect assemblages. In S. Needham, *Excavation and salvage at Runnymede Bridge, 1978: the Late Bronze Age waterfront site*, 277-325. British Museum, London.

Robinson, M. A., 2000. Coleopteran evidence for the Elm Decline, Neolithic activity in woodland, clearance and the use of the landscape. In, A. S. Fairbairn (ed.) *Plants in Neolithic Britain and beyond. Neolithic studies group seminar papers* **5**, 27-36. Oxbow Books, Oxford.

Sadler, J. P., Buckland, P.C. & Rains, M., 1992. Bugs: an entomological database. *Antenna.* **16**, 158-166.

Sernander, R., 1908.. On the evidence of Post-glacial changes of climate furnished by the peat mosses of northern Europe. *Geologisk Förenings Stockholm Förhandlingar* **30**: 465-478.

Sinka, K. J. & Atkinson, T. C., 1999.. A mutual climatic range method for reconstructing palaeoclimate from plant remains. *Journal of the Geological Society, London* **156**: 381-396.

Skidmore, P., 1985.. *The biology of the Muscidae of the World.* Dordrecht, W. Junk.

Southwood, T. R. E. & Henderson, P. A., 2000. Ecological Methods (3rd ed.). Blackwell Science, Oxford.

Taylor, K. C., Lamorey, G. W. Doyle, G. A., Alley, R. B., Grootes, P. M., Mayewski, P. A., White, J. W. C. & Barlow, L. K., 1993.. The 'flickering switch' of climate change. *Nature* **361**: 432-436.

Wagner, P. E., 1997. Human Impact or Cooling Climate? The "Little Ice Age" and the beetle fauna of the British Isles. In A.C.Ashworth, P.C.Buckland & J.P.Sadler (eds.) *Studies in Quaternary Entomology - An Inordinate Fondness for Insects. Quaternary Proceedings* **5**, 269-276.

White, J. W. C., 1993.. Don't touch that dial. *Nature* **364**, 186.

Figures

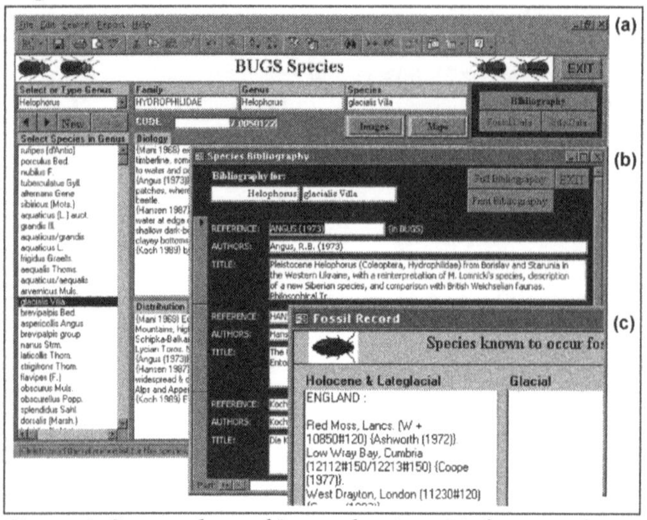

Figure 1. Screen shots of Bugs, showing: (a) the main data screen; (b) bibliography for the selected species; and (c) fossil record for the selected species.

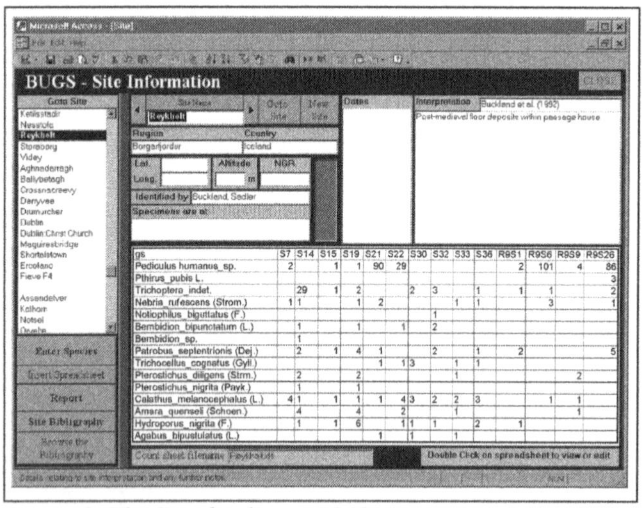

Figure 2. The Site database with the species count sheet for "Reykholt" selected (Buckland et al. *1992). Species names are listed down the left of the count sheet, and samples displayed in columns thereafter. Double clicking opens the sheet in MS Excel ©.*

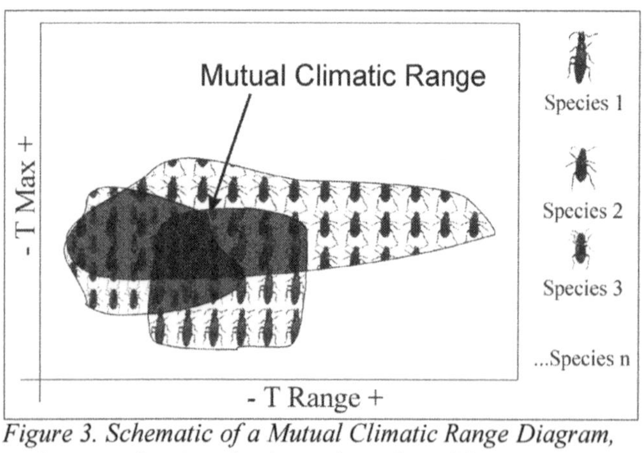

Figure 3. Schematic of a Mutual Climatic Range Diagram, in this case showing the thermal overlap of three species. TMax is the mean temperature of the warmest month of the year, which the species will tolerate, and TRange the difference between this and TMin, the mean of the coldest month. The Mutual Climatic Range itself is the area of most

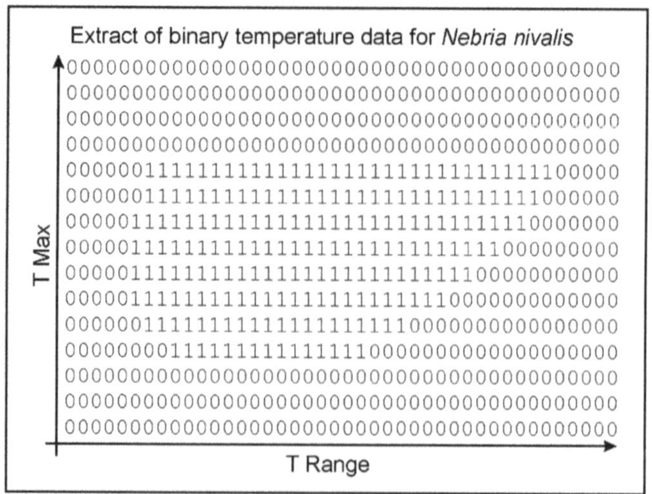

Figure 4. Binary temperature envelope data from the MCR Birmingham program. Each digit represents 1°C , of a grid with the dimensions 0 to 38°C for TMax and 5 to 65°C for T Range.

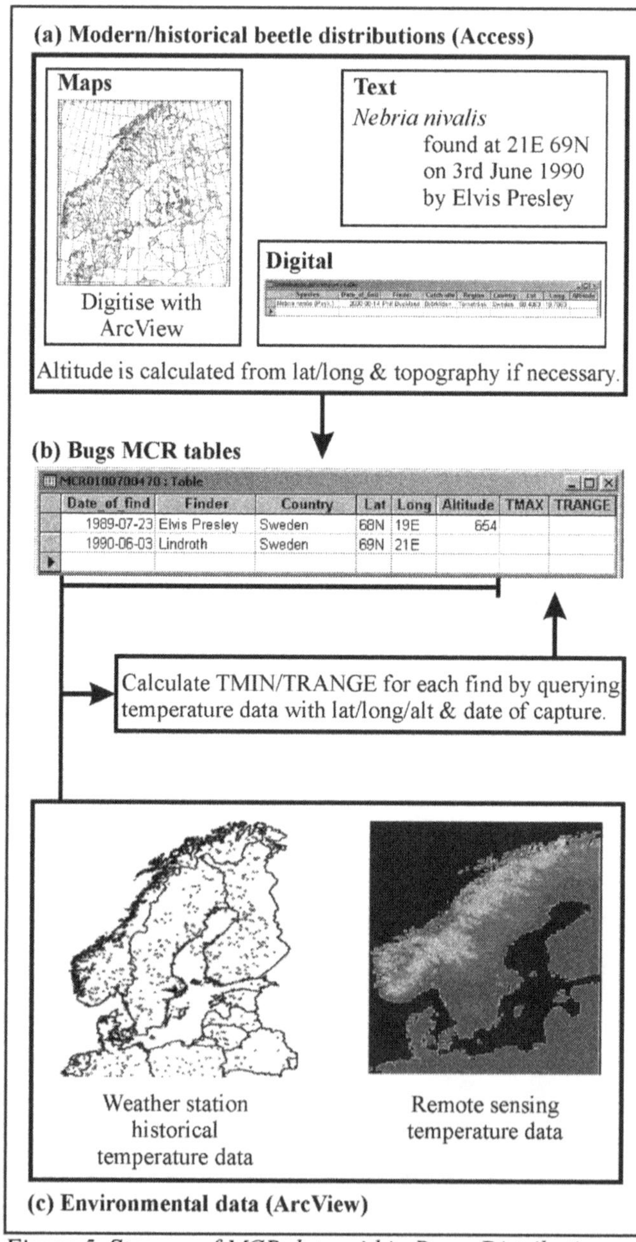

Figure 5. Storage of MCR data within Bugs. Distribution data is (a) digitised/entered into (b) Bugs MCR tables, and then (c) ArcView's Spatial Analyst used to calculate TMAX

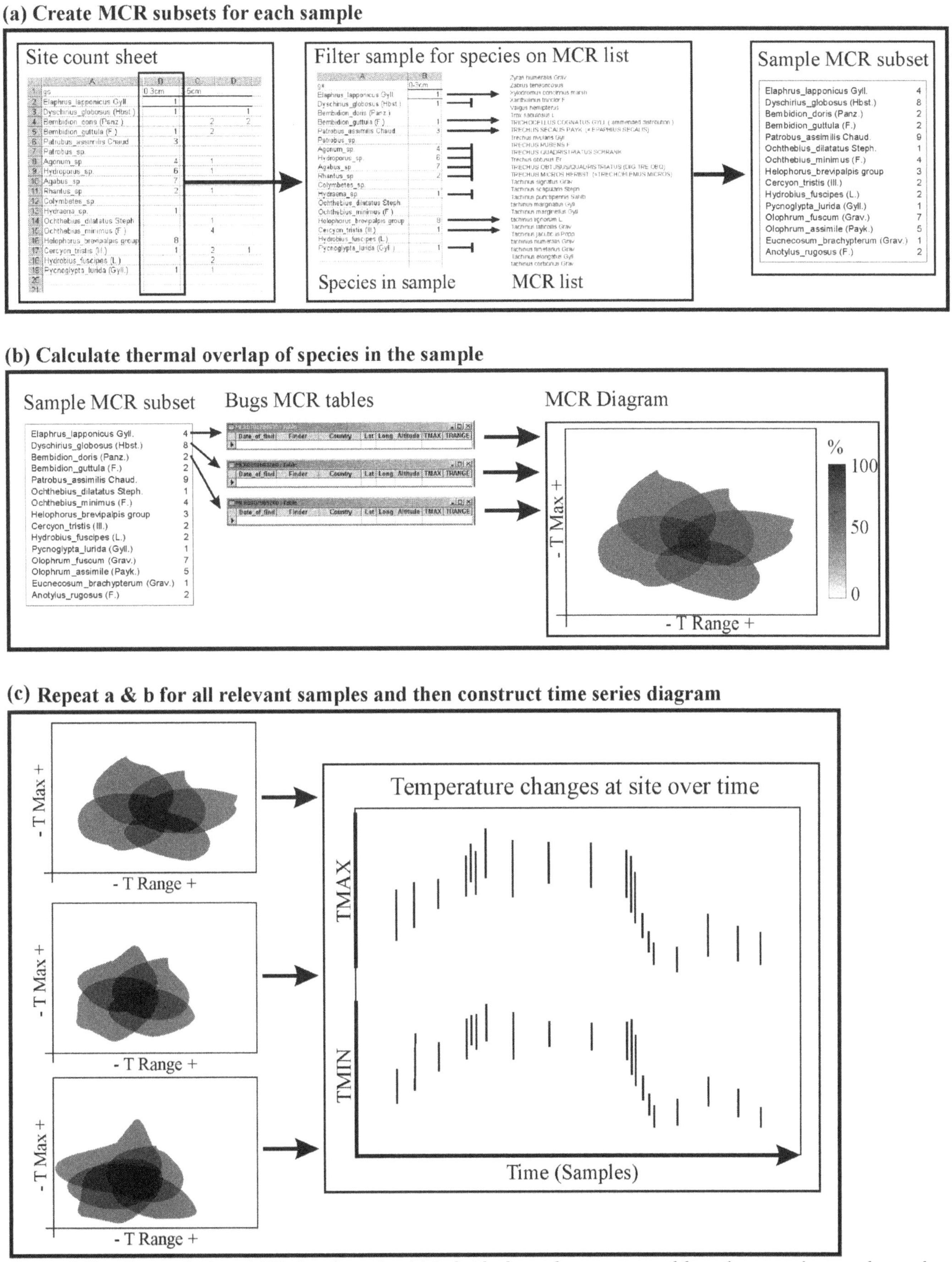

Figure 6. How Bugs calculates MCR data for a site. (a) Individual samples are extracted from the count sheets and parsed against the list of MCR viable species in Bugs, to produce a Sample MCR subset. (b) Each species in the MCR subset has an associated MCR table which is used to calculate thermal envelopes, which are then overlaid to produce the MCR diagram.. (c) The calculations are performed for every sample in the count sheet, it is up to the researcher to judge the relevance of the results. If the count sheet represents a chronological/stratigraphic sequence then a composite time series diagram can be produced.

Pattern recognition applied to Rock Art

Diego Diaz
Universidad de Buenos Aires
Nicolás Vila 470 21 C
Buenos Aires, Argentina 1421
Phone: +54-11-4381-5933 - E-mail: didibart@hotmail.com

Damián Castro
Universidad de Buenos Aires
17525 NE 40th Street Appt D320
Redmond, WA 98052, USA
Phone: +1-425-8618679 - E-mail: damianc@microsoft.com

Abstract: This paper was made in the context of a project that also develops a DA (database application), aimed at storing a large part of Argentinean rock art sites data.

This database is an analytical approach to that subject, and it is the second stage of the mentioned project, that began with a bibliographical and geographic data store. That software is currently in use and is part of the PROINDARA project that is supported by the University of Buenos Aires. The goal of this second stage, NA (neural application), is to add classification and tools to the database. Pattern recognition is among the capabilities that we are adding at the analysis stage.

Our conclusion is that, bearing in mind that the quality of our working pattern populations is very important, while we are storing images in the databases, we will also have to improve our aim to achieve good initial classifications.

Key words: Neural, Rock Art, Argentina, Database, Application

The database application

This presentation is an overview of our current works in Rock Art within the University of Buenos Aires PROINDARA project. We are working in two areas: Database Application DA and Neural Application NA. The first is concerned with the organization of the data and the second is concerned with meaning of that data.

The aim of the project is then, to provide informatic solutions that could fit the needs of the rock art researcher. The first stage was to develop a database system inside Database Application DA to load geographic information about rock art sites. This stage was called "descriptive database" and was focused on the political division of the world with tables like countries, states, counties, etc. to help the researcher to locate the rock art remains quickly and in situ, and to acquire the geographical information of the site. DA also has tables that can store coordinates to easily export this data to a GIS. This descriptive database also contains tables with information about authors and articles, that the user can link with the sites that are referred to in these articles.

At a second stage we are developing another database system of DA to load analytical information about the image. This stage is called "analytical database" and is focused on the description of the set of elements, in a formal and technical way. The contextual information is related to the images' position in the archaeological site. To analyze this data we use traditional queries (written in SQL language) and some tools taken from the IA like neural networks and data mining.

From an epistemological view, the computer can help anthropologists to do their jobs in two ways: the analytical and the synthetic. Firstly we consider the amazing computer ability to calculate in all of its aspects (mathematical and logical) and secondly, we consider the relatively new ability to make exploratory analyses and suggest new hypotheses. As regards the analytical aspect, our DA is not restricted to aristhotelical logic. And this is not trivial since very few people know about these other kinds of logics, even in the academic world. These logics are an output of the logic revolution of Frege, Russell, Whitehead, etc. at the beginning of the XX century. It was demonstrated, in Prof. Reynoso's work (tesis inédita), the capability of PROLOG to write his own rules in a metalanguage sense, or similarly, the ability of PROLOG to implement any kind of logic (like modal, deontic, fuzzy, etc.); and this was the key to prove the versatility of these logics in the adecuation to different kinds of anthropological problems. In the synthetic aspect we include tools like neural networks, as our NA, and simulation models among others. This tools deals with empirical aspects of the investigation, generating hypothesis or testing in virtual time different kinds of simulations. A great tip of the synthetic issue is the fact that the actual output is more important than the process which generates this output; in the sense that, for example, in simulation models, the rules may always be the same, but the output has a huge range of possibilities with minimal changes in the input data. The epistemological literature distinguished two stages in the scientific work: the discovery context and the justification context. The first is related with the process where the hypotheses arises and with the moment where the researcher interacts with the empirical data; the second is related with the validation of this empirical data and with the consequences that these facts bring to the theory that rules the research. The parallel between analytical aspect and justification context and synthetic aspect and discovery context is almost obvious. The inference we extract is that there are some informatic tools to help the scientific in the justification context and there are some others tools to help them in the discovery context. This is just an idea and should not be considered as the final say in the matter. This idea could bring a bit of order in the scientific work, but all of us know that reality is much more complex than the idea behind it.

The rock art area provides a fertile field for computer jobs. Between image management and geographic information the range is very wide. And in all of them the researcher can use a lot of tools in the epistemological sense. From an analytical perspective we develop a geographic and a bibliographic database system as a part of DA. Based on a relational model (model that has a wide spread in the scientific and commercial worlds) we design the different tables and the different relations between them to answere queries for the geographical location of the site and for bibliographic information related with the sites. The central unit in this model is the Sites table (Fig. 1).

The table stores the following information: Country, State, County, Site, Absolute Height, Localization (this is a memo field where stored data about certain references to locate the site like "to the west of Lechuza River", etc, and in future releases we expect to implement full text queries on it), Longitude, Latitude, Kind of Site (i.e., cavern, wall, exposed, etc.) and Image (this field relates with another table that stores the n images of the site that the researcher could access). We don't consider the ecological and spatial context in our original design by the great offer of the GIS available today so we focus on product portability. The universal measures we use, like Longitude and Latitude, the versatility of the product base (MSAccess '97) to import and export data in almost all kind of file format (including the most power relational database engine like SQLServer, Oracle, Informix, etc.) and the fact that we store the image reference, not the image itself in the database, (allowing the researcher to access the image without opening the database) garantees the application portability.

The Articles table (Fig. 2) stores the following information: Article, Author (this table relates with the AuthorsXArticles, a table that covers the large amount of relationships between Articles and Authors), Site (as the relationship between Sites and Articles are many as well, we have another table to cover it: ArticlesXSites), Article Title, Publication Title, Collection (i.e. if it is a journal), Edition, Edition Place, Edition Year and a Topographic Index (that help to locate in the physical library of the I.C.A. (Instituto de Ciencias Antroplógicas de la Universidad de Buenos Aires) the required Article). At this time the system stores just the name of the Article, not the article itself. In future releases we expect to load the article itself and let the archaeologist implement full text queries. To gain this objective we must export our product base (MSAccess '97) to a most powerful engine like SQLServer which in its last release has introduced this new feature.

The second stage of the DA, the "analytical database" (named in this way because it stores formal and technical information about the draws and not for its epistemological connotations) is focused in the Element; a single unit in the draw, i.e.: in a herd of Llamas, a single Llama. Another important table is the Element Set, i.e.: the herd of Llama, that keeps a foreign key relationship with Element and provides a range from 1 to n elements in the element set.

The Element table (Fig. 3) stores the following information: Site, Sector, Topographic Unit, Element Set, Element, Relative Position (with some site references like a crevice, panel, etc.), Sizes (of the element), Module, Attitude, Expression and Image. We have another set of tables related to Element where we load data about formal and technical attributes of each element. These tables are: General Formal Type, Master Formal Type, Detail Formal Type, Formal Type by Element, General Technique Type, Master Technique Type, Detail Technique Type, Technique Type by Element, where the user can create up to 3 levels of attributes of each type.

The Element Set table stores the following information: Site, Sector, Topographic Unit, Element Set, Relative Position, Image and Articulation (This field relates this table with ArticulationXElementSet where we keep the relationship between elements in the Element Set with regard to any kind of Articulation i.e.: opposite, rotation, etc.).

From a synthetic point of view we expect to implement in the future, data mining techniques to DA, to give the researcher a new source of hypothesis. This relatively new technique is part of a wider area known as Knowledge Discovery in Databases (KDD), that involves data warehousing and artificial learning among others. These tools focus on the evolution of data along the time or another continuous variable, extracting hidden knowledge from databases. These knowledge banks could be retrieved with standard techniques like those written in ANSI SQL, but it would be time-consuming to implement it with the standard techniques. The information extracted from database brings to archaeologists new configurations that can suggest new hypotheses.

The neural application

One of the sections of the project is the implementation of the neural application (NA) to make classifications of the image data from the different sites. This module is the first implementation of the synthetic methods mentioned earlier. Even though NA is partly dependent on the DA, in the sense that the images will primarily be located there, the interoperation procedures are not yet developed. At this experimental stage, we are feeding the neural application externally, mostly to simplify the testing process.

The NA has two main sections. On one hand, the problem section, which is in charge of administrating the pattern storage.

On the other hand the Neural Wizard that creates and trains the neural network. But there is an important interaction between them because depending on the type of network the user chooses, the information stored in the database will be interpreted differently by the program.

The problem section has a tree-like view to represent the hierarchy where the problem itself is represented by the root, with the predefined classes on the second level and the samples for each of these on the third level. When the program, after the training, identifies a pattern as belonging to a class, it puts it under the corresponding class node. There are slight changes in this behavior when the Kohhonen Learning vector is applied: all the patterns are reclassified and may change their affiliation to a class state. In fact, in future versions, we are thinking of using an "Unknown" system class to store the data before the Kohonen learning vector. The screen shot in figure 4 shows what the tree looks like.

Because the actual images are stored through DA, in NA only the binary black and white representation exists, and is stored as an array of bytes. It is expected to develop a color version in the future, though it represents an important trade off regarding the memory needed to represent color. However, each point in the image is represented by two bytes, so theoretically 65536 different values might be allocated. Hence, a color scale of this quantity at least, will be supported without losing performance or without needing additional memory.

The neural section of NA was designed as a wizard that guides the user through the processes of creating and training the network.

The first step is when the user decides what the name will be, to identify the set of classes and patterns in the database-referred above as the "problem"- and to select a physical place on the disk where the file will be stored. In this step, you can also select how big the pattern matrix will be.

The second step is to decide the type of the neural network. The alternatives are Back Propagation, Learning Vector and Probabilistic. Back Propagation and Probabilistic are both used to classify new patterns when a predefined set of classes with their sample patterns are already known. Learning Vector (the Knohonen network) is used when the classes are not known. In those cases the network classifies the patterns using a class quantity as a goal. It is expected that similar patterns will belong to the same class and the network may classify the pattern in less classes than the goal. It will never find more classes than that.

The options apply to the types. The compression network is used to reduce the sizes of the input matrix before the training begins. This feature was proposed by Sejnowski et al. in 1991. When Vanilla is selected, a normal back propagation algorithm is used. Momentum option applies a change to the algorithm: new weight changes of neuron connection in each separate training step do not totally affect the final weights. They will affect some percentage instead. In this way, we can prevent the "oscillation" of the network, where the class of last trained pattern is always the winner.

Bidimensional neighborhood applies to Learning Vector type and Normalized to the probabilistic. (Fig. 5)

The following step is the network topology (Fig. 6). Here, we must decide the quantity of layers and neurons. For the Learning Vector there are no internal layers and the output layer neurons is the maximum quantity of classes to classify. In the Probabilistic there is only one internal layer and is equal to the quantity of patterns. The Back Propagation network permits decide freely the internal layers and their neurons.

Finally, we have to decide the training parameters: how many cycles, how much error is allowed, the learning rate, etc. A cycle means a propagation through all the layers. The error is how far is the actual output from the correct one (of course, this is only valid for supervised learning)

The first experiments

We are working with one of the rock art sites included in the DA. The site and its analysis were presented in "Congreso Internacional de Arte Rupestre", Tarija, Bolivia September of 2000, by María Isabel Hernandez Llosas et al.

There is one concept we extract from the Methodology section of this paper in order to organize the training experiments of the neural network: element. This means that we are isolating the paintings from the context and we are not using the special units, even though they are relevant to the taxonomies that might be created after the training. But it would add an overhead to input matrix and we choose, for this very first approach, work only at the lowest level.

We trained our network using three classes of elements: abstract forms, zoomorphic and anthropomorphic. Figure 7 shows some patterns we used.

We used the Back Propagation method. We added a set of three patterns per class and trained the network with different combinations of cycles and learning rates.

One thing that happened from the beginning was the oscillation. The last class of the last pattern trained was always the winner. To avoid this situation, we decrease the learning rate. This means more training time. In the unique case we successfully recognize a pattern of the network, it took almost 24 hours of training.

Of course this was in made in a desktop machine and not in a two or more processors server, but we think that we can optimize this process in some way. There is no other alternative, in fact, because considering the increase in time when a pattern is added, a minimal quantity of them, like 50, may take weeks to be trained.

Future perspectives

There are several tasks that we have to make, besides the optimization of our code and the improvement of the application capabilities. The more important is the writing of our own neural procedures, optimized for this specific kind of problem. Even though the user interface and the database code were developed by our team, we used an external code library, which implemented the neural network algorithm. So it would give us more versatility if we can optimize the neural code for this specific problem.

Another important task is to implement genetic algorithms to search for the optimal network for a specific set of patterns. The genetic algorithm might be very powerful for this kind of optimization and may considerably reduce the learning time to successfully produce recognitions and classifications.

References

Hernández Llosa, María Isabel. 2000. "Los grabados de campo morado", Congreso Internacional de Arte Rupestre", Tarija, Bolivia

Hertz, John. 1991.*Introduction to the theory of neural computation*, Redwood City: Addison Wesley.

Figures

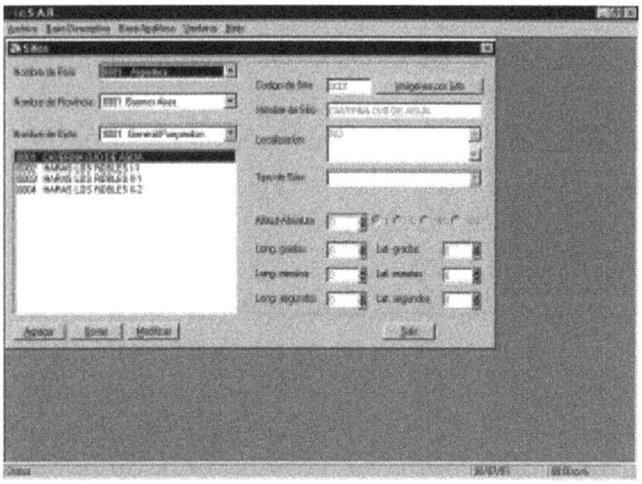

Figure 1. Sites table

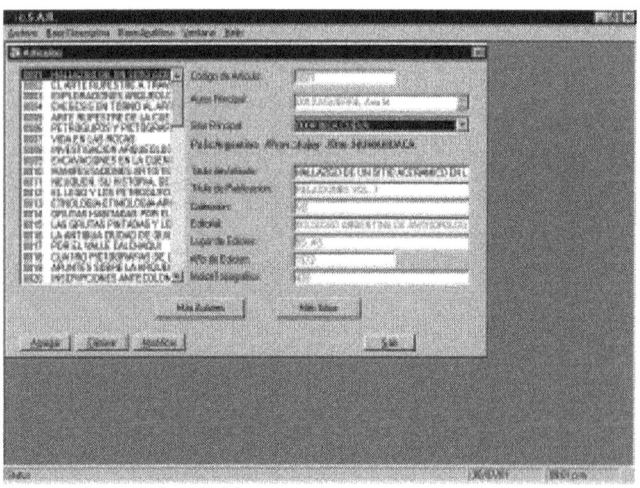

Figure 2. Articles Table

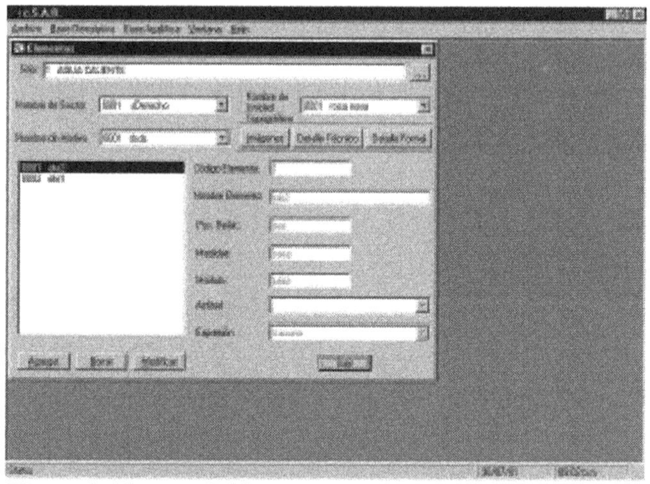

Figure 3. Element Table

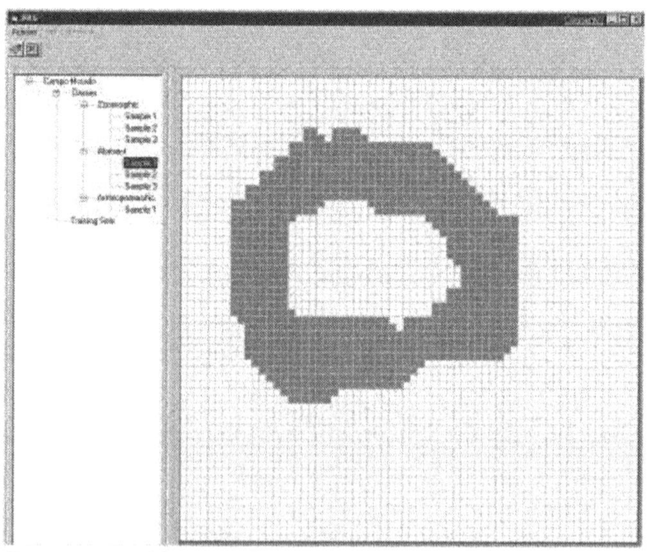

Figure 4. Problem window, where classes and patterns are defined

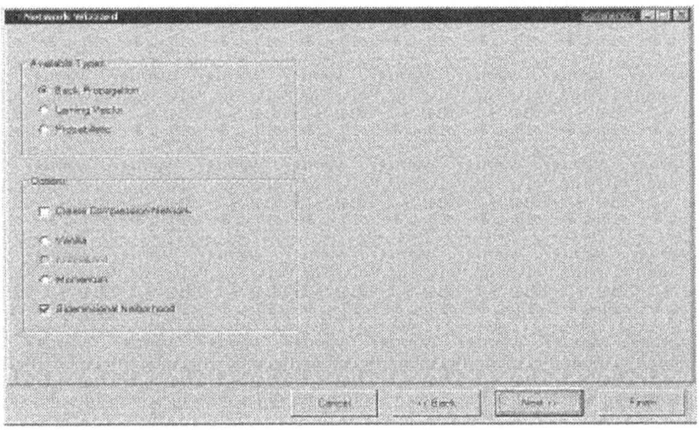

Figure 5. Second Step: neural network type

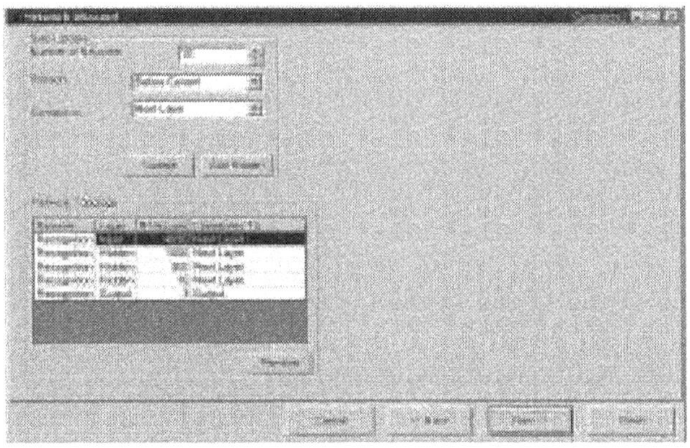

Figure 6. Neural network topology window

Figure 7. Compo Morado photograph (detail)

Creating of a database for prehistoric sites: which are the goals, the strategy and what means to put in place?

Pierre Corboud
Dpt of Anthropology and Ecology
Geneva University
Case postale 511
CH-1211 GENEVE 24, Switzerland
Phone: +41 22 702 69 89 - Fax: +41 22 300 20 46 - E-mail: Pierre.Corboud@anthro.unige.ch

Abstract: Nowadays, computer science allows the creating of archaeological database to enable the gathering, updating, and management of all kinds of information necessary for the research and synthesis in the field of archaeology. However, in view of the past experiment in the seventies, one must be very cautious and be aware that the «archaeological data banks» must be plentiful, maintained, and exist for years before giving any profitable results. We wish to discuss the problems of the data pertaining to archaeological sites, especially those from prehistoric time. The archaeological site constitutes the ideal unit of description and study to deal with theoretical and methodological aspect of the creation of a relational database. The relationship between the geographical chronological and documentary entities imposes the use of very precise options in function of the chosen goals. We intend to show what could be the line of reasoning behind those options with the help of some real world examples and an ongoing project.

Introduction

The database from archaeological sites, and particularly those of prehistoric sites require the organization and implementation of a specific type of archaeological database. Indeed, the constitution of a database of archeological sites can be justified according to extremely varying goals and concerns. It is, for instance, the process of gathering and laying out of information relating to sites from the same geographical area, belonging to the same cultural phase, in order to study the diffusion of a culture. It could also be the assembly of all the data relating to the sites from one county or from one province in order to manage and protect the saved remains. Thus, to simplify, one can distinguish between the goals relating to safeguard of the patrimony and those directed rather towards the scientific research. Sometimes, these two categories are more or less the same, but in other cases, the options chosen in the creation of such a database are relatively divergent, especially when it comes to choosing and organizing the various fields. We won't go on developing this opposition, but rather we will attempt a systematization of the theoretical and strategic steps, which will lead to a management or research tool adapted to the needs of the users. Moreover, it is essential that this tool guarantee a certain perennity of the data recorded in the base. Indeed, the lifetime of a database is a very important criteria when it comes to its creation out of the necessity to facilitate the unavoidable upgrades and to allow and easy evolution towards other database management systems. It is not useful, here, to mention the software packages and applications relating to managing such databases. Currently, the technical evolution of the hardware and the software is so fast and unforeseeable that it is not desirable, at least in the first phase of elaboration, to create a database according to a precise computer science environment. It is, however, preferable to seek to ensure the structural solidity of the data model, according to the environment of the research or the management of the objects to be described in this system. This last remark is certainly obvious to all of you, but it is not useless to point out it ...

As always in archaeology, like in the majority of the scientific fields, the success of a project depends mainly on the adequacy between the objectives, the characteristics of the available data and the environment of the research in the broad sense.

Accordingly, the principal goal for which the database of the sites will be created is essential in the definition of the various documentary modules and the relations to be established between them, especially when it comes to the input modules and output products. By documentary module we mean all the fields or headings of a data table, which share the answer to the same type of question.

The basic model

Independently of the choice of the goals and the definition of the environment of the research, the structure of the various documentary modules can be summarized in the definition of the basic model (fig. 1). This basic model is intended to answer the three following questions :
1) Where is the site?
2) When was it occupied?
3) What are its intrinsic characteristics?

This basic model thus gathers, in the shape of three modules of information, the essential data or interpretation, which one can expect to find in the description of an archeological site. On the other hand, this information does not give any details on the source of the data used for the description of the site. It is, in fact, more a synopsis of the main fields and proposed interpretations.

Such a model would present very little interest within the framework of a data management in the short or long term. Indeed, each update would require a return to the primary data, which are not contained in the table but in external documents, and which, is not always referenced.

This defect, fatal for the long-term survival of such a system, must absolutely be corrected by the implementation of an extended model, which gives reference to the sources, or, still better, which contains the sources of the information needed for the descriptions or the interpretations relating to the archeological sites.

The extensive model (fig. 2)

Indeed, one cannot consider that an archaeological data (for example a layer of habitat belongs to a specific culture) is an Ḥabsolute factḤwhose validity cannot be questioned. Each interpretation or conclusion is the result of an assessment based on a set of information and documents. This assessment could be modified several times by the appearance of a new data or simply the intervention of a more qualified expert could modify. There too, it shows evidence, which is too seldom taken into account in the most widespread documentary applications.

Thus, the extensive data model should integrate the categories of the following information to allow for a easy return to the sources :

A) References to the articles, works, reports of excavations, maps, graphical documentation, etc. from which the 'archaeological facts' were extracted ;

B) The essential information textually extracted from these sources, or more precisely the transcription of the quoting of articles or works which provide significant data, whose interpretation could be ambiguous or which represent useful information on the history of the research and interpretations of the site. In this category we will retain for example the citations of texts who mention the locations of formerly spotted sites, but whose localization is vague or unknown.

Generally, it is always preferable to transcribe an original citation rather than to interpret it and translate it according to our system of today's reference. The interpretation in a homogeneous system, for example in a typo-chronological system of common use, must intervene later on and in another documentary area accompanied by a reference to its author. We shall come back later about this rule which could be named 'traceability principle of data and interpretation.'

Historic of knowledge

This last remark brings us in another field of documentary analysis which is historical knowledge on an archeological site, as much as to the physical operations realized on this site, the results obtained or as to subsequent interpretations. In this type of module, we find the information relating to discovery of the site, to the studies carried out (survey, excavations, analyses of artifacts, etc.) from which it is possible to identify the results and interpretations and, finally, old interpretations resulting from these various interventions.

Thus, the incorporation of these new documentary modules in a extended data model should allow the update of such a database permanently whatever the type of objective one considers for the database of sites, without requiring a new document search, but simply by supplementing the references to the sources and by correcting out-of-date interpretations and conclusions.

Importance of the relation to the bibliographical references and to the documents

With the exception of the ground, all the articles, reports and works that mention the observations, constitute the principal documentary source of an archeological site and the interpretations gathered on this site since its discovery. One of the essential characteristics of the archaeological research, in opposition to certain exact sciences (such physics and biology) is to not be able to neglect the writings of the former authors. Indeed, the archaeological observation – in particular the excavation – must destroy its object of study to describe it and understand it. This remark is even more relevant for the sites not excavated, described in the past, but partially or completely destroyed by the action of nature or the man. In this situation, the mention of the bibliography of reference related to a site described is an essential and mandatory element in any database of archeological sites. Moreover, only the relation between each work of this bibliography, the page or the figure, will make it possible to retrieve the source of information which made it possible to answer the questions included in the basic database model.

The articulation of the site database with a bibliographical reference table, which includes also the gray literature, is a central element of this type of applications. This goes as well for all the non-bibliographical documentation (maps, drawings, collections of archaeological artifacts etc,) the mention of which is necessary as well as the indication of their location.

Relation with the collected archaeological artifacts

Independently of the bibliographical references to the archaeological artifacts collected or observed on the site, the mention of the artifacts preserved in the museums is essential to discuss the dating of the various levels and successive occupations. In these applications, the goal of the relation artifacts–site is not to carry out a museographic inventory of all the artifacts collected and attributed to each of the sites of the database. Such an objective is not only utopian, but also especially inadequate in this context. It is rather preferable to mention the sets of artifacts preserved in each depository, sorted by their categories and their typo-chronological dating.

Before treating relations between the various modules and their logical articulation, it is necessary to open a parenthesis to define the significance, which we give to an archeological site, which is the unit of elementary description.

Definition of the elementary unit of description

In the applications which are used as example here, the unit of elementary description corresponds to the definition of the archeological site as physical, topographic and logical entities, whose geographical and chronological limits can be traced with more or less precision. It is for example about a surface of ground on which the existence of a Neolithic village is known. There, later, a necropolis of the Bronze Age was built. In this case, the unit of description is the topographic layout, which include all of the vestiges of these two periods and these two types of sites (fig. 3). The limits of this definition are obvious. They prohibit to easily distinguish the characteristics from each identified archaeological level and to describe for each occupation the archaeological artifacts discovered and the structures observed. In this case, this limitation is desired, because the objective is to treat a whole of sites and not a whole of occupations belonging to the same site. Moreover, the attribution of several occupations to an archaeological site is an interpretation and not a fact. Thus, the various chronological horizons recognized as well as the characteristics of those horizons should be described in a secondary table belonging to each site's record.

It should be noted that the conservation of a site is not necessary to its description according to such a definition. A destroyed or displaced site, for which we have enough precise information to locate the old position of it, the dating and the characteristics will have as much its place in our database that a site well preserved, studied or not.

Some principles of relations between the documentary modules and the fields

Before approaching the various logical structures between the units of information of the database of the sites, it is useful to propose and explain some principles, which govern the construction of these relations (fig. 4).

The first principle could be named 'recognition of the data sources'. It is a question of allowing, in all the cases, the access to the primary data, be it bibliographical, graphic or constituted by collections of artifacts. This principle is rather easy to apply as long as the documents and artifacts concerned are catalogued in known systems of inventories (libraries, documentary archives, museum's catalogues, or photographic records.)

The second principle is that of the 'traceability of interpretations'. Its strict application is more difficult, because, in this view, each term in each field should mention its author, the date and the data used for this interpretation. Practically, this principle can be simply solved by mentioning the successive dates and responsibilities for the data entries and the corrections and by admitting that the last expert has read again the whole record for the site and takes the responsibility for its contents.

The third principle, which is corollary of the first two, is to guarantee the logical and graphic distinction, in the screens of the system of database management, between the modules and the fields which contain raw data resulting from the documents and observations, and those which contain interpretations or expertise. Such a distinction is essential as much for the people who enter the data than for those which question them. One could name this principle 'separation of the raw data and interpretations'.

From the various data modules quoted and of the three principles stated here, it is possible to build a data model adapted to the goals that we fixed ourselves, by limiting to the maximum the number of tables.

Logical structure of the various documentary modules, models simplified

This model takes into account only the primary data to be integrated in an archeological site's database (fig. 5). Moreover, this model is in conformity with the recommendations of the CIDOC (CIDOC 1995) One can thus distinguish various modules :

- The topographic module, in the table of sites, which includes the fields, which provide the localization administrative and geographical of the site, with the mention of the sources from which the co-ordinates were established.
- The bibliographical module, which gathers the extent of the references to the texts and documents relating to the site, as well as the quotations chosen in these texts.
- The historical module, which resume the succession of observations and studies listed for this site.
- The module of dating, with all information, not interpreted which makes it possible to attribute the occupations to a culture or a given period.
- Lastly, the module of interpretation, which presents a synthesis of all the significant data and proposes a chronological attribution of the occupation of the site.

A sample application : the ARCHEODAT database

At this point we are through with the theoretical considerations on the archeological site's database, in order to briefly present a concrete application, which started in Western Switzerland twenty years. At its conception, the goal of this project was to gather all the informative data resulting from an extensive archaeological survey on the littoral sites of the Lake of Geneva, Morat and Neuchâtel, and to integrate the results of this operation. It is now a management tool on archaeological knowledge about a set of sites, as well as a referential database, allowing the access to the original data, which made it possible to elaborate this knowledge.

The archeological sites that we consider here were occupied between the middle of Neolithic period and the end of the Bronze Age, between 4000 and 800 years before Christ. None was excavated completely, but all were the subject of observations or small surveys, which made it possible to gather a certain number of information on their situation, their dating and their nature. One of the characteristics of this project was that the majority of the studied sites had been discovered between 1854 and 1900, particularly for the south bank of Lake of Neuchâtel and the Lake of Morat. For the majority they were not the objects of studies or observations during this period. As well as the recent observations carried out on the ground by our team, the old data thus occupy a significant place.

With regard to its structure and its interface, the ARCHEODAT database underwent several transformations, in order to regularly adapt it to the new management tools of computer sciences, but also to the evolution of the questions and goals of our project. Currently, after having developed the last application with the MS-Access software, we are in the process of transferring all the data in an Oracle database, which permits an access by the Web, far beyond the limits of the limited local network of the University of Geneva.

The simplified structure of the data of ARCHEODAT is as follows, with the articulation between four principal tables, which are the table of the sites, the table of the bibliographical references, the table of the 14C dating and that of the dendrochronological dating. In detail, the fields are distributed in various modules, which are :

- The topographic module, with the topographic coordinates and the references to the source of these coordinates.
- The bibliographical module, which is in fact the table of the bibliographical system of management of our department.
- The module of the ancient citations.
- The module of the historic of the studies, which is also a reminder of the main results of recent observations.
- The module of the 14C dating, which proposes the standard dates, but also the calibrated dates.
- The module of the dendrochronological dating, essential to treat sites preserved in wetland.
- The module of synthesis and interpretation of the data, which summarizes it.

Lastly, as a conclusion with this presentation, homage should be paid to a Swiss prehistorian, who was perhaps despite himself, the inventor of the database of archeological site. Frederic Troyon was, indeed, the first to circulate in the ḤcantonḤof Vaud a questionnaire about the antiquities and archaeological richness, in order to collect the data necessary to the protection and the study of the sites of his land. This endeavor has enabled the creation of the first archaeological map of the ḤcantonḤof Vaud, published in 1874. This is the prototype of our current archaeological and GIS database (Troyon 1858 and 1860, Bonstetten 1874)

Also, during a stay in Sweden between 1845 and 1846 F. Troyon had the merit of proposing this questionnaire to the Royal Prince of Sweden, to be used as model for the archaeological inventory of the archeological sites of the Kingdom of Sweden. This marked an historical and initial point of the management and the study of the archaeological patrimony of Sweden.

References

Bonstetten (G. de). 1874. *Carte archéologique du canton de Vaud.* Toulon : C. Mihière.

CIDOC-ICOM. 1995. *Projet de norme documentaire internationale pour les sites archéologiques. Draft international core data standard for archaeological sites and monuments*. Paris : ICOM.

Troyon (F.). 1858. Statistique des Antiquités de la Suisse Occidentale. *Indicateur d'antiquités suisses*, 2, juin, 25-29.

Troyon (F.). 1860. *Habitations lacustres des temps anciens et modernes*. Lausanne : G. Bridel. (Soc. d'histoire de la Suisse romande, Mémoires et documents ; 17).

Figures

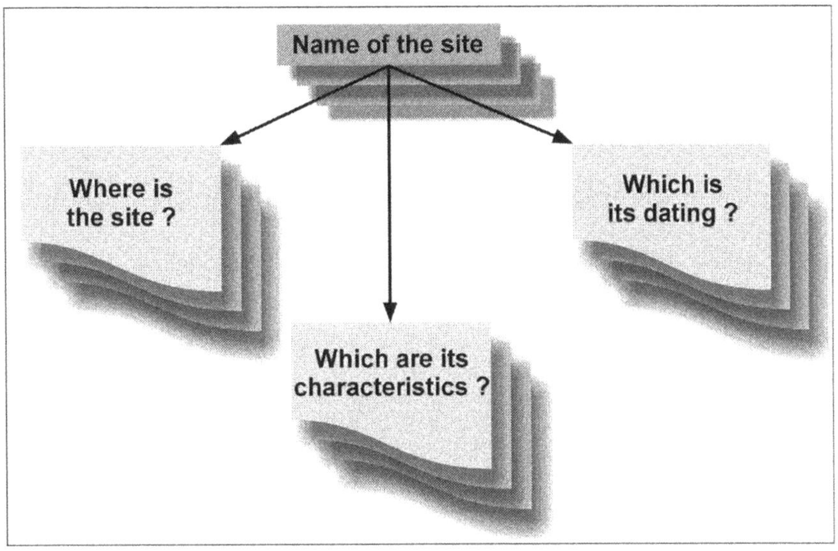

Figure 1.

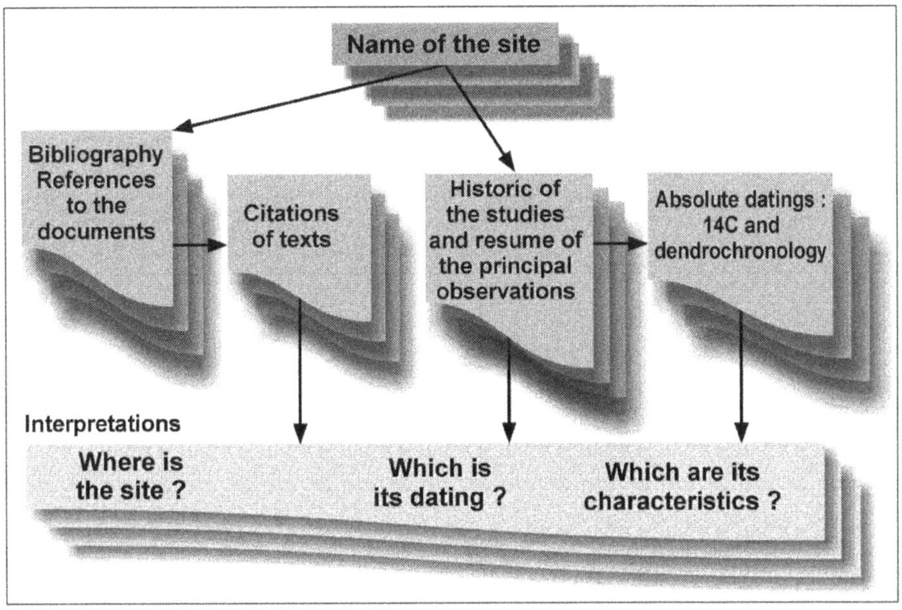

Figure 2.

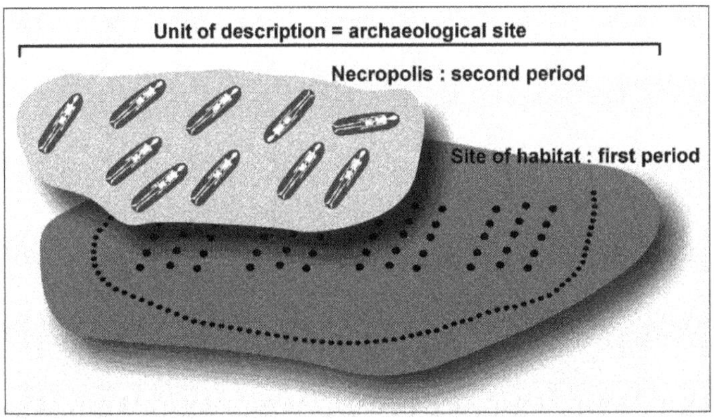

Figure 3.

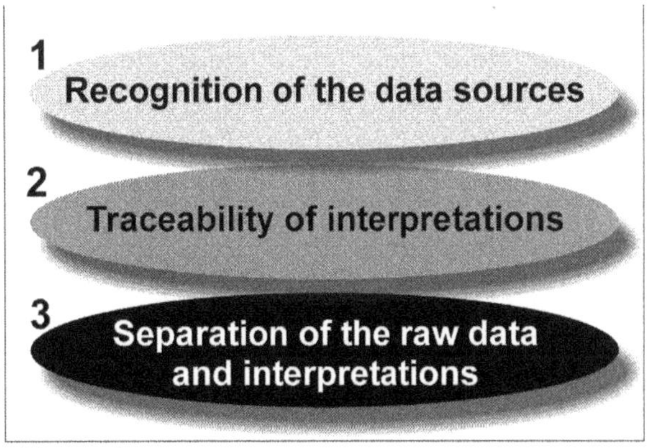

Figure 4.

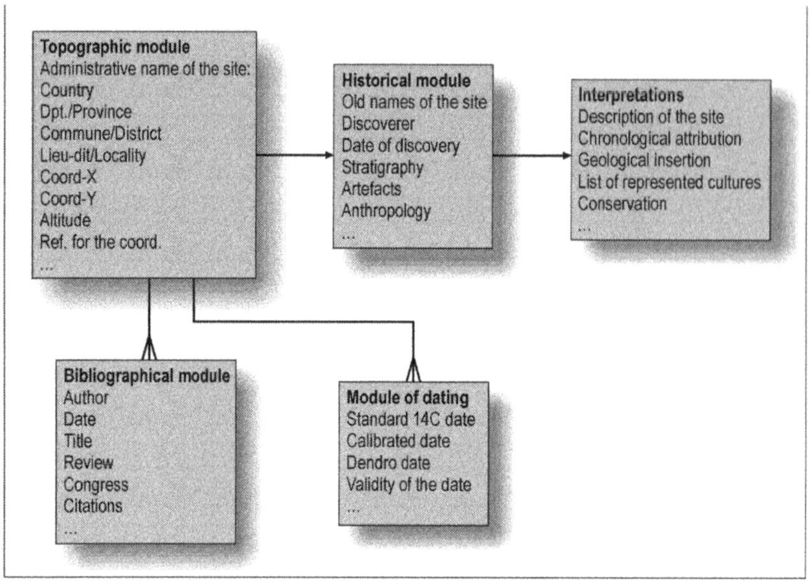

Figure 5.

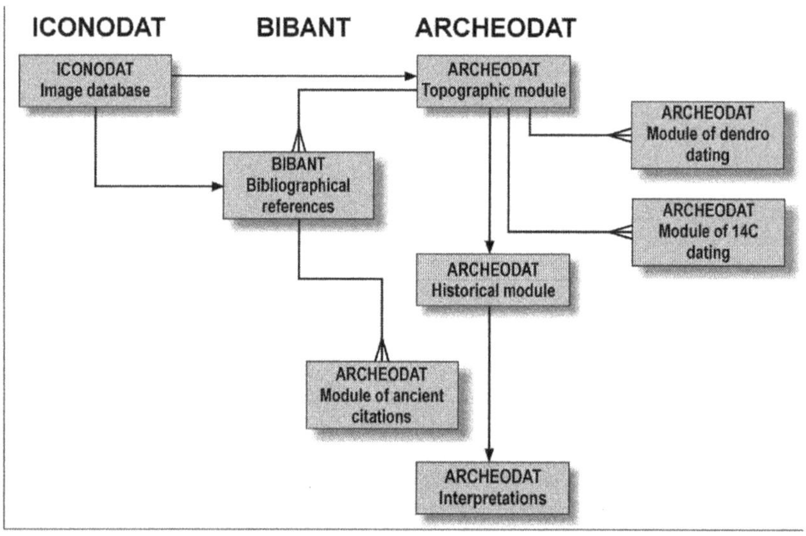

Figure 6.

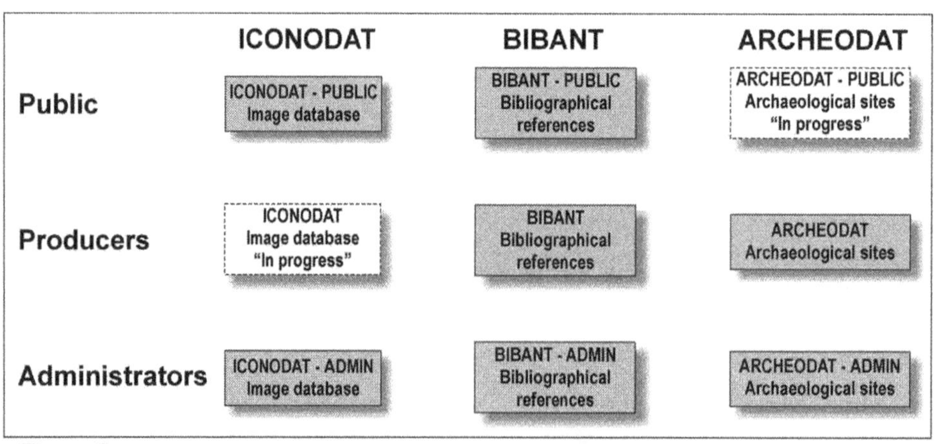

Figure 7.

The Computer catalogue of the Kunstkammer museum collections and perspectives of an Internet-shared anthropological database

Yuri Osintsov
Peter the Great Museum of Anthropology and Ethnography (Kunstkammer)
3 University Emb.
St.-Petersburg, 199034, Russia
Phone: +7-812-3280025 - Fax: +7-812-3280811 - E-mail: Y.V.Osintsov@mae.nw.ru

Yuri Chistov
Peter the Great Museum of Anthropology and Ethnography (Kunstkammer)
3 University Emb.
St.-Petersburg, 199034, Russia
Phone: +7-812-3232719 - Fax: +7-812-3280811 - E-mail Y.Chistov@pobox.spbu.ru

Dmitriy Gerasimov
Peter the Great Museum of Anthropology and Ethnography (Kunstkammer)
3 University Emb.
St.-Petersburg, 199034, Russia
Phone: +7-812-3280025 - Fax: +7-812-3280811 - E-mail dger@dg6388.spb.edu

Abstract: The Computer catalogue of the Kunstkammer Museum is part of a large museum database that has been in development in the Museum of Anthropology and Ethnography (Kunstkammer) since 1996. The two major purposes of the project are: 1) organizing of museum storage routine and museum documentation processing on an up-to-date level and 2) providing free access via Internet to scientific and inquiry information.

We provide Internet users with information about collections in cultural and physical anthropology stored in the Museum. Attributive information for the materials, such as archaeological sites and cultures, historical periods, dating, geographical locality is also available.

The web part of the computer catalogue is based on dynamic web pages. Multi-aims and multi-conditions search through the database is available. The purposes of creating a museum computer catalogue, project achievements and perspectives of its development, as well as system architecture and data layout are special topics for discussion in this paper.

At present information on approximately 400 archaeological, 571 anthropological, and 200 ethnographical collections and subcollections is available for database users. The Russian version of the Kunstkammer database is accessible in Internet on http://www.mae.nw.ru/maedb or from the Museum WEB-site: http://www.kunstkamera.ru.

Key words: Museum computer catalogue, data base, internet, anthropology, archaeology.

Introduction

Collections kept in the Peter the Great Museum of Anthropology and Ethnography (Kunstkammer) present ethnological, archaeological and physical anthropological materials from all over the world. Museum storage consists of about 6,5 thousand collections that contain about 700 thousand pieces. These collections are important for researches in cultural and physical anthropology and are also very attractive for amateurs and everyone interested in human origin and culture.

The developing of a computer technology and Global Computer Network allows the organisation of information about the Museum's collections on an up-to-date level and to give access to this information to anyone interested in it via Internet. Thus, researchers in all branches of anthropology gain an opportunity to check if there are any materials for their research, to get some information about them to decide if it is necessary to work with these collections personally and to get some reference information or even some characteristic of the material that they can use in scientific work. The Russian version of the Kunstkammer Database is now available at http://www.mae.nw.ru/maedb. This Internet source is also very useful for students in anthropology and anyone interested in the origin and history of mankind.

Purposes

There are three purposes of creating the Database of Kunstkammer collections – research tool, storage system and educational database of cultural and physical anthropology.

Science

As a research tool the Database provides opportunity to search and select materials for certain research purposes and therefore it should consist of information important for researchers, including description of specific features for each kind of materials, such as technological and typological features of flint objects and ceramic, measurements of skulls, etc. It has already been mentioned that Kunstkammer collections present materials in three branches of anthropology such as archaeology, ethnology and physical anthropology, and the Database contains specific descriptions of any kind of collection.

It will be also possible to make use statistic, such as measuring of technological indexes of prehistoric flint industries or to use the data for multivariate statistical analyses in physical anthropology.

Storage

Since the Kunstkammer museum was founded in 1714, the organization of its storage system is quite complicated – materials from different places can be kept in the same collection or different collections can consist of materials from the same place (e.g. from the same archaeological site). Thus, it was necessary to reflect this complicated structure in the Database.

Education

As an educational database in anthropology it will consists of a large number of articles about archaeological periods, cultures, different ethnicities, etc., presented by collections in the Database.

Data layout

During the development of the project there were created separate schemas for description of archaeological, ethnological and physical anthropological collections. Then the schemas were merged and expanded, and the Museum storage system was integrated in the Database. As a result we tried to get the universal data layout for description of materials in all branches of anthropology (fig. 1).

Each record in the Database has mandatory fields: Identifier, Subordination identifier, Name and Additional info. These attributes uniquely identify any object in the database and allow to build a hierarchical structure tree. Indeed, a hierarchical tree reflects the structure of any typology anyone uses, i.e. a hierarchical tree is the core element of any scientific systematization. Such structure allows us to create procedures, libraries and client applications that can operate with any type of objects stored in the database.

All other record fields are optional in relation to data access, but serve to describe a certain object's peculiarity.

The whole scheme can be divided into the following parts or schemes: common information (common), documents (docs), information about any persons mentioned in the Database (persons), information about any organizations or organizational units (orgs), museum receipts information (receipts), museum storage system (storage), archaeological sites (sites), collections' and objects' data (sets).

Common scheme

Table common.chronology
There are two schemes currently presented in the table: relative and absolute chronology. Archaeological periods in the relative chronology scheme have no beginning and end dates because it is impossible to find universal dating for them. Absolute dating scheme consists of dating of events such as excavation dates or the existence period of an archaeological site. In the latter case period limits are the earliest and the latest possible dates of a site's existence.

Mandatory fields: Name – name of period or event; Info – description of a period or background for event dating.

Optional fields: Date_start; Date_end

Table common.locality
This table accumulates information about spatial location of object, such as an archaeological site.

Mandatory fields: Name – name of unit or site; Info – description of the characterized area

Optional fields: Northlimit – northern latitude; Eastlimit – eastern longitude; Southlimit – southern latitude; Westlimit – western latitude.

There are three schemes in this table – geographical regions, administrative units and absolute coordinates.

Table common.cultures
This table is factual about archaeological cultures or ethnicities.

Mandatory fields: Name – name of archaeological culture or ethnicity; Info – description of archaeological culture or ethnicity.

Table common.materials
Describes materials museum pieces objects are made of.

Mandatory fields: Name – material name; Info – material description

Persons' scheme

Table persons.list
This table consists information about all persons referred to in the Database.

Mandatory fields: Name – full name; Info – additional info about the person.

Optional fields: Surname – family name; Fstname – first name; Sndname – middle name; Dates_id – dates of life, reference to the table Common.chronology; Address – mail address; Phone – phone number; Fax – fax number; Email – e-mail address; Url – uniform resource locator, e.g. internet personal pages.

Orgs scheme

Table orgs.types
Types of organizations or units listed in the table.

Mandatory fields: Name – organization type; Info – type description.

Table orgs.profiles
Organization field of activity kept in the table.

Mandatory fields: Name – organization profile; Info – profile description.

Table orgs.tree
All organizations related to collections are kept in this table.

Mandatory fields: Name – name of the organization; Info – additional information about the organization.

Optional fields: Abbr_name – abbreviation of the organization's name; Manager_id – the manager of the organization's unit, reference to the table Persons.list; Address – address of the organization; Phone – phone numbers; Fax – fax numbers; Email – e-mail addresses; Url – uniform resource locator, e.g. internet site of the organization.

Docs scheme

Table docs.types
This table serves to control over search for any type of documents.

Mandatory fields: Name – document type; Info – description of the document type.

Optional fields: Viewer – reference to the viewer to see a document if there is a digital version; Editor – reference to the programme to edit a document if there is a digital version.

Table docs.tree
Information about all documents related to any Database objects, such as stored documents or bibliography, etc. is kept in this table.

Mandatory fields: Name – the name of the document, Info – the description of the document

Optional fields: Type_id – type of the document, such as book, manuscript, photography, audio-record etc., references to the table Docs.types; Author_id – the author of the document, references to the table persons.list; Publisher_id – publishers (if any), references to the table Orgs.tree; Date – date of publishing; Abstract – abstract of the document with hyperlinks to related documents; Path – place of the document's storage; Url – hyperlink to the Internet-version of the document.

Receipts scheme

Table receipts.types
In this table such types of collection acquiring as buying, presenting, expedition, etc. are described.

Mandatory fields: Name – receipt type; Info – definition of the receipt type.

Table Receipts.list.
Any set of objects acquired by the Museum at the same date, from the same person or organization receives a unique number in the Main Museum Incoming Book. This set could be gathered by different persons, e.g. the same incoming set could originate from different archaeological sites, but we mean that this set could be collected by some person or organization. As we have said, there are several ways of collections' acquisition (receipt types) – a collection could be presented, sold, replaced from a different organization, obtained by the Museum's expeditions, etc.

Mandatory fields: Name – the number in the Main Museum Incoming Book; Info – any information on a collection's acquisition.

Optional fields: Date – the acquisition date; Type_id – receipt type, reference to the table Reciepts.types; Org_id – the organization that materials came from, reference to the table Orgs.tree; Collector_id – the person who gathered these materials, a reference to the table Persons.list; Supplier_id – the person who delivered a set of objects to the Museum or who is responsible for moving materials from another organization, a reference to the table Persons.list.

Storage scheme

Table storage.types
This table contains information about types of places of the collection's storage. The hierarchical tree allows describing different types of storage as an exposition or museum fund.

Mandatory fields: Name – type of storage; Info – definition of the type.

Table storage.tree
This table enumerates museum storage system up to the shelf or box level.

Mandatory fields: Name – name of the storage place; Info – additional information.

Sites scheme

Table sites.types
This table contains information about types of sites. It is especially important in archaeological sites' description (e.g. tumulus, crypt, tomb, cave burial, etc.)

Mandatory fields: Name – type of a site; Info – definition of the type.

Table sites.tree
A site is a very important Database object. It's not simply an archaeological site. We use this term to describe common origin of a set of objects. It is possible to describe site as a dot in a three-dimension space where the first axis is location, the second – time, and the third – cultural or ethnical identity. Archaeological site is the clearest example of this conception – materials from a certain archaeological site reflect certain cultural tradition in a certain period of time from a certain place. In the case of a multi-layer site each occupation layer should be described as a separate site. Hierarchical structure also allows to describe the whole site in a separate record.

The conception of a site is also very useful when describing an ethnographical collection. In this case certain sites reflect materials on a certain ethnical or social group that existed or exists in a certain time in a certain area.

Spatial and temporal location of a site could be described in different schemas – geographical regions, administrative units, and geographical coordinates, absolute and relative chronology. For some objects' sets collected several ages ago our knowledge is limited. But if we know something about their origin we should create a new site record. We call it a "ghost-site". For example, in a collection there are objects from East Siberia – that is all we know about them. But we make a site record – with the Geographical region attribute – East Siberia.

Mandatory fields: Name – site name; Info – short description of a site with a hyper-reference to a hyper-text document with description of the site and some images, such as photographs, maps, cross-sections, etc. These documents contain information of educational significance.

Optional fields: Period_id – the period of a site's existence in the relative chronology schema, a reference to the table Common.chronology, scheme relative chronology; Absdate_id – the period of a site's existence in absolute dates from the earliest possible to the latest possible dates and the background of absolute dating, reference to the table Common.chronology, scheme absolute chronology; Geo_id – geographical region where the site is located, a reference to the table Common.location, geographical scheme; State_id – the administrative unit in which the site is located, a reference to the table Common.location, administrative scheme; Coord_id – geographical coordinates of a site and a short text description of the site's location, a reference to the table Common.location, scheme geographical coordinates.

Sets scheme

Table sets.types
This table keeps information about types of sets, e.g. collection, subcollection, composed object, etc.

Mandatory fields: Name – types of set; Info – definition of the type.

Table sets.profiles
This table contains information about profile of sets, including attribute definition for different types.

Mandatory fields: Name – profile of sets; Info – definition of the profile.

Table sets.tree
The basic object in the database is collection – it's a set of objects kept in the museum with a certain museum registry number. As a "perfect" collection we imply a set of subjects from a certain site, which was obtained in a certain period of time by a certain person (e.g. materials of archaeological or ethnological expedition from one field season). As a rule materials with the same incoming number form the same collection. But sometimes materials with the same incoming number could be separated in several collections (e.g. physical anthropological and archaeological materials from the same archaeological site are kept in different funds).

At the same time, if materials from different archaeological sites come into the Museum at one time and receive the same incoming number, they could be stored under the same collection number. But the hierarchical tree allows making sub-collection records – each one for different site. In such case link_id identifier of subcollection record refers to collection record. Collections from the same site or from a certain area of ethnological research also could be divided according to materials the items are made of. It also makes more sense as it is preferable to keep objects of different materials in different conditions, that means in different storage places.

Following the Database logic, each collection item with its own item number is described as a low-level subcollection, and sometimes there could be complex objects (e.g. composite tool) and it is necessary to describe each part of such object in the database as lower-level subcollection. It should also be mentioned that sometimes several subjects (e.g. several flakes or pot-shreds or different bones of one skeleton) could be registered under the same item number. Identifying each collection with certain organization or organization department allows to describe in the Database a collection not only from the Kunstkammer museum, but also from any museum or institution that possesses anthropological collections.

Mandatory fields: Name – collection number or any name of subcollection (things); Info – information about storage or any collection movement as participation in exhibitions or laboratory studies etc.

Optional fields: Receipt_id – incoming number of materials in collection, reference: to the table Receipts.list; Dep_id – Museum or organization where collection is stored or research department, which is responsible for the collection, reference to the table Orgs.tree; Date – the date of a collection's registration. Recorder_id – the person who registered the collection, reference to the table Persons.list; Keeper_id – the person who is responsible for the collection, reference to the table Persons.list; Subj_count – number of objects in the collection;

Item_count – number of items in the collection; Material_id – material that the object is made of, reference to the table Common.materials; Site_id – archaeological site or certain area where ethnographical or physical anthropology collection was obtained from, reference to the table Common.sites; Extractor_id – the person who obtained the collection (e.g. the chief of an expedition), reference to the table Persons.list; Extract_date – period when the collection was obtained, reference to the table Common.chronology; Type_id – type of the set; Profile_id – type of objects according to typological scheme or collection profile, reference to the table Sets.profile; Storage_id – place of collection or subcollection storage such as exhibition showcase or shelf, reference to the table Storage.tree.

Table sets.docs
One document can be related to many Database objects or one object can be related to many documents. This structure is reflected in the table Docs.tree.

Fields: Doc_id – Document name, reference to the reference table Docs.tree; Set_id – the object that a document is related to.

Table sets.attributes
Describes objects in collections according to typological schemes, given in Sets.profiles. A certain type of objects needs a certain description template. In the same collection objects from the same site could be of different profiles – for example, ceramic and flint in archaeological collections. In the Database we can support description of each type in its own scheme.

Mandatory fields: Name – name of the attribute; Info – additional information; Profile_id – reference to the attribute definition; Set_id – reference to the object in the table Sets.tree.

System architecture

For technical support of all of the declared purposes we chose a client-server system architecture. The server side is divided into two subsystems – the databank and the set of stored procedures for access. The client side consists of access libraries, granting high level interface to the server stored procedures, and the client applications themselves.

In order to keep strict control over access to the data, enforce data integrity and data access speedup direct manipulations in the databank were denied – only operations with data going through the stored procedures are allowed.

Client access library presents two levels. The low one is the extension of the ADO library. It consists of the following objects: "Connection" – database connection and interface for stored procedures; "Node", which represents the node of the data tree, and "Nodes", which is collection of "Node" objects; "Field", which represents attribute of any record with attribute description, and "Fields" – collection of "Field" objects, that is classic database record.

ActiveX controls make a higher level with objects "Tree" and "Properties". The first is designed to navigate through the hierarchical tree of records and the second – to display and operate above the records' attributes.

Thus, the system architecture can be presented as a stack: ActiveX controls – ADO Extensions – Stored Procedures – Databank.

Using such system architecture with conjunction of the sets of the mandatory fields in any database record automatically gives us the following results: 1) the system is invariant for adding new objects and object attributes; 2) the total search through the database is available because all records have the set of standard attributes; 3) as data access is going through the stack, it is easy to convert inner protocols to standards ones, such as HTTP, LDAP.

Achievements

So far we have reached the following achievements: the server level is completed, a data management tool and a web page generator are created.

The data management tool looks like Windows Explorer and has a similar operating logic. It supports navigation through tree-organized records, records adding and editing.

The page generator is built using ASP technology and allows navigation and search. Three types of search are available: SEARCH BY ABSOLUTE DATES, total SEARCH BY NAME of the objects of any type and MULTIPURPOSE SEARCH, which reflects relations between objects. In advanced or MULTIPURPOSE SEARCH any type of data that has a name in the Name field of record, such as archaeological sites, cultures, periods, geographical, administrative regions, etc., could be the aim of a search. The user can choose the purpose of the search and then any kind and non-limited amount of search conditions. E.g., it is possible to find all archaeological cultures that existed on the territory of the European part of Russia in the Neolithic period, or geographical regions where a certain archaeological culture was spread. The only limit for such searches is that sites or regions to be found should be presented in materials stored in the Kunstkammer and described in the Database. It may well be true for the reason that all low-level database tables are related with each other through collection tables.

The created system is time shared and multi-user. According to the user's access permission it is possible for several users at the same time to read, add or edit data via local network or via Internet.

Some limits are still not overcome: at present all applications exist only in the Russian version, no access restriction is supported, and thus some fields with internal Museums' information are closed for web-users.

At present information about 400 archaeological, 571 anthropological, and 200 ethnographical collections and subcollections is available for database users.

Perspectives

The system described above was designed not only as a catalogue of Kunstkammer collections but also as a number of scientific, research and educational data bases connected with this catalogue. For instance, together with this project there was developed a project of creation of a Russian Craniological Data Bank that encompasses materials from almost all anthropological museums of Russia. Most probably, in the future there will be developed a number of other data bases connected with the catalogue of Kunstkammer collections (such as a bibliography of publications about the Museum's collections, etc.)

At present European and American anthropologists and archaeologists are working on organizing several global projects that cannot be realized without creation of big specialized databases. The developed technique, in our opinion, opens brilliant opportunities for the creation and administrating of such resources in the global computer network.

The architecture of the project also make it possible to create a powerful education and reference resource on anthropology and archaeology of the peoples of the world. The authors of the project did their best to provide the attribution of the Museum's collections with references, including gazetteer, reference books, illustrations, hyperlinks to other projects in the net, etc.

Speaking about the technical maintenance and upgrowth of our museum Internet-based computer catalogue, it should be mentioned that we have the following for the nearest future: 1) developing multi-language support; 2) integrating XML technology in Web application and access to the Museum directory by LDAP protocol.

Integrating an XML technology allows us to provide the created scheme of description of anthropological collections in standard XML-Schema terms. It also makes it possible to use convenient schemes granted by other providers. XML-documents in combination with style tables allow to use personal settings in web layout depending on the user's preferences, access tool and communication bit rate.

The ideology of data operating protocol, which is used in the present database is practically very close to LDAP ideas. In the nearest future we plan to remake and expand our protocol to conform to the LDAP standards. It will allow us to open access to the Kunstkammer Database for outer users.

As a result, we would be able to organize access to data and metadata by standard protocol, which supports references between directory servers and standard methods to operate with these references. In perspective, it will make it possible to organize a Common Museum Network. Using this protocol, any Network member will be able to create a personal client application layout for a certain (scientific or other) need. So, anyone interested in cooperation is welcome. We are looking for partners.

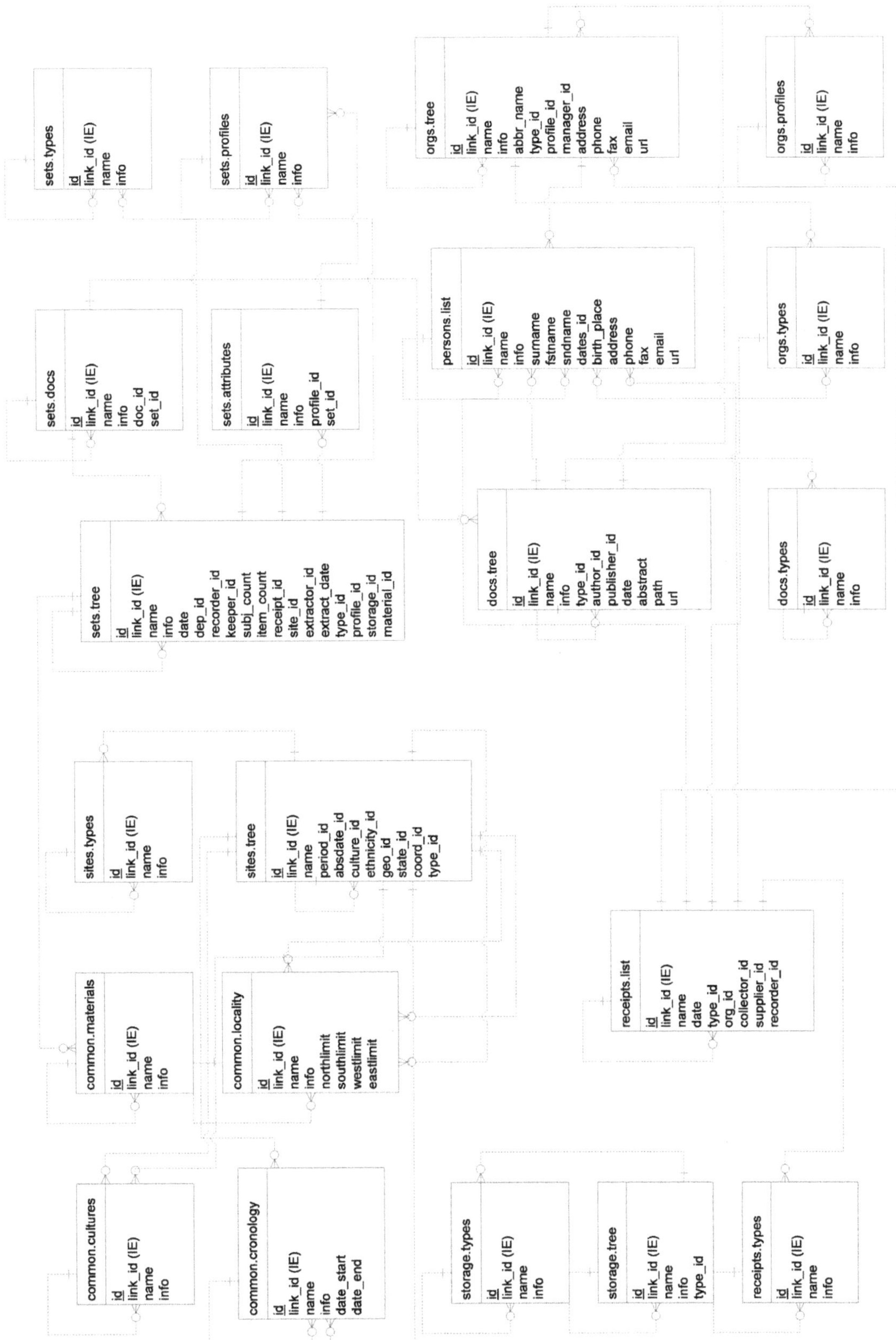

Figure 1. Data layout of the Kunstkammer Computer Catalogue.

Reclaiming Old Data: The Wasden Site Research Project

E. S. Lohse
Idaho State University
Idaho Museum of Natural History
PO Box 8096
Pocatello, ID 83209
E-mail: lihesrne@isu.edu

J. Anderson
Idaho State University
Idaho Museum of Natural History
PO Box 8096
Pocatello, ID 83209

Abstract: The Wasden site is a significant Paleoindian site in southeastern Idaho, USA. Excavated between 1961-1979, this large collapsed lava tube has preserved evidence of prehistoric people hunting mammoth and other extinct Pleistocene species dated c.11,000 B.P., and two separate bison kills involving at least 60 animals dated c.8000 B.P. This significant site has never been fully analyzed, nor a final report published. Individual papers and theses have been written but the data has never been integrated in a comprehensive, searchable electronic database. This paper presents an overview of our efforts to reclaim and recast archaeological data to allow research on an important forty year old site excavation. Our results will be available online and on CD-ROM, with applications in GIS, and ACCESS. The CD-ROM is scripted in Macromedia Director and contains all documentation, written reports and papers, artifact and site images.

Keywords: data reclamation; databases; GIS; multimedia

Data Defined

Archaeological excavations yield a wide range of information, encoded in variable constructions, measurement systems, and file formats. The problems for data compilers multiply as time between excavation, analysis, publication and compilation increases. This paper addresses our efforts to reclaim "old data" from an archaeological site first excavated almost forty years ago. At that time, all excavation records and subsequent analyses were preserved only as handwritten documents and scribbled notes. No standard excavation recording system was used. Two principle excavators worked on the site from 1963-1978. The usual labor force consisted of amateur volunteers and fieldschool students.

This is not a novel concern, as a glance at the literature on data management reveals (e.g., Andresen and Madsen 1996; Cheetham and Haigh 1992; Clubb and Lang 1996; Hadizlacos and Stoumbou 1995; Hansen 1993; Lamprell et al. 1995; Murray 1995; Stewart 1996). One important issue is the development of standards for data definition, maintenance and dissemination (e.g., Beargrie and Greenstein 1998; McCartney, Robertson and Cowgill 2000; RCHME 1993; Stewart 1995; Van Leusen et al. 1996; Wise and Miller 1997). Another is description of incomplete data in a standard relational database (e.g., Eiteljorg 2000).

Our practical difficulties are compounded by dealing with a significant archaeological site excavated and analyzed over two decades beginning some forty years ago. Our goal is data reclamation and then integration of this information in a standard database that will be archived. We have a number of specific concerns:

Need to archive archaeological data from a significant site
Development of a web site for online dissemination of information
Development of a comprehensive museum database that incorporates Wasden data in a regional database
Use of digital imaging for recording archaeological information and for enhancing future research

Wasden '60s-70s: Data Transformed

The Wasden site consists of a collapsed lava tube on the Snake River Plain, southeastern Idaho, which was occupied over the past eleven thousand years by prehistoric hunters (Figure 1). The lowest levels contain Folsom projectile points and mammoth, bison and camel bones. Just above, is a major concentration of bison bone representing several discrete entrapments, butchering and processing events that occurred about eight thousand years ago. Excavation at Wasden began in 1964 as an archaeological dig of the Upper Snake River Prehistoric Society and the Idaho Museum of Natural History. Three separate adjacent caves were explored: Owl Cave or Wasden (10BV30), Coyote Cave (10BV31) and Dry Cat Cave (10BV32). All three lie at an elevation of 5000' a.s.l. Work at the Wasden site from 1964 to 1974 proceeded solely on volunteer labor. From 1975-1978 work was conducted by Idaho State University archaeological field schools, funded by a National Science Foundation grant (SOC75-10340). The site deposits exceeded fifteen feet in depth, consisted largely of aeolian deposition, and were finely stratified. The site yielded a large collection of microfaunal and macrofaunal remains, collected in three stratigraphic columns for identification and bagged by excavation level. The focus of research from 1975-1978 was to excavate deposits in the alcove on one side of the bisecting trench and define an association of fluted stone points and extinct Pleistocene elephant, camel and horse remains. The Wasden site was the first find of butchered elephant remains in a stratified rockshelter in the Northern Intermountain West, USA.

The Wasden site overhang is roughly circular with a diameter of approximately 22 meters at the dripline. Before excavation, the southern portion was filled with sediments sloping down to the floor of the cave from the south. The northern portion of the cave created a sheltered area about 18 meters wide and 4-16 meters back from the dripline. Deposition consisted of fine laminae of wind-blown sediments and water-deposited sediments broken by periodic episodes of rockfall from frost-weathering of the laminar basalts constituting the walls and overhang. Stratigraphy is marked by the presence of strata indicative of clear climatic shifts and accompanying changes in basic geomorphological processes affecting the site. The master stratigraphic profile (Figure 2) shows major stratigraphic features of the site sequence within the overhang. A sequence of ice-wedge casts is bracketed by two radiocarbon dates of about 8160 B.P. and 7750 B.P. A redeposited layer of Mazama Ash is dated about 6600 B.P. The eastern half of the cave had been excavated down to heavy roof-fall resting on bedrock. That roof-fall slopes down from the front to the rear of the cave. The overlying loess becomes correspondingly deeper toward the rear of the cave as well (Butler 1969). The remains of elephant, bison and camel bone were recovered from roof-fall blocks on the floor at the rear of the alcove.

In the 1965 excavation the grid system was set up using two-meter excavation test units, with a North-South axis and an East-West axis. As such the grid system was labeled with a three character name, eastern side of the site carrying an "E" designation, western side of the site carrying a "W", followed by a number (x-axis) and a letter (y-axis). The E/W lines are still marked on the north ceiling of the alcove. The excavation concentrated in the eastern section of Owl Cave, and generally proceeded down in five centimeter arbitrary excavation levels. Recovered deposits were passed through 3/8" screen. The later, 1975-77 excavations proceeded within the same grid system but varied in thickness of levels removed and the screen used.

Butler (1971) published a number of articles describing the bison remains recovered from the general site deposits. Those on the modern site surface are *Bison bison*. Those from lower loess deposits are too fragmentary to identify to species and too infrequent to constitute a good sample. The bison bone bed dated about 8000 B.P., however, yielded the disarticulated remains of more than 60 individual bison. The morphology of skulls and horn cores ranged from modern *Bison bison* to *Bison antiquus*. Butler concluded that the bison population comprised yearlings and adults of both sexes. The cave deposits also produced plentiful microfaunal remains including rabbits (*Sylvilagus*) and pocket gophers (*Thomomys*). *Sylvilagus idahensis* thrives in sagebrush areas and numbers are highly correlated with shrub density. The Northern pocket gopher is more common along streams and feeds on roots and stems of grasses. Guilday's (1969) study indicated pronounced shifts in total numbers of small mammals over time. Mean numbers of individuals are higher before 6600 B.P. and lowest between 3500-5500 B.P., a period correlated with the Altithermal, and intervals of dry and warm climate. There was also a shift in the ratio of pocket gophers to pygmy rabbits about 6600 B.P. Prior to that time, pocket gophers wre more numerous than pygmy rabbits. These shifts indicate a shift from grass to sagebrush vegetation coincident with the climatic change to a warmer, drier climate post-5500 B.P.

The Wasden site is one of the most significant archaeological sites in the western United States. The cave shows evidence of Native American use from about eleven thousand years ago to the early historic period (Butler 1978, 1986; Lohse 1994). A major occupation is the large bison kill dated about 8000 B.P. It seems that the site was used by prehistoric hunters at this time to drive bison into the overhang where they were dispatched and butchered. Discrete activity events like this are rarely documented in the archaeological record and examination of these assemblages of bones and artifacts will offer dramatic insights into prehistoric lifeways. The lower, less well defined mammoth kill was the focus of an incomplete dissertation by Susanne Miller. Her final report, dated 1/29/80, described these prehistoric events as workshops of extensively broken and modified bones from mammoth, bison and camel. To date though these assemblages remain incompletely studied and reported on. A number of papers and posters were presented by Miller on selected aspects of the butchering assemblages but no comprehensive published summary is yet available. Manuscripts on site environmental reconstruction were completed but not published. The associated megafauna populations have been studied and characterized (Butler, Gildersleeve and Sommers 1971; Miller 1982, 1989).

These collections and associated documentation are housed at the Idaho Museum of Natural History. They have received little attention from researchers because they remained in storage without a comprehensive catalog. Our current work focuses on accurately compiling the original site record to create a detailed map of the 8000 year old bison bone bed in a GIS environment . This first required reorganization of the Wasden site collection. The majority of the artifacts and faunal specimens had been stored in hundreds of cardboard boxes in a number of locations within the museum (Figure 3). Miller's dissertation research and the specimens she worked with had been stored in locked cabinets in the Paleontology Division's storage area. Materials analyzed by Butler had been stored in cabinets in the Anthropology Division. We set up a project room in the Anthropology Division and all artifacts and faunal specimens were taken there to be arranged on steel shelving. Graduate and undergraduate students, enrolled in lab courses, spent a year extracting specimens from boxes, entering these in a master catalog, and organizing specimens on the shelving.

Attendant paper documentation, including excavation records, analysis notes, correspondence, manuscripts, and published articles were gathered together in the Wasden project room and placed in lockable fire files. These were initially gathred from files in the Southeastern Archaeological Repository, Anthropology Division, but as work proceeded, documents were also recovered from Miller and from the amateur members of the original excavation crews. Documentation held and catalogued by the Museum was primarily that produced by the work of and the Idaho State University archaeology fieldschool crews.

No standard excavation record keeping system had been developed over the two decades of Wasden site excavation, and notes consisted mostly excavators' field notebooks, artifact and specimen cards, and maps and profiles drawn in the field. No comprehensive inventory had been made and our students carefully combed published papers and manuscripts to cross-reference field notes and analysis records. Problems encountered were myriad and included redundant specimen numbers, redundant or contradictory provenience notations, conflicting field and lab records, and missing level or excavation square notebooks. The most complete documentation was that supplied for the published work of Butler on the 8000 year old bison bone bed and that developed by Miller for her research on the 11,000 year old mammoth level. We quickly learned that other aspects of the site were equally interesting and important. We also learned that some important parts of the documentation record were missing or suspect.

Reclamation Strategy

Reclaiming the Wasden site data is a significant addition to our knowledge of regional prehistory. It adds to our understanding of Paleoindian and Early Archaic hunting societies on the North American continent in general. Several points are paramount in reclaiming this site record.

Our current IMNH database design could not simply be imposed on the Wasden collection (cf. Lohse 1996; Lohse and Sammons 1998, 1999). In many cases, information was missing or recorded on disparate scales of measurement. Bringing the Wasden collection into an electronic format facilitates further analysis of this important site. Analysis of the well defined bison kill levels supplies a compelling case study emphasizing the value of using digital imaging in database construction, analysis, interpretation and information sharing (Anderson 2001). Publication of the analysis of the entire Wasden site assemblage will deliver information on a very significant paleoindian site that has yet to be published in any detail yet it is listed in summaries of Idaho prehistory and in introductory textbooks on North American prehistory. Putting selected elements of this project online will lead to significant collaborative efforts by regional and national experts interested in Paleoindian and Early Archaic hunting cultures.

Reclamation Methodology

The Wasden paper database, consisting of 4 artifact catalogs and 4 faunal catalogs were converted into a Microsoft Access relational database format (Figure 4). The new digital version can be queried, and can be placed into statistical software. This digital database was then inventoried against the actual collection of specimens to look for glaring errors, such as errors in classification. It was also inventoried to check for missing artifacts. The resulting inventories list particular classes of artifacts, and partially record associations with cultural and natural features. The database does not include small faunal remains, which were apparently never catalogued though these number in the thousands.

Data reclamation is a time and money consuming process. The Wasden catalogs were placed into two tables in the master Wasden database, which also includes tables that are set for particular analyses, such as Funcan (Functional) and Techan (Technological) for future stone tool analysis (Lohse 1993:33-64, 1996; Lohse and Sammons 1998). The current artifact database consists of twenty fields for over 1600 records. The faunal database consists of ten fields for 8000 records, reflecting almost exclusively the large faunal remains. The old specimen catalogs were entered into our Microsoft Access 2000 tables. Data entry was done by students and took over two hundred hours. The data were then proofed for accuracy. The edited database was placed into SPSS by converting the Access database into a dBase file, which required an additional twenty hours to put into a useable format by converting fields from a string to a numeric format. Only some of the fields were in a standard format, which would have enabled a software solution for conversion. Fields, such as grid square, feature, and layer, were in a standard form. Problems were encountered when these variables were given as ranges, reflecting the inaccuracy of the collection methods. Some of the fields, such as the explanation of skeletal element in the faunal database, required hand entry of code numbers that would allow for nonparametric correlation testing and basic exploration statistics to be performed on the Wasden material. The skeletal elements were grouped into fifty different classes based on how they were originally coded in the specimen catalogs. In total, close to three hundred hours were needed simply to put the older Wasden material into a useable format. This does not count the hours required to create the original paper databases. Nor do these hours reflect additional statistical analysis, entry into GIS ArcView, the basic stone tool analysis, any formal analysis of bone, or a complete inventory of the Wasden material.

The many practical problems ranged from large to small. Even when the data are in a consistent format, there are errors, from excavation, cataloging, analysis and simple changes in handling and storage over several decades. Things are coded incorrectly. There are errors in data transfer within the original paper catalogs, and then again in the transfer from paper to digital format. We found redundant data to be helpful, as when a grid unit highlight the error even if the analysts cannot with assurance reconstruct the provenience. Without a standard system for excavation, there were inconsistencies in the way that people recorded the data in the field and in the lab. Variables recorded in the paper database as depth, BMD, and BOD, are different depth measurements taken from different site datums. Excavators did not measure from a standard corner in the excavation unit, and this has compounded our problems in putting the master stratigraphic profiles together.

Data Production

Transfer of the paper record to a standard electronic format is essential for effective manipulation of the Wasden site collection.

Entry into a GIS Environment

A GIS, by definition, is the linking of descriptive database information to visual representation, normally a map or a digital image, as in the case of raster GIS data. As such there are several ways to bring such information into a useable GIS. For the Wasden project we brought information into the GIS database using three different approaches.

The first approach was to construct a GIS of the grid system used in the Wasden excavation using ArcInfo. The information was coded using Microsoft Wordpad and appropriate ArcInfo syntax to create a text document. The Wordpad text document was then brought into ArcInfo and using the generate command a GIS layer was developed. The database created when the GIS layer was generated was then amended to bring additional information to the database, such as the grid unit designation.

The second approach used in reclamation of the Wasden data was using the constructed Microsoft Access database saved as a dbase file with a .dbf extension. ArcView uses dbase as its relational database core and is only able to bring in files with the .dbf extension. The file was then used to generate a database within ArcView. The database in ArcView was then used to create a point theme, representing the artifact distribution within the Wasden site. This event theme was then saved as an ArcView file.

The final approach was digitizing the maps using a Numonics Grid Master digitizing pad and ArcView GIS to reproduce the paper maps that were created during the original excavations at Wasden. These paper maps had been stored rolled, which necessitated that they be flattened. These original documents were copied, and the copies used for the digitizing. The original documents were thereby protected from possible damage. The photocopying introduces some level of error, but this error level, roughly 1%, was considered to be acceptable. In fact the error level, with most of the maps at a scale of 1:10, fell within the acceptable level for the program to register the map. ArcView accepts a registration error of .004 in the unit of measurement. The database was then amended to include additional information about the polygons that were created representing rock fall and faunal material (Figure 5).

ArcView lends itself to the presentation of complex data, by displaying it in a visual format. The presentation of the Wasden data has been accomplished in several ways. Using the Layout module in ArcView allows themes to be manipulated for hardcopy presentation and export in a variety of digital formats, i.e. .jpg, .tif, etc. This allows a variety of options for the display of GIS data. For the Wasden project themes and complete views were exported for use in an online environment. The Wasden site is one of the first archaeological sites in the western United States to be online in a useable GIS format as part of the "Digital Atlas of Idaho."

Imaging Analysis

Recording and analysis of the artifacts and specimens from the Wasden site is enabled by use of sophisticated digital imaging hardware and software. CCTV cameras attached to binocular microscopes with fiber optic illumination allows generation of high quality, high magnification digital images. Digital data is manipulated through ImagePro Plus software allowing analysis of pixel patterns, exact quantification of those patterns, entry into standard relational databases, and easy statistical manipulation. Artifacts from the Wasden site show clear evidence of Paleoindian hunting and butchering patterns (Figures 6, 7). Perhaps more importantly, storage of the live, information rich digital images and links to databases holding numerical information to the level of a single pixel, ensures accurate dissemination of information.

Online Presentation of Reclaimed Data

The Digital Atlas of Idaho is a GIS based interactive production housed on the Idaho Museum of Natural History server, and links web sites of the Idaho Geological Survey and the Idaho State University College of Education. The Digital Atlas was funded by a multi-year award of the Idaho State Board of Education Technology Inventive Grant, and is a cooperative venture with principal investigators from Idaho State University, Boise State University and the University of Idaho. Data sets dealing with anthropology, archaeology, geography, history, surface and subsurface geology, seismic activity, hydrology, and vertebrate and insect ecology are linked using ArcView GIS and Visual Basic Map Objects Graphical User Interface (GUI).

Future Work

The databases are intended to compile information about southern Idaho for distribution to Idaho's Universities and the Idaho Virtual Campus, as well as elementary and secondary schools. Teaching modules are being created for elementary and secondary educators.

Archaeological data is currently being added to the Digital Atlas. Specifically the digital atlas archaeology section has data layers with links to survey and excavation data and provides digital images of diagnostic specimens and sites online. Part of the Digital Atlas contains links to the Anthropology Collections database at the Idaho Museum of Natural History, with thumbnail images of those collections housed at the Idaho Museum of Natural History. These images and the database itself are placed online as they become available. The Digital Atlas archaeology sections include regional overviews and culture history sections. The Atlas, because it provides information for multiple levels of users, becomes an introduction to archaeology and Idaho prehistory for elementary and secondary students, it gives elementary and secondary educators teaching material, and it also becomes an accessible repository of information for all interested in Idaho Archaeology and Idaho prehistory.

The Wasden project has become a template of how to place, or not to place information online. The strategy employed has been to divide the presentation of sites into relatively easy to produce modules. These modules are self-contained autonomous pieces. Being self-contained, these pieces are interactive with the entire online presentation when placed into the web site structure. This allows the web page to flow easily while pieces are added, and allows the web site a completed look although it is under continuous construction.

For example with the Wasden site the original work focused on and ultimately published on two occupations, one from roughly 8,000 BP and the other from roughly 11,000 BP, within the site. These two occupations are the natural beginning point to create the online Wasden site. After the data reclamation and GIS construction discussed earlier, a textual overview, or narrative for the site was constructed. Artifact record shots were taken and the constructed GIS maps were brought in as JPEGs as well as placed in ArcView format for download. ArcExplorer is available for download, allowing even those without access to standard GIS software to view some of the GIS mapping.

References

Anderson. J., 2001.Old Databases, New Information: A perspective of data recovery from the Wasden Site. Paper presented at the Sixty-Sixth Annual Society for American Archaeology Conference April18-22, 2001. New Orleans, Louisiana.

Andresen, J, and Madsen, T., 1996. IDEA – the Integrated Database for Excavation Analysis. In Kamermans, H. and Fennema, K. (eds.), Interfacing the Past: Computer Applications and Quantitative Methods in Archaeology CAA95, Vol. I, pp.3-14. Leiden: Institute of Prehistory.

Beargrie, N. and Greenstein, D., 1998. A Strategic Policy Framework for Creating and Preserving Digital Collections. In Guidelines for Digital Preservation. Arts and Humanities Data Service. http://ahds.ac.uk.manage/framework.htm

Butler, B.R., 1969. More Information on the Frozen Ground Features and Further Interpretation of the Small Mammal Sequence at the Wasden Site (Owl Cave), Bonneville County, Idaho. *Tebiwa* 12(1):58-63.

Butler, B.R.,1971. The Origin of the Upper Snake River Country buffalo. Tebiwa 14(2):1-20.

Butler, B.R.,1978. *A Guide to Understanding Idaho Archaeology*, 3rd ed. Special Publication of the Idaho Museum of Natural History. Pocatello.

Butler, B.R., 1986. Prehistory of the Snake and Salmon River Area. In "Great Basin," Warren L. d'Azevedo (ed.), pp.127-134, *Handbook of North American Indians* Vol.11, William Strutevant (gen. ed.). Washington, D.C.: Smithsonian Institution.

Butler, B.R., Gildersleeve, H. and Sommers, J. 1971. The Wasden site bison: Sources of morphological variation. In A.H. Stryd and R.D. Smith (eds.), *Aboriginal man and environments on the Plateau of North America*, pp.126-152. University of Calgary Archaeological Association, Calgary.

Cheetham, P. N., and Haigh, J. G. B., 1992. The archaeological database new relations? In Lock, G. and J. Moffett (eds.), Computer Applications and Quantitative Methods in Archaeology, BAR International Series, S 577, pp.7-14.

Clubb, N. D., and Lang, N. A. R., 1996. A strategic appraisal of Information Systems for archaeology and architecture in England past, present and future. In Kamermans, H. and K. Fennema (eds.), Interfacing the Past: Computer applications and quantitative methods in archaeology CAA95, pp. 51- 72. Leiden: Institute for Prehistory.

Eiteljorg, H., 2000. If we preserve the files , who will use them? Paper presented at the 66th annual Society for American Archaeology Conference March 2000, Philadelphia, PA.

Guilday, J., 1969. Small mammal remains from the Wasden site (Owl Cave), Bonneville County, Idaho. *Tebiwa* 12(1):41-45.

Hadizlacos, T. and Stoumbou, P.M., 1995. Conceptual Data Modelling for Prehistoric Excavation Documentation. In Kamermans, Hans and K. Fennema (eds.), Interfacing the Past: Computer applications and quantitative methods in archaeology CAA95, *Analecta Praehistorica Leidensia* Vol. I, pp. 21-30. Leiden: Institute of Prehistory.

Hansen, H. J., 1993. European archaeological databases: problems and prospects. In Andersen. J., T. Madsen and I. Scollar, Computing the Past: Computer applications and quantitative methods in archaeology CAA92 pp. 229-237.

Lamprell, A., Salisbury, A., Chambers, A., and Stoddart, D., 1995. An Integrated Information System for Archaeology Evidence. In Kamermans, Hans and K. Fennema (eds.), Interfacing the Past: Computer applications and quantitative methods in archaeology CAA95, *Analecta Praehistocia Leidensia* Vol. I, pp. 37-42. Leiden: Institute of Prehistory.

Lohse, E.S., 1993. Manual for Archaeological Analysis: Field and Laboratory Analysis Procedures. Department of Anthropology Miscellaneous Paper No. 92-1 (revised). Pocatello: Idaho Museum of Natural History.

Lohse, E.S.,1994. The southeastern Idaho prehistoric sequence. *Northwest Anthropological Research Notes* 28:137-160.

Lohse, E.S., 1996. A computerized descriptive system for functional analysis of stone tools. *Tebiwa* 26(1):3-66. Online at lohserne@isu.edu.

Lohse, E.S., 1998. *A manual for archaeological analysis: Field and laboratory procedures.* Archaeological Survey Miscellaneous Paper 98-1 (revised). Online at lohserne@isu.edu.

Lohse, E.S. and Sammons, D., 1998. A computerised data base for lithic use-wear analysis. In Proceedings of the 1997 Computer Applications in Archaeology Conference, V. Gaffney (ed.). *BAR International Series* S750. CD-ROM.

Lohse, E.S. and Sammons, D., 1999. *Digital Stones*: A guide to analysis of stone tools. Beta CD-ROM. Laboratory for Digital Imaging, Idaho Museum of Natural History. Online at lohserne@isu.edu.

McCartney, P., Robertson, I. and Cowgill, G., 2000. Using Metadata to Address Problems of Data Preservation and Delivery. Paper presented at the 66th annual Society for American Archaeology Conference March 2000, Philadelphia, PA.

Miller, S.J., 1982. The archaeology and geology of an extinct megafauna/fluted point association at Owl Cave, the Wasden site, Idaho: a preliminary report. In *Peopling of the New World*, Ballena Press Anthropological Papers No. 23, Ericson, Taylor, and Berger (eds.), pp81-95. Los Altos: Ballena Press.

Miller, S.J., 1989. Characteristics of mammoth bone reduction at Owl Cave, the Wasden Site, Idaho. In *Bone Modification*, Bonnichson and Sorg (eds.) pp.381-393. Orono: Center for the Study of the First Americans, Institute of Quaternary Studies, The University of Maine.

Murray, D.M., 1995. The management of archaeological information: A strategy. In Wilcock, J.D. and K Lockyear (eds.), Computer Applications and Quantitative Methods in Archaeology 1993, BAR International Series 589, pp. 83-87.

RCHME, 1993. *Recording England's past: A data standard for the extended national archaeological record.* London: Royal Commission on the Historical Monuments of England.

Stewart, J., 1995. Transforming the management and access of museum collections with IT. In Stevens, T. and V. James (eds.), The Future for Europe's Past, Proceedings of COMETT Workshops 3..37-3.40. Swan: Swansea Institute of Higher Education on behalf of the EU COMETT Programme.

Stewart, J., 1996. Where is the knowledge we have lost in information? In Fahy, A. and W. Sudbury (eds.), Information: The hidden resource, museums and the Internet, Proceedings of the Seventh International Conference of the MDA, pp. 419-423, Cambridge: MDA.

Van Leusen, M., Champion, S., Lizee, J., and Plunkett, T., 1996. Towards a European Archaeological Heritage Web. In Kamermans, H. and K. Fennema (eds.), Interfacing the Past: Computer applications and quantitative methods in archaeology. CAA95, pp. 507-514. Leiden: Institute for Prehistory.

Wise, A. and Miller, P., 1997. Why Metadata Matters in Archaeology. *Internet Archaeology* 2. http://intarch.ac.uk/journal/issue2/wise_index.html.

Figures

Figure 1. Views of the Wasden site: Above, aerial view showing Owl Cave (arrow indicator), Coyote Cave to north and Dry Cat Cave to the south (USDA Photograph; Butler 1968: Fig 4); Below, view of the site from the west before excavation began in 1965 (Butler 1968: Fig. 5).

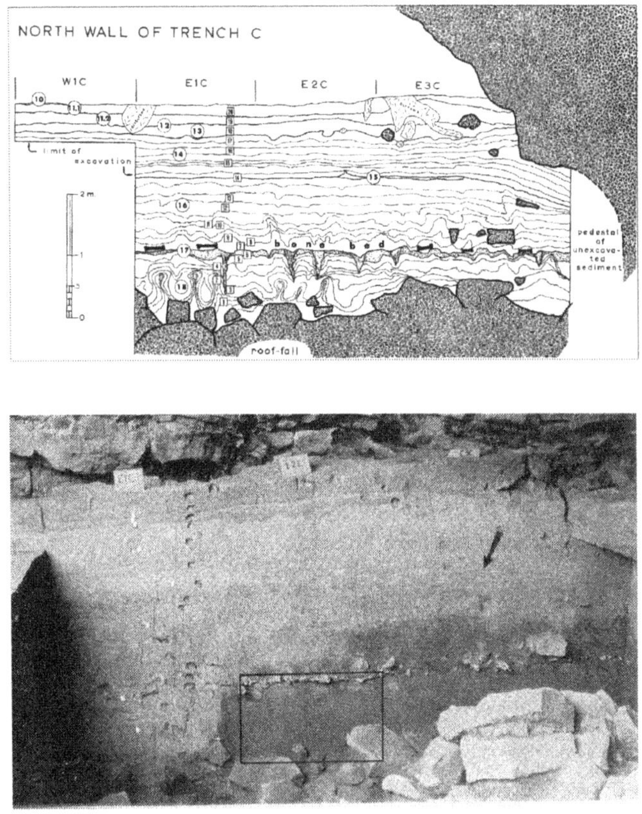

Figure 2. Wasden site stratigraphy. Top, stratigraphic drawing of north wall of Trench C. Numbers indicate strata recognized in the field. Distortion of strata lines below Layer 14 represent cryoturbation of bedding planes. Ice-wedge casts are shown below Layer17. Note the prominent bone bed indicated as Layer 17. Butler (1971:Fig.7). Bottom, photograph of north wall of Trench C. Arrow points to identified Mazama Ash Layer 17. Box shows ice-wedge casts below the bison bone bed radiocarbon date about 8100 B.P. Butler (1968:Fig.8a).

Figure 3. Sorting Wasden site specimens in the project room before data entry and authentication.

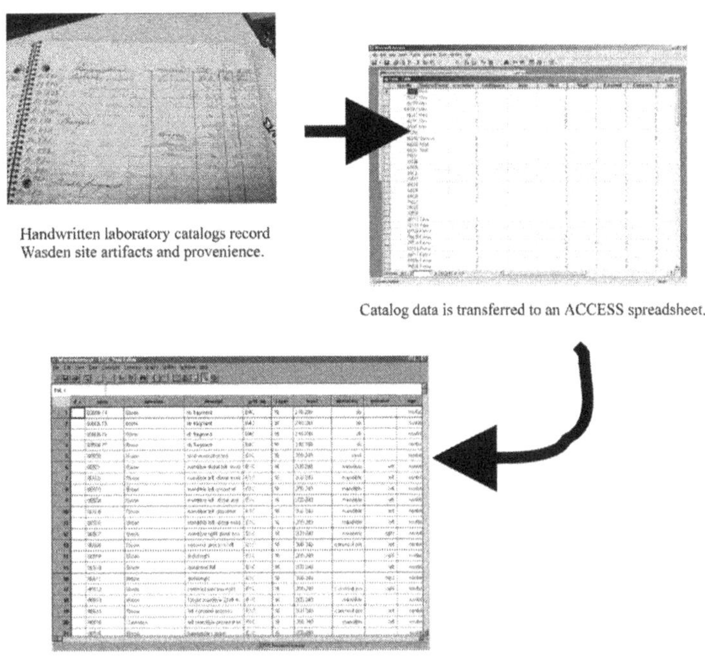

Figure 4. Data Records and Data Entry: Variable formats.

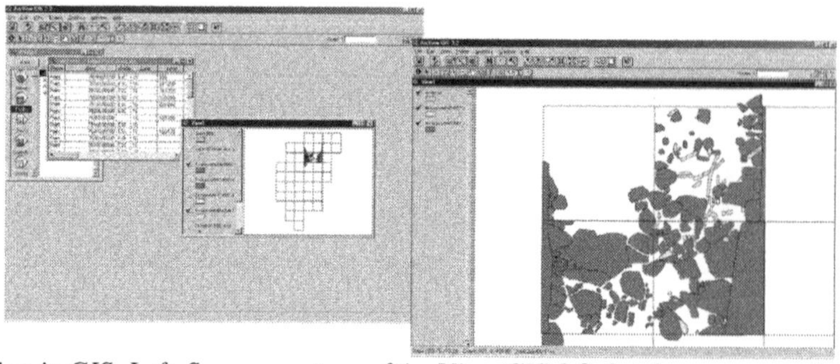

Figure 5. Data representation in GIS. Left, Screen capture of ArcView GIS 3.2 database screen: Attributes of Layer 16 database.dbf; right, Screen capture of plan map in ArcView GIS 3.2 of level 647-657 b.s.d, Unite E4E-E5E, E4F-E5F, Owl Cave

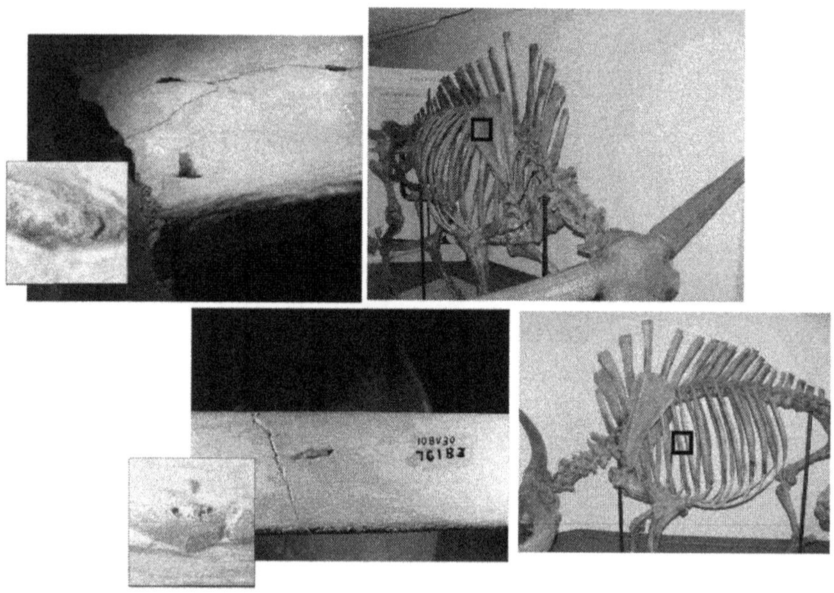

Figure 6. Date Interpretation: Forensic evidence of projectile angle of penetration using Bison scapula and rib recovered from units E3C and E1B at the Wasden site in the Layer 16 "Bison Bed".

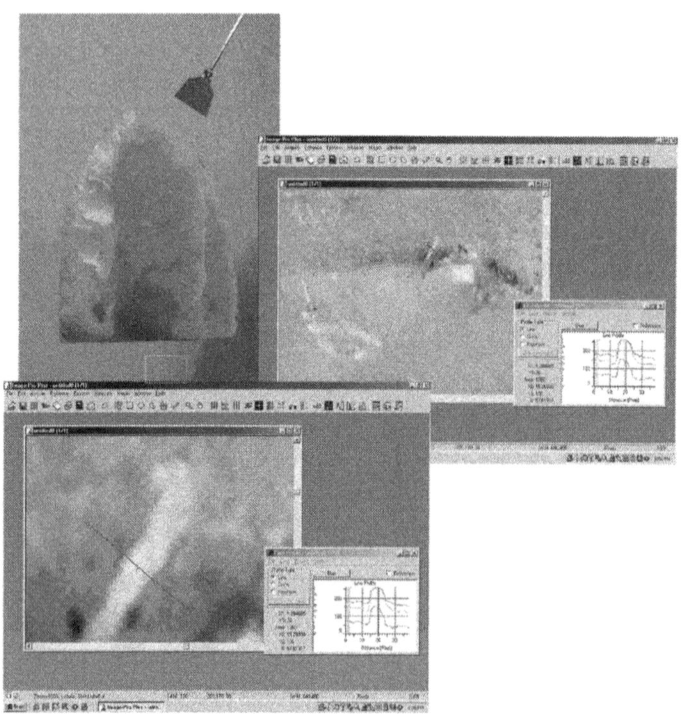

Figure 7. Image Analysis: Microscopic examination of a Folsom point fragment recovered from unit E6G at 6.34m b.u.d., Wasden site, in the Layer 18 "Mammoth Layer". Upper left, record shot with scale; Center right, magnified view of sinew fragment at 60X; Center left, magnified view of the same sinew Fragment at 200X. Graphs display measurements of spectral values that are downloaded to standard relational databases.

The Archaeology-Palaeobotany-Palynology Database on the Palaeolithic, Mesolithic and early Neolithic sites of the Former USSR Area

Yuri Stepanov
Peterlink Web L.L.C.
7a Chapygina st.
St.-Petersburg, Russia
Phone/Fax +7(812) 3275343 - E-mail: stepanov@petelink.ru

Galina Levkovskaya
Institute of History of Material Culture Russian Academy of Science (IHMC RAS)
18 Dvortzovaja emb.
St.-Petersburg, 191186, Russia
Fax: +7(812) 3116271 - E-mail: info@gml.spb.ru

Mikhail Anikovitch, Nikolai Anisutkin, Elena Beliaeva, Vladimir Shumkin,
Veronika Stegantzeva and Vladimir Timofeev
Institute of History of Material Culture Russian Academy of Science (IHMC RAS)
18 Dvortzovaja emb.
St.-Petersburg, 191186, Russia
Fax: +7(812) 3116271

Galina Sinitsyna
Institute of History of Material Culture Russian Academy of Science (IHMC RAS)
18 Dvortzovaja emb.
St.-Petersburg, 191186, Russia
Fax: +7(812) 3116271 - E-mail: galina@AS6238.spb.ru

Anastassia Bogolubova
The State Russian Museum
Department of Modern Electronic Technologies
4 Inzenernaya st.
St.-Petersburg, 191011 Russia
Phone: +7(812) 3134145 - E-mail: anb@peterlink.ru

Andrey Stegantzev
Geological faculty
St.-Petersburg State University
7 Universitetskaya emb.
St.-Petersburg, Russia
E-mail: pogantsev@mail.ru

Abstract: The Database of Paleobotanical (charcoals, woods, seeds etc.) and Palynological finds from the archaeological sites of the former USSR area contains some data on the early and middle Palaeolithic (BARPP-A), six archaeological and ecological stages of the upper Palaeolithic – Mesolithic (BARPP-B1–BARPP-B6) periods and the first information on the early Neolithic period.

The Internet address of the database is following: http://www.peterlink.ru/~stepanov/gml.

Key words: Database, Palynology, Palaeobotany, Pollen, Charcoal, Palaeolithic, Mesolithic, Neolithic

Introduction

The Database concentrates the Palaeobotanical and Palynological information on Stone Age sites from different regions of the former USSR area that is dispersed in a large number of mainly small articles, published in the issues on different sciences (archaeology, botany, geology, palaeogeography, palaeoclimatology, palaeopedology, geophysics, etc.) and often in unknown publisher houses of small towns, museums, etc. This information was accumulated through the joint work of palaeobotanist and palynologist (Levkovskaya), the specialists on Internet or computer technologies who for some years were involved in archaeological expeditions (Stepanov, Bogolubova , Stegantzeva, Stegantzev), and specialists on the Palaeolithic (Anikovitch, Anisutkin, Beliaeva) and Mesolithic–Neolithic–Bronze ages (Shumkin, Sinitzina, Timofeev) archaeology of different FSU areas. The database is located in St.-Petersburg: partly in some personal computers of scientists at the Institute of History of Material Culture Russian Academy of Science and partly on the Peterlink web server: http://www.gml.spb.ru/ BARPP-C/.

The general characteristics of the Database materials

The Archaeology-Palaeobotany-Palynology Database includes materials on the early and middle Palaeolithic (BARPP-A), on six chronological, archaeological and palaeogeographical periods of the upper Palaeolithic and Mesolithic (BARPP-B) and on the Neolithic (BARPP-C). The upper Palaeolithic–Mesolithic database contains data on the following chronological, archaeological and palaeoecological periods: BARPP-B1 (43,000–37,000 BP) — the initial upper Palaeolithic, extremely dry and cold climatic stage; BARPP-B2 (36,000–25,000 BP) — the early upper Palaeolithic, the late Valdai megainterstadial; BARPP-B3 (24,000–20,000 BP) — the middle upper Palaeolithic, the period between megainterstadial and maximum of late Pleistocene glaciation in Europe and Siberia; BARPP-B4 (19,000–15,000 BP) — the late upper Palaeolithic, maximum of late Pleistocene glaciation in Europe and Siberia; BARPP-B5 (14,000–10,000 BP) — the final upper Palaeolithic, late glacial climatic period; BARPP-B6 — the Mesolithic (early Holocene). BARPP-C contains some data on the early Neolithic (the development of this part of BARPP is scheduled for the year 2001).

The general principles of the BARPP organization have been published (Levkovskaya et al. 2000; 2001). Part of the databases A, B and C information is placed on the Peterlink webserver. The system has the following technical characteristics: Intel — based PC; Free BSD 4.2 stable; Russian Apache 1.3.17; programming language — Perl 5.00503.

13 maps presented in the BARPP show the geographical location of all early Palaeolithic (18), middle Palaeolithic (51) and some upper Palaeolithic (111), Mesolithic (85) and early Neolithic sites with palaeobotanical or palynological information that are known in the FSU. Parts of these maps are published (Levkovskaya et. al. 1999). Data not only on archaeological layers but on archaeologically sterile sediments of the geological sections and sites are presented in the BARPP. The palynological information dominates the information about the finds of the macro remains of the plants. The database includes palaeobotanical and palynological information in the form of author's pollen diagrams, floristic tables and SEM-micrographs. It contains geological profiles, palaeoclimatic curves, different dating (palaeomagnetic, C14, ESR, TL), European Pollen Database (EPD) text forms with about 90 items describing each key site (geographical location, bibliography, addresses of the researches, etc.) and the additional text forms with information on archaeological complexes, climatic stratigraphy, information about the plant marco-remain finds, etc.

An especially important part of BARPP presents the data (partly not yet published) on the best stratified archaeological sites such as the key early and middle Palaeolithic group sites of Korolevo (Transcarpathia) and Kudaro (South Osetia), the middle and late Palaeolithic group sites of the Middle Dniestr basin (Ketrosy, Korman 4, Molodova 1, 5) and Anuj river at Altai (Denisova, Anuj 2, Ust-Karakol 1), the upper Palaeolithic group from the Don river (Kostenki group sites), Mesolithic–Neolithic group from Eastern Latvia (Lubana lowland) and from Upper Volga (with tens of archaeological sites), etc. The correlation of Palaeolithic archaeological sites with global climatic stages (with the Greenland Pleistocene isotope O16/O18 curve) based on palynological data is given in the BARPP. Such correlation transforms local archaeological sites or local key archaeological areas to the world standard geoarchaeological territories. The BARPP contains some new data (Levkovskaya 1999; 2000) on palaeoecology of transition periods (middle/upper Palaeolithic, Mesolithic/Neolithic) and some new SEM-information on plant-gathering.

The information from the openly accessed (Internet) and closed parts of the BARPP-A, BARPP-B and BARPP-C could be sent in English or in Russian to interested persons. Requests should be sent to the following address: info@gml.spb.ru

Conclusion

The BARPP needs international financial support that will permit its improvement (to translate all materials in English, to improve some graphs, etc.). It is important to organize the European Archaeology-Palaeobotany-Palynology Database by the International group of Scientists. The international database will be useful for European archaeologists, palaeobotanists and palaeoecologists.

Acknowledgements

BARPP-A, BARPP-B and BARPP-C have been organized thanks to the Russian Foundation for Humanitarian Research financial support (grants: N 96-00-00425a of 1996–1997, 98-01-0031a of 1998–2000 and 01-01-00269a of 2001–2003).

The materials of the present article were presented at the CAA 2001 conference on Gotland thanks to the CAA Organization Committee grant. One of the co-authors (A.Stegantzev) visited Gotland thanks to this grant. The co-authors are grateful to CAA Organization Committee for financial support and are especially grateful to Professor Göran Burenhult and coordinator Johan Arvidsson for their practical help in the organization of this important scientific contact.

References

Levkovskaya, G. 1999. *Palaeobotanical Database for the Archaeological Sites of the Former USSR Area and It's Meaning for Studying the Problem of Transition from the Middle to Upper Palaeoilithic* // Specificity of the Eastern Europe Upper Palaeolithic. Abstracts of the International conference. St.-Petersburg. Pp. 39–40.

Levkovskaya, G., N.Anisutkin, E.Beliaeva, Yu.Stepanov, A.Bogolubova. 1999. *Distribution Maps of the Palaeolithic Sites Studied by Pollen and Plant Macrofossil Analyses (Database for the Former USSR Area)* // Proceedings of the 5th European Palaeobotanical and Palynological conference (EPPC). Acta Palaeobotanica. Supplement 2. Cracow, Poland. Pp.561–565.

Levkovskaya, G.M., M.V.Anikovitch, N.K.Anisutkin, E.V.Beliaeva, V.I.Beliaeva, Yu.V.Stepanov, A.N.Bogolubova. 2000. *The Archaeology-Palaeobotany-Palynology Database for the Palaeolithic Epoch of the Former USSR Area* // Archaeological News 7. St.-Petersburg. Pp. 380–383. In Russian.

Levkovskaya, G. 2000. *Palaeoecological Stresses of the Middle–Upper Palaeolithic transitions at Transcarpathians, Russian Plain, Caucasus, Altai* // 6th Annual Meeting European Association of Archaeologists. Abstracts. Lisbon. Portugal. Pp. 97–98.

Levkovskaya, G., M.Anikovitch, N.Anisutkin, E.Beliaeva, A.Bogolubova, V.Stegantzeva, Yu.Stepanov, V.Timofeev, G.Sinitzina, V.Shumkin. 2001. *The Internet Archaeology-Palaeobotany-Palynology Stone Age Database for the Former USSR Area* // Information Technology in Biodiversity Research. Proceeding and Abstracts of the 2nd International Symposium. St.-Petersburg. Pp. 100-102.

Never under-estimate the power of a model

Clive Robert Orton
University College London Institute of Archaeology
31-34 Gordon Square
London WC1H 0PY, UK
Phone: +44 (0)20 7679 4749 - Fax: +44 (0)20 7383 2572 - E-mail: c.orton@ucl.ac.uk

Abstract: Statistical models used in archaeology tend to be either implicit and very simple, or complicated, formal, and associated with the use of specialized software. The case is made here for the explicit use of relatively simple models, as a valuable aid to archaeological interpretation. The argument is illustrated by reference to the problem of interpreting the quantities of pottery discarded at a Romano-British pottery production site. The specific question is "how many firing seasons does this pottery represent?" Conventional approaches suggest a range of possible answers that is wider than that given by external evidence (from consumption sites), but the use of successively more refined models can narrow the range to a point where it provided useful interpretative information. The sensitivity of this approach to the initial assumptions is also discussed.

Key words: Statistical models, confidence intervals, archaeological interpretation.

Introduction

The uses of statistical models in archaeology tend to fall into one of two extreme categories. One the one hand, we have the informal, often implicit, role of a model as a basis for the use of a particular statistical test or analytical technique. For example, if we use a t-test to compare two sets of data, we are making implicit assumptions about the Normality of distributions, equality of variances, etc., which together constitute a model of the behaviour of the variables concerned. If we decide (for example) to take logarithms of the data before undertaking an analysis, we are modifying our model to try to bring it more into line with real life. All this takes place so automatically that we scarcely recognize that a model is involved.

At the other extreme, we have the formal models that are needed if we are to employ more specialized statistical techniques. Until recently, these were rare in archaeology, but they are now becoming increasingly common through the growing use of sophisticated analytical packages such as *Oxcal* (Bronk Ramsey 1995) and *Bcal* (Buck *et al* 1999). This increase can be linked to the use of a Bayesian approach to statistical analysis, which requires a more formal model-building stage. Here the archaeologist is forced to be very explicit and detailed about their model of (for example) the stratigraphic relationships between their samples, if they are to benefit from the analytical power that such techniques can bring.

Somewhere between the two lies an area where simple statistical models can be used in the interpretation of relatively simple datasets. The simple 'back of envelope' calculations that archaeologists may use to try to interpret the quantities of material from a site are frequently inconclusive, because the margins of error on their figures are so wide that the final answer is almost meaningless, or could be arrived at by other means. By building a simple but realistic model of the situation, the limits may be reined in to a point at which they may actually become useful. Just such a situation is presented below – the 'Highgate' problem.

Background

The site known as Highgate Wood is a small Romano-British pottery production site, located about 8 km north-west of the Roman city of *Londinium* (Brown and Sheldon 1969). The main phase of production, which is dated to *c.* AD 90 to 140, consists mainly of necked jars, bowls, and beakers in a grey sandy fabric, known as *Highgate Wood C* (Brown and Sheldon 1974). Excavations in the late 1960s and early 1970s revealed six pottery kilns of this phase, together with what felt like an enormous quantity of 'waster' pottery fragments. Post-excavation work showed them to weigh well over a tonne in total.

In recent years discussion has focussed on what these quantities might mean in terms of the organization of pottery production on the site – how much pottery was produced, how it fitted in to the broader picture of *Londinium*'s pottery supply, etc. An initial attempt to quantify this problem (Tyers 1997) yielded inconclusive results, but provided the parameters for a more precise analysis.

The 'Highgate' problem

Evidence from *Londinium*, in the form of Highgate Wood pots found in dated contexts, suggests that this phase of production at Highgate Wood lasted for about 50 years (Tyers 1997: 9). On reflection, one tonne or so of wasters is not a lot to show for 50 years' potting, raising questions about the intensity of the use of the site, e.g. was it in use every year? To answer them, we need to estimate the number of firing seasons represented by the waster pottery, and to compare this estimate with the overall life-span of the site.

We approach this by noting that the total number of waster pots must be the product of:
the number of firing seasons,
the number of weeks per season,
the number of firings per week,
the number of pots per firing, and
the 'waster rate', i.e. the ratio of pots spoilt to pots fired.
The number of waster pots can be estimated from the site catalogue, while estimates for the other factors can be obtained from experimental and ethnographic evidence (see below).

Number of waster pots

The sums of the raw values from the catalogue are (Tyers 1997: 2):

weight	1304 kg,
eves	1295 eve,
rim sherds	13,347,

implying a brokenness statistic of just over 10 rim sherds per pot.

The two extreme interpretations of these figures are:

1. each rim sherd represents a single pot; about 90% of each waster pot has either not been recovered or not been recorded. Therefore, about 13,300 pots are represented.
2. each pot represented is substantially complete, but it has not been possible to reconstruct them because of their abraded condition, their dispersal across several contexts and the large quantity of pottery overall. Therefore, about 1300 pots are represented.

Both of these extremes seem unlikely, and the actual number of pots represented on the site must lie somewhere between the two (Tyers 1997: 3).

An alternative approach is through the weight of the vessels, since whatever the problems of measuring the eves of small and abraded rim sherds, all the pottery was weighed and this should be more reliable (Tyers 1997: 5). A regression analysis of the weights and rim diameters of complete or near-complete pots at Highgate (Orton, unpublished) suggests that the average weight of a waster jar is about 650 g, and of a beaker is about 600 g. An overall average of 650 g is estimated, to allow for other forms (such as bowls) which may be slightly heavier. This compares with a figure of about 500 g per eve from sites in *Londinium* (Tyers 1997: 7); it may be that larger pots are more prone to failure in firing than smaller ones. If we accept an average of 650 g per pot, the 1304 kg translates into about 2000 pots. This is again a minimum figure, as it assumes that the pots at Highgate are virtually complete.

An additional complication is that there appears to be under-recording of the Highgate pottery in 'Layer 2' (i.e. from elsewhere than the dumps, kilns and ditches), compared to the non-local pottery (Tyers 1997: 7-8). Tyers (*ibid.*) calculates that adjusting for this would increase the total weight of pottery from the site to about 2000 kg. This would raise the upper and lower limits of the number of pots represented to 3000 and 20,000 respectively. Arguing on the basis that some cross-joins (i.e. sherds from the same pot in different contexts) can be expected, Tyers (1997: 8) reduces these limits to 2500 and 4500 pots.

Number of weeks per season

Tyers (1997: 9) assumes that the firing season was confined to the period from June to September, giving a maximum of 20 weeks in round figures. He takes a span of four weeks as the minimum.

Number of firings per week

Experimental firings suggest that the firing itself would take the best part of one day and the cooling the best part of another. Taking into account the time taken to load and unload the kiln, Tyers (1997: 9) assumes that a maximum of two firings per week could be achieved. His minimum is one firing per week.

Number of pots per firing

The kilns constructed during the kiln experiments (Anon 1972; 1973) were modelled on those of the Roman potters, and it was estimated that a full load would comprise *c.* 180 to 200 pots. This may be an under-estimate as a temporary superstructure might have allowed a larger load than the permanent dome used in the experiments. Tyers (1997: 9) therefore suggests 250 and 150 as the upper and lower limits on the firing capacity of the kilns.

Waster rate

The first Highgate kiln experiments achieved a waster rate of 78%! The report on the experiment suggests a 20% waster rate, based on the comments of an experience modern potter and comparison with experiments in firing Roman pottery elsewhere (Anon 1973: 59). The more experienced Roman potters, familiar with their tools and environment, should have achieved a lower figure. Tyers (1997: 8) therefore suggested a lower limit for the waster rate of 5%, and an upper limit of 20%.

The maximum and minimum values for each of these factors are given in Table 1.

Number of firing seasons

The lower limit for the number of seasons is found by dividing the lower limit for the number of pots by the upper limits of all the other factors, i.e.
2500/(20 x 2 x 250 x 0.2) = 1.25 seasons.
The probable upper limit for the number of seasons is found by dividing the probable upper limit for the number of pots by the lower limits of all the other factors, i.e.
4500/(4 x 1 x 150 x 0.05) = 150 seasons,
and the absolute upper limit would be
20000/(4 x 1 x 150 x 0.05) = 667 seasons.
Since these are, strictly speaking, kiln-seasons, any figure greater than 50 implies that not only did firing take place every year, but more than one kiln was in operation for at least part of the time. We are therefore no nearer the answer to the question of the frequency of use of the kilns.

A statistical approach

Model 1

The initial approach provides us with a simple arithmetical model, namely

$N = S.W.F.P.R$ (1)

where:

N = number of waster pots,
S = number of firing seasons,
T = number of weeks per season,
F = number of firings per week,
P = number of pots per firing,
R = waster rate.

As we have seen above, applying maximal and minimal values to each variable leads to a very wide range of possibilities for the 'unknown' variable S.

To make further progress, we have to model the variability of the factors concerned, as well as the relationship between them. At its simplest, we could construct a model (model 1) from the assumptions that:

1. each variable has a lognormal distribution,
2. the lower and upper limits given above represent confidence intervals at chosen probability levels (the absolute upper limit for N is not used because it is an absolute, not a confidence, limit),
3. each variable is independent of all the others.

Then equation (1) can be rewritten as

$\log(S) = \log(N) - \log(T) - \log(F) - \log(P) - \log(R)$ (2)

with

$\mathrm{var}(\log(S)) = \mathrm{var}(\log(N)) + \mathrm{var}(\log(T)) + \mathrm{var}(\log(F)) + \mathrm{var}(\log(P)) + \mathrm{var}(\log(R))$ (3)

The expected values of each logged variable, and their variances, can be calculated by assuming that the logged upper and lower limits are the logged means ± k standard deviations, where k is chosen to give the required probability level. It is possible to choose different probability levels for different variables (perhaps reflecting different levels of confidence in the assigned limits), but for the time being we choose a common level across all variables.

Three levels were chosen: a 'conventional' one of 95% (k = 1.96), a 'strict' one of 99% (k =2.576) and a 'relaxed' one of 90% (k = 1.645). We also need to specify the confidence limits for S; here a 'conventional' 95% confidence interval has been chosen. The results are shown in Table 2.

Model 2

Model 1 has been criticized on the grounds that F (number of firings per week) cannot be lognormal, but must have a Bernouilli (either/or) distribution because it can only take the values 1 and 2. This raises an interesting question about the nature of the variability that we are modelling – is it variability in past events (e.g. some weeks there was one firing and some weeks two), or uncertainty in our knowledge (e.g. we know there was likely to be between one and two firing per week, but cannot be more precise), or both? This criticism also draws attention to problems of modelling the number of firings in a season by multiplying the number of 'active' weeks by the number of firings per week. For example, if work had taken place on an n-day firing cycle (where n `•7), this would create correlations between successive values of F, which would therefore not be independent.

To overcome this, a more flexible model is suggested (*model 2*), in which T (weeks per season) and F (firings per week) are replaced by two new variables L (total length of season in days) and M (interval in days between successive firings). The effect is to focus all the chronological variability within a season in a single variable (M). The model thus becomes

$N = S.(L/M).P.R$ (4)

i.e. $S = N.M/L.P.R$

Using the same assumptions as for model 1, we then obtain

$\log(S) = \log(N) + \log(M) - \log(L) - \log(P) - \log(R)$ (5)

with

$\mathrm{var}(\log(S)) = \mathrm{var}(\log(N)) + \mathrm{var}(\log(M)) + \mathrm{var}(\log(L)) + \mathrm{var}(\log(P)) + \mathrm{var}(\log(R))$ (6)

We next need to consider appropriate limits for L and M. Starting from Tyers' maximum of 20 firing weeks (140 days) per season as the mean calendar length of a season (assuming no time is 'lost'), a range of from 90 to 160 days might be reasonable as a confidence interval for L. For M, Tyers gives an absolute minimum value of 3 days, so a lower bound of 4 days for the confidence interval might be reasonable. The maximum value is more difficult to assess; assuming a maximum of 7 consecutive days' 'down-time' would give an upper bound of 11 days. However, this leads to a mean of about 18 firings per season, compared to only about 13 under model 1. To maintain the same number of firings per season under model 2 would require an upper limit of 22 days (i.e. a maximum 'down-time' between firings of 18 days, which seems rather high). Using bounds of 4 and 11 days for the confidence interval of M leads to table 3; bounds of 4 and 22 days give results very similar to table 2, as might be expected.

Archaeological conclusion

Under model 1, the number of firing seasons is likely to lie between about 4 and 40, depending on the reliance that is placed on the limits assigned to the other variables. This is a reduction on the span of possible outcomes implied by the original analysis (about 25% of the original span of 150 years). Under model 2, with its rather more intensive firing regime, the upper limit of S is reduced to about 20/30 seasons and the span of possible outcomes to about 30% of the original. In either case, it is unlikely that there was a firing every year of the site's existence. This is an important conclusion, because it contradicts the common assumption that 'fixed' kilns are not associated with 'itinerant' potters.

Statistical discussion

Apart from improving our estimate of the 'dependent' variable (in this case, S), this approach also yields useful diagnostic information about the 'independent' variables. The variances of the logged variables show their 'importance', i.e. the relative contributions that they make to the variability of the dependent variable. The orders of importance (greatest to least) under the two models are:

model 1	model 2
T	R
R	M
F	*L*)
N	*N*)
P	P

These show that, if we want to improve our estimate, we should concentrate on reducing the variability of the 'seasonal' factors (T and F or M and L), and the waster rate R, rather than the 'potting' factors (N and P). For example, even if we knew the value of P exactly, it would make very little reduction in the variability of our estimate of S.

We also need to consider the possible effects of mis-specifying the model. This has two aspects: the shapes of the distributions of the variables and the assumption that the variables are uncorrelated.

Distribution of the variables

The calculations of the means and variances from the ranges are based on the assumption that all have lognormal distributions, which was made purely for computational convenience. We therefore need to take a closer look at the variables and assess the likely form of the distribution of each. It seems reasonable to assume that N and P have normal distributions, but given the sizes of the standard errors (= standard deviation/mean) of these variables, the difference between normal and lognormal will not be great. R seems more likely to be genuinely lognormal, reflecting a potter's wish to minimize the waster rate, and the occasional higher values that are likely to occur. The distribution of F has already been discussed above. T is more difficult; one might suppose that there was a 'natural' firing season, reduced by a random number of inclement weeks, which can be expressed by $T = 22 - T_L$, where T_L has a lognormal distribution. This approach is not followed up here, because of the decision to replace F and T by L and M in moving from model 1 to model 2. Since L is the overall length of the firing season, a normal distribution seems most appropriate, but for the same reason as for N and P, a lognormal approximation does not seem unreasonable. M seems much more suited to a lognormal distribution, as skewness is likely to be a notable feature of its distribution.

Thus, although much more work could be done on finding the most appropriate distribution for each variable, the use of the lognormal seems reasonable as a first step.

Correlations between variables

The formulae for var(log(S)) – equations (3) and (6) – carry an implicit assumption that the 'independent' variables are not correlated with each other. The formula if this assumption is not made is of the form

$$\mathrm{var}(A - B - C) = \mathrm{var}(A) + \mathrm{var}(B) + \mathrm{var}(C) - \mathrm{cov}(A, B) - \mathrm{cov}(A, C) + \mathrm{cov}(B, C)$$

(Stuart and Ord 1994: 351).

This means that, under model 1, positive correlations between N and any of the other variables will *decrease* var(S), while positive correlations between the other variables will *increase* var(S). Since N is an estimate made by archaeologists from a fixed body of data, it is not likely to be correlated with other variables. There are no obvious correlations between T, F, P and R, except that certain potting strategies (e.g. a need to produce a given number of pots in a season) might suggest a negative correlation between T and F. This would have the effect of reducing var(log(S)), but as it is only one correlation amongst many possible ones, the effect is not likely to be great.

A similar argument would hold for model 2, when we should consider possible correlations between L, M, P and R. The possible negative correlation between T and F has disappeared, and there seems to be no reason for replacing it by one between L and M.

There is thus no reason to suppose that the model is so seriously mis-specified as to cast doubt on the likely values of S, and the outcome seems to be robust.

General discussion

The use of a simple stochastic model has here made the difference between an outcome that tells us nothing that we did not already know from other sources of evidence, and an outcome which, although still quite vague, makes a significant contribution to the interpretation of the site. Such a use could be equally valuable in many other interpretative situations. The model can be very easily implemented as a spreadsheet, which allows rapid 'what if' calculations based on different values of the variables. It is a potentially valuable to the archaeologist's toolkit.

Acknowledgements

I am grateful to Paul Tyers for asking me to look at this problem, and for his forbearance for the errors made *en route* to this response to it. Mike Baxter and Ioulia Papageorgiou commented on earlier versions of this paper.

References

Anon. 1972. The Horniman Museum Kiln Experiment at Highgate Wood – Part 1. In *London Archaeol* 2, no. 1, 12-17.

Anon. 1973. The Horniman Museum Kiln Experiment at Highgate Wood – Part 2. In *London Archaeol* 2, no. 3, 53-59.

Bronk Ramsey, C. 1995. Radiocarbon Calibration and Analysis of Stratigraphy: the Oxcal Program. In *Radiocarbon* 37, 425-430.

Brown, A. E. and Sheldon, H. L. 1969. Early Roman Pottery

Factory in North London. In *London Archaeol* 1, no. 2, 38-44.

Brown, A. E. and Sheldon, H. L. 1974. Highgate Wood: the Pottery and its Production. In *London Archaeol* 2, no. 9, 222-231.

Buck, C. E., Christen, C. E. and James, G. N. 1999. Bcal: an on-line Bayesian radiocarbon calibration tool. In *Internet Archaeol* 7 (http://intarch.ac.uk/journal/ issue7/buck)

Stuart, A. and Ord, K. J. 1994. *Kendall's Advanced Theory of Statistics Volume 1 Distribution Theory* (6th ed.) London: Edward Arnold.

Tyers, P. 1997. *How much pottery is there at Highgate?* unpublished discussion document.

Tables

	lower limit	probable upper limit	absolute upper limit
pots represented	2500	4500	20,000
weeks per season	4	20	
firings per week	1	2	
pots per firing	150	250	
waster rate (%)	5	20	

Table 1: maximum and minimum values for each of the factors

	Probability level			
	90%	95%	99%	Original limits
Lower limit	3.3	4.2	5.6	1.25
Probable upper limit	56.1	44.7	33.7	150

Table 2: 95% confidence limits on the value of S (number of firing seasons) under various assumptions about the confidence levels implied by the limits assigned to other variables, under model 1.

	Probability level			
	90%	95%	99%	Original limits
Lower limit	3.0	3.6	4.5	1.25
Probable upper limit	31.0	25.6	20.3	73

Table 3: 95% confidence limits on the value of S (number of firing seasons) under various assumptions about the confidence levels implied by the limits assigned to other variables, under model 2.

Deriving ancient foot units from building dimensions: a statistical approach employing cosine quantogram analysis

Jari Pakkanen
Dept of Classics, Royal Holloway, University of London
Egham, Surrey TW20 0EX, UK
Phone: +44–1784–443211 - Fax: +44–1784–439855 - E-mail: j.pakkanen@rhul.ac.uk

Abstract: It is generally held by scholars of Greek metrology that the length of a foot standard can be derived from the dimensions of building elements. The Erechtheion at Athens is used as a case study to demonstrate that besides architectural measurements and inscriptional evidence, statistical analysis should also be used in the process. The method is based on D. G. Kendall's cosine quantogram analysis, and the result validity is assessed by Monte Carlo simulation. In the second part of the study an artificial data set based on the Arsenal in Piraeus is used to study the effects of noise on quantogram analysis. The main results of the study are that based on building block dimensions the length of the Erechtheion foot unit can be established as c. *324 mm and that a uniform noise-level of ±10 mm does not cause difficulties in identifying units of quarter-foot or larger in quantogram analysis.*

Key words: Greek architecture, cosine quantogram analysis, Monte Carlo simulation, kernel density estimates, metrology, foot unit, Erechtheion, Arsenal in Piraeus

Introduction

It is very often illuminating to get an external evaluation of a problem, and ancient metrology is no exception. N. R. J. Fieller has recently written a survey on some statistical questions in archaeology, and on metrology he writes as follows: "[T]here is a substantial literature on the subject which is largely contained in the more informal publications of local archaeological societies and it is in the main by 'amateur archaeologists'. Measuring archaeological constructions is an enjoyable holiday activity. Sadly, although many of these apply some version of the quantal techniques [...], most fall into obvious pitfalls in selecting data" (Fieller 1993:286). However, in the context of Greek architecture, it is possible in some cases to solve the question of data selection on the basis of inscriptional evidence: this is not a new idea, German archaeologist W. Dörpfeld used it to calculate the length Greek foot units at the end of the 19th century (Dörpfeld 1890), but it should be combined with a thorough statistical evaluation. In general, metrological analyses of Greek architecture have not employed statistics to any great depth[1] and the whole procedure of deriving foot unit lengths from architectural dimensions is in need of further study.

One of the buildings Dörpfeld used in his metrological studies was the late fifth-century BC Erechtheion on the Athenian Acropolis (Dörpfeld 1890:168–170). The first part of this paper presents a restudy of Dörpfeld's definition of the Erechtheion foot standard based on a cosine quantogram analysis of the blocks named in the inventory carried out on the incomplete building in 409/8 BC (*Inscriptiones Graecae* I[3] 474). The quantogram results are compared against the inscription dimensions, and different data models can be used to test the reliability of the analysis. Also, the Erechtheion data provides an opportunity to study the effect of small sample size on quantogram analysis: as often in archaeological contexts, the number of relevant measurements is limited.

In the second part of the paper I will use another Attic inscription to produce a simulated data set (*Inscriptiones Graecae* II[2] 1668). Parts of the fourth-century BC Arsenal in Piraeus, designed by Philon, were recently discovered near the ancient harbour installations, but the main features of the building had already previously been reconstructed based solely on the inscription.[2] The reason for using an artificial data set is that the *exact* length of the foot unit behind the measurements is then known: the effect of different levels of noise on the reliability of the quantogram analysis can therefore be evaluated.

Method

The method of this paper is based on D. G. Kendall's cosine quantogram analysis for detecting a quantum of an unknown size in a data set; after the initial analysis the result validity is evaluated by using Monte Carlo simulations (Kendall 1974). The effect of simulation distributions in the second stage of the study has been questioned by P. R. Freeman (1976),[3] which is why I will use several different data models to test whether the results are dependent on modelling. I will also introduce kernel density estimate (KDE) distributions as possible models to be used in the Monte Carlo simulations.[4]

In the case of Greek architecture the quantum hypothesis is that a building dimension X can be expressed in terms of an integral multiple M times a quantum q plus a small error component e:

$$X = Mq + e. \qquad (1)$$

The error may equally well be due to ancient building execution or modern measurement, and it should be significantly smaller than q.

To find out whether dimension X can usefully be given in terms of quantum q, X is divided by q and the remainder (or error) e is analysed: the closer to 0 or q it is, the better the fit. The formula for calculating the amount of clustering around q is

$$f(q) = \sqrt{2/N}\sum_{i=1}^{n}\cos(2\pi\epsilon_i/q), \qquad (2)$$

where N is the number of building dimensions. The higher the score $f(q)$, the higher is the probability of q being the right candidate for the quantum (fig. 1); Monte Carlo simulations can then be used to assess whether the function peak is high enough to be considered a 'true quantum'. In the simulations random data sets are created from non-quantal distributions, and these are then analysed in the same way as the original data to determine whether peaks as high as or higher than the original regularly arise from these distributions.

The Erechtheion

The block dimensions of the Erechtheion inscription are listed in table 1. Instead of using Dörpfeld's measurements I have used a more recent monograph on the building (Dörpfeld 1890:169–170; Paton & Stevens 1927).[5] There are some discrepancies in the data, but I have intentionally retained them: the most obvious is the length of the frieze block which is given in the inscription as 4 feet while the measured average length for the east frieze block is 1.94 m, which must be 6 feet. As we see from column 4 of table 1, the reported block dimensions in the inscription are not overly precise. There are several possible explanations: if the inventory was only needed for recognising the blocks which required further work, taking measurements to the closest quarter-feet or in some cases to the nearest round number of feet could have been thought sufficient; or, if the surveyors were working with the original building specification which also listed the block dimensions, these might be quite different from the final sizes of the blocks finished only at the building site. The length of the Erechtheion foot unit is usually given as 326–328 mm, but this is calculated on the basis of only a few dimensions (Dörpfeld 1890:171; Dinsmoor 1961:358).

In fig. 1 is presented the cosine quantogram of the Erechtheion data with the quantum score $f(q)$ is plotted against q. There are two clear peaks, the first at 162.0 mm and the second almost exactly twice the first at 324.2 mm. Since the analysis makes no *a priori* assumption on the quantum length, or actually even its existence, it is very interesting that the correspondence with the inscription data is as good as it is. The height of the peak at 324 mm is 3.67 and at the half-foot mark 162 mm only a fraction less, 3.65.

The results of Monte Carlo simulations employing several different data models are collected in table 2; plots of four distribution models are presented in fig. 2. For each distribution 1000 simulations were run. The results are quite similar except for the first simulation employing the bootstrap method. The reason why bootstrap should not be used in quantum analysis is that multiple replication of an observation in the resampled data set very often emphasizes the height of the maximum quantum peak, which is evident when the bootstrap score is compared with other simulations (cols. 2–3 in table 2): the 5% significance level for the peak is very high at 5.12 and 85% of bootstrap simulations produced a higher peak than the Erechtheion score of 3.67. The significance level of 3.50 of the uniform distribution is a fraction more than the distribution models on lines 3–7. This is most likely due to a slight over-representation of long measurements produced by the model. The F distribution, the histogram distribution and the kernel estimate distributions all produce very similar results with the 5% significance level range as 3.34–3.41.[6] Therefore, the choice of simulation distribution does not have an effect on the interpretation of the results *in this case*. However, more extensive tests with other data sets should be carried out to determine if they lead to the same conclusions and whether kernel density estimation provides a general solution for the choice of distribution model. Since the results of the five last models are as close as they are, the simulations can be combined to obtain more accurate values of 5000 runs (line 8. in table 2): the score for 5% significance level is 3.36, and the Erechtheion peak height of 3.67 is significant at the level of 1.7%. It is very important that even without inscriptional evidence on the length of the unit, we would have clearly accepted the quantum at 5% significance level.

Philon's Arsenal in Piraeus

The data set used in the computer simulations to study the effect of noise on quantogram analysis is based on 18 Arsenal inscription measurements which could very likely be discovered in a modern excavation of a Greek building, even if it was not well preserved: these are all dimensions of the building plan or of individual blocks. They, and their length in feet given in the inscription, are listed in table 3. The calculated size in millimetres (col. 3) is based on the above determined Erechtheion average foot length of 0.3245 m; actually, any other length within the range of supposed Greek foot units would have served the analysis equally well. One of the measurements, the width of the wall orthostate block, is given in the inscription with the accuracy of one-sixteenth of a foot,[7] all others in terms of quarter-, half- or a round number of feet.

Fig. 3 presents a summary of the Arsenal simulations. Adding a *uniform* noise of ±10 mm to the building dimensions does not have any effect on the quantogram analysis: the top two KDE plots in fig. 3 show that all 50 simulations picked the quarter-foot mark of *c.* 81 mm as the quantum (top left) and all had a very high maximum peak score of 4 or more (top right). The second set of simulations demonstrates that a noise of ±20 mm has very little effect on the length of detected quantum, but peak scores are now clearly lower than in the first set. The vertical line at 3.5 indicates a peak score which most often will be recognized as significant in Monte Carlo simulations, depending of course on the data: only 22 of 50 simulations produced a peak of 3.5 or higher. A noise-level of ±30 mm is enough to collapse the peak at quarter-foot and make the half-foot mark of c. 162 mm the mode of the distribution; 29 simulations peak in the region of 81 and 162 mm, and only a few produce significantly high peaks. Addition of a noise of ±40 mm and ±50 mm gradually diminishes the proportion of correct quanta being detected while there is very little change in the height of maximum peaks produced in the simulations.

The next obvious question is that what kind of 'noise' can be expected in measurements of Greek buildings. J. J. Coulton has made the following observation: "It would appear that in many buildings a variation of about 0.01 m. between 'identical' elements was considered acceptable for the smaller elements, while variations of several centimetres might occur in the larger ones. [...] We may [...] presume that discrepancies between the model and the design were at least as great as, and perhaps greater than, the variations of 'identical' parts from the mean sizes"(Coulton 1975:94). The published dimensions of the Arsenal give some indication of the unit length and variation of the building execution, but they are problematic, as is evident from Table 4.[8] In the case of the pillar stylobate width there is quite likely a quarter-foot discrepancy between the design and the execution. Luckily, such a discrepancy would not effect quantogram analysis since the modified length would be recognized as 3.5 feet instead of 3.25 feet. However, it might be advisable to avoid the longer measurements where larger errors are likely to occur, since as we see in fig. 3, a noise-level of ±10 mm does not create any problems for detecting the correct quantum.

Conclusions

The cosine quantogram analysis of the building block dimensions suggests that the length of the Erechtheion foot standard was *c.* 324 mm. The difference in length compared to the previous proposals of 326–328 mm suggests that a more thorough metrological analysis of the building blocks should be carried out. However, it is quite encouraging that even though the Erechtheion sample size is small with $n = 19$, quantogram analysis can produce significant results when the building inscription is used to select the appropriate dimensions. In general, provided that a sufficient number of blocks were executed relatively accurately in multiples of quarter-feet, using block dimensions rather than larger measurements, such as the width and length of the building, will more likely result in discovery of an architectural foot unit.

Software

I have implemented the computer programs used in the cosine quantogram analyses, Monte Carlo simulations, and kernel density estimations on top of Survo MM, the Windows version of the statistical program; very warm thanks are due to S. Mustonen for providing a copy of the program. C. C. Beardah's MATLAB routines were used for calculating the optimal window widths of the kernel density estimates (see Beardah & Baxter 1999 for a recent bibliography on KDE).

Endnotes

[1] R. C. A. Rottländer's work should be mentioned here as an exception; see e.g. Rottländer 1996.
[2] The inscription is dated to 347/6 BC. For earlier reconstructions of the building, see e.g. Jeppesen 1958, 69–101. On the remains of the building, see Steinhauer 1994, 1996; Rottländer 1997.
[3] Freeman's method of calculating posterior distributions is very closely connected to Kendall's quantogram analysis, as has been demonstrated by B. W. Silverman 1976.
[4] For a recent bibliography on KDE in archaeology, see Beardah & Baxter 1999, 168–169.
[5] When several measurements are reported, the average is given in Table 1. I have omitted Dörpfeld's 'Eckquader' because his identification of the block cannot be correct: it was only discovered in later excavations and it is very fragmentarily preserved; see Paton & Stevens 1927:51, 330 n. 2. The length of the Kariatid porch roof block is given for the visible part: only the upper surfaces needed redressing and these blocks were in place when the inventory was carried out; Paton & Stevens 1927:pl. 27.
[6] The step used in the quantogram analysis of the simulations was 0.25 mm; to test that the step is adequately small, two control simulations of 500 runs using a step of 0.05 mm and STE and DPI-2 KDE distributions were executed: there are no discrepancies between results of the larger and smaller step values.
[7] The reason for this is that the width is one dactyl or fingerbreath wider than the normal wall block width of $2^1/_2$ feet.
[8] The published width of 18 m for the building is not without problems since it results in a reconstruction where the side walls and pillars are not placed at the centres of discovered foundations; see Steinhauer 1994:fig. 40. The centre- and sidenave widths are also dependent on the building width.

References

Beardah, C. C. & M. J. Baxter. 1999. Three-dimensional data display using kernel density estimates, in J. A. Barceló et al. (eds), *New Techniques for Old Times. Computer Applications and Quantitative Methods in Archaeology. Proceedings of the 26th Conference, Barcelona, March 1998.* BAR International Series 757, Oxford: Archaeopress, 163–169.

Coulton, J. J. 1975. Towards Understanding Greek Temple Design: General Considerations, in *Annual of the British School at Athens* 70, 59–99.

Dinsmoor, W. B. 1961. The basis of Greek temple design: Asia Minor, Greece, Italy, in *Atti del settimo congresso internazionale di archeologia classica* 1. Rome: L'Erma di Bretschneider, 355–368.

Dörpfeld, W. 1890. Metrologische Beiträge, in *Mittheilungen des Keiserlich Deutschen Archäologischen Instituts, Athenische Abteilung* 15, 167–187.

Fieller, N. R. J. 1993. Archaeostatistics: old statistics in ancient contexts, in *Statistician* 42, 279–295.

Freeman, P. R. 1976. A Bayesian analysis of the Megalithic yard, in *Journal of the Royal Statistical Society* A 139, 20–35.

Jeppesen, K. 1958. *Paradeigmata. Three Mid-Fourth Century Main Works of Hellenic Architecture Reconsidered.* Aarhus: Aarhus University Press.

Kendall, D. G. 1974. Hunting quanta, in *Philosophical*

Transactions of the Royal Society of London. Mathematical and Physical Sciences A 276, 231–266.

Paton, J. M. & G. P. Stevens. 1927. *The Erechtheum*, Cambridge (Mass.): Harvard University Press.

Rottländer, R. C. A. 1996. Studien zur Verwendung des Rasters in der Antike II, in *Jahreshefte des Österreichischen Archäologischen Instituts in Wien* 65, 1–86.

Rottländer, R. C. A. 1997. Die Maßeinheit des Schiffsarsenals des Philon im Piräus, in *Bulletin antieke beschaving* 72, 105–114.

Silverman, B. W. 1976. Discussion of Dr Freeman's paper, in *Journal of the Royal Statistical Society* A 139, 44–45.

Steinhauer, G. 1994. Die Skeuothek des Philon, in W. Hoepfner & E.-L. Schwandner (eds), *Haus und Stadt im Klassischen Griechenland*, München (2nd ed.): Deutscher Kunstverlag, 44–50.

Steinhauer, G. 1996. La découverte de l'Arsenal de Philon, in Harry Tzalas (ed.), *Tropis* 4. *4th International Symposium on Ship Construction in Antiquity*, Athens: Hellenic Institute for the Preservation of Nautical Tradition, 471–479.

Tables

1.	2. *IG* I^3 474 (feet)	3. Measured dimension (mm)	4. Length of foot unit (mm)
Wall block L	4	1300	325
Wall block W	2	664	332
Wall block H	$1\frac{1}{2}$	490	327
Epikranitis L	4	1300	325
Epikranitis W	3	912	304
Epikranitis H	$1\frac{1}{2}$	491	327
Architrave L	8	2608	326
Architrave W	$2\frac{1}{4}$	770	342
Architrave H	2	633	317
Kariatid porch roof block L	13	3890	299
Kariatid porch roof block W	5	1648	330
East frieze block L	4	1940	485
East frieze block W	1	285	285
East frieze block H	2	617	309
North frieze block W	1	298	298
North frieze block H	2	683	342
Geison block L	4	1297	324
Geison block W	3	960	320
Geison block H	$1\frac{1}{4}$	284	227
			x 323.3

Table 1. Erechtheion, Athens. Building dimensions.

1.	2. (q), = 5%	3. , (q) = 3.67
1. Bootstrap	5.12	85%
2. Uniform distribution	3.50	2%
3. *F* distribution	3.34	2%
4. Histogram distribution	3.37	2%
5. KDE, STE, $h = 286.9$	3.35	1%
6. KDE, DPI-2, $h = 320.3$	3.36	2%
7. KDE, Normal scale, $h = 403.6$	3.41	1%
Combined models 3.–7., $n = 5000$	3.36	1.7%

*Table 2. Erechtheion, Athens. Cosine quantogram analysis of building blocks; results of the Monte Carlo simulations (*n = *1000 for each run).*

1.	2. *IG* II^2 1668 (feet)	3. Calculated dimension (mm)
Building W	55	17848
Doorpost L	10	3245
Doorpost W	2	649
Door W	9	2921
Centre-nave W	20	6490
Side-nave W	$12\frac{1}{4}$	3975
Euthynteria block L	4	1298
Euthynteria block W	3	974
Euthynteria block H	$1\frac{1}{2}$	487
Corner euthynteria block L	$4\frac{3}{4}$	1541
Orthostate block L	4	1298
Orthostate block W	$2\frac{9}{16}$	831
Orthostate block H	3	974
Wall block W	$2\frac{1}{2}$	811
Wall block H	$1\frac{1}{2}$	487
Pillar W	$2\frac{3}{4}$	892
Pillar stylobate L	4	1298
Pillar stylobate W	$3\frac{1}{4}$	1055

Table 3. Arsenal, Piraeus. Simulated inscription data (1 foot = 0.3245 m).

1.	2. (feet)	3. Measured dimension (m)	4. Length of foot unit (mm)
Building W		18	
Steinhauer (between walls)	55		327.3
Rottländer (at euthynteria)	$55\frac{1}{2}$		324.3
Axial distance of pillars		3.50	
Steinhauer	$10\frac{3}{4}$		325.6
Rottländer	11		318.2
Jeppesen	$11\frac{1}{4}$		311.1
Centre-nave W	20	6.55	327.5
Side-nave W with pillar	15	4.90	326.7
Pillar stylobate L	4	1.30	325.0
Pillar stylobate W	$3\frac{1}{4}$	1.15	353.9

Table 4. Arsenal, Piraeus. Building dimensions and foot unit lengths (based on Jeppesen 1958, Steinhauer 1994 and Rottländer 1997).

Legends

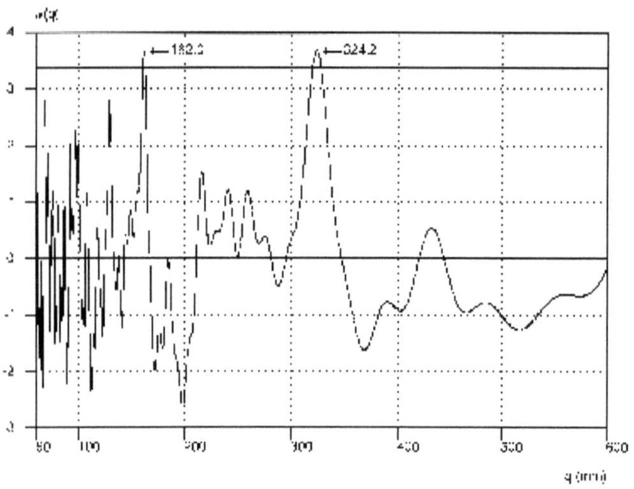

Figure 1. Erechtheion, Athens. Cosine quantogram of building block measurements.

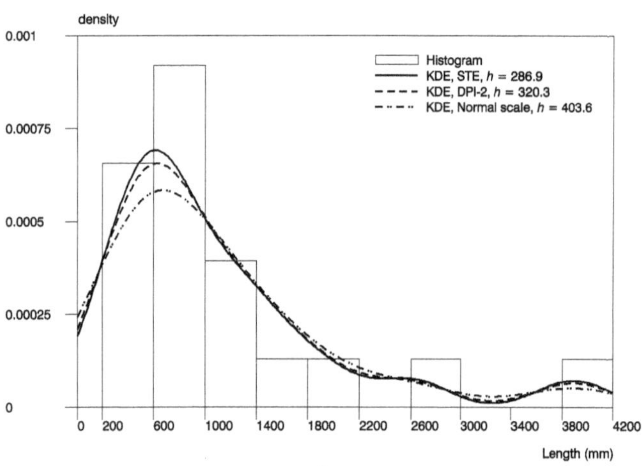

Figure 2. Erechtheion, Athens. Four of the distribution models used in the cosine quantogram Monte Carlo simulations of building block dimensions.

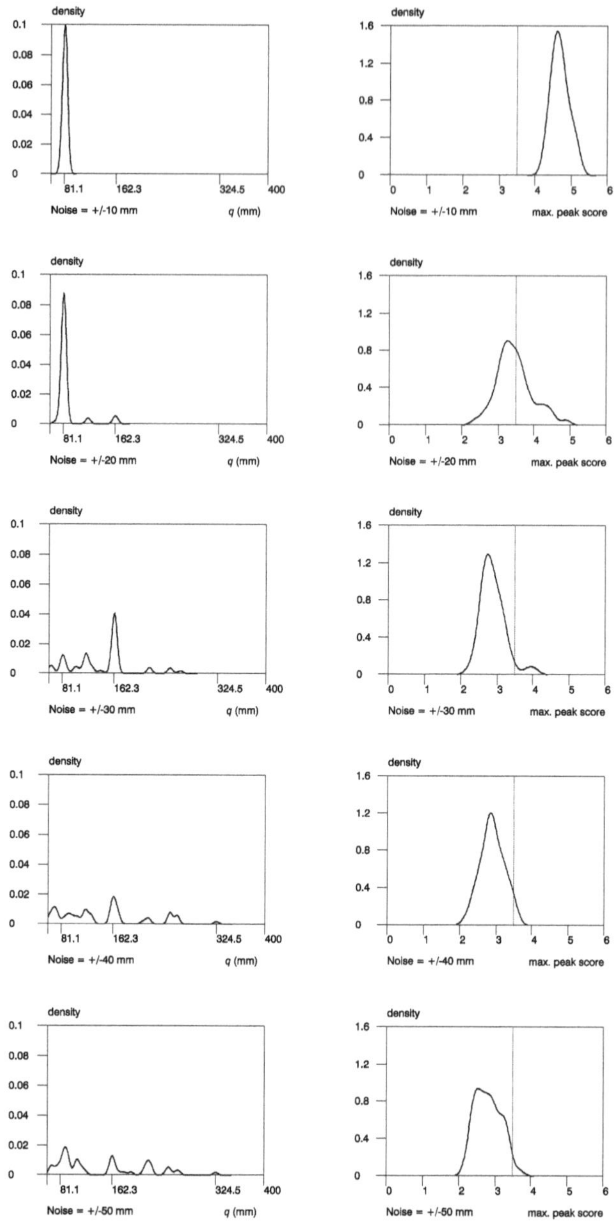

*Figure 3. Arsenal, Piraeus. Kernel density estimates of the effect of noise on the reliability of the quantogram analyses of simulated inscription data (*n = *50 for each run). Length of maximum quantum (on the left, band-width* h = *4) and maximum peak score (on the right,* h = *0.14).*

A Genetic Algorithm problem solver for Archaeology

Carlos Reynoso
Urquiza 65 depto. 35 (1215)
Buenos Aires, Argentina
Phone/Fax: +54-11-4865-9630 - E-mail: billyr@microsoft.com

Edward Jezierski
Universidad de Buenos Aires
E-mail: edjez@microsoft.com

Abstract: Generic algorithms are based on the mechanics of natural selection and genetics. They proved to be an important methodology in the development of search and machine-learning methods, but its relevance to archaeology is still to be tested. This paper consists in a basic introduction to the subject, and the presentation of a "workbench" program (to be distributed for free) implementing a general problem solver. The workbench requires a definition of the problem and a specification of how to solve it. A problem is defined in terms of data types (which specify how to decode alleles to phenotypical values), variables (of the specified data types) which define the genotype of an individual, and an evaluation function to maximize or minimize (the tool accepts a script, a series of IF-THEN-ELSE rules, or a compiled object). A solution configuration consists of which operators to use with what activation probabilities, a selection policy, a fitness mapping policy, an initial population size and a cut condition. The result produced is a collection of individuals with phenotypical and genotypical values that optimize the fitness/evaluation function until the cut condition is fulfilled. Archaeologists are invited to submit their problem definitions in order to test them.

Keys words: modeling – archaeology – genetic algorithm – genetic computing – simulation.

Genetic Algorithms Refresher

Genetic Algorithms (GAs) were initially developed by John Holland at the University oh Michigan in the seventies (Holland 1975). In the eighties they were combined with neural networks, resulting in a paradigm sometimes called "neural Darwinism". In the nineties it was usual to implement GAs in computer programs; today we talk of Genetic Programming, instead of GA. Anyway, GAs are used in search & optimization problems. Given a problem space in which to search and/or optimize, GAs will mimic natural evolutionary patterns to some degree in order to find individuals that meet a desired goal. Some of the most distinctive characteristics of GAs are:

- Genetic Algorithms are a stochastic, yet structured search (they are not a variation of Montecarlo approaches). Under the operators considered, GAs can be proven to converge to the solution.

- GAs tune the search based on a population of search points rather than only one search point. The position of a search point depends on the whole previous search population, not just a previous search point.

- Genetic Algorithms perform the search with no knowledge of the actual structure of the solution space but rather perform the search in a space which is a representation of the solution space

The following are common terms used when discussing Genetic Algorithms:

Population: A set of individuals at a given time,
Individual: A particular solution to a problem (it may not be the best solution),
Phenotype: The values that make up a solution,
Genotype: The 'genetic material' of an individual that can be transformed into a phenotype.

A comparison between natural and GA terminology is in order:

Natural	Genetic algorithm
Chromosome	String
Gene	Feature, character or detector
Allele	Feature value
Locus	String position
Genotype	Structure
Phenotype	Parameter set, alternative solution, point (in the solution space)
Epistasis	Non linearity

GAs typically implement three operators: reproduction, crossover and mutation.

The characteristics outlined above and their underlying mathematics make Genetic Algorithms especially suitable for problems that:

- Have nonlinear search spaces – while GAs do searches in a space, the search is in a space that defines a representation of the solution, not the solution itself. This effectively decouples the solution space from the search space. For example, a genetic algorithm trying to find the highest number between 0 and 7 (obviously 7) could try to solve the problem in terms of finding the arrangement of 3 bits that yields the highest numbers. The GA in this case is unaware of the fact each bit sequence represents a natural number. It just asks an evaluator – what is the 'fitness score' for an individual whose genotype is (whose genes say) "*101*"? The phenotype ("5" in this case) has no incidence over how the GA acts on the genes.

- Are highly polymodal – many points in the problem space may yield good solutions. This results to some extent from the point above. Going back to the example, in terms of search complexity, 0 and 1 are at the same Hamming distance than 0 and 4 – thus a search will have little tendency to fall in 'peaks' or 'valleys' of the problem space. This is also enhanced by the fact that the search is performed in terms of populations, not single points.

GAs have many other attributes whose relative importance depends on the application at hand.

Following is a quick summary of a generic Genetic Algorithm process. This will help understand the design and implementation processes that had to be undergone to do problem solving with Genetic Algorithms.

Process to use GAs

After much experience in working with Genetic Algorithms we have found that the following is a simple yet effective representation of the process needed to use them:

1) Define a Problem
2) Define a Solution
3) Run the Solution

These phases are outlined in detail below.

Define the Problem

When presented with a problem to be solved it is first important to understand the freedoms and restrictions of the problem, and the expected outcome.

Take for example a simple problem:
Find X such that X^2 is the highest, with x being natural numbers between –2 and 5

From here we can infer that:
- We want to search (or optimize - a premise for GA use)
- The search is done on a number of X's,
    - The Xs are samples on a continuous dimension,
    - The dimension is bound between –2 and 5 and has 7 samples,
- We are searching for a result of X^2,
- We want to maximize the result.

For most cases, the following approach – implemented in our Genetic Algorithm workbench – covered the definition of most problems we were encountered with during a number of research activities.

1) Identify Data Types for the Problem
    a. Identify type – Continuous or categorical
    b. Identify data type values – classes or bounds & samples
2) Identify variables that belong to certain data types
3) Express a transformation that will express fitness in terms of those variables and other context information. This can be done by
    a. A generic function, script, program or object
    b. A set of rules of the form: *if [condition] then [delta$_T$] else [delta$_F$]*
4) Express the desire to maximize or minimize the fitness function.

An individual is then merely a representation of a solution. This representation is not done in terms of the problem variables but in terms of a problem independent alphabet. For ease of understanding in this paper we will assume that it is a binary alphabet (of 1's and 0's). For simplicity we will also assume the alphabet is known a priori and is constant. In the example above, an individual could be defined as three 1's and 0's: (-2=000, -1= 001, … , 4=110, 5=111).

Define Solution

We call a "Solution" a set of parameters that will tell a Genetic Algorithm how to operate on a problem. Typical solution settings include:

1) A population size (we will assume single populations of fixed sizes)
2) Genetic operators, with associated activation probabilities and operator-specific parameters
3) Objective-Fitness transformation expressions, which will determine how a result in the objective function (e.g. X^2) maps to actual 'individual' fitness.
4) Cut conditions (which tell the GA runtime to stop the optimization process)

Run A Solution

Running a solution entails creating a population of individuals based on the problem definition and repeatedly using a process such as selection, operation and evaluation to create new populations of better individuals – until a cut condition is met. The actual processing of the run is done by a generic GA runtime.

The basic operation of the runtime is a loop that consists of generating new individuals based on fitness, and evaluating them. Operators cause genetic material to change as new individuals are created. The loop ends when a cut condition is met.

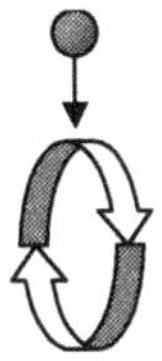

Create a population, then...

Evaluate the new individuals of the population,

Based on fitness, select individuals to reproduce

and create a new generation.

As they reproduce, apply operators that modify the genetic structure of the new individuals.

The three-step process and structure described above was implemented in a Genetic Algorithm workbench.

The GA Workbench

Using this simple yet flexible structure the workbench developed has been used to resolve problems such as:
"Solve a 100+ city blind traveling sales person (bTSP) problem"
"Train a neural network robot controller to find all lights in a maze – avoiding walls"
"Tune performance of any distributed computer application"

The workbench mainly consists of:

- A Windows user interface that lets users define problems, solutions, and control & monitor runs
- A runtime implemented as a series of COM objects which have the fundamental GA algorithms and some COM interfaces for user-defined data types, operators, cut conditions, user interface extensions etc
- A library of commonly used data types, operators, cut conditions, and fitness mappings.

This workbench saves problems as *.GAp* files and solutions as *.GAs* files. These are structured storage files that can be browsed using the data objects in the runtime.

The individual evaluation can be expressed as a VBScript function. This allows the user to enter a program with any desired complexity for the evaluation, even creating custom or 3rd party remote COM objects on other computers.

Using the GA Process and Workbench for a particular problem

1) Defining the problem
   a. Defining data types of the problem parameters
   b. Defining variables
   c. Defining an Evaluation of Performance
      i. Defining how to apply the configuration to a subject system
      ii. Testing the simulation/system with the configuration
      iii. Obtaining a 'performance' number
2) Defining a solution's parameters (population sizes, etc.)
3) Running the GA

Sample screen showing variables, their data type and their description, as entered for a particular application (Fig. 1)

Sample screen where a user may enter his own problem to solve. The tool provides syntactic helpers so that the user does not need to master the programming language, and many parameters for the system are visually accessible.

This screen shows the monitoring of a genetic algorithm run, with increasing fitness values (Fig. 2).

GAs and Archaeology

As observed, GA terminology usually has to do with rather outmoded mendelian ideas. We are fully aware of it. On the micro level, the scientific understanding of DNA structure has advanced a lot, and that terminology is no longer used. Today, genes and their behavior bear relatively little resemblance to simple parameter setting. Anyway, conventional GA and GP really work, and the new, more accurate metaphors (and their computational implementations) are still under construction.

GAs were occasionally used in social sciences. In 1979 R. G. Reynolds developed a GA-like adaptation schema to model prehistoric hunter-gatherer behavior. In 1981, Smith and DeJong used GA search to assess calibration of population migration. In 1985, R. Axelrod simulated behavioral norms using GA. The present paper is not an implementation of an archaeological problem, but an invitation to test a paradigm and the offering of a tool.

Why are we going to use GA in archaeology? Well, in the first place they are widely used in a multiplicity of domains, such as mathematics, engineering, political science, medicine, pattern recognition. After all, in non-analytic systems, semantics does not matter: algorithms are blind, and meaning-independent. There are relatively few problem structures, and they are the same across all disciplines. Besides, GA behaves especially well when processes are at stake, and archaeological problems generally are related to processual and diachronical issues. We think GA will be helpful to understand the underlying structure of an archaeological problem as a generic problem:

- Does your archaeological reasoning really qualify as a problem?
- Do you know enough about your subject matter in order to formulate a problem, or something essential is missing?
- How good a problem is it? Is it trivial? Is it solvable?
- What does a GA implementation of your archaeological problem look like?

GAs are specially suitable for problems having nonlinear search spaces and highly polymodal problems. They help to asses the solution process as a search procedure, and today we have a deep knowledge of them. Furthermore, they help to automate the invention process, which involves factors such as intuition, creativity, abduction. Once upon a time, anthropology and archaeology developed the first evolutionary theories and concepts outside biology. With GA, in a sense, evolution theory comes back home.

References

Axelrod, R., 1985. *Modeling the evolution of norms.* Paper presented at the American Political Science Association Meeting, New Orleans.

Holland, J., 1975. *Adaptation in natural and artificial systems.* Ann Arbor: The University of Michigan Press.

Goldberg, D., 1989. *Genetic algorithms in search, optimization & machine learning.* Addison Wesley, Reading.

Koza, J., 1992. *Genetic Programming: On the programming of computers by means of natural selection.* Cambridge: MIT Press.

Koza, J. et al, 1999. *Searching for the impossible using genetic programming.* Proceedings of the Genetic and Evolutionary Computation Conference, Orlando. San Francisco: Morgan Kaufmann.

Reynolds, R. G., 1979. *An adaptive computer model for the evolution of plant collecting and early agriculture for hunter-gatherers in the valley of Oaxaca, Mexico.* Unpublished doctoral dissertation. Ann Arbor: University of Michigan Press.

Smith, T. and K. A. DeJong, 1981. Genetic algorithms applied to the calibration of information driven models of US migration patterns. *Proceedings of the 12th Annual Pittsburgh Conference on Modeling and Simulation*, pp. 955-959.

Figures

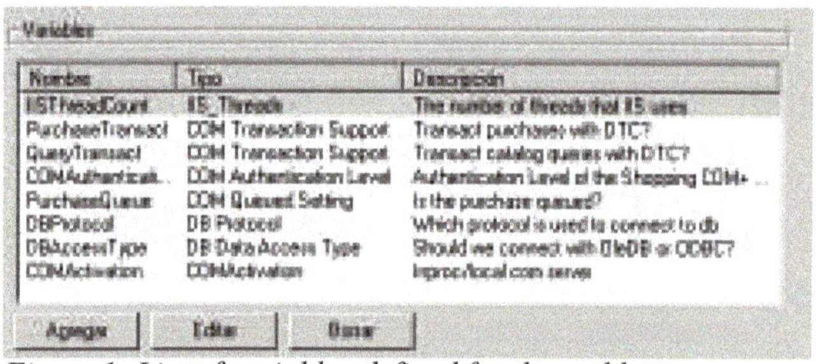

Figure 1: List of variables defined for the problem.

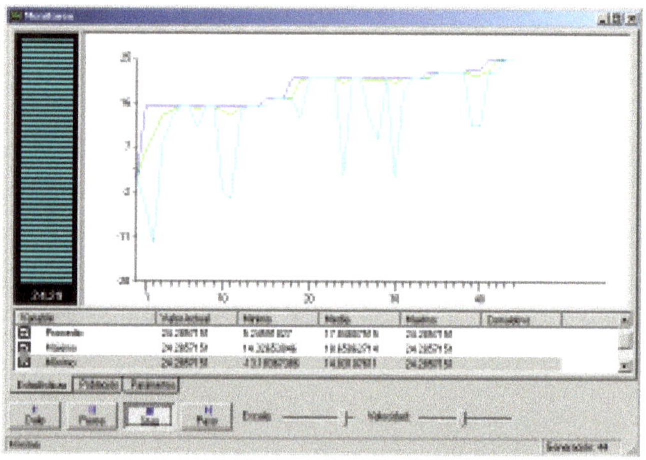

Figure 2. The image shows the GA Workbench console during a sample run. The image is provided for illustrative purposes. Screenshots were not obtained the experiment. The chart below displays maximum, minimum and average fitness for all individuals in the population. Note how the average tends to the maximum, and how the minimum can vary so much from the maximum, while the maximum is quite stable.

Visualising the Neolithic transition in Europe

Thembi Russell
Department of Archaeology
University of Southampton
Southampton SO17 1BJ
England, UK
E-mail: mtr@soton.ac.uk

James Steele
Department of Archaeology
University of Southampton
Southampton SO17 1BJ
England, UK
E-mail: tjms@soton.ac.uk

In this short paper we will discuss methods of visualization and analysis of the radiocarbon record of a large-scale diffusion process. We will use the spread of the European Neolithic as a case study.

Since the 1960s, when Grahame Clark published a map of the spatial distribution of early Neolithic radiocarbon dates (Clark 1965; Figure 1), many people have analysed and interpreted this dimension of the archaeological record using map visualizations and statistical techniques. There is still substantial disagreement in the literature regarding mechanisms of diffusion of the cultural elements of early agricultural strategies (adoption-diffusion, demic diffusion with interbreeding, or population replacement). Perhaps these debates can never be resolved simply by analysing the chronological record of this diffusion process. Nevertheless, the quality and quantity of radiocarbon data available continues to make this case study an ideal test bed for techniques of data mapping and of the statistical characterization of spatial patterns.

In a recent project funded by an AHRB grant to Stephen Shennan and James Steele, a new radiocarbon database of the late Mesolithic and early Neolithic record in Europe was compiled (Gkiasta *et al*., in press). The database contains fields identifying each site, its location and cultural affiliations, and details of published radiocarbon dates relating to late Mesolithic or early Neolithic stratigraphic units. The database is archived on the Web with the AHDS data service in York, England (http://www.ads.ahds.ac.uk). In the course of its compilation it became clear that associating dates with specific cultural elements (pottery, individual domesticates, etc.) would require extensive consultation of primary site publications and archives, since the consolidated date lists and Dating Lab records that were being used usually only gave a broad cultural affiliation. Nevertheless, it still seemed worthwhile to analyse these new data as a meaningful record of a cultural process.

In an initial attempt to visualise the new database, we replicated Clark's map in which early Neolithic sites were assigned to 1200 year 'bins' on the basis of their uncalibrated modal radiocarbon age (Figure 2). In comparison with Clark's dataset (containing 53 dates), it is clear that ours (using 508 dates) is restricted geographically to Europe and excludes sites in Turkey, the Fertile Crescent and North Africa. The spatial distribution of sites in the 4000-5200 bp range is greatly expanded, and is no longer restricted to the central European loess belt. In other respects, however, the big picture remains substantially similar to that mapped by Clark 35 years ago.

In the early 1970s, Ammerman and Cavalli-Sforza (1971) derived some rate estimates for the diffusion of agriculture (seen as a travelling wave) using regression techniques (Figure 3). Time was measured using the modal uncalibrated radiocarbon ages of 54 early Neolithic sites, and space was reduced to the single dimension of geodesic distance from Jericho (assumed to be the location of origin of the diffusion).

Repeating this major axis regression analysis with our larger early Neolithic dataset (508 uncalibrated radiocarbon dates) gives substantially similar results (Figure 4). Ammerman & Cavalli-Sforza (1971) found an average diffusion rate from an assumed origin in Jericho of about 1 km/year, and found a high value for r (0.89) in their sample of 53 Neolithic sites - suggesting that this rate was quite representative of the process generally. A similar analysis using sites indexed in the new database yielding the major axis equation:

Y (km) = 8240 - 0.89 X (yrs bp).

This suggests that the overall rate of spread is ~1.1 km/year and that the mean departure time from Jericho was ~9,250 bp (uncal.). In this case linear regression of the two variables produces a correlation coefficient, r = 0.74. In other words, with the larger data set now available the mean rate of spread is similar to that observed by Ammerman & Cavalli-Sforza, although the dispersion around that rate is somewhat greater (Gkiasta *et al*., in press). We can note, at this point, one major problem with such an approach: the need to specify a single origin for the diffusion process.

Ammerman and Cavalli-Sforza (1984) subsequently mapped the diffusion in two dimensions, using spatial interpolation techniques and a larger data set of 106 early Neolithic sites (Figure 5). This technique does not require us to specify an origin to the diffusion, which can now be inferred from the orientation of the gradient in site age, apparent in the isochrons that are fitted at 500 year intervals.

We also repeated this analysis, using the interpolation (r.surf.idw, interpolation using the 12 nearest neighbours) and contour fitting (r.contour) commands in the GRASS GIS package (Figure 6). Our own visualization indicates a less simple spatial structure than that depicted by Ammerman and Cavalli-Sforza. It is likely that this is partly because in each case, all available sites were used (as opposed to those which were the earliest in each quadrat of a sampling grid) – and we have a much larger database (once again using 508 uncalibrated radiocarbon years). But we must also recognize the methodological limitations of this technique. Interpolation error will be large in regions with sparse or no sites to use as control points, and the interpolation has been continuous across sea as well as land (although the sea has subsequently been masked off). This technique therefore risks giving a misleadingly coherent impression of the spatial structure of any such dataset. As a control we repeated this exercise using our method (interpolation in GRASS GIS) with Ammerman and Cavalli-Sforza's original dataset of 106 Neolithic dates as the input data. Figure 5 (Ammerman and Cavalli-Sforza's isochron map) and Figure 7 (created using our method and Ammerman and Cavalli-Sforza's original dataset) show very similar results.

We have been experimenting with other map visualizations and statistical analyses (see also Glass *et al.* 1998), and here we report for the first time on two of them. First, we have experimented with geographically-weighted regression (Fotheringham, Brunsdon and Charlton 2000, 2001). This technique allows us to detect local variation in trends in large-scale spatial datasets, by weighting each control datum in a regression analysis inversely to its distance from a specified point location. In Figure 8 we have plotted results of a locally-weighted linear trend analysis for the ages of early Neolithic sites, for each of a regular grid of points. We have used projected co-ordinates, and we have only used the earliest sites in each 60km–by-60km quadrat of a sampling grid. The orientations of the arrows show the direction in which sites get younger. The lengths of the arrows scale to the rate of spread. The colours of the arrows indicate the strength of the locally-weighted linear trend (where red is a well-fitting model, and blue is a poorly-fitting model). Obviously we don't believe that the diffusion of agriculture was taking place out in the North Atlantic! It is merely a convention that vector field diagrams should plot values on a regular grid in this way. But what this shows us is that we can only pick up significant regional trends in regions where our dataset is full, rather than sparse. This technique therefore combines the best elements of the previous two (as used by Ammerman and Cavalli-Sforza), but avoids their weaknesses. We can observe local variation in rate and direction of diffusion, and its statistical strength. We do not need to assume any single origin, and we can see the regions where the model fits well and the regions where it fits badly.

Geographically-weighted regression also allows us to vary our characterization of 'local' (Figure 9). By increasing the bandwidth, we expand the window over which trends are detected until, in the limit (here, bandwidth=2000, Figure 9d), the weighting has no discernible effect and we simply see the major linear trend of the dataset as a whole. Unsurprisingly, it indicates an origin somewhere in the Fertile Crescent.

All these techniques require us to treat radiocarbon dates as point values. However, after calibration, the irregular nature of such dates (seen as probability distributions) makes using the mode or the median rather problematic, as we can see from this example (Figure 10 - the date of a Neolithic site at Pfyn, Switzerland). Our second new visualization technique tries to take account of this problem. We have used a technique similar to that used in the 'ShowTime' animation (which depicts pollen records of the changing distributions of major plant taxa in late glacial and postglacial North America - http://www.ngdc.noaa.gov/paleo/softlib.html).

We have divided the calibrated probability distributions of all our Mesolithic (207 radiocarbon dates) and Neolithic site dates (508 radiocarbon dates) into 100-year 'bins', and plotted these on a series of maps (at 100-year intervals - see Figure 11 for examples). The size of each of the dots is scaled to the probability that the site was occupied in each 100-year interval. We have then stacked the maps up as an animation, with Mesolithic sites plotted in red and Neolithic sites plotted in blue. The animation is available on the Web, as a PowerPoint file (http://www.soton.ac.uk/~mtr/Europe.ppt). In some ways this animation tells the story much more immediately than does any of the preceding methods. However, it does this by addressing the part of our brains that can discern pattern in dynamic visual input, and if we are to work with explicit models then we still need to find ways of translating such perceptions of pattern into verbal form or into simple summary statistics.

Continent-scale models of the Neolithic transition have been less popular with many archaeologists in recent years, because they subsume regional variation and because they often confound observations of pattern with interpretations of process. Better interpretation of process will require not just subdivision of the radiocarbon dataset into more specific sets of cultural associations, but also predictive modelling of the patterns expected as a result of different mechanisms (population expansion, adoption-diffusion, cultural convergence). In this paper our emphasis has been solely on the observation of pattern. The two new techniques represent efforts to meet some of the criticisms of continent-scale modelling, by incorporating statistical uncertainty and regional variation into the visual characterization of a continent-wide diffusion process.

Acknowledgements

We gratefully acknowledge the assistance of Stephen Shennan, the project's lead P.I., and of Marina Gkiasta, the research assistant at UCL who compiled the database. This work has been funded by the AHRB and by the University of Southampton.

References

Ammerman, A.J. and Cavalli-Sforza, L.L. 1971. Measuring the rate of spread of early farming in Europe. *Man n.s.* 6: 674-688.

Ammerman, A.J. and Cavalli-Sforza, L.L. 1984. *The Neolithic Transition and the Genetics of Populations in Europe.* Princeton: Princeton University Press.

Clark, J.G.D. 1965. Radiocarbon dating and the expansion of farming culture from the Near East over Europe. *Proceedings of the Prehistoric Society* 31: 57-73

Fotheringham, A.S., Brunsdon, C., and Charlton, M.E. 2000. *Quantitative Geography.* London: Sage

Fotheringham, A.S., Brunsdon, C., and Charlton, M.E. 2001. *Geographically Weighted Regression.* http://www.ncl.ac.uk/geography/GWR/

Gkiasta, M., Russell, T., Shennan, S. and Steele, J. In press. Origins of European agriculture: the radiocarbon record revisited. *Antiquity*

Glass, C., Steele, and Wheatley, D. 1999. Modelling spatial range expansion across a heterogeneous cost surface. In *Procs. CAA 97, Birmingham. BAR Int Series* 750: 67-72. Oxford: Archaeopress.

Figures

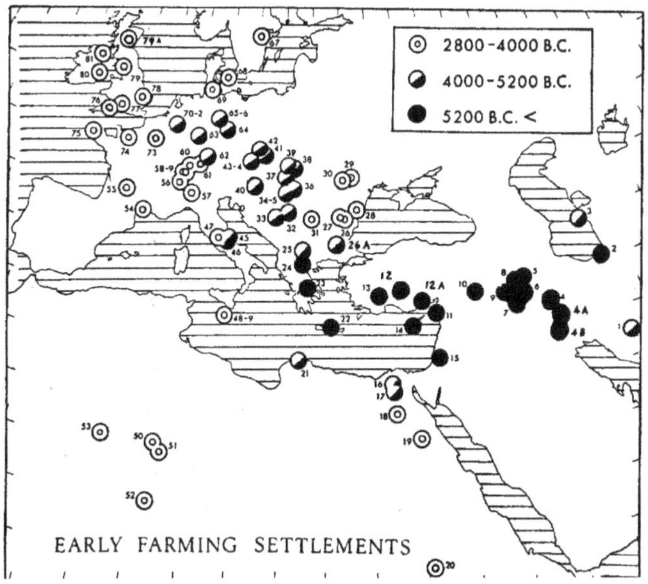

Figure 1. The distribution of early farming sites (Clarke 1965:46). Clarke uses 53 early Neolithic uncalibrated radiocarbon dates in years BC. Sites are grouped together into 1200 year interval bins.

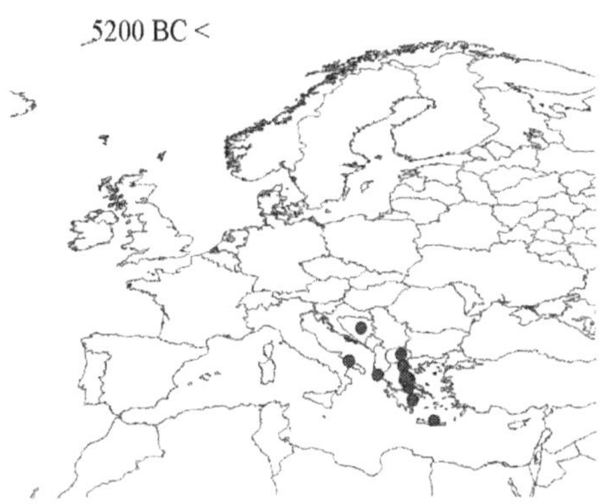

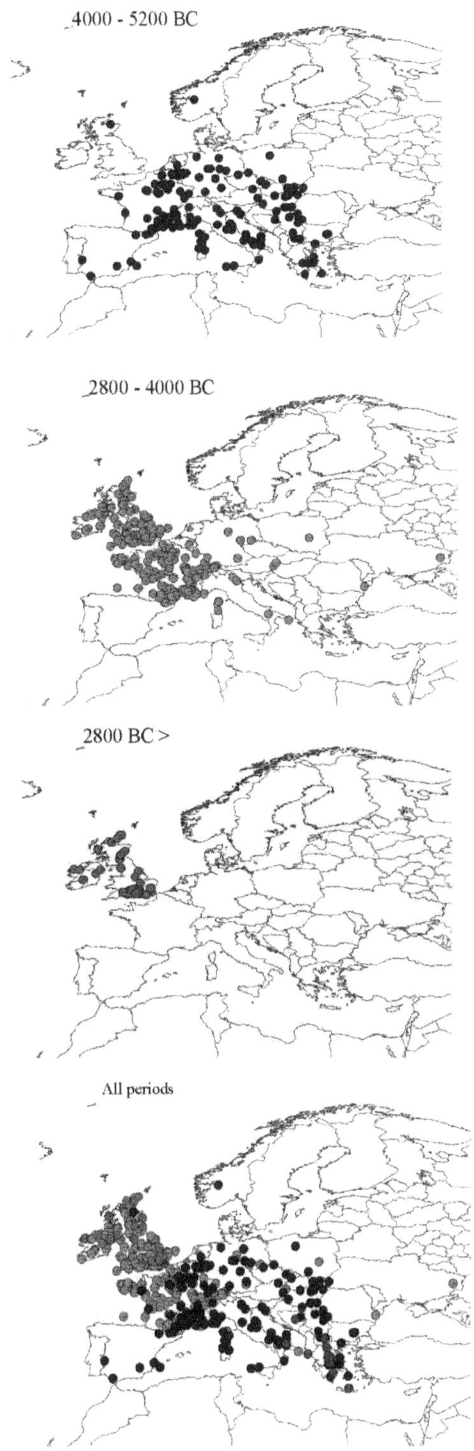

Figure 2: Clarke's method reproduced using 508 early Neolithic dates in uncalibrated radiocarbon years BC.

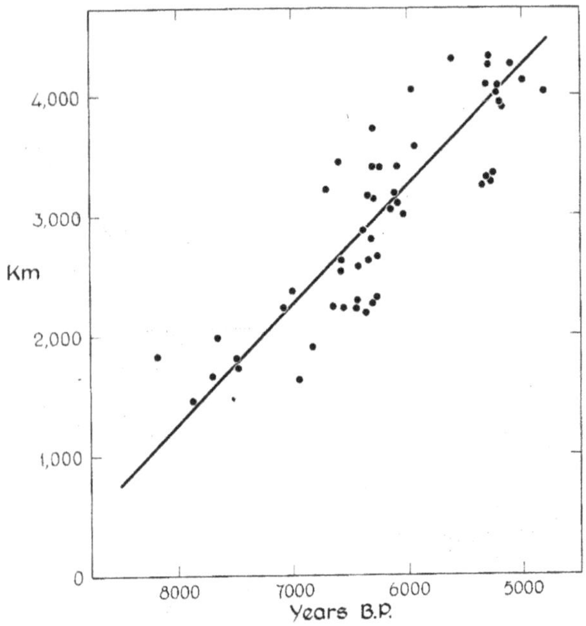

Figure 3. Regression analysis of the age of early farming sites with their distance from an origin in Jericho (Ammerman & Cavalli-Sforza 1984:53). This uses 53 uncalibrated radiocarbon dates.

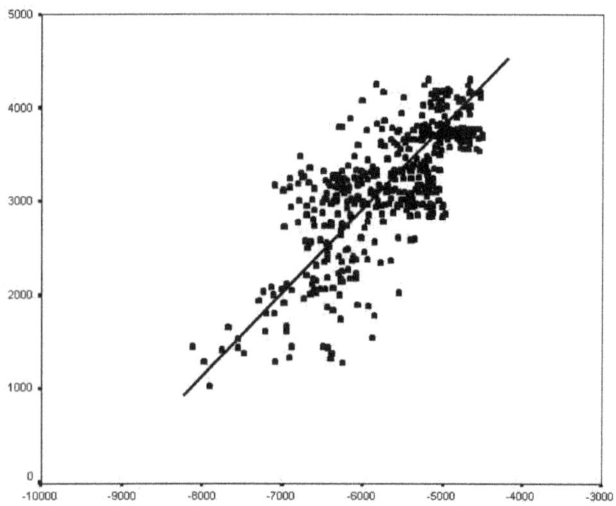

Figure 4. Regression analysis of geodesic distance from an assumed origin at Jericho (kms, y-axis) with the site age (years bp, x-axis). 508 uncalibrated radiocarbon dates (for early Neolithic sites) are used in the analysis.

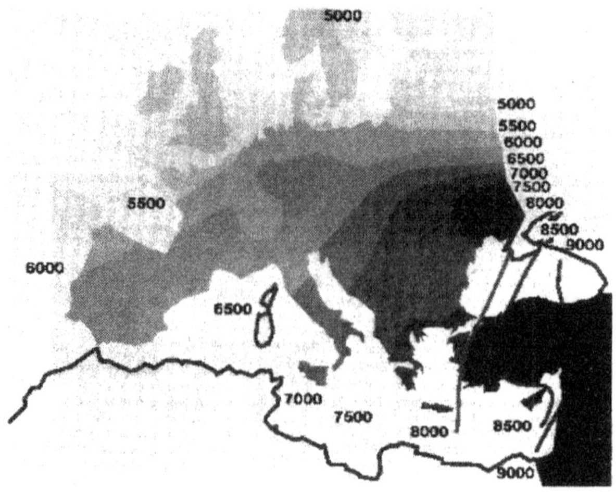

Figure 5. Isochron map of the spread of the first farmers to Europe (Ammerman & Cavalli-Sfroza 1984). 106 sites are used and dates are uncalibrated years BP. Isochrons are drawn at 500 year intervals.

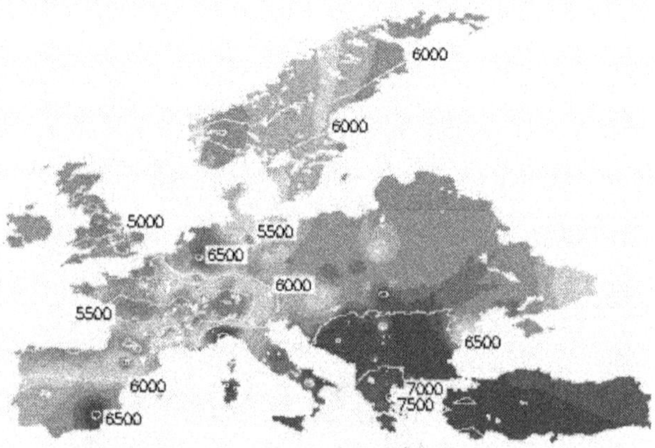

Figure 6. Isochron map for the distribution of early Neolithic sites using 508 radiocarbon dates. Dates are uncalibrated years BP. Isochrons are created at 500 year intervals.

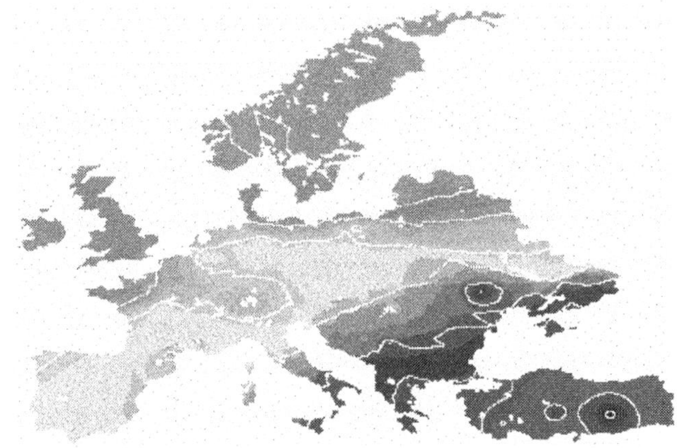

Figure 7. A control map. This isochron map is generated using our method of interpolation and contour fitting coupled with Ammerman and Cavalli-Sforza' a data (106 uncalibrated radiocarbon dates). The difference between this Figure and Figure 5 lies in the methods of interpolation and contour fitting, as the input data is the same in both cases.

Figure 8. Results of a locally-weighted linear trend analysis for the ages of early Neolithic sites, for each of a regular grid of points. We have used projected co-ordinates, and we have only used the earliest sites in each 60km–by-60km quadrat of a sampling grid. The orientations of the arrows show the direction in which sites get younger. The lengths of the arrows scale to the rate of spread. The colours of the arrows indicate the strength of the locally-weighted linear trend (where red is a well-fitting model, and blue is a poorly-fitting model). Bandwidth = 400.

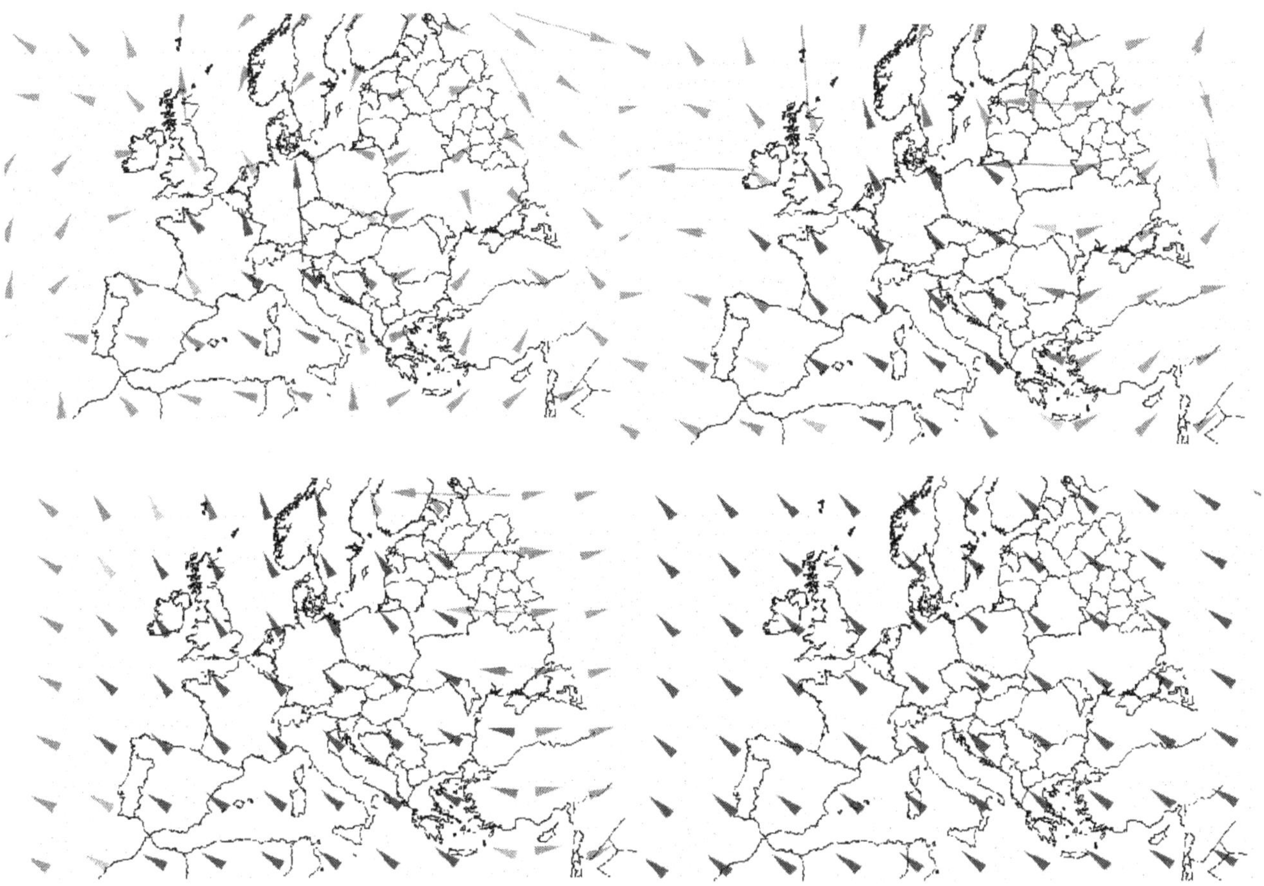

Figure 9. Results of a locally-weighted linear trend analysis for the ages of early Neolithic sites, for each of a regular grid of points. Key as for Figure 8. Bandwidth = 400, 650, 950, 2000.

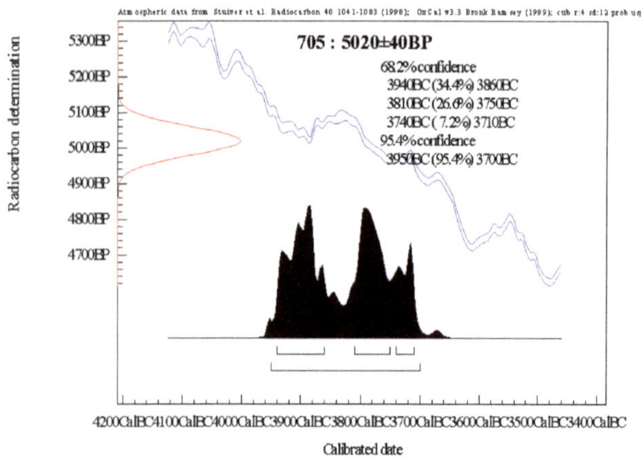

Figure 10: The calibrated probability curve for the Neolithic site of Pfyn in Switzerland to show the bimodal shape of the calibrated radiocarbon date.

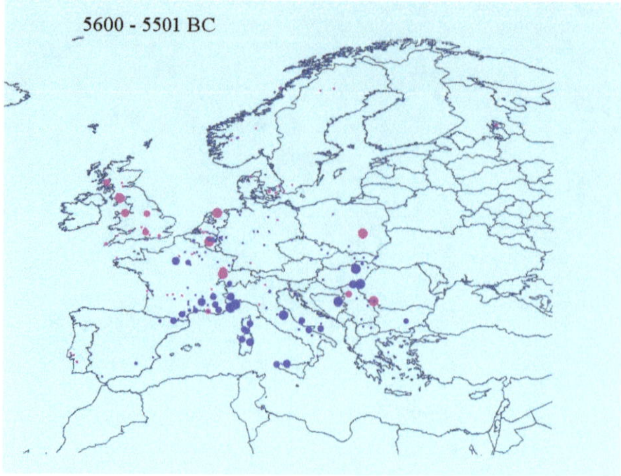

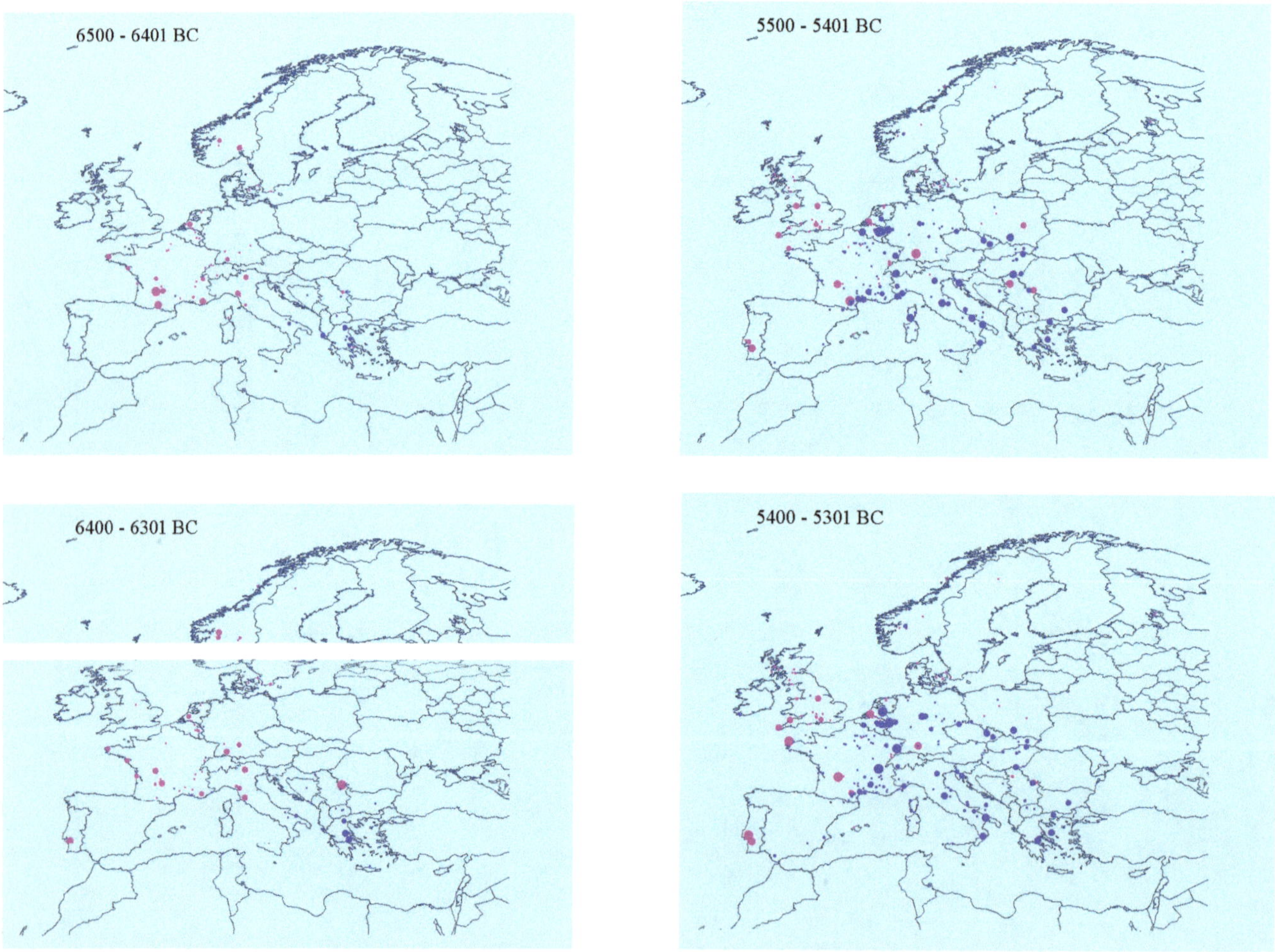

Figure 11. The distribution of early Neolithic sites. The fluctuation in the probability that the site is occupied at each 100 year interval (as calculated from the area under the calibrated probability curve of the radiocarbon date) is reflected in the size of the circular symbol marking a site's location. The probability that a site is occupied at a given time varies from less than 10 % (smallest symbol area) to greater than 80% (largest symbol area).

Archaeological databases: what are they and what do they mean?

Robert Schlader
Idaho State University
Idaho Museum of Natural History
344 north 15th Ave Apt. 11B
Pocatello, Idaho, 83201
Phone: +1-208-478-2642 - E-mail: schlrobe@isu.edu

Abstract: In order to understand what an archaeological database is and what it means we have to do a few things first. We need to discuss what a database really is. Once done with that, we need to discuss what makes a "good" database; specifically what we need to do to ensure that the information we distill from the data contained therein is both good and useful. Following this we take a brief look at the ways in which databases are currently being used in archaeology. Then we will look at a fairly new approach to archaeological databases, which is attaching an archaeological database to a Geographic Information System (GIS) for a better method of creating and displaying our information. A problem exists with this however; there is a problem with META-DATA. We will look into this problem and then dissolve it. When all this is complete, we will be at a spot where we can begin to (finally) talk about the theories driving the use of databases in archaeology. When this is complete we will have a closer fit for what databases are, and what they mean for archaeology.

Key words: Databases, GIS, METADATA, DBMS, American Archaeology

Introduction

Archaeologists require a method to organize and store their data. One method commonly used is to record everything into a series of notebooks and draw maps. Another method is to create and fill an electronic database. Both can be used to access the data stored in them, but only one can re-organize that data on the fly. This is only one benefit of using a database. Currently most archaeologists spend little to no time actually thinking about their databases. Most archaeologists will do no more than consider what few pieces of information they feel they need to look at to answer their research question (personal communication with Richard Holmer, November 2000). This lack of thought about proper data design has left archaeology lagging behind the rest of the academic world (Carroll, 2000). This lack of consideration about proper data design has been the cause of a major difficulty among groups who are attempting to share information, especially over the internet (Eiteljorg, 1997). When we consider this, it begins to look strange that archaeologists really don't understand what databases are. What archaeologists do understand are the theories of materialism, functionalism, processualism, post-processualism, and other "-isms." What we will find is that, even though not aware of it, archaeologists utilize these theories in their databases, and that they tend to design their databases to exploit whichever theory they happen to be using to ask their research questions.

Welcome to databases

A database is a shared, integrated computer structure that houses collections of end user data (raw facts that are of interest to the end user) and Metadata ("data about data" through which the data are integrated). In a sense, a database resembles a very well organized electronic filing cabinet in which powerful software, known as a Database Management System (DBMS), helps manage the cabinet's contents (Coronel, 1997, p. 4). Information is the key to our modern world; even the world of archaeology, and information is constructed by the proper manipulation of raw data. Databases not only store the raw data that we collect, they allow us to manipulate that data in very complicated ways (Codd, 1974, 1990). We can re-configure the way our data is displayed on the screen, the way it is printed out, and even the way it is stored on the computer. Through these manipulations we create the information that we need to answer our research questions, as well as appease those that hold the purse strings of our budgets (Date, 1995).

Making a database work correctly

The currently popular method for creating a good database is to create a database using a Relational DBMS (Coronel, 1997). What allows relational databases work are a few simple concepts. These concepts are: keys, controlled redundancy, entity integrity, and referential integrity (Coronel, 1997). The way that the data is manipulated in the database is through the use of "keys." Primary keys uniquely identify some piece of data (which is known as entity integrity)(Coronel, 1997), and foreign keys link one primary key to another by placing a copy of one table's primary key into another table (which is known as referential integrity)(Coronel, 1997). This "copied key" allows the two entries in two different tables to refer to one another, and thus a relationship is created between the two tables (and hence the name relational database)(Coronel, 1997). By only allowing small parts of the tables to be copied the redundancy amidst the data is reduced to a controllable level (strikingly enough this is called controlled redundancy)(Coronel, 1997).

And now for something completely different: how archaeologists use databases

In archaeology there are two main reasons for having a database. One is for the long-term storage of data, this is better known as a data warehouse or an electronic archive (Eiteljorg, 1997). The other is for the express use of researchers and other active users of the data, better known as a "user group." It is a fact that most researchers maintain independently designed, widely varying databases for the express purpose of answering their specific research question (Eiteljorg, 1997).

The differences between the types of databases are not necessarily a problem until that database is finished being actively used. When the data is no longer in use, where does it go? What happens to it? Most researchers simply save a copy of their data on whatever medium they find handy, send that copy to a museum (if in fact they do anything with the copy), and that's the end of that (Robinson, 2000).

There are a number of data warehouses available other than museums. One is the Digital Archiving Pilot Project for Excavation Records, which is better known as DAPPER (Robinson, 2000). Another is Eiteljorg's Archaeological Data Archive Project or simply ADAP (Eiteljorg, 2000). Both of these digital archives will take a user database and maintain it for long-term storage and re-use.

GIS: a new perspective for information dissemination

There are a small number of new technologies that have come into their own recently that are of great interest to archaeologists concerning data. One of these new technologies is the Geographic Information System, or GIS, but what is a GIS?

In essence a GIS can be said to be a database that contains some sort of spatial information that then can be used to create a map for display, manipulation, and presentation (my definition). Where this comes in to play is that most archaeologists are collecting spatial data about the objects they uncover. This spatial information not only (potentially) covers large areas, but also may extend over many layers of the site. Due to this glut of spatial information, GIS becomes a godsend when it comes time to map out object distributions and densities as well as creating a series of overlays to account for site depth. What this does is allow for identification of spatial patterns within a site, and will (hopefully) facilitate future regional analysis and intersite comparisons of archaeological data (Moyes and Awe, 2000).

METADATA: the glue that holds a database together

One key to the way that GIS software functions is through the use of METADATA. In a standalone database the metadata for the database consists of two elements: the data dictionary, and the data definition language.

> The *data definition language* is the part of the DBMS that allows the user to set up a new database, to specify how many attributes there will be what the types and lengths or numerical ranges of each attribute will be and how much editing the user is allowed to do. This establishes the *data dictionary*, a catalog of all of the attributes with their legal values and ranges (Clarke, 1997).

What most GIS software does is to extend this simple set of metadata to include other information as well. This additional data is of a more general nature than that contained in the database's data dictionary (Weibel, 1995). The extended metadata contains information about the nature of the database, the location of the data in the database, the existence of similar information, what data the database contains, how you can view the data in the database, and who it is that holds and maintains the data (Wise and Miller, 1996) among other things.

The extended set of metadata allows us to do three things that the data dictionary does not allow. First it allows us to find out what the data set covers without having to look at the data. Second, it allows us to locate data faster. Third, it allows us to group the data into larger groups based on the similarity of the dataset (Wise and Miller, 1996). What this allows us to do, once the metadata is filled out, is narrow our search to exactly the elements in the database that we need to look at without actually having to sort through the data.

The true benefits of using metadata in this way only become evident when one is attempting to get information out of databases from different locations (specifically online databases), or different databases (such as an excavation database versus a data warehouse). In essence, the more abstract the metadata the faster we can narrow our searches and find the information we are looking for.

So, if metadata is used to speed up the process of narrowing down a search for information, what kinds of problem(s) can there be with that? There are really two problems with metadata. First there is the problem of a consistent set of fields, i.e. each database has the exact same set of attributes (date entered, type of object, name of person entering the data, etc.) in their metadata (Wise and Miller, 1996). Second there are the allowable entries within each field in the metadata, i.e. whether the "type of object" field will allow "Desert Side-notched", "DSN", the number "3" which codes for Desert side-notched, or any other of many different names for this kind of object (Wise and Miller, 1996).

The first of our two problems with metadata is a real problem that can only be solved by establishing an agreed upon structure. There is a lot of work being done currently to solve this problem (Wise and Miller, 1996; Cathro, 1997; Weibel, 1995). The second of our two problems is in fact a very old problem that probably will never be solved. This old problem is the problem of typology that has plagued archaeology for centuries. With the current state of DBMSs all we need to do is simply hardcode a lookup table that contains all the possible entries for any typological conflict that may occur for any object classification.

Unfortunately the proponents of each individual typology refuse to come together in agreement for what to call these objects for a variety of different, and always-personal reasons. Once we come to terms that most of the fighting concerning a standard structure for archaeological metadata is in fact a battle over typology we can then, hopefully, come to a final agreement on the much needed metadata standard. Once this standard is agreed upon and implemented there will no longer be a problem with metadata.

What databases really mean: Archaeological theory through databases.

Now that we know what databases are, why they are the way they are, what neat things we are beginning to do with them, and the major problems we are having with them, we can finally get to what are the driving theories behind the use of databases in archaeology.

Most archaeologists truly don't think much about what theory is driving their use of databases; they just use them as one of a suite of tools to answer their main research question. What they don't realize is, however, that they are in fact following one of three fundamental theoretical perspectives for databases. The three theories that I have identified that archaeologists use are: first, a Functionalist or Marxist approach that looks at the data as being nothing more than a representation of the material being described; second, a cognitive approach that gives the raw data to the observer in a form that the observer can then compile in his or her own way so the observer can build his or her own interpretation; and third, a stylistic or artistic approach that strives for greater clarity or aesthetic appeal in the presentation of the dataset as a means of providing visual support to the arguments based on that dataset.

These three approaches differ on one very important aspect. This aspect can be boiled down to how the researcher uses the database to construct his or her interpretations of the dataset. Someone using the Functionalist or Marxist approach views databases as either a means to an end, i.e. an analysis of the information that can be created through the manipulation of the data contained in the database (which doesn't differ in the slightest from the way a paper database is viewed), or as an electronic substitute for the actual objects themselves that then can be manipulated to help create an acceptable story. The key to this approach is the speed with which the final analysis can be created. A paper database requires much more time to "sift" for the juicy tidbits of data that are needed to make a compelling story. Rapid returns of analysis are dictated by the time constraints of contract archaeology and Cultural Resource Management in the US, and this then, requires the use of a method other than paper for their database. The financial concerns of not running over budget while still producing a thorough analysis dictates a methodology that requires the shortest amount of manipulation of data, and a Functionalist or Marxist approach fits this nicely.

The cognitive approach offers an alternative to the Functionalist or Marxist approach that some in the US and around the world are turning to. This approach normally provides a short overview of the data contained in the database, and then provides access to the entire dataset so that the dataset can be manipulated by anyone. For Cultural Resource Management in the US this allows the contract archaeologist a method of providing the desired information to their employers without having to go through the expense of actually doing the analysis, as well as providing a kind of "buffer" between the researcher and the financial managers. For others around the world this method is a way that they can expand their research by allowing multiple perspectives to be brought to bear on the dataset. It is felt that with multiple perspectives looking at the same data at the same time that a better view of the information contained within the dataset will arise. As one can imagine, most archaeologists using the cognitive approach to their databases are providing the interface to their data online via a web accessible design.

The third approach I am calling the presentational approach, though I feel that term is somewhat misleading. This approach is one in which the database is used in a manner that helps to create a better public presentation of the researchers conclusions, not simply as the data repository from which the information that makes up the conclusions were drawn. For obvious reasons most archaeologists who utilize GIS databases use this approach when creating the high quality maps that accompany their reports. By creating a better visual representation of the information contained in the database than a chart or graph provide, more information is conveyed with fewer typed words. This then in turn allows the researcher to spend more time on the actual results of his or her research than on the explanation of what the data contains. The drawbacks of this approach are high dollar costs, and time spent in the creation of the presentation. To utilize the presentational approach one must purchase the software and hardware that will allow the presentation to be created, which can be quite expensive. Then a considerable amount of time must be spent in training and the use of the software, and production of the actual final products. Once complete the rapid production of information rich visual representations of the data makes it all worthwhile.

Conclusion

What we find when looking at this is that most archaeologists don't consider their database as having any real bearing on the theories they use to develop their conclusions about their data. We have seen that this is not the case at all. A well-designed database can aid the archaeologist develop an analysis of his raw data that may not have appeared other wise. In fact most archaeologists develop their databases based directly on the desired end product that they are pursuing. Unfortunately for archaeology most of the end product being produced in the US is for cultural resource management, funded directly by the state and federal agencies that are only interested in how fast a project can be completed. This has caused archaeologists in the US to stop asking questions that require data design beyond a simple count of artifacts. Once archaeologists begin to recognize that they have been unconsciously constructing their databases biased on quite solid theoretical foundations they should be able to fine-tune their use of databases to generate more information than ever before.

References

Anderson, D. G. and Faught, M. K. 1998. *A North American Paleoindian Projectile Point Database.* http://www.anthro.fsu.edu/special/paleo/paleoind.html

Baker, Tony. 1999. *Digital Crabtree: Computer Simulation of Folsom Fluting.* Online publication. http://www.ele.net/algor/crabtree.htm

Barker, Richard. 1990. *CASE Method Entity Relationship Modeling* (Addison Wesley Publishing Company,), clothbound; ISBN 0-201-41696-4.

Baumann, Steven M. 1999. *Integrating GIS and Cultural Resources Databases for Archeological Site Monitoring.* Journal Info Crm: cultural resource management. Volume 22, Number 933. SICI 1068-4999(1999)22:9L.33:IGCR;1-

Burrough, P. A. 1986. *Principles of Geographic Information Systems for Land resource Assessment.* Oxford: Clarendon Press.

Carroll, M. 1999. *Preserving Archaeological Digital Data: Report of the NCPTT Working Group.* From *SAA Bulletin*, vol. 18, no. 5.

Cathro, Warwick. 1997. *Metadata: An overview.* National Library of Australia. http://www.nla.gov.au/nla/staffpaper/cathro3.html.

Clarke, K. C. 1995. *Analytical and Computer Cartography.* 2nd edition. Upper Saddle River, NJ: Prentice Hall

Codd, E. F. 1974. *The Relational Approach to Data Base Management: An Overview*, Third Annual Texas Conference on Computing Systems.

Codd, E. F. 1990. *The Relational Model for Database Management* (Addison-Wesley Publishing Company,), 538 pages; clothbound; ISBN 0-201-14192-2

Coronel, Rob. 1997. *Database Systems: Design, Implementation, and Management. Course* Technology, a Division of International Thomson Publishing. Massachusetts.

Clarke, Keith. 1997. *Getting Started With Geographic Information Systems.* Prentice-Hall, Inc. Simon & Schuster / A Viacom Company. Upper Saddle River, New Jersey.

Crane, Gregory. 1991. *Hypermedia and the Study of ancient culture.* Journal Info Ieee computer graphics and applications. JUL 01 1991 v 11 n 445

Date, C. J. 1995. *An Introduction to Database Systems* (Addison-Wesley Publishing Company,), 839 pages; clothbound; ISBN 0-201-54329-X.

Duecker, K. J. 1979. *Land Resource Information Systems: a review of fifteen years experience.* In *Geo-Processing*, vol. 1, no. 2, pp. 105-128.

Eiteljorg, Harrison. 1997. *Electronic Archives.* http://intarch.ac.uk/antiquity/electronics/eiteljorg.html.

Eiteljorg, Harrison. 2000. *If We Preserve the Files, Who Will Use Them?* http://csa.brynmawr.edu/saa/saa-adap.html

EPA. 1997. *EPA Scientific Metadata Standards Project.* http://www.lbl.gov/~olken/epa.html.

Gilman P. 1997. *Securing a Future for Essex's Past. ArcUser* Magazine, http://www.ersi.com/news/arcuser/1099/essex.html

Heyworth, M. P., Richards J., & Winters J., Eds. 2000. *Internet Archaeology: Where Next?* http://csa.brynmawr.edu/saa/saa-ia.html

Hodder, I. 1999. *The Archaeological Process: An Introduction*

Lohse E.S. and D. Sammons. 1997. *DIGITAL STONES: A Basic Introduction to the Analysis of Stone Tools.* Beta version CD-ROM.

Lohse E.S. and D. Sammons. 1997. *A computerized Database For Lithic Use-Wear Analysis.* BAR Conference, 1996.

Lohse E.S. and D. Sammons. 1997. *Metalanguage or Metabaggage: Archaeological database constructions in cyberspace.* Paper presented at the 63rd annual SAA meeting, Seattle, March 25-29, 1998.

McCartney, P., Robertson I., & Cowgill G. 1997. *Using Metadata to Address Problems of Data Preservation and Delivery: Examples from the Teotihuacan Data Archiving Project.* http://csa.brynmawr.edu/saa/mccartney.html

Moyes, H. and Awe, J. 2000. *Spatial Analysis of an Ancient Cave Site. ArcUser* Magazine, http://www.ersi.com/news/arcuser/1000/cave.html

Robinson, D. 1997. *Digital Archiving Pilot Project for Excavation Records (DAPPER)* http://csa.brynmawr.edu/saa/dapper.html

Star, J. and Estes J. E. 1990. *Geographic Information Systems: An Introduction.* Upper Saddle River, NJ: Prentice Hall.

Weibel, S. 1995. *Metadata: The foundations of resource description.* D-Lib Magazine, July. http://www.dlib.org/dlib/July95/07weibel.html.

Wise, A. and Miller P. 1995. *Why metadata matters in archaeology.* Internet Archaeology 2. http://intarch.ac.uk/journal/issue2/wise_toc.html.

Wolle, A. 1997. *Çatalhöyük: Excavations of a Neolithic Anatolian Höyük* http://catal.arch.cam.ac.uk

The Use of Permutations to Explain the Hackness Cross Tree Rune Inscription

Richard Sermon
Gloucester Archaeology Unit
Barbican Road
Gloucester GL1 2JF
England
E-mail: richards@glos-city.gov.uk

Abstract: All that remains of the Hackness Cross are two damaged stone fragments of an 8th or 9th century Anglo-Saxon monument. The cross fragments bear five inscriptions, two coded or cryptic inscriptions, and three in Latin which refer to an abbess named Oedilburga. The first of the cryptic inscriptions is written in what appears to be a form of Ogham, while the second is written in a combination of Anglo-Saxon Runes and Tree Runes. In 1994 I gave a paper at the CAA conference held at Glasgow University on the use of computer generated permutations in attempting to decipher the cryptic inscriptions (Sermon, 1995, 253-257 and 1996, 101-111). This work interpreted the Ogham inscription as a dedication in Old Irish which translated 'Cross to King Jesus, rock of help from Angus', and the standard Anglo-Saxon Runes as an Old English anagram meaning 'Oedilburg knew me'. However, no interpretation was offered at that time for the Tree Rune inscription. Recently a more critical re-examination of the Runic inscription has been carried out, where only those characters that could be read with a high degree of confidence were accepted. As a result one of the permutations of the Tree Rune inscription was found to contain the personal name 'Bosa' who was bishop of York from 678 to 705 AD, and a contemporary of Abbess Hilda who founded the monastery at Hackness.

Introduction

The monument consists of two damaged stone fragments from an 8th or 9th century Anglian cross, presently located in the south aisle of St Peter's church at Hackness in North Yorkshire (see Figure 1). The stones were discovered some time before 1848 in an outhouse on the former monastery site at Hackness Hall, and appear at some time to have been used as a gatepost. The fragments are from the top and the bottom of the cross shaft, and together stand to a height of 1.5 metres. However, the original height of the monument would have been something like 4.5 metres. The stones are decorated in relief with vine scroll, interlace, the feet of two beasts, and what is presumably the head of Jesus. In its original form the Hackness cross would have been equal to the well known examples at Bewcastle and Ruthwell. The cross fragments also bears five inscriptions, two coded or cryptic inscriptions, and three in Latin which all refer to an abbess named Oedilburga (see Table 1).

Oedilburga was presumably abbess of the monastery at Hackness that according to Bede (Sherley-Price 1968, 249) was founded by Abbess Hilda of Whitby in 680 AD; *That same night it pleased Almighty God to make her death (Hilda's) known by means of a vision in a monastery some considerable distance away, which she had founded that same year at Hackness (Hacanos).* Hilda's successor to the monastery at Whitby was Abbess Aelffled, who is mentioned in the *Life of St Wilfrid* by Eddius Stephanus (Webb and Farmer 1983, 171), when she visits her dying brother King Aldfrith of Northumbria at Driffield in 705 AD; *This verbatim account was given us by trustworthy witnesses, among whom were Abbess Aelffled, herself a king's daughter and Abbess Aethilberg (Aedilberg).* These events all fit into place when we realise that Hackness lies on route between Whitby and Driffield. It would therfore make perfect sense for Aelffled to have broken her journey at Hackness and from there on to have been accompanied by Aethilberg who is almost certainly the same person as Oedilburga commemorated on the cross.

The first of the cryptic inscriptions is written in what appears to be a form of Ogham, an alphabet developed in Ireland in the 4th century AD. In my previous work this was interpreted as a dedication in Old Irish which translated 'Cross to King Jesus, rock of help from Angus'. The second cryptic inscription is written in a combination of standard Anglo-Saxon Runes (see Figure 2) and more unusual Tree Runes.

The Runic Inscription

The Runic inscription is located on the east side of the upper fragment and consists of 15 Anglo-Saxon Runes, 35 Tree Runes and three Latin letters, combining to form a six line inscription (see Figure 3). In my previous work the first two lines of the inscription (+ E M B D W Œ / G N L G U I Œ R) were interpreted as an Old English anagram containing the name Œdilburg, in keeping with the three Latin inscriptions which also commemorate Oedilburga (see Table 2). Here the words are spoken by the cross, in a familiar Anglo-Saxon literary device known from the poem *Dream of the Rood.* Parallels also exist for Runic anagrams, including a number of riddles from the from the *Exeter Book.* The three Latin letters ORA at the end of the Runic inscription were interpreted as an abbreviation for the Latin *orate* or pray.

The Tree Runes employed here are thought by Page (1973, 64-66 and 1999, 83-86) to be a form of *Hahalruna*, which are described in the 9th century *Isruna* Tract, and are similar to Norse runes from Maes Howe on Orkney. The Runic alphabet (futhorc) is split into four groups of eight letters. Each Rune is then represented by a vertical stemline, with the number of arms to the left indicating the group in which the Rune occurs and those to the right indicating its position within that group, thus giving rise to an alphabet of 32 letters (see Figure 4). However, due to the fragmentary state of the inscription, the actual reading of each Tree Rune is rather more problematic. If we compare the number of arms on each Tree Rune that have been identified by Haigh (1875, 349-391), Collingwood (1927, 59-61) and Brown (1930, 52-75) there are some significant differences (see Table 3). Nevertheless, this comparison reveals broad agreement for the Tree Runes identified in the second line, and would suggest that this is the best preserved part of the inscription. Following a recent amd more critical re-examination of the inscription and published photographs (Lang, 1991, Illustration 463), only eight of the Tree Runes could be safely identified with a high degree of confidence (see Table 4).

As well as uncertainty about the correct reading of each individual Tree Rune, attempts to decipher the inscription have been further complicated by uncertainty about the order in which the Tree Rune letter groups were employed. If we divide the Anglo-Saxon Runic alphabet into four groups of eight letters, we then need to establish the order in which these letter groups correlated to the Tree Rune letter groups. This gives us a total of 24 different possible permutations, which have now been generated for the eight legible characters using a computer program. Having examined all the permutations, the majority were largely unintelligible, however, number 19 appeared to contain the name Bosa (see Table 5). For some time it has been suggested that the first group of Tree Runes (1/1 to 1/8) may correspond to the final group of eight letters in the Anglo-Saxon Runic alphabet (Derolez, 1954, 140-142), given that both groups appear to comprise the least used characters. It is therefore interesting to note that this is also the case with permutation 19 (see Figure 5), and that it is paralleled by the Maes Howe inscription where the Runic letter groups are used in reverse order.

According to Bede (King, 1930, 131), a monk named Bosa was raised at Abbess Hilda's monastery in Whitby, and later became bishop of York from 678 to 686 AD, and from 691 until his death in 705 AD; *In short, we have since seen five from the same monastery, afterwards bishops, and all these men of singular worth and holiness, whose names are Bosa, Aetla, Oftfor, John and Wilfrid. Of the first we have said before, that he was consecrated bishop of York.* Bosa would certainly have known about the monastery at Hackness, being a daughter house of Whitby, and is likely to have been held in high regard by these communities, which both lay within his diocese. We should also bear in mind that the early church in Northumbria was a close knit community, and that Hilda, Bosa, Aelffled and Aethilberg would all have known one another (see Table 6).

Conclusions

The legible Tree Runes number only eight out of 35, for which we have 24 possible readings or permutations. Finding that one of these permutations contains a known personal name is by itself a significant result. However, when we consider that this is not just any name but that of Bosa bishop of York, a contemporary Abbesses Hilda, Aelffled and Aethilberg (Oedilburga), it seems unlikely to be just a coincidence. In my previous work on the Hackness Cross I had concluded that there was probably little more that could be done on the Tree Rune inscription, given its fragmentary state. However, over the following years I have found it very difficult to put the problem down. While re-reading some of the previous papers on the cross, I realised that there was considerable agreement as to the reading of the second line of Tree Runes. With this in mind I decided to look again at computer output from my earlier work. This has lead to the present interpretation (see Table 7), which I believe fits both the historical and cultural context of the monument.

References

Brown, G B. 1930. *The Arts in Early England VI*, London

Collingwood, W G. 1927. *Northumbrian Crosses of the Pre-Norman Age*, London

Derolez, R. 1954. *Runica Manuscripta*, Brugge

Haigh, D H. 1875. The Monasteries of St Heiu and St Hild, *YAJ*, 3

King, J E. ed 1930. *Bede, Historical Works*, II, Leob Classical Library

Lang, J. 1991. *Corpus of Anglo-Saxon Stone Sculpture*, III, British Academy

Page, R I. 1973 and 1999. *An Introduction to English Runes*, London

Sermon, R. 1995. in Huggett J and Ryan N, eds, *CAA 1994*, BAR S600, 253-257

Sermon, R. 1996. The Hackness Cross Cryptic Inscriptions, *YAJ*, 68

Sherley-Price, L. ed 1968. *Bede, A History of the English Church and People*, London

Webb, J F & Farmer, D H. ed 1983. *The Age of Bede*, London

Tables

Table 1. Hackness Cross Latin Inscriptions

+ REL(IGI)OSA ABBATISSA OEDILBVRGA ORATE PR(O NOBIS)
+ Religious Abbess Oedilburga pray for us

(OEDI)L(BVR)GA SEMPER TENENT MEMORES COMMV(NITATE)S TVAE TE MATER AMANTISSIMA
Oedilburga your communities hold you always in memory most loving mother

OEDILBV(RGA) BEATA A(D S)EMPER T(E REC)OLA(NT)
Blessed Oedilburga always may they remember you

Table 2. Interpretation of Standard Anglo-Saxon Runes

Reconstruction	+ ŒDILBURG GNŒW ME
Old English	+ Œðilburg cneow me
Interpretation	+ Oethilburg knew me

Table 3. Comparison of Previous Tree Rune Readings

Line 1										
Haigh	4/8	-/-	-/-	-/-	-/-	-/-	-/-	-/-	-/-	-/-
Collingwood	3/8	-/-	-/-	-/-	-/-	-/-	-/-	-/-	-/-	-/-
Brown	4/8	2/4	3/4	4/1	3/-	-/-	3/3	-/-	-/-	-/-
Line 2										
Haigh	3/2	2/4	3/8	2/1	2/5	2/4	3/2	-/-	-/-	-/-
Collingwood	4/2	2/4	3/8	1/1	2/5	2/4	3/1	-/-	-/-	-/-
Brown	4/2	2/4	3/8	1/1	3/5	2/4	3/1	1/1	-/-	-/-
Line 3										
Haigh	3/2	-/-	-/-	-/-	-/-	3/8	3/2	2/4	3/3	-/-
Collingwood	3/1	-/-	-/-	-/-	-/-	3/6	3/2	3/3	3/3	-/-
Brown	3/2	1/1	-/-	-/-	4/6	3/8	3/3	3/5	4/4	-/-
Line 4										
Haigh	-/-	-/-	-/-	-/-	-/-					
Collingwood	-/-	-/-	-/-	-/-	-/-					
Brown	1/1	1/1	2/3	-/-	3/8					

Table 4. Legible Tree Runes

Line 1	4/8	-/-	-/-	-/-	-/-	-/-	-/-	-/-	-/-	-/-
Line 2	4/2	2/4	3/8	1/1	3/5	2/4	-/-	-/-	-/-	-/-
Line 3	3/2	-/-	-/-	-/-	-/-	-/-	-/-	-/-	-/-	-/-
Line 4	-/-	-/-	-/-	-/-	-/-					

Table 5: Tree Rune Program Output

01	q	ae j d f l j	b	.....
02	d	b j q f io j	ae	.....
03	q	ae m s f e m	n	.....
04	s	n m q f io m	ae	.....
05	d	b ea s f e ea	n	.....
06	s	n ea d f l ea	b	.....
07	q	ae o d h l o	b	.....
08	d	b o q h io o	ae	.....
09	q	ae m w h r m	u	.....
10	w	u m q h io m	ae	.....
11	d	b ea w h r ea	u	.....
12	w	u ea d h l ea	b	.....
13	q	ae o s t e o	n	.....
14	s	n o q t io o	ae	.....
15	q	ae j w t r j	u	.....
16	w	u j q t io j	ae	.....
17	s	n ea w t r ea	u	.....
18	w	u ea s t e ea	n	.....
19	d	b o s a e o	n	.....
20	s	n o d a l o	b	.....
21	d	b j w a r j	u	.....
22	w	u j d a l j	b	.....
23	s	n m w a r m	u	.....
24	w	u m s a e m	n	.....

Table 6. Hackness Timeline (614-714 AD)

614	Hilda is born.
649	Hilda is appointed abbess of the monastery at Hartlepool.
654	Aelffled is born.
657	Hilda founds a monastery at Whitby, where Aelffled and Bosa are trained.
678	Bosa is appointed bishop of York, following the expulsion of Wilfrid.
680	Hilda founds a daughter monastery at Hackness.
680	Hilda's death is seen in vision at Hackness.
680	Aelffled is appointed abbess of Whitby.
705	Aelffled visits her dying brother king Aldfrith, accompanied by Abbess Aethilberg.
705	Death of Bosa.
714	Death of Aelffled.

Table 7. Interpretation of Runic Inscription

Reconstruction	Translation
+ Œdilburg	+ Oethilburg
gnoew me	knew me
d	d . . . .
Bosa eo	Bosa eo . . . .
n	n . . . .
. . . . . ora(te)	. . . . . pray

Figures

Figure 1. The Hackness Cross

Figure 3. Hackness Runic Inscription

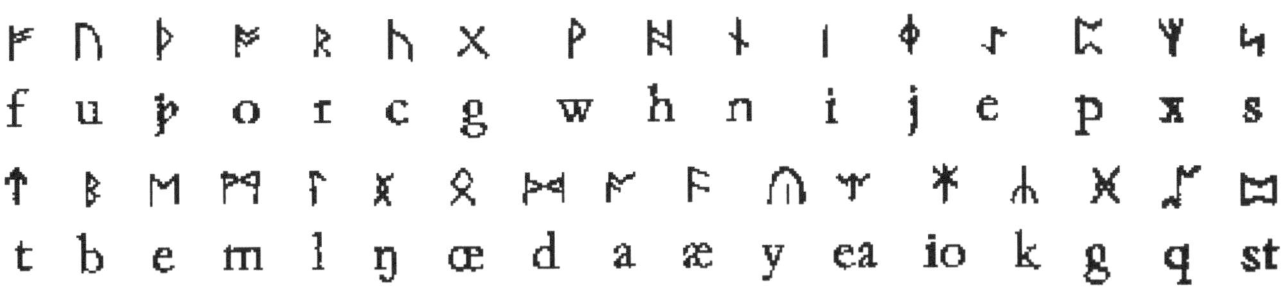

Figure 2. Anglo-Saxon Runic Alphabet

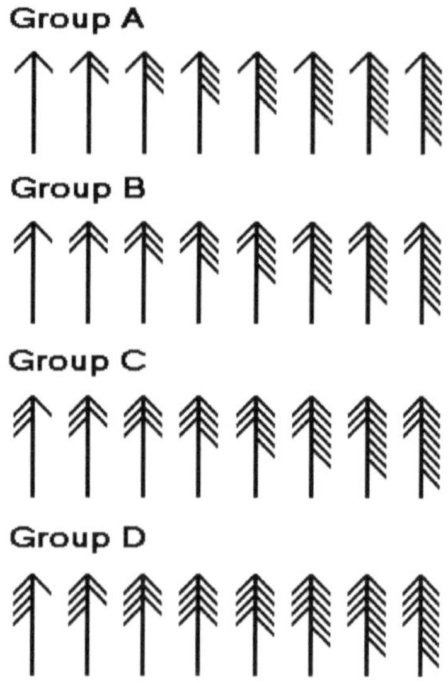

Figure 4. Tree Rune Letter Groups

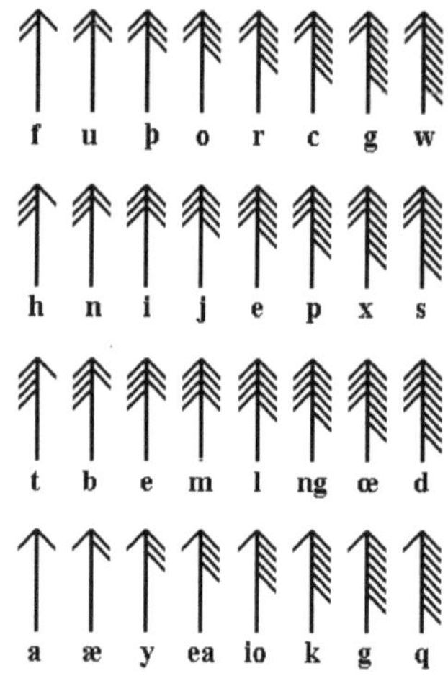

Figure 5. Tree Rune Alphabet Reconstruction

The computer as a tool in a dialectical approach to data processing

Maria Shinoto
Institut für Ur- und Frühgeschichte
Marstallhof 4
69117 Heidelberg
http://www.uni-heidelberg.de/institute/fak8/ufg

Abstract: For a study of sites and artefacts in a small region of South Japan, the author developed a combination of photography and a relational data base that provides tools for data input, analysis, output and manipulation. New results from analysis in the course of an investigation can change research design within the data base continuously. At the same time the structure is kept consistent. The most efficient part of the data base in this sense is the 'artefact area' with its flexible treatment of artefacts, attributes and classes. The present article focuses on this aspect.

The hard- and software equipment is shortly introduced. Basic theoretical ideas about observations, properties and attributes and their significance for classification shall explain the structure of the 'artefact area' of the data base with respect to statistical, archaeological and technological standards. Two representative examples of application from the case study in Japan are presented.

Key words: relational data base, normalization, continuous adaptation of research design, classification, numerical taxonomy, cladistics, consistency, site, artefact, pottery, Japan.

Introduction

The problem

Collection, analysis and categorize, visual and verbal documentation of data or pictures of artefacts and sites is one basis of our discipline. Although there is a natural sequence in these activities, each step is related with the others. For example, the "analysis" step both influences the later step of ordering and acts as a feedback mechanism on the earlier step of collection. This is the reason for a systematic research problem: how can we adapt our research design to our knowledge, which is inevitably changing in the course of a research project?

Up to now research projects have followed more or less either of two principles. The researcher can start with knowledge from earlier studies in the field and go on, intuitively adapting research strategies to what seem to be new insights or knowledge. From a systematic point of view, this is scientifically not satisfying, although talented researchers have gained great results with studies of this kind. The other, scientifically more acceptable approach is to start with a well defined research design based on knowledge from earlier studies, then collect data, and finally analyze the data in a separate step. A well known epistemological problem becomes more severe than in the "intuitive" approach, since research in a preset framework can give new insights into details, but results of a new quality will be difficult to produce. What is more, the researcher has to wait until the end of the study to understand whether data collection did give new insights or was at least valid.

During a study in South Japan, I encountered this problem when collecting site and artefact data from a small region in order to investigate the settlement activities during several centuries. The study (Shinoto in prep.) is referred to as the "case study" in this article. In the course if the study a solution to the problem had been developed which is mainly based on a data base that controls a continuous modification of the data structure and the research design - both during data collection and afterwards - on a scientifically acceptable basis. This concept has parallels in epistemological ideas like hermeneutics and dialectics, in the realm of physics the idea of self-consistency is utilized in a similiar fashion, in statistics it would be the Bayesian approach. On the one hand these are "handy slogans", easy to misunderstand, but on the other hand they can serve as a starting point for a discussion about whether this approach is appropriate in archaeological research or not (see Shinoto in prep.). For the present article it should be sufficient to summarize that permanent evaluation of research design on the basis of new data is not only most fruitful but also well accepted in other fields of science.

This article cannot present the whole system. Instead I shall focus on a central part of it, the collection and modification of data about artefacts, in order to clarify the idea proper and the task of the computer with examples from the case study. In other words: the system serves a wide range of purposes, but this article focuses on the problem of artefact classification.

Classification background

During the last 40 years, methods based on numerical taxonomy - a form of classification that was predominant in biology during the sixties of the last century - have become predominant in archaeological classification (e.g. Clarke 1962, Clarke 1978 (1968), Hill & Evans 1972, Hodson 1966, Kampffmeyer et al. 1988, Stehli & Zimmermann 1980; see also Dunnell 1971, Whallon 1972, Whallon & Brown (eds.) 1982). At the same time, traditional approaches (Vossen 1970) continued to be used or were further systematized (Adams & Adams 1991).

Adaptation of numerical taxonomy from biology to archaeology was thought to lead to a scientifically satisfying alternative to traditionally more intuitive approaches. But in the late sixties and the seventies, numerical taxonomy lost its dominance in biology and was nearly replaced by another classificatory method called cladistics, which in its original form works without statistical analysis. Both methodological trends have their advantages and disadvantages, they have undergone substantial changes in biology since then (Mayr 1990, Panchen 1992). It is not as easy to transform the ideas of cladistics to archaeology as it was with numerical taxonomy, though it is a very stimulating task. A first attempt concerning methodological and theoretical aspects will be published in another paper (Shinoto in prep.).

For the present paper it should be sufficient to maintain that numerical taxonomy and its developments are not the only scientific way to archaeological classification, there are different models of at least equal scientific standard. The alternative approach deduced from biology needs a clear definition of attributes that can be utilized for classification and their close observation. A clear idea about what is the explanans and what is the explanandum in the classification is inevitable. Evaluation of the observations may involve statistics but does not necessarily do so.

For the artefacts of the case study - mainly pottery of rough making called Narikawa pottery - classifications already exist. Two classifications show a more traditional approach (Ikehata 1980, Tatara 1981) the other is more systematic though it is not a numerical taxonomy (Nakamura 1987). All classifications are based on profile description and cover more than 400 years divided into 4 to 5 stages. To achieve a more detailed classification with information about more than just chronology, the case study had to start with new ideas about research design and new views on the pottery. Therefore, systematic studies on pottery (e.g. Dohrn-Ihmig 1976, Gardin 1976 (1985), Kampffmeyer et al. 1988, Karstens 1994, Orton, Tyers & Vince 1993, Sablatnig & Menard 1996, Shepard 1955 (1980), Traunecker 1981) were adapted to the case in South Kyûshû in order to get a more vivid idea about the ancient society (Shinoto in prep.).

Prerequisites

This article does not comment on the basic rules for structuring data in statistics (see e.g. Shennan 1988 (1990) or any introduction to descriptive statistics). It does not explain the terminology of relational data base theory, the customary drawing of relations among tables, nor does it explain the rules for normalization (see e.g. Adams & Beckett 1997 or any introduction to relational data bases).

System features

Photographs

In the case study, photographs were made instead of drawings. This was one important factor for developing new ideas about the material, mainly pottery of rough making, known as "Narikawa-pottery". Drawings of artefacts can be integrated into the system for documentation and further analysis. On the one hand, photographs are closer to the original, but not a good basis for structured analysis because of their redundancy of individual information. On the other hand, drawings rarely offer new insights since they are standardized with respect to certain traditions, only those features are recorded that are of interest in the framework of this tradition.

The photographs of the case study became a source for new insights and an understanding of the pottery that could not be gained with traditional drawings. The redundancy of individual information was avoided to a certain degree by taking photographs from standardized positions so that photographs of different objects from the same position could be selected in the data base and visually compared. To achieve more than this information, additional spontaneous photographs were taken where it seemed to be of interest (Shinoto in prep.).

Database

The relational data base is designed in order to record information from different regions, sites, documentation in literature, excavations, artefacts and the places where artefacts are stored. The relation consists of several tables which refer to "areas of content" (fig. 1). "Log tables" are not yet implemented; they shall create text data of each data manipulation and structural changes - which is only roughly done by hand at the moment. One area is concerned with the location of the site. Information about administrative units from hamlet up to the national state can be stored in several tables. These are related to tables with information about the natural geography like traditional landscapes, river systems or mountain ranges. A table for maps and automatic mapping may be added in the future and serve as a GIS module together with the analysis tools.

The central "site" table of the "site area" is related to the location tables. Informations related to sites are divided into further tables concerning excavation campaigns and excavation units like layers, artificial units like excavation grids or anthropomorphous house pits etc. - The "artefact area" is related to these excavation units and further divided into attribute and class tables and tables that store pictures or other non-verbal information about the individual find. - The "time" area is concerned with relative and absolute dating.

Information about storage of artefacts could be related to the artefact table - in order to record information about storage for each artefact individually. For practicability, I related the storage area to the site table. The mass of artefacts is stored at the same place, in the case of an exception this can be mentioned separately in a "comment data field". The same solution was chosen for the documentation tables, which store information about notes, protocols, articles or books. It may make sense to relate these two areas to excavation campaigns in the future rather than to the whole site.

The system is designed for international research. Proper names are stored, sorted and searched in the characters of the original language and additional columns (data fields in the actual record of a table) for transcription into Latin characters and for alternative namings. This feature was designed for the situation in Asia. But since anywhere in the world research most often covers several countries, and since places are named differently by different groups, this may be a helpful feature elsewhere as well.

The equipment

The system was developed with standard software that is available on a Macintosh. It can be used on the Windows platform with the same software, which is unfortunately not the case with Linux. When the project started, analogue photography with scanning offered the better quality particularly as far as colour management is concerned. For any new case study to be started nowadays, digital technology should be used in order to reduce cost and increase speed and quality of the research (details in Shinoto in prep.). The whole equipment is easy to transport and can be used on site.

The data base is the centre of the system, it serves data storage, manipulation, analysis. In addition, via ODBC or data export it provides data for more sophisticated analysis. In the future these may be integrated into the software package. The system was developed with the DBMS "4th Dimension", starting with version 3.5. At the moment it is under further development using version 6.58, while the latest available version is 6.71.

This article shall present the ideas as independent from the underlying software as possible. The final aim is to create a software package that is compiled as a stand-alone application which does not need any programming by the end user. When working with the program, the user creates one separate data file for his individual case, independent of the subject of the study. - The aim of this article is to explain the basic ideas of the system in order to enable the reader to create a similiar solution with his own DBMS.

First normalization of artefacts and attributes

One may create columns for attributes in the artefact table. But archaeological practice and the theory for relational data base design shows that this is inappropriate. Using the rules for normalization, one can create a structure for artefacts and attributes that is archaeologically sensible and practicable. - It is important to distinguish attributes and attribute states (Clarke 1978 (1968)) in this context, where each attribute represents a generic term with the attribute states as its concrete manifestations, varying from artefact to artefact.

When designing only one table (fig. 2 a) for artefact and attributes together, different fields will stay empty for each record. If artefacts of different kinds are recorded, like stone tools and pottery, each record has to offer attribute fields for both kinds. In every record the fields concerned with the other kind of artefact will stay empty. In addition, artefacts are not all preserved well. Pottery in particular is most often found as sherds from different vessel parts. Therefore attributes referring to vessel parts that are not preserved cannot be recorded.

To keep the database slim, one may exclude the columns for attributes from the artefact table, creating separate tables for each attribute instead (fig. 2b). If a certain attribute state is observed for an artefact, a record is created in the respective table. Since every attribute requires a table, the data base structure has to be changed for different cases.

The solution in fig. 2c is more elegant. Unique artefact data like ID numbers are recorded in the "artefact table", attribute states are stored in one "attribute states table", regardless the attribute concerned. Several attribute states are related to and specified by one attribute which is recorded in the "attribute table". This forms an n:1 relation between the tables for attribute states and attributes. The number of attribute states is most often limited. Therefore, one attribute state can be assigned to several artefacts, forming a 1:n relation. But at the same time an artefact most often shows more than one attribute, thus forming a 1:n relation with the "attribute states table". Together, both artefact and attribute state form a n:m relation. - In real life, this n:m relation requires an additional table in order to form a 1:n and a m:1 relation as fig. 7 will show.

This data base structure needs no change in the course of research, because attributes and attribute states can be added as records by the user rather than added as tables by the programmer. However, attributes are poorly structured and attribute states are difficult to sort and search.

The same ideas apply to what is called "classes" in this article and may be referred to as "types", "groups" or even "chronological stages" in other contexts. A classification is similar to attributes with the classes being the equivalent for the attribute state. The class name can be stored in the attribute table like an attribute. In many cases, several classification systems exist for the same groups of artefacts, and if we do not wish to favour one, we have to store several class names for one artefact. Again, all these can be stored in the "attribute state table". - When speaking of "attribute tables" and "attribute state tables" in this article, classifications and classes are included.

The nature of attributes

The model proposed is too simple for both archaeological practice and computer technology. The archaeologist as well as the computer needs structured data to work with. In the following sections I will discuss the nature of archaeological observations, the significance of observations for archaeological classification, the nature of variables in statistics as well as the rules for normalization and the basis for encoding data in a computer. This shall lead to a model of an intelligent data base structure that is not limited to a certain case. In other words: the software package shall be applicable to any new project without new programming.

Observations: attributes, properties and classification

Data to be recorded are first of all observations, structured or unstructured. Observations may form attributes or properties. Both may be significant for classification or not as long as they show a certain structure (fig. 3).

Occasional features are recorded as unstructured observations. As soon as they are observed more or less regularly, they should be reconsidered, structured and changed into a structured observation. Observations may concern attributes or properties. Attributes are those features that are formed by intention or without intention as a by-product of intentional work in the course of the artefact production. Thus, they may reflect the intended usage, fashions or stilistic traditions, technological standard and availability of resources - to mention just a few matters that are of interest for classification. The origin of properties is not limited to the stage of production, they can also occur as a result of usage or deposition in the earth. If they occur during the stage of production, they do not reflect intentions of the producer but unintended matters like faulty production. In most cases properties are observed occasionally. If they occur frequently, particularly as usage marks, they may become of interest for classification as well. - In the case study, a certain vessel form is used for cooking in open fire and for storage. Usage marks like soot on certain vessel parts help to understand the usage of the individual pot (e.g. Uno 1999). Comparison of pots with soot and those without helps to find functional differences reflected in attributes that were overlooked before. Besides distinguishing attributes and properties one may distinguish between observations that are significant for classification and those that are not. Of course, unstructured, accidentally occuring observations cannot be significant for classification, but they are nevertheless worth being recorded in the data base and considered for interpretation.

Only few observations reflect the intention of the producer immediately and are as such significant for classification. Some structured attributes are too detailed. In these cases, the intention either has to be deduced from several specific observations, forming a "deduced attribute" or it has to be reconstructed by unifying and simplifying several detailed observations to one "simplified attribute". Two examples may clarify the idea (fig.4).

Fig. 4 shows sherds from five pedestaled bowls, the bottom of bowls from below in the three examples above and two feet below. Two bowls on top still show a clay ball in its centre that functioned as a connection between bowl and foot. So we can record the observation "clay ball" as well as a deduced attribute "adherence technique" with its attribute state "with a clay ball". The bowl in the centre shows a hole instead of the ball, so for the same vessel part we cannot record the observation "clay ball", but the same deduced attribute state "with a clay ball" for the attribute "adherence technique". For classification, only the deduced attribute is of importance. It builds a bridge to another part of the same vessel form: the two examples below show the feet of pedestaled bowls with a hollow on top. It is the trace of the clay ball that connected bowl and foot. Therefore, in case of the feet we observe a third morphological attribute that represents the same deduced, significant attribute for classification. As for the producer of these vessels, she (several studies let us suppose that they were women) made the clay balls intendedly, holes and hollows may be the unintended side effect, but these observations also reflect her intention to connect both vessel parts with a clay ball.

Beside deducing a significant attribute from several observed and unsignificant attributes, simplification is another important method - as the example "temper" may show. In the course of material collection, one may count and measure temper grains and record their colour or shapes and so on. But this may not lead to a significant attribute for classification, since most often it was not the intention of the producer to add a certain proportion of grains of certain colour and size or shape. Temper was chosen from available resources, perhaps mixed or sieved and added in certain amounts to the clay. It is the task of the archaeologist to find the regularities in the many insignificant but detailed attributes to form groups that reflect the intention of the producer who wanted to produce a certain kind of fabric. Even these simplified temper groups may be too detailed to be significant for classification, since the final intention was to form a certain fabric. The correlation of temper groups with other features of the fabric, like porosity, may be investigated to finally form a significant attribute "fabric".

The last example indicates another basic problem. How do we know that simplification is at its end and that the attribute is significant? Classes are further simplifications of attributes, the border between attribute and class is fluid. As soon as attributes from different realms like "decoration" and "technology" or as soon as properties are integrated, one may speak of classes. Classes may vary with respect to the purpose and nature of the classification, one artefact may refer to several classes at the same time (Shinoto in prep.).

Structure in archaeology, statistics and data bases

Attributes and attribute states have their counterparts in statistics: the variable with categories or values. "Sherd colour" or "rim diameter" are attributes in the realm of archaeology and variables in the realm of statistics. "Red" or "21 cm" would be examples of affiliated attribute states in archaeology, called categories or values in statistics. This structure can be transformed to the structure of relational data bases. This process is called normalization, the structure is reflected in a table for the attribute and another table for the attribute states (fig. 5).

The form of structured attributes varies, as the example from sherd colour and rim diameter above has shown. The variation is structured; like statistical variables they can be on nominal, ordinal, interval or ratio scale. In the case study, I did not encounter attributes on interval scale, for several reasons they should be treated together with attributes on ratio scale. Attribute states on different scale have their counterparts in different data field types of the columns of a data base table (fig. 6).

Observations without statistical scale are best recorded in text fields or alternatively in picture fields or BLOBs. Attribute states on ratio scale are best encoded in data fields of real type, although integers are another possible solution. Attribute states on nominal scale are best encoded in alphanumeric data field types. Attributes on ordinal scale have to be treated different. In the present study, these attribute states are presented in the layouts as alphanumerical variables like "small", "medium", "large", but recorded as an integer "1", "2" or "3" in the respective data field, encoded automatically by the program. This solution is helpful when sorting items with respect to their rank or analyzing them with methods that need data on ordinal scale. While the user can name the attribute without regard to the alphabetical order, the computer is always informed about the rank from the affiliated integer value.

Towards final normalization

The above considerations lead to a new design of the "artefact area" in the data base (fig. 7).

Unstructured observations are stored in the "attribute states (observations) table" and related to keywords in the "attribute without scale table" that may lead to structured attributes in the future. If not, the keywords help to get a better overview of the unstructured observations by searching and sorting. Each keyword forms a record with definitions and comments. Observations on ratio scale are another simple case in terms of normalization. The definition of the attribute is stored in the "ratio scale attribute table", the related 1-table to the "attribute state table". Since attribute states on ratio scale have no previously defined values, each value is stored individually in the "attribute state table", which is again the related n-table to the "artefact table".

The situation is more difficult with attributes on nominal and ordinal scale. In contrast to observations without scale and on ratio scale, here the names and values of the attribute states have to be defined before data input starts. In these cases the values of the attribute states are recorded in the "attribute state table" at the same time when the attribute proper is defined in the "attribute table". Each attribute state forms one record with additional definition.

It has become clear from the above explanations that it is necessary to define attributes and record them before recording of attribute states for the actual artefacts starts. When recording attribute states for a certain artefact in the next stage of research, the user chooses the respective attribute from a drop down list. The computer creates this list from the actual "attribute tables" with regard to the scale of the attribute. If the attribute chosen is on nominal or ordinal scale, the user can choose the attribute state from another list. This list is created from the related records in the "attribute states tables". After the attribute state is chosen, the data base creates a new record in the "attribute states list table" which is the related n-table to the "artefact table". For attributes without or on ratio scale the user can record individual informations and values directly in the "attribute state tables". In other words, the information about the attribute states of a certain artefact is stored in the tables "attribute state list" for nominal and ordinal scale, in the tables "attribute state" in the other cases.

Let us return to the definition of attributes. Ratio scaled data need a unit of measurement which is set in the "attribute table" beside the definition. Values concerning size often consist of up to three dimensions which should not be divided into separate attributes. Therefore the definition consists of three dimensions. This is an important feature for irregularly formed attributes as well, where the dimensions can represent minimum, maximum and normal extension. It was a very important feature in the case study, where rim diameter of the same pot could vary between 27 and 33 cm, to mention just one example.

All tables directly related to the artefact show an additional column called "parameter". This is necessary to explain where the respective attribute is observed. As for diameter, one could form several attributes for rim diameter, neck diameter etc., but in this data base another approach was chosen. "Diameter" would be the attribute, "rim" would be the parameter. Parameters form lists with different values. They are defined at the stage of attribute definition in the "attribute table" and stored in BLOB type fields (binary linked objects). This structure is a good solution as regards the demands of the computer, but it weakens the stringent structure of the archaeological attributes. The problem is controlled by the software when attribrutes are sorted, searched and modified.

Up to now I spoke of attributes alone. All this applies to properties and classifications as well, classification and property being the equivalent to the attribute, class and property state being the equivalent to the attribute state. They are stored in the same tables. Classifications are always structured and on nominal scale as shall be explained below. So all information referring to classes is stored in the tables for attributes on nominal scale (fig. 7). Unstructured properties are stored together with the attributes without scale, structured properties are most likely to be stored in the same tables like attributes on nominal scale. Considering the origin of structured properties (fig. 3), cases for properties on higher scales should be rare. To distinguish classes, attributes and properties, there is another column in the "attribute tables" called "specification" (fig. 7). The user has to decide whether he is creating a new attribute, a class or a property.

The dialectical approach: modifying observations

As explained above, observations have to be modified from time to time in order to achieve significant data for archaeological work. Observations can be individual, very detailed and precise in the beginning, deduction or simplification of many precise observations naturally lead to more vague and global data on a lower statistical level, finally on nominal scale (fig. 8). Although statistical analysis on higher scale offers more possibilities for sophisticated analyses, the final and significant attributes on nominal scale are the bases for further archaeological research, for interpretation of artefacts in social, functional, historical or other contexts.

The process of modification needs to be done carefully in order to keep the data base consistent. Fig. 9 shows the steps from the decision to interrupt the process of data collection in order to evaluate the data and re-design the research down to the step of applying the new structure. Fig. 10 shows the next steps that are necessary to carefully modify the data already collected before new data can be collected with regard to the new structure or rules.

To speak in terms of a dialectical approach, the structure during data collection in the beginning is the "thesis". A new idea emerges during data collection, which lead to an "antithesis". This is the hypothesis that can be investigated with statistical or other methods. Therefore, a population in the statistical sense has to be defined (Sahner 1997, Orton 2000) - it can vary with regard to certain vessel parts, artefacts found in certain excavation units and so on. The data base creates a selection that represents the population. Analyses may utilize the whole population. But it may be appropiate to work with random samples that are easy to select in a data base with dialogues where the user can set the parameters. If analyses show that the hypothesis has a high probability, the user has to define in which way the old attributes have to be modified. This can be a simple grouping of many old attribute states to few new attribute states, but it may concern more complex rules. In the case study, the rules were recorded manually and modification was performed manually. It is another task for the future to implement recording and using these "modification rules" in the data base. Finally the new attribute is recorded in the "attribute table". It represents the "synthesis", to complete the comparison with dialectics.

The next step is to adapt the old records to the new research design (fig. 10). For this, all records containing the old attribute have to be selected. They are either deleted and replaced by new records or new records are recorded in addition, and the old records are kept for documentation. This is recommended, since most often the old records offer detailed information that is lost in the new, simplified or deduced attributes.

The flow chart in fig. 10 can be applied to any software or programming language. It shows how each old record is checked by the software, that an equivalent new record is created and that the old record is either deleted or marked as an old record. Deleted records are stored as text in another table and can be saved as text files from time to time. Marking a record as "old" has the effect that - in principle - it cannot be used for another analysis again.

This is an important feature in order to keep research consistent. For example an analysis which uses the attribute "fabric" should not use the attribute "porosity" if this is already part of the "fabric" definition. A contrary example may be the combination of vessel volume and vessel profile. Rim diameter may be used to calculate the volume, but it may be necessary for description of form or profile as well.

This problem concerns the "modification rules" mentioned before. It was controlled manually in the present study. To automatize the process, attributes are now being integrated in a familiy structure with "attribute genealogy". Attributes in the same family cannot be used to form a new level or "generation" of deduced or simplified attributes in the same family. As for the rim diameter in our example, this attribute would refer to two attribute families, and this would allow it to be used twice.

Examples from the case study

Two examples (Shinoto in prep.) may clarify the ideas. The first example concerns "fabric" and shows the emergence of a significant attribute out of non structured descriptions via detailed attributes up to extremely simplified, significant attributes. The second example shows the emergence of an attribute "surface treatment" from spontaneous, "virtual" piling of photographs based on intuition and leading to the definition of an attribute on nominal scale.

Fabric of the pottery in the case study is so irregular that it could not be classified in earlier studies. Detailed observations of features like colour or opacity of temper grains, grain size, porosity and so on (Orton, Tyers, Vince 1993) were recorded in many site reports (e.g. Kagoshima daigaku bunkazai... 1995) but never led to a better understanding of the fabric. In the present study, attributes concerning fabric were observed in an unstructured manner in the beginning. After a while, the observation records were checked for regularities or possibly significant features. These were now transformed to several attributes on varying statistical scales. For testing their validity, the artefacts with unstructured fabric observations were reexamined, the new attributes being recorded for each artefact. When it turned out that the modifications worked, new artefacts were recorded with this structure. The first result showed that sherd colour, feel, grain size distribution, form and quantity of pores and temper proportion seemed to be of importance. In the following sequences, the character and correlation of these features were investigated and simplification went on. Finally, three fabric groups with subgroups could be formed. These proved valid in terms of archaeology, since they show correlation with time and vessel form (Shinoto in prep.) and mineralogically, since several methods of analysis confirmed the groups and could explain their making (Shinoto in prep., Hoffbauer & Shinoto 2000, Shinoto & Hoffbauer 2000).

Fig. 11 shows some of the analyses in the course of the study. The simple table on the left groups artefacts with respect to attributes that seemed of importance in the beginning. The diagramme on the top right shows the relation of temper quantity and the distribution of grain size in the sherds. It was achieved with a quick and easy application of ideas based on fuzzy control rather than measuring and counting. The last table shows the correlation of colour - that was simplified from Munsell Soil Color Charts to three significant groups - with the temper groups that emerged out of the diagramme above. These examples and the other analyses were done with simple but efficient methods, the samples were random samples between 40 and 200 items. The sequential, self-consistent approach on small scale proved more successful than the traditional attempts of the last decades. An example of different nature describes the emergence of surface treatment groups.

The surface of the pottery shows a structure called "hakeme" which is made with various kinds of wooden spatulae (fig. 12). Japanese research on this technique achieved knowledge about how certain patterns were achieved (Sahara 1986). For several years, the width of each spatula and of the incisions as well as the number of incisions and the direction of the original spatula movement was recorded carefully in the excavation reports. Still, these efforts did not lead to any further conclusion and are abandoned from observation nowadays. During data collection for the case study, I got the impression that certain patterns correlate with time in the later period, but exactly like in Japanese research, recording data in detail did not lead to the formation of significant attributes. Recording of spontaneous, intuitive impressions seemed to be more appropriate to understand surface treatment than numbers and exact definitions. Instead of recording data on site, I later checked the records for photographs with details of surface treatment which seemed to be characteristic or interesting for some reason. These photographs were "piled up" in a "photo stacks table" which is an additional table in the artefact area of the data base.

Piling up photographs is a virtual act. By clicking on a button in the artefact table, a record in the "photo stacks table" is created with information about the artefact and picture. A keyword has to be chosen from a pop up menu that is created from a list. The keyword list represents the "stacks", it can be modified at any time. In order to create a more intuitive user interface, "piling up" photographs by drag and drop into several sections of a virtual desk represented by another input layout shall be implemented in the future. However, the less intuitive layout served well in the case study.

From time to time, the stacks can be checked. Are the records in one stack varied, should they be divided into more and smaller stacks, or should several stacks be unified to one larger? Grouping may not be convincing in any way, so records from one pile have to be moved to another. All this can be done spontaneously. Each action is recorded in a "log table", which makes the process transparent later. In the course of this process, a clearer idea about what makes the difference between the groups emerges. When the groups become convincing intuitively, the stacks or keywords can function as category names for attributes on nominal scale, and attribute records can be created for each artefact. Description of each attribute state will not be one single word or value, but a longer description and discussion. Although vague "understanding" of the emerging attribute may be intuitive, this understanding has to be expressed in a clear definition in order to become an attribute.

In the case of surface treatment, this procedure was extremely helpful, surface treatment forms groups that show correlation to time. "Hakeme" became an important factor in structuring the latest chronological phase of the pottery, which may cover up to 200 years and cannot be structured by vessel profile alone.

Summary and prospects

The system presented here is based on researches of the last 40 years which treat problems of data base design, classification in general and in archaeology and systematic approaches to pottery research. References on the cognitive aspect of classification, which becomes important in the "dialectical" approach derived from cladistics, are not mentioned (see e.g. Shinoto in prep.). The data base's basic structure was created before material collection in South Japan started in 1997, but development continued in the course of material collection and analysis. The first analyses were made with exported data and spreadsheet software, implementation of most common analyses and diagramme tools is going on.

The case study (Shinoto in prep.) proved that the consistency of a material collection can be maintained during research with continuously changing research design. More important than that, these changes proved to be most appropriate in understanding the material of the actual study. New insights were achieved that would not have been achievable by dividing research in two separate steps of data collection and data analysis. The flexible approach to research design offers a more vivid image of the ancient society.

The system is not limited to the case study but can be applied to different research problems in other regions and times.

Acknowledgements

I would like to thank the company AS Verlag & Consulting GmbH, where I could acquire the necessary computer knowledge and which provided me with hardware equipment. AGFA Gevaert AG provided me with slide films and advice in Japan, and the All Nippon Airways AG transported the heavy equipment to Japan and back. The Institute for Egyptian Studies at the University of Bonn gave the permission to test the system with their collection of prehistoric Egyptian pottery, my thanks in particular to Christina Regner. The study has greatly benefited from discussions with people from other disciplines, thanks to Martin Engel and Dr. Giovanni Galizia. Zoltán Böröcz, who commented on a draft of this paper. The views presented in this paper and any mistakes contained therein are the sole responsibility of myself. I gratefully acknowledge the community of Fukiage and the academic institutions in Kagoshima which supported the research most openly and generously.

References

Adams, D., Beckett, D., 1997. *Programming 4th Dimension. The Ultimate Guide.* Foresight Technology.

Adams, W.Y., Adams, E. W., 1991. *Archaeological Typology. A dialectical approach to artifact classification and sorting.*

Buck, C.E., Cavangh, W.G., Litton, C.D. 1996. *Bayesian Approach to Interpreting Archaeological Data.* Chichester et al., John Wiley & Sons.

Clarke, D. L., 1962. Matrix Analysis and Archaeology with particular reference to British Beaker Pottery. 371-382, *Proceedings of the Prehistoric Society 28.*

Clarke, D. L. 1978 (1968). *Analytical Archaeology.* 2nd edition revised by Bob Chapman 1978. London.

Dohrn-Ihmig, M. 1976. Die Anwendung statistischer Prüfverfahren bei der Stilanalyse und bei der Erkennung von Typen am Beispiel verzierter bandkeramischer Gefässe. 1-25, *Praehistorische Zeitschrift 51.*

Dunnell, R. C., 1971. *Systematics in Prehistory.* New York.

Gardin, J.-C. et al., 1976 (1985). *Code pour l'analyse des formes de poteries. Paris, Éditions du Centre National de la Recherche Scientifique.* 1985, dépôt légal initial 1976.

Hill, J. N., Evans, R.K., 1972. A model for classification and typology. 231-273, Clarke, D. L. (ed.) *Models in Archaeology*. London.

Hodson, F. R., 1966. Searching for the structure within multivariate archaeological data. 90-105, *World Archaeology 1*.

Hoffbauer, R., Shinoto, M., 2000. Joint archaeological and mineralogical research on pottery production in Ancient Japan. 989-992, Rammlmair D., et al. (eds.). *Applied Mineralogy in Research, Economy, Technology, Ecology and Culture. Proceedings of the sixth International Congress on Applied Minearalogy ICAM 2000, Göttingen*. Rotterdam, Brookfield.

Ikehata, Kôichi,1980. Narikawa-shiki doki no saibun hennen shian. (Proposal for a chronology of the Narikawa pottery.) 1-41, *Kagoshima Kôko 14*. Kagoshima.

Kagoshima daigaku maizôbunkazai chôsa shitsu (ed.), 1995. *Kagoshima daigaku maizôbunkazai chôsa shitsu nenpô 9-10*. (Annual reports of the Section for Investigation in Archaeological Properties at the University of Kagoshima 9-10). Kagoshima.

Kampffmeyer, U., Zamperoni, P., Teegen,W.-R. , Graca, L., 1988. *Untersuchungen zur rechnergestützten Klassifikation der Form von Keramik*. Helmut Ziegert (ed.): Arbeiten zur Urgeschichte des Menschen 11. Frankfurt a.M., Bern, New York, Paris

Karstens, K., 1994. *Allgemeine Systematik der einfachen Gefässformen. Münchener vorderasiatische Studien*. Münchener Universitätsschriften 16; Philosophische Fakultät 12. München, Wien.

Mayr, E., 1990. Die drei Schulen der Systematik. 263-276, *Verh. Dtsch. Zool. Ges. 83*.

Nakamura, Naoko, 1987. Narikawa-shiki doki saikô. (Re-thinking Narikawa pottery.) 57-76, *Kadai kôko 6*. Kagoshima.

Orton, C., 2000. *Sampling in Archaeology*. Cambridge manuals in archaeology.

Orton, C., Tyers, P., Vince, A. 1993. *Pottery in archaeology*. Cambridge manuals in archaeology.

Panchen, A. L., 1992. *Classification, evolution and the nature of biology*. Cambridge. Reprinted 1994, digital reprint 1999.

Sablatnig, R., Menard, C., 1996. Computer based Acquisition of Archaeological Finds. The First Step towards Automatic Classification. 429-446, Moscati, P., Mariotti, S. (eds.), *3rd Intl. Symposium on Computing and Archaeology, Rome, Vol. 1*.

Sahara, M., 1986. "Nendo kara yakiage made." (From paste to firing.) 27-41, Kanaseki & Sahara (eds.) *Yayoi doki I*. Tôkyô, Yuzankaku.

Sahner, H., 1997. *Schliessende Statistik*. Statistik für Soziologen 2. Stuttgart, 4th ed.

Shennan, S., 1988 (1990). *Quantifying Archaeology*. Edinburgh University Press. (Reprinted with minor corrections 1990.)

Shepard, A. O. 1956 (1980). *Ceramics for the archaeologist*. Washington D. C., Carnegie Institution of Washington. (Manuscript submitted 1954, published 1956, reprint 1980.)

Shinoto, M., Hoffbauer, S., 2000. Kagoshima ken Fukiage chô Tsujidôbaru iseki no Narikawa shiki doki no kôbutsugakuteki kenkyû. (Mineralogical studies on the Production Process of the Narikawa Pottery from the Tsujidôbaru Site in Fukiage, Kagoshima prefecture." 91-104, *Jinruishi Kenkû – (Journal of the Society of Human History) 12*.

Shinoto, M., (in prep.) *Neue Forschungen zur Kofunzeit in Südjapan. Die Siedlungskammer um Tsujidôbaru und ihre Keramik*. Arbeitstitel Dissertation Heidelberg.

Stehli, P.; Zimmermann, A. 1980. Zur Analyse neolithischer Gefässformen. 147-177, *Archäo-Physika 7*.

Tatara T. 1981. Narikawa-shiki doki no kentô. (Investigations on the Narikawa pottery.) 89-116, *Kagoshima Kôko 15*. Kagoshima.

Traunecker, C. 1981. Code Analytique de Profils de Céramique de l'Ancienne Égypte. 49-77, Arnold D. (ed.) *Studien zur altägyptischen Keramik*. Mainz.

Uno, Takao. 1999. Kofunjidai chû/kôki ni okeru shokki, chôrihô no kakushin. Ritsuryôseiteki shokki yôshiki no kakuritsu katei. (Innovations in Food Vessels and Cooking Styles in the Middl and Late Kofun Period. The Process of Establishing Food Vessel Styles of Ritsuryô Society.) 25-42, *Nihon Kôkogaku 7*.

Vossen, R. 1970. Klassifikationsprobleme und Klassifikationssysteme in der Amerikanischen Archäologie. 29-79, *Acta Praehistorica et Archaeologica 1*. Berlin.

Whallon, R. 1972. A new approach to pottery typology. 13-33, *American Antiquity 37-1*.

Whallon, R., Brown, J. A. (eds.). 1982. *Essays on Archaeological Typology. Kampsville Seminars in Archeology 1*. Evanston/ Illinois.

	1996/97	2001	Request for future research
Computer	PowerBook 5300 (48 MB/2 GB)	PowerBook G3/400 (320MB/6GB)	notebook with any processor architecture
OS	MacOS 7.5	9.04	UNIX based OS
Photographs	Nikon FM	=	digital, 48 bit, more than 7 Mega-pixels; add-on for camera control from inside the data base software
	135 mm	=	=
	55 mm Macro	=	=
Paper scans	Ricoh FS2	Heidelberg 1450	-
	Cirrus 2.0	Linoscan 6.x	-
Slide scans	Nikon LS2000*	"	-
	LaCie SilverScan 4.x*	"	-
Digital imaging	Photoshop 3.0	Photoshop 4.0	-
CAD (maps/GIS, processing of artefact data)	MiniCad 5.0	VectorWorks 8.5	CAD-add-on or data exchange module for CAD packages independent of the software
DBMS	4th Dimension 3.5	version 6.56	newest version or SQL data base
Statistics	4th Dim, Spreadsheet software		add-on for advanced analyses in the data base or dynamic data exchange

*Table 1. Equipment used at the beginning of the study, at present and requests for future development. (= same as earlier; " same as above; - not necessary; * bought after material collection in 1998).*

Figures

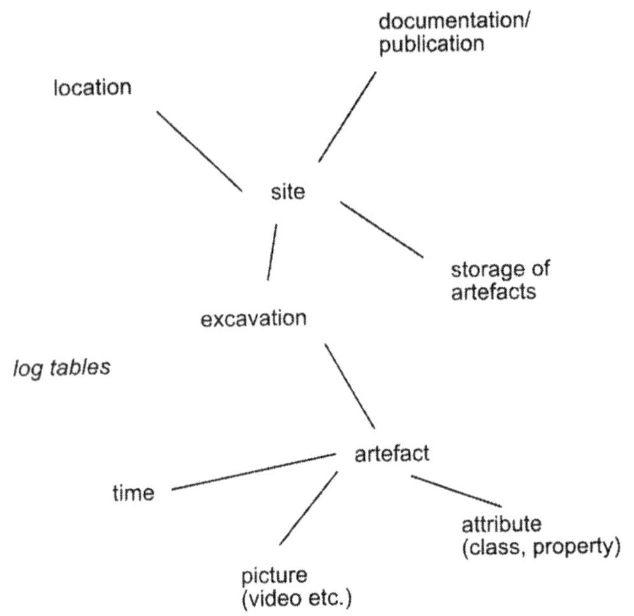

Figure 1. Overview of the data base structure. Each term represents an "area", i.e. a normalized structure of tables that store informations about the respective topic. Lines indicate a relation, where the central table of one area relates the area with the central table of another neighbouring area. At the moment, the data base consists of about 40 tables in 9 areas.

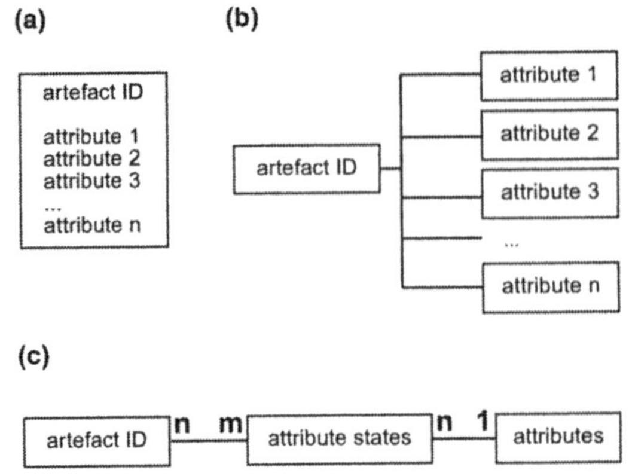

Figure 2. Examples for simple organization of artefact and attribute state relations, the starting point to develop the relation in Fig. 7. - (a) Flat file solution with individual fields for each attribute in order to record one attribute state if it is observed. (b) Each attribute is assigned to one specific table. If an artefact shows a certain attribute state, one record is created in the respective table. New attributes require an additional attribute table. (c) Relation where artefacts, attributes and attributes states are assigned to specific tables. For each artefact several attribute state records are created in the" attribute states table".

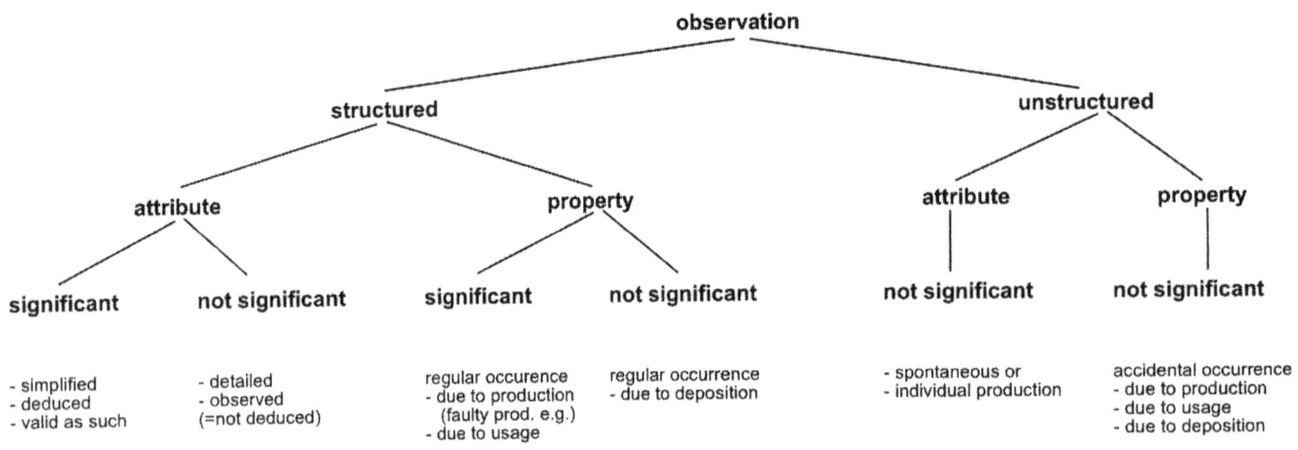

Figure 3. Classes of observations and their significance for systematic artefact classification.

Figure 4. Attribute "adherence technique" deduced from several attributes at dufferent vessel parts.

archaeology	statistics	relational database (normalization)
attribute	variable	table "attribute"
attribute state	category / value	table "attribute state"
artefact	unit	table "artefact"

Figure 5. Structural parallels between archaeological attributes, statistical variables and a normalized relation.

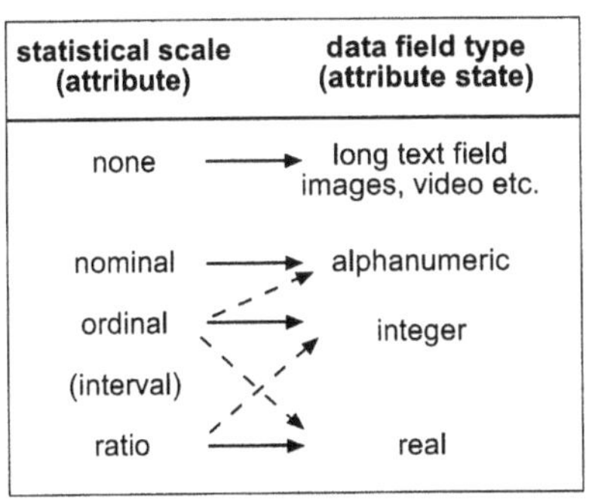

Figure 6. Assigning attributes on a certain statistical scale to data field types.

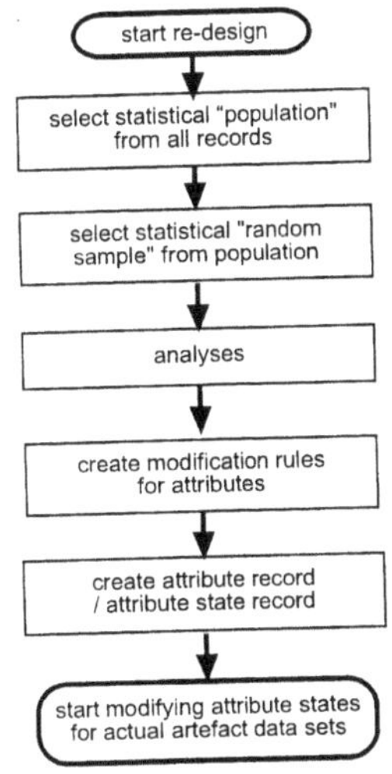

Figure 9. Sequence of steps that lead to modification of the research design based on new knowledge from new data.

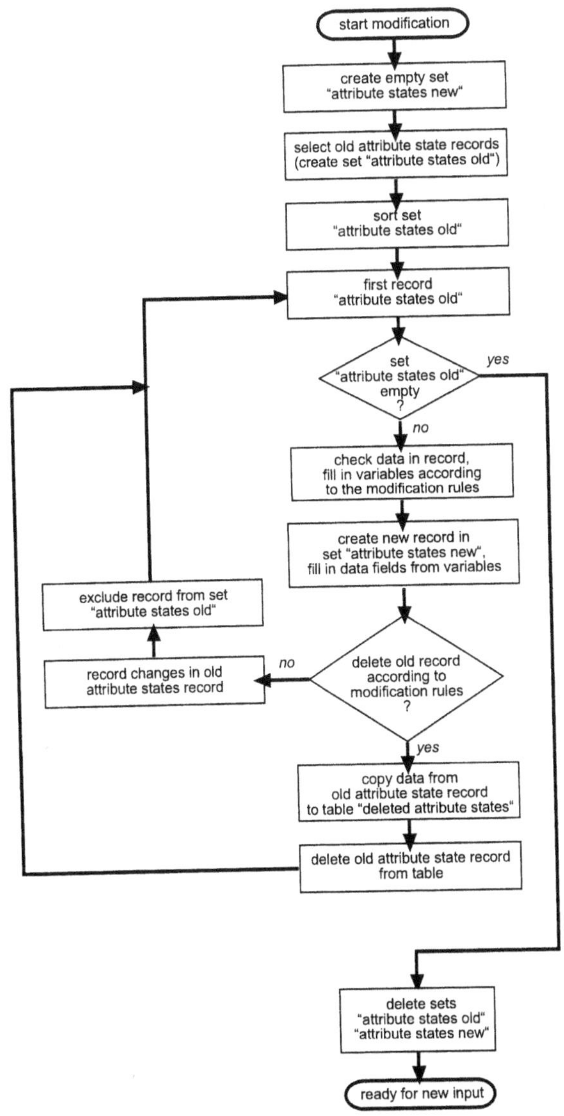

Figure 10. The process of modifying attribute states and property states of exsiting records after definition of the new attribute, property or class. Sets represent the selection. Deleting a selection of records means deleting the actual records whereas "sets" means deleting the representation without effect on the records.

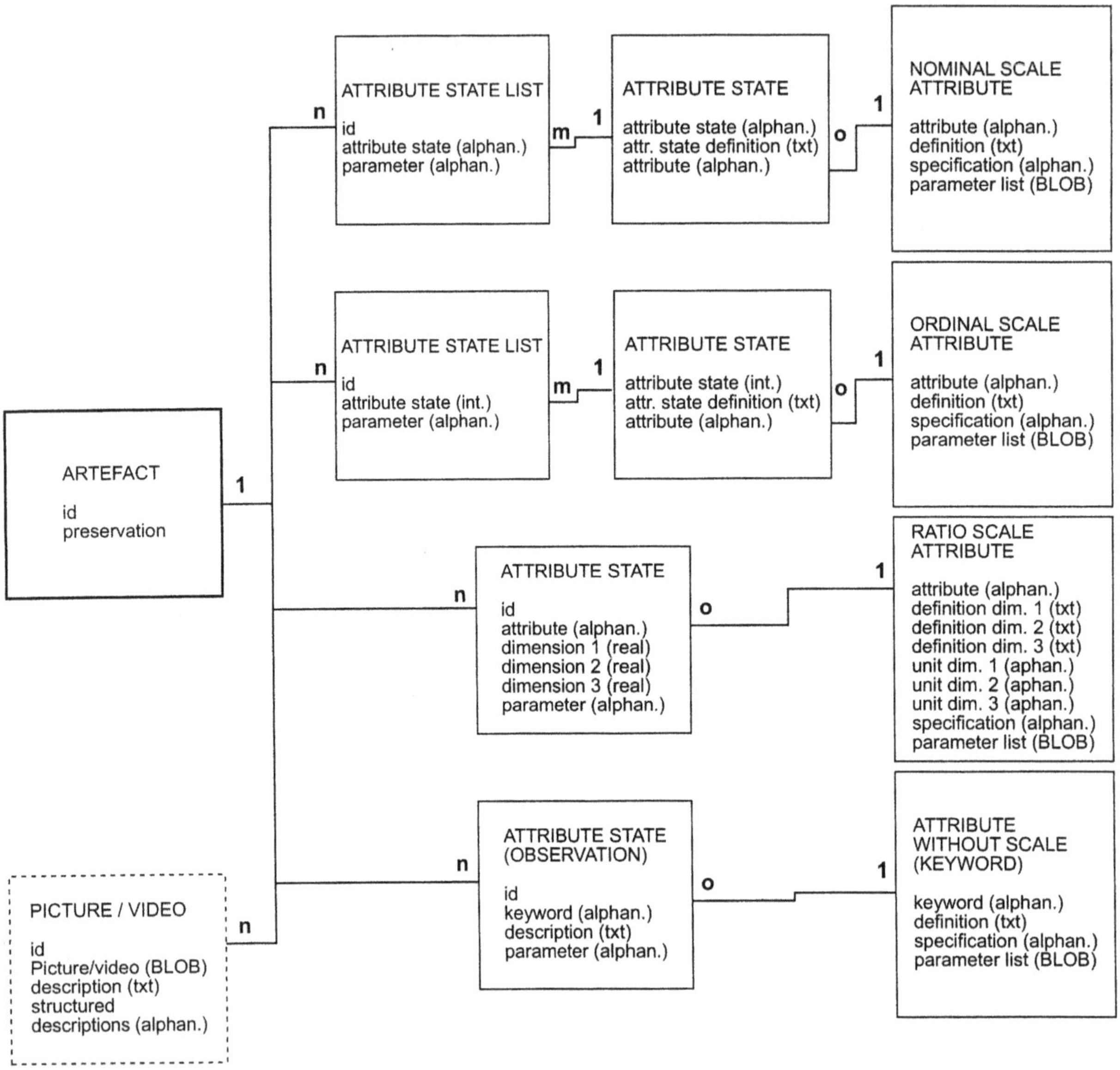

Figure 7. Tables to store observations, properties, attributes and classes (represented by the term "attribute") in the artefact area of the database. Pictures can be understood as nonverbal observations beside documentation, therefore they are included into this diagramme. – "l" and "n,m,o" represent the respective sides of a 1:n relation. The n-part of the relation is represented with different letters to show comparable levels.

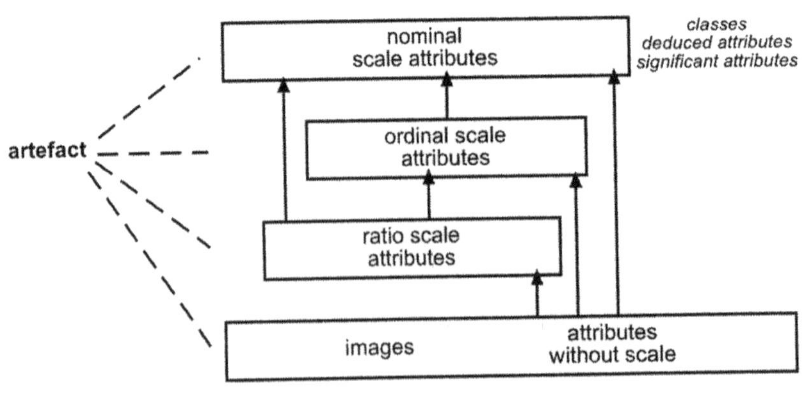

Figure 8. The sequence of scales in forming few, vague but significant attributes and classes from many precise, individual properties or attributes.

Probe	Farbe Braun	Sand	km	Menge w	n/m	e	ü	hauptsächliche Partikelgröße Kst	SK	K	M	G	SG	ÜG
112		a	a					a						
881		a	a					a						
825		a	a					a						
815		a	a					a						
229		a	a					a						
234		a	a						a					
236		a	a						a					
705		a	a						a					
843		a		a					a					
931		a		a					a					
826		a		a					a					
853	a			a					a					
1064	a			a					a					
821	a			a						a				
873	a			a							a			
332	a			a							a			
726	a			a							a			
735	a			a							a			
706		a		a							a			
709		a			a				a					
356	a				a					a				
736	a				a						a			
422	a				a						a			
592	a				a						a			
811	a				a						a			
195	a				a						a			
340	a				a						a			
331	a				a						a			
182	a				a						a			
949	a				a						a			
1098	a				a						a			
829		a			a						a			
1043		a			a						a			
832		a			a							a		
208		a				a					a			
186		a				a					a			
198		a				a					a			
371		a				a					a			
824		a				a					a			
836	a					a						a		
779		a				a						a		
385		a				a						a		
830		a				a						a		
810	a						a			a				
831		a					a				a			
797		a					a				a			
831		a					a				a			
797		a					a				a			
816		a					a						a	

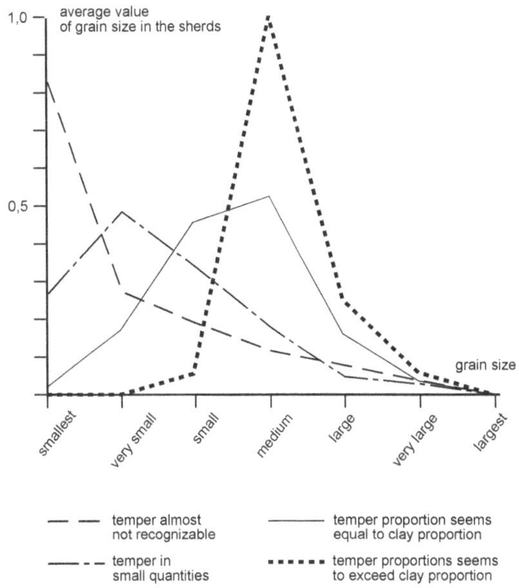

	fine (fine)	(coarse)	medium	coarse	*sum*
light yellowish brown with red	8	4	4		*16*
"dirty" brown with red		7	11	2	*20*
pale yellowish brown without red				13	*13*

Figure 11. Some analyses on random samples that led to the three fabric groups. Left: table of observations that were structured from unstructured observations; n=50, 1 sample without data. Top right: correlation of grain size distribution and temper quantitiy leading to temper groups; n=200, 7 samples without data. Bottom right: correlation between temper groups and colour groups simplified from detailed observation with colour charts; n=50, 1 sample without data.

Figure 12. Some characteristic examples of surface treatment of pottery in the case study.

Papers in the Posters and Workshops categories

Presentation of a Database for funerary Analysis and Proposals for its Increments and Developments

Francesca Fulminante
E-mail: fulminante@tiscalinet.it

Abstract: This note presents a database application for funerary analysis.It shows the database structure, some examples of the analysis which can be obtained with a such structured database and some proposals for its increment and development.

Introduction

This brief note presents a database application of funerary analysis of early Iron Age cemeteries of Latium Vetus.

The database is aimed at studying and reconstructing the evolution of late Bronze Age and early Iron Age Latin society and better understanding the well known phenomenon of Orientalizing princely tombs and their predecessors in a period which is crucial for the development of proto-urban and urban centre in middle-thyrrenian Italy[1].

This research constituted my thesis dissertation[2], and has been conducted in the tradition of the studies of the so-called Archeology of Death[3], paying a particular debt to the theoretical approaches of Processual and Post-Processual Archaeology (Contextual Archaeology)[4].

On this occasion it might not be possible to refer to the historical implications and ramifications of this study, which demonstrates the precocious rising of social differentiation in early Iron Age Latin communities, as it is intended to present the methodological approach and its computer applications.

Database structure

The database system has been created for *Windows* 95 using the *Access* program in the version of *Office* 97 *Professional*. It consists of two related data-sets:

1) The first, BURIALS, concerns the dead and their tombs. It comprises these fields: site, ID-tomb, period and dating of the context; sex and age of the dead; funerary ritual, orientation and structure of the tomb, eventual presence of *semata*, enclosures or *tumuli*; modality of deposition of the corpse or of the cremated remains, modality of deposition of the funerary equipment; spatial relationship between the tombs; at the end, fields devoted to the registration of eventual traces of cult related to the dead (objects, pottery, animal or faunal remains etc.).

2) the other, EQUIPMENT, concerns the objects accompanying the dead in the tomb and includes the following fields: pottery-class or material of the object, definition, type, miniature or normal size, local or imported production, position of the object in relation to the corpse in inhumation-tombs or in cremations which imitate inhumations.

The two related data-sets are accessible through the same data-entry mask (Figure 1).

Data analysis examples

Using *Queries* in *Access* or exporting data in *Excel* and correlating them in *Pivot* tables it is possible to present different analysis of the data:

1) diagrams showing the correlation between variables such as, for example, the age of the dead and the length of the ditch-grave (Figure 2);

2) diagrams and curves of demography (Figure 3);

3) systematic tables correlating different ritual aspects of the burials to the sex and the age of the individuals (Figure 4):
   - A) funerary ritual;
   - B) orientation of the tomb;
   - C) structure of the tomb;
   - D) presence or absence of external *semata*, enclosures or tumuli;
   - E) modality of deposition of the corpse or of the burned rests;
   - F) elements indicating ritual activity or veneration of the dead;

4) diagrams and curves of wealth which might be considered as a sign of social status (Figure 5).

Proposals for further developments of the database system

The database presented is quite simple and was elaborated to study edited tombs of Latium Vetus in the Iron Age period. However it could be improved using any of the following (Figure 6):

1) being enriched with GIS applications in order to include spatial analysis;

2) being added and related to other more analytical data-sets not only to record but also to classify archaeological finds (pottery and other equipment-objects), anthropological and faunal remains etc.;

3) being associated with sites data-sets in order to study sites related to their cemeteries and compare archaeological evidence from different contexts. It would be interesting, for example, to compare pottery and other finds from different contexts (funerary, domestic or religious) considering both the functional and the symbolic aspects of the objects;

4) evolving into a software for both studying edited burials and documenting new excavated necropolis.

End notes

[1] For the formation of urban society in middle-tyrrenian Italy see A. Carandini, *La nascita di Roma. Dei, Lari, eroi e uomini all'alba di una civiltà*, Roma 1997 and A.M. Bietti Sestieri, *The Iron Age Community of Osteria dell'Osa. A socio-political development in central thyrrenian Italy*, Cambridge 1992 with their rich bibliography. The two authors shows different points of view on the problematic subject.

[2] F. Fulminante, "*Princely tombs" in Latium Vetus between the end of early Iron age and the beginning of Orientalizing Age*, thesis dissertation in Archeology and greek-roman Art History, Faculty of Letter, University of Rome "La Sapienza", November 1998. The study was conducted under the guide of prof. Andrea Carandini and prof. Gilda Bartoloni of the University "La Sapienza" of Rome and prof. Alessandro Guidi of the University of Verona. My special gratitude is due to prof. Paolo Carafa of the University of Calabria (Cosenza).

[3] I. Morris, *Burial and Ancien Society. The rice of the Greek City-State*, Cambridge 1987 and I Morris, *Death-Ritual and Social Structure in classical Antiquity*, Cambridge 1992.

[4] I. Hodder, *Reading the past*, Cambridge 1986, and I. Hodder, W. Proucel (edd), *Contemporary Archaeology in Theory*, Oxford Cambridge 1996; with particular regard to funerary archeology a useful summary of the story of the studies in .M. Cuozzo, *Prospettive teoriche e metodologiche nell'interpretazione delle necropoli: la post-processual archaeology*, in *AION(archeol)*, n.s.3, 1996, p.1-38.

Figures

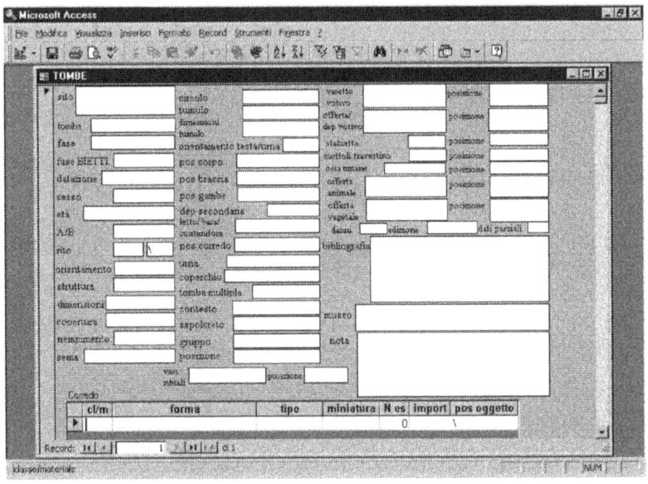

Figure 1. Data-entry mask which describes the burials with the dead and their equipment

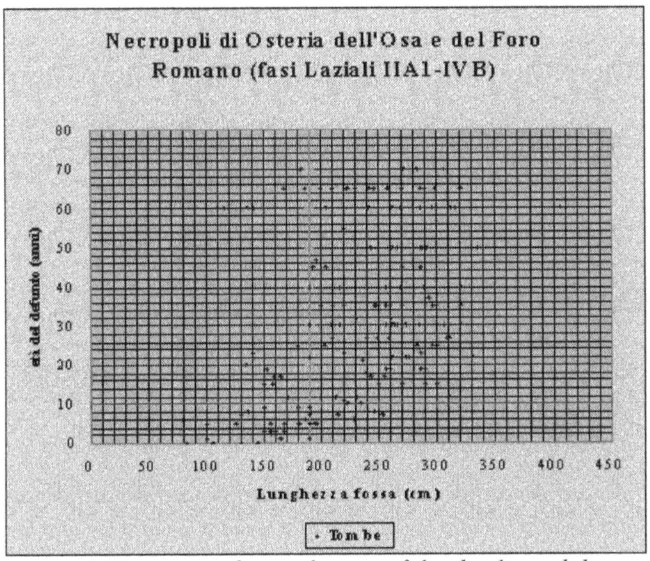

Figure 2. Diagram relating the age of the deaths and the length of the ditch-grave

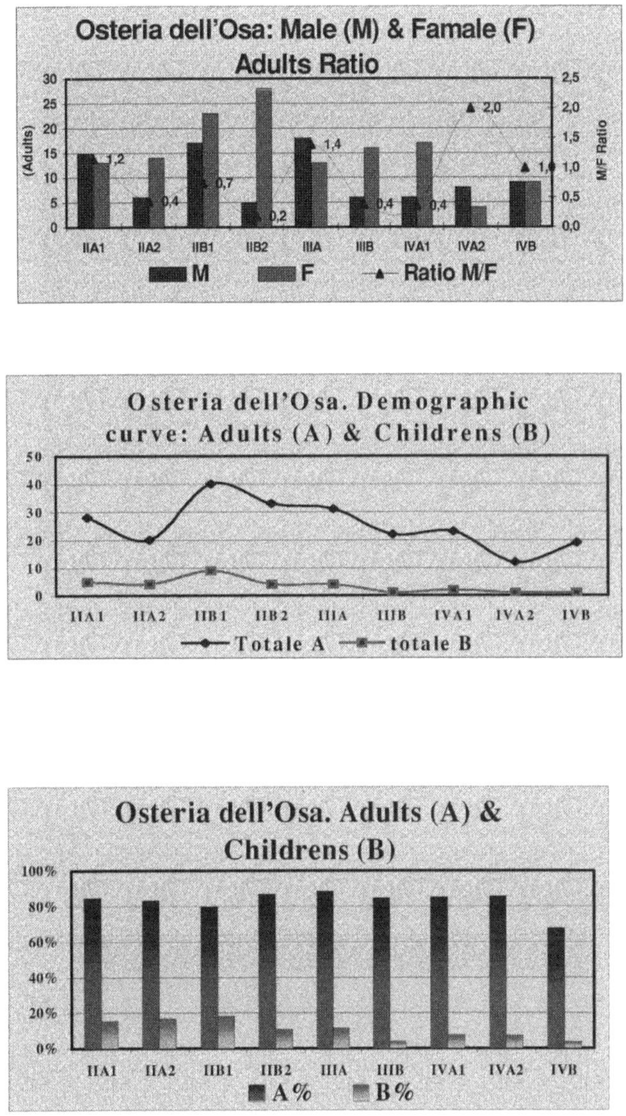

Figure 3. Examples of diagrams and curves of demography

sito	Gabii
contesto	NE
sepolcreto	Osa
fase BIETTI	IIA1

Conteggio di a/b	sesso				
	A		B		Totale
rito	F	M	F	M	
C	0%	69%	0%	0%	33%
I	100%	31%	100%	100%	67%
Totale comp	100%	100%	100%	100%	100%

LEGEND:
A = Adult B = Children
F = Female M = Male \ = Unknown
C= Cremation I= Inumation
P= Pit PR= Offset-pit
F1 = dich-grave < 2 m
F2 = dich-grave > 2 m & < 3 m
ST = filling of tufa-stones
ST-P = filling of tufa-stone along the perimeter of the grave
sarc le = wood coffin

Conteggio di tomba					a/b	sesso			
					A		B		Totale
struttura	copertura	riempimento	urna	letto/bara	F	M	F	M	
PR	ST	ST-P	\	dolio-capanna	0%	6%	0%	0%	3%
		Totale ST-P			0%	6%	0%	0%	3%
		ST	olla ovoide	dolio-cop LT	0%	6%	0%	0%	3%
			capanna	dolio-cop disco	0%	19%	0%	0%	9%
				dolio	0%	19%	0%	0%	9%
		Totale ST			0%	44%	0%	0%	21%
Totale PR					0%	50%	0%	0%	24%
P	ST	ST-P	capanna	dolio	0%	6%	0%	0%	3%
		Totale ST-P			0%	6%	0%	0%	3%
Totale P					0%	6%	0%	0%	3%
F2	ST	ST-P	\	sarc le	23%	0%	0%	0%	9%
				\	38%	19%	0%	0%	24%
		Totale ST-P			62%	19%	0%	0%	33%
		ST	\	\	38%	13%	0%	0%	21%
		Totale ST			38%	13%	0%	0%	21%
Totale F2					100%	31%	0%	0%	55%
F1	ST	ST-P	capanna	\	0%	13%	0%	0%	6%
			\	sarc le	0%	0%	50%	0%	3%
				\	0%	0%	0%	100%	6%
		Totale ST-P			0%	13%	50%	100%	15%
		ST	\	\	0%	0%	50%	0%	3%
		Totale ST			0%	0%	50%	0%	3%
Totale F1					0%	13%	100%	100%	18%
Totale complessivo					100%	100%	100%	100%	100%

Figure 4. Example of a systematic table regarding rite and tombs-structure

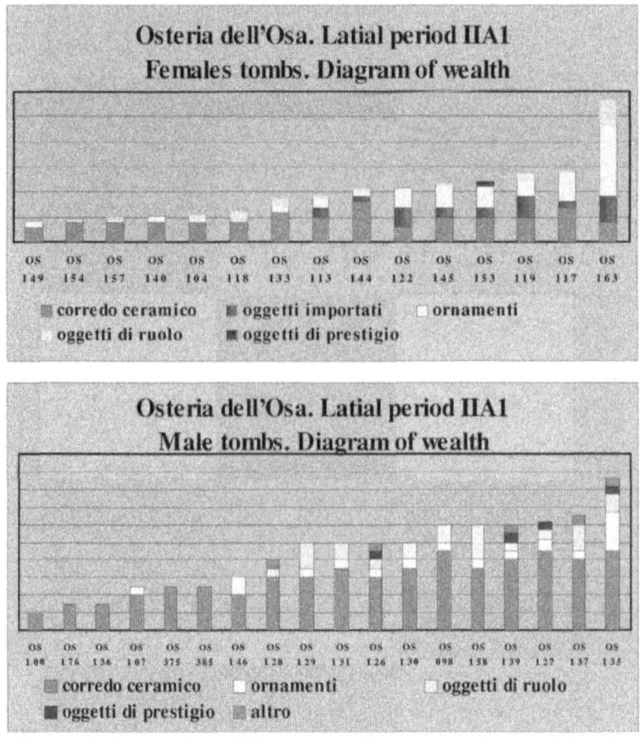

Figure 5. Examples of diagrams illustrating the wealth of the tombs

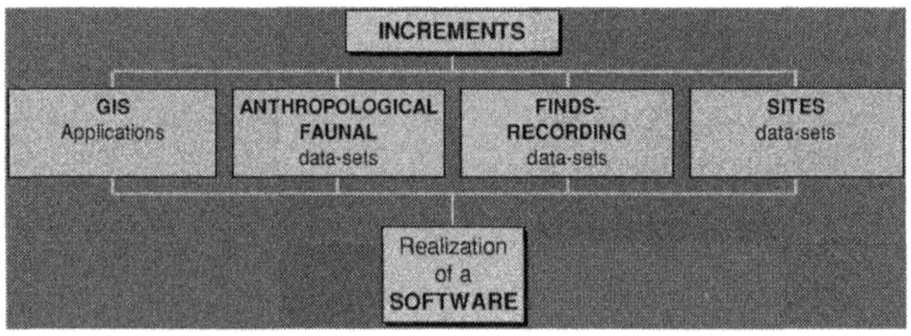

Figure 6. Schematic proposal for further developments

A GIS Application for the study on water supply and draining system in the ancient capital cities in Japan

Akihiro Kaneda
Independent Administrative Institution National Research Institute for Cultural Properties
Nara National Cultural Properties Research Institute
2-9-1, Nijo-cho, Nara-shi, 630-8577, Japan
Phone: +81-742-34-3931 - Fax: +81-742-40-4041 - E-mail: akaneda@nabunken.go.jp

Susumu Morimoto
Independent Administrative Institution National Research Institute for Cultural Properties
Nara National Cultural Properties Research Institute
2-9-1, Nijo-cho, Nara-shi, 630-8577, Japan
Phone: +81-742-34-3931 - Fax: +81-742-40-4041 - E-mail: susumu@nabunken.go.jp

Abstract: The ancient capital cities in Japan had straight streets running east to west and north to south. These streets divided the city into a checkerboard pattern of regular blocks.

There was a drainage ditch on each side of the street. In cities, these ditches were essential for the daily life. The Heijo capital was the principal capital in the eighth century AD. The Nara National Cultural Properties Research Institute started the archaeological and historical investigations into the Heijo capital in 1955. There are many data of drainage ditches at the present time. We must utilize these data to study the ancient cities.

Based on the above, this paper aims to introduce a current status of our project. At present, we are carrying out preparation of basic data and experiments with model areas.

Key words: GIS, ancient city, water supply and drainage system, Heijo capital, TIN, draining ditch.

Introduction

The ancient capital cities in Japan were formed based on the cultures of East Asia. These cities were the center of Japanese political, economic, ritual, and cultural life. Straight streets running east and west and north and south divided the city into a checkerboard pattern of regular blocks.

Such cities were constructed in the Kinki region of Japan (fig.1). Representative cities are Fujiwara capital (684-710 A.D.), Heijo capital (710-784 A.D., situated in the current city of Nara), Nagaoka capital (784-794 A.D.) and Heian capital (794 A.D.-, situated in the current city of Kyoto).

A street in any of these cities had a ditch on each side of the street. The ditches worked as a drainage through which runoff rain water and contaminated household water would flow. In cities, these drainage channels were essential for the city to function as a place for people to live in.

It is also notable that various archeological finds, such as wooden tablets, wooden implements, pottery, clay objects, roof tiles and natural features have been found in or near these ditches.

The ditches are being studied from various viewpoints. Studies include analysis of written materials by historians, examination by historical geologists, and excavation by archaeologists. The information obtained from excavations is particularly significant in that it increases our knowledge of such cities.

The enormity of excavation records has made it difficult to utilize a large amount of data obtained by means of the conventional analytical methods. It is now essential to introduce a new method that will integrate the accumulated research data.

It is possible to reconstruct a draining system based on data regarding both the plan position and the altitude from excavation. It is also possible to compare the overall system of the city that may appear based on a macroscopic perspective and the actual models that may be created based on individual examples.

To that end, a means that enables integration and analysis of the accumulated spatial information is necessary. A geographic information system (GIS) would appear to meet these needs.

Based on the above, research is currently underway with a main focus on the drainage ditches of Heijo Capital (fig.2), the most famous city during the Nara period (710-784A.D.).

The goals of the research are as follows:

1. Collect the yields from excavations conducted to date based on a uniform standard and use them as a basis for future research.

2. Build a spatial database using a GIS, and clarify the city plans and actual designs of water supply and drainage systems of ancient cities.
3. Contribute to research in various disciplines through the organization and publication of the collected data.

While the research is still underway, we would nonetheless appreciate the opportunity to present an interim report and exchange opinions with various persons.

We believe that interdisciplinary collaboration is possible with regard to this research not only with experts in archeology, history and historical geology, but also with various other scientists, including experts in information science, urban sociology, civil engineering, historical metrology and hydrology.

Direction of research

This research is being carried out in a three-pronged direction.

(1) Development of water supply and drainage system analysis system

A system that accepts data input, carries out interpolation-based generation of three-dimensional data, and analyzes water flow must be developed. AutoCADMap2000 and IDRISI32 will be used. In addition, because a TIN model needs to be created, AutoCAD Land Development Desktop will be introduced.

(2) Collection of basic data relating to the water supply and drainage system of Heijo capital

Plan and altitude data on the drainage ditch within the city will be extracted from the results of excavation. Adjustments are made to the data in order to incorporate it into the analysis system. A digital elevation model (DEM) will be prepared from the existing 1/1000 scale Heijo capital topographical map created by the National Research Institute for Cultural Properties, so that it may serve as a basic topographical map.

(3) Comparison of water supply and drainage systems in ancient cities in Japan and East Asia

Information will be collected regarding cities in East Asia, particularly Japan, Chinese Continent and Korean Peninsula, to study the well-planned nature of the ancient cities from the perspective of water supply and drainage systems.

Current status

Preparation of basic data on a drainage ditch.

153 pieces of basic data on a drainage ditch has been successfully collected from data derived from an investigation conducted by the National Research Institute for Cultural Properties. Data is being extracted from data obtained from an investigation conducted by the Nara City Board of Education. Data possessed by other organizations will be reviewed as well. The data is stored in a computer as it is collected (fig.3).

In order to analyze the entire water system in the Heijo Capital, data will be collected regarding the top and bottom edges of both banks and the deepest points of the drainage ditches extracted at given intervals. Observation of some areas of the cross-section of the stratum provide important information that provides clues regarding changes to the ditches because it sheds some light on the burying and remodeling of them.

The incorporation of such information is given high priority.

Experiments with model areas

In developing the analysis system, two areas were chosen as model areas from among the collected drainage ditches data, and the contents of the necessary data and the format thereof are being reviewed and input support modules are being developed while the model areas are used for specific input. Many of the records from excavations indicate the altitude at any given plan point, so we are considering the creation of models using TIN for input purposes.

An experimental model of SD6400, a drainage ditch located on the west side of Eastern First Avenue (Higashi Ichibo-oji), will be presented here (fig.4-5).

Based on the models, we plan to study the direction of the water flow, confluence and forking, and changes in the flow amount due to precipitation.

Preparation of DEM

The 1/1000 scale topographical map that serves as the basis for the Heijo Capital DEM contains many symbols, and some of the contour lines are missing. Therefore, contour lines are manually extracted and the missing areas are supplemented. This map is read in using a scanner, and converted into vector data after skeletonization (fig.6). It is finally converted into mesh data. This method has already been used in regard to archaeological sites in Japan. The extraction of contour lines has been completed, and we are now determining the method by which to digitize the data.

We believe that such data that has been collected and accumulated is useful not only in archaeological research but also in research and applications in various many other fields of study. The results of our research will be published in the form of books and data compilations.

In addition, we plan to advance our research regarding the development plan and the actual construction of the streets based on the data accumulated through this research.

Conclusion

This project has just begun. We would appreciate advice and guidance from various experts. Excavations of ancient cities are progressing on a daily basis, and new finds are continually being uncovered. We hope to pursue our research while trying to find a way to use these finds. We also hope that this system can comprise a system able to incorporate new excavation data and reflect them in the research.

Acknowledgements

This research constitutes entitled 'A GIS application for the study on water supply and draining system in the ancient capital cities in Japan' (subject number 11309014; representative researcher: Ikuo Tanabe; fellow researchers: Ken'ichi Kobayashi, Kazuto Inoue, Susumu Morimoto, Kenkichi Ono, Tsuyoshi Ozawa, Yoshiharu Nakajima and Akihiro Kaneda) subsidized by Grant in Aid for Scientific Research that commenced since 1999.

References

Ozawa, T., 1997. Fujiwara: The Establishment of an Ancient City. Quarterly of Archaeological Studies, vol.44-3:52-71. (Printed in Japanese)

Kaneda, A., 2000. A study of water-use systems in ancient cities by use of GIS. Papers and proceedings of the Geographic Information Systems Association, vol.9:429-432. (Printed in Japanese)

Legends

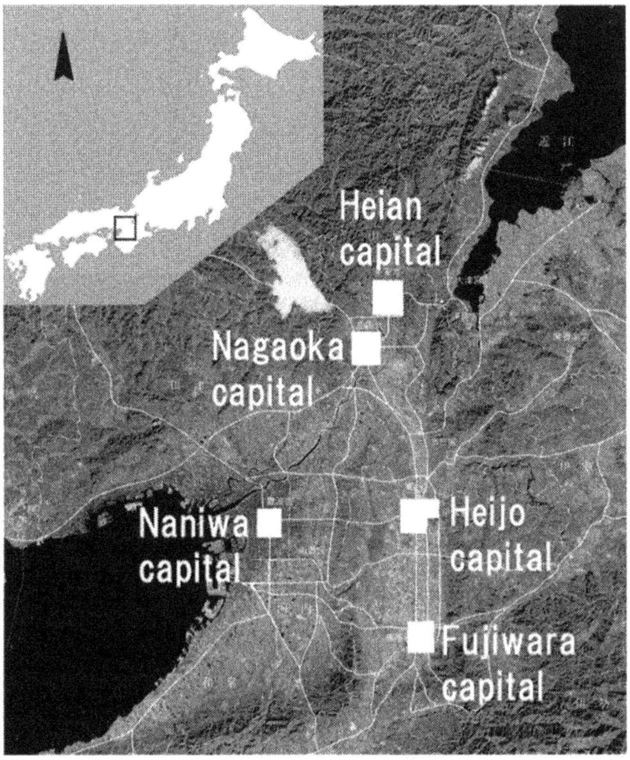

Figure 1. The ancient capital cities in Japan.

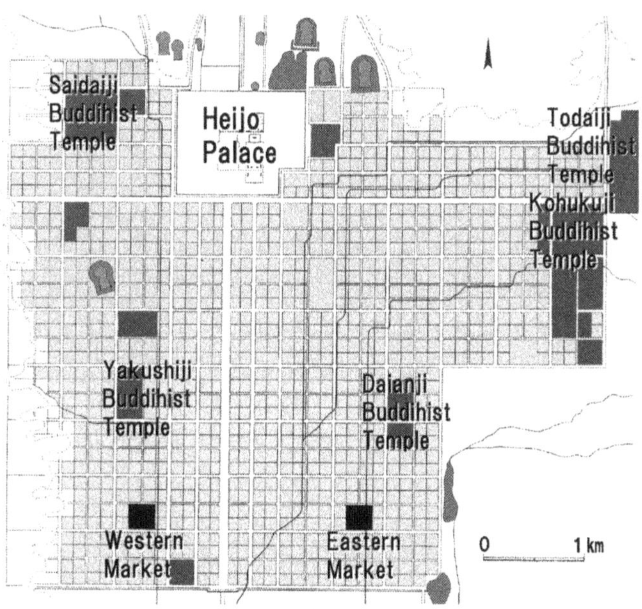

Figure 2. Plan of Heijo capital.

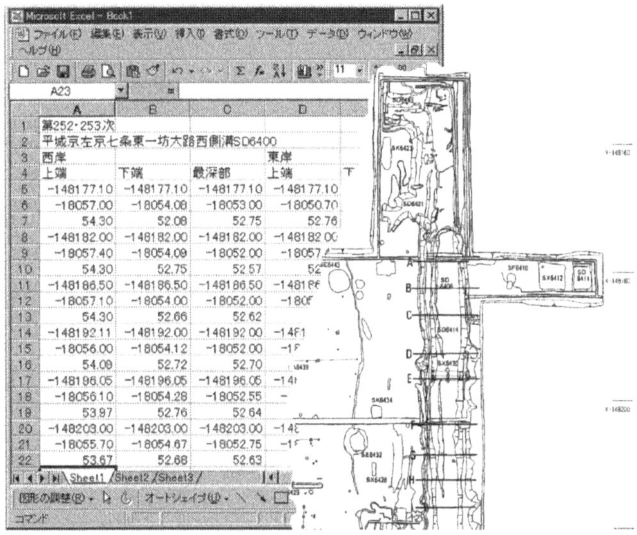

Figure 3. Collection of basic data on a draining ditch.

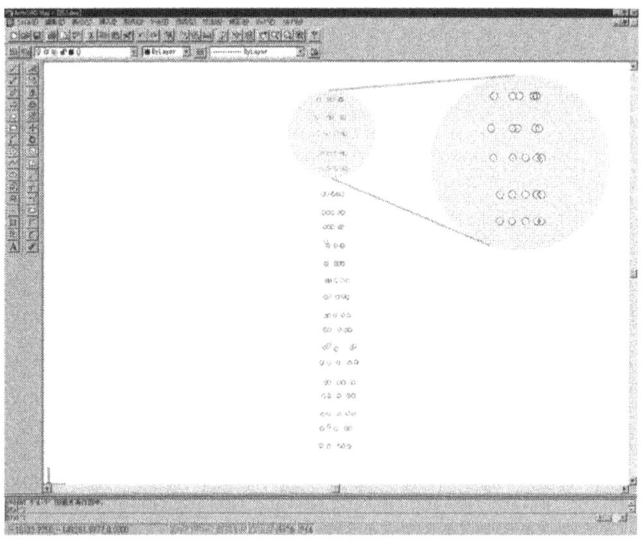

Figure 4. Showing data points as to a draining ditch.

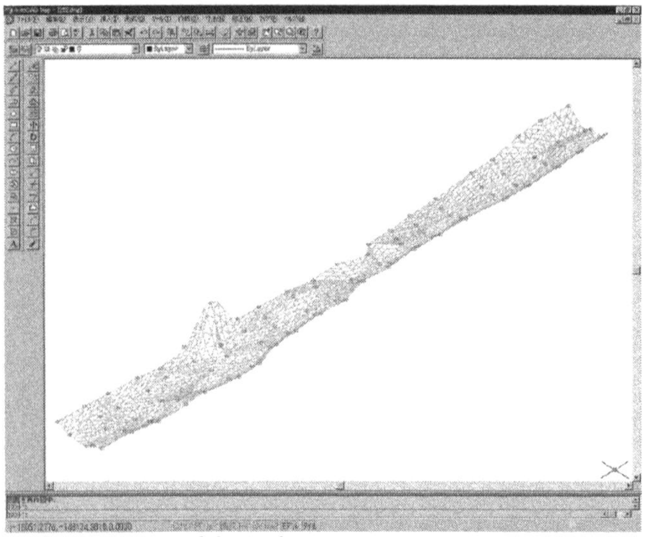

Figure 5. TIN modeling of SD6400.

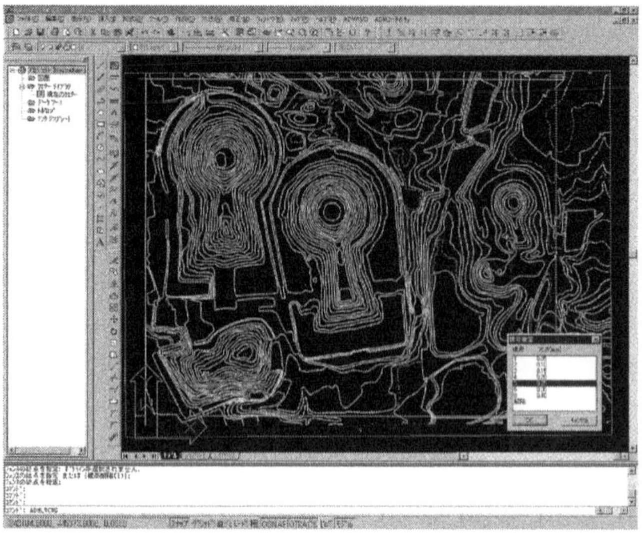

Figure 6. Preparation of DEM

Bjäre and Bowland: Computer Applications in *European Pathways to Cultural Landscapes*, a Culture 2000 programme

Jenny Nord Paulsson
Malmö Kulturmiljö
Box 406
201 24 Malmö
Sweden
Phone: +46 40 344261 - E-mail: jenny.paulsson@malmo.se

Graham Fairclough
English Heritage
23 Savile Row
London, W1S 2ET
UK
Phone: + (0) 20 7973 3124 - Fax: + (0) 20 7973 3111 - E-mail: graham.fairclough@english-heritage.org.uk

Abstract: The European Pathways to Cultural Landscapes project is funded by the European Union as part of the Culture 2000 initiative. It is a network of twelve projects in ten countries working to expand understanding, and appreciation of the European cultural landscape, and public access to it using new methods and new technology. The project is still in the first of its three years, and this paper therefore sets out intentions more than achievements. It describes two of the projects that concern very different landscapes: Bjäre in Sweden and Bowland England, where the focus is on web-based virtual pathways and historic landscape characterisation using GIS and related applications to capture subjective perceptions as well as 'objective' facts.

Key words: Cultural landscape, Historic Landscape Character, European Pathways to Cultural Landscapes, Culture 2000, GIS, QTVR

Background to the EPCL

European Pathways tothe Cultural Landscapes is a 3-year multi-national programme with EU funding through the Culture 2000 Programme. It is designed to study, understand and promote the cultural landscape of ten countries in keeping with the aims of the European Landscape Convention. Two cities in Germany are the leader partner (Albersdorf) and the administrative co-ordinator (Aschaffenburg). There are 12 separate national projects, representing most parts – but unfortunately not the far south – of Europe, from Ireland to Estonia and from Sweden to Italy (fig 1).

The background of the EPCL programme is to be found in several parallel developments. First of all the programme grew out of an earlier EU-funded project entitled European Cultural Pathways (ECL),(ECP), to which EPCL is a follow-up. ECLECP was a project with the aim to create trails and paths through archaeological areas. There were five partners. This earlier project made contacts that opened up the possibility for another European project. When ECP ended in 1999 all partners felt that a continuation and also an expansion of the co-operation were desirable. Additional partners were found, and in February 2000, the 12 partners of the new project gathered at Kilkenny in Ireland to found a permanent network that would concentrate on research and communication of cultural landscapes in order to achieve sustainable management for the future.

Another parallel development can be found in the common frustration that appeared in different organisations all through Europe that Cultural Landscapes have few workable management tools or little protection in law. The recently-published European Landscape Convention is a step forward, and a timely development that might lead us to a positive management for landscapes in the future.

The network established at the meeting in Kilkenny successfully applied for funding within the EU culture 2000 program and the EPCL programme began in December 2000. We sought Culture 2000 funding for 918,000 euros, which with matching funds was at the 1.5m ceiling for Culture 2000 Action 2 programmes. In January 2001 formal notification was received of a grant of 858,427 euros, which will be supported by c600,000 euros from national contributions raised separately by the twelve national projects.

The programme has three-fold objectives: the understanding, communication and sustainable management of cultural landscape as part of the cultural diversity of Europe, taking into account the European Landscape Convention, even though it is not yet in force. It will focus on new methods and new technology, notably GIS and internet communication. It will find common solutions across the range of projects, but part of the objective is to recognise and work within the diversity that is characteristic of Europe and that is fully represented by the 12 national projects. Their study areas include a typical range of landscape types, from coastal to high alpine regions, from dry lands to wet lands, from lowlands to highlands, from rural to urban areas.

The programme is primarily concerned with the understanding and assessment of the cultural landscape throughout Europe, and the requirements of the European Landscape Convention. It is designed to explore how to increase democratic access – intellectual, virtual and physical – to the European cultural landscape, which is the major emphasis of the Convention. It will foster exchange of information and expertise between the partners, identify best-practice in methodological terms, and explore new technologies. It will have a particular focus on the management and conservation of the cultural landscape and its contributions to local and regional diversity and identity within Europe.

The project has two levels. The European-wide core of the project, co-ordinated from Aschaffenburg, provides for seminars, staff exchanges, web-sites, and the publication of a book, leaflets and educational material. Within this there are twelve separate national projects, each with distinctive agendas, differing methods and diverse approaches but all sharing common goals and fully complementing each other (fig 1).

The 12 projects are:

Prachensko, Czech Republic
Funen, Denmark
Bowland and the Lune Valley, England
Kaali, Estonia
Untamala, Finland
Albersdorf, North Germany
Spessart, South Germany
Irish Midlands, Eire
Paneveggio, Italy
Bjare,Bjäre, Sweden
Halland, Sweden
Arfon, Wales

A common internet-site, *www.pcl-eu.de*, is currently under construction. More information about the common project as well as links to the national projects will be found there.

The national projects chose as their study areas places that are recognised cultural landscapes but that are also vulnerable, faced with further change and threats to their identity that need to be managed sustainably and carefully, and that are remote in different senses, particularly in terms of being little visited and under-funded.

This paper describes the EPCL project through two of the national projects: one of the two Swedish Projectsprojects and the English project. Because work has only recently started, and will not be completed until 2004, these sections of this paper are necessarily accounts of intentions rather than reports of achievement. They outline what is planned, and future papers at CAA will be able to report on outcomes.

The Swedish project – the Bjäre Peninsula – is located in the north-west of Skåne, the southernmost part of Sweden. The area was also participating in the former ECP project. In Sweden the concept of Historic Landscape Characterisation (see below) is still rather new, therefore the national project will try to adapt the English methodology into a Swedish context. The partners are universities, museums and associations concerned with archaeology and heritage management.

The English project – Bowland and Lune Valley – is located in Lancashire in NW England. Here it is partly a follow-up to a recently completed Historic Landscape Characterisation county-wide project (see below). As well as meeting ECPL objectives, it will therefore additionally take forward a national agenda for improving the understanding and conservation of the landscape, notably into stakeholder issues. With English Heritage and Lancashire County Council, the partners are scientific institutes and academies, universities, museums and local associations concerned with archaeology and heritage management.

Goals and frameworks

Through gaining a better knowledge of cultural landscapes and their development through time the EPCL project wishes to achieve a sustainable future management where the prehistoric and historic dimensions will be able to exist together with modern developments, and where cultural tourism could be developed. A wider understanding and appreciation of the cultural heritage may also contribute to a sense of community value and local pride - as well as a wish to care for it alongside the modern developments that are expected to increase in different senses in the near future.

One important theme that EPCL will work with is *Historic Landscape Characterisation* (HLC). The main objectives of HLC cover the following areas:
- understanding: summarising current knowledge about the historic dimension of the present-day landscape,
- public awareness: new ways of involving the public
- archaeology: helping to direct future research
- land-use planning: providing information for controlling developments and managing landscape change
- agricultural and land management: providing advice on priorities for conservation and archaeological expenditure to achieve sustainable land-use.

The EPCL programme adopts the European Landscape Convention definition of landscape, which is; "*an area, as perceived by people, whose character is the result of the action and interaction of natural and/or human factors*". The Convention applies this definition to all parts of a country's territory, to urban as well as rural areas, to everyday as well as outstanding landscapes, to degraded as well as well-preserved places.

All the different projects in the EPCL will use a similar philosophy of landscape; some projects will explore historic landscape character, partly based on the English approach to HLC (Fairclough this volume). Much of its precise methodology and definitions might have to be modified and adapted, however, to be useful in their new national contexts. GIS (*Geographical Information System*) will also be used in a variety of different ways by all projects conducting the HLC work, as well as those carrying more conventional archaeological techniques of data collection and analysis - some of the 12 projects will be principally concerned with studying past environments and landscapes.

For the purpose of characterising the present day cultural landscape (as opposed to other types of landscape-based archaeological work) the EPCL national projects are likely to follow these principles:

i) A focus on present-day not past landscape
ii) An emphasis on time rather than space as the principal attribute of cultural landscape, and on ways of capturing this within spatial computer systems
iii) An approach, facilitated by GIS, that aims to be comprehensive not selective, describing landscape as the sum of all its parts not just areas with especially high levels of archaeological site survival
iv) Reflecting the dynamic rather than static character of the landscape: the 'living landscape' concept, a recognition and acceptance (/celebration) of change;
v) Interest in pattern and process more than merely sites or monuments
vi) Using area not point data, patterning not places, landscape not sites
vii) Employing generalisation not detailed recording; thus establishing a general description of an area to highlight its own character and (more importantly for EPCL) commonality as well as diversity at European level.
viii) Recording perception (leaning on the Convention's phrase "... as perceived by people ...") and recognising that interpretation not record, ideas not facts comprise landscape, which is seen as an idea not a thing. Even though it consists of physical objects, landscape is constructed by our minds and emotions from a combination of interpretation, perception and association. An important EPCL goal is to discover methods of collecting public perceptions and of capturing and understanding them through computer applications.
ix) Working in desk-based ways, with limited new field work, bringing information that already exists into new perceptions at landscape scale.
x) Treating the work as a process, with provisional rather than definitive results, provoking as many questions as answers: all historic landscape characterisation is provisional.

Some of these principles fit straightforwardly into GIS and computer applications – (v) and (vi) for example. Others will require innovation in GIS (eg the emphasis on time and change – (ii) and (iv)); others again present real challenges, such as how to capture perception under (viii) and the uncertainty inherent in (x).

The Bronze Age Landscape of the Bjäre Peninsula

The Bjäre peninsula is situated in the north-west of Skåne, the southernmost county of Sweden, and has a well preserved prehistoric heritage, mainly consisting of Bronze Age graves and places with cup-marks.Bjäre is a rather small area, the peninsula measures about 10 x 10 km, and here more than 500 mounds from the BronzeAge are known and almost as many stone-settings from the late Bronze Age and early Iron Age, as well as about 300 places with cup-marks. Peculiarly enough there are almost no visible traces from other prehistoric periods and most of the known settlement sites consist of undated scatters of flint in the surface.

The landscape of today in the area consists of a rather open landscape with fields for grazing as well as arable fields. The predominant boundary featuresin this landscape are the stonewalls that were made in connection with the agricultural shifts according to the Land Reform law of 1807. The overall picture of the landscape of today is much affected by the shifts, since they implied a comprehensive redistribution of farmland. The principle behind them were that small patches of land should be put together to form larger and more efficient fields. The old common grazing land was also divided and put under the plough. The layers of visible remains of human activity in the landscape may then be summarised as a well-preserved funeral landscape from the Bronze Age period around which the farming landscape of later periods has developed.

In 1997, through the non-profit association Bronstid, the area became the leading partner in ECP, *European Cultural Paths*. The objectives in this project was to make the cultural heritage in the landscape more accessible as well as to gain more knowledge about it. This was the starting point of a great deal of work that still is going on in the area. The partners in the new EPCL project of the Bjäre peninsula are the University of Lund, the agricultural University of Alnarp, the Regional Museum in Kristianstad and the association Bronstid, organisations that also represent the structure of the whole project : Research, Understanding, Communication and Management.

Understanding

Historic Landscape Characterisation will be a considerable part of the research in the Bjäre project. The above mentioned principles will be borne in mind but also some effort will be made to be able to reconstruct the cultural landscape of the Bronze Age period which stands out so predominantly in the cultural heritage of the area. The creation of HLC maps within a GIS system that will take into account landscape change, time of change, sort of change, and of course a characterisation of the present day landscape. The HLC of Bjäre will be conducted as a pilot project (together with the slightly different Halland project) and a big effort will be put in to modify the English HLC method so that it could work in a Swedish context.

One of the national themes for the Bjäre project is to take account of present day vegetation. We will take a look at places where vegetation from a well managed grazing land typical from the time before the shifts is still preserved. The mounds still have this kind of vegetation and one check with a pollen analyses from the filling of a mound and its former ground level below it shows that the vegetation today is not very different from the past vegetation. In the project another five pollen analysis in mounds are planned to confirm this. This means that the vegetation is considered not only as rare species but in fact also as a common cultural heritage which gives the landscape a vivid history.

In the work to gather the known information about the landscape we will use several computer applications. For theThe HLC and the connected database ArcView and Access will be used. Into the database allwill be created in *ArcView* and *Access*. All available information will be put into the database in different tables that will be connected to different layers in ArcView (see table 1). In this system all available information shouldabout the area will be put; for example prehistoric heritage, vegetation, HLC and historical maps and amaps. Table II shows the first outline of the HLC table. One main goal is that the municipalitywill use it for planning as well as developersearly in their planning work, andalso the agricultural authorities of course. And will be able to use this material early in their planning work. The database will of course also be very useful for education and research the material will be very research. One of many important tasks is to explore the possibilities in the GIS-based HLC for useful.

The database outlinepredicting models of archaeological site distribution, location and survival in relation to later landscape history.

Communication

Communication is an important as well as very difficult within the EPCL. When the database is ready to be used, it is expected to be available through different modes.mediums. Through public computers at the library for example as well as on a CDRomCD-Roms and on the Internet, - but then maybe in a slightly lighter version. Communication also means a two-way dialogue,it is not only for the project to give information from gatheringwill not only be a provider of information in different archives and sources as well as from the research made,- we also need the public to giveus information to the project about their cultural landscape. In the landscape convention, as mentioned above, the definition of "landscape" is; *an area, as perceived by people, whose character is the result of the action and interaction of natural and/or human factors,* this means that we need to take into account the local appreciation of the landscape and an increased democratic process as well as access is definitely needed. To be able to achieve this the Internet as well as public computers will be used, also seminars, meetings and local projects are being planned.

One of the outputs in the project is a Virtualvirtual pathway. In the former ECP project two physical trails were created in the area, which is why in EPCL we would like to concentrate on virtual pathways. The work with this pathway has started already and will be in the form of a photographic tour through the landscape by the means of *Quick Time Virtual Reality* (QTVR). By using a program called VR Worx we can make 360 degrees panoramas which can be connected to one another as well as to maps and still photos through hotspots. The QTVR is a very vivid way to show a landscape both through the internet and CD-Roms. It is also very useful as a documentary piece, as well as it is a way to bring the landscape into the office when dealing with for example landscape archaeology.

The local non-profit association Bronstid have made a reconstruction of a Bronze Age house, called *Bovigården* (the Bovi farm), around this a Bronze Age centre is being built-up and it is already used as a place where schoolchildren are having their pre-historic education in a very vivid way. Bovigården will play an important part in the local communication.

Management

To achieve a sustainable future management of the cultural landscape, you need above all interest, knowledge and communication. Throughthe EPCL programme we believe that we can achieve a better situation in the future when it comes to these things.

In all the EPCL's goals for the Bjäre project are as follows. Through gaining a better knowledge of the cultural landscape and its development through time the EPCL project wishes to achieve a sustainable future management where the prehistoric and historic dimensions will be able to exist together with modern developments,developments. Already today, because of the former ECP project as well as the new-started EPCL project, we can notice that the communication with the local planning authorities has improved which is very promising for the future. For this purpose the GIS based material will be and where cultural tourism could be developed. A wider understanding and appreciation of the cultural heritage may also contribute to a sense of community value and local pride, -as well as a wish to care for it alongside the modern developments that are expected to increase in different senses in the near future.
vital.

Products

Beside the GIS based material and the Virtual Pathway the Bjäre project will also make contributions to the Overall Programmes Book, produce local leaflets and a book about the local cultural landscape and its development. The outcome of the HLC methodology in a Swedish context will be evaluated in a method report.

The Cultural Landscape of the Forest of Bowland and the Lune Valley

The Bowland/Lune Valley project covers an area of North Lancashire (NW England) of nearly 1000 sq km. It unites two very different but historically closely connected areas. The Lower Lune is a wide-floored valley with evidence for long occupation and successive land-use from pre-Roman Iron Age to the modern day; Roman forts, medieval villages, and several layers of field systems are the most visible remains. The Bowland Forest is essentially an un-enclosed upland mass of grit-stone moor-land, with medieval and later dispersed upland settlements, still occupied, focussed around deep valleys containing small market centres. There is a long history of successive intake and enclosure at the moor's edge. A rich prehistoric resource lies largely hidden and unknown beneath blanket peat; in the medieval period, the area supported large vaccaries – ranch-like estates – but the post-medieval landscape is one of small farms punctuated by large high status houses and estates. Everywhere, however, place-names of pre-Roman, anglo-saxon and Scandinavian types signpost earlier phases of landscape history, and there is industry: stone quarries, lime kilns and textiles mills exploiting the fast flowing streams.

One of the interesting aspects of the project is that there are large urban populations very close to this remote rural area on its west and south margins. Their use of Bowland as a weekend recreational area is therefore a significant aspect of its cultural significance, and one which the EPCL project hopes to consider and map.

The whole area is included within one of the county-wide historic landscape characterisation (HLC) projects that since 1994 have been promoted by English Heritage, the national agency concerned with the management of the historic environment (Fairclough, this volume, notably fig 1 and table 3). HLC is a GIS map-based technique designed to produce a generalised understanding of the historic and archaeological dimension of the present-day landscape, and one major aim of the Bowland and Lune project is to explore the existing Lancashire HLC in more detail at a local level, examining the flexibility of GIS to absorb more and more local and diverse perceptions. The area has traditionally been under-studied by archaeologists, especially in relation to landscape, and its cultural landscape is under-appreciated both publicly and amongst archaeologists. It is an area under threat of change that requires better-informed and more effective approaches to management.

HLC serves a variety of uses, such as education, research, land-management and CAP-linked agricultural environmental incentive schemes, spatial planning and environmental impact assessment. All these have a place in the ECPL project. In particular, the new project will look at better ways of making GIS information easily available and accessible to a wide range of different users.

The principal outputs of the Bowland /Lune project will be:

- intellectual access to the landscape through a virtual trail using new technology similar to that described above in relation to Bjäre, and detailed GIS mapping of the landscape, developed from the existing HLC and drawing in many other data sets, and to be made available through libraries, the web and CDRoms;
- two physical trails through the cultural landscape, with sign-posting, information boards and guide leaflets; and
- conservation and management guidance for use by local government spatial planners, land managers and the general public; this will be underpinned by the GIS data and the results of public opinion surveys and interviews with farmers; school contacts participation and liaison; in particular the project will look for ways to draw public perceptions into the GIS database.

There are three main stages in the project, corresponding to EPCL's main strands:

Stage 1: Understanding

Collecting together known information and extending knowledge about historic landscape, by

a) Data collection (through consultation with experts, trawling data sets such as the SMR and defining and mapping sites, historic landscape character and place-names; sub-dividing / reworking HLC Types, detailed mapping of sample areas on GIS)
b) Fieldwork (including rapid assessment, validating HL Types, public opinion and participation surveys)
c) Analysis and evaluation (including measurement against criteria such as visibility, potential, sensitivity, opportunity)

Stage 2: Communication

Transmitting information and ideas in various forms to public audiences, by

a) creating two physical pathways through the landscape
b) creating at least one web-based virtual pathway / portal into the landscape
c) other publicity (leaflets etc)
d) participation in public meetings

Stage 3: Management

Using the results of GIS analysis and public participation to produce

a) landscape management guidelines
b) Site and setting management prescriptions
c) Integration other targets (eg agri-environmental grants)
d) Supplementary Planning Guidance
e) Sites and Monuments Record enhancement

The project will produce reports of several types :

i) contributions to the Overall Programmes Book
ii) local leaflets and book
iii) a method report, and technical guide to the GIS databases produced
iv) an academic journal paper on methodology etc

The project will address many broad themes that have relevance at a local, national and international level:

- the accuracy and applicability of HLC methodology, and its relationship to other, mainly site-based, types of data; it will explore ways of using GIS to convert point data and distributions into generalised area data. A major aspect will be the relationship between broad historic landscape character on HLC and individual sites and monuments, and in particular how to use GIS-based HLC for predictive modelling of archaeological site distribution, location and survival in relation to later landscape history.
- the character of historic dispersed settlement, and the relationship between settlement pattern, historic land use, natural resources and historic landscape character; again this will necessitate novel applications of GIS data capture and handling to extend HLC methodology.
- the assessment of time-depth and evidence for change (and continuity) in the landscape), and using HLC and the SMR to define visibility, sensitivity and perception in today's landscape
- defining the rarity and vulnerability of historic assets, and historic components within the landscape, leading to the establishment of management priorities and their integration with other programmes of work. It should be possible to find new ways of using GIS to deliver

- management plans in interactive form, with feedback mechanisms to allow the results of action, and the effect of development, to update strategies.
- promoting access to cultural landscape through GIS and Internet applications, and engaging local communities, including seeking reactions and changes to expert-led assessments and characterisations. In the long term, if it is successful, this will probably be the most significant outcome of EPCL.

Conclusion

This paper has discussed only two of the twelve national projects within the Culture 2000 European Pathways to Cultural Landscapes programme, but it has emphasised both their commonality and their differences. EPCL's main concern is to understand, to communicate and to find ways of managing the very different cultural landscapes that are represented in the 12 national projects within the whole programme that are spread across ten countries. Work has only just started this year and will continue until 2004, and further similarities and differences will doubtless emerge from the national and regional cultural differences across the EPCL network. In this work, GIS-based Historic Landscape Characterisation (HLC) as used in England but modified to suit different landscape types, will play an important part, but each of the slightly different national projects will also have their own special themes and questions to be answered.

Bjäre and Bowland, the two national projects presented in this paper, are very dissimilar areas, with distinctive archaeologies and histories, and quite different landscapes. Yet (like all twelve areas that will be studied during ECPL) the two areas share a 'European-ness' that is defined by overall similarity as well as by national and regional differences. Both the people who modified their environments in the past, thus bequeathing to us our landscapes, and the people today who 'create' landscape by understanding, perceiving and recognising it, all shared a common European culture heritage and common human responses. So much of these considerations are intangible and subjective, however, even when they are based on the material remains of the past, and it is how to use computer applications for such areas of understanding and perception that is one of ECPL's biggest interests.

Tables

Access tables	ArcView layers
1. HLC	1. HLC (polygon)
2. Vegetation inventory	2. Vegetation (polygon)
3. Site and Monuments record (pol)	3. Site and Monuments record (pol)
4. Site and Monuments record (dots)	4. Site and Monuments record (dots)
5. Site and Monuments record (lines)	5. Site and Monuments record (lines)
6. Inventory of flora	6. Inventory of flora (dots)
7. Rare and threatened species	7. Rare and threatened species (dots)
8. Areas with protection	8. Areas with protection (polygon)

Table 1. The first database outline

Polygon	*Code*	*Area*	*Dating*	*Interpretation*	*Validity*	*Comment*	*Date*	*Signature*
Unique number	Change	Place	Time of change	Type of landscape	...of the characterisation	Metadata		

Table 2. A first trial edition of the HLC table

European Pathways to Cultural Landscapes

Figures

Figure 1. Map showing location of Bjäre and Bowland and the other ten EPCL projects.

Ancient Mantineia's Defence Network Reconsidered Through a GIS Approach

S. Topouzi
Laboratory of Geophysical - Satellite Remote Sensing & Archaeo-environment
Institute of Mediterranean Studies – Foundation of Research & Technology (F.O.R.T.H.)
Melissinou & Nik. Foka 130
PO. Box 119
GR-74100 Rethymnon, Crete, Greece
Phone: +30831-56627 - Fax: +30831-25810 - E-mail: topouzi@students.phl.uoc.gr, or topouzi@yahoo.gr

A. Sarris
Laboratory of Geophysical - Satellite Remote Sensing & Archaeo-environment
Institute of Mediterranean Studies – Foundation of Research & Technology (F.O.R.T.H.)
Melissinou & Nik. Foka 130
PO. Box 119
GR-74100 Rethymnon, Crete, Greece
Phone: +30831-56627 - Fax: +30831-25810 - E-mail: asaris@ret.forthnet.gr

Y. Pikoulas
Department of History
Archaeology and Social Anthropology
University of Thessaly
Argonafton & Filellinon str.
GR- 38221 Volos, Greece
Phone: +30421-74000 - E-mail: ypikoul@eie.gr

S. Mertikas
Laboratory of Geodesy & Geomatics Engineering
Department of Mineral Resources Engineering
Technical University of Crete
GR-73100 Chania, Crete, Greece
Phone: +30821-37429 - Fax: +30821-37476 - E-mail: mertikas@mred.tuc.gr

X. Frantzis
Laboratory of Geodesy & Geomatics Engineering
Department of Mineral Resources Engineering
Technical University of Crete
GR-73100 Chania, Crete, Greece
Phone: +30821-37429 - Fax: +30821-37476 - E-mail: frantzis@mred.tuc.gr

A. Giourou
Laboratory of Geophysical - Satellite Remote Sensing & Archaeo-environment
Institute of Mediterranean Studies – Foundation of Research & Technology (F.O.R.T.H.)
Melissinou & Nik. Foka 130
PO. Box 119
GR-74100 Rethymnon, Crete, Greece
Phone: +30831-56627 - Fax: +30831-25810 - E-mail: giourou@students.phl.uoc.gr

Abstract: Ancient Mantineia, located in the centre of Peloponnese, Greece, consists of an elliptical fortification wall, almost 4 km long containing 10 gates and 126 towers. Since the 19th century, a number of watchtowers have been located at the surrounding slopes of the Mantineian plain and the existence of a defensive network has been suggested.

GIS analysis has been used specifically for studying the defensive network of the ancient settlement and its potential coverage. The gates of the fortification walls and the watchtowers were mapped through high precision GPS units. These data were superimposed on a DEM created by the digitization of 20m elevation lines covering a region of 1000 Km^2. Geological maps, aerial and SPOT satellite imagery were also incorporated to the GIS analysis.

Viewshed analysis was performed in order to determine the extent and degree of intervisibility between the watchtowers situated in the nearby hills and mountains and the gates of the city walls. The data were correlated to the results of previous surface surveys that had located ancient roads and paths. Cumulative viewshed analysis was also employed to suggest the number of the defence constructions that were able to control a specific area. In this way, a thematic map of the potential defensive region has been created, suggesting either possible gaps or candidate archaeological sites that may have been used to complete the defence system of the region. The results of the analysis were superimposed on a SPOT Panchromatic image draped on the 3D DEM. Finally, a fly-over simulation in the area of interest was produced serving as a visual verification of the above analysis.

Keywords: Mantineia, defence network, GIS, GPS, viewshed analysis.

Introduction – Aim of the study

The technological advancement of computers has proved to be of great help in Archaeology. New methods and technologies improve and upgrade the traditional approaches in archaeological research. Upgraded computer systems have been systematically used for image processing, creation of databases, multi-variable analyses of archaeological and environmental data, three-dimensional modeling and virtual reality.

During the last years, Geographic Information Systems, designed to collect, store and analyze objects and phenomena with a geographic variable, have been widely used in archaeological applications. Such an application is the study of the Ancient Mantineia's defence network, in Arcadia, Peloponnese, Greece (Topouzi, et al 2000).

Ancient Mantineia, located in the center of Peloponnese, Greece, consists of an elliptical fortification wall, almost 4 km long, containing 10 gates and 126 towers. Since the 19th century, a number of watchtowers have been located at the surrounding slopes of the Mantineian plain (Loring 1895; Fougères 1898; Lattermann H., 1913; Frazer 1965) and the existence of a defensive network has been suggested (Pikoulas 1990-1; 1995).

The term "network" suggests a carefully planned defensive system, consisting of watchtowers with intervisibility between one another. Such intervisibility must also exist between the watchtowers and the city's fortification walls. Previous studies have not been able to identify the effectiveness and extent of this defensive network. Performing a viewshed analysis based on the digital elevation model of the area was considered as the appropriate method to provide some complementary information regarding the type and structure of this network.

A number of other analyses, such as slope and aspect, were performed in addition to the viewshed analysis. In the future, the corresponding results could be used in order to create a predictive model, capable of suggesting new, archaeological sites, unknown until now.

The study area – The defensive structures

The ancient city of Mantineia was located within three valleys of central Arcadia, Peloponnese, and was surrounded by mountains. The elevation varies from 624m to 640m.

Habitation in the area begun during the Early Helladic period. During the early historical years, Mantineia was divided in 5 demes (Strabo, VIII 3.2), the names of whichare not known. The classical city bears two chronological phases, one before 385 B.C., when it was destroyed by Agissipolis, king of Sparta, and one after 370 B.C., when it was reconstructed. The remnants of the city's fortification walls belong to the second phase. They consist of a stone foundation, built according to the irregular polygonal masonry system, over which a brick wall, now destroyed, was raised. The walls' shape is elliptical, having a perimeter of 3942m. The N-S axis measures 1349m and the E-W axis measures 1087m.

Almost equally spaced towers and 10 gates (fig.1). From each gate a road started, connecting Mantineia with its neighboring cities.

Five watchtowers (fig.1) and a small fortress lie at the surrounding slopes. The watchtowers are similar to one another, built from local limestone according to the irregular polygonal masonry system. Their dimensions are almost 5´5m and it has been suggested that they should have been about 8m in height (Pikoulas 1990 – 1). The small fortress of the town Nestani is built of the same material and masonry system. All of them control important road passes.

Methodology of the GIS construction

The methodology used to create the specific GIS project consisted of the following steps : Digitization, GPS positioning, Digital Elevation Model Construction, Remote Sensing, Georeference and Database Construction.

Digitization

The 20m elevation contours, datum points, roads, rivers, wells, modern settlements and the railroad network were digitized from 84 topographic maps of the Greek Army Geographic Service (scale 1:5000). The Root Mean Square (RMS) error was kept under 0.005, providing a very good accuracy in the digital transformation of the maps. The total area covered by the topographic maps was nearly 1000 km^2. In addition two geological maps (scale 1: 50000) were digitized and a database containing the geological formations was constructed. A geological formation (attribute), represented by a different color and pattern, was assigned to every closed polygon.

GPS positioning

All the defensive constructions were accurately positioned using two sub-cm accuracy Global Positioning System units. During the five days of fieldwork, static relative geodetic positioning was used for mapping the architectural monuments of the region. The technique involved the placement of a continuous tracking receiver at a fixed and known position and the use of a second rover receiver for measuring at the sites of interest. The coordinates of the base receiver were determined with high accuracy by measuring to GPS satellites for a period of one hour at a nearby army datum point. The data from the roving receiver were corrected by comparing computed pseudoranges of the base station with the measured pseudoranges from the rover receiver.

The coordinates of the four corners of each watchtower were determined (fig.2). One point was chosen for each gate of the city's walls in order to position them on the walls' digitized plan (fig.3). A total of 38 points were established, including the army datum points. Processing of the GPS data was conducted daily, so as to evaluate the coordinates and their accuracy. Point coordinates were determined in both the satellite reference system (i.e. the World Geodetic Reference System-84, WGS'84) and the national system of EGSA'87 (Greek Geodetic System 1987) used by the National Cadastre of Greece. After completing the processing, the coordinates were exported and stored in a spreadsheet (Excel format) in order to import and use them in the database of the Geographic Information System. The accuracy of the measurements was kept below 10 centimeters.

Digital Elevation Model Construction

The digital elevation model (DEM) of the area was created in TNTmips software, using the 20m contours combined with digitized topographic points. The processing method used was that of the Minimum Curvature, a method with the optimum result when combining contours and points. The vector format elevation values produced a 60m pixel size DEM. The two-dimensional and three-dimensional DEM (fig. 4) were further enhanced through the processing of the histogram.

Remote Sensing

One SPOT 4 panchromatic and four multi-spectral (XS) images were also used to create pseudochromatic images of the region (fig.5). Supervised image classification techniques were applied in an effort to identify areas of archaeological interest, together with other environmental characteristics, with unsatisfactory results. In addition, the panchromatic image has been used as a background layer, over which all other data have been superimposed. Besides satellite images, forty-eight air photos (scale 1: 15000) taken in 1980, have been used to create a photo mosaic (fig.6).

Georeference

Since the aforementioned data (GPS coordinates, satellite images, aerial photo and topographic maps) were in different reference systems, it was necessary to proceed with the registration of them to a common system. The Greek Geodetic Reference System 1987 (EGSA '87 was chosen for the specific purpose. A polynomial transformation was used for transforming the topographic maps from HATT (a previously used reference system in Greece) to EGSA '87. Afterwards, the transformed road cross-sections were used in order to georeference the satellite images, the aerial photos and the geological map.

Database construction

Microsoft Access was used for the design and the construction of the database to be used for the registration, storing and searching of the archaeological information. It consisted of 12 tables which included various types of information, such as the name and the type of the sites, chronology, information regarding excavations and surveys in the area, photos, plans and bibliography. The database was later exported in DBF format to be implemented in the GIS analysis.

GIS analysis and results

After completing the construction of the GIS it was possible to perform a number of analyses in order to determine the existence of a defensive network in the Mantineia area.

Viewshed analysis

Viewshed analysis was considered as the appropriate method to provide some complementary information regarding the level and structure of the defense network. As it was mentioned above, the digital elevation model (DEM) of the wider area of interest was created by digitizing 84 topographic maps (scale 1:5000 and 20m elevation contours). A DBF containing the GPS coordinates of the watchtowers and the gates of the city's walls was superimposed on it. An additional height of 8m for the elevation data was allowed for each one of the gates of the wall of the ancient city of Mantineia, as well as for the watchtowers, in order to take into account the height of the structures. Visibility parameters were set from -90^0 to 90^0 for the vertical field of view and 0^0 to 360^0 for the horizontal field of view, defining visibility control towards every direction (Smith 2000). Similarly, the view distance was set to a maximum value, in order to cover the entire DEM. The results, for each site, were encoded as a binary raster in which visible cells have a value of 1 and those without visible contact have a value of 0. The synthesis of the binary viewshed maps was made either in the form of a logical union (multiple viewshed) (fig.7) or in the form of cumulative viewshed maps (Wheatley 1995) (fig.8). A cumulative binary raster was created to represent the visibility of the wall gates, the visibility of the watchtowers and the intervisibility between both of them.

The results of the viewshed analysis confirmed the previous assumption that the defensive constructions in the Mantineian plain formed a defence network in combination with the city's fortification walls. It is suggested that the location of the structures was chosen according to their intervisibility, in order to cover possible gaps in their visual communication (fig.9).

After completing the viewshed analysis and having generated the three-dimensional DEM, it was possible to overlay the georeferenced SPOT panchromatic image (10m resolution), the corresponding viewshed analysis results and the location of the archaeological sites (fig.10). It was concluded that the watchtowers were built in order to control areas with ancient road passes, covering gaps in places, which were not directly visible from the city's wall gates. Finally, using the above-mentioned data and serving as visual verification, a three- dimensional fly-simulation was created by setting a route, passing over the defensive structures.

Slope and Aspect Analyses

The slope analysis is the measure of surface steepness and it can be calculated in either degrees or percentage values, with the option to rescale the resulting values to span the full 8-bit data value range (0-255). The aspect analysis is a measure of slope orientation and it refers to the direction that the slope faces (Smith 2000). Both analyses were conducted using the DEM. The raster cell size was identical to the DEM cell size (60m.) and the direction of the sun was set to 90° (vertical), being tested to have the best visual result. The slope values are represented in degrees. A value of 0° represents a surface with no slope and 90° represents a vertical plane. Aspect output cell values describe the direction that the slope inclines relative to north. The TNTmips software, which was used to perform the slope and aspect analysis, can create an 8-bit output raster by scaling the 0 to 360 degree range of a compass to the 8-bit data space, using 0 to 240 for the compass direction (0 = north, 60 = east, 120 = south, 180 = west) and 254 indicating no slope (i.e. a flat region).

It was calculated that the city's wall gates were built on almost flat ground (slope 0-6°), a result that was expected since the city is in the middle of a plain. The watchtowers and the small fortress were built on the top of hills with maximum slope value 39°, surrounded by much steeper slopes, indicating the defensive nature of the sites (fig.11).

Conclusions

The GIS has served as a tool to validate the existence of the defence network in the plain of Mantineia and the extent of its control. It was made obvious that the defensive structures in the area were constructed following careful planning from the city of Mantineia, in order to control passes and roads, critical for its defence in an eventful chronological period.

Further exploitation of the above results, combined with environmental and archaeological data, could produce a predictive model, for detecting sites of potential archeological interest and a useful tool for the management of the cultural resources of the region.

References

Fougères G., 1898. Mantinée et l'Arcadie orientale, Paris.

Frazer J.G., 1965. Pausanias's Description of Greece, vol.IV, N.York.

Lattermann H., 1913. Nestane und das Argon Pedion, AA 26 (1913): 395–428

Loring W., 1895. Some ancient routes in the Peloponnese, JHS 15(1895): 25-89.

Pikoulas Y.A., 1990 – 1, Purgoi: Diktuo, chrissi, apories kai erotimata, Prosegisi me ta dedomena tis Argolidas, Arkadias, Lakonias, Horos 8 – 9 (1990 – 1) : 247 – 257.

Pikoulas Y.A., 1995. Odiko diktuo kai amuna, Athens.

Smith R.B., 2000. Modeling Watersheds and Land Surfaces, Getting Started booklet of the TNT-mips Software of Microimages Inc., Lincoln, Nebraska.

Topouzi S., Soetens S., Giourou A., Sarris A., 2000. The Application of Viewshed Analysis in Greek Archaeological Landscape, 6th Annual Meeting of the European Association of Archaeologists, 10th-17th September, Lisbon, Portugal.

Wheatley D., 1995. Cumulative Viewshed Analysis: a GIS-based method for investigating intervisibility, and its archaeological application, in Archaeology and GIS : A European Perspective, G.Lock – Z.Stancic (eds.), London : 171 – 185.

Figures

Figure 1 : One of the gates of Mantineia's walls (left) and the watchtower of St. Tryphon (right)

Figure 2 : Positioning the NE corner of the watchtower at St. Tryphon

Figure 3 : GPS positioning of a gate of the Mantineia's fortification walls

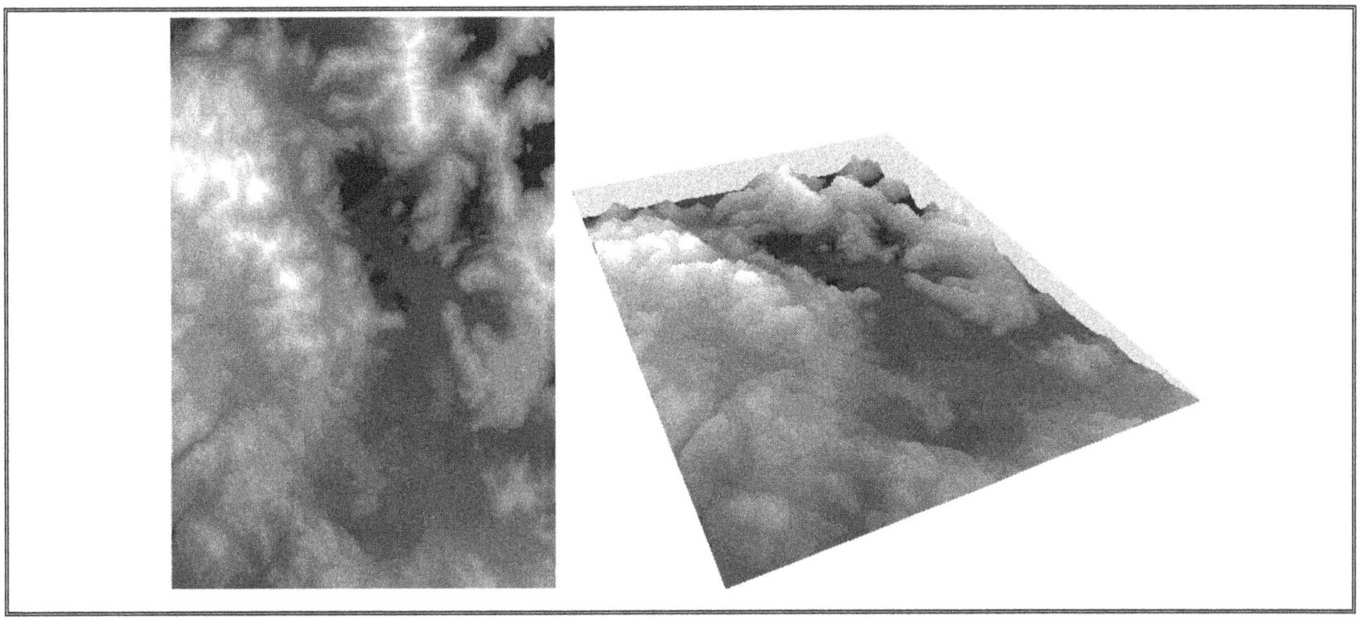

Figure 4 : 2D (left) and 3D (right) Digital Elevation Model (DEM) of the Mantineia plain (auto normalize contrast)

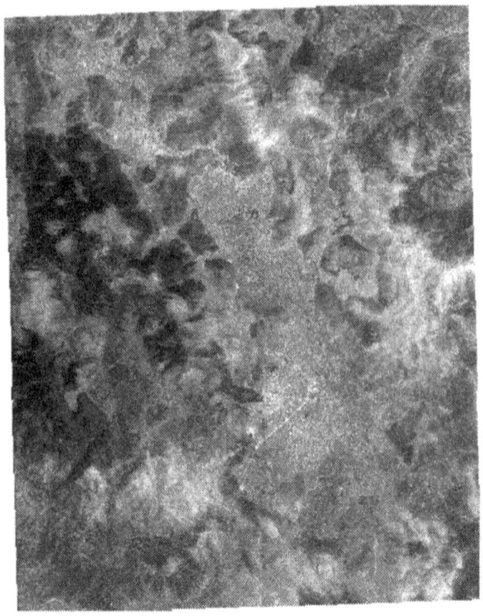

Figure 5 : Pseudochromatic image of the wider Mantineia area. Combination of SPOT panchromatic, XS1 and XS4 images

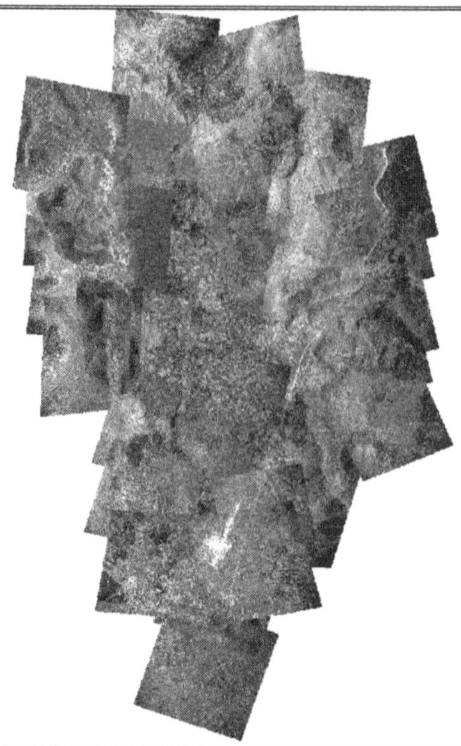

Figure 6 : Aerial image mosaic

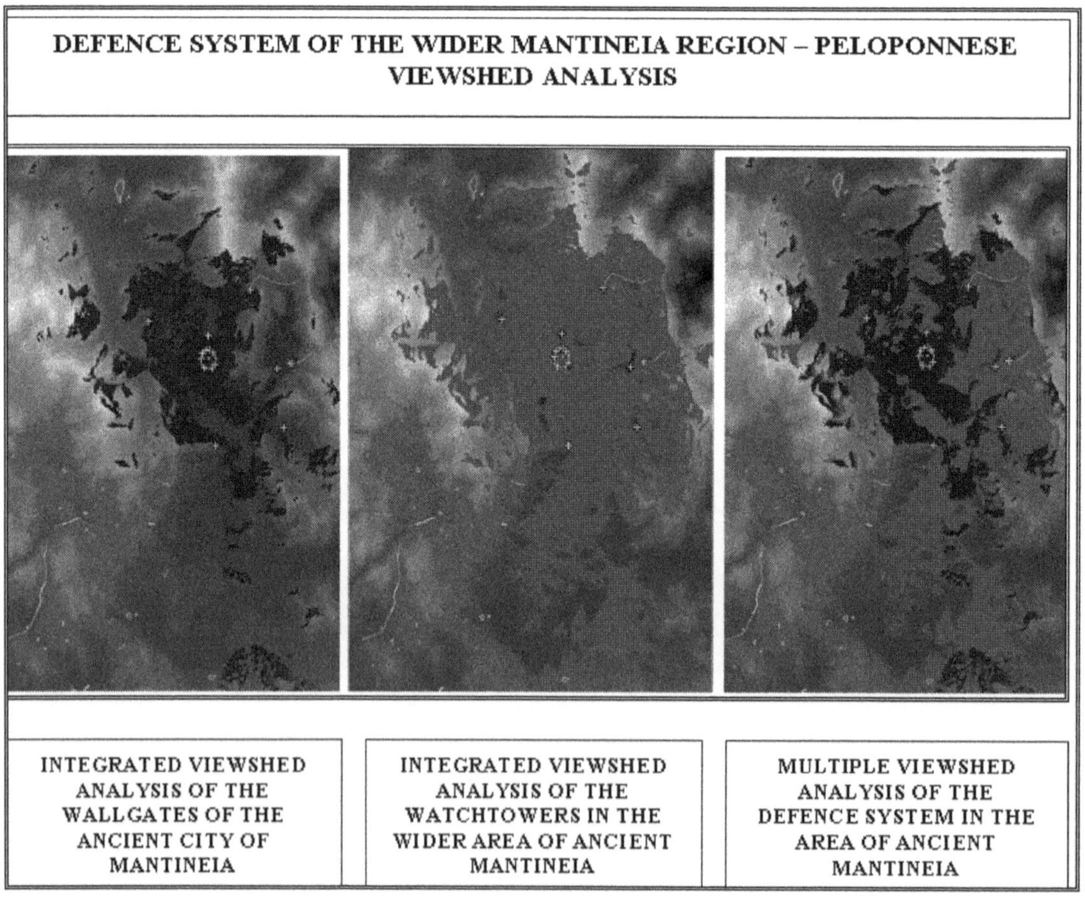

Figure 7: Single and multiple viewshed analysis of the defensive network of the Mantineia plain

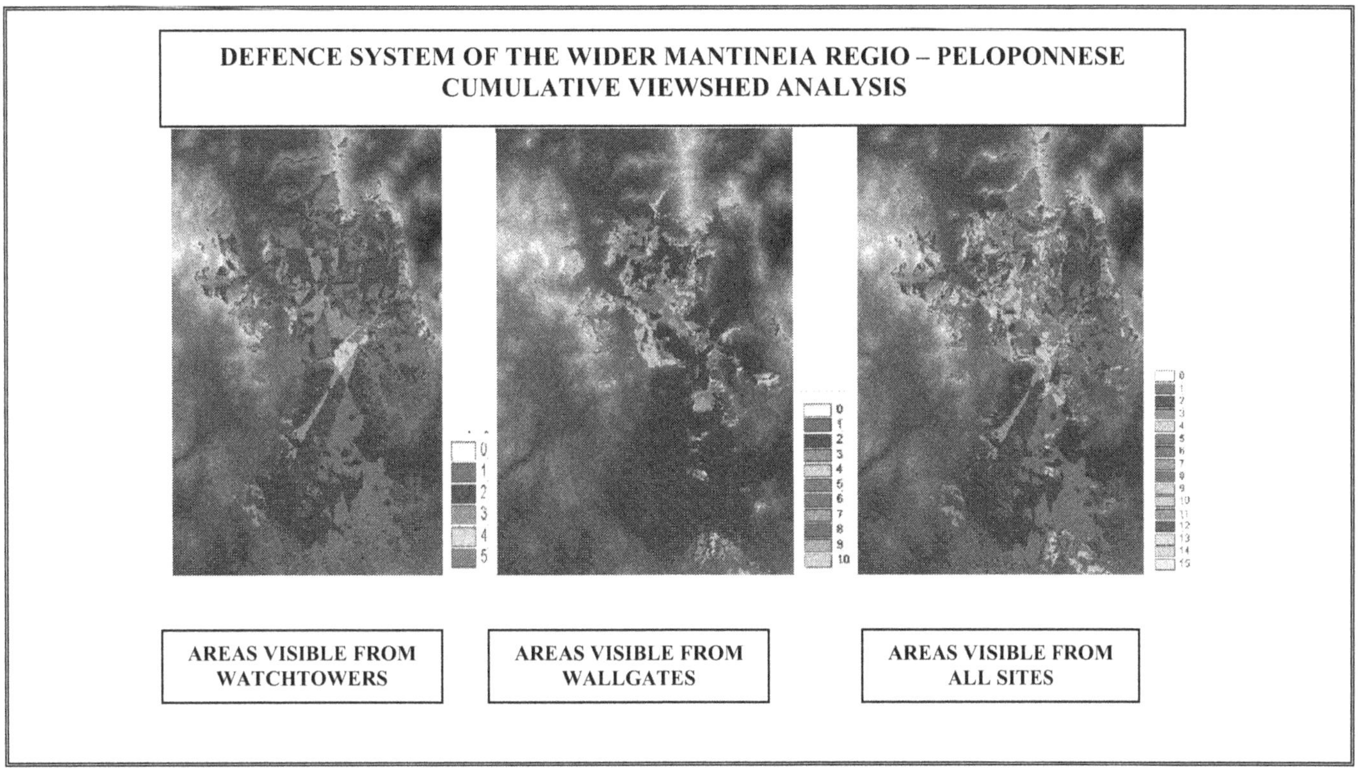

Figure 8 : Cumulative viewshed analysis of the defensive network of the Mantineia plain

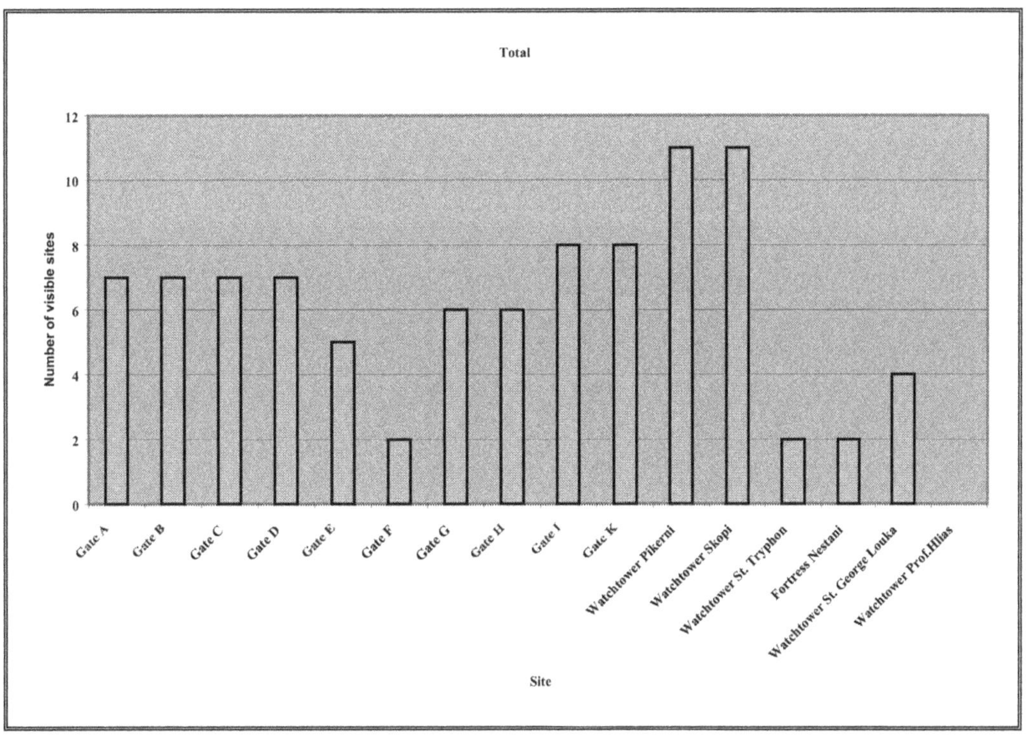

Figure 9 : Chart representing the total number of the visible sites from each defensive construction

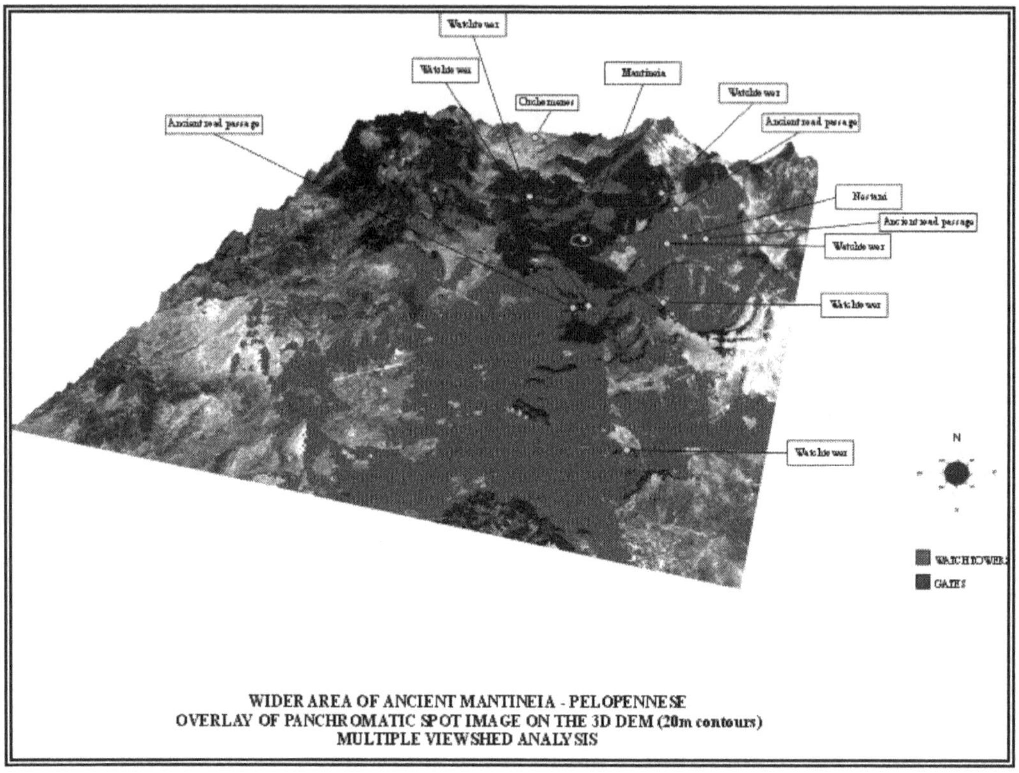

Figure 10 : Overlay of the results of the multiple viewshed analysis of the defensive network of the Mantineia plain, the outline of archaeological features and the SPOT PAN imagery on the 3D DEM.

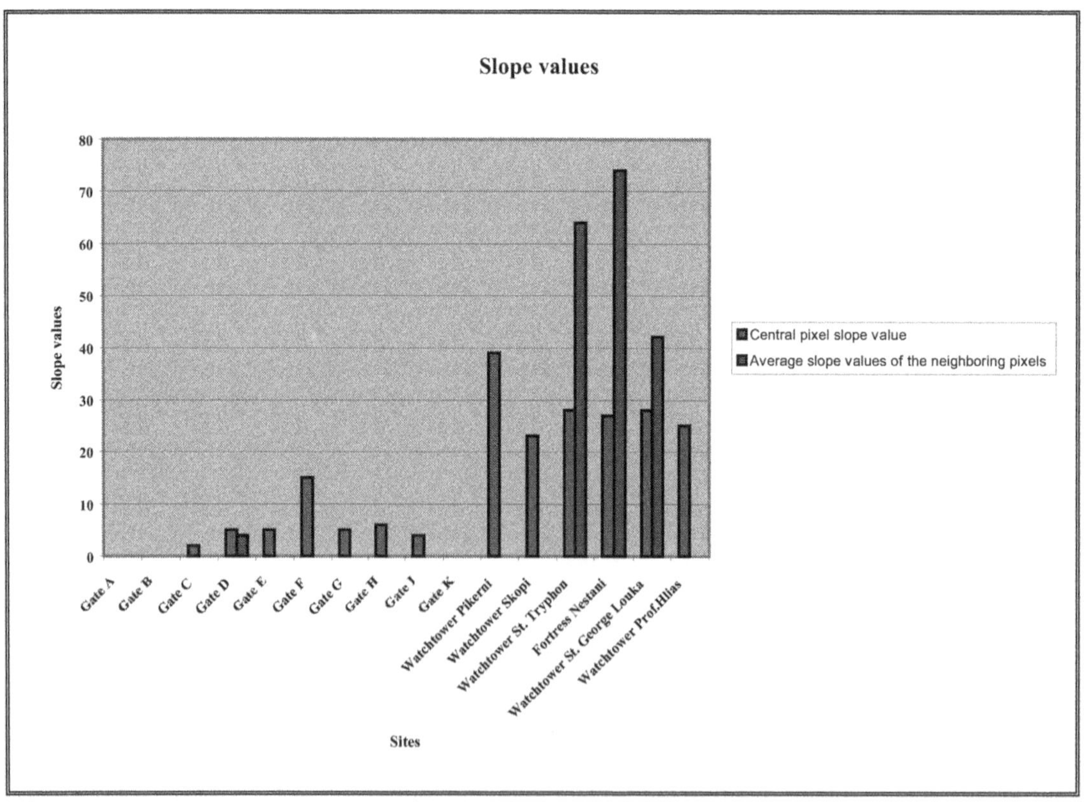

Figure 11: Chart representing the slope values for all the defensive sites

Laser Scanning and Digital Close Range Photogrammetry for Capturing 3-d Archaeological Objects: a Comparison of Quality and Practicality

Athanasios Velios
Royal College of Art / Victoria and Albert Museum Conservation programme
Sculpture Conservation
Victoria and Albert Museum
London, SW7 2RL, UK
Phone: +44 20 7942 2134 - Fax: +44 20 7942 2092 - E-mail: t.velios@vam.ac.uk

John P. Harrison
Department of Earth Science and Engineering
Imperial College
London, SW7 2BP, UK
Phone: +44 20 7594 7348 - Fax: +44 20 7594 7444 - E-mail: j.harrison@ic.ac.uk

Abstract: One of the commonest problems in the conservation of big and heavy fragmented objects is the estimation of the original relative position of the fragments (known as the matching problem). Help on the matching problem is obtained using specialized software for the estimation of the original position of every fragment. Prior to this the geometry of the fragments must be stored on the computer. We consider two non-contact capturing methods: laser scanning and digital close range photogrammetry. Laser scanning has been found to be rapid in use producing a dense, irregular point cloud that describes the object, stored in a digital file. Close range photogrammetry is slower, resulting in sparse measurements on a much more regular pattern. Data is stored on film first, and then converted to a digital file. We show how this intermediate step offers advantages related to the potential for future refinement of the model. We suggest the combined use of laser scanning for quick and easily usable models together with photogrammetry for future reference records.

Key words: laser scanning, photogrammetry, matching, Propylaia, recording, capturing

Introduction

Stone archaeological objects are often discovered in fragments, as breakage due to mechanical trauma is likely to have occurred at some time since their creation. The conservation of these objects involves the time-consuming process of the estimation of the original position of each fragment, in order to perform reconstruction of the overall shape of the object. This problem becomes more complex when dealing with heavy and large objects, as these are difficult to mechanically handle. Although modern computer software can help in this process by allowing virtual movement and matching of the fragments, prior to this virtual manipulation digital recording of the fragments is necessary.

In this paper, two remote methods for digitally recording objects — 3D laser scanning using hand-held scanners and digital close range photogrammetry — are compared on the bases of their suitability for making permanent high quality records of the objects and assisting in their virtual reconstruction. For this comparison, a cornice from the Propylaia of the Acropolis of Athens was used as a case study (figs 1-2).

Photogrammetric capturing process used at the Propylaia

When photogrammetrically capturing objects that possess only one scientifically important face (e.g. engraved objects), two cameras placed opposite the face of interest are usually used. There are many examples of the use of photogrammetry for this purpose, as this methodology has been used for the recording of buildings, objects and monuments (Bähr 1988, Cuny 1981, Masygan 1993, Milella 1997 and Rivett 1977). However, in the case of the cornice, a complete 3D model (comprising all sides of the object) is necessary, as the criteria for the matching of fragments prior to virtual reconstruction use a combination of characteristic features located on all sides of the object. In contrast to the previous case, the authors were unable to locate any previous references describing photogrammetric set-ups for the complete capturing of ancient objects. Indeed, there are only a few references describing similar procedures in other fields (Müller 1982). This is so because:

a) much more complex set-ups are needed for the complete capturing of the object (an example to demonstrate this is described later);

b) photogrammetric tools are developed mainly for the earth-mapping market, and close range photogrammetry, being a secondary market, tends not to have the necessary funds to support the research on its own (Bähr 1988); and

c) the rapid recent development of laser scanning has overtaken the development of close range photogrammetric techniques because of certain advantages that we will describe later.

In our work, the set-up for photogrammetrically capturing the Propylaia's cornice (fig. 4) involved photographing the object in 12 pairs of stereo-images (i.e. 24 photographs in total). These photographs were taken from 12 elevated positions around the object, without the object being moved. Elevated positions were selected in order to allow simultaneous capturing of both the sides and top of the object (the top is scientifically important, and if a lower elevation was used, it would not be visible). Twelve camera positions were selected (three on each side) so that sufficient overlap of the image data would exist to permit complete measurement of the object.

Equipment

The reference points for measuring the object were given by a rectangular metallic frame. An important factor in on-site photogrammetric measurement of objects is portability of the equipment, and because of this, the frame is designed to be readily disassembled and reassembled, and made of aluminium for lightness. The frame is also made to be adjustable, in order to encompass objects of various sizes. This is achieved through the bars of the frame being made of three, concentric square cross-sectioned adjustable elements, accurately positioned relative to each other with the aid of locating pins that pass through drilled elements in each bar (fig. 3). Hence, by changing the position of the locating pins, the size of the device can be changed from a minimum of 120cm × 120cm × 120 cm up to a maximum of 300cm × 300cm × 300cm. The bars of the frame are also equipped with rectangular targets (12 each bar, 144 in total) that are used as photogrammetric control points (fig. 3). Accurate fabrication of the frame means that the positions of the targets relative to each other are readily computed on the basis of the dimensions of the frame's components, and as a result high-quality photogrammetric control easily achieved.

A semi-metric ROLLEIflex 6006 camera (film size 6cm × 6cm) equipped with a reseau plate back and a 40mm lens was used for the photography. This equipment was selected for the following two reasons:

a) It is much simpler to use, and is smaller lighter and cheaper, than a fully metric camera. As a result, there is much less risk involved in its use than with a fully metric camera.

b) Because the available space around the object for setting up the equipment was limited, a wide-angle lens had to be used, and the only calibrated wide-angle lens available was for a ROLLEIflex camera.

Calibration

The frame was measured using terrestrial surveying techniques, and the relative positions of the control points estimated with a three-dimensional spatial accuracy of 1mm. However, it was recognized that the assembly – disassembly process would introduce errors that could only be quantified through re-measurement of the frame, which negates the frame's main purpose. The impracticability of constantly re-measuring the frame led us to decide that monitoring of the error would be done during the data analysis stage, using the photogrammetric software.

Photography

Figure 4 shows the positioning of the equipment around the object. Subsequently, the photogrammetric models produced from every stereo-pair were aligned and scaled according to the fixed control points of the frame, so that they shared the same co-ordinate system and could then be combined in the same 3D space to produce the complete model.

The photography took place in a single day, starting at 7 a.m. and finishing at 6 p.m. This same eleven-hour period was used for the laser scanning process, in order that a comparison could be made between the techniques in terms of the amount of data obtainable in a given time. A total of 96 photographs were taken. Photogrammetry usually uses a colour positive (i.e. slide) film, due to its superior photographic properties over colour negative film. However, these films have limited exposure latitude, which means that accurate setting of exposure time on the basis of light meter readings is required. With this work, as the object was located outdoors, there were substantial changes to the light throughout the day. This indicated that constant adjustments to the exposure times on the basis of light meter measurements would be required, and this would dramatically increase the time taken for the photography. As a result, we chose to adopt a colour negative film with wide exposure latitude, and used only a limited number of exposure times determined from a few light measurements. The film used was FUJICOLOR Superia 100, with an ISO film speed of 100 and a diffuse RMS granularity value of 4.

Data Processing – Resulting models

The films were scanned using a PhotoScan 2000 scanner (distributed in the U.K. by ZI Imaging U.K.). The scanner was used at its maximum resolution of 7¼m, in order for us to be able to take full advantage of the photogrammetric potential of the film, and the size of the resulting digital image was therefore about 8500 pixels by 8500 pixels. The software used for the processing of each pair of digital images in order to produce a digital elevation model was ImageStation 2000 (developed by ZI Imaging). Usefully, after photogrammetric orientation of the models using the co-ordinates of the control frame, the software provided us with reports on the estimation of the accuracy of the measured control points. In this way it was possible to assess the accuracy achieved while building the stereo-models, and this is shown in table 1 for two objects of different shape.

The error values given in the table describe the mean estimated error of the control points of the frame. It is expected that the corresponding values for the measured points on the object will be smaller than this. This is because the object appears towards the centre of each image, whereas the frame appears towards the edges, and the photographic errors due to camera and lense distortions increase way from the image centre.

The digital elevation models from each pair of photographs were combined to produce a complete model of the object as a single entity. The resulting model consisted of a group of points (generally referred to as a point cloud) with X, Y, Z co-ordinates in the frame's co-ordinate system. The minimum distance between two points on the object's surface was about 5mm, which means that, although the position of each point was estimated with the accuracy shown above, the low density of the points results in surface features located between any two points not being detected. Clearly, closer point spacing is desirable, but the point spacing obtained is that which resulted from the image matching algorithms used in the process, rather than being specified *a priori*. As the performance of these algorithms is affected by the quality of the images, the low point density is thought to have occurred because of:

a) low picture quality on the medium, exemplified by the extreme diversity in lighting conditions causing a loss of data; and
b) the scanning of the negatives, with user-defined colour settings possibly affecting the quality of the resulting digital image and hence limiting the available data.

Production of the digital elevation model is a time consuming process. Although several software packages include algorithms for the automatic calculation of the co-ordinates of the points in such a model, it is usually the case that the automatic process needs to be monitored by a user who intervenes when necessary. The production of the models for the two objects took about 8 hours of systematic work to complete.

Laser scanning capturing process

Laser scanning is a method for capturing objects that is extensively used in engineering. There have been several applications of this technology to the recording of the cultural heritage, beginning in the 1980s (Ahmed 1983, Taylor 1987) and with very good results in more recent years (Larson 1994, Levoy 1999). However, there are few applications with specific orientation to conservation. In 1992, Baribeau et al. (Beribeau 1992) described the use of a laser scanning system for the evaluation of the damage caused to paintings by vibrations. More recently, Geary (Geary 2001) has reported on the use of laser scanners to capture the geometry of polychrome sculpture. We conclude that there is little guidance available in the published literature for the use of laser scanning in conservation.

Equipment

The scanning of the objects here was performed using a laser scanning device ("Model Maker H", available in the U.K. from 3D Scanners Ltd., and in Greece from FTS Ltd.) mounted on a precision articulated arm (the "Bronze" arm produced by FARO). The theoretical accuracy of the equipment is about 0.5mm (Faro 1999), and a simple calibration process (see below) confirmed this accuracy. Nevertheless, given the crystal structure of the pentelic marble used in the building of the Propylaia (Tanoulas 1994), it is possible that greater error could be introduced by the refraction of the laser beam in the crystals, as pentelic marble can be transparent up to 15mm from its surface (Orlandos 1994: 83). However, given the deterioration of the material and the patina, which has made the surface opaque, this error is not considered to be of significant magnitude here.

Calibration

The calibration process of the scanner involves the scanning of an object with known dimensions (usually a cube). The resulting digital model, when compared to the dimensions of the real object, provides all the information for the calibration of the instrument. This calibration is then done automatically within the control software. However, the accuracy of the system in any given application depends on both the user and the object being scanned, and therefore the actual model accuracy obtained is not known.

Scanning

A critical component of the scanning process is for the camera embedded within the scanning device to detect the laser beam reflected from the object. Direct sunlight on the marble would therefore make the process impossible, and because of this the scanning was done indoors (fig. 5).

The nature of the scanning instrument and the support to the object meant that at any one set up there was at least one surface of the object that was hidden and therefore could not be scanned. To overcome this problem, after the visible sides had been scanned, metallic registration marks were placed on the object (with water-soluble glue) and scanned separately. The object was then rotated such that the registration marks and the previously hidden surfaces were visible to the scanner, and the registration marks scanned again (fig. 6) before being removed. In this way, sufficient data were obtained to allow the results from the scans of the various surfaces to be linked through co-ordinate transformation.

It was found possible to undertake both the calibration process and the scanning in a single day: Work commenced at 8 a.m. and was concluded at 6 p.m.

Data Processing – Resulting models

As with photogrammetry, the output of the scanner is a point cloud, with each point defined by a triplet of values corresponding to X, Y, Z co-ordinates in 3D space. Because of the scanner's potential for collecting 3600 points per second (300 points per frame, 12 frames per second) the output files can contain vast numbers of co-ordinate values and are usually extremely large. As a result, post processing of these files involves the filtering of the points, such that coincident points are removed. The software which accompanies the scanner (or other commercial software packages) is capable of such basic filtering of point clouds.

Comparison of the techniques

Accuracy – Quality of the models

From inspection of the figures of the models (fig 7-10)[1] and table 2, it is readily apparent that the number of points in the photogrammetric model (8,907) is much smaller than that in the scanned model (479,374). It is therefore obvious that a lot more information is exported (not captured — see the later section on Ethics) by the scanner than by the photogrammetric software. The effect of this is clearly seen in the photogrammetric model where, because of the sparse points, small-scale details cannot be seen on the final model.

However, an issue of possibly greater importance than the quantity of data is their quality (i.e. accuracy). The software used in the photogrammetric method can provide us with information, for each stereo-pair of photographs, that includes the estimation of the error both for every control point and for the overall accuracy of the model. On the contrary, similar information is not available when exporting data from the scanning software. The only indication of the accuracy of the scanned data is the performance data provided by the manufacturer of the instrument. As described earlier, prior to scanning the instrument is calibrated, but this only shows whether or not the equipment at least reaches the accuracy stated by the manufacturer. We can assume that the accuracy of the scanned data equals this, but as the quality of the scanning process depends on both the experience of the user and environmental conditions, it is possible that the stated accuracy is not always the accuracy of the model.

Hence, judging from the resulting models of the two methods, laser scanning gives the greater number of measurements (more points and improved surface definition) and probably (but not necessarily) better accuracy than the photogrammetric system. However, it is not possible for the accuracy of the scanning system to be determined as robustly as that of the photogrammetric system.

Time

The work reported on here was a preliminary investigation, rather than a highly optimized production process. As a result, we can see with hindsight how simple improvements to our methods and equipment would reduce the capturing time. For example, with the photogrammetric work, had the object been placed closer to the ground, then the use of a stepladder for the photography would have been avoided and as a consequence the amount of time needed for this greatly reduced.

However, the time needed for photogrammetrically capturing extremely large objects is almost the same as the time needed for capturing small objects, as the number of images required is the same. In contrast, the scanning of extremely large objects would dramatically increase the time of the recording process, since the area that needs to be scanned will be substantially greater. For such large objects it is likely that scanning will be an impractical solution.

As far as post-processing is concerned, in the case of the cornice it is obvious that laser scanning software performed its tasks much more quickly (~2 hours) than the photogrammetric software (~8 hours). Again, here, one has to consider the case when it is extremely large objects being captured. The post-processing of the data for scanning could be extremely slow and demanding of computational resources. On the other hand, photogrammetric software runs equally fast no matter what the object size is, as its speed depends on the size of the digital image that holds the information, and not the object itself. There is, of course, the related issue of accuracy of the measurements and their spatial distribution, in that small images of large objects will be processed quickly, but at the expense of quality of the final digital elevation model.

Practicality

It appears that most issues regarding practicality of the equipment are down to personal prejudice. For example, both sets of equipment are portable and can be easily carried on site, and both methods depend on the ambient lighting conditions. It is likely though that some users would feel happier with one set of equipment than the other for purely personal reasons.

One issue of practicality that is important is that of the sensitivity and the robustness of the equipment in terms of tolerance to physical abuse. Both laser scanning cameras and ordinary cameras are precision instruments that require careful handling, and although the laser scanning camera felt much the more delicate, the authors are not aware of any research undertaken to compare the physical tolerance of the equipment. Although such work would probably be pleasurable to undertake, it would without doubt be expensive!

Finally, we must consider practicality in terms of the personnel requirements. Laser scanning is the more orientated to an automatic approach for the production of a digital model by non-specialist personnel, in that the training needed for successfully using the scanner does not presuppose any technical knowledge apart from that of the use of a computer generally. On the other hand, in order to be able to produce a model using photogrammetry, specialized knowledge of the subject is needed, along with sufficient experience of modern equipment and processes.

Ethics

According to the international charter of Venice (Vrieze 1964: Article 16), keeping accurate records of objects is an essential part of the conservation process. In laser scanning the collection of the data ends as soon as scanning has been completed, and no additional information can be extracted from the object unless a new scan takes place at a higher resolution. On the other hand, with photogrammetry an uninterpreted and essentially complete representation of the object is recorded on a medium (i.e. film), and the information is extracted after the recording using an appropriate technique. Hence, the inevitable improvements in these techniques will make additional information available for extraction in the future. Permanent records of the object are kept on the film, and this is why careful photography is important.

Matching of fragments

In terms of using these digital elevation data, or point clouds, for extracting useful information for the matching of the fragments, we can make two points. Firstly, in the context of architectural elements, we believe that the point density achieved by photogrammetry is sufficient to capture identity patterns of the objects. This can be useful when large models (as result from laser scanning) would rather be avoided for speed of computer processing. Secondly, Papaioannou et al. (Papaioannou 2001) state that the better the definition of the fractured surface, the more reliable the results of the matching, in which case laser scanning may be preferable. Our investigations are continuing to determine whether photogrammetric data may be appropriate for measuring the roughness of the fractured surfaces, and whether this can assist in the matching of fragments.

Discussion

The comparison of these techniques was based on very specific equipment and software that, according to the authors' experience, combine good quality with accessibility. Although better equipment (and probably software) will undoubtedly exist, it was not accessible to use. As a result, although a more extensive comparison of the methodologies would have been useful, it would have had a rather theoretical character.

One of the major problems we encountered during the recording of the objects was the capturing of the hidden side. We consider that further research is necessary on at least two aspects of this problem. Firstly, a simple process for accurately registering the point clouds that arise from different sides of the same object during laser scanning, without having to attach any registration marks to the object, is urgently required. Although such registration is currently feasible through software, the authors are not aware of any reference to its reliability or accuracy. Secondly, and similarly, simple techniques are required that will allow alteration of the photogrammetric set-up such that all surfaces of an object can be captured, without the object having to be moved.

Lack of standardisation during post-processing of the data is also a significant problem. Several parameters exist that have to be optimally selected for the best results in terms of accuracy, and used consistently. We depended on the skills and expertise of specialist personnel employed by the various equipment suppliers for this; such a solution is impractical at large.

Finally, the issue of introducing high quality photogrammetric control in a simple fashion is one that requires urgent attention. We chose to use a precision-built frame, but due to its cubic shape, distortions can easily occur during disassembly and reassembly. If a control frame is to be used, development of a simple and robust system that is both dependably accurate and able to surround large objects is required.

Conclusions

Laser scanning gives high definition digital models of objects, quickly. However, no improvement to these data can occur later on, and no permanent uninterpreted record of the object exists. Current photogrammetric platforms in combination with medium format photography are unable to produce such dense models, although if new software packages and film scanners are introduced in the future, more information can potentially be extracted. The use of photography does, however, mean that a permanent uninterpreted record of the object exists.

Consequently, permanent accurate records that adhere to the principles of the international charters of conservation, can only be kept using photogrammetry. However, because today's laser scanning technology is much more efficient in the immediate production of models that are useful to practical conservation, we suggest that the combined recording of objects with both photogrammetry and laser scanning be implemented.

Acknowledgements

The authors thank Professor Alan Cummings (Royal College of Art) for his support and the Departments of Geomatic Engineering at University College London, and Engineering and Surveying at the National Technical University of Athens for providing access to their equipment. Finally, we gratefully acknowledge the Committee for the Restoration of the Acropolis Monuments for granting access to the objects. This research was funded by the State Scholarship Foundation of Greece.

End notes

[1] In the figures of the scanned model, only one quarter of the total data are displayed, as images of the point cloud in full resolution are totally black due to the point cloud's density.

References

Ahmed F. A., 1983. *A Laser Scanning Plotter for Measurement of Archaeological Objects,* Journal of Field Archaeology, Boston University, Volume 10, Number 2, pp.236 – 238.

Bähr H.P., 1988. *The use of Photogrammetry in the Analysis of Deformation, Stable-Unstable? Structural Consolidation of Ancient Buildings,* Centre for the conservation of Historic Towns and Buildings, Leuven University Press, pp.157 – 166.

Beribeau R., Rioux M., Godin G., 1992. *Recent Advances in the Use of a Laser Scanner in the Examination of Paintings,* Restoration 1992, United Kingdom Institute for Conservation.

Cuny F.C., Mugnier C.J., 1981. *Close Range Photogrammetric Analysis of Earthen Buildings Under Seismic Loading,* International Workshop Earthen Buildings in Seismic Areas, May 24-28, Volume 1, pp. 131-138.

Faro Technologies Inc., 1999. *Acceptance and Accuracy*

Certification, Issue 8 (revised August 2000).

Geary A., 2001. *Computer Related Imaging in Conservation: The Visualisation in Three-Dimensions of the Original and Present Appearances of European Polychrome Sculpture Using Laser Scan Data, Presented in Virtual and Enhanced Reality Environments,* PhD by Thesis, Royal College of Art.
Larson J., 1994. *The Use of Laser Scanners for Recording and Reproducing Sculpture,* Conservation News, United Kingdom Institute for Conservation, Number 54, pp. 10.

Levoy M., 1999. *The Digital Michelangelo Project,* Second International Conference on 3D Digital Imaging and Modelling, Ottawa, Canada, October 1999.

Masygan R.J., 1993. *The Use of Multi-Image Close-Range Photogrammetry for Recording Historic Architectural and Archaeological Sites,* 10th Triennial Meeting, Washington, DC, USA, ICOM Committee for Conservation. Volume 1, pp.195 – 197.

Milella N., Fiore N.G., 1997. Study of Structural Interventions and Changes Affected Altamura Cathedral (XII Century) by Means of Digital Photogrammetric Techniques, 4th International Symposium on the Conservation of Monuments in the Mediterranean, 6-11 May 1997, Volume: 2.

Müller B.G., 1982. *Zur wahl der aufnahmeanordnung fuer ingenieur-photogrammetrische nahbereichsaufnahmen (On the determination of the camera orientation for close range photogrammetry),* Vermessungswesen und raumordnung, Volume 5, pp. 217-237.

Orlandos A., 1994. *Construction Materials of the Ancient Greeks and their Application.* Athens: Deytera publishing.
Papaioannou G., Karabassi A., Theoharis T., 2001. *Virtual Archaeologist: Assembling the past,* IEEE Computer Graphics and Applications, Issue: 2, pp.53-59.

Rivett L., 1977. *Photogrammetry – Its Potential Application to Problems in Australian Archaeology,* Conservation of Rock Art, Perth, Institute for the Conservation of Cultural Material, September 1977, pp.15 – 21.

Tanoulas T., Ioannidou M., Moraitou A., 1994. *Study for the Restoration of the Propylaea.* Athens: Ministry of Culture, Committee for the Preservation of the Acropolis Monuments, Vol. 1.

Taylor J.M. et al., 1987. *Applications of a Laser Scanner to the Recording and Replication of Museum Objects,* 8th Triennial Meeting, Sydney, ICOM Committee for Conservation.

Vrieze (de), P. L., et al, 1964. *The Venice Charter.* Venice: ICOMOS.

Tables

	Fragment No 1 (highly elongate)	Fragment No 2 (approximately cubic)
Extent X	1m	0.4m
Extent Y	0.25m	0.25m
Extent Z	0.5m	0.45m
Error	±0.003m	±0.0021m

Table 1. Estimated error in photogrammetric measurements compared to the size of the object.

	Photogrammetry	Laser scanning
Extent X	1,050mm	1,040mm
Extent Y	245mm	261mm
Extent Z	481mm	472mm
Number of Data Points	8,907	479,374

Table 2. Size of the models (mm) and number of points.

Figures

Figure 1. Elongate fragment of the cornice.

Figure 2. Cubic fragment of the cornice.

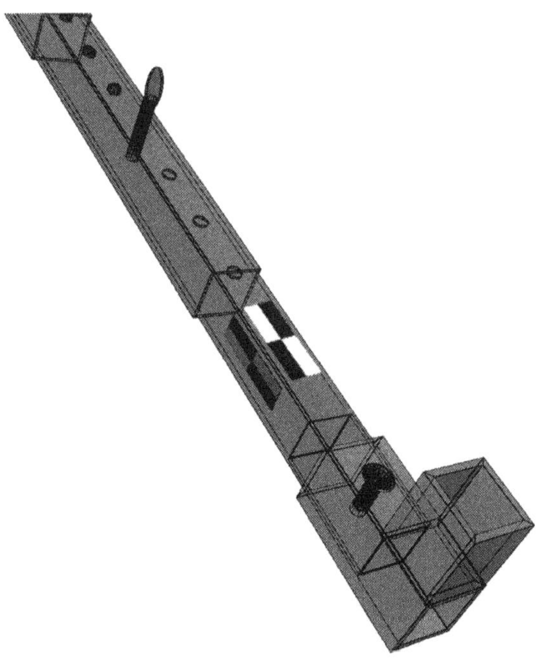

Figure 3. Extension pins and targets.

Figure 4. Photogrammetric set-up.

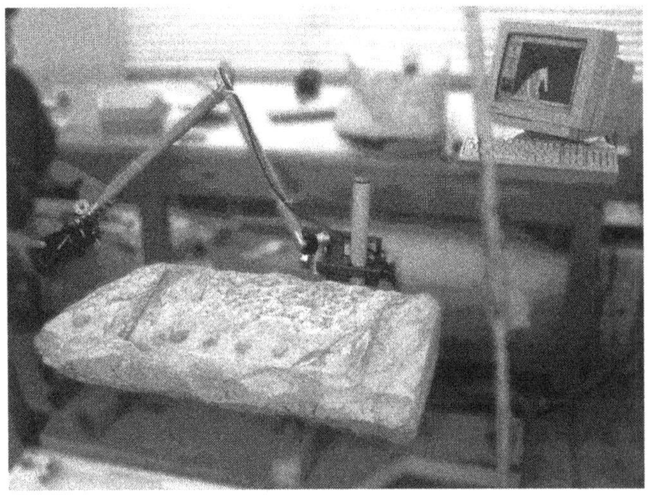

Figure 5. Scanning of the cornice.

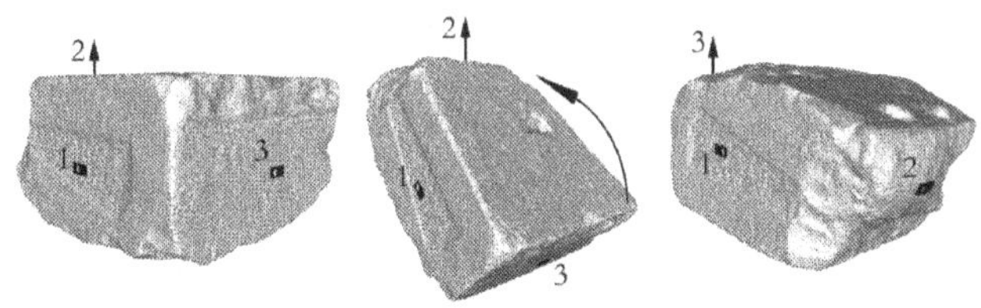

Figure 6. Registration marks.

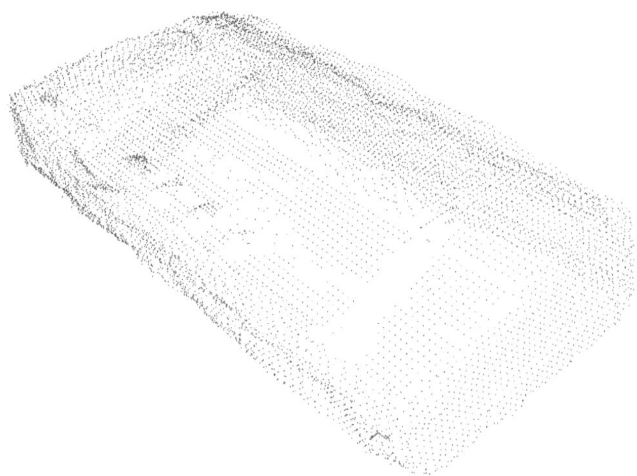

Figure 7. Photogrammetric point cloud.

Figure 8. Scanning point cloud.

Figure 9. Photogrammetric shadowed surface.

Figure 10. Scanning shadowed surface.

www.ingramcontent.com/pod-product-compliance
Lightning Source LLC
La Vergne TN
LVHW072246300926
849160LV00018B/1602